MICROPROCESSORS AND DIGITAL SYSTEMS
SECOND EDITION

DOUGLAS V. HALL

Instructor, Portland Community College, Portland, Oregon
Design Engineer, Hall Electronics Consultants

Gregg Division
McGraw-Hill Book Company

New York Atlanta Dallas St. Louis San Francisco Auckland Bogotá
Guatemala Hamburg Johannesburg Lisbon London Madrid Mexico Montreal New Delhi
Panama Paris San Juan São Paulo Singapore Sydney Tokyo Toronto

Sponsoring Editor: Paul Berk
Editing Supervisor: Iris Cohen
Design and Art Supervisor: Judith Yourman
Production Supervisor: S. Steven Canaris

Cover Photograph: Richard Megna, Fundamental Photographs

Library of Congress Cataloging in Publication Data

Hall, Douglas V.
  Microprocessors and digital systems.

  Includes bibliographical references and index.
  1. Microprocessors. 2. Digital electronics.
I. Title.
TK7895.M5H34 1983   621.3819′5   82-20878
ISBN 0-07-025552-0

Microprocessors and Digital Systems, Second Edition

      6 7 8 9 0 SEMBKP 8 9 0 9 8 7 6

ISBN 0-07-025552-0

CONTENTS

Preface   v

1 PROTOTYPING AND THE USE OF TEST EQUIPMENT   1

Prototyping   1
Digital Voltmeters and Logic Probes   6
Oscilloscopes   6
Other Test Equipment   15

2 DIGITAL LOGIC GATE CHARACTERISTICS AND INTERFACING   17

Analog versus Digital   17
Binary Code: A Base-2 Number System   17
Basic Logic Gates   19
Logic Family Characteristics   21
Input Signal Conditioning and Signal Sources   41
Output Signal Handling   45
Interfacing Logic Families to One Another   48
Interfacing Logic Gates to Simple Displays and
   Relays   50
An Important Point   51

3 COMBINATIONAL LOGIC, MULTIPLEXERS, CODES, AND ROMs   54

Combining Logic Gates   54
Synthesizing Simple Logic Gate Circuits   66
Switches, Multiplexers, and Demultiplexers   68
Binary-Based Codes   74
Code Converters   79
Read-Only Memories   86
PLAs, FPLAs, FPGAs, and PALs   91

4 FLIP-FLOPS, COUNTERS, AND REGISTERS   96

Latches and Flip-Flops   96
Asynchronous or Ripple Counters   104
Synchronous Counters   111
Registers and Shift Registers   121

5 READ/WRITE MEMORIES   135

Random Access Memory   135
Shift Register Memories   146
Magnetic Core Memory   148
Magnetic Tape Storage   149
Magnetic Disk Data Storage   150
Floppy Disk System   152
Electron Beam Accessed Memory   155
Access Time versus Storage Capacity   155
Testing Semiconductor Memories   155

6 D/A AND A/D CONVERTERS   161

Digital-to-Analog Converters   161
Analog-to-Digital Conversion   166
Applications   172

7 DIGITAL ARITHMETIC   178

Addition   178
Subtraction   181
Arithmetic Logic Unit   185
Binary Multiplication and Division   187
Fixed-Point Numbers   190
Floating-Point Numbers   191

8 MICROPROCESSOR STRUCTURE AND PROGRAMMING   194

General Organization and Program Flow in a Digital
   Computer   194
Microcomputer CPUs: The 8080A and 8085A
   Microprocessors   197
Assembly Language Programs for the 8080A and
   8085A   199
The 8080A/8085A Instruction Set   199
Writing Assembly Language Programs   209

9 8080A/8085A SYSTEM HARDWARE AND TIMING 222

An 8080A Microprocessor System   222

An 8085A Microprocessor System   228

Address Decoding   235

Port Decoding: Memory-Mapped I/O and Direct I/O   237

Address Decoding for an 8085A System   238

Microcomputer Timing Parameters   239

Wait, Hold, and Halt States   241

Interrupts   243

8085A Transition State Diagram   245

Troubleshooting an 8080A or 8085A Microcomputer System   246

The Zilog Z80   249

One-Chip Microcomputers   250

10 THE MC6800 MICROPROCESSOR AND MICROPROCESSOR EVOLUTION 256

MC6800 Microprocessor   256

A Simple MC6800 System   272

6800 Interrupts   275

Troubleshooting a 6800-Based Microcomputer System   278

Descendants of the MC6800   279

The MOS Technology 6500 Microprocessor Family   281

Summary and Comparison of Currently Available Microprocessors   286

11 INPUT AND OUTPUT INTERFACING 291

Simple, Polled, and Interrupt I/O   291

Handshake Input and Output with a Microcomputer   296

Interfacing Keyboards to Microcomputers   302

Interfacing Seven-Segment Displays to Microcomputers   305

Power Circuits   314

D/A Converter Interfacing and Applications   315

A/D Converter Interfacing and Applications   316

Common Bus and Data Communication Standards   318

Interfacing a CRT Display Monitor to a Microprocessor   337

Microprocessor Peripheral Interface Trends   342

12 PUTTING IT ALL TOGETHER 345

A Microprocessor-Based Smart Scale   345

An MC6800-Based EPROM Programmer   365

Overview of Industrial Process Control   377

An SDK-85-Based Process Control System   379

13 PROTOTYPING AND TROUBLESHOOTING μP-BASED SYSTEMS 391

Building and Debugging a Prototype   391

Testing, Debugging, and Linking Program Modules   394

Building, Testing, and Debugging a Prototype without a Development Board and Monitor Program   398

Production Processes   400

Production Test and Field Service Tools and Techniques for Microprocessor-Based Instruments   402

14 HIGHER-LEVEL LANGUAGES 407

Overview   407

Systems and Programs Used to Write and Execute High-Level Language Programs   408

Programming in BASIC   410

Programming in FORTRAN   418

Programming in Pascal   425

Trends in Development of Microcomputer-Based Products   432

APPENDIX

A. State Diagrams and Algorithmic State Machine Charts   435
B. Explanation of New Logic Symbols   437
C. Data Sheets   451
    Intel 2716 PROM   452
    Zilog Z80/Z80A CPUs   457
D. Votrax SC-01 Speech Synthesizer—Phoneme Chart   465

INDEX   467

PREFACE

The explosive growth of electronics technology and the electronics industry has made it very difficult for any one textbook, or even any sequence of courses, to cover all the information an electronics technician will require throughout a career. Therefore, two skills are particularly important to the technician who wishes to succeed in the field and grow with it. First is the ability to read and interpret manufacturers' data books, application notes, and service manuals. Second, and this has been supported by much industry feedback, is that electronics technicians need to be fluent in the use of test equipment and to understand systematic approaches to troubleshooting.

This second edition of *Microprocessors and Digital Systems*, and its correlated lab manual, *Experiments in Microprocessors and Digital Systems*, will help you to develop these skills in the area of digital and microprocessor systems.

This book uses a spiral approach. That is, a topic is introduced in as much detail as is required for basic understanding, and then each time the topic is met again, more depth is added.

To help develop skill in the use of test equipment, the text starts with a chapter on the use of test equipment, such as logic probes, signal generators, and oscilloscopes, commonly used to troubleshoot digital circuitry. The use of logic analyzers is described in Chaps. 5 and 13. Chapter 13 also discusses the use of development systems, in-circuit emulators, and signature analyzers.

To help develop the skill of systematic troubleshooting, this book uses a systems approach throughout. An electronics person needs to know not only the function of components, but also how the components are connected together and how they function as a system. A digital clock, a frequency counter, and a successive approximation A/D converter are discussed in detail as examples of digital systems. A microprocessor-based scale, EPROM programmer, and industrial controller are described in detail as examples of microprocessor-based systems. Systematic prototyping and/or troubleshooting methods for these systems are explained.

To help develop skill at interpreting manufacturers' literature, this text uses many current data sheets and explains the parameters found on these. The new IEEE/IEC dependency notation logic symbols recently added to digital data books are described and explained. A standard TTL data book such as the Texas Instruments *TTL Data Book for Design Engineers*, the National Semiconductor *Logic Databook*, or the Signetics *Logic-TTL Data Manual* should be used in conjunction

with the chapters on digital systems to gain further familiarity with data books. Latest editions of the Intel *MCS 80/85 Family User's Manual*, the Intel *Components Catalog*, and the Motorola *Microprocessors Data Manual* should be used in conjunction with the chapters on microprocessors in this book.

On the basis of the premise that it does little good for a technician to know the theory of a circuit without also knowing how to connect and test the circuit, this book starts with a detailed discussion of the use and misuse of test equipment. Following this, Chap. 2 presents binary numbers, basic gates, a "nuts and bolts" comparison of most common logic families, interfacing, signal conditioning, and problems commonly encountered in digital logic.

Chapters 3 through 7 proceed from simple combinational logic to code conversion, ROMs, counters, registers, RAMs, A/D converters, and arithmetic logic units. With the completion of these chapters, all the components used in a microcomputer will have been thoroughly covered.

Starting with the general organization of a digital computer or microprocessor system, Chap. 8 moves quickly into the structure and assembly language programming of a specific microprocessor, the 8080A/8085A. Experience has shown that detailed instruction with one or two specific processors is more effective for teaching technicians than a generalized approach using a hypothetical or generic device. When you become thoroughly familiar with the structure, programming, and problems of one or two microprocessor families, it is relatively easy to learn other families as needed. The spiral study approach is continued in Chap. 9 with a detailed discussion of the hardware of 8080A/8085A-based systems. Chapters 8 and 9 together give a solid understanding of basic 8080A/8085A-based systems. Chapter 9 also briefly discusses the architectures and instruction sets of the Z80 and 8048 microprocessors.

In order to acquaint you with a second major microprocessor family, Chap. 10 describes the structure, instruction set, and hardware of a 6800-based microprocessor system, and compares them with the 8080A/8085A system. The software learning spiral continues with the introduction of indexed addressing modes. Chapter 10 also discusses the architecture and instruction set of the 6502 microprocessor and then compares three of the commonly available 16-bit processors.

Since most applications of a microprocessor require it

to interface to the outside world, Chap. 11 analyzes a wide variety of input and output devices and methods, including keyboards, LED displays, CRTs, and A/D converters. The initialization and applications of several programmable peripheral devices are described. Each manufacturer, such as Intel and Motorola, has a thick book of available programmable peripheral devices. There is no space here to describe all of these, but if you learn the procedure for initializing a few specific devices, you can get the details for others from the manufacturers' data books as needed.

Chapter 12 combines these I/O devices and the microprocessors discussed in the previous chapters into three examples of complete microprocessor-based products: an 8085A-based "smart" scale, a 6800-based EPROM programmer, and an 8085A-based industrial controller. The circuits, flowcharts, and assembly language programs of these products are analyzed. Then, in Chap. 13, they are used to teach troubleshooting of LSI systems. Both hardware and software troubleshooting techniques are developed. Enough specific information is given so that with a relatively inexpensive microprocessor development board, such as the Intel SDK-85, the Motorola MEK6800D2, or the Motorola MEK6800D5, you can single-step through a program discussed in the text and observe address, data, and control bus signals. If a logic analyzer is available, a simple test loop program portion can be run and observed.

Chapter 14 gives an introduction to high-level programming languages. Decrease in the per-bit cost of memory and the high cost of debugged assembly language programming has led to much programming being done in more time-efficient high-level languages. The major program for a system can be written in a high-level language and assembly language modules inserted only as needed for initializing programmable devices, time-dependent sections, or interrupt service subroutines. BASIC is used for most of the examples in this chapter because of its widespread use, and because of its availability in several inexpensive home computer systems. FORTRAN and Pascal are also introduced.

I wish to express my profound thanks to the following people: Marybelle B. Hall, my wife, who critiqued my writing, assembled the manuscript from countless fragments, and made other important contributions too numerous to mention; Fred Mann of Texas Instruments, Inc., for his help with dependency notation logic symbols; Professor John L. Morgan of DeVry Institute of Technology in Irving, Texas, for his thorough review of the manuscript; my colleagues at Portland Community College: David M. Hata, instructor and COOP work experience supervisor who made many suggestions and served as a pipeline for feedback from industry, Dr. Richard Morris who reviewed Chaps. 6 and 14, and William O'Dell who reviewed Chaps. 1–3; friend Michael Hudson who helped build and test the industrial controller in Chap. 12; friends Rosemary Griffeth and Melody Reasoner who typed the manuscript and became computer-literate in the process.

Finally, I wish to thank my many students and ex-students working in the electronics industry who try to keep me tuned in to their needs.

I would very much appreciate hearing from users of this book—any comments, criticisms, or suggestions you might have would be most helpful.

Douglas V. Hall

CHAPTER 1

PROTOTYPING AND THE USE OF TEST EQUIPMENT

At some time in your career as an electronics technician or student, most likely you will have to build and test some digital circuits. This may be for prototyping or for modification of existing equipment. Success of the circuit depends not only on its design but also on the care put into laying it out and constructing it. Quickly putting a pile of parts together in a "bird's nest" configuration usually produces a circuit that is unlikely to work, difficult to troubleshoot, and undependable in use. Therefore, the first topic to consider in the building of circuits is the proper construction of *a prototype* or *modification circuit.*

At the conclusion of this chapter, you should be able to:

1. Choose a material on which to construct a prototype.

2. Lay out and build the circuit neatly, using color-coded wiring and effective bypassing.

3. Test basic logic states by using an oscilloscope, a DVM (digital voltmeter), or a logic probe.

4. Use a pulse generator or time-mark generator as a signal source.

PROTOTYPING

BOARD SELECTION

Many materials are available on which to build prototype or modification circuits. The material chosen will depend on the required life and use of the circuit. For laboratory use or other short-term applications where a permanent circuit is not needed and operating fre-

quency is below 10 MHz, solderless prototyping boards such as those available from Global Specialties Corporation (see Figure 1-1), E & L Instruments, Inc., and others are appropriate. Individual sections snap together to form as large an area as is needed. Socket sections are used to mount integrated circuits (ICs) and other components, as shown in Figure 1-4. As shown in Figure 1-1a, the four holes that are laterally next to each hole into which each pin of the IC would be inserted, are connected electrically to the pin. This scheme allows four wires or component leads to be connected to each pin of the IC. Bus strip sections for power and ground often are supplied as separate sections (see Figure 1-1b). In the bus strips, two independent conductors run the length of the strip. These sections may be interspersed between socket sections as needed. For example, two bus strips could be put in to give ground, +5, −12, and

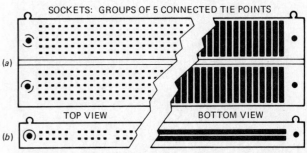

FIGURE 1-1 *(a)* Solderless proto board IC socket strip section showing internal construction. *(b)* Solderless proto board bus strip section showing internal construction. *(Courtesy Global Specialties Corporation.)*

1

+12 V. Since these boards are easily reusable, their long-term cost is low.

For more permanent work, such as a part of a system prototype or a field modification of equipment, universal printed circuit boards such as those available from Douglas Electronics or Vector Electronic Company, Inc. (Figure 1-2) are more applicable. These boards come in many styles and sizes. They are predrilled to accommodate standard components (such as ICs or transistors) or sockets, and many have standard edge connectors. Power and ground buses are supplied on the boards. Connections are made either by soldered jumper wires or by installing wire wrap sockets and using wire wrap jumpers. Wire wrap jumpers are connections made by tightly wrapping the stripped end of the wire around a pin on the wire wrap socket with a wire wrap tool or gun (see Figure 1-3a). If properly done, as shown in Figure 1-3b, these connections are as dependable as soldered connections.

(a)

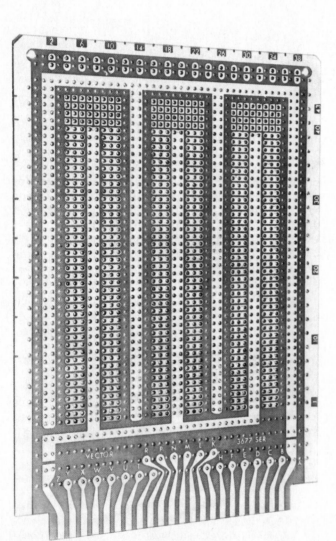

FIGURE 1-2 A universal PC board. *(Courtesy Vector Electronics Company, Inc.)*

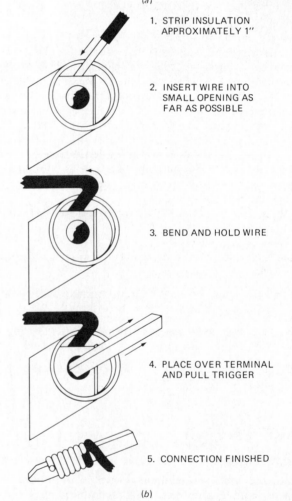

1. STRIP INSULATION APPROXIMATELY 1″

2. INSERT WIRE INTO SMALL OPENING AS FAR AS POSSIBLE

3. BEND AND HOLD WIRE

4. PLACE OVER TERMINAL AND PULL TRIGGER

5. CONNECTION FINISHED

(b)

FIGURE 1-3 *(a)* A wire wrap gun is used to make jumpered connections. *(b)* How to make wire wrap connections. *(OK Machine and Tool Corporation.)*

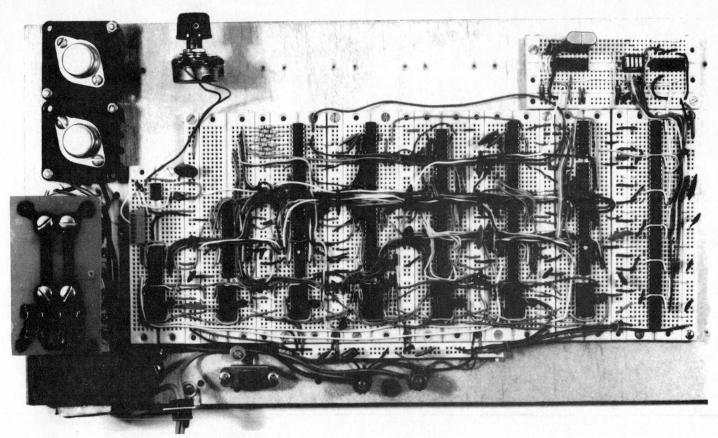

FIGURE 1-4 Solderless proto board construction of an MSI frequency counter.

Once the material on which to build the circuit is chosen, study the schematic to determine the number of ICs and other parts. From this you can determine the required area for the board. Leave ⅛ to ½ in [3.2 to 12.7 mm] between ICs for wiring, so that the ICs can be easily removed if necessary. A standard 14- or 16-pin IC usually requires about 1 in² [6.5 cm²] of board space. Also remember to leave extra space so that one or two ICs may be added to revise the circuit later without redoing the entire board. Then the appropriate universal PC (printed circuit) board can be acquired or the necessary prototyping board assembled. When you are assembling a solderless prototyping board, it is a good idea to screw the sections on a piece of plywood or rigid plastic. This gives the unit some structural integrity and prevents breakage of the locking tabs on the sections. Terminal binding posts can be added to the board for power and ground connections (see Figure 1-4).

CIRCUIT LAYOUT

Careful planning of the circuit layout simplifies wiring, minimizes errors, and makes troubleshooting easier. Overall circuit layout should parallel, as much as possible, the layout on the schematic, assuming the schematic is orderly! If it does not, try to arrange the circuit for a logical signal flow. This helps anyone looking at the board to easily find sections of the circuit and trace signals through it. Also, if a printed circuit board is to be laid out for the circuit later, much of the preliminary work will have been done already.

Wherever possible, all ICs should be pointed in the same direction to reduce the chance of one being put in backward. This also makes it easier to keep track of pin numbers when you are wiring or troubleshooting.

WIRING THE CIRCUIT AND EFFECTIVE BYPASSING

COLOR-CODED WIRING   An easy way to reduce wiring errors and aid in troubleshooting a circuit is to color-code all wiring. Multipair communications cable, such as that used by telephone companies, can be stripped of its outer jacket to yield a large variety of color combinations. For solderless prototyping board, however, make sure the wires are not stranded and are 22 or 24 gage. Larger wire or leads stretch the internal contacts, and smaller wire may not always make dependable contact. Strip off about ¼ in [6.4 mm] of the insulation for solderless prototyping board connections. For wire wrap construction, a special 30-gage wire is used.

If possible, build and test one section of a circuit at a time to simplify debugging. Keep resistor and other component leads at a minimum to prevent accidental short-circuiting. Keep connecting wires short and route them around ICs and sockets so that a defective IC can be replaced without removing and, often incorrectly, replacing the wires. Capacitors and other components should be mounted so the values are easily readable. Resistors should be mounted with their color codes reading from left to right or from top to bottom so that it is easier to spot errors (such as inserting a 15-Ω resistor instead of a 1-MΩ resistor).

As each connecting wire or component is added, trace the corresponding section on the circuit schematic, using a yellow or other color *transparent* marking pen to keep track of what has been completed. The schematic is still visible and shows you where to start when you come back after a weekend. When a section has been completely wired, check each IC pin by pin, from pin 1 to the highest number, and then from the highest number pin back to pin 1, making sure every pin is accounted for and connected to the proper point or points. While this may seem laborious, the time spent in careful construction and checking is well repaid by having more circuits work the first time and by having fewer blown ICs.

TREE WIRING   When you are working on solderless prototyping board, run a separate wire from the terminal binding posts, where power enters the board, to each of the individual power bus strips. Also, run separate ground connections to each ground bus strip from the ground binding post (see Figure 1-4). With the addition of proper bypass capacitors, this configuration decouples or isolates the power-supply buses from one another so that voltage transients or noise generated in one section is prevented from affecting others. This multiple wiring also ensures that all sections get the same supply voltage. A single wire jumping from bus strip to bus strip ("daisy chaining") will give a lower voltage to the section at the end of the line. For small circuits this probably won't matter. For larger circuits it will.

DECOUPLING AND BYPASSING   Decoupled wiring is a good habit to develop. A 25- to 100-μF electrolytic capacitor and a 0.1μF capacitor should be connected between the power binding post and the ground binding post, or between the power and ground printed circuit traces where they enter the board, to filter out voltage transients. The obvious question is, Why bother to put a 0.1-μF capacitor in parallel with one which is 250 to 1000 times larger? The answer is that every capacitor, because of its leads and internal construction, has some series inductance. This series inductance together with the capacitance forms a series-resonant circuit at some frequency. Basic ac (alternating current) theory shows that below this resonant frequency the net reactance is capacitive, decreasing to a minimum at resonance. Above the resonant frequency the net reactance is inductive, increasing as the frequency rises. Because of the increasing inductive reactance above the resonant frequency, the capacitor becomes less and less effective

at short-circuiting or filtering out high-frequency voltage transients produced by logic gates changing states. The resonant frequency of a 100-μF capacitor with 1 in of lead on each end is about 80 kHz, assuming 20×10^{-9} H of inductance per inch of wire and $f_r = 1/(2\pi\sqrt{LC})$. At frequencies much above this, the 100-μF capacitor is not an effective bypass filter. Although not an effective bypass at low frequencies, a parallel 0.1-μF capacitor with 1-in [2.5-cm] lead length extends the bypassing to 2.5 MHz. A combination of the two capacitors thus provides filtering over a wide range of frequencies.

To minimize the inductance of bypass capacitors, and thereby improve their performance, the leads should be kept short. For example, reducing the lead length of the 0.1-μF capacitor to ¼ in [6.4 mm] on each end extends the effective bypassing from 2.5 to 5 MHz. For TTL (transistor-transistor logic) and CMOS circuits, inexpensive 0.1 or 0.01-μF bypass capacitors should be placed between power and ground leads next to each IC or at least next to every other IC. For higher-frequency bypassing, ECL (emitter-coupled logic) or Schottky TTL circuits need some 100- to 1000-pF capacitors in addition to the 0.1 and 0.01-μF capacitors.

INITIAL TESTING

After carefully wiring the circuit, checking component polarities, etc., use an ohmmeter to check that there are no shorts between power supplies and ground. Then apply power and look and listen for violent signs of circuit discontent, such as smoke, pops, or fizzes. If you see or hear any of these, quickly turn off the power and do a quick check for excessively hot components by touching them lightly with your finger. Care is required because many ICs and electrolytic capacitors, when inserted backward, quickly get hot enough to give a nasty burn! Also look for miswiring, incorrect resistor values, and solder bridges which often cause noise or hot components.

A hot electrolytic capacitor is either internally short-circuited or, more likely, placed in the circuit with the polarity reversed. Some of the "epoxy blob" tantalum electrolytic capacitors require close examination to find the polarity markings. Also, the markings are occasionally wrong. Once it has been reverse-polarized, an electrolytic capacitor usually is destroyed and should be replaced. Large electrolytics, when reverse-polarized, can actually explode with a sound equal to that of a firecracker. If you are applying power for the first time to a circuit containing large electrolytics, you should use a 120-V autotransformer or Variac to slowly bring up the circuit to full supply voltage. This allows the dielectric in the capacitors to build up and gives you time to find any hot capacitors before they build up enough internal pressure to pop.

The most likely causes of a hot IC are backward placement in the socket or reversed supply polarity as a result of a wiring error. Another cause could be an output short-circuited to V_{cc} or ground. In some logic families, connecting the outputs of two ICs will cause them to get hot if one output is trying to go high while the other is trying to go low. Reversed ICs should be replaced.

If visual inspection does not reveal an apparent initial problem, an ohmmeter can be used to find possible short-circuits or components. When such obvious problems are eliminated, power up again and check with a DVM to verify that all power-supply voltages are present. Short-circuits causing loading down of the power supply can be found quickly with a current-tracing probe such as the Hewlett Packard 547A (see Figure 1-5) or the General Radio Bughound. Since the current probes respond only to changing current, the power-supply connection is broken, and the power-supply input to the circuit is pulsed with a low-impedance pulse source such as the Hewlett Packard 546A logic pulser. The brightness of the 547A current tracing probe tip light indicates the amount of current flowing. Move the tip of the probe along a power-supply wire or trace that shows the highest current. When the short-circuited point is passed, the light will go out.

If you still have a hot IC and there seems to be no wiring error, short circuits, or incorrect supply voltages, *turn off the power* and replace the IC. When you are removing ICs from sockets or prototyping boards, use a small screwdriver (or IC remover) to lift one end and then the other until the IC pops out. If you try to pull out ICs with your fingers, you are liable to end up with a perforated thumb and bent IC leads. Continued overheating, when power is reapplied, may mean a missed wiring error or an error on the schematic used to build the circuit. Get a data sheet for the IC involved, and check that the pin numbers on the schematic are correct. When you work with TTL, it is very easy to get used to wiring ground to pin 7 or 8 and V_{cc} to pin 14 or 16. However, a common TTL latch, the 7475, has V_{cc} on pin 5 and ground on pin 12. The "normal" supply pins, 16

and 8, are both outputs. Watch for this kind of error, and be aware that designers sometimes change an IC in a design, but forget to change the pin numbers on the schematic to correspond.

FUNCTION TESTING AND TROUBLESHOOTING

When the circuit passes the smoke and heat tests, consider which tests to perform to determine whether the circuit is operating properly. What signals must be applied? What outputs are expected? How can these inputs be produced? What instruments are needed to measure these inputs and outputs? Without carefully planned tests and competent use of these instruments, you can waste a tremendous amount of time in poking, probing, and general witch-hunting. Try to approach problems systematically. The rest of this chapter introduces the operation of common test equipment. Later chapters discuss more complex test equipment as the circuits requiring them are introduced.

To start, let us assume that you are testing the sample conditional circuit shown in Figure 1-6a. If you remember that the gates shown are NAND gates, and that in

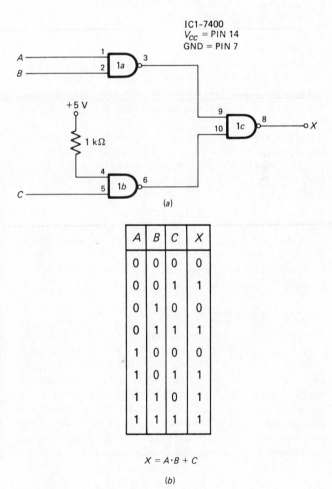

A	B	C	X
0	0	0	0
0	0	1	1
0	1	0	0
0	1	1	1
1	0	0	0
1	0	1	1
1	1	0	1
1	1	1	1

$$X = A \cdot B + C$$

(b)

FIGURE 1-6 (a) Digital circuit. (b) Truth table.

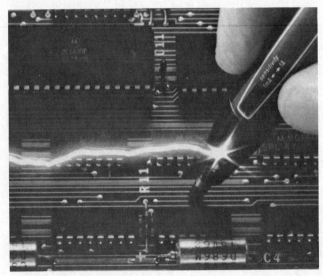

FIGURE 1-5 Current tracing probe HP547A. *(Hewlett-Packard Company.)*

positive logic the output of a NAND gate is high when any input is low, then you can construct a truth table for the circuit, as shown in Figure 1-6b. (You may recognize that the output expression for this circuit is $X = A \cdot B + C$.) If this is a discrete circuit, you can put jumper wires from input to ground or V_{cc} to establish the input conditions. If the circuit is part of a larger circuit, you will have to determine the input conditions by using an oscilloscope or a logic probe. Then the actual output can be compared with the output predicted from the truth table.

Logic voltage levels should be checked directly on the IC pins. Try to set up more than one input combination from the truth table and then check both the signal-source pin and the signal-destination pin for proper logic levels. If the output is not as predicted, do not automatically assume the IC chip is bad. For example, in the circuit shown in Figure 1-6a, check the signal at pin 3 and pin 9 of IC1 to see that both are at the same level. Likewise, check pins 6 and 10. When you have eliminated cold-solder joints, solder bridges, and miswiring, and the circuit still does not operate as predicted, *then* try replacing the IC chip and recheck levels.

DIGITAL VOLTMETERS AND LOGIC PROBES

DVMS

Since the input and output signals for the sample circuit in Figure 1-6a are just dc (direct current) levels, there are several instruments you can use to check them, such as digital voltmeters, logic level probes, and level testers.

A digital voltmeter will tell you very accurately whether an output is a legal high or a legal low. (As shown in Chap. 2, the actual numerical value varies with the logic family.) Many DVM probes, however, are not physically suited to probing on ICs. While you are looking at the readout to get a value, the probe may slip and short-circuit adjacent IC pins. Also the continuous looking back and forth from circuit to readout is tiring. Some DVM manufacturers have overcome these problems by making small, hand-held units with the readout in the probe.

LOGIC PROBES

For quick determination of logic levels when the precision of the DVM is not required, you can use a *logic probe* such as the HP545A or the Continental Specialties Corporation LP-1 (see Figure 1-7). These probes have one LED (light-emitting diode) that comes on if the input is high and another that comes on if the input is low. Voltage comparison levels in the probe can be adjusted to correspond to those of common logic families, and most logic probes also have provisions to indicate the presence of a train of pulses or a single pulse. The logic probe concept has been extended to multipin, clip-on level testers with one LED for each pin. These work well if they are externally powered. If they take power from the IC being tested, however, the 200-mA current drain required to light as many as 15 LEDs may disrupt the circuit. Also, some pins on chips such as the 74121 monostable multivibrator will be loaded down by these testers. For these pins an oscilloscope should be used.

OSCILLOSCOPES

An oscilloscope also can be used to check the logic levels in a simple conditional circuit as above. However, its real strength lies in in-depth analysis of dynamic logic circuits where you need to measure pulse width, estimate frequencies, detect brief unwanted states, or compare the timing of two or more signals. Since an oscilloscope often is your main test instrument, you should take the time to learn the functions of all the controls on the particular oscilloscope you will be using. This enables you to use the full power designed into the instrument. Some timing problems, for example, may be made visible only by the use of delayed sweep. In this chapter and throughout the book, general use of this and many other oscilloscope functions is discussed. Consult the manufacturer's manual for operational details of your individual oscilloscope.

GETTING A TRACE ON A TRIGGERED SWEEP OSCILLOSCOPE

When you first turn on a triggered sweep oscilloscope, it is quite likely that no trace will appear on the screen. To get a visible trace, turn up the intensity, set the triggering on automatic, and adjust the horizontal and vertical position controls to the centers of their ranges. If the oscilloscope has several sweep modes (delayed, delayed triggered, etc.), select main sweep and then adjust the trigger level or trigger stability control until a display is obtained. If a trace is still not visible, press the beam-finder button. This control compresses the range of the vertical and horizontal position controls to make a trace visible anywhere in their range. Once you find the trace, the horizontal and vertical position controls can be used to center it. If your oscilloscope does not have a beam finder, you will have to find the trace by slowly rotating the vertical and horizontal position controls.

For best display on the screen, reduce the intensity to a point where the trace is clearly visible, but well below the point where the beam is burning a hole in the phosphor coating on the screen. Lower intensity allows you to focus the display to a finer and more accurate trace and prolongs the life of the CRT (cathode ray tube). An astigmatism control, if present, is adjusted until the vertical and horizontal portions of a display pulse are both brought into focus at the same time by the focus control. If a horizontal trace is not parallel to the horizontal axis on the screen, find the trace rotation control and correct the problem. The trace rotation control may be on the rear of the oscilloscope.

FIGURE 1-7 Logic probe HP545A. Note also the glomper clip on an IC.
(Hewlett-Packard Company.)

ALTERNATE OR CHOPPED SWEEP

After a stable trace has been obtained, the vertical volts/division and horizontal time/division settings appropriate for the signal to be displayed can be selected. *Always make sure the variable volts/division and time/division controls are in the calibrated position.* If you wish to display two signals on the screen with a single beam, you must select either chopped or alternate display mode.

In the alternate mode, one sweep of the beam across the screen displays the channel 1 signal; the next sweep displays the channel 2 signal. A dc voltage, added to each of the displays by adjustment of the vertical position controls, separates the waveforms from one another on the screen. For viewing high-frequency signals where sweep speeds are above about 1 ms/division, your eye will not detect the alternations, and the two displays will appear to be present at the same time. For viewing low-frequency signals, which require slow sweep speeds, the alternate mode will produce a blinking display which is very difficult to analyze.

A stable display of two low-frequency signals requires the chopped mode in which the beam rapidly switches, or chops, back and forth between the two channels during each sweep across the screen. The beam traces first a segment of the channel 1 waveform, then a segment of the channel 2 waveform, then another segment of channel 1, and so on across the screen. Since this chopping is faster than your eye can detect (50 to 100 kHz), the displays again appear to be continuous and present at the same time. The chopped mode cannot be used for high-frequency signals because chopping the waveform may adversely affect the apparent switching time of observed waveforms. For example, a signal on channel 1 may change levels while the beam is displaying channel 2. The display will not reflect this change, however, until the beam swings back to display channel 1. You will have no way of knowing precisely when the level change occurred.

If your oscilloscope is a multicompartment type, then a similar control near the time/division switch will select chopped or alternate displays between the left and right amplifier compartments. The above comments concerning the choice of chopped or alternate displays also apply to this selection.

DC OR AC COUPLING

Direct-current coupling connects the input directly to the vertical amplifier so that any input voltage is amplified and produces a display deflection. If the oscilloscope is set with a vertical scale of 2 V/division and 5 V dc is applied to the input, the trace will move up 2½ divisions [5 V × 1 division/(2 V)] from its no-signal position. If the polarity of the input is reversed, the trace will move down 2½ divisions from its no-signal position. In this coupling mode then, the oscilloscope can be used as a

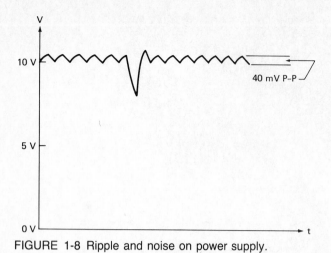

FIGURE 1-8 Ripple and noise on power supply.

crude (3 to 5 percent accuracy) dc voltmeter. This is sufficient for determining whether voltage supplies are present and whether a signal line is at a legal high or a legal low level. For example, a gate that is loaded too heavily may still show a pulse on its output even though the pulse may not be reaching a legal low or high level. With dc coupling, the oscilloscope will quickly show this problem. A ground position usually is given on the coupling selector to aid in calibration.

When the input to the scope is switched to ac coupling, a capacitor is inserted in series with the signal lead. This capacitor blocks any dc component of the incoming signal and allows only the ac or changing portion of the signal to be displayed. A useful place to use ac coupling would be when you are looking for noise spikes or ripple on a power supply, as in Figure 1-8. The power supply is at 10 V dc with a 40-mV ac ripple signal superimposed on the dc. If dc coupling is used, you cannot expand the display enough to study the 40-mV ac signal without the 10-V dc causing the display to go off the screen. The ac coupling blocks the dc, so that the 40-mV signal appears centered on zero on the screen. The display of the 40-mV signal can then be expanded to fill the whole screen for detailed analysis.

PROBES AND PROBE COMPENSATION

To get the best possible signal to the oscilloscope and least load the circuit under test, probes are usually used. The most common are the ×1, ×10, and ×100 passive voltage probes. The ×10, for example, indicates that the signal appearing at the input of the oscilloscope is one-tenth as large as the signal appearing at the probe tip. Oscilloscope probes are considerably more than a piece of wire with a fancy handle, and they must be adjusted to match the individual input on which they are being used to get the best results.

The input impedance of a typical mid-frequency oscilloscope appears as a parallel combination of a 1-MΩ resistor and a capacitor of 10 to 50 pF connected to

ground (see Figure 1-9a). If you connect the input of the oscilloscope directly to a logic circuit, the 1-MΩ resistor probably will not cause any problem, but the 10 to 50 pF of added capacitance may load the circuit under test enough to change its timing and cause a malfunction. A solution to this problem is to use a ×10 or ×100 probe. Refer to Figure 1-9b. The 9-MΩ resistor and the 1-MΩ input impedance of the oscilloscope form a 9:1 or ×10 voltage divider at dc. As the frequency increases, however, the input capacitance of the oscilloscope shunts the 1-MΩ resistor and decreases the amplitude of the signal at the input. Without $C1$, the bandwidth of the probe is only a few kilohertz. If $C1$ is adjusted so that the time constants of the two networks are equal ($R1C1 = R2C2$), the input to the scope will be one-tenth of the input to the probe tip for all frequencies. The total input impedance of the probe will go down from its dc value of 10 MΩ as the frequency increases, but the maximum capacitive load placed on the circuit by the probe will always be less than 10 pF, and usually less than 5 pF.

Most oscilloscope manufacturers have made it easy for you to compensate the probe to obtain the best frequency response, or, in other words, adjust $R1C1 = R2C2$. Connect the probe output to the channel of the oscilloscope where it will be used, and the probe tip to the calibrator output of the oscilloscope. Set the vertical volts/division and the horizontal time/division controls to show one cycle of the calibrator waveform filling most of the screen. Then by either turning a trimmer capacitor or, on some Tektronix probes, loosening the locking nut on the probe and rotating the probe barrel, adjust for the best square corners on the displayed calibrator waveform. Too little $C1$ will give rounded corners on the waveform; too much $C1$ will give overshoot or peaked corners (see Figure 1-9c and d). Remember, the probe compensation must be adjusted with the probe connected to the oscilloscope input on which it will be used. If a probe connected to the calibrator waveform gives a small amplitude display such as that in Figure 1-9f, very likely $R1$ is open in the probe or the center conductor of the probe cable is broken. The differentiated display is produced by capacitive coupling of the signal through $C1$ or through the capacitance formed by the ends of the broken center conductor.

The amplitude accuracy of the probe and vertical amplifier also should be checked by using the calibrator output. When you do this, make sure the variable volts/division control is in the calibrated position. If the display amplitude is appreciably in error, consult the oscilloscope manual for proper adjustment sequence.

Since it is difficult to connect oscilloscope probes to an IC chip without short-circuiting adjacent pins, an IC clip or "glomper" clip is very helpful (see Figure 1-7). These clip onto an IC like an old-fashioned spring clothespin and have short leads extending from each IC pin to which an oscilloscope probe can be connected. You can reduce the chance of short-circuiting adjacent test clip pins with the probe tip by bending every other pin on the test clip down about 45°.

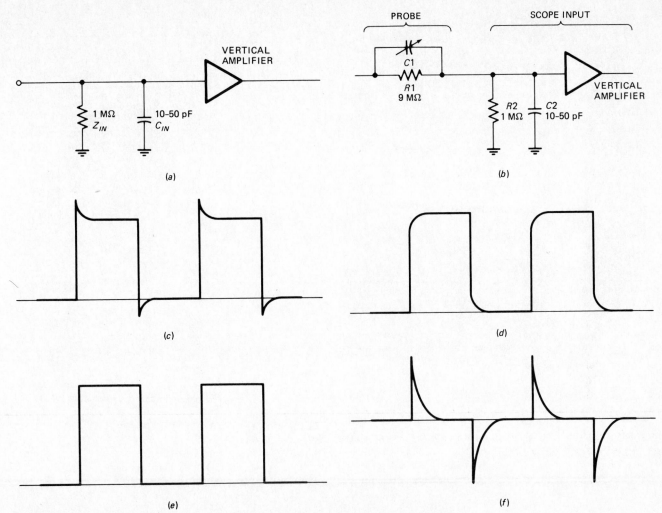

FIGURE 1-9 *(a)* Equivalent input impedance of midfrequency oscilloscope. *(b)* Equivalent circuit for ×10 oscilloscope probe and oscilloscope input. *(c)* Oscilloscope waveform with too much *C*1 in probe adjustment. *(d)* Oscilloscope waveform with too little *C*1 in probe adjustment. *(e)* Waveform with correct probe adjustment. *(f)* Possible oscilloscope waveform with center wire of oscilloscope probe broken, or internal resistor open.

TRIGGER CONTROLS

In the preceding discussion of oscilloscope use, the triggering and sweep controls are described only briefly. For many applications only a few of these controls need be used. However, knowledge and proper use of all these controls make many elusive problems visible. Systematic signal tracing is usually more effective than randomly replacing ICs until, by chance, a defective one is found. The following discussion will, where a specific example is necessary, refer to a common laboratory oscilloscope plug-in, the Tektronix 5B42 delaying time base (see Figure 1-10). If you don't have one of these units, it should be relatively easy to find the corresponding controls on the oscilloscope or time base plug-in you do have.

TRIGGER SOURCE First consider the source of triggering. For most applications you will use the internal trigger source. This means that a small portion of the signal from the vertical amplifier is "picked off," amplified, and shaped to produce the trigger that starts the sweep of the beam across the CRT. On the 5B42 this is done by pushing in either the LEFT or RIGHT button in the trigger source column to determine which vertical amplifier compartment supplies the trigger. When you are trying to display signals of different frequencies such as counter outputs, it is usually best to select the lowest frequency signal as the source of the trigger. This gives a stable display and ensures that the phase relationship of the displayed waveform is correct.

When you want to look at a small signal that has a

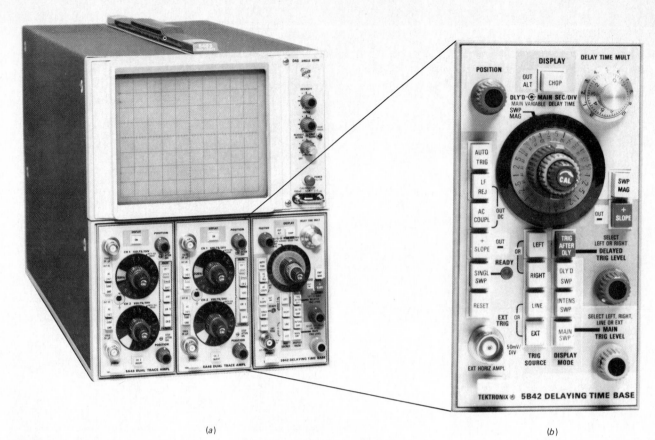

FIGURE 1-10 *(a)* Tektronix 5440 mainframe. *(b)* Tektronix 5B42 delaying time base.
(Courtesy of Tektronix, Inc.)

frequency related to the ac power-line frequency, such as the ripple on a dc power supply, the line source often will give a stable display. Line source picks off 60 Hz from the oscilloscope power transformer and uses this to produce the oscilloscope trigger.

External trigger source is used when you want to start the sweep at a time determined by an external condition rather than by a transition of the signals being displayed. For example, if you have a three-output circuit and you want to compare the *A* and *B* outputs only when the *C* output makes a low-to-high transition, connect the *A* and *B* signals to the vertical amplifiers and the *C* signal to the external trigger input. Select external trigger and set the trigger slope selector to +. The beam will be triggered to start a display when the *C* output makes a low-to-high change, and the CRT will display those states of *A* and *B* present after that transition. Usually at least a 100-mV signal is required for the external trigger input. For logic signals where the amplitude is large, a ×10 probe on this input will minimize capacitive loading on the gate output. For more extensive circuits you can get an external trigger from a device called a *word recognizer*, which produces a trigger only when up to 32 inputs match switch-selected low or high states.

TRIGGER INPUT COUPLING   The common types of trigger input coupling are dc, ac, ac low-frequency reject, and ac high-frequency reject. The names are almost self-explanatory. Specifically dc coupling is used for triggering on very low frequencies or slowly changing signals such as ramps. In spite of its name, dc coupling can be used up to the full bandwidth of the oscilloscope. And ac coupling is for triggering on signals above 50 Hz and removing the effect of any dc components on triggering, such as when you are looking at a ripple on a power supply. If, however, unwanted 60-Hz noise or hum is present on a triggering signal, use the ac low-frequency reject coupling. This coupling attenuates any signal components below a few kilohertz. If high-frequency noise is producing difficult triggering, then the ac high-frequency reject can be used to attenuate signal components above about 50 kHz. These filters affect only the trigger coupling; they do not affect the CRT display.

TRIGGER SLOPE AND LEVEL CONTROLS   Trigger slope and trigger level controls are often confused, but really they are quite simple. Selecting a + on the slope control establishes that the beam will be triggering on a low-to-high transition, and selecting − on the slope control starts the sweep on a high-to-low transition. The level control determines at which voltage during the transition triggering will take place. The two controls are independent. It is quite possible to trigger on a positive slope and a negative level, or vice versa. You can demonstrate this by displaying a 1-kHz sine wave so that it vertically fills about three-quarters of the screen. Center the level control, select + slope, and set for internal trigger source. The display should show the positive-going portion of the sine wave at the left edge of the screen, starting at about the zero crossing point. As you rotate the level control, the starting point (or trigger point) for the display should be movable from the negative peak of the sine wave up to the positive peak. You can switch to − slope and see the display start with a negative-going portion of the sine wave and observe the effect of the level control of this display. When you use an external trigger, the slope switch works as before, and the level control will adjust the trigger point over a range of at least −1.5 to +1.5 V. Many better-quality oscilloscopes have an internal delay line in series with each vertical amplifier. This delays the signal long enough for the sweep to get started so that you can see displayed the leading edge of the pulse on which you triggered.

TRIGGERING MODES   In the automatic triggering mode, the sweep generator free-runs so that a trace is always visible on the screen even with no input signal to the oscilloscope. When a signal is applied, this signal together with the level or stability control establishes triggering. Automatic triggering is convenient for checking dc levels and simple signals because it gives you a ground reference trace and does not require you to readjust the trigger level. However, for complex waveforms and higher frequencies, the normal or nonautomatic triggering mode gives best results. A high-frequency sync mode, if present, is used to trigger on signals too small or too high in frequency for the normal triggering circuits. In this mode a stable display is obtained by adjusting the free-running frequency of the sweep with the level control to have an integral, submultiple frequency of the displayed frequency. This is similar to the mode used to synchronize old-fashioned, free-running oscilloscopes.

The single-sweep mode is used to photograph events that occur only once or at very low rates. When single sweep is selected, a reset will light the ready light. The next trigger pulse starts a single sweep of the beam, the ready light goes out, and all further pulses are ignored unless the sweep is reset.

Some oscilloscopes have a *trigger holdoff* control. This is used to get a stable display of waveforms that do not repeat symmetrically when the trace reaches the right side of the screen. Normally, when the oscilloscope completes its sweep across the screen, it will trigger on the next active edge of the input signal to start the next sweep. The trigger holdoff control can be adjusted to make the trigger recognition circuit wait until the desired active edge of the signal appears. Then each sweep of the beam will display the same section of the input asymmetrical waveform.

SWEEP MODES

For most oscilloscope applications the main or normal sweep is sufficient. However, some specific examples show the importance of magnified, delayed, and triggered delayed sweep modes. Figure 1-11a shows a simple control clock waveform for a digital instrument. (To get a stable display of an irregular waveform such as this, sometimes it is necessary to adjust the trigger holdoff control or varible time/division.) The horizontal scale is 5 ms/division, and the section from A to A′ represents one control cycle (pattern repeats at A′).

Superficially, the observed waveform looks exactly like the waveform shown in the instrument's service manual. The symptom shown by the instrument, however, is that an operation which is supposed to be started by the low-to-high transition after point B actually is started at time B. This leads to the suspicion that a very narrow unwanted pulse or "glitch" may be present at point B. But if the glitch is only 10 or 15 ns wide, it will not be visible with the 5 ms/division sweep setting. Using a faster sweep speed such as 2 ms/division will move point B off the screen, where it is not visible.

MAGNIFIED SWEEP   Another attempt to expand the region of interest is to use the sweep magnifier. This expands the trace in the horizontal direction by a factor of 5 or 10. The horizontal position control then can move the trace across the screen until the region of interest is displayed. For the above problem, however, the magnification is not nearly enough to display the suspected tiny glitch. Delayed sweep can probably expose the problem.

DELAYED SWEEP   First, observe that there are two time/division controls, two slope controls, and two level controls. One set of controls is for the main sweep, and the other set is for the delayed sweep. (Refer to Figure 1-10 if you don't have a suitable oscilloscope available.) A stable display as in Figure 1-11a is obtained by using the main sweep controls. The delay time multiplier is set to the center of its range, and the delayed time/division selector is set to a sweep faster than on the main sweep, 1 μs/division, for example. When the intensified sweep is engaged, a section of the displayed trace appears brighter or intensified. The width of this intensified region corresponds to the delayed sweep, time/division setting. You can rotate the delayed sweep, time/division control and observe the width of the intensified zone change. The position of the intensified portion along the trace is controlled by turning the delay time multiplier control. When the delayed sweep button is depressed, the intensified portion expands to fill the entire width of the screen. Some oscilloscopes allow both the main

PROTOTYPING AND THE USE OF TEST EQUIPMENT

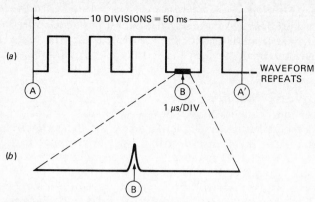

FIGURE 1-11 *(a)* Control clock oscilloscope waveform. *(b)* Glitch at point *B* shown by use of delayed sweep.

FIGURE 1-12 Video waveform.

sweep view and the expanded or delayed sweep view to be seen at the same time. Since the two time bases are independent, you can set the delayed time base to faster and faster sweeps and thus get greater magnification of the region of interest. The intensified section is moved across the *B* region of the trace by the delay time multiplier until the presence or absence of a glitch is obvious (see Figure 1-11*b*).

TRIGGERED DELAYED SWEEP A further enhancement of delayed sweep is *triggered* delayed sweep. In this mode, the delayed sweep is triggered at a level transition selected in the main sweep display. With delayed sweep only, the delay time multiplier moves the intensified portion smoothly across the entire screen. With triggered delayed sweep, the intensifier or delayed segment can only start a + or − edge or transition selected by the slope selector switch of the delayed sweep time base. If a positive slope is selected, the intensified section jumps from one positive edge to the next positive edge of the displayed waveform as the delayed time multiplier is rotated.

When you examine a waveform such as that shown in Figure 1-12, triggered delayed sweep often gives less display jitter when large magnification is required because the sweep is triggered by a specific transition rather than just being triggered by the end of the delay interval.

EXTERNAL HORIZONTAL AMPLIFIER
X-Y SWEEP MODE

For some applications, such as determining the transfer function of a logic gate, you need to have the horizontal axis of the oscilloscope display represent voltage rather than time. This is done by rotating the time/division control to the amplifier position or otherwise selecting the external horizontal amplifier input. This is sometimes referred to as the *X-Y mode* because the vertical axis now corresponds to the *y* axis and the horizontal axis corresponds to the *x* axis in standard graphing terminology. The horizontal position of the display on the

screen is proportional to the amplitude of dc or ac signals applied to the external horizontal input, just as signals applied to the vertical input affect the vertical display. A positive input voltage to the horizontal amplifier produces deflection to the right as you face the CRT. Some oscilloscopes have variable horizontal gain, but many have a fixed horizontal deflection of 50 mV/division.

You may have used the *X-Y* mode in an ac theory laboratory for making Lissajous figures to demonstrate sine wave phase relationships. Here we show how to use this mode for analyzing a logic gate transfer curve, which is discussed further in Chap. 2. Figure 1-13*a* shows the setup for displaying the output voltage versus an applied input voltage for a TTL inverter such as a 7404. The output of the inverter is connected to the vertical input, and the vertical deflection is set to 0.5 or 1 V/division. The inverter input sweep signal is connected to the external horizontal input of the oscilloscope with a ×10 probe. If the oscilloscope has a fixed horizontal deflection of 50 mV/division, then the probe and the oscilloscope together give an equivalent horizontal deflection of 500 mV/division. Since the input voltage range of interest is 0 to +5 V (5000 mV), the 10 horizontal divisions of 500 mV/division will provide a good scale on which to display the input. If power and signal are applied to the inverter, the display on the oscilloscope should resemble Figure 1-13*b*, which is called the *input-to-output transfer curve* of the gate. For the input sine wave at zero, the output of the inverter is high so the display is at point *A*. As the sine wave increases in the positive direction, the trace moves to the right. When the input voltage passes a threshold voltage of about 1.3 V, the output of the inverter begins to drop and reaches a minimum as the trace moves to point *B*. As the sine wave input swings from its positive peak back to zero, the curve is retraced. The zener diode protects the input of the gate from being damaged, if the output of the generator is accidentally set too high, by limiting the input to the gate to 5 V. The negative half-cycles of the sine wave have no effect because the zener diode prevents the input of the gate from going any more negative than about 0.7 V, but the positive half-cycles will keep tracing out the curve to give a continuous display. If the display is calibrated and started at the left edge of the screen, the threshold of crossover voltage can be determined from the screen. A point to remember when you use the external horizontal input is that its bandwidth is usually only 1 or 2 MHz. Many scopes having two vertical input channels allow the channel 2 vertical input to be used as the input for the horizontal component of an *X-Y* display. This is advantageous because then both the vertical and horizontal of the display are calibrated and controlled by the volts/division controls.

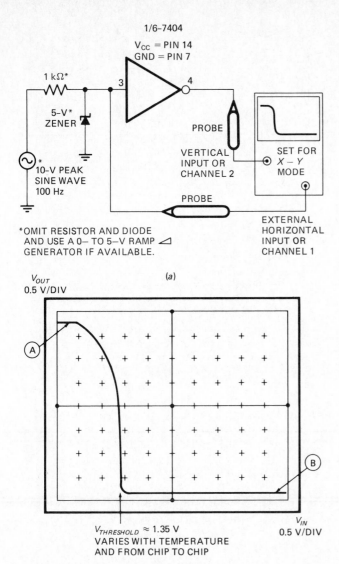

1/6-7404

V_{CC} = PIN 14
GND = PIN 7

1 kΩ*

5-V*
ZENER

*
10-V PEAK
SINE WAVE
100 Hz

3

4

PROBE

VERTICAL
INPUT OR
CHANNEL 2

SET FOR
$X - Y$
MODE

PROBE

EXTERNAL
HORIZONTAL
INPUT OR
CHANNEL 1

*OMIT RESISTOR AND DIODE
AND USE A 0– TO 5–V RAMP
GENERATOR IF AVAILABLE.

(a)

V_{OUT}
0.5 V/DIV

A

B

$V_{THRESHOLD}$ ≈ 1.35 V
VARIES WITH TEMPERATURE
AND FROM CHIP TO CHIP

V_{IN}
0.5 V/DIV

(b)

FIGURE 1-13 (a) Setup for showing transfer curve of a logic gate. (b) Transfer curve of a TTL 7404 inverter.

z AXIS

The z axis refers to modulating the intensity of the trace. The external intensity or z axis input is usually on the rear panel of the oscilloscope. For a typical Tektronix oscilloscope, a negative 5-V signal applied to this input will turn off the beam if the front panel intensity control is set to show a trace. A positive 5-V input will turn on the beam if the intensity control is set below the point where a trace would be visible. This feature can be used to intensify portions of a trace for emphasis similar to the way in which the intensified sweep mode internally brightens a trace. Another application involves applying synchronized vertical and horizontal sweeps to the vertical and horizontal inputs to produce a raster display, as is done in a television set. Video information applied to the z axis input modulates the intensity of the beam as it sweeps, to produce a picture or graphics display.

STORAGE OSCILLOSCOPES

In our discussion of oscilloscope use, we referred to a standard dual-trace, triggered oscilloscope. For many applications this type is adequate. When viewing very low frequency or single-shot signals with this type of oscilloscope, however, you see a bright dot moving slowly across the screen to trace the waveform rather than seeing the whole waveform at once as you do for higher frequencies. The moving-dot display is difficult to analyze unless you have a photographic memory. One solution to the problem is to use a camera with the single-sweep mode as previously described. Another solution is to use an oscilloscope with a special CRT that retains a visible "trail" where the beam has swept. Thus the display is stored. Depending on the type of storage oscilloscope, the display may be kept from seconds to days. An erase control permits removing the display at any time. For low-bandwidth, slow-sweep displays, storage is directly on the phosphor of the CRT. For high-bandwidth, fast-sweep displays, the storage is on meshes behind the CRT phosphor. Some common storage oscilloscopes divide the screen into an upper half and lower half which may be written on or erased separately. This lets you display and store the waveform of a circuit, modify the circuit, and then compare a new waveform with the original. Very high sweep rate storage oscilloscopes such as the Tektronix 7834 can be used to capture and measure nanosecond-width pulses that happen only once or at a very low repetition rate.

Some storage oscilloscopes first convert the analog input signal to a series of digital (binary-coded) values. The digital values can be stored in a semiconductor memory and then read out over and over to produce a display on the CRT. The digital values also can be easily processed by a microcomputer to calculate risetimes, periods, frequencies, and peak voltages. Then these values can be displayed on the screen.

The operation of the vertical amplifier, time base, sweep modes, and triggering of a storage oscilloscope is the same as previously described for a standard triggered sweep oscilloscope. However, since there are several types of storage oscilloscopes, you will have to consult the manual of your specific unit for write, store, and erase procedures. These are the most important points to remember for preserving the storage screen: Keep the intensity down, avoid repeated use of the same area of the screen, minimize storage time, and do not use the store mode unless necessary.

SAMPLING OSCILLOSCOPES

Another area in which a standard oscilloscope is not applicable is for displaying high-frequency signals above 350 MHz or their equivalent-pulse risetimes of less than 1 ns. If you want to display one cycle of a 1-GHz sine waving filling a 10-cm horizontal by 8-cm vertical screen, a beam sweep speed of 20,000 cm/μs is required. At this speed the beam is moving across the screen so fast that it does not leave a visible trace. You

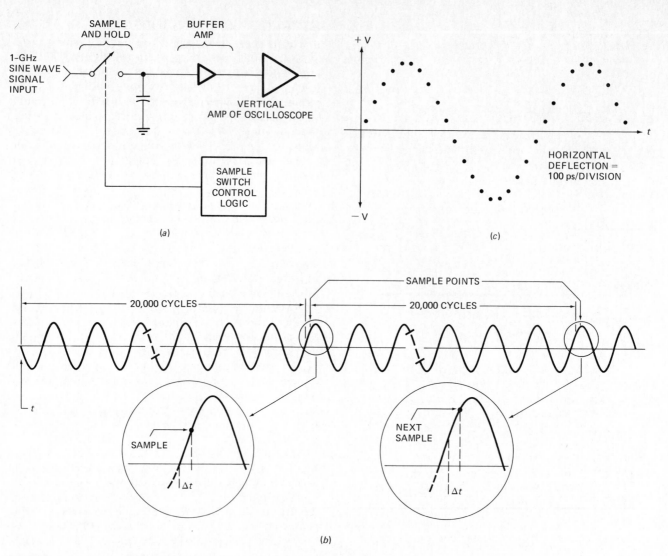

FIGURE 1-14 *(a)* Sample and hold circuit input for a sampling oscilloscope. *(b)* A 1-GHz sine wave shows sample points. *(c)* Sampling oscilloscope display.

may have noticed that with any oscilloscope, as you increase the sweep speed, you have to increase the intensity to keep the trace visible. Another problem with trying to display these high-frequency signals with a standard or real-time oscilloscope is the extreme bandwidth required for the vertical amplifier.

For regularly repetitive signals these problems are solved by the use of a sampling oscilloscope. Rather than the usual real-time view, where the display is traced directly by an input signal, a sampler creates a display with a series of point values assembled from different waves. Here is how this is accomplished.

The input to a sampling oscilloscope is a *sample and hold circuit*, which can be represented by a switch and a capacitor as in Figure 1-14a. When the switch is closed, the voltage on the capacitor charges to follow the value of the input. When the switch is opened, the capacitor holds the voltage present at that time. This voltage is

amplified by the high-impedance buffer amplifier and the vertical amplifier to make one point of the display. An additional point is created each time the switch closes and opens. Typically the switch closes and opens at about a 50-kHz rate, or about once for every 20,000 input cycles if a 1-GHz sine wave is being displayed. A sample is being taken from only one of every 20,000 input cycles, but since the cycles are all identical, this is acceptable. Now, if each sample were taken at the same time in a cycle, the oscilloscope would get the same value for each sample, and the display would simply be a straight horizontal line. If sample 1 is taken at time t (see Figure 1-14b), sample 2 is taken 20,000 cycles plus a small time Δt later, and sample 3 is taken 20,000 cycles plus Δt after that, then the sample point appears to step along the input waveform in increments of Δt and gives display points for the entire waveform (see Figure 1-14c). The horizontal time base of a sampling oscilloscope is calibrated to give equivalent time in picoseconds/division or nanoseconds/division; therefore, risetimes or frequencies can be determined easily from the display.

DIGITAL READOUT OSCILLOSCOPES

Some oscilloscopes such as the HP1725A have a built-in digital multimeter for measuring ac/dc volts, ohms, and amperes. The digital readout of the DMM (digital multimeter) is also used to give the time in nanoseconds or microseconds between two intensified dots on the CRT display. This feature lets you easily measure risetimes or delay displays.

OTHER TEST EQUIPMENT

PULSE GENERATORS AND TIME-MARK GENERATORS

The use of a pulse generator as a signal source usually is quite straightforward, but there are a few precautions. Before you connect a pulse generator to a circuit, first use an oscilloscope on dc coupling to make sure the offset and amplitude controls on the pulse generator are set to give legal levels for the logic family you are using. (Refer to Chap. 2 for the legal levels of the common logic families.) An accidental overvoltage from a low-impedance source such as a pulse generator can easily destroy the input of a logic gate. Second, if the pulse generator has a 50-Ω output impedance, use 50-Ω cable such as

RG-58 to carry the signal from the generator to the circuit, and then connect a 50-Ω resistor from the cable shield to the center conductor where it connects to the circuit. The reasons for this are discussed in Chap. 2 under the heading "Transmission Lines." Third, when you turn off power, the pulse generator should be turned off at the same time as or before the logic circuits to prevent current flow through logic chips in the wrong direction.

Time-mark generators are much more accurate pulse generators which put out narrow pulses at precisely timed intervals. A crystal oscillator ensures the accuracy of the timing. These time marks are used for calibrating the time bases of oscilloscopes and other equipment.

LOGIC-STATE ANALYZERS

When you analyze digital circuits that have many output lines, such as a random access memory, it is advantageous to be able to display and compare eight or more signals at the same time. This is one of the functions of a *logic-state analyzer*. In its simplest form, a logic-state analyzer is just a device that chops or alternates between eight or more input channels to produce an oscilloscope trace for each. A word recognizer allows you to trigger when a specific word or pattern of 1s and 0s is present on the input lines. Logic analyzers also may display the information from these channels as a string of 1s and 0s, as a plot of points, or as hexadecimal words. Most logic analyzers contain a continuously recording memory that permits display of data which occurred before the trigger. In Chap. 5 we discuss the use of logic analyzers for analyzing specific circuits.

REVIEW QUESTIONS AND PROBLEMS

1. What are the advantages and disadvantages of solderless prototyping boards?

2. When might you choose to use a universal printed circuit board to build a circuit on?

3. Why is a 0.1-μF capacitor put in parallel with a 25-μF capacitor to bypass the power-supply leads entering a printed circuit board?

4. What problems does daisy-chain wiring cause?

5. Why should the circuit layout follow the schematic layout as much as possible?

6. What size and type of wire should be used with solderless prototyping board?

7. How can you keep track of where you left off when wiring a circuit from a schematic?

8. Why is color-coded wiring important?

9. Describe the initial testing procedure for a newly built circuit.

10. What is the advantage of the logic probe over most DVMs for checking logic levels?

11. *a.* What are likely causes of a hot electrolytic capacitor?
    b. What are likely causes for a hot IC?

12. Describe the use of a logic probe. On which circuits should they not be used?

13. What is the difference between chop and alternate sweep modes? Which mode is used to view two high-frequency signals?

14. Is ac or dc input coupling best for viewing logic signals?

15. Figure 1-15 shows the schematic of a simple transistor amplifier stage. What type of coupling would you use to display:

    a. The bias voltage on the base?

    b. The input signal appearing at the base?

    c. The output signal on the collector?

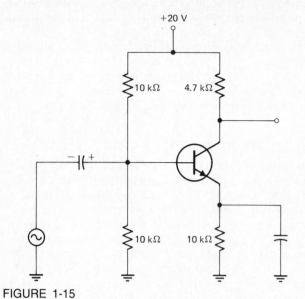

FIGURE 1-15

16. Why are probes used on oscilloscopes inputs?

17. *a.* How are oscilloscope probes adjusted for best frequency response?

    b. With a ×10 probe connected to its calibrator output squarewave, an oscilloscope displays the waveform in Figure 1-16. What is the problem, and how do you correct it?

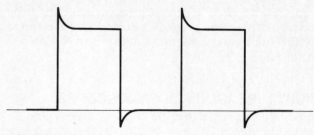

FIGURE 1-16

18. What is the difference between the trigger level control and the trigger slope control?

19. What is the use of single sweep?

20. Describe the use of intensified and delayed sweep.

21. In the *X-Y* display mode, what quantities are represented by the two axes of the oscilloscope display?

22. *a.* What types of signals are best displayed by a storage oscilloscope?

    b. Why should the intensity of the beam be kept low?

23. *a.* For what types of measurements are sampling oscilloscopes required?

    b. What is the principle of operation of a sampling oscilloscope?

24. Why should the amplitude and offset of a signal generator output be checked with an oscilloscope before the signal generator is connected to a circuit?

25. What is the frequency of best bypassing for a $0.001\text{-}\mu\text{F}$ capacitor with a total of ½ in [12.7 mm] of leads?

REFERENCES/BIBLIOGRAPHY

Using Your Oscilloscope Probe, Application Note, Tektronix, Inc./A-3051-1.

Allison, Gordon (ed.): "Teknique—Understanding Oscilloscope Triggering Controls," *Tekscope*, vol. 6, no. 5 (November/December 1974), pp. 11–13.

Global Specialities Corporation Breadboard and Test Equipment for the Professional and Hobbyist! (catalog), Global Specialties Corporation, 1982.

Bechtold, Jim (ed.): "Troubleshooters in Action," *Bench Briefs—Service Information from Hewlett-Packard*, September–October 1977, pp. 1–5.

Tektronix Products Catalog, Tektronix, Inc., 1982.

Tektronix 5A18N Dual Trace Amplifier Instruction Manual, Tektronix, Inc., May 1976.

Tektronix 5440 Oscilloscope Instruction Manual, Tektronix, Inc., April 1976.

Tektronix 5B42 Delaying Time Base Instruction Manual, Tektronix, Inc., June 1976.

Tektronix 5111 Storage Oscilloscope Instruction Manual, Tektronix, Inc., August 1976.

CHAPTER 2

DIGITAL LOGIC GATE CHARACTERISTICS AND INTERFACING

In Chap. 1 we discussed circuit construction techniques and the use of common test equipment. After some introductory material, in Chap. 2 we discuss basic logic gates and their input-output characteristics. Using the knowledge of test equipment you gained in Chap. 1, you can measure and compare these characteristics. This will help fix them in your mind and give you more experience with test equipment.

After completing this chapter, you should be able to:

1. Describe the difference between an analog and a digital quantity.

2. Convert decimal numbers to binary and binary numbers to decimal.

3. Write the truth tables for AND, OR, NAND, NOR, inverter, and exclusive OR logic gates.

4. Use manufacturer's data sheets to determine the input voltages and currents, the output voltages and currents, and the propagation delay times for common logic gates.

5. Given the schematic of the internal structure of a common logic family IC gate, predict the output state for high or low inputs.

6. Draw and build the circuits to properly condition or produce input signals for common logic families.

7. Draw and build the circuits to interface common logic families to one another or to other devices such as displays and relays.

ANALOG VERSUS DIGITAL

Electronics involves two basic kinds of quantities: analog and digital. *Analog* quantities can have any value.

For example, the output voltage of the circuit in Figure 2-1a can be adjusted to any of an infinite number of values between 0 and 5 V. Most natural quantities such as temperature, pressure, and time are analog. A sine wave is an example of an analog signal. A moving-coil voltmeter is an analog instrument, since the needle can point to an infinite number of positions.

A *digital* quantity or signal can have only certain fixed values. The output voltage of the circuit in Figure 2-1b, for example, has only two possible values: 0 V if the switch is open and 5 V if the switch is closed. If its risetimes and falltimes are considered instantaneous, a square wave is an example of a digital signal. As its name implies, a digital voltmeter is a good example of a digital instrument. The display can indicate only certain fixed values such as 1.95 V, 1.96 V, etc. Also, it uses digital circuitry to compute or measure the voltage under test.

Calculators and most computers work with digital signals which have only two values such as 0 and 5 V. The question that immediately comes to mind is, How can a calculator store in its memory a number such as 327 using only two voltage levels? The answer is that the calculator or computer stores and manipulates numbers in a *binary*-coded form.

BINARY CODE: A BASE-2 NUMBER SYSTEM

REVIEW OF DECIMAL SYSTEM

To learn the structure of the binary number system, the first step is to review the familiar decimal, or base-10, number system. Figure 2-2a shows a decimal number with the value of each placeholder or digit expressed as a power of 10. The digits in the decimal number 5346.72

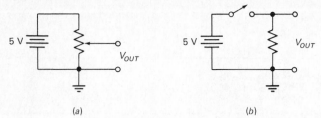

FIGURE 2-1 Circuits illustrating analog versus digital. (a) Analog output circuit. (b) Digital output circuit.

tell you that you have 5 thousands, 3 hundreds, 4 tens, 6 ones, 7 tenths, and 2 hundredths. The number of symbols needed in any base number system is equal to the base number. In the decimal number system then, there are 10 symbols, 0 through 9. When the count in any digit position passes that of the highest symbol, a carry of "1" to the next digit is produced and the digit rolls back to zero. A car odometer is a good example of this.

A number system can be built by using powers of any number as placeholders or digits, but some bases are more useful than others. It is difficult to build electronic circuits that can store and manipulate 10 different voltage levels, but relatively easy to build circuits that can handle two levels. Therefore, a *binary*, or *base-2*, number system is used.

BINARY SYSTEM

Figure 2-2*b* shows the value of each digit in a binary number. Each binary digit represents a power of 2. A *binary digit* is often called a *bit*. Note that digits to the right of the *binary point* represent fractions used for numbers less than 1. The binary system uses only two symbols, 0 and 1. Therefore, in binary you count as follows: 0, 1, 10, 11, 100, 101, 110, 111, 1000, etc.

To convert a binary number to its equivalent decimal number, multiply each digit by the decimal value of the digit and just add these. The binary number 101, for example, represents $1 \times 2^2 + 0 \times 2^1 + 1 \times 2^0$, or

$$
\begin{array}{cccccc}
5 & 3 & 4 & 6 & . & 7 & 2 \\
10^3 & 10^2 & 10^1 & 10^0 & & 10^{-1} & 10^{-2}
\end{array}
$$

(a)

$$
\begin{array}{ccccccccc}
1 & 0 & 1 & 1 & 0 & . & 1 & 1 \\
2^7 & 2^6 & 2^5 & 2^4 & 2^3 & 2^2 & 2^1 & 2^0 & 2^{-1} & 2^{-2} \\
128 & 64 & 32 & 16 & 8 & 4 & 2 & 1 & \frac{1}{2} & \frac{1}{4}
\end{array}
$$

(b)

FIGURE 2-2 Number systems and value of placeholders. (a) Decimal. (b) Binary.

$4 + 0 + 1$. This, of course, is equal to 5 in decimal. For the binary number 10110.11 you have

$$N = 1 \times 16 + 0 \times 8 + 1 \times 4 + 1 \times 2$$
$$+ 0 \times 1 + 1 \times 0.5 + 1 \times 0.25 = 22.75$$

To convert a decimal number to binary, there are two common methods. The first is simply a reverse of the binary-to-decimal method just discussed. For example, to convert the decimal number 21 to binary, first subtract the largest power of 2 that will fit in the number. For 21_{10} the largest power of 2 that will fit is 16, or 2^4. Subtracting 16 from 21 gives a remainder of 5. Put a 1 in the 2^4 digit position, and see whether the next lower power of 2 will fit in the remainder. Since 2^3 is 8 and 8 will not fit in the remainder of 5, put a 0 in the 2^3 digit position. Then try the next lower power of 2. In this case, the next is 2^2, or 4, which *will* fit in the remainder of 5. Therefore, a 1 is put in the 2^2 digit position. When 2^2, or 4, is subtracted from the old remainder of 5, a new remainder of 1 is found. Since 2^1, or 2, will not fit into this remainder, a 0 is put in that position. A 1 is put in the 2^0 position because 2^0 is equal to 1, and this fits exactly into the remainder of 1. The result shows that 21_{10} is equal to 10101 in binary. The conversion process is somewhat messy to describe, but easy to do. Try converting 46_{10} to binary. You should get 101110.

The second method of converting a decimal number to binary is shown in Figure 2-3. Divide the decimal number by 2, and write the quotient and remainder as shown. Divide this quotient and following quotients by 2 until the quotient reaches zero. The column of *remainders* will be the binary equivalent of the given decimal number. Note that the *MSD (most significant digit)*

$$227_{10} = \underline{\quad ? \quad}_{\text{Binary}}$$

			Least Significant Binary Digit ↓					
2)227	=	113	R1	×	1	=	1
2)113	=	56	R1	×	2	=	2
2)56	=	28	R0	×	4	=	0
2)28	=	14	R0	×	8	=	0
2)14	=	7	R0	×	16	=	0
2)7	=	3	R1	×	32	=	32
2)3	=	1	R1	×	64	=	64
2)1	=	0	R1	×	128	=	128

↑ Most Significant Binary Digit

227 Check

$$\therefore 227_{10} = 11100011_2$$

FIGURE 2-3 Conversion of a decimal number to binary and a binary number to decimal.

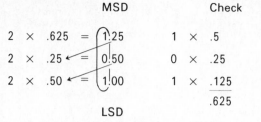

MSD				Check	
2	×	.625	=	1.25	1 × .5
2	×	.25	=	0.50	0 × .25
2	×	.50	=	1.00	1 × .125
					.625

LSD

FIGURE 2-4 Conversion of a decimal number less than 1 to binary.

HIGH	LOW
1	0
TRUE	FALSE
+5 V	0 V
ON	OFF
ASSERTED	UNASSERTED

FIGURE 2-5 Some terms or names for positive logic digital states.

is on the *bottom* of the column and the *LSD (least significant digit)* is on the *top* of the column. You can demonstrate that the binary number is correct by reconverting from binary to decimal, as shown in the right-hand side of Figure 2-3.

You can convert decimal numbers less than 1 to binary by successively multiplying by 2 and recording carries until the quantity to the right of the decimal point becomes zero, as shown in Figure 2-4. The *carries* represent the binary equivalent of the decimal number, with the *most significant digit* or *bit* at the *top* of the column. Decimal 0.625 equals 0.101 in binary. For decimal values that do not convert exactly, as this one did (the quantity to the right of the decimal never becomes zero), you can continue the conversion process until you get the number of binary digits desired.

At this point it is interesting to compare the number of digits required to express numbers in decimal with the number needed to express them in binary. In decimal, one digit can represent 10 numbers, 0 through 9; two digits can represent 10^2, or 100, numbers, 0 through 99; and three digits can represent 10^3, or 1000, numbers, 0 through 999. In binary, a similar pattern exists. One binary digit can represent two numbers, 0 and 1; two binary digits can represent 2^2, or 4, numbers, 0 through 11; and three binary digits can represent 2^3, or 8, numbers, 0 through 111. The pattern, then, is that N decimal digits can represent 10^N numbers, and N binary digits can represent 2^N numbers. Eight binary digits can represent 2^8, or 256, numbers, 0 through 255.

BASIC LOGIC GATES

Previously we stated that most computers and digital circuits use binary code to represent all numbers and quantities. This is done because it allows each digit to be represented by only two symbols, 0 and 1. In digital circuitry these two symbols are referred to as *states*. Figure 2-5 shows some of the terms or conditions commonly used to describe these two states.

The basic building blocks of digital circuitry are *logic gates*. They are called this because they manipulate the binary data in a logical way. Some examples will clarify this.

AND GATE

Figure 2-6a shows a simple circuit with two switches in series. The light will be on only if switch A is on *and* switch B is on. This is referred to as the *logical AND function* and is represented by the logic gate symbol in Figure 2-6b.

TRUTH TABLE   A simple way to show the output state for each possible input state combination is by a *truth table*. Figure 2-6c shows the truth table for a two-input AND gate. Since there are two binary input variables, A and B, there are four possible input combinations. It is important when you write the truth table for a logic gate or digital circuit to always show the output for all possible input combinations. An easy way to do this is to write the input states in *binary counting sequence* as shown in Figure 2-6c. For a three-input gate there are eight possible input combinations. They are simply the binary numbers 000, 001, 010, 011, 100, 101, 110, and 111. An AND gate can have any number of inputs.

BOOLEAN EXPRESSIONS   Truth tables are a valid way to show the response of logic gates, but they are somewhat awkward and space-consuming. A more compact way of expressing the logic function of a gate uses a *boolean algebra* expression. Boolean algebra is similar to, but not quite the same as, ordinary algebra. The boolean algebra expression for the AND gate in Figure 2-6 is $Y = A \cdot B$. The dot between A and B indicates the AND function. Often the dot is omitted, and the expression is simply written as $Y = AB$. An AND gate can have

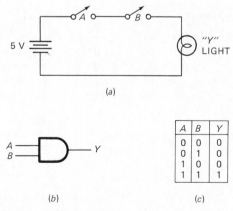

FIGURE 2-6 AND function representations. *(a)* Switch equivalent circuit. *(b)* Logic gate symbol. *(c)* Truth table.

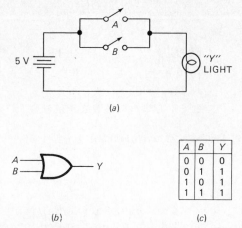

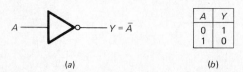

FIGURE 2-7 OR function representations. *(a)* Switch equivalent circuit. *(b)* Schematic symbol. *(c)* Truth table.

any number of inputs. The expression for a three-input AND gate is $Y = A \cdot B \cdot C$, or $Y = ABC$.

OR GATE

Another common logic gate is the *OR gate.* Figure 2-7a shows a switch-equivalent circuit for an OR gate. The light will be on if switch *A or* switch *B* is on. Figure 2-7b shows the schematic symbol and Figure 2-7c shows the truth table for a two-input OR gate. The boolean algebra expression for an OR gate is $Y = A + B$. The plus sign indicates the OR function. Note that for the $A = 1$ and $B = 1$ case in the truth table, this sign does not mean the same as it does for ordinary algebra. An OR gate also can have any number of inputs. The boolean expression for a three-input OR function is $Y = A + B + C$.

INVERTER

Another useful logic gate is the *inverter,* or *NOT gate.* Figure 2-8a shows the schematic symbol and Figure 2-8b the truth table for an inverter, or NOT gate. The output is simply the opposite, inverse, or *complement* of the input. The boolean expression for this is $Y = \overline{A}$. The bar over the *A* indicates inversion and is read "*A* not."

NAND GATE

AND gates and OR gates can be combined with inverters to form other basic gates. An AND gate can be combined with an inverter to form a NOT AND, or *NAND*, gate. Figure 2-9a illustrates this using schematic symbols, and Figure 2-9b shows the truth tables for AND and

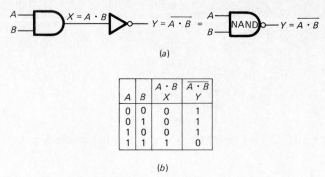

FIGURE 2-9 Deriving the NAND function. *(a)* Schematic symbols. *(b)* Truth tables.

NAND gates. The outputs for the NAND gate are just the inverse of those for the AND. The bubble on the output of the schematic symbol indicates this. The boolean expression for a NAND gate is $Y = \overline{A \cdot B}$, or just $Y = \overline{AB}$. An important point to remember when you are troubleshooting circuits containing NAND gates is that *the output of a NAND gate is high (1) whenever any input is low (0).*

NOR GATE

An OR gate with an inverter on its output forms a *NOR* gate. Figure 2-10a shows the schematic symbol for a NOR gate. Figure 2-10b shows the truth tables for an OR gate and a NOR gate. The boolean expression for a NOR gate is $Y = \overline{A + B}$. Note that for this gate the output is low (0) if any input is high (1).

XOR AND XNOR GATES

Last of the basic gates are the *exclusive OR (XOR)* and *exclusive NOR (XNOR)*. Figure 2-11a shows the schematic symbol for exclusive OR and Figure 2-11b the symbol for the exclusive NOR gates. As shown by its truth table in Figure 2-11c, the exclusive OR gate might be called a disagreeable gate because its output will be high (1) only if the two inputs disagree with each other. The boolean expression for an exclusive OR can be written as $Y = A \cdot \overline{B} + \overline{A} \cdot B$ or $Y = A \oplus B$.

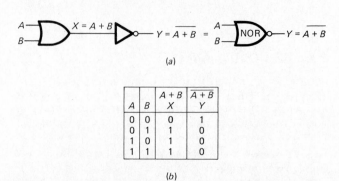

FIGURE 2-10 Deriving the NOR function. *(a)* Schematic symbols. *(b)* Truth tables.

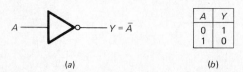

FIGURE 2-8 Inverter or NOT gate. *(a)* Schematic symbol. *(b)* Truth table.

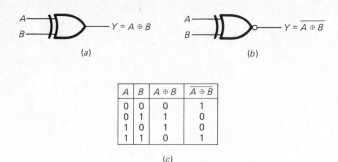

A	B	$A \oplus B$	$\overline{A \oplus B}$
0	0	0	1
0	1	1	0
1	0	1	0
1	1	0	1

(c)

FIGURE 2-11 XOR and XNOR functions. (a) XOR schematic symbol. (b) XNOR schematic symbol. (c) Truth tables.

An exclusive NOR gate might be called an agreeable gate, or equality detector, because its output will be high (1) if the two inputs agree with each other. The boolean expression for an exclusive NOR is $Y = AB + \overline{A}\overline{B} = \overline{A \oplus B}$.

LOGIC FAMILY CHARACTERISTICS

In the preceding sections of this chapter we introduced the basic logic gates. In Chap. 3 and following chapters we show how these gates are used. In the following sections of this chapter we discuss the characteristics of commonly available families of logic gate devices.

TTL LOGIC GATE CHARACTERISTICS

The first logic family to be discussed is *TTL (transistor-transistor logic)*. Throughout the late 1960s and the 1970s the standard 7400 version of this family and its variations accounted for a very large share of the digital IC market. It is a mature logic family. In other words, many different functions are available, and manufacturers know how to make these chips with high yields. Therefore, the cost is low. As the industry evolved toward higher level of integration (putting more gates and functions on a single chip), TTL could not be used because it required too great a chip area per gate and dissipated too much power (heat) per gate. To overcome these problems, *large-scale integrated circuits* have mostly utilized *MOS* (metal-oxide semiconductor), *CMOS*, or I^2L devices. However, much TTL logic is still encountered, and so you should thoroughly understand it.

TTL IC PACKAGES AND NUMBERING Usually TTL gates are found in *dual in-line packages*, or *DIPs*. The package may have 14 to 24 pins depending on the logic functions contained in it. Figure 2-12 shows the pin diagram for a DM5400/7400 which contains 4 two-input positive logic NAND gates in a single 14-pin package.

Most TTL family parts are identified by a number of the form AA74XXX. The AA represents a letter or letters,

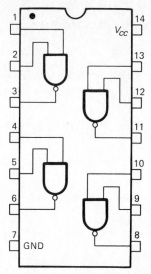

FIGURE 2-12 Pin diagram for DM5400/7400 quad two-input positive logic NAND gate.

such as DM, that are a code for the name of the company which manufactured the device. For example, DM represents National Semiconductor, SN represents Texas Instruments, and S represents Signetics. The XXX stands for a two- or three-digit number that identifies the function of the part. For example, the 7402 contains four two-input NOR gates, the 7430 contains one eight-input NAND gate, and the 74193 contains a 4-bit binary counter. Consult a TTL data book for a list of available functions.

The IC package also may have stamped numbers that indicate the date of manufacture and batch number.

A TTL DATA SHEET Table 2-1 shows a typical data sheet for a 5400/7400 TTL gate. V_{CC} is the power-supply voltage required to operate the entire device or IC. Input and output voltages are the logic level signals for each gate in the device. The first parameters to look at on any data sheet are the maximum ratings and operating conditions. The absolute maximum V_{CC} of 7 V warns you that if you have a variable power supply, you should adjust its voltage to the center of the operating supply voltage range, +5 V, before connecting it to TTL devices. An accidentally applied supply voltage of more than 7 V may destroy the device.

Under operating conditions, we see the ranges of supply voltage and ambient temperature over which the manufacturer guarantees the performance of the circuit. The 5400 series is functionally the same as the 7400 series, but as you can see, the 5400 series is guaranteed to operate over a wider range of supply voltages and ambient temperatures.

According to the data sheet, the maximum signal input voltage in the positive direction is +5.5 V. The maximum input voltage in the negative direction (input diode clamp voltage) is −1.5 V. To make sure you do not exceed these ratings, always use an oscilloscope on dc

TABLE 2-1
DM5400/DM7400, DM5410/DM7410, DM5420/DM7420, DM5400 DATA SHEET

Absolute Maximum Ratings

V_{CC}	7.0 V
Input voltage	5.5 V
Storage temperature range	−65 to +150°C
Fan-out	10
Lead temperature (soldering, 10 s)	300°C

Operating Conditions

	Min	Max	Unit
Supply voltage V_{CC}			
DM54XX	4.5	5.5	V
DM74XX	4.75	5.25	V
Temperature T_A			
DM54XX	−55	+125	°C
DM74XX	0	70	°C

ELECTRICAL CHARACTERISTICS*

PARAMETER	CONDITIONS	MIN	TYPICAL	MAX	UNITS
Input diode clamp voltage	$V_{CC} = 5.0$ V, $T_A = 25°C$, $I_{in} = -12$ mA			−1.5	V
Logic 1 input voltage	$V_{CC} = $ Min	2.0			V
Logic 0 input voltage	$V_{CC} = $ Min			0.8	V
Logic 1 output voltage	$V_{CC} = $ Min $V_{in} = 0.8$ V, $I_{out} = -400\ \mu A$	2.4			V
Logic 0 output voltage	$V_{CC} = $ Min $V_{in} = 2.0$ V, $I_{out} = 16$ mA			0.4	V
Logic 1 input current	$V_{CC} = $ Max $V_{in} = 2.4$ V			40	μA
Logic 1 input current	$V_{CC} = $ Max $V_{in} = 5.5$ V			1	mA
Logic 0 input current	$V_{CC} = $ Max $V_{in} = 0.4$ V			−1.6	mA
Output short-circuit current†	$V_{CC} = $ Max $V_{in} = 0$ V, $V_0 = 0$ V DM74XX	−20		−55	mA
	DM54XX	−18		−57	mA
Supply current, logic 0‡	$V_{CC} = $ Max $V_{in} = 5.0$ V		3	5.1	mA
Supply current, logic 1‡	$V_{CC} = $ Max $V_{in} = 0$ V		1	1.8	mA
Propagation delay time to logic 0, t_{PD0}	$V_{CC} = 5.0$ V, $T_A = 25°C$, $C = 50$ pF		8	15	ns
Propagation delay time to logic 1, t_{PD1}	$V_{CC} = 5.0$ V, $T_A = 25°C$, $C = 50$ pF		13	25	ns

*Unless otherwise specified, min/max limits apply across the −55 to +125°C temperature range for the DM54XX and across the 0 to 70°C range for the DM74XX. All typicals are given for $V_{CC} = 5.0$ V and $T_A = 25°C$.
†Not more than one output should be short-circuited at a time.
‡Each gate.
SOURCE: National Semiconductor.

coupling to check that the signal levels from a pulse generator or other signal source are between 0 and +5 V.

Next examine the electrical characteristics section of the data sheet for the voltage level that a TTL device will accept as a legal logic 1, or high, and the voltage it will accept as a legal logic 0, or low. The minimum logic 1 input voltage $V_{IH,MIN}$ of 2.0 V means the manufacturer guarantees that the gate will recognize an input voltage of 2.0 V or more as a high and use it to determine the output state of the gate. The maximum logic 0 input voltage $V_{IL,MAX}$ of 0.8 V indicates that any input voltage less than or equal to 0.8 V is recognized by the gate as a legal low. If the input signal voltage is between 0.8 and 2.0 V, the output logic level is *indeterminate*, so input signals should avoid or pass quickly through this region.

On the output of the gate, the given minimum logic 1 output voltage $V_{OH,MIN}$ of 2.4 V guarantees that an output high will always be equal to or greater than 2.4 V. Typically an output high will be 3 V or greater. Note that the minimum *output* high voltage is 0.4 V higher than the minimum voltage required for an *input* high on a following gate (see Figure 2-13a). This 0.4-V difference is referred to as a *noise margin*. It ensures that a small noise transient on a connecting line cannot change the state of the next gate. A transient has to be greater than a 400-mV peak before it can swing the input of the next gate below the 2.0-V input high minimum. A similar noise margin of 400 mV exists between the maximum logic 0 *output* voltage $V_{OL,MAX}$ of 0.4 V and the maximum logic 0 *input* voltage $V_{IL,MAX}$ of 0.8 V (see Figure 2-13a). As is shown later, excessive loading or incorrect input signals can eliminate one or both of these noise margins and give erratic operation.

Another useful way of showing the input and output voltage levels for a logic gate is by a transfer curve. This is simply a graph of output voltage versus input voltage. Figure 2-13b shows the transfer curve for a TTL inverter. Figure 1-13 in Chap. 1 shows how this graph for a TTL gate can be displayed on an oscilloscope.

Observe in Figure 2-13b that as the input voltage increases above the $V_{IL,MAX}$ of 0.8 V, the output voltage drops below the necessary $V_{OH,MIN}$ of 2.4 V. The V_{in} region of the graph where V_{out} is changing most steeply is called the *threshold voltage*. It is about 1.35 V. For a V_{in} of less than 2.0 V, the V_{out} rises above the desired $V_{OL,MAX}$ of 0.4 V. From this graph you can easily see why the input voltage range of 0.8 to 2.0 V is referred to as indeterminate. For this input range the output will not be a legal high or a legal low.

To discuss the input and output current specifications for a TTL gate, we must first define source current and sink current. Figure 2-14 shows the output of a TTL gate connected to input of a following gate. The resistors and diodes shown represent the equivalent internal circuitry of the inputs and outputs of typical TTL gates. If the output of gate 1 is low, as shown in Figure 2-14a, then a positive (conventional) current will flow from the +5-V V_{CC} supply, through gate 2, to gate 1. The current then flows through gate 1 to ground. An output is said

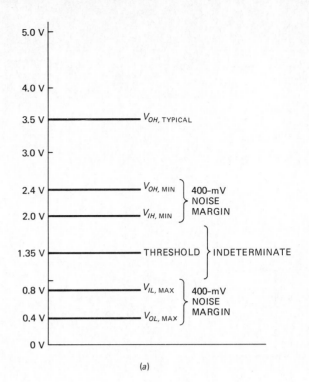

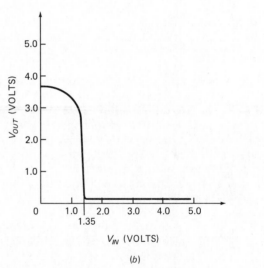

FIGURE 2-13 (a) TTL voltage levels and noise margins. (b) Transfer curve for a TTL inverter.

to *sink* current if it creates a current path from the input of a following gate to ground or to a negative supply. The output of gate 1 in Figure 2-14a, then, sinks current when it pulls the input of gate 2 low. Table 2-1 shows that the maximum logic 0 input current ($I_{IL,MAX}$) for a TTL gate is −1.6 mA. This means that for a TTL input to be pulled low, the output connected to it may have to sink as much as 1.6 mA to ground. On the data sheet a minus sign on a current indicates that conventional current is flowing *out of* the indicated device pin. A plus sign on a current indicates that conventional current is flowing *into* that device pin.

Figure 2-14b shows a gate with a high output connected to the input of a following gate. Conventional

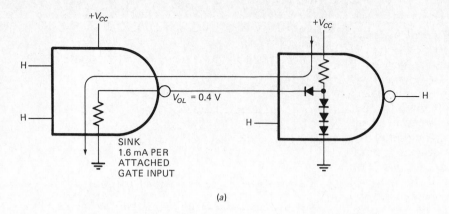

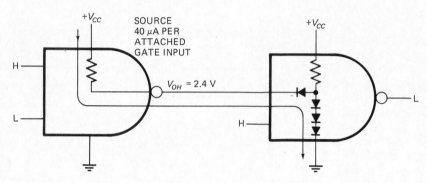

(a)

FIGURE 2-14 Output sink and source current. *(a)* Output low sinks 1.6 mA maximum from next gate input. *(b)* Output high sources 40 μA maximum to next gate input.

current flows from the +5-V supply, through gate 1, and out to the input of gate 2. Then the current flows through gate 2 to ground. An output is said to *source* current if it creates a current path from the positive supply to the input of the following gate. When the output of gate 1 is in the high state, it sources the current required to pull the input of the following gate to a high. Table 2-1 shows that the maximum logic 1 input current ($I_{IH,MAX}$) for a TTL gate is 40 μA. Therefore, the output of gate 1 may have to source as much as 40 μA when pulling the input of gate 2 high. In a later section we explain why the input low current is so much greater than the input high current.

A look in Table 2-1 at the conditions following the logic 1 output voltage and logic 0 output voltage shows that outputs are designed to source 400 μA in the high state ($I_{OH,MIN}$) and sink 16 mA in the low state ($I_{OL,MIN}$).

Fan-out is the maximum number of gate inputs that can be connected to a single gate output without preventing the output from reaching legal high and legal low voltage levels. Fan-out is equal to either I_{OL}/I_{IL} or I_{OH}/I_{IH}, whichever calculation produces the lower number. For this gate I_{OL}/I_{IL} = 16 mA/1.6 mA = 10, and I_{OH}/I_{IH} = 400 μA/40 μA = 10. A standard TTL output, therefore, can source and sink enough current to drive as many as 10 standard TTL inputs. If you connect more than 10 gates to an output, the saturation voltage of the gate's output current sinking transistor will increase above the 0.4-V maximum specified by the manufacturer for a load of 10 gates. Increasing this voltage decreases the noise margin for all the connected inputs

and makes them more susceptible to noise transients. In circuits where an input is connected to ground with a resistor (such as the Schmitt trigger or oscillator discussed later in this chapter), the value of the resistor must be low enough to keep the input voltage under or near 0.4 V. Dividing 0.4 V by 1.6 mA gives a value of 250 Ω for this resistor. However, for circuits that are not expected to encounter worst-case operating conditions, 470 Ω is a commonly used value.

The next parameter to consider is the *supply current*, or total current drawn from the power supply to operate each gate. Notice that a gate producing a logic *low* output requires 3.0 to 5.1 mA, while a gate producing a logic *high* output requires only 1.0 to 1.8 mA. When the gate switches states, the change in required supply current is a cause of transients on the V_{CC} line. When you are initially estimating the power requirements for a circuit, use the maximum supply current value for each IC and double it. This gives a margin of safety and makes power available for additional circuitry you may need.

The final parameter, *propagation delay*, is the time needed for a change on an input of a device to cause a change on the output. Figure 2-15 shows how propagation delay and several important pulse parameters are defined. The propagation delay time is measured from the 50 percent point of an input transition to the 50 percent point of the corresponding output transition. If the output is making a transition from a low to a high, the propagation delay time is identified as T_{PLH} or T_{PD1}. If the output transition is from a high state to a low, the propagation delay time is identified as T_{PHL} or T_{PD0}.

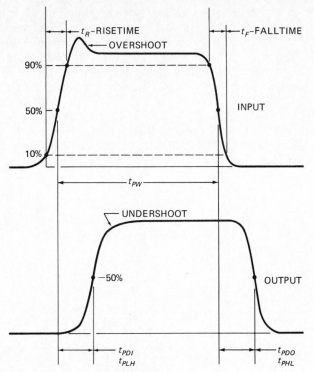

FIGURE 2-15 Pulse parameters.

Note in Table 2-1 that the T_{PD1} and T_{PD0} are different for a TTL gate. The reason is explained in a later section of this chapter.

Other pulse parameters to observe in Figure 2-15 are risetime, falltime, and *pulse width*. *Risetime* T_R is usually defined as the time it takes a pulse to go from 10 percent of its amplitude to 90 percent. *Falltime* T_F is the time needed for the pulse to go from 90 to 10 percent. Occasionally the 20 and 80 percent points are used for T_R and T_F. Risetime and falltime also may be called *transition time low to high*, T_{TLH}, and *transition time high to low*, T_{THL}, respectively. The width of a pulse, T_{PW}, usually is specified as the time between the 50 percent points of the pulse transitions.

SUMMARY OF TTL GATE CHARACTERISTICS A standard 7400 series TTL gate requires a power-supply voltage between 4.75 and 5.25 V. It requires a supply current of a few milliamperes.

To pull a TTL input low, you must apply a voltage between 0.0 and +0.8 V. The device or circuit used to pull an input low must be able to sink as much as 1.6 mA of current to ground.

To pull a TTL input high, you must apply a voltage between +2.0 and +5 V. The driving device or circuit must be able to source as much as 40 μA.

The output of a standard TTL gate can sink or source enough current at the right voltage levels to drive the inputs of 10 other standard TTL gates. To indicate this ability, the gate is said to have a fan-out of 10.

The guaranteed output high voltage of 2.4 V is 0.4 V greater than the required input high voltage of 2.0 V. This gives a 0.4-V noise margin between the output of a driving gate and the input of a receiving gate. The same noise margin exists for the output low maximum of 0.4 V and the maximum input low voltage of 0.8 V.

It takes 10 to 25 ns for a change on the input of a TTL gate to cause a change on the output. This time is called *propagation delay*.

TTL INTERNAL CIRCUITRY: THEORY OF OPERATION

In the previous section we discussed the characteristics of a standard TTL gate without regard to its circuitry. In this section we discuss the workings of a standard TTL gate and explain the origin of the previously described parameters.

BIPOLAR TRANSISTOR AS A SWITCH AND AN INVERTER Logic gates are essentially electric switches. Integrated-circuit logic gates use bipolar or MOS transistors to make these switches. Figure 2-16 shows the circuit for an NPN bipolar transistor switch that functions as an inverter. If a logic high voltage of +5 V is applied to the input, the base-emitter junction of the transistor will be forward-biased and the transistor turned on. The base will be at about 0.7 V. This leaves 5 V − 0.7 V, or 4.3 V, across the 10-kΩ base resistor. Therefore, a current of 4.3 V/10 kΩ, or 0.43 mA, will flow into the base. This base current will cause a much larger current flow from the collector to the emitter of the transistor. The ratio of collector current to base current is called *beta* (β), and typically it is about 100. This means that the collector current will be 100 times the base current. For the circuit shown, the base current of 0.43 mA would theoretically produce a collector current of 43 mA. However, a current of 43 mA through the 1-kΩ collector resistor would produce a voltage drop of 43 V across it (43 mA × 1 kΩ = 43 V). This is clearly impossible with a 5-V supply. What actually happens is that the collector current increases until the voltage drop across the collector resistor pushes the collector voltage down to within a few tenths of a volt of that on the emitter. For the example shown, the collector will be at about 0.1 V, which is a logic low. In this condition the transistor is said to be *saturated*. This term comes from the fact that the base region of the transistor is saturated with charges because there is not enough voltage on the collector to pull out all the

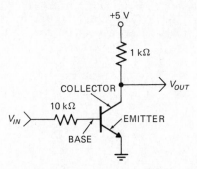

FIGURE 2-16 NPN bipolar transistor switch/inverter.

available charge. The important point here is that a saturated bipolar transistor has a voltage of a few tenths of a volt between its collector and its emitter. For the circuit in Figure 2-16 then, a logic high on the input produces a logic low on the output.

If the input of the circuit in Figure 2-16 is tied to ground, a logic low, the base-emitter junction will not have the 0.7 V required for it to be forward-biased. Therefore, the transistor will be off. With no base current there is no collector current. With no collector current there is no voltage drop across the 1-kΩ collector

resistor. The collector of an "off" transistor floats to whatever voltage is applied. Therefore the collector will be at +5 V, which is a logic high. A logic low on the input turns off the transistor and produces a logic high on the output. The circuit acts as an inverter. The transistor in this circuit is either off or saturated.

TTL NAND GATE CIRCUITRY
Figure 2-17 shows the internal circuitry of a standard TTL NAND gate. To analyze the operation of this circuit, assume the *B* input is tied high to +5 V, so that it has no effect. Assume also

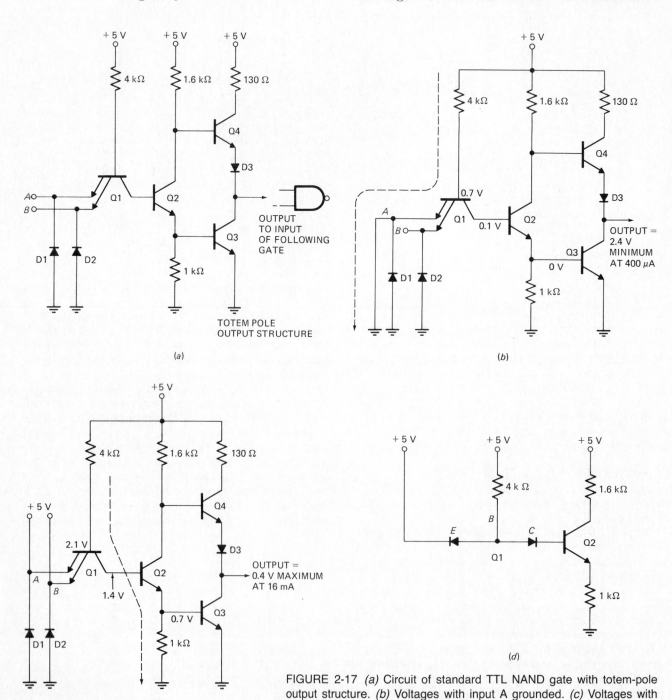

FIGURE 2-17 *(a)* Circuit of standard TTL NAND gate with totem-pole output structure. *(b)* Voltages with input A grounded. *(c)* Voltages with input A high. *(d)* Ohmmeter view of Q1.

that the A input is grounded or at a logic low state, as in Figure 2-17b. Grounding the A input gives a path for base current to flow through $Q1$ and turn on $Q1$. The base of $Q1$ is at V_{BE} above ground, or about +0.7 V. This leaves 4.3 V across the 4-kΩ resistor. With Ohm's law you can calculate the base current to be about 1.1 mA. Since $Q1$ has no source of collector current, 1.1 mA of base current will pull $Q1$ into saturation, and its collector voltage will be only about 0.1 V above its emitter voltage. Since the emitter is grounded, the collector of $Q1$ and the base of $Q2$ are about 0.1 V. This is much less than the 0.7 V required to turn on $Q2$, so $Q2$ is off. With no current flowing through $Q2$, no voltage is developed across the 1-kΩ resistor, and the base of $Q3$ is at ground. Therefore, $Q3$ is also off and has no effect on the output, since the collector of an off transistor acts as an open switch.

Now that you have followed the emitter path of $Q2$ to the output, go back and follow the collector path of $Q2$ to the output. Since $Q2$ is off, its collector has no effect on the base of $Q4$. $Q4$ is supplied with base current by the 1.6-kΩ resistor connected to +5 V. The maximum output load current that $Q4$ would have to supply in this state is the 0.4 mA, or 400 μA, needed to drive the inputs of up to 10 load gates. The small base current needed to produce this output develops little voltage drop across the 1.6-kΩ resistor. The base of $Q4$ is near 5 V. The output is a V_{BE} drop of 0.7 V plus a diode drop across $D3$ of 0.7 V below this, or typically about 3.6 V. For standard TTL, an output voltage greater than 2.4 V is a legal logic high level. We see, then, that a low or ground on the A input makes the output high. You can trace currents and voltages to show that the B input is equivalent to A in function. Therefore, the output of this TTL NAND gate is high if A is low, if B is low, or if both A and B are low.

Next consider the effect of making input A high (see Figure 2-17c). Again assume the B input is tied high to +5 V. With input A also at a positive logic high, or tied to +5 V, there is no path for $Q1$ base-emitter current to flow, so $Q1$ is off. However, the ohmmeter view of an NPN transistor in Figure 2-17d shows an available current path to the base of $Q2$ through the 4-kΩ resistor and the collector-base diode of $Q1$. This path supplies sufficient base current to $Q2$ to saturate it. $Q2$ supplies enough drive to $Q3$ to turn it on and to drive it into saturation. The voltage at the output now is the voltage across a saturated transistor—a few tenths of a volt—which is a logic low.

Now go back and follow the other path to see how the high on the A input affects $Q4$. The emitter of $Q2$ is at 0.7 V, as set by V_{BE} of $Q3$. Since $Q2$ is saturated, its collector is only about 0.1 V above this, or about 0.8 V, with respect to ground. This is considerably less than the 1.4 V necessary to turn on the series combination of the $Q4$ base-emitter junction and $D3$. Therefore $Q4$ is off and has no effect on the low output. The purpose of $D3$, in fact, is to ensure, as much as possible, that $Q4$ is never on at the same time as $Q3$. If $Q3$ and $Q4$ were both on for very long, a large current would flow directly from

V_{CC} to ground, heating the chip and wasting power. Even with this *totem-pole* output structure using $D3$, there is a brief instant each time the output changes states at which both transistors are on. This allows a short surge of current through the on transistors to ground, which produces voltage spikes or transients on the V_{CC} line. Unless you have sufficient bypass capacitors, as discussed in Chap. 1, these transients may disturb other circuits.

The purpose of the input clamp diodes $D1$ and $D2$ in Figure 2-17 is explained in a later section of this chapter on input signal conditioning.

SUMMARY OF TTL NAND GATE CIRCUIT OPERATION
If any one of the inputs is low, then $Q1$ is conducting as a transistor, $Q2$ and $Q3$ are off, and the output is pulled high by $Q4$. If all the inputs are high, the collector-base junction of $Q1$ functions as a diode. This turns on $Q2$ and $Q3$ to pull the output to a logic low. This follows the boolean expression for a NAND gate, $Y = \overline{A \cdot B}$.

IMPORTANT POINTS TO REMEMBER IN BUILDING CIRCUITS USING TTL
A TTL input left open acts as a high because there is no base-emitter current path through it for $Q1$. In practice, unused TTL AND or NAND gate inputs should be tied to V_{CC} with a 1-kΩ resistor. The 1-kΩ resistor protects the gate input from any large voltage spikes on the V_{CC} line. Unused OR or NOR gate inputs can be tied directly to ground.

Standard TTL outputs should never be connected together. If you try tying together two standard totem-pole output structures as in Figure 2-18, an often fatal fight will occur between the outputs if one output tries to go high and the other tries to go low. You can see in Figure 2-18 that this creates a low-impedance connection from V_{CC} to ground through $Q3$, $D4$, and $Q5$. Some manufacturers include 130-Ω resistors in series with the collectors of $Q4$ and $Q5$ to limit short-circuit current. These resistors will save the chip if only one gate output in the chip is accidentally short-circuited to ground. You can't count on the resistor being there, so never connect totem pole outputs together.

Bypass capacitors, typically 0.01 to 0.1 μF, should be connected between V_{CC} and ground to filter out V_{CC} transients caused by gates switching states.

Now that you are familiar with the theory of operation of the internal circuitry of a TTL gate, the previously mentioned difference between the propagation delay times T_{PLH} and T_{PHL} can be explained. During a low output state $Q2$ and $Q3$ (see Figure 2-17c) are in saturation. Before these transistors can turn off and change the state of the output, the charges stored in the base region during saturation must be drained off. The time needed to do this is referred to as *storage time*. Since storage time is not a problem during transistor turn-on, $Q3$ and $Q4$ turn on faster than they turn off, contributing to a difference in propagation delay times. This difference is typical for most members of the TTL family and can cause timing problems on circuits such as the

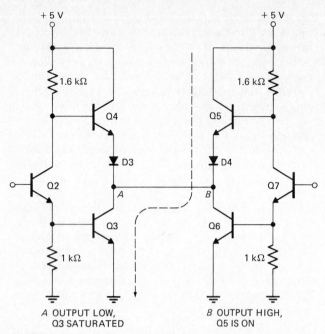

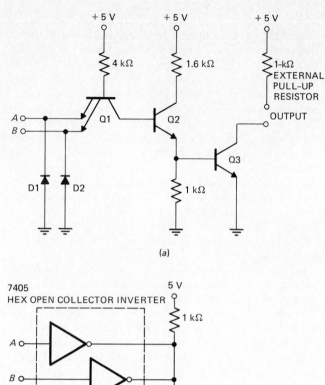

FIGURE 2-18 Low-impedance path produced by tying together two TTL totem-pole outputs.

inputs of a fast digital-to-analog (D/A) converter. If you connect more than 10 gate inputs on an output, you not only reduce the noise margins as discussed earlier, but also increase the propagation delay times by adding more capacitance which must be charged or discharged during output changes. This explains why you should not exceed the fan-out limit for a TTL gate.

OPEN-COLLECTOR TTL Some standard TTL circuits are available with an *open-collector* output structure instead of the totem-pole output we discussed. The open-collector circuit replaces the upper output transistor and diode with an *external* resistor to V_{CC} (see Figure 2-19a). When the lower transistor ($Q3$) is off, this *pull-up* resistor brings the output to a high.

The advantage of the open-collector circuit is that, without having to use another gate, several outputs can be tied together to give a *wired-AND* function, as shown in Figure 2-19b. For this circuit the output will be high only if all the inputs are low. Note two alternative symbols that indicate this wired function.

The disadvantages of open-collector logic are that it is noisier and slower than standard totem-pole output and difficult to troubleshoot. If one of several outputs is internally short-circuited low, the only ways to find the culprit are to replace chips one by one or to trace the current to the bad output with a current tracing probe.

THREE-STATE OUTPUT LOGIC Open-collector logic mostly has been replaced by *three-state* or *Tri-State* output logic. (Tri-State is a registered trademark of National Semiconductor Corporation.) Three-state logic has the normal TTL low and the normal TTL high output states plus a high-impedance, floating output state created by turning off both output transistors with a separate *enable* input (see Figure 2-20a). An external

FIGURE 2-19 (a) Circuit of standard TTL NAND gate with open-collector output. (b) Six-input NOR gate made with 7405 open-collector inverters.

signal applied to the enable input controls whether the output is active or in the "don't care" floating state. Many three-state outputs can be connected on a common bus line, but only one output can be enabled at one time. You can select which output you want to hear from by enabling only the output of that gate. As shown in Figure 2-20b, data also can be selectively routed off the bus line with three-state gates. If the output of gate A is enabled, then gate A will determine the data on the bus line and that data will be available at the output of D, E, or F, whichever is enabled.

TTL SUBFAMILIES

Now that the important parameters of a logic family have been discussed and the standard 5400/7400 TTL family has been analyzed, we describe and compare the

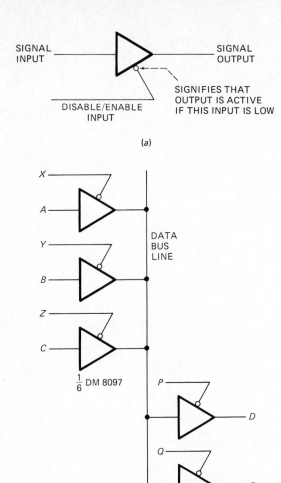

SIGNAL INPUT

SIGNAL OUTPUT

SIGNIFIES THAT OUTPUT IS ACTIVE IF THIS INPUT IS LOW

DISABLE/ENABLE INPUT

(a)

DATA BUS LINE

$\frac{1}{6}$ DM 8097

(b)

FIGURE 2-20 *(a)* Three-state output buffer. *(b)* Three-state output buffers used to selectively send data on or off a data bus line.

parameters of several other common logic families. Table 2-2 summarizes these parameters and comparisons.

A term often used to compare efficiencies of logic families is the *speed-power product.* The lower the speed-power product is, the more energy-efficient the family. Although manufacturers often use units of picojoules for the speed-power product, you can use the data in Table 2-2 to compare the speed-power products of various logic families by simply multiplying the gate delay by the power per gate.

The first families discussed in the following sections are those derived by modification of the standard TTL circuit.

HIGH-POWER AND LOW-POWER TTL

If all the resistor values in the basic circuit of Figure 2-17 are reduced, more current flows and the circuit switches

twice as fast. This modification is the basis of the 74H00/54H00 TTL family, known as *high-speed,* or *high-power, TTL.* As you can see from Table 2-2, devices in this family have a shorter propagation delay than regular TTL, but they use much more power per gate. The 74H00 family is essentially obsolete and mostly has been replaced by Schottky TTL, the 74S00 family discussed later.

For applications where the short propagation delay or high speed of standard TTL is not required and less power dissipation is desired, the *low-power TTL* 74L00/54L00 family was developed. Increasing the value of all the resistors in the basic circuit of Figure 2-17 decreases the power by a factor of 10 and the speed by a factor of about 3 to produce this family. As you can determine from Table 2-2, each 74L00 low-power TTL output can drive a maximum of two standard TTL inputs or 10 low-power inputs. For most applications, CMOS logic or low-power Schottky TTL has replaced low-power TTL.

SCHOTTKY TTL

In their eternal quest for speed, integrated-circuit designers sought a way of overcoming the storage time problem of the saturated transistors in the standard TTL gate. The solution came in the form of a unique diode made with a metal-semiconductor junction. This *Schottky diode,* as it is called, has a forward voltage drop of about 0.3 V and turns on or off very fast because it has no storage time problem. Refer to Figure 2-21*a.* A Schottky diode connected between the base and collector of a transistor goes into conduction when the collector voltage of the transistor drops about 0.3 V below the base voltage. This clamps the collector voltage at about 0.4 V and prevents the transistor from going into saturation. Since the transistor is not saturated and the Schottky diode has no storage time, now both can be turned off very rapidly.

In ICs the Schottky diode is made as part of the transistor. Figure 2-21*b* shows the schematic symbol for a *Schottky diode–clamped transistor.* By making most of the transistors in the IC gates Schottky diode–clamped and using a slightly different circuit, the *Schottky TTL* 74S00/54S00 family was created. The members of this family use less power than the corresponding parts in the high-power TTL family, but they are at least twice as fast as high-power TTL and 4 times as fast as regular TTL. However, the fast switching and short risetime edges of Schottky TTL can cause problems unless connections between gates are kept short and the power supply is well bypassed. We discuss this further in a later section on transmission lines.

LOW-POWER SCHOTTKY

Increasing the value of the resistors in Schottky TTL gate circuits converts them to *low-power Schottky TTL,* a family optimized for minimum propagation delay, fast switching, and low power. The 74LS00/54LS00 family is slightly faster than standard TTL but requires only one-fifth the power. Thus its speed-power product is 5 times better (lower) than standard TTL. Improved versions of low-power Schottky such as the Texas Instrument's 74ALS series have decreased the typical T_{PD} from 8 to 4 ns and the

TABLE 2-2
COMPARISON OF COMMON LOGIC FAMILY CHARACTERISTICS

LOGIC FAMILY	OPERATING SUPPLY VOLTAGE MIN	MAX	MINIMUM LOGIC 1 INPUT VOLTAGE	MAXIMUM LOGIC 0 INPUT VOLTAGE	MINIMUM LOGIC 1 OUTPUT VOLTAGE	MAXIMUM LOGIC 0 OUTPUT VOLTAGE	MAXIMUM LOGIC 0 INPUT CURRENT	MAXIMUM LOGIC 1 INPUT CURRENT	MINIMUM LOGIC 0 OUTPUT CURRENT	MINIMUM LOGIC 1 OUTPUT CURRENT	TYP. PROP. DELAY T_{PHL}	T_{PLH}	MAXIMUM COUNTER FREQUENCY	FAN-OUT	TYPICAL POWER DISSIPATION PER GATE	TYPICAL t_r 10-90%	SPECIFIC DEVICE
7400	4.75	5.25	2.0	0.8	2.4	0.4	1.6 mA	40 µA	16 mA	400 µA	8 ns	13 ns	35 MHz	10	10 mW	15 ns	7400
5400	4.5	5.5															5400
74H	4.75	5.25	2.0	0.8	2.4	0.4	2.0 mA	50 µA	20 mA	500 µA	6.2 ns	5.9 ns	50 MHz	10-74H	22 mW	7 ns	74H00
74L	4.75	5.25	2.0	0.8	2.4	0.4	0.18 mA	10 µA	3.6 mA	200 µA	30 ns	60 ns	3 MHz	10-74L	1 mW	30 ns	74L00
74S	4.75	5.25	2.0	0.8	2.7	0.5	2.0 mA	50 µA	20 mA	1 mA	5 ns	5 ns	125 MHz	10-74S	19 mW	3 ns	74S00
74LS	4.75	5.25	2.0	0.8	2.7	0.4	0.4 mA	20 µA	8 mA	400 µA	8 ns	8 ns	45 MHz	20-74LS	2 mW	15 ns	74LS00
74ALS	4.5	5.5	2.0	0.8	2.5	0.4	0.2 mA	20 µA	10 mA	400 µA	5 ns	4 ns	32 MHz 74ALS192	6-74 20-74ALS	1 mW	—	74ALS00
PECL III OR MECL III	-4.7	-5.7	-1.095	-1.485	-0.980	-1.600	MIN 350 µA	350 µA	40 mA	40 mA	1 ns	1 ns	1000 MHz	DC → 63 MECL III	60 mW	1 ns	MC1662
MECL 10,000	-4.7	-5.7	-1.105	-1.475	-0.980	-1.630	MIN 0.5 µA	265 µA	50 mA	50 mA	2 ns	2 ns	200 MHz	DC → 92 MECL	25 mW	3.5 ns	MC10101
I²L	1	15	0.7	0.4	V_{CC}	0.4	Injector current	Injector current	20 mA	Zero without resistor to V_{CC}	Adjustable 25-250 ns	—	—	Depends on injector current	6 nW-70 µW	—	—
Low-threshold PMOS	$V_{CC}=+5$ V ∓ 5% $V_{DD}=-9$ V ± 5% $V_{GG}=V_{DD}$		3.0	0.65	3.5	$V_{CC}-6$ V MOS OR 0.45	1 µA	1 µA	1.6 mA @ 0.45 V	0.2 mA @ 3.5 V	—	—	1 MHz	—	—	—	1702A ROM
+12, +5, -5 NMOS	$V_{CC}=+5$, $V_{DD}=+12$, $V_{BB}=-5$		3.3	0.8	3.7	0.45	±10 µA	±10 µA	1.9 mA @ 0.45 V	150 µA	—	—	—	—	—	—	8080A
NMOS +5 ONLY	4.5	5.5	2.0	0.8	2.4	0.4	10 µA	10 µA	2.1 mA @ 0.4 V	100 µA	—	—	—	1-74 5-74LS	—	—	2102 RAM
4000 A +5 V CMOS	3	15	⅔V_{CC}	⅓V_{CC}	V_{CC} - .01	0.01	10 pA	10 pA	.12 mA @ .5 V	.12 mA @ 4.5 V	50 pF 180 ns	50 pF 125 ns	1 MHz CD4040	AC=3-10 CMOS	Depends on operating frequency	175 ns	CD4011
4000B +5 V CMOS	3	18	⅔V_{CC}	⅓V_{CC}	V_{CC} - .01	0.01	±1 µA	±1 µA	0.4 mA @ 0.4 V	1.6 mA @ 2.5 V	50 pF 160-320 ns	50 pF 210-420 ns	2.5 MHz 4520	DC=∞ AC=3-10 CMOS	—	100 ns	CD4081B
4000B +15 V CMOS	3	18	⅔V_{CC}	⅓V_{CC}	V_{CC} - .01	0.01	±1 µA	±1 µA	3 mA @ 1.5 V	3 mA @ 13.5 V	50 ns	65 ns	6 MHz 10 V 4520	3-10 CMOS	—	40 ns	CD4081B
74C00 +5 V CMOS	3	15	3.5	1.5	4.5	0.5	5 nA	5 nA	0.4 mA @ 0.4 V	0.36 mA @ 2.4 V	50 pF 90 ns	50 pF 90 ns	2 MHz 74C90	3-10 CMOS 2-74L00	—	100 ns	74C00
74HC00 +5 V CMOS	3	6	3.5	1.5	4.95	0.05	1 µA	1 µA	4 mA	10 mA	15 ns	15 ns	25 MHz	10 74LS	—	20 ns	74HC00

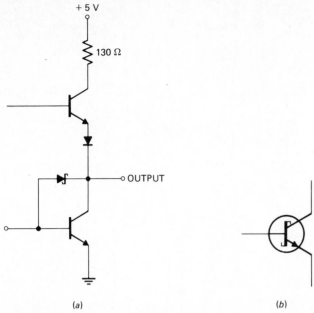

(a)   (b)

FIGURE 2-21 (a) Schottky diode clamp to prevent TTL output transistor collector from reaching saturation. (b) Schottky diode–clamped transistor schematic symbol.

power dissipation from 2 to 1 mW. This gives an even better speed-power product. The 74LS and 74ALS families are rapidly replacing standard 7400 series TTL because of their obvious speed and power advantages.

Another advantage of low-power Schottky is that the propagation delay times for a low-to-high and a high-to-low transition are more nearly *matched*. Also, low-power Schottky is a handy *interface* between CMOS and standard TTL. As explained later in the chapter under CMOS, most CMOS will not directly drive standard TTL although they will drive low-power Schottky. Then the low-power Schottky can drive two standard TTL inputs, which is usually sufficient. Many logic functions not found in standard TTL are now available in low-power Schottky TTL.

POINTS TO REMEMBER IN WORKING WITH TTL GATES:

1. Tie unused inputs to V_{CC} with a 1-kΩ resistor or directly to ground.

2. The outputs of TTL gates should not be connected together. Exceptions to this rule are open-collector and three-state TTL.

3. The maximum V_{CC} is 7 V.

4. The maximum signal input is −1.5 to +5.5 V.

5. There is a maximum fan-out of 10 gates within each TTL family. Fan-out is 20 in 74LSXX and 74ALSXX families.

6. Install a 0.1- or 0.01-μF bypass capacitor between V_{CC} and ground next to each IC or as often as is feasible.

7. Connecting leads should not be greater than 12 to 14 in [30.5 to 35.6 cm] for standard TTL and 3 to 4 in [7.6 to 10.2 cm] for Schottky TTL.

STILL MORE SPEED— EMITTER-COUPLED LOGIC

Emitter-coupled logic, or ECL, is another group of logic families that utilizes a very different circuit from that of the TTL families to achieve very short propagation delays. Figure 2-22a shows the circuit of a typical ECL gate of the MECL 10,000 family. Note the supply voltages of ground and −5.2 V. The simplest gate in these families is an OR-NOR gate rather than the NAND gate of the TTL families. Refer to Fig. 2-22a and mentally remove Q2, Q3, and Q4. You can see that Q1 and Q5 have their emitters coupled together to form a simple differential amplifier. The Q7 and Q8 transistors provide low impedance emitter follower outputs from the collectors of the differential amplifier. Q6 supplies a stable reference V_{BB} of −1.3 V to Q5, one side of the differential input amplifier. If the bases of Q1 and Q5 were at the same voltage, −1.3 V, then the two transistors would share the emitter current equally. When the base voltage on Q1 is pulled low to the −5.2-V V_{EE}, no current will flow through Q1 and its collector voltage will go up to give a high on the NOR output. The increase in current through Q5 will drop its collector voltage to give a low on the OR output. As you can see from Table 2-2, a minimum output high level is −0.980 V and a maximum output low level is −1.630 V. Internal resistors R_P tie the bases of unused inputs low to −5.2-V V_{EE}, so they do not affect the output. If input A is now switched from a low state to a high state of −0.980 V, current flow will switch from Q5 to Q1, since the base of Q1 is now at a higher voltage than the base of Q5. Q5 turning off will raise the OR output to a high level, and Q1 turning on will drop the NOR output to a low level. Raising any one of or all the inputs to a high level will cause the OR output to go high and the NOR output will go low. Emitter pull-down resistors, typically 510 Ω, are tied from each output to −5.2 V to equalize the output risetimes and falltimes.

As was shown for TTL, a useful way of illustrating graphically the input and output voltage characteristics and noise margins for ECL is by a transfer curve. Figure 2-22b shows the transfer curves for the ECL MC10102 quad two-input OR-NOR gate IC. At first glance, the number of different voltage parameters may seem confusing; but if they are taken one at a time, these can be clarified.

VOLTAGE PARAMETERS The $V_{IH.MAX}$ of −0.810 V (at 25°C) means that at an input voltage any more positive than this, the manufacturer will not guarantee the output voltage of the device to be in the proper range for the input of a following device. A $V_{IL.MIN}$ of −1.850 V indicates the most negative input low voltage for proper output levels to a following gate. The $V_{IHA.MIN}$ and

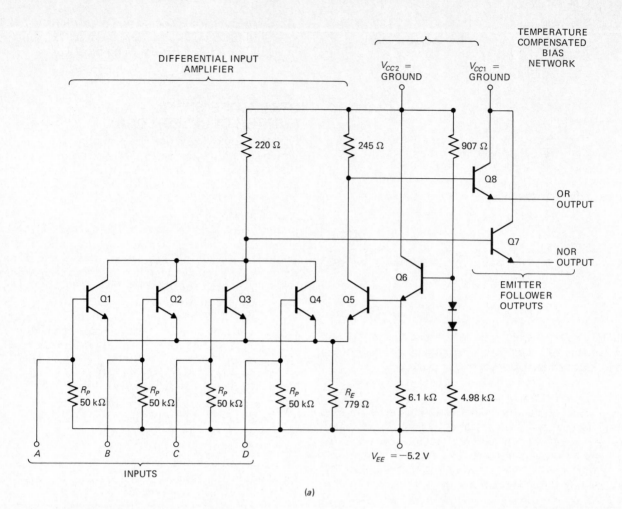

(a)

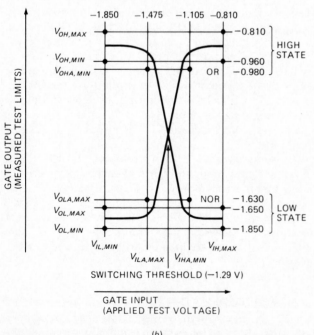

(b)

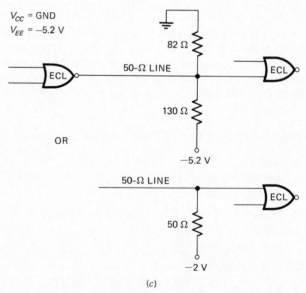

(c)

FIGURE 2-22 *(a)* Typical emitter-coupled logic gate of the Motorola MECL 10,000 family. *(b)* ECL OR-NOR transfer curves. *(Courtesy Motorola.)* *(c)* Terminations for ECL 50-Ω transmission line.

$V_{ILA,MAX}$ specifications indicate how close an input voltage can get to the switching threshold without putting the output at a voltage not legal for a following gate. The $V_{OHA,MIN}$ and $V_{OLA,MAX}$ specifications indicate the worst-case output voltages with the input at the worst-case voltages nearest the switching threshold ($V_{IHA,MIN}$ and $V_{ILA,MAX}$).

NOISE MARGINS

The noise margin for the low state is defined as $V_{ILA.MAX} - V_{OLA.MAX}$. By using these values from either Figure 2-22b or Table 2-2 MECL 10,000 series, the low-level noise margin is computed as

$$-1.475 \text{ V} - (-1.630 \text{ V}) = 0.155 \text{ V}$$

The noise margin for the high state is defined as $V_{OHA.MIN} - V_{IHA.MIN}$. For the parameters shown in Figure 2-22b or Table 2-2 MECL 10,000 series, the high-level noise margin is calculated as

$$-0.980 \text{ V} - (-1.095 \text{ V}) = 0.115 \text{ V}$$

Compare the ECL noise margins with those for TTL (0.400 V). ECL is designed with lower noise margins than TTL because its lower output impedance makes it less susceptible to noise.

The very short propagation delay time of emitter-coupled logic is the result of several design tricks. First, currents are kept high and output impedance is kept low so that stray and circuit capacitance can be quickly charged and discharged. The low-impedance output also gives gates a large fan-out, the ability to drive long transmission lines, and less chance of picking up noise on signal lines. Second, the logic swing is kept small, less than 1 V typically, so capacitance has to be charged and discharged less than in other logic families. Third, storage time problems are avoided by setting the output voltage levels and differential amplifier current so that none of the transistors ever goes into saturation.

The disadvantages of ECL are its relatively high power dissipation, its non-TTL compatible voltage levels, and its fast risetime transitions which require any connecting wire over a few inches to be treated as a transmission line. To avoid voltage reflection, which is explained in the section on transmission lines, such lines must be terminated with a resistor equal to the value of the characteristic impedance of the line. Figure 2-22c shows that a 50-Ω ECL transmission line can be properly terminated with either a 50-Ω resistor to -2 V or, the Thevenin equivalent, an 82-Ω resistor to ground and a 130-Ω resistor to -5.2 V. Since the -5.2-V supply is ground for an ac signal, the 82- and 130-Ω resistors appear in parallel to give an equivalent resistance of 50 Ω. The second scheme avoids needing a separate -2-V supply, but uses more power. The 510-Ω output pull-down resistors are not needed if a termination is used.

The Motorola Emitter-Coupled Logic III (MECL III) family is very powerful because of the very high frequencies at which it is capable of operating, but the requirement of a termination for any connecting wire over 1 in [2.5 cm] long makes it difficult to work with. In 1968, Motorola introduced the MECL 10,000 family, which kept short gate propagation delays, but increased the output risetimes from the 1 ns of MECL III to about 3.5 ns. This kept most of the speed and increased the maximum unterminated line length to 8 in [20.3 cm]. An improved version of MECL 10K, the MECL 10KH se-

ries, has reduced the t_{PD} to 1 ns. The MECL 10K is often used in high-speed computers. A MECL 10K family microprocessor, the 10,800 set, is also available.

I²L

Another bipolar transistor logic approach is current injection logic, or *integrated injection logic* (I²L). This type of logic is not used for discrete gate ICs as are the previous families, but is used in ICs containing thousands of gates such as a complete digital watch circuit or a complete microprocessor such as one version of the Texas Instruments SBP9989. The circuit of the basic I²L gate is shown in Figure 2-23. Q1 and Q2 are current sources, the value of which is programmed by an external resistor and the applied V_{CC}. This injector current, as it is called, can be set anywhere from 10 nA to 100 μA. If input A is open, or high, the Q1 injector current will flow through the base-emitter junction of Q3. This current drives Q3 into saturation and gives an output low voltage of about 50 mV on its collector. If a low of 50 mV or ground is applied to input A, the injector current will be steered away from the base of Q3, Q3 will be off, and its collector will just float high to whatever voltage is connected to it. When the output is connected to the input of another I²L gate as in Figure 2-23, the high-level output voltage will be the 0.7-V V_{BE} of Q5. Since Q3 and Q4 are in parallel, if either the A or the B input is high, the output at the junction of their collectors is low. This is the positive logic NOR function.

There are several advantages to I²L logic. First, the simplicity of the gates and their low power make it possible to make many gates in a small chip area. Second, since the currents are constant, no switching transients are produced on the V_{CC} line as with TTL or CMOS. Third, by programming the injector current, you can

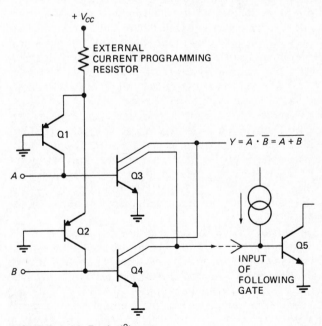

FIGURE 2-23 Basic I²L gate.

vary the propagation delay and the power dissipation over a very wide range. An injector current of 10 nA gives a propagation delay of 100 μs, and an injector current of 100 μA reduces the propagation delay to about 25 ns. For slow applications such as a digital watch circuit, you can cut the injector current way down and thus use very little power. Another advantage of I^{2}L is that it can be made easily on the same chip with bipolar analog circuits such as op-amps.

A disadvantage of I^{2}L is that it requires one more step during its manufacturing process than MOS, which is its main competitor in the LSI (large-scale integration) market.

MOS LOGIC

All the logic families discussed so far use bipolar transistors in their circuitry. The families to be discussed next use enhancement-mode MOS (metal-oxide semiconductor) transistors.

REVIEW OF MOS TRANSISTORS   To refresh your memory concerning MOS transistors, Figure 2-24 shows the structure, conductance curves, and schematic symbols with proper voltage connections for the common types of MOSFET. The structure of the first of these, the enhancement-mode type, is shown in Figure 2-24a. The N+ regions connected to the source and drain are just heavily doped N-type silicon in which most carriers are electrons. A P-type substrate is electrically insulated from the control gate by a layer of silicon dioxide and a layer of silicon nitride which protects the silicon dioxide from contamination. As the device is shown, there is no conducting path from the source to the drain because the substrate, source, and drain form two back-to-back PN junctions. At room temperature the P substrate will contain some free electrons produced by thermal agitation as well as the "holes" produced by doping. If the substrate is tied to the source and a positive voltage applied to the gate, these free electrons will be attracted to the edge of the substrate near the gate. The attracted electrons form a conducting electron or N channel (labeled induced channel on Fig. 2-24a) between the source and drain. Figure 2-24b shows that the more positive the gate is made, the more drain current flows through the device. Negligible drain current flows until a threshold voltage $V_{GS.TH}$ of typically 2 to 3 V between gate and source is reached. Note that the schematic symbol in Figure 2-24c shows the channel as a broken line which reminds you that no channel exists unless the gate is made positive, to *enhance* the conduction; hence the name enhancement-only (E-only) MOSFET.

The structure of the depletion-enhancement MOSFET shown in Figure 2-24d is identical to the enhancement-only MOSFET except that an N-type channel is diffused between the N+ source and drain regions. This means that even with zero voltage applied between the gate and source substrate, a conducting channel exists between source and drain current will flow. If the gate is made more positive with respect to the source, free electrons

are attracted from the substrate and the channel is "enhanced," so more drain current flows. If the gate is made more negative, electrons are repelled from the channel region. This "depletes" the channel and raises its resistance, so less current flows. The conductance curve in Figure 2-24e illustrates the effect of gate source voltage on drain current for this D-E, or depletion-enhancement MOSFET. The schematic symbol in Figure 2-24f shows the channel as an unbroken line, reminding you that the channel exists as a ready-made part of the device, before any gate voltage is applied.

E-only MOSFETs and D-E MOSFETs also are easily made as P-channel devices by changing the N regions to P and the P regions to N. The polarity of the applied voltages also is reversed. For a P-channel, E-only MOSFET, the source is connected to a positive supply and the drain to a more negative one. To get a P-channel, E-only device to conduct, the gate is made more negative than the source.

Because the input of a MOS transistor is electrically isolated, almost no input current is required and the input acts as only a capacitive load to the output of a preceding gate.

The extremely high input impedance of MOS-based logic creates severe handling problems because the thin layer of oxide isolating the gate is easily ruptured by static electricity. An attempt to protect the device is made by including protective circuitry on each input such as shown in Figure 2-24g. Diodes D1 and D2 will conduct if the input voltage goes 0.7 V above V_{SS}. D3 is a sharp-knee zener with a breakdown voltage of about 35 V, and it will conduct if an input voltage goes 35 V positive from V_{DD}. Since the gate oxide breakdown voltage usually is between 100 and 200 V, this circuit should protect the oxide, and it does up to a point. However, the static voltage generated by walking across a carpet on a low-humidity day may be 10 times as much as the protection circuit is capable of shunting. The protection circuitry cannot be improved without degrading the performance of the gate, so care in handling is required.

GUIDELINES FOR HANDLING MOS DEVICES   You may tend to become very careless in handling MOS. Then on a low-humidity day when you are wearing a nylon lab coat over a wool shirt, every MOS circuit or chip you touch seems to be bad or marginal. The static voltage generated by your nylon coat rubbing on the wool shirt has turned you into a "MOS killer." Rather than develop a bad reputation, follow these simple MOS handling guidelines:

1. Integrated circuits should be stored in *conductive foam* (usually dull black or pink) or *aluminum foil.* Never put the chips in white styrofoam, which is an excellent static generator.

2. *Ground to an earth ground* (the round prong of all properly wired outlets) *your soldering-iron tip, your metal bench, and the chassis of any test equipment used.*

3. *In a low-humidity, high-static condition, connect your wrist to earth ground with a length of wire and a 1-MΩ series resistor.* The resistor prevents you from getting a fatal shock if you touch a voltage source, but allows static to drain off.

4. *As with all integrated circuits, do not remove or insert MOS ICs with the power on.*

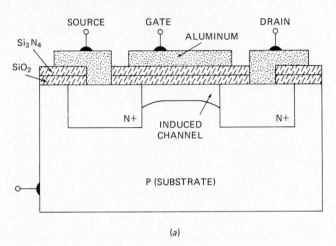

(a)

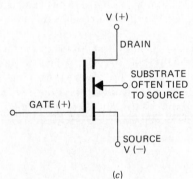

(c)

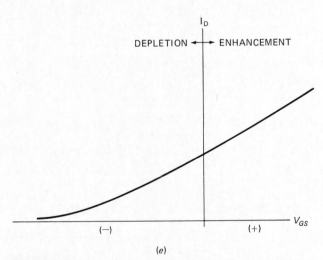

(e)

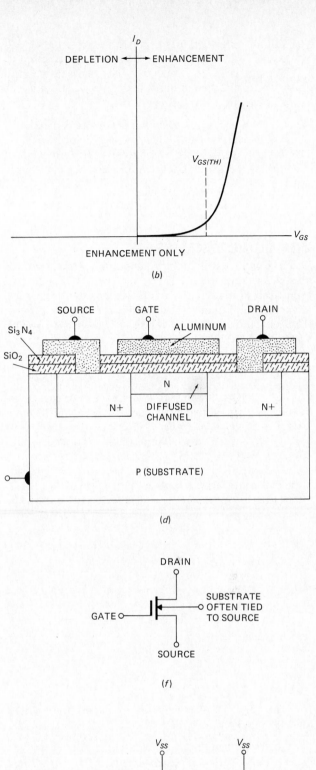

(b)

(d)

(f)

(g)

FIGURE 2-24 *(a)* E-only MOSFET structure. *(b)* E-only MOSFET transconductance curve. *(c)* E-only MOSFET schematic symbol. *(d)* D-E MOSFET structure. *(e)* D-E MOSFET transconductance curve. *(f)* D-E MOSFET schematic symbol. *(g)* Typical MOS input static protection circuitry.

5. *Do not apply signals to an input with the circuit power off.* With no V_{CC} present, a low-output-impedance signal generator may forward-bias the protection diodes and force enough current through to destroy them.

6. *Connect all unused input leads to V_{CC} or V_{DD},* whichever keeps the desired logic for the gate. Never leave them open, because open inputs accumulate static and float to unpredictable logic levels.

7. *Put resistors in series with any MOS inputs connected to points off a printed-circuit board.*

MOS LOGIC FAMILIES

Because gates using MOS transistors are low in power dissipation and require little chip area, MOS logic has been used in many LSI circuits such as shift registers, memories, and microprocessors. Common variations of MOS logic include PMOS, NMOS, CMOS, DMOS, HMOS, VMOS, and SOSMOS. CMOS is the only family in the group which has available as discrete chips simple gate functions as in TTL.

PMOS   PMOS, the first of this group to be mass-produced, uses enhancement-mode, P-channel MOS transistors to form gates. Figure 2-25*a* shows the circuit of a basic high-threshold P-channel gate. Note the typical supply voltages of ground, −13 and −27 V. In some versions the drain of Q3 is tied to V_{GG} to eliminate one supply. The negative supply is then referred to as V_{GG}. Q3 in this circuit is simply a resistive or current source

load for Q1 and Q2. In other words, Q3 is designed so that its ON-resistance is greater than the combined R of Q1 and Q2 when they are both ON. If input A or B is at a voltage high near V_{SS}, then Q1 or Q2 is off and the output is pulled down to a low voltage near V_{DD} by Q3. When inputs A and B are both pulled to a low voltage near V_{DD}, both Q1 and Q2 conduct. This connects the output to the positive supply through the small resistance of the two "on" transistors, to give an output high level near V_{SS}. In positive logic this is a NOR gate: $\overline{A \cdot B} = Y = \overline{A} + \overline{B}$. Input logic levels required for higher threshold PMOS are 0 V and a voltage near V_{DD}.

LOW-THRESHOLD PMOS   To continue our discussion of PMOS, early high-threshold versions of PMOS required high-voltage swing drivers to interface signals from TTL. Later, low-threshold versions of PMOS use silicon gates instead of metal gates for the internal transistor. This improves the speed and gives better TTL input and output compatibility. Typical power-supply voltages for low-threshold PMOS are V_{CC} of +5 V, V_{DD} of −5 V, and V_{GG} of −12 V; or V_{CC} of +5 V and V_{DD} and V_{GG} of −12 V. Silicon-gate PMOS operating on either of these voltage combinations can drive one standard TTL load. Most low-threshold PMOS devices can accept input logic levels of +5 V and 0 V or +5 V and −12 V.

NMOS   PMOS was the first MOS family because P-channel processing had fewer problems with contamination than N-channel processing. The higher speed and lower chip area per transistor possible with N-channel devices led manufacturers to develop NMOS devices

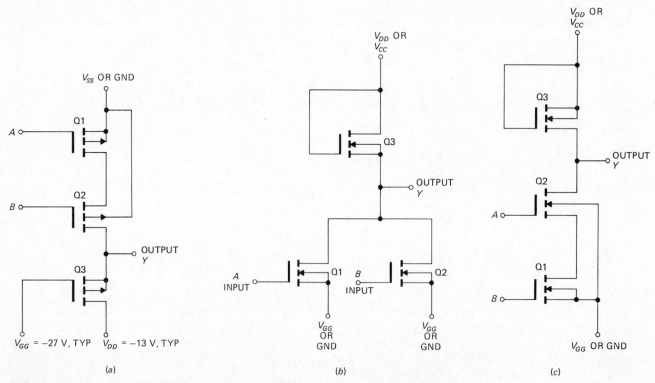

FIGURE 2-25 *(a)* Basic PMOS gate. *(b)* NMOS NOR gate. *(c)* NMOS NAND gate.

as soon as processing technology permitted. Most of the present MOS memories and microprocessors employ some variation of N-channel MOS. Figure 2-25b and c shows the circuit of a basic NMOS NOR gate and NAND gate. For the NOR gate (Figure 2-25b), Q3, an N-channel enhancement-mode FET, is simply a load resistor or current source. If inputs A and B are at a voltage near V_{GG}, then Q1 and Q2 will be off and the output will be pulled high by Q3. If either the A or B input is pulled to a voltage near V_{DD}, then Q1 or Q2 will turn on and pull the output low to V_{GG}. This satisfies the expression for a positive logic NOR gate, $Y = \overline{A + B}$. You can analyze the circuit in Figure 2-25c to see how it functions as a positive NAND gate.

Early LSI NMOS circuits such as the 8080A microprocessor use a V_{CC} of +5 V and a V_{BB} bias supply of −5 V. They also use a V_{DD} supply of +12 V to improve the speed of internal circuits. The outputs are TTL voltage-compatible and can sink enough current to pull a TTL input to a low state. Later NMOS circuits, such as the 2102 RAM, use silicon-gate technology and a push-pull output structure to provide sufficient drive for a single TTL load, while using only a single +5-V supply.

When troubleshooting systems containing PMOS and early NMOS, you are likely to find a wide variety of different supply voltages. We cannot detail here all the possible levels and currents. Table 2-2 gives the input-output levels, etc., for low-threshold P-channel silicon-gate MOS operating on +5 and −12 V and N-channel silicon-gate MOS operating on +5-, −5-, and +12-V supplies. Also shown are the operating levels for silicon-gate NMOS which operates on +5 V and ground. Fortunately the semiconductor manufacturers have now concentrated on making most MOS circuits operate on a single 5-V supply.

VMOS, DMOS, AND HMOS

VMOS, DMOS, and HMOS are structural variations of NMOS processing that produce circuits with much less propagation delay. Figure 2-26a, b, and c compares the structures of standard NMOS, VMOS, and DMOS. VMOS derives its name from its V-shaped structure and the fact that current flows vertically from source to drain, rather than horizontally as in the standard NMOS. VMOS transistors, because of their low capacitance and high speed, have shown promise as radio-frequency power amplifiers as well as in LSI logic circuits.

DMOS shortens the effective channel length to decrease propagation delay, by double-diffused doping in the gate region. In Figure 2-26c this is represented by the N− region under the gate which is not present in the standard NMOS. HMOS or high-performance MOS achieves short propagation delays by scaling down proportionately all the dimensions of the N-channel MOS transistors on the chip. The only difficulty with HMOS seems to be that optimum performance requires a supply voltage less than the *de facto* standard of +5 V. All these versions are theoretically capable of reaching ECL speed while using much less power and chip area than ECL.

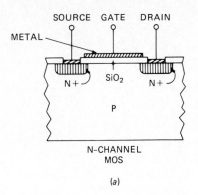

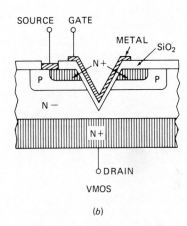

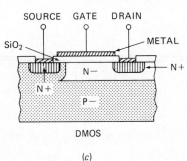

FIGURE 2-26 *(a)* NMOS structure. *(b)* VMOS structure. *(c)* DMOS structure. *(Electronic Design.)*

CMOS

At about the same time as P-channel technology was being developed for LSI circuits, C, or complementary, MOS was used to produce a logic family with logic functions comparable to those found in TTL, but with the low-power dissipation typical of MOS devices. CMOS circuits use much more chip area than PMOS, but they are faster, have input and output characteristics more compatible with common logic families, and can run on a single supply voltage between 3 and 15 V. Figure 2-27a and b shows the circuits for basic CMOS NOR and NAND gates. Note the use of complementary N- and P-channel enhancement-mode MOS transistors which gives the family its name. In Figure 2-27a, a basic

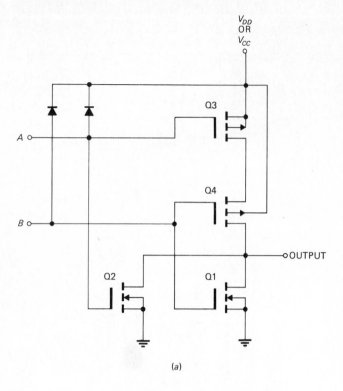

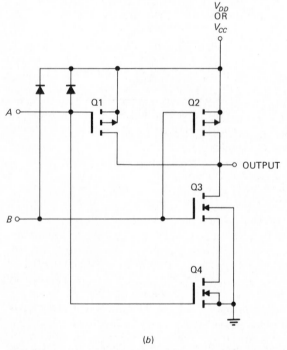

(a)

(b)

FIGURE 2-27 (a) CMOS NOR gate structure. (b) CMOS NAND gate structure.

CMOS NOR gate, if input A is at a high level near V_{DD} (V_{CC}), then the P-channel $Q3$ will be off and the N-channel $Q2$ will be on. $Q2$ then pulls the output to a low level near ground. A high on B input will hold off $Q4$ and turn on $Q1$, also pulling the output low. If both the A and B inputs are at a low level near ground, $Q1$ and $Q2$ will be off and $Q3$ and $Q4$ will be on. This pulls the output to a

high level near V_{DD} (V_{CC}). Since the output is low with A, B, or both A and B high, this circuit performs the NOR function. You can use similar analysis to show how the CMOS circuit in Figure 2-27b performs the NAND function. There are four common series of CMOS: the original 4000A series; an improvement of this series referred to as the 4000B series; the Fairchild 4500 series; and the National 74C00 series, which is similar in characteristics to the 4000B series but has the same logic function and pin numbers as the corresponding TTL chips.

I/O CHARACTERISTICS AND NOISE MARGIN Now that the basic structure of CMOS gates has been discussed, the input and output characteristics are described. Table 2-2 shows the parameters for the 4000A series, with 5-V V_{CC}: the 4000B series, with a 5-V supply and with a 15-V supply; the 74C00 series with a 5-V supply; and the new 74HC00 series with a 5-V supply. (*Note:* Regarding the confusing variety of subscripts used for supply voltages on MOS devices, the positive supply for CMOS devices such as the 74C and 74HC series which use *only* a *+5-V* supply is usually referred to as V_{CC}. On other CMOS or NMOS devices the positive supply is often called V_{DD}, and the negative supply or ground is referred to as V_{SS}. For the most part, the +5-V supply has become known as V_{CC} throughout the industry, regardless of whether it is connected to a drain, source, or collector.) Since the input of a CMOS gate is a MOS transistor, the input impedance is very high, and only about 10 pA of input current is required in either a high or a low state. In actual integrated circuits, diode networks as shown in Figure 2-24g are included on each input to help prevent static breakdown. Total input capacitance is 5 to 7 pF.

If you have a CMOS gate and a dual-trace oscilloscope, you can use the *X-Y* mode of the oscilloscope as described in Chap. 1 to determine the input-output transfer curve (see Figure 2-28a). The worst-case input voltage levels are about ⅓V_{CC} for voltage input low maximum, and about ⅔V_{CC} for voltage input high minimum. The threshold voltage is about ½V_{CC}. The worst-case output voltage levels are approximately equal to V_{CC} minus 10 mV for a high level and V_{SS} or ground plus 10 mV for a low level. You can draw a diagram such as that in Figure 2-28b to show that these input and output levels give effective noise margins of ⅓V_{CC} for a high and ⅓V_{CC} for a low level. For CMOS operating on a 5-V supply, the noise margin is about 1.7 V, or more than 4 times the noise margin of TTL. However, since the MOS gates have higher input and output impedances, their connecting wires pick up noise more easily, and in reality the noise rejections of 5-V CMOS and of TTL are roughly comparable. At higher operating voltages the CMOS has a noise-rejecting advantage.

CMOS creates noise spikes on the V_{CC} line during each output transition because of a current surge from V_{CC} to ground for the instant when both output transistors are on. Bypass capacitors help cure this problem.

CMOS is often said to be TTL-compatible, but this is somewhat misleading. One problem is that some CMOS

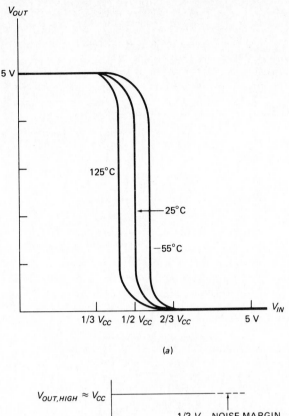

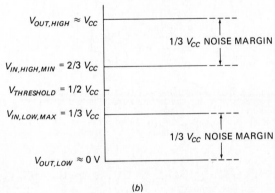

FIGURE 2-28 (a) CMOS transfer curves. (b) CMOS levels and noise margins.

discussed in greater detail, but the example above introduces you to some of the factors involved.

PROPAGATION DELAY TIMES The several hundred nanosecond propagation delays of CMOS operating on a 5-V supply limit its use to applications of a few megahertz or less. Increasing the supply voltage to 15 V decreases the propagation delay to under 100 ns and allows CMOS to be used for somewhat higher frequencies. The propagation delay and risetime of CMOS are lengthened a great deal by capacitive loading. With a capacitive load of 15 pF or three gate inputs, a typical B-series gate has a maximum t_{PD} (propagation delay time) of 50 to 75 ns. If the load is increased to 50 pF, or 10 gate inputs, then t_{PD} increases to a maximum of 360 ns. The long, variable t_{PD} of CMOS can cause timing problems in some circuits.

POWER DISSIPATION One of the main advantages of CMOS is its very small power dissipation when it is operated at low frequency. A typical CMOS gate dissipates only about 10 μW when operating at 1 kHz with a 5-V supply. You may wonder why operating frequency is mentioned along with power dissipation. The reason is that both of the main sources of power dissipation in CMOS increase as frequency of operation increases. The first is the current flowing from V_{CC} to ground through the output transistors when they are both on for an instant during switching. At higher switching frequency these transistors are both on more often, and the average power dissipation increases. The other main source of power dissipation is the current that flows to charge and discharge the capacitance of a gate input. When an input is changed from a low to a high state, current must be supplied through the upper P-channel output transistor to charge the input capacitance up to a voltage near V_{CC}. When the output goes low, this input capacitance discharges to ground through the lower N-channel output transistor. Higher-frequency operation means charging and discharging input capacitance more often, thereby increasing the average current to ground and the average power dissipation. As you can see in Figure 2-29, which shows the typical power dissipation per gate versus input frequency for several common logic families, above 1 MHz the power dissipation of 5-V CMOS is greater than that of low-power Schottky TTL. Above 5 MHz, 15-V CMOS uses more power than even ECL. Often CMOS is used in low-frequency, low-power, battery-operated instruments.

CMOS TRANSMISSION GATES AND THREE-STATE OUTPUTS Standard CMOS outputs cannot be tied together for the same reason that totem-pole TTL outputs can't be tied together. If one output is trying to go high and the other low, a direct-current path to ground results. With CMOS powered at 10 V or more, this is usually fatal to one or both chips. For applications requiring several gate outputs to be tied to a common data bus line, and only one output selected at a time, three-state output CMOS devices are used. As with TTL, the three output states are high, low, and floating or don't

doesn't have enough output current to drive TTL. The output current drive of CMOS operating on 5 V, as you can see in Table 2-2, is only 0.4 mA at an output low voltage of 0.4 V. This is enough to drive two low-power TTL inputs or one low-power Schottky input, but much less than the 1.6 mA required for a standard TTL input. To drive standard TTL from CMOS, you have to use a buffer such as the 4050. Another problem with CMOS-TTL compatibility is possible inadequate voltage levels. On the input of a CMOS gate operating on a 5-V supply, a minimum of 3.6 V is required for a legal input high. Manufacturers of TTL guarantee an output high of only 2.4 V for the standard totem-pole output. The *typical* TTL output of 3.5 V cannot be depended on to pull the CMOS input to a high level unless an external 10-kΩ resistor is connected from the TTL output to V_{CC}. Later in this chapter the interface of different logic families is

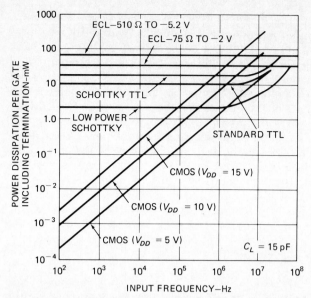

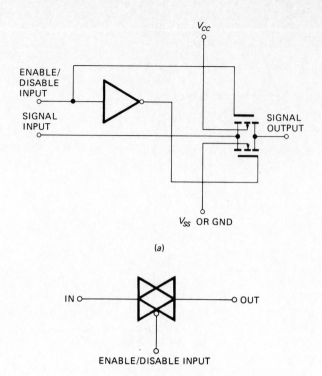

FIGURE 2-29 Typical power dissipation versus input frequency for common logic families. *(Fairchild Low Power Schottky Data Book.)*

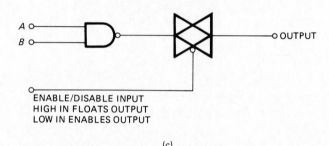

FIGURE 2-30 *(a)* Simple CMOS transmission gate. *(b)* Schematic symbol. *(c)* Transmission gate used to produce three-state output on standard gate.

care. Some CMOS use a unique circuit called a *transmission gate* to produce the three-state output (see Figure 2-30a). If both the N- and P-channel transistors in this circuit are turned on, then the output is connected to the input by the low channel resistance of the parallel transistors. The output follows the input high or low. If both transistors are off, they appear as open circuits or an open switch, and the output floats. The inverter in the transmission gate gives the opposite polarity signals necessary to control the N- and P-channel transistors. Figure 2-30b shows the schematic symbol for a transmission gate. A low on the enable/disable input turns on the output, and a high on the enable/disable floats the output. Figure 2-30c shows the connection of a transmission gate onto a standard CMOS structure to give a three-state output. Transmission gates are essentially logic-controlled switches. They are useful for multiplexing analog or digital signals onto a common line and are discussed in Chap. 3. They are available four or eight to a package, for example, the 4016 CMOS quad bilateral switch.

SILICON-ON-INSULATOR CMOS

A high-performance version of CMOS is made by building the CMOS transistors on an insulating substrate. An insulating substrate reduces capacitance and allows operation up to 50 MHz or more. Some devices use sapphire as the substrate and are referred to as SOS (silicon on sapphire). Other devices use silicon dioxide as the substrate and are referred to simply as SOI (silicon-on-insulator) devices. SOI CMOS is used mostly for LSI circuitry such as memories and microprocessors.

POINTS TO REMEMBER WHEN WORKING WITH CMOS:

1. Observe static electricity precautions as with any MOS logic.

2. Make sure all unused inputs are tied to V_{CC} or ground.

3. Turn off low-impedance signal sources at the same time as or before turning off CMOS circuits.

4. Do not attempt to drive standard TTL with a standard CMOS output.

5. Do not tie two CMOS outputs together unless they are both three-state, and then make sure they are not both enabled at the same time.

6. Minimize capacitive loading on outputs to keep t_{PD} as short as possible.

7. In higher-frequency circuits, watch for timing problems caused by the relatively long t_{PD} of CMOS.

DEVELOPING NEW LOGIC TECHNOLOGIES

As semiconductor processing has evolved, new technologies have been sought to increase the density of circuitry on chips and to make logic circuits operate at high frequencies. Some companies have been working on multilevel logic that can carry more information per gate by having up to 10 output voltage levels rather than just two as does traditional logic. This means that many fewer gates need be put on an integrated circuit such as a calculator chip.

Hewlett-Packard has produced gates with GaAs (gallium arsenide) FETs and Schottky diodes which have a propagation delay of less than 100 ps. This is 5 to 10 times faster than the fastest ECL. Superconducting Josephson junction gates with a propagation delay of about 10 ps have been produced. Such developments are leading to ultra-high-speed computers and logic families which operate up to several gigahertz.

INPUT SIGNAL CONDITIONING AND SIGNAL SOURCES

INPUT SIGNAL REQUIREMENTS

In testing or troubleshooting digital logic, a common source of problems is improper input signals. The requirements of a signal source for digital logic are as follows:

1. Output voltage high level is solidly above the minimum input high voltage, but not greater than $V_{IN,MAX}$, which will break down the input.

2. Output voltage low level is well below the maximum input low voltage, but above the $V_{IN,MIN}$ (voltage less than $V_{IN,MIN}$ may break down the input).

3. Output of signal source can sink or source sufficient current to maintain proper logic levels under load.

4. If the signal source is a switch, only a single transition should be input, not a pulse from each bounce of the switch contacts.

5. If the signal source is a clock or pulse source, the risetime of the pulses should be fast enough that the output of the driven gate does not oscillate or stay in its active region, where both output transistors are conducting, long enough to overheat the chip.

For the first three requirements, use Table 2-2 for the specific values. One source of problems is signal genera-

tors whose output is centered on zero. When the pulse output amplitude is set large enough to drive the gate input to a logic high level, the negative swing of the pulse will exceed the maximum input voltage rating of the chip. For TTL, a voltage more negative than about 0.7 V will forward-bias the input clamp diodes. A current of 10 mA or more through these diodes will probably destroy them. The static-protection diodes on the input of MOS or CMOS circuitry can suffer the same fate if an input signal swings too far negative or if the signal generator is left on when the chip power supply is turned off. In a later section of this chapter we discuss getting proper-level signals from one logic family to another.

SWITCHES AND SWITCH DEBOUNCING

When you use switches as signal sources, the bouncing of switch contacts may last several milliseconds. As you can see in Figure 2-31a, the bouncing produces a half-dozen or so output pulses rather than a single low-to-high transition as desired. The TTL input pulls high between each contact bounce. A storage oscillo-

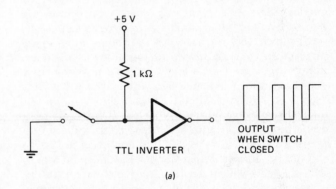

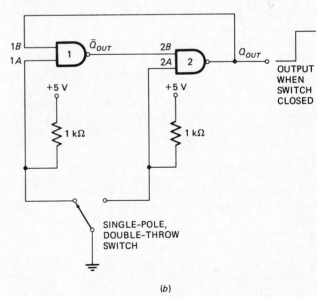

FIGURE 2-31 (a) Output of TTL inverter showing effect of switch bounce. (b) Simple switch debounce circuit.

scope can be used to demonstrate this bouncing. Figure 2-31b shows a simple circuit for producing debounced output transitions. If the switch initially ties input 1A to ground, then the $\overline{Q}$ output of NAND gate 1 and the 2B input of gate 2 are high. Since the 2A input is pulled high by the resistor to +5 V, both inputs of gate 2 are high and the Q output is low. When the switch is thrown over to the other position, input 1A goes high and input 2A goes low. The low on 2A gives a high to the Q output and the 1B input. Since both inputs of gate 1 are now high, its $\overline{Q}$ output is now low. This low, applied to the 2B input of gate 2, holds its Q output high regardless of the bouncing on the 2A input. A complementary ($\overline{Q}$) output is available from the output of gate 1. For applications using many switches, several switch debouncers are available in a single IC package such as the 74279. Switches on microprocessor data inputs can be debounced by a debounce routine in the program.

COMPARATORS

Signals too small to drive logic can be amplified to logic levels with comparators such as National's LM311 or LM219, which have TTL- and CMOS-compatible outputs. For high-frequency applications up to 300 MHz, the Advanced Micro Devices AM685 has ECL outputs and less than 6-ns propagation delay. Comparators are essentially operational amplifiers in an open-loop mode, so they have very high gain. When the negative signal input of the comparator is tied to ground as a reference voltage and a small ac signal is applied to the positive signal input as in Figure 2-32a, the output will go to a logic high level each time the input signal goes a few millivolts above ground. Each time the signal input drops below the ground reference voltage, the output drops to a logic low level. With pin 8 connected to *ground*, the output of the LM319 will produce the standard TTL logic voltage swing from +5 V to 0 V, rather than the typical open loop op-amp output voltage swing from +V to −V (+5 V to −5 V). A comparator, then, detects and gives an output that indicates whether the input signal is above or below the reference input voltage. The reference voltage can be any voltage within the operating range of the particular device, not just zero. Therefore, comparators also can be used to detect and give a logic level indication when the voltage output of a pressure or temperature transducer passes a reference voltage set for maximum safe boiler operating conditions, as in Figure 2-32b.

SCHMITT TRIGGERS

Another problem is created when an input signal has the proper voltage levels but the risetime and falltime of the input signal are longer than about 1 μs for TTL, or 5 μs for some CMOS inputs. Noise on the input signal may cause the output to oscillate high and low if the input voltage stays in the threshold region too long. On ac-coupled inputs, as in some flip-flops, a slow transition may not inject enough energy to produce the desired internal transitions. A cure for slow transition time is the Schmitt trigger, a circuit that uses positive feedback to speed up output risetimes and falltimes.

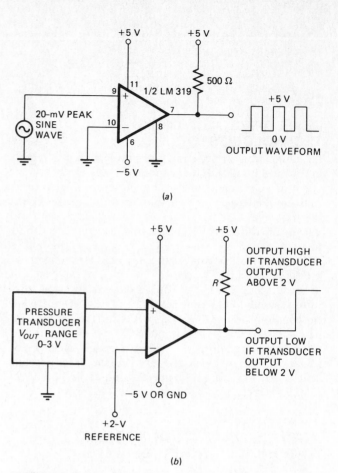

FIGURE 2-32 (a) Comparator used to produce logic level signals from small sine wave input. (b) Comparator used to detect when pressure of a boiler exceeds safety limit.

Schmitt-trigger circuits have built-in hysteresis, which means different thresholds for positive-going input transitions than for negative-going transitions. For TTL, the positive-going input transition threshold is typically 1.7 V, but on a negative transition, the output will not change until the input gets all the way down to 0.9 V. Therefore, a small noise voltage on the input cannot cause the output to oscillate during a slow input transition. These Schmitt-trigger circuits are available in TTL as the 7413 or 7414 and in CMOS as the 4584 or 4093. For single applications such as producing a logic level clock from the 60-Hz power line, a homemade Schmitt trigger as shown in Figure 2-33 can be used. Positive feedback to speed up the slow transition of the rectified 60-Hz sine wave is supplied by the 4.7-kΩ resistor from output to input. If you use CMOS inverters for the circuit, multiply the two Schmitt trigger resistor values by 10.

SIMPLE LOGIC CLOCKS OR PULSE SOURCES

For testing or driving digital circuits, often it is handy to have a compact pulse source which can be built right on the prototype.

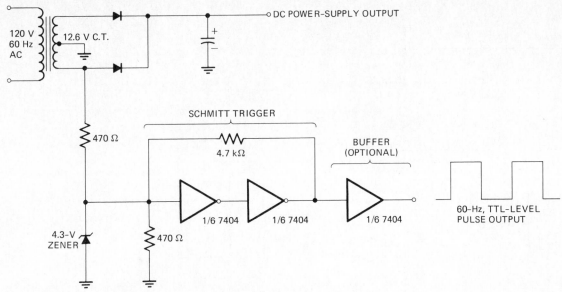

FIGURE 2-33 Simple Schmitt trigger conditioner used to produce 60-Hz clock from ac line.

555 TIMER AS AN ASTABLE MULTIVIBRATOR PULSE SOURCE

When powered with +5 V and ground, a 555 timer IC, in the circuit of Figure 2-34a, will produce a continuous train of pulses, which are directly TTL- and CMOS-compatible.

The circuit operation is described as follows. When power is applied, the output on pin 3 is high. The timing capacitor from pin 2 to ground is charged through R_A and R_B in series, until the voltage on the capacitor is equal to $\frac{2}{3}V_{CC}$. An internal comparator senses when the capacitor reaches this value, changes the previously high output to a low state, and turns on a transistor which discharges the capacitor through R_B alone to ground. Another internal comparator detects when the capacitor has discharged down to $\frac{1}{3}V_{CC}$, switches the output back to a high state, and turns off the discharge transistor to start the charge part of the cycle again. Frequency of the output pulses can be determined from the formula

$$f = \frac{1.44}{(R_A + 2R_B)C}$$

or from the chart in Figure 2-34b. To use the chart, select the frequency you want on the horizontal axis, follow this frequency line upward until it intersects an available capacitor value, and then read the value for $R_A + 2R_B$ from the nearest diagonal line. For example, for a frequency of 100 Hz and a capacitor of 0.1 μF, $R_A + 2R_B$ is nearly equal to 100 kΩ. If you want an approximately square wave output, then make R_A much smaller than R_B. For the above example, you might use an R_A equal to the minimum value for R_A of 1 kΩ and an R_B of 47 kΩ. Since the duty cycle or time the output is high divided by the total time of one cycle is equal to $(R_A + R_B)/(R_A + 2R_B)$, the output is only about 0.5 percent in error for these values. The maximum $R_A + R_B$ is

about 6.6 MΩ, and the minimum R_A or R_B is 1 kΩ. Resistor R_B can be replaced with a variable resistor and a fixed 1-kΩ resistor in series to give variable-frequency output. Timing capacitor value can vary from a few hundred picofarads to 1000 μF or more. For high-value capacitors, use very low-leakage types. Bypass capacitors are always required from V_{CC} to ground and from pin 5 to ground.

555 TIMER AS A MONOSTABLE MULTIVIBRATOR PULSE SOURCE

The 555 timer IC can be used also as a monostable multivibrator which will give an output pulse only when triggered by a high-to-low trigger input on pin 2 of the circuit, as in Figure 2-35a. The trigger width should be less than the desired output-pulse width. The time high for the output pulse is equal to $1.1RC$. Figure 2-35b gives a design chart for the 555 used as a monostable. If you need pulse high times less than 2 or 3 μs, use a 74121 or 74122 TTL monostable multivibrator. Trigger input leads on monostables should be kept short to avoid false triggering caused by induced noise pulses.

CRYSTAL OSCILLATORS

For applications requiring higher-frequency or more stable frequency pulses than the 555 is capable of, usually we turn to crystal-based oscillators. These are available as complete packaged units that plug into a standard IC socket. But if you don't have one of these, you can use one of the circuits in Figure 2-36a, b, or c. In the TTL version of Figure 2-36a, the resistor voltage dividers bias the inverters in their active region, about 1.35 V, where their gain is highest. Since each inverter gives a phase shift of 180°, the output of the second inverter is in phase with the input of the first. At its series resonant frequency the crystal supplies positive or in-phase feedback to produce oscillations at this frequency, 1 MHz for the circuit

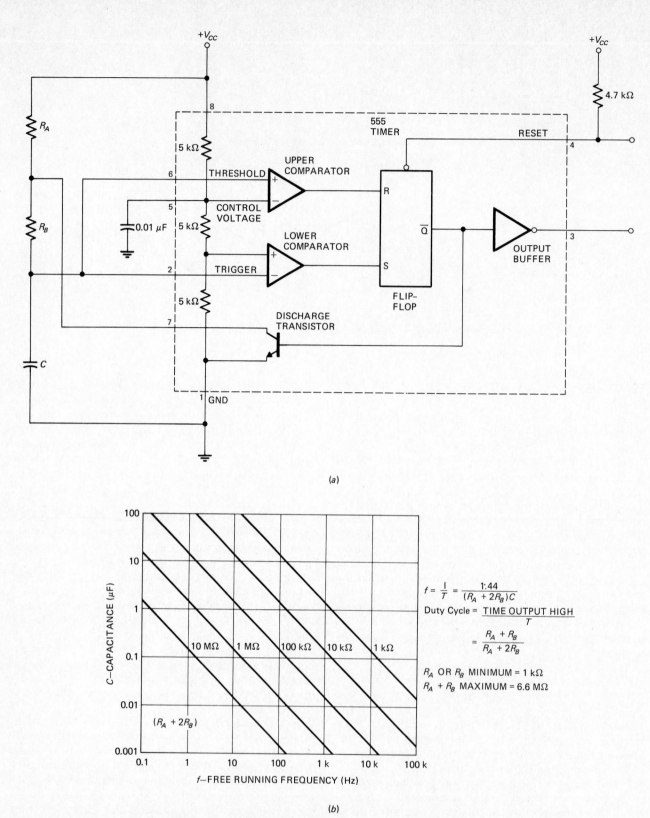

FIGURE 2-34 (a) LM555 timer IC used as a TTL or CMOS pulse source. (b) Chart for determining R and C values for 555 astable. (National Semiconductor.)

shown. Some adjustment of the resistors in each divider may be necessary to make sure the inverters are biased near the center of their active region. Low-impedance crystals work best, and it is a good idea to use a third inverter as a buffer on the output so that changes in loading do not affect the oscillator frequency. Lower frequencies can be derived from the stable 1-MHz source with digital dividers, as is discussed in Chap. 4.

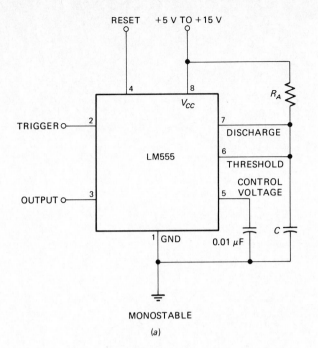

MONOSTABLE

(a)

TIME OUTPUT HIGH = 1.1RC

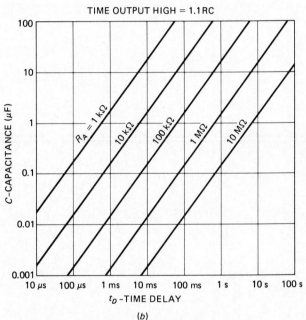

(b)

FIGURE 2-35 *(a)* LM555 timer used as a monostable multivibrator or pulse stretch. *(b)* Chart for 555 monostable R and C values. *(National Semiconductor.)*

The CMOS oscillator circuit in Figure 2-36b uses a single inverter and the parallel resonant mode of the crystal because, when it is operated on a +5-V supply, the propagation delay of two CMOS inverters in series is too long to produce a 1-MHz oscillator. R_F biases the gate output into its active region. A buffer gate on the output of the oscillator is again recommended.

The oscillator circuit in Figure 2-36c uses a standard MECL 10,000 series NOR gate to produce a high-frequency oscillator with ECL output levels. Active region

biasing is developed by the 1.6- and 5.1-kΩ resistors. As before, the crystal supplies positive feedback to produce oscillations. The 510-Ω resistors are pull-down resistors on the MECL emitter follower outputs.

OUTPUT SIGNAL HANDLING

For short-distance connections between the output of one gate and the input of the next, a simple jumper wire or printed-circuit board trace will deliver a signal with no problem. However, as the distance between the driving and driven gates increases beyond a certain length, different for each logic family, the connecting wire, coaxial cable, or trace must be recognized as a transmission line and treated as such. If it is not treated as a transmission line, reflection problems may cause extra input signal transitions.

TRANSMISSION LINES

A transmission line will appear to the driving gate as a load equal to the characteristic impedance of the line Z_o. Z_o is equal to $\sqrt{L_o/C_o}$, where L_o is the inductance per foot of length and C_o is the capacitance between the two conductors per foot of length. Since L_o and C_o both contain "per foot," this cancels out. Z_o then is the same value whether the transmission line is 5 m or 5000 m long! For a wire above ground or two open wires, Z_o will be between 30 and 600 Ω. Coaxial cables are manufactured with standard fixed impedance values such as 50, 62, 75, 93 Ω, etc. The Z_o of a common coaxial cable, RG-58AU, is 50 Ω. Uniform, low-impedance transmission lines are created on printed-circuit boards by leaving the copper foil covering on one side of the board as a ground plane, where possible, and by controlling the width of the signal trace on the other side. The fiberglass-epoxy material serves as a dielectric between the two conductors. For example, a 0.106-in [0.27-cm] wide trace creates a 50-Ω transmission line on standard 0.062-in [0.16-cm] fiberglass-epoxy board with ground plane. Multilayer printed-circuit boards have the ground plane as a middle layer, to allow transmission lines on both top and bottom of the board. The velocity of a signal along a transmission line is dependent on the dielectric material and is between 6 and 8 in/ns [15.2 to 20.3 cm/ns]. (For reference, the velocity of light is about 11.8 in/ns [30.0 cm/ns].)

If you put a pulse on one end of an infinitely long transmission line as in Figure 2-37a, the pulse would just propagate along the line forever. If the line is not infinitely long and the end away from the source is open, the pulse will pile up charge, or voltage-double, at the high-impedance, open end of the line and send an in-phase, reflected pulse back to the source (see Figure 2-37b). The way it does this is as follows. Note in Figure 2-37b that, at the open end, the reflected pulse adds to the incoming pulse for the duration of the incoming pulse to produce a voltage that is double the amplitude of the incoming pulse. (It is as if the reflected pulse

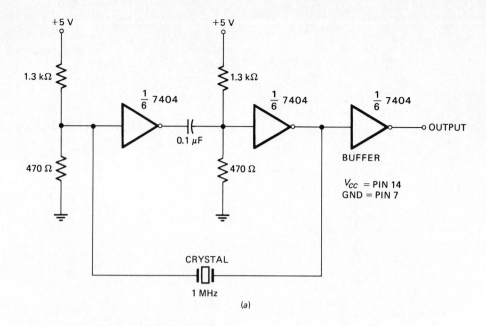

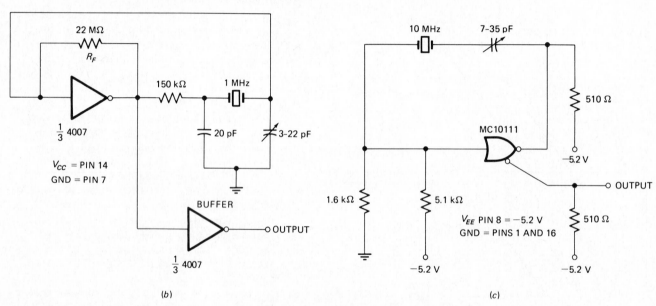

FIGURE 2-36 *(a)* TTL crystal oscillator. *(b)* CMOS crystal oscillator. *(c)* ECL crystal oscillator. *(Courtesy Motorola.)*

climbed up over and rolled down the back side of the incoming pulse, and then traveled back down the transmission line to the source in the same phase and amplitude as it left.) If the end of the line away from the source is short-circuited, a reflected pulse is produced that is 180° out of phase with the pulse sent out (see Figure 2-37*c*). This reflected pulse *must* be 180° out of phase because the *sum* of incoming and outgoing voltages must equal *zero* at the *short-circuit*. A reflected pulse will go back to the source. If the output impedance of the gate is not equal to the impedance of the line, the pulse will reflect again to the end of the line. Figure 2-37*d* shows the effect this bouncing back and forth has

on the pulse waveform seen at the receiving end of a cable with an open end. As you can see, the waveform overshoots and then rings across the logic threshold voltage several times instead of the single low-to-high transition desired. Figure 2-37*e* shows the waveform seen at the end of a line, which has a low impedance connected across it. As you can see, the first part of the low-to-high transition undershoots and may or may not make it above the threshold voltage.

A simple way to prevent these reflection problems is to trick the pulse going out the line into thinking that it sees an infinite-length line. You do this by connecting across the end of the line a resistor equal in value to the

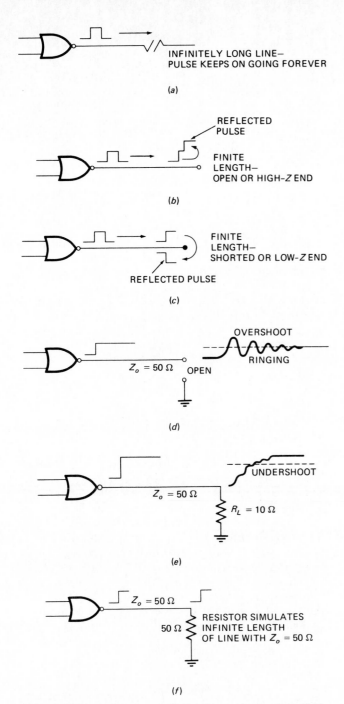

FIGURE 2-37 *(a)* Pulse applied to transmission line. *(b)* In-phase reflection of pulse from open end of transmission line. *(c)* Out-of-phase reflection from short-circuited end of transmission line. *(d)* Signal observed at open end of a transmission line showing overshoot and ringing after several reflections. *(e)* Signal observed at end of transmission line terminated with *R* less than *Z*. *(f)* Signal observed at end of transmission line terminated with *R* equal to *Z*.

characteristic impedance of the line, as shown in Figure 2-37*f*. The resistor dissipates the energy of the pulse and prevents reflections. This method is referred to as

parallel termination, and, where possible, it is the best to use. For example, when using a pulse generator with a 50-Ω output, you should connect to it with 50-Ω coaxial cable and terminate the end of the cable with a 50-Ω resistor. To summarize: The output of a line terminated with a resistance more than Z_o will show overshoot; the output of a line terminated with a resistance less than Z_o will show undershoot; but a line terminated with a resistance equal to Z_o will transmit a pulse without reflection problems.

Now that you know a little bit about transmission lines, your next question probably is, How long a connecting wire can I get away with before I have to worry about all that? The answer is that you will have no problem if the output risetime of the gate is 3 or 4 times the propagation delay for the length of the line. For TTL, with a 15-ns risetime, you can have a line with 4- to 5-ns delay. Since the signal travels about 7 in/ns [17.8 cm/ns], you can use a wire over ground plane of 28 to 35 in [0.7 to 0.9 m] without problems. Without ground plane, open-wire length should be no more than 12 or 14 in [30.5 to 35.6 cm]. For 5-V CMOS with several hundred nanosecond risetimes, the distance becomes 20 or 30 ft [6.1 to 9.1 m]. But for MECL III with its 1-ns risetimes, the maximum open-line length before problems drops to about *1* in [2.5 cm]. Fortunately, ECL outputs can drive parallel termination resistors. As discussed previously, ECL operating on −5.2 V can be parallel-terminated for 50-Ω lines either by connecting a 50-Ω resistor from the line to −2 V or with a divider of 82 Ω to ground and 130 Ω to −5.2 V. The second method works because the voltage divider produces about −2 V with a source or Thevenin impedance equal to 82 Ω in parallel with 130 Ω, or about 50 Ω.

Now the purpose of clamp diodes *D*1 and *D*2 on the inputs of a TTL gate in Figure 2-17*a* can be explained. A high-to-low pulse hitting a standard TTL input without the clamp diodes will see a low impedance and be reflected back and forth to give a waveform as shown in Figure 2-38. As you can see, the waveform may cross the threshold to a high again before finally settling down to a low. The clamp diodes *D*1 and *D*2 of Figure 2-17*a* pre-

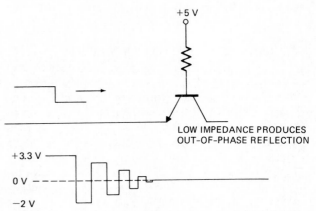

FIGURE 2-38 Signal received at TTL input without clamp diodes.

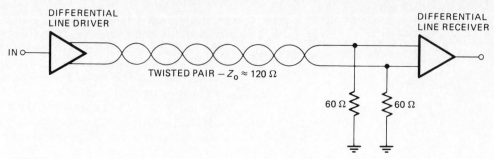

FIGURE 2-39 A differential line driver and receiver are used for data transmission.

vent the TTL inputs from going any more negative than about 1 V. Below this voltage the clamp diodes conduct to drain energy from the pulse and reduce the amplitude of the reflection to the point where it doesn't come near the threshold voltage upon its return.

LINE DRIVERS AND RECEIVERS

The standard outputs of most common logic families such as CMOS and TTL cannot drive the resistive load of a parallel termination at the end of a line. To send signals from these families over long distances, it is best to use specially designed output drivers such as the Signetics N8T13. These drivers accept TTL input levels and put out enough current to drive 50-Ω or higher impedance lines with a termination resistor at the end. The corresponding line receiver is the Signetics N8T14. The line receiver has a high input impedance, so it does not load the termination.

When signals are transmitted a long distance in very noisy environments, differential line drivers with complementary outputs such as the 75110 are used to drive a line of two wires twisted together. The signal is received by a differential line receiver as in Figure 2-39. Since both wires of the connecting twisted pair pass through the same environment, noise is induced equally in each wire. This common mode noise is rejected by the differential input of the receiver. Note that the two terminating resistors are in series across the twisted pair line, so their sum is equal to Z_o of the line.

POWER AND GROUND PROBLEMS

The two main ground problems encountered are IR voltage drops in long wires and noise induced in lines by combining logic grounds with ac power grounds. The first is illustrated in Figure 2-40. The IR voltage drops in the wire leave only 4 V across the ICs on board 2, which, if they are TTL, is not enough to operate them properly. This is another illustration of daisy-chain wiring which was warned against in Chap. 1. The problem is eliminated by branched wiring or producing the +5-V supply with a voltage regulator on each board and providing separate heavy-gage grounds from each board to a single common ground point (see Figure 1-4).

The problem of noise induced in signal grounds by ac noise currents can be eliminated by making sure that the entire system is grounded to the ac safety ground at a single point. Refer to the circuit in Figure 2-41. Unless the connection between the signal ground and ac safety ground in instrument 2 is broken at point X, noise voltages generated in the safety ground will appear on the signal ground also. Excessive noise on the signal ground can disrupt the operation of logic gates in instrument 2.

INTERFACING LOGIC FAMILIES TO ONE ANOTHER

When interfacing to different logic families, you must consider both the voltage levels and the current requirements of each family. Table 2-2 will help you with this.

TTL SUBFAMILY INTERCONNECTIONS

Interconnection among different TTL subfamilies is simple because the input and output voltage levels are all compatible. Care must be taken not to exceed the fan-out limits. Remember, the 74LS00 family and the 74L00 family can drive only two standard TTL inputs. When you have a load made up of gate inputs from different families, add up the maximum input low currents

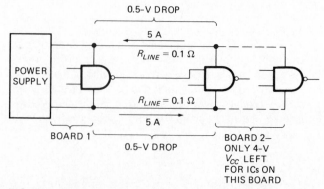

FIGURE 2-40 How IR voltage drops cause logic malfunction.

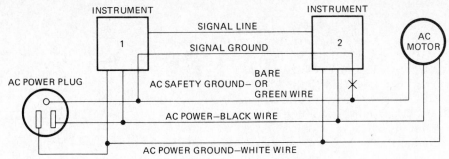

FIGURE 2-41 Illustration of potential ground-loop problem.

to make sure the total does not exceed the output low current-sinking capability of the driving gate. For example, to see whether a 74LS00 output can drive a load of one 7400 gate and five 74LS00 gates, use Table 2-2 to find the maximum logic 0 input currents. For the 7400 this is 1.6 mA, and for the five 74LS00 gates this is 5×0.36 mA, or 1.8 mA. The total input load current is 3.4 mA, which is less than the 4 mA that the 74LS00 output is capable of handling. Therefore, the 74LS00 can drive these gates.

TTL TO CMOS TO TTL

When you connect TTL families to CMOS operating on 5 V, a 10-kΩ pull-up resistor usually is required on the output of the TTL to make sure the output reaches a legal high level for the CMOS. Interface of CMOS outputs to TTL family inputs requires careful consideration of the input low current drive required by the TTL and the output low drive capability of the particular CMOS circuit. For example, the CD4000A NOR gates operating on a 5-V supply may sink only their minimum (worst-case current) of 0.4 mA in an output low state. This would be enough to drive two low-power TTL inputs or one low-power Schottky TTL input. However, the CD4011AE NAND gates with a 5-V supply may sink only their minimum output low current of 0.12 mA, which would not be enough to drive even one low-power TTL input low. Little problems such as this may be overlooked in a prototype and not show up until an instrument goes into production, where it is found that only those ICs which happen to have a little more drive will work in a circuit. This demonstrates the circuit designer's rule: What works for one chip once in a while, will not work for all chips all the time.

The 4000B, the 74C00, and the 4500 series CMOS are all standardized to drive one low-power Schottky TTL or two low-power TTL loads. If you are in doubt or need more drive, use buffers such as the CD4049 inverting or CD4050 noninverting, which come six to a package and can each drive up to two standard TTL inputs. To interface in either direction between CMOS operating on +10- or +15-V supplies, and TTL or CMOS operating on 5-V, translators such as the 74C901 hex inverting or 74C902 hex noninverting buffers can be used.

TTL TO PMOS TO TTL

High-threshold PMOS output voltage swing is typically from ground to some negative voltage around −13 V. Newer low-threshold PMOS outputs swing from a +5-V high to a −12-V low. To connect from TTL or 5-V CMOS levels to high-threshold PMOS levels, you can use a level translator/buffer such as the National Semiconductor DS7800. Some early low-threshold PMOS circuits have data inputs that will accept TTL levels, but have clock inputs that require external drivers to supply a full +5- to −12-V swing. Later low-threshold PMOS has inputs that can all be driven by TTL with a pull-up resistor or by 5-V CMOS, and outputs that can drive one standard TTL input. A PMOS output will try to pull down to −12 V, but TTL input clamp diodes or CMOS static diodes prevent the output from dropping any more negative than about −1 V. Since a PMOS output cannot sink enough current to harm the diodes, this is acceptable.

NMOS TO TTL TO NMOS

Most NMOS, whether powered on +5 V and ground or +5 and −12 V, has inputs designed to accept standard TTL or 5-V CMOS levels. If the TTL is driving any other TTL along with an NMOS input, a pull-up resistor to +5 V is required to make sure an input high is high enough to give a good noise margin. The outputs of most NMOS will drive one standard TTL input, four low-power Schottky TTL inputs, or eight low-power TTL inputs.

TTL TO ECL TO TTL

TTL, CMOS, or NMOS to ECL level conversions are easily done with the 10124 quad TTL-to-ECL translator. TTL outputs can be connected directly to the translator inputs, but since these inputs each require sinking as much as 3.2 mA, a buffer is required between a CMOS or NMOS output and a translator input.

For conversion from ECL to TTL, CMOS, or NMOS levels, the 10125 quad ECL-to-TTL translator may be utilized. If you are driving a mixture of TTL and NMOS, a 10-kΩ pull-up resistor to +5 V on the output of the translator ensures a proper MOS high level.

TTL TO I²L TO TTL

Since the output structure of I²L is similar to an open-collector TTL, a resistor can be tied from the output to +5 V to give NMOS, CMOS, or TTL output levels. If the injector current is low, then the output transistor will not sink enough current to drive TTL directly. A CMOS buffer makes a good connecting link. TTL, NMOS, or CMOS outputs will drive I²L inputs directly, but a series current-limiting resistor must be used to prevent excessive current flow through the base-emitter junction of the input transistor. Most I²L LSI circuits have inputs and outputs directly compatible to TTL and CMOS.

INTERFACING LOGIC GATES TO SIMPLE DISPLAYS AND RELAYS

LEDs

Many times you may want to have a visible indication of the output state of a logic gate without having to connect an oscilloscope or a DVM. If properly connected, a simple light-emitting diode (LED) can do this. An LED is a diode, usually made from gallium arsenide phosphide, that emits visible light when current is passed through it in the forward direction. The color of the light given off is determined by the material and may be red, orange, yellow, or green. The forward voltage drop across the diode varies from about 1.5 V for a red LED at 20 mA to about 3 V for a green LED at the same current.

Standard TTL, Schottky TTL, or high-power TTL can supply enough drive current to light an LED. If you wish to indicate when the output of a gate goes high, the circuit of Figure 2-42a will do this. To indicate a low output, the circuit of Figure 2-42b may be used. Current-limiting resistors are needed to prevent overheating of the gate output or the LED.

To drive an LED from low-power Schottky TTL, low-power TTL, CMOS, or NMOS, a handy way is to use a driver transistor, as shown in Figure 2-43a or Figure 2-43b. Note current-limiting resistors on the base of the driver transistors. The value of the collector resistor is determined by subtracting the voltage drop across the LED (about 1.5 V for a red LED at 20 mA) and saturation voltage of the transistor (say, 0.5 V at 20 mA) from the 5-V V_{CC}. The resultant voltage divided by the desired LED current gives the resistor value. Using the above values of 20 mA and a resistor voltage drop of 3 V gives a resistor value of 150 Ω. The value of the base current-limiting resistors is not critical; 1 to 10 kΩ is appropriate. It is important to be aware that a gate output which is driving an LED directly (Fig. 2-42) *usually will not reach legal logic levels*. The transistor-driven LED indicator, however, will *not* load the logic circuit output.

LAMPS

Small low-voltage incandescent lamps can be driven by the circuits in Figure 2-43a or b. With 6- or 4.5-V lamps, the collector current-limiting resistor can be left out.

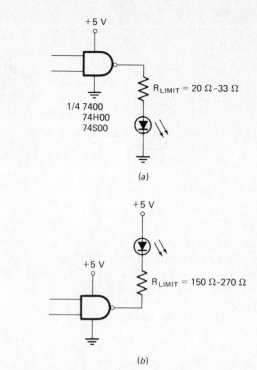

FIGURE 2-42 *(a)* Standard TTL gate connected to light LED when output is high. *(b)* Standard TTL gate connected to light LED when output is low.

A circuit such as that in Figure 2-43a can be used to drive a neon lamp. These lamps require 55 to 150 V to turn on; the collector-emitter and collector-base breakdown voltages of the driver transistor must be greater than this applied high voltage. The collector current-limiting resistor must be increased to a much higher value, usually around 100 kΩ.

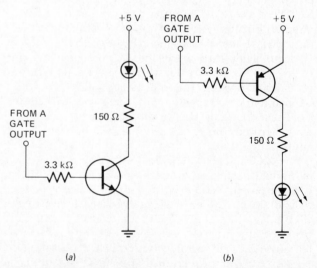

FIGURE 2-43 *(a)* Transistor LED driver for detecting high output. *(b)* Transistor LED driver for detecting low output.

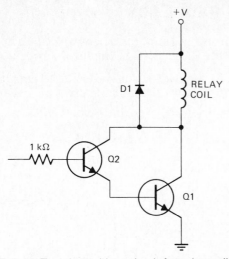

FIGURE 2-44 Transistor driver circuit for relay coil.

MECHANICAL RELAYS

When you want to control larger direct currents up to a few amperes with logic level outputs, the transistors in the circuits of Figure 2-43a and b can be replaced by Darlington transistors. For control of alternating currents or still larger direct currents, use either a mechanical or solid-state relay. When a control current is passed through the coil of a mechanical relay, an electromagnet is formed which pulls some heavy-duty switch contacts open or closed. The circuit of Figure 2-44 can be used to drive a low-voltage relay coil, needing up to 200-mA coil current, from TTL, CMOS, or NMOS logic levels. Diode D1 across the relay coil is absolutely necessary to prevent the inductive "kick" or back (electromotive force) voltage, produced when current through the coil is shut off, from destroying the transistors. Since the relay contacts are just mechanical switches, they can be used to turn on 120- or 240-V ac lights, heaters, motors, etc., which cannot be controlled by simple transistor circuits.

SOLID-STATE RELAYS

Mechanical relays have the problem that the contact points arc, then eventually burn out and fail to make contact. This and other relay problems are solved by employing solid-state relays that use a Triac switch rather than contact points to switch the large currents. Since the switching is done inside a semiconductor device, there is no arcing. Also, many solid-state relays have an internal zero-voltage detector, whose circuitry makes sure the Triac switch is triggered only at the zero-crossing points of the 120-V ac power input. Zero-point switching minimizes the radio-frequency interference (RFI) created by rapidly turning large currents on or off. Most solid-state relays have TTL-compatible inputs, and some have the input circuitry optically coupled to the triac control circuits. In these optically coupled circuits, the input current lights an internal LED. Light from the LED drives the base of a phototransistor connected to the Triac switch trigger circuitry. Since light transmits the signal, optical coupling completely separates or isolates the input and control circuit voltages from each other. Figure 2-45 shows a block diagram of a solid-state relay with both optical coupling and zero-point switching.

Optical couplers are available as separate units, such as the Monsanto MCT277. They can be used whenever you want to transmit a logic signal but keep the voltage supplies separate.

AN IMPORTANT POINT

In this chapter, you have been introduced to the basic operation, characteristics, and uses of several common logic families. Because of the vast amount of information represented, and because some of the parameters shown in Table 2-2 for comparison purposes vary a great deal among family members, this chapter is just a start. You should obtain manufacturers' data books and application notes for any specific logic family with which you will be working. Much of the material in these data books and application notes should be understandable to you now.

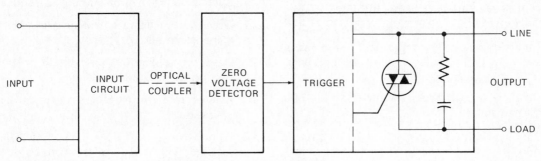

FIGURE 2-45 Solid-state relay with optical coupling. *(DEC.)*

REVIEW QUESTIONS AND PROBLEMS

1. State whether each of the following quantities or devices is analog or digital:
   a. Speed of a car
   b. Energy levels of electrons around an atom
   c. Time
   d. Number of water molecules in the ocean
   e. Socket ratchet
   f. Transformer output voltage

2. Convert the following decimal numbers to binary:
   a. 10
   b. 72
   c. 4096
   d. 0.704
   e. 0.35

3. Convert the following binary numbers to decimal:
   a. 110
   b. 1010101
   c. 1111
   d. 0.1101
   e. 0.011

4. Write the truth table and the boolean expression for the following gates:
   a. Inverter
   b. Two-input AND
   c. Three-input OR
   d. Two-input NAND
   e. Four-input NOR
   f. Two-input exclusive OR
   g. Two-input exclusive NOR

5. Complete the following:
   a. The output of an AND gate with one input tied low will be at a logic _____.
   b. The output of a NOR gate with both inputs tied low will be at a logic _____.
   c. The output of an exclusive OR gate with one input low and the other high will be at a logic _____.

6. Draw a top view of a typical 16-pin dual in-line package IC. Label pin numbers. How do you recognize pin 1 on a DIP device?

7. a. For standard TTL give the maximum V_{CC} voltage and the maximum input voltage with V_{CC} at 5 V.
   b. What is the difference between the 7400 series TTL and 5400 series TTL?

8. Give the worst-case input high and low voltages, output high and low voltages, and typical propagation delay for a standard TTL gate.

9. Define noise margin. What are the noise margins for a TTL gate?

10. What is the maximum current required to produce an input low on a standard TTL input? How much current is required for an input high?

11. What is the fan-out of a standard TTL gate? What are the effects of exceeding this fan-out?

12. A logic high of 3.6 V is applied to the circuit shown in Figure 2-46. What logic level will be present at the output?

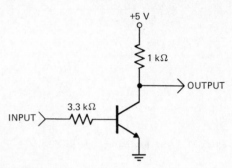

FIGURE 2-46

13. What is the purpose of the diode in the totem-pole output structure of the TTL?

14. Why should two totem-pole outputs not be connected together?

15. Why are bypass capacitors needed between V_{CC} and ground near TTL circuits? What are good values of bypass capacitors to use?

16. What are the advantages and disadvantages of open-collector outputs?

17. What are the three states of three-state output TTL? What are three-state outputs used for?

18. Why is Schottky TTL much faster than standard TTL?

19. Use Table 2-2 to answer the following questions:
   a. How many standard TTL inputs can one 74LS series output safely drive?
   b. How many 74ALS inputs can one 74ALS output safely drive?
   c. How many 74ALS inputs can one 74C00 output safely drive?

20. What are the typical supply voltages for ECL? Why do ECL gates have such a short propagation delay? Describe a major problem of ECL connections.

21. Describe the operation of an I^2L gate. What are some advantages of I^2L?

22. Describe the handling precautions of MOS transistors or integrated circuits.

23. Draw a circuit for a CMOS inverter and describe its operation.

24. What is a major advantage of CMOS at low frequency? Why does this advantage disappear at higher frequencies?

25. What should be done with unused TTL inputs? With unused CMOS inputs?

26. Why should the amplitude of a signal generator be checked with an oscilloscope before it is connected to a logic gate?

27. Show a simple circuit that can be used to produce a debounced switch signal.

28. Describe the operation of a comparator.

29. Why is a Schmitt trigger required to condition some input signals?

30. When should a crystal oscillator be used as a signal source instead of a cheaper source such as the 555 timer circuit?

31. Show a waveform produced by a pulse at the end of a transmission line terminated with a resistance greater than the characteristic impedance of the line.

32. What value of resistor should be used to terminate a 75-Ω transmission line?

33. What is a proper termination for a 50-Ω transmission line from ECL operating on a −5.2-V supply?

34. Why should power ground and the signal ground be joined at only one point?

35. When you are interfacing TTL to CMOS, why is it a good idea to use a pull-up resistor on the output of the TTL?

36. Why will most CMOS not drive standard TTL directly? What are two methods of properly interfacing from CMOS to standard TTL?

37. What is required to interface ECL to TTL?

38. With a circuit diagram show how an LED can be used to indicate a low output state on a standard TTL gate.

39. Show a circuit that can be used to drive a 100-mA relay coil from a CMOS or TTL gate. Why must a reverse-biased diode always be placed across the relay coil?

40. What are the advantages of solid-state relays over standard mechanical relays?

REFERENCES/BIBLIOGRAPHY

Signetics Analog Manual—Applications/Specifications, Signetics Corporation, 1976, sec. VI, "Timers," pp. 105–127.

Logic Data Book, National Semiconductor, 1981.

Linear Integrated Circuits (handbook), National Semiconductor, February 1975.

COS/MOS Digital Integrated Circuits (Selection Guide/Data Application Notes), 1975 RCA Solid State DATA BOOK Series.

Interface Integrated Circuits (handbook), National Semiconductor, October 1975.

Signetics Data Manual: Logic, Memories, Interface, Analog, Microprocessor, Military, Signetics Corporation, 1976.

CMOS Data Book, National Semiconductor, 1978.

MECL System Design Handbook, 1st ed., Motorola Semiconductor Products, Inc., compiled by the Computer Applications Engineering Department: William R. Blood, Jr. (author); Jon M. DeLaune and Jerry E. Prioste (contributors); Edmund C. Tynan, Jr. (ed.), October 1971.

Solid State Relays, RS Series (data sheet), Catalog no. ISC-109, IDEC Systems and Controls Corporation, Santa Clara, CA.

Morris, Robert L., and John R. Miller (ed.): *Designing with TTL Integrated Circuits* (prepared by the Applications Staff of Texas Instruments Incorp.), Texas Instruments Electronics Series, McGraw-Hill, New York, 1971.

Signetics Digital Linear MOS Applications (handbook), sec. 7, "MOS Applications," Signetics Corporation, 1974.

Low Power Schottky and Macrologic TTL, Fairchild Semiconductor, Components Group, Fairchild Camera and Instrument Corp., 1975.

McMOS Handbook—Products, Characteristics, Applications, Motorola Semiconductor Products, Inc., 1974.

Blakeslee, Thomas R.: *Digital Design with Standard MSI & LSI,* Chap. 10: *Nasty Realities II,* "Noise and Reflections," Wiley, New York, 1975, pp. 256–278.

Intel Data Catalog 1982, Intel Corporation, 1982.

Allan, Alan: "Noise Immunity Comparison of CMOS versus Popular Bipolar Logic Families," *Application Note AN-707,* Motorola Semiconductor Products, Inc., 1973.

"Field Effect Transistors in Theory and Practice," *AN-211A Application Note,* prepared by Applications Engineering, Motorola Semiconductor Products, Inc.

CHAPTER 3

COMBINATIONAL LOGIC, MULTIPLEXERS, CODES, AND ROMs

In Chap. 2 we introduced basic logic gates and discussed the input and output parameters of common logic families. This chapter shows how logic gates can be combined to form digital systems such as a furnace controller, multiplexer, or code converter.

Upon completion of this chapter, you should be able to:

1. Predict the output of basic digital circuits in positive logic, negative logic, or mixed logic.

2. Use boolean algebra to simplify logic expressions.

3. Use a Karnaugh map to simplify up to a four-variable logic expression.

4. Draw circuits for simple gates used as switches.

5. Draw circuits for a multiplexer and a demultiplexer used to transmit several signals on a single wire.

6. Given the appropriate tables, convert a number in one code to the equivalent number in another code.

7. Show how exclusive OR gates are used to generate and detect parity.

8. Given a specific ROM (read-only memory), determine the number of address and enable lines required.

9. Define the timing and other parameters found on a ROM data sheet.

10. Describe the programming and erasing of a typical EPROM or EEPROM.

11. Describe the differences between a ROM and a programmable logic array (PLA).

COMBINING LOGIC GATES

Digital logic circuits may be represented by logic diagrams, truth tables, or boolean algebra expressions. It is important to learn to work with all three forms. Figure 3-1 shows the three forms for a common gate circuit. Given the logic diagram in Figure 3-1a, the truth table is constructed by tracing the effect of each possible input combination through the circuit. The input combination of $A = 0$, $B = 1$, and $C = 1$, for example, produces a 0 on the output of the upper AND gate and a 1 on the output of the lower AND gate. The 0 from the upper AND and the 1 from the lower AND cause a 1 on the Y output of the gate. A circuit such as this is easy to trace if you remember that for an AND gate the output will be 0 or low if any input is 0 or low. The boolean expression for a circuit can be derived directly from the logic diagram by writing the expression for each section and combining them as shown in Figure 3-1a. The output of the upper AND gate is $A \cdot B$. The output of the lower AND gate is $B \cdot C$. These terms are ORed together so their final output Y is equal to $A \cdot B + B \cdot C$.

To produce the boolean algebra expression from the truth table, first write a term for each input combination that produces an output 1 or *true* in the truth table. The circuit in Figure 3-1, for example, produces an output 1 for inputs of $A = 0$, $B = 1$, and $C = 1$. In boolean algebra this term would be expressed as $\overline{A} \cdot B \cdot C$. Note that all three variables are included in the term. To form the complete boolean expression, the individual terms are ORed (summed). For the circuit in Figure 3-1 the result is

$$Y = \overline{A} \cdot B \cdot C + A \cdot B \cdot \overline{C} + A \cdot B \cdot C$$

as shown below the truth table. The expression is said

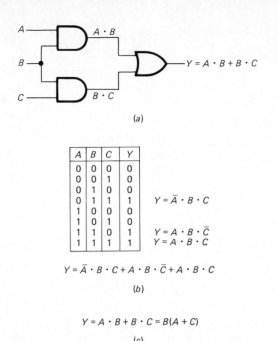

$$Y = \bar{A} \cdot B \cdot C$$

$$Y = A \cdot B \cdot \bar{C}$$
$$Y = A \cdot B \cdot C$$

$$Y = \bar{A} \cdot B \cdot C + A \cdot B \cdot \bar{C} + A \cdot B \cdot C$$

(b)

$$Y = A \cdot B + B \cdot C = B(A + C)$$

(c)

FIGURE 3-1 Combinational logic circuit. *(a)* Schematic. *(b)* Truth table. *(c)* Output expression.

to be in a *sum-of-products* form. This result is correct, but it is not the same as that derived directly from the logic diagram, and it is not the simplest boolean expression that describes the function of the circuit. Later we show that the two expressions are equivalent. The boolean expression derived either from the logic diagram or the truth table will seldom be in its simplest form. Two common methods of simplifying logic functions are boolean algebra and Karnaugh mapping.

BOOLEAN ALGEBRA

Figure 3-2 shows the rules for boolean algebra. Some of these may seem strange at first, but you can see from either a truth table or a logic diagram for each that they are valid. Note that some of these rules are not the same as they are for ordinary algebra. The rules are valid for any number of variables, but for simplicity they are shown here with the minimum number for each rule.

Rule 1, $A \cdot 0 = 0$, simply states that if any input of a multiple-input AND gate is tied low, the output will be low regardless of the logic state on the other input(s). This is called *disabling* the gate and sometimes called *masking* because any signal on the A input will not get to the output. You can see this in the AND/NAND truth table of Figure 3-2.

The NAND function also is shown in the truth table for the first rule because it is used so commonly. For a NAND gate the output will be high if any input is low.

Rule 2 indicates that if one input of a two-input AND gate, or all but one input of a multiple-input AND gate, is tied high, the output will be the same as the remain-

ing input. This is easily seen from the truth table in Figure 3-2. A high on all but one input, therefore, *enables* an AND gate to pass the signal on A to the output.

A NAND gate with all inputs but the signal input A tied high will function as an inverter.

Rule 3 indicates that if the inputs of an AND gate are tied together, the output follows the input. This configuration is sometimes referred to as a *buffer* and is shown in the truth table as the case $A = 0$ and $B = 0$ and the case $A = 1$ and $B = 1$.

You can also determine from the truth table for rule 3 that a NAND gate with its inputs tied together functions as an inverter. Remember, however, that in practice tying together inputs is not desirable because of the added capacitive load this places on a driving gate.

Rule 4 shows that if a variable is ANDed with its complement, the output is always low. Because one of the inputs is always low, the output is low.

Rules 5, 6, 7, and 8 show the corresponding results for OR/NOR gates. Note that two-input OR/NOR gates are enabled by a low on one input and disabled by a high on one input. If the inputs of an OR gate are tied together, the output follows the input. A NOR gate with its inputs tied together functions as an inverter. If a variable is ORed with its complement, the output is always high because one input is always high. Remember that a high into any input of an OR gate produces a high on the output.

Rule 9 states that two inversions of a logic signal cancel. A 0 passed through one inverter produces a 1. This 1 passed through another inverter produces the original 0 again.

Rules 10 and 11 indicate that the inputs of a gate are equivalent and interchangeable.

Rules 12 and 13 show that the order in which variables are ANDed or ORed doesn't matter. For example, if A is ORed with B and the result ORed with C, this will give the same result as ORing B with C and ORing that result with A.

Rules 14 and 15 illustrate the *distributive properties* of boolean expressions. Rule 14 is equivalent to factoring in ordinary algebra. Rule 15 is not so obvious, but you can think it through as follows. The first expression says that if $A = 0$, both B and C have to be 1 to produce an output 1. If A is a 1, then B and C are "don't cares" because a 1 ORed with anything produces an output 1. For the second expression, if $A = 0$, then 0 ORed with B reduces to B, and the $A = 0$ ORed with C reduces to C. Therefore, the output will be 1 only if B and C are both 1 as before. If A is 1, then B and C are "don't cares" because $A = 1$ makes the $A + B$ term equal to 1 and the $A + C$ term equal to 1 (see rule 6). Since $1 \cdot 1 = 1$, the output will be a 1. This discussion has shown simply how the rules of boolean algebra can be used to reduce or alter the form of logic expressions.

Rules 16 and 17 are commonly used *reductions*. To prove rule 16, first factor A out of both terms to give $A(1 + B)$. Then since $1 + B$ is equal to 1 (rule 6), this reduces to $A \cdot 1$, which is equal to A (rule 2). To prove

1. $A \cdot 0 = 0$

2. $A \cdot 1 = A$

3. $A \cdot A = A$

4. $A \cdot \overline{A} = 0$

5. $A + 0 = A$

6. $A + 1 = 1$

7. $A + A = A$

8. $A + \overline{A} = 1$

9. $\overline{\overline{A}} = A$

10. $A \cdot B = B \cdot A$

11. $A + B = B + A$

12. $(A \cdot B) \cdot C = A \cdot (B \cdot C)$

13. $(A + B) + C = A + (B + C)$

14. $A(B + C) = AB + AC$

15. $A + BC = (A + B)(A + C)$

16. $A + AB = A$

17. $A + \overline{A} B = A + B$

De Morgan's Theorem

18. $\overline{A \cdot B} = \overline{A} + \overline{B}$

19. $\overline{A + B} = \overline{A} \cdot \overline{B}$

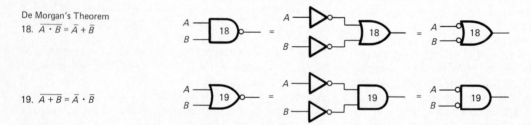

SIGNAL A	CONTROL B	AND Y	NAND $\overline{Y}$
0	0	0	1
0	1	0	1
1	0	0	1
1	1	1	0

SIGNAL A	CONTROL B	OR Y	NOR $\overline{Y}$
0	0	0	1
0	1	1	0
1	0	1	0
1	1	1	0

FIGURE 3-2 Boolean algebra equivalents.

rule 17, use rule 15 to expand the left side of the equation as follows:

$$A + \overline{A}B = (A + \overline{A}) \cdot (A + B)$$

Since $A + \overline{A} = 1$ (rule 8), this reduces to just $A + B$.

Before going on to the last two rules we show another example of how these rules are used to simplify logic expressions. Figure 3-1 shows the logic diagram and the truth table for a common logic circuit. As explained earlier, the boolean expression derived directly from its logic diagram is

$$Y = A \cdot B + B \cdot C$$

The boolean expression derived from the truth table is

$$Y = \overline{A}BC + AB\overline{C} + ABC$$

Using the boolean rules, you can show that these are equivalent as follows. Starting with

$$Y = \overline{A} \cdot B \cdot C + A \cdot B \cdot \overline{C} + A \cdot B \cdot C$$

factor AB out of last two terms, to give

$$Y = \overline{A} \cdot B \cdot C + A \cdot B \cdot (\overline{C} + C)$$

NAND   NOR $\overline{A + B} = \overline{A} \cdot \overline{B}$ AND OR

A	B	$A \cdot B$	$\overline{A \cdot B}$	$A + B$	$\overline{A + B}$	$\overline{A}$	$\overline{B}$	$\overline{A} \cdot \overline{B}$	$\overline{A} + \overline{B}$
0	0	0	1	0	1	1	1	1	1
0	1	0	1	1	0	1	0	0	1
1	0	0	1	1	0	0	1	0	1
1	1	1	0	1	0	0	0	0	0

$$\overline{A \cdot B} = \overline{A} + \overline{B}$$

FIGURE 3-3 Truth table demonstrating De Morgan's theorem.

Since $\overline{C} + C = 1$ (rule 8), the expression reduces to

$$Y = \overline{A} \cdot B \cdot C + A \cdot B$$

Factoring B out of the two remaining terms gives

$$Y = B \cdot (A + \overline{A} \cdot C)$$

By using rule 17 this can be reduced to

$$Y = B(A + C)$$

Multiplying this as shown in rule 14 gives

$$Y = A \cdot B + B \cdot C$$

which is the same as the expression derived directly from the logic diagram.

Rules 18 and 19 are usually referred to as *De Morgan's theorem*. They are very useful in analyzing and simplifying logic circuits. Rule 18 states that inverting the inputs to an OR gate produces a NAND function. In other words, a NAND gate can be thought of as an OR gate with inverted inputs. You can prove this by comparing the $\overline{A \cdot B}$ and $\overline{A} + \overline{B}$ columns in Figure 3-3.

Figure 3-2 also shows this theorem in logic symbol form. The inverted input OR symbol is often used to represent a NAND gate in logic diagrams. Figure 3-4*a* shows a schematic using the symbol to represent one-quarter of a 7400 quad two-input NAND gate IC. Drawing the schematic in this way makes it easier to determine the output expression and truth table. Since the output inverting bubbles of IC1a and IC1b are canceled by the inverting bubbles on the inputs of IC1c, you can see that the output is just $A \cdot B$ ORed with $C \cdot D$. The output expression is

$$Y = A \cdot B + C \cdot D$$

This is a common method of producing the OR function by using only positive logic NAND gates. The circuit of Figure 3-4*b* uses the same parts as that in Figure 3-4*a* and has the same truth table, but its output expression is not as immediately obvious as it was for the circuit using the inverted-input OR representation of IC1c.

Rule 19 states that a NOR gate is equivalent to an AND gate with inverted inputs. You can prove this by comparing the $\overline{A + B}$ and $\overline{A} \cdot \overline{B}$ columns in the truth table of Figure 3-3. Figure 3-2*b* also shows this theorem in logic symbol form. In Figure 3-5 the inverted AND representation of the CMOS 4001 NOR gate IC2c makes it easy to see that the output expression is

$$Y = (A + B) \cdot (C + D)$$

The inverting bubbles on the outputs of IC2a and IC2b are effectively canceled by the inverting bubbles on the input of IC2c. The technique shown in this example is often used to produce the AND function using only positive logic NOR gates.

Another useful way of thinking of De Morgan's theorem when you are using it to simplify boolean expressions is as follows. Splitting or joining the bar over the terms of a boolean expression changes the sign between

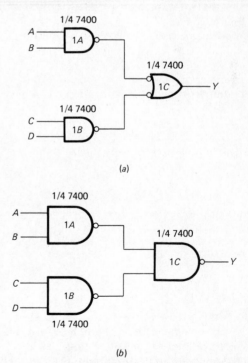

FIGURE 3-4 Logic circuit using positive logic NAND gates. *(a)* Inverted input OR representation of NAND. *(b)* Standard all-NAND schematic.

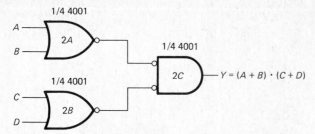

FIGURE 3-5 Use of inverted-input AND symbol to simplify tracing NOR gate circuit logic.

the terms. For example, splitting the bar over $\overline{A \cdot B}$ changes the sign to give $\overline{A} + \overline{B}$, which is equivalent according to rule 18. Joining the bar over $\overline{A} \cdot \overline{B}$ gives $\overline{A + B}$, which is equivalent according to rule 19. Here is how this trick is used to simplify boolean expressions. The expression derived directly from the three-NAND-gate circuit in Figure 3-4b is

$$Y = \overline{(\overline{A \cdot B}) \cdot (\overline{C \cdot D})}$$

Splitting the top bar and changing the sign between terms give

$$Y = \overline{(\overline{A \cdot B})} + \overline{(\overline{C \cdot D})}$$

Since double bars over any term cancel, this reduces to

$$Y = A \cdot B + C \cdot D$$

which is the same result as that determined directly from Figure 3-4a.

As another example, the boolean expression

$$Y = \overline{(\overline{A + B}) + (\overline{C + D})}$$

can be simplified by splitting the top bar to give

$$Y = \overline{(\overline{A + B})} \cdot \overline{(\overline{C + D})}$$

Again the double bars cancel to give

$$Y = (A + B) \cdot (C + D)$$

In conclusion, then, a NAND gate is equivalent to an OR gate with inverted inputs ($\overline{A \cdot B} = \overline{A} + \overline{B}$) and a NOR gate is equivalent to an AND gate with inverted inputs ($\overline{A + B} = \overline{A} \cdot \overline{B}$).

SUMMARY OF LOGIC GATE CONVERSIONS

To help you remember them, Figure 3-6 summarizes the gate transformations that can be done by adding inverters to the output, inputs, or both outputs and inputs of a gate. Adding an inverter only to an output transforms horizontally in the diagram, such as from an AND to a NAND or from a NOR to an OR. Adding inverters to only the inputs transforms along a diagonal in the diagram.

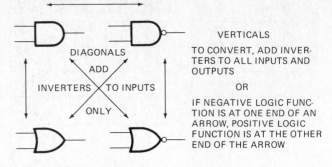

ADD INVERTER TO OUTPUT
TO CONVERT EITHER DIRECTION

DIAGONALS
ADD
INVERTERS TO INPUTS
ONLY

VERTICALS

TO CONVERT, ADD INVERTERS TO ALL INPUTS AND OUTPUTS

OR

IF NEGATIVE LOGIC FUNCTION IS AT ONE END OF AN ARROW, POSITIVE LOGIC FUNCTION IS AT THE OTHER END OF THE ARROW

ADD INVERTER TO OUTPUT
TO CONVERT EITHER DIRECTION

FIGURE 3-6 Summary of logic gate conversion.

An OR gate with inverted inputs functions as a NAND gate ($\overline{A} + \overline{B} = \overline{A \cdot B}$), and an AND gate with inverted inputs functions as a NOR gate ($\overline{A} \cdot \overline{B} = \overline{A + B}$). Note that the arrows are double-ended. This means that transformations in either direction are valid. These diagonal arrows summarize De Morgan's theorem.

The vertical arrows in Figure 3-6 represent the gate transformations performed by adding inverters to all the inputs and outputs of a gate. The boolean expression for a NAND gate with inverted inputs and an inverted output is

$$Y = \overline{\overline{\overline{A} \cdot \overline{B}}}$$

The long double bars cancel, so that

$$Y = \overline{A} \cdot \overline{B}$$

can be expressed as

$$Y = \overline{A + B} \qquad \text{(De Morgan's theorem)}$$

These vertical arrows show the correspondence between positive logic functions and negative logic functions, which is the topic of the next section. Another important point of the diagram in Figure 3-6 is that it shows how any logic function can be created by using just one kind of gate, a NAND for example, and inverters. Often tricks such as this are used by circuit designers to reduce the number of different ICs required in a design and to avoid wasting gates. For example, if only one NOR gate is needed, instead of putting in a 7402 quad two-input NOR gate IC, which contains four positive logic NOR gates, a designer might use a leftover NAND gate and three leftover inverters to produce the NOR function.

NEGATIVE LOGIC

So far in this book, for simplicity we have dealt with only *positive logic* conventions in which the more positive output voltage of a gate is a logic 1 and the more negative output voltage is a logic 0. ECL, early PMOS, and some computers often use *negative logic,* in which these conventions are reversed. In negative logic, a 0 is the most positive voltage, and a 1 is the most negative. For high-threshold PMOS, then, a negative logic 1 may be −13 V and a 0 may be ground. For TTL a negative logic 1 or high is 0.4 V, and a negative logic 0 or low is 2.4 V.

Positive Logic NAND

A	B	Y
0.4 V = L	0.4 V = L	2.4 V = H
0.4 V = L	2.4 V = H	2.4 V = H
2.4 V = H	0.4 V = L	2.4 V = H
2.4 V = H	2.4 V = H	0.4 V = L

$Y = \overline{A \cdot B}$

Negative Logic NOR

A	B	Y
0.4 V = H	0.4 V = H	2.4 V = L
0.4 V = H	2.4 V = L	2.4 V = L
2.4 V = L	0.4 V = H	2.4 V = L
2.4 V = L	2.4 V = L	0.4 V = H

$Y = \overline{A + B}$

(a)

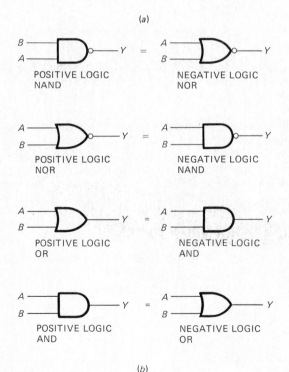

(b)

FIGURE 3-7 Negative and positive logic equivalents. *(a)* Truth tables. *(b)* Schematic symbols.

A gate such as one of the 7400 positive logic NAND gates still responds to input voltages in the same way. But since the logic level definitions are reversed, the gate will perform a NOR function in a system using negative logic conventions. You can use the truth tables in Figure 3-7a to show this. The upper part of the truth table shows the voltage levels and the equivalent logic levels for the familiar positive logic NAND gate. The lower part of the table shows the same input voltages with their negative logic levels. Focusing on the logic states for the lower part of the table, you can see that the output is low if any of the inputs is high. You should recognize this as a NOR gate function. A *positive logic NAND gate* performs the *NOR function in negative logic.* Similar analysis will show that a *positive logic NOR gate* performs the *NAND function in negative logic.* Likewise, a positive AND gate is equivalent to a negative logic OR gate and vice versa, as shown in Figure 3-7b. The vertical lines in Figure 3-6 indicate the equivalent gates in positive and negative logic. These vertical lines also indicate that you can convert, for example, a NAND gate to a NOR gate by inverting all inputs and outputs. In system schematics using entirely negative logic, the schematic symbols used are those for the functions of the gates. A gate actually may be one-quarter of a 7400, but in a negative logic schematic it will be shown as a NOR gate. Figure 3-8a and b shows the schematic and truth table for a circuit using negative logic. Once the gates are represented as their proper negative logic functions, the analysis is the same as for a positive logic circuit. Pure negative logic systems are seldom encountered; but if you do have to troubleshoot one, remember to equate the proper voltage levels with logic high and logic low.

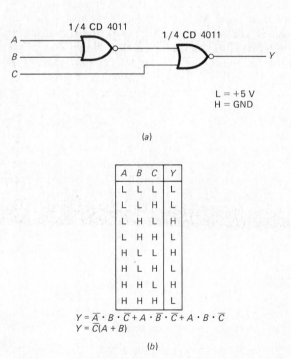

(a)

A	B	C	Y
L	L	L	L
L	L	H	L
L	H	L	H
L	H	H	L
H	L	L	H
H	L	H	L
H	H	L	H
H	H	H	L

$Y = \overline{A} \cdot B \cdot \overline{C} + A \cdot \overline{B} \cdot \overline{C} + A \cdot B \cdot \overline{C}$
$Y = \overline{C}(A + B)$

(b)

FIGURE 3-8 CD4011 positive logic NAND gate used as NOR gate in negative logic. *(a)* Schematic. *(b)* Truth table.

59

COMBINATIONAL LOGIC, MULTIPLEXERS, CODES, AND ROMS

MIXED LOGIC

In a system using positive logic conventions, the *active, asserted,* or *true* level for signals is a voltage high. Data books usually name a device according to its positive logic function. The familiar 74LS00, for example, contains four gates which each perform the positive logic NAND function. Figure 3-7 shows a voltage truth table for one of these gates. If the active (asserted) level of the inputs and the output is 2.4 V, you can see that the output function Y is the NOT function of $A \cdot B$, or

$$Y = \overline{A \cdot B}$$

In a system using negative logic conventions, the active, asserted, or true level for signals is a voltage low. From the voltage truth table in Figure 3-7, then, you can see that a gate in this 74LS00 device performs the NOT function of $A + B$, or

$$Y = \overline{A + B}$$

In system schematics using mixed logic conventions, signals with voltage high, active, asserted, or true levels and signals with voltage low, active, asserted, or true levels are both present. A letter or symbol after each signal name on the schematic indicates whether the asserted level for that signal is a voltage high or a voltage low. An active high signal is indicated by −H, .H, −1, or no symbol. KEY-H, KEY.H, KEY-1, or just KEY are ways in which an active high signal called KEY might be represented. An active low signal is indicated by −L, .L, −0, /, or *. BELT-L, BELT.L, BELT-0, BELT/, and BELT* are ways in which an active low signal called BELT might be represented. On a given schematic, only one set of symbols is used to indicate whether a signal is asserted low or asserted high. Most logic designers and system schematics now use mixed logic conventions. As you will see, using this mixed notation makes it very easy to determine the overall logic functions directly from the schematic. For the most part, this convention is a logical extension of the symbol equivalencies shown in the discussion of De Morgan's theorem.

Figure 3-9a shows how a 74LS00 gate might be represented in a mixed logic system schematic. In mixed logic, this device is thought of as an AND gate with active or asserted high inputs and an active or asserted low output. If the A input is asserted (H) and the B input is asserted (H), then the output will be asserted (L). This, then, is mixed logic, because the asserted level for the inputs is high (no bubbles) and the asserted level for the output is low (bubble present).

As demonstrated by De Morgan's theorem and Figure 3-3, a positive logic NAND gate can be represented as an OR gate with inverted inputs. The mixed logic schematic symbol for this is shown in Figure 3-9b. Bubbles on the inputs indicate that the asserted level for the inputs is low. The absence of a bubble on the output indicates that the active level for the output is a high. This symbol, then, says that the output will be asserted (H) if the A input is asserted (L) or the B input is asserted (L).

Figure 3-10 shows the two ways of representing several common types of logic gates by using mixed logic symbols. A positive logic NOR gate, such as found in a 74LS02, is thought of as an OR gate with active or asserted high inputs and an active or asserted low output. It also may be represented as an AND gate with active or asserted low inputs and an active or asserted high output. Note that an inverter simply changes the asserted level of a signal. Figure 3-10 should be for you a review of De Morgan's theorem. In the next section, we show how these symbols are utilized with active high and active low signals in a mixed logic system.

ANALYZING MIXED LOGIC CIRCUITS

Figure 3-11 shows a simple mixed logic circuit that you might have to analyze in a system. The active (asserted) level of the A-L signal is changed by the inverter to A-H. This signal is ANDed with B-H. The bubble on the output indicates that the output will be active (asserted) low. The way to describe this circuit is: If A is asserted (L) and B is asserted (H), the output will be asserted (L). The function would be written

$$Y - L = (A\text{-}L) \cdot (B\text{-}H)$$

Figure 3-12 shows another mixed logic circuit for you to analyze. From the symbol you can see immediately that the output function is an OR and the output is asserted high. The A-L signal is active low and is going into an active low input. Therefore, it will appear in the output expression as just A-L. The B-H signal is going into an active low input, so it will appear complemented in the output expression. In other words, the B-H signal will have to be *not* asserted in order to assert the active low input of the 74LS00. Some schematics put a slash on the signal line to indicate this. Then the output expression is

$$G\text{-}H = (A\text{-}L) + (\overline{B\text{-}H})$$

The output will be asserted (H) if A is asserted (L) or if B is not asserted (L).

Still another example of mixed logic is shown in Figure 3-13. BACK will be asserted (L) if XACK is asserted (L) or AACK is asserted (L), and ADEN is asserted (L).

A final example of mixed logic is shown in Figure 3-14. The output will be the sum (OR) of three terms. The bubble on the output of the 74LS32 symbol and the

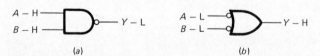

FIGURE 3-9 Mixed logic equivalents for positive logic NAND gate. *(a)* AND representation. *(b)* OR representation.

FIGURE 3-10 Mixed logic names and symbols for common logic gates.

TTL NUMBER	POSITIVE LOGIC NAME	MIXED LOGIC NAMES AND GATE SYMBOLS	
74LS00	QUAD 2–INPUT NAND	AND GATE (WITH ACTIVE–LOW OUTPUT)	OR GATE (WITH ACTIVE–LOW INPUTS)
74LS02	QUAD 2–INPUT NOR	AND GATE	OR GATE
74LS04	HEX INVERTER	INVERTER	INVERTER
74LS08	QUAD 2–INPUT AND	AND GATE	OR GATE (WITH ACTIVE–LOW INPUTS AND OUTPUT)
74LS10	TRIPLE 3–INPUT NAND	AND GATE	OR GATE
74LS20	DUAL 4–INPUT NAND	AND GATE	OR GATE
74LS32	QUAD 2–INPUT OR	AND GATE (WITH ACTIVE–LOW INPUTS AND OUTPUT)	OR GATE

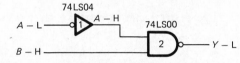

FIGURE 3-11 Mixed logic AND gate circuit.

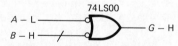

FIGURE 3-12 Mixed logic OR gate circuit.

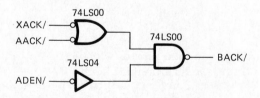

FIGURE 3-13 Example of mixed logic circuit.

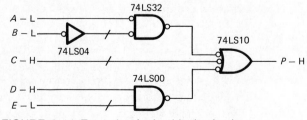

FIGURE 3-14 Example of mixed logic circuit.

bubble on the input of the 74LS10 symbol can be thought of as canceling, so the term for this section is $A\text{-}L \cdot B\text{-}L$. The C-H signal is applied to an active low input, so its term is complemented in the output expression. This term is then $\overline{C\text{-H}}$. For the 74LS00 section, the bubble on its output cancels the bubble on the 74LS10 input, so the term for this section is $D\text{-H} \cdot \overline{E\text{-L}}$. The total expression, then, is

$$P\text{-H} = (A\text{-L} \cdot \overline{B\text{-L}}) + (\overline{C\text{-H}}) + (D\text{-H} \cdot \overline{E\text{-L}})$$

Analyzing mixed logic schematics is very important in troubleshooting modern digital systems. More practice problems are given at the end of the chapter.

NEW SYMBOLS FOR MIXED LOGIC SCHEMATICS

To clearly indicate that mixed logic conventions are being used in a system schematic, IEC (International Electrotechnical Commission) and IEEE (Institute of Electrical and Electronic Engineers) have adopted a new set of symbol standards. (See App. B.) As shown in App. B, these standards actually use rectangular shapes for all logic devices, with symbols to indicate the particular logic function performed by the device. An ampersand (&), for example, is used for the AND function. The symbol ≥1 represents the OR function. In schematics using mixed logic, these standards replace the familiar inverting bubble used in pure positive or pure negative logic schematics with a triangle wedge. Figure 3-15a shows the new standard logic symbol for the 74LS00 quad two-input positive NAND gates. Figure 3-15b

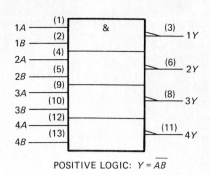

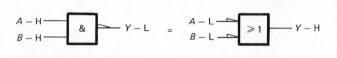

POSITIVE LOGIC: $Y = \overline{AB}$

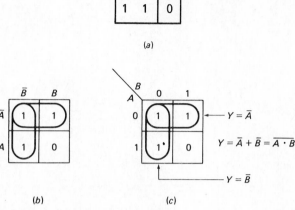

(a)

(b)　　　　(c)

FIGURE 3-15 (a) Dependency notation symbol for quad two-input positive logic NAND gate. (b) Equivalent symbols for single mixed logic NAND gate. (c) Hybrid symbols for mixed logic NAND gate.

FIGURE 3-16 Karnaugh mapping of a NAND gate. (a) Truth table. (b) Karnaugh map with letter variables. (c) Karnaugh map with 1s and 0s.

shows the two ways in which a gate in this device can be represented.

Since these standards are quite new, most existing mixed logic schematics use traditional shape symbols and bubbles on inputs and outputs. For clarity, in this book (the second edition) we use the traditional AND, OR, and inverter shapes, but we show wedges on inputs and outputs as in Figure 3-15c to indicate that mixed logic is being employed.

FURTHER EXAMPLES OF SIMPLIFICATION OF LOGIC FUNCTIONS

In previous sections we showed how logic expressions can be simplified by using boolean algebra. By factoring and substituting you may be able to get a reduced expression. The only problem with this approach is that, just as when you try to prove a theorem in geometry, one wrong turn can have you wandering around in circles forever. A more dependable and easier-to-remember method is *Karnaugh mapping*.

KARNAUGH MAPPING

Karnaugh mapping is a graphic method of plotting the truth table for a logic expression and then extracting a simplified expression from the plot. Karnaugh maps (K-maps) can be used for up to eight input variables, but their use will only be discussed for up to four input variables. It is currently more practical and common to simplify expressions of five or more variables by use of computer programs.

DEFINING THE K-MAP STRUCTURE

The K-map is a two-dimensional form of a truth table, consisting of a square or rectangular array of adjacent cells or squares. Figure 3-16a, b, and c shows the truth table and two ways of drawing the Karnaugh map for a simple two-input NAND gate.

There is one square in the map for each possible combination of the input variables. Two variables give four squares, three variables give eight squares, and four variables give sixteen squares. Therefore a boolean function of n variables requires a K-map consisting of 2^n cells or squares. See Figure 3-17a and b for examples of 3- and 4-variable expressions and truth tables and their corresponding K-maps.

Each square inside the map is at the intersection of an externally labeled column and row. So each cell or square has a unique address which is specified by the intersecting row's label and column's label.

Rows and columns are labeled using a binary sequence in which only one digit changes at a time. Figure 3-17, for example, shows the AB sequence as 00, 01, 11, 10. (A later section on additional kinds of binary codes describes this as the Gray code sequence.) This labeling produces a map where the full address (row label plus column label) of any square will differ by no more than one digit from the address of any adjacent cell. Also, note in Figures 3-16 and 3-17 that for boolean expressions of "n" input variables, "r" of those variables are the row labels, and the remaining n minus r variables are column labels. No more than two variables are used for any column or row address label.

The logic state in each box is the output state for the input variables at the top of its column and at the left of

A	B	C	Y
0	0	0	0
0	0	1	0
0	1	0	1
0	1	1	1
1	0	0	0
1	0	1	0
1	1	0	1
1	1	1	0

$Y = \bar{A}B\bar{C} + \bar{A}BC + AB\bar{C}$

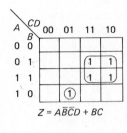

$Y = \bar{A}B + B\bar{C}$

(a)

A	B	C	D	Z
0	0	0	0	0
0	0	0	1	0
0	0	1	0	0
0	0	1	1	0
0	1	0	0	0
0	1	0	1	0
0	1	1	0	1
0	1	1	1	1
1	0	0	0	0
1	0	0	1	1
1	0	1	0	0
1	0	1	1	0
1	1	0	0	0
1	1	0	1	0
1	1	1	0	1
1	1	1	1	1

$Z = A\bar{B}\bar{C}D + BC$

$Z = \bar{A}BC\bar{D} + \bar{A}BCD$
$+ A\bar{B}\bar{C}D + ABC\bar{D}$
$+ ABCD$

(b)

FIGURE 3-17 *(a)* Example of a 3-variable expression, truth table, and corresponding K-map. *(b)* Example of a 4-variable expression, truth table, and corresponding K-map.

its row. In other words, a 1 or 0 is written in each square (cell) as determined by evaluating the original boolean function using the input conditions which are specified by the row and column address. For example, the box in the lower right-hand corner of Figure 3-16b and c indicates that the output is low if A is high and B is high.

SIMPLIFYING THE K-MAP
Now that you know what a K-map is, the next question is: How does one get a simplified expression from it?

1. First construct the K-map and enter the 1's (and 0's) as described above.

2. Next examine the K-map for *adjacent* 1's. Note that two cells are *adjacent* if their addresses differ by no more than *one* digit. Therefore diagonal squares are *not* adjacent. However, the *opposite extremes (edges)* of any row or column *are* adjacent. *Circle all single 1's which have NO adjacent ones in any direction.*

3. The remaining ones will be looped (circled, enclosed) in groups of twos (pairs), fours (quads), and eights (octets) according to the following rules.
   a. These groupings (loops or enclosures) are integer powers of 2 (2^n).
   b. The groupings will be square or rectangular in shape.

c. The enclosures (loops) must be as large as possible.
d. Every 1 must be included in at least one loop but a 1 may be used in more than one loop if it allows that loop to be a larger 2^n group.
e. There should be as *few* loops as possible, but enough loops to include *all* 1's at least once.
f. Loops may wrap around the borders of the map because borders have adjacent addresses.

4. Each loop or enclosure will be translated into a product term for the final simplified output expression.
   a. The address digits that change as one moves from square to square within a loop are ignored (not used) in the product term.
   b. The address digits that do *not* change as one moves from square to square within a loop are needed (used) in the product term. Constant 1 terms are high (uncomplemented) in the product. Constant 0 terms are low (complemented) in the product.

   For example, if you look at the circled highs (1's) in the two boxes to the right of $\bar{A}$ in the map of Figure 3-16b, you can see that for A low, the output is high, whether B is low or B is high. Therefore, the state of B input doesn't matter, as long as A is low. To obtain the partial output expression for this situation, you leave out the variables whose input state doesn't matter and write down the input state of the variable which gives the high output: $Y = \bar{A}$. Now, if you look at the circled highs in the column under $\bar{B}$, you can see that it doesn't matter what state A is in as long as B is low: $Y = \bar{B}$. Note that a 1 in the map can be included in more than one term, and that all 1's must be included in at least one term of the final expression.

5. Finally, form the OR of all the terms. Referring to Figure 3-16b again, since all the output highs are now accounted for, you can see that the output is high if A is low or if B is low, or the complete output expression is written as:

$$Y = \bar{A} + \bar{B}$$

Note that you could have used Figure 3-16c in the same way to obtain this output expression. This expression $(Y = \bar{A} + \bar{B})$ still doesn't look like the expression for a NAND gate unless you remember from De Morgan's theorem that an OR gate with inverted inputs is equivalent to a NAND gate:

$$\bar{A} + \bar{B} = \overline{A \cdot B}$$

If you did not recognize the truth table of Figure 3-16a as that of a NAND gate and you extracted the sum-of-products or so-called *minterm* expression directly off the truth table, you would get:

$$Y = \bar{A} \cdot \bar{B} + \bar{A}B + A\bar{B}$$

63

COMBINATIONAL LOGIC, MULTIPLEXERS, CODES, AND ROMS

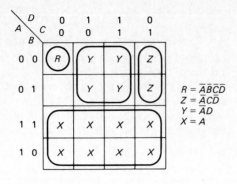

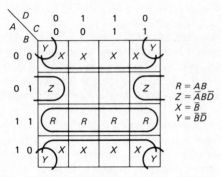

$$R = \overline{A}\overline{B}C\overline{D}$$
$$Z = \overline{A}C\overline{D}$$
$$Y = \overline{A}D$$
$$X = A$$

$$R = AB$$
$$Z = \overline{A}B\overline{D}$$
$$X = \overline{B}$$
$$Y = \overline{B}\overline{D}$$

FIGURE 3-18 Some possible groupings of Karnaugh maps used to reduce the number of variables in a term or an expression.

The boolean algebra identities can be used to reduce the expression to its familiar form, but the Karnaugh map method is direct. In cases with more input variables, this becomes even more obvious.

Now examine an example with three input variables in Figure 3-17a.

1. After construction of the K-map, it is examined for single (no adjacency) 1's.

2. Since none are found, all pairs that are only pairs (a 1 adjacent to only one other 1) are looped. Note that two pairs are looped and also that the loops overlap.

3. Since at this point all 1's are accounted for, we are ready to write the product terms. From the map you can see that when A is low and B is high, C can be either low or high so you can write the term $\overline{A}B$. For the other loop you can see that when C is low and B is high, A can be either low or high, so you can write the term $B\overline{C}$.

4. Finally these two terms are ORed together to form the final output expression:

$$Y = \overline{A}B + B\overline{C}$$

The four-variable Karnaugh maps in Figure 3-18 will now be used to demonstrate how grouping adjacent boxes together properly reduces the number of variables in an expression. A four-variable map, as mentioned before, requires sixteen boxes. If an expression R has only a single 1 in the box labeled R, you can see that it takes all four variables to give an expression $R = \overline{A}\overline{B}\overline{C}\overline{D}$ for the single box R. If the map for an expression has 2 "ones" in adjacent boxes, as the Zs in Figure 3-18a, then only three variables are needed to describe the term, since one of the variables can be either a "one" or a "zero." In the example shown, B doesn't matter and $Z = \overline{A}C\overline{D}$. If four adjacent "ones" can be grouped together horizontally, vertically, or in a block as shown, two variables are eliminated. The expression for Y in the example shown is simply $Y = \overline{A}D$. Grouping eight boxes together as shown for the X expression removes three variables, and the resultant expression is just $X = A$ for the example shown in Figure 3-18.

Since the Karnaugh map can be considered to be wrapped around a cylinder in both the vertical and horizontal directions, Figure 3-18b shows some other ways in which boxes can be grouped to simplify the output expression. Remember, all "ones" on the map must be included in a term of the output expression and for the simplest output expression, the groups should contain the largest possible number of ones in groups of proper size (powers of 2).

For an example illustrating the K-map simplification steps for a 4-input variable expression or truth table, refer to Figure 3-17b.

1. The K-map is constructed as described previously, and 1's are entered into the appropriate squares.

2. Next the map is examined for single 1's (no adjacencies). In Figure 3-17b one "single" is found and looped.

3. The map is next examined for pairs that are only pairs which cannot be included in a larger grouping. In Figure 3-17b no isolated pairs are found.

4. Since this is a 4-input variable map, we now look for a largest possible grouping of 8 (octet), in order to enclose any remaining unlooped 1's. In this case there is no octet.

5. Since there are still unlooped 1's, next we look for groupings of 4 (quads), which may include other previously looped 1's as well as unlooped 1's. We find one grouping of 4, and loop it.

6. The last looping step is to loop any remaining pairs if you still have any 1's that were not enclosed by previous steps. For this Figure 3-17b map example, all 1's have already been looped.

7. Now we write the product terms for each loop formed. The term for the single is $\overline{A}B\overline{C}D$. The quad loop has B high and C high, while A and D are either high or low and therefore not used in the term. So we obtain the product BC.

8. Finally the product terms are ORed to obtain the simplified output expression $Z = \overline{A}B\overline{C}D + BC$.

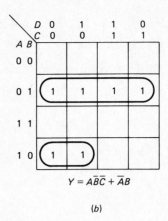

A	B	C	D	Y
0	0	0	0	0
0	0	0	1	0
0	0	1	0	0
0	0	1	1	0
0	1	0	0	1
0	1	0	1	1
0	1	1	0	1
0	1	1	1	1
1	0	0	0	1
1	0	0	1	1
1	0	1	0	0
1	0	1	1	0
1	1	0	0	0
1	1	0	1	0
1	1	1	0	0
1	1	1	1	0

$$Y = \overline{A}B\overline{C}\overline{D} + \overline{A}B\overline{C}D + \overline{A}BC\overline{D} + \overline{A}BCD + A\overline{B}\,\overline{C}\,\overline{D} + A\overline{B}\,\overline{C}D$$

(a)

FIGURE 3-19 (a) Truth table requiring simplification of output expression. (b) Karnaugh map and simplified output expression. (c) Simplified circuit built with NAND gates and inverters.

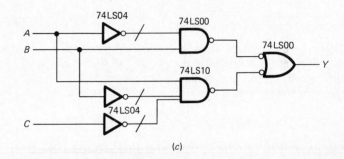

$$Y = A\overline{B}\,\overline{C} + \overline{A}B$$

(b)

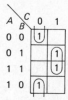

(c)

Two final simplification examples are shown below to illustrate the differences between drawing a K-map directly from a *truth table* and drawing a K-map directly from a *boolean expression*.

TRUTH TABLE TO K-MAP

Figure 3-19a shows a truth table and the unsimplified boolean expression for it. Figure 3-19b shows the Karnaugh map for this function which contains a 1 for each output 1 term in the truth table. Four "ones" across the center of the map can be grouped as shown to give the term $\overline{A}B$. The two "ones" on the bottom of the map can be grouped to give the term $A\overline{B}\,\overline{C}$. These two terms are summed (ORed) to give the complete sum-of-products expression:

$$Y = A\overline{B}\,\overline{C} + \overline{A}B$$

This expression is much easier to implement with gates than the original expression which contained 6 terms. Figure 3-19c shows how the simplified expression can be built with only three NAND gates and inverters.

BOOLEAN EXPRESSION TO K-MAP

To simplify a logic function directly from the boolean expression, first draw and properly label a Karnaugh map for the required number of variables. Then for each term in the boolean expression, put a 1 in the corresponding square of the Karnaugh map. The simplified expression can then be extracted from the Karnaugh map as shown above.

As an example, consider the boolean expression:

$$Y = \overline{A}\,\overline{B}\,\overline{C} + \overline{A}BC + ABC + A\overline{B}\,\overline{C}$$

It contains three variables so an 8-box Karnaugh map is required. Figure 3-20 shows the required map with a

FIGURE 3-20 Karnaugh map used to simplify a three-variable expression.

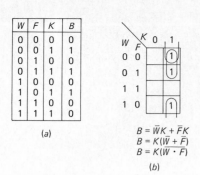

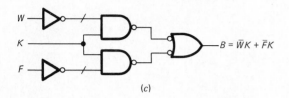

$$B = \overline{W}K + \overline{F}K$$
$$B = K(\overline{W} + \overline{F})$$
$$B = K(\overline{W \cdot F})$$

(b)

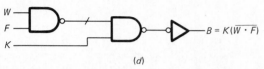

(c)

(d)

FIGURE 3-21 Karnaugh maps are used to synthesize an automobile seat belt warning logic circuit. (a) Truth table. (b) Karnaugh map. (c) Sum-of-products circuit. (d) Alternate, further simplified circuit.

"one" in each box which corresponds to a term of the original boolean expression. The boxes for $\overline{A}\overline{B}\overline{C}$ and $A\overline{B}\overline{C}$ can be looped together as shown to give $\overline{B} \cdot \overline{C}$. The other two terms can be looped together to give $B \cdot C$. The result then is the sum of these terms or

$$Y = \overline{B} \cdot \overline{C} + B \cdot C$$

which incidentally can also be written as

$$Y = \overline{B \oplus C} \quad (B \text{ exclusive NOR } C)$$

SYNTHESIZING SIMPLE LOGIC GATE CIRCUITS

Analyzing logic circuits is often easier if you understand how they are used to do real world jobs. This section shows how two such jobs, an automobile seat belt warning buzzer and a home gas hot-air-furnace controller, are done with logic gates.

AUTOMOBILE SEAT BELT WARNING SYSTEM

The automobile seat belt warning system functions as follows. The manufacturer wishes a buzzer to sound for 15 s when there is weight on the driver's seat, the key is on, and the seat belt is not fastened. Also, to get double

usage out of the buzzer and sensors, it is desired to have the buzzer sound when there is no weight on the seat and the key is in the ignition. This serves as a warning to prevent the driver from getting out of the car and leaving the keys in the ignition. The logic can be wired into the ignition switch so that it works in only the off and on positions of the switch, not in the accessory position. This connection prevents the buzzer from sounding when a person just wants to sit in the car and listen to the radio on a moonlit night.

The variables for this system are W, the weight in the seat; F, the state of having the seat belt fastened; K, the state of the key being on or off (not in accessory position); and B, the buzzer. Figure 3-21a shows a possible truth table for this circuit. A 1 in the truth table represents the on condition for each variable. Figure 3-21b shows the Karnaugh map and the simplified expression produced from it. The sum-of-products expression from the Karnaugh map can be implemented with three NAND gates and two inverters, as shown in Figure 3-21c. If, however, the sum-of-products expression is factored and a De Morgan's theorem substitution is made, as shown in Figure 3-21b, then the result can be built with only three NAND gates, as shown in Figure 3-21d. This shows how a combination of Karnaugh maps and boolean algebra is used to reduce the number of IC packages required for a particular circuit.

GAS HOT-AIR-FURNACE CONTROLLER

The first task in figuring out a system such as this is to define the inputs and outputs. As you can see in Figure 3-22a, the controller has four inputs and two outputs. For the input from the thermostat T, a high means the room temperature is above the temperature setting, and a low indicates the room temperature is below the setting. A low from the safety temperature limit sensor S indicates the bonnet, or metal-encased airspace around the furnace, is at a safe operating temperature; but a high says the bonnet is getting too hot and the gas should be turned off. A high from the pilot safety sensor P indicates the pilot is on and will ignite the main burner when turned on. A low from the pilot safety sensor means that the pilot is off and the gas should not be allowed to turn on. (This safety precaution usually is implemented by the red button you have to hold down for one minute when lighting the pilot.) The bonnet temperature sensor B gives a high output when the air in the furnace above the burner has heated to the temperature at which the circulating fan will turn on. The two outputs are the control for starting the fan F and the control for turning on the gas-valve solenoid G. In both cases, high (1) is on.

Figure 3-22b shows a possible truth table for these inputs and outputs. At first glance, this truth table looks complicated, but a closer look reveals that only two input combinations give a high for the GAS ON output:

$$G = \overline{T} \cdot \overline{S} \cdot \overline{B} \cdot P + \overline{T} \cdot \overline{S} \cdot B \cdot P$$

You could build the circuit for this function with gates

as it stands, or you can try to simplify the expression by one of the following methods.

SIMPLIFICATION BY AN INTUITIVE APPROACH
If you compare the two truth table lines that give a high G

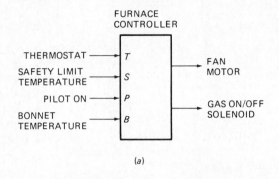

(a)

INPUTS				OUTPUTS		
T	S	B	P	$F1$	G	$F2$
L = COLD	L = OK	L = COLD	L = OFF	L = OFF	L = OFF	L = OFF
L	L	L	L	L	L	L
L	L	L	H	L	H	L
L	L	H	L	H	L	H
L	L	H	H	H	H	H
L	H	L	L	L	L	L
L	H	L	H	L	L	L
L	H	H	L	H	L	H
L	H	H	H	H	L	H
H	L	L	L	L	L	L
H	L	L	H	L	L	L
H	L	H	L	L	L	H
H	L	H	H	L	L	H
H	H	L	L	L	L	L
H	H	L	H	L	L	L
H	H	H	L	L	L	H
H	H	H	H	L	L	H

(b)

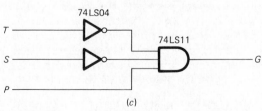

(c)

FIGURE 3-22 Logic gate furnace controller. *(a)* Block diagram. *(b)* Truth table. *(c)* Circuit for gas control.

output, you can see that the T, S, and P inputs are the same. One combination has B low, and the other has B high. Apparently, then, you don't care about the input state of B for your expression. All you care about is $\overline{T} \cdot \overline{S} \cdot P$ in your expression for G, so you can simplify to

$$G = \overline{T} \cdot \overline{S} \cdot P$$

SIMPLIFICATION BY USING BOOLEAN ALGEBRA
Starting with the original expression for G above, you can factor $\overline{T} \cdot \overline{S} \cdot P$ out of the combined expression to give

$$G = \overline{T} \cdot \overline{S} \cdot P(B + \overline{B})$$

Since $B + \overline{B}$ is equal to 1, the expression simplifies to

$$G = \overline{T} \cdot \overline{S} \cdot P$$

which is the same as your intuitive result above. This final expression states that you want the gas on when the room temperature is low, the bonnet temperature is safe, and the pilot is on. This can be built with gates as shown in Figure 3-22c.

The expression for FAN ON is a little more complicated, but it can be described as follows. First, collect all the terms for input combinations that give a high output:

$$F1 = \overline{T} \cdot \overline{S} \cdot B \cdot \overline{P} + \overline{T} \cdot S \cdot B \cdot \overline{P} \\ + \overline{T} \cdot \overline{S} \cdot B \cdot P + \overline{T} \cdot S \cdot B \cdot P$$

All terms contain $\overline{T} \cdot B$, so you can factor this out of each term to give

$$F1 = \overline{T} \cdot B(\overline{S} \cdot \overline{P} + S \cdot \overline{P} + \overline{S} \cdot P + S \cdot P)$$

Since the parentheses contain all the possible combinations of $S \cdot P$, the expression in the parentheses simplifies to 1, and the overall expression simplifies to

$$F1 = \overline{T} \cdot B$$

This expression simply says that you want the fan on when the room temperature is low and the bonnet is hot.

If you want the fan to stay on and cool down the bonnet after the thermostat goes high and shuts off the gas, you can add four more H's to the fan column as in the $F2$ column of Fig. 3-22b. The resultant expression for $F2$ is

$$F2 = \overline{T} \cdot \overline{S} \cdot B \cdot \overline{P} + \overline{T} \cdot S \cdot B \cdot \overline{P} \\ + T \cdot \overline{S} \cdot B \cdot \overline{P} + T \cdot S \cdot B \cdot \overline{P} \\ + \overline{T} \cdot \overline{S} \cdot B \cdot P + \overline{T} \cdot S \cdot B \cdot P \\ + T \cdot \overline{S} \cdot B \cdot P + T \cdot S \cdot B \cdot P$$

All the terms contain B, so you can factor out B to give

$$F2 = B(\overline{T} \cdot \overline{S} \cdot \overline{P} + \overline{T} \cdot S \cdot \overline{P} \\ + T \cdot \overline{S} \cdot \overline{P} + T \cdot S \cdot \overline{P} + \overline{T} \cdot \overline{S} \cdot P \\ + \overline{T} \cdot S \cdot P + T \cdot \overline{S} \cdot P + T \cdot S \cdot P)$$

67

COMBINATIONAL LOGIC, MULTIPLEXERS, CODES, AND ROMS

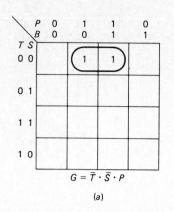

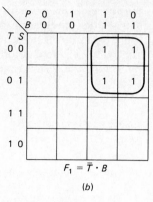

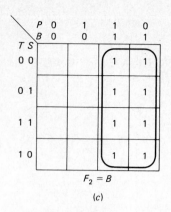

$G = \overline{T} \cdot \overline{S} \cdot P$

(a)

$F_1 = \overline{T} \cdot B$

(b)

$F_2 = B$

(c)

FIGURE 3-23 Karnaugh maps are used to simplify gas and fan expressions for furnace controller. *(a)* Gas. *(b)* Fan 1. *(c)* Fan 2.

Since the parentheses contain all the possible combinations of $T \cdot S \cdot P$, one term in parentheses will always equal 1, and you can replace the entire expression in parentheses with a 1. In other words, you don't care about the state of the T, S, and P inputs. The whole expression reduces to

$$F2 = B$$

which is even easier to build than $F1 = \overline{T} \cdot B$. No gates are needed. You can simply run a wire from the B input to the FAN MOTOR relay. The result simply says, "Turn on the fan any time the bonnet is hot."

SIMPLIFICATION BY USING KARNAUGH MAPPING
If you compare the number of gates that would be required to implement the first expression for FAN ON with that for the final result, the reason for simplifying is obvious. Now go back and try a Karnaugh map on the truth table for the furnace controller in Figure 3-22b, so that you realize that variables don't always have to be labeled A, B, C, etc. Construct the gas output expression map for yourself; then compare your results with Figure 3-23a. From the circled pair of 1's you can immediately see that the output expression is

$$G = \overline{T} \cdot \overline{S} \cdot P$$

Now try doing a map for the first fan expression, and compare your results with Figure 3-23b. The four circled 1's in Figure 3-23b immediately give you a simple output expression of

$$F1 = \overline{T} \cdot B$$

A map for the $F2$ expression is shown in Figure 3-23c. Since eight boxes can be taken together, the expression is

$$F2 = B$$

The use of Karnaugh maps is demonstrated further in a later section of this chapter on code converters, but first we discuss some other interesting applications of basic gates.

SWITCHES, MULTIPLEXERS, AND DEMULTIPLEXERS

In this section we show how you can use gates as logic-controlled switches, how many lines of data can be

time-multiplexed on a single wire, and how any logic expression with up to five input variables can be built with one simple IC package.

TIMING DIAGRAM VIEW OF SIMPLE GATES AND NAND GATE SWITCHES
Up to this point, truth tables have been used to describe and analyze logic gate circuits. For logic circuits with constantly changing inputs, another useful way to represent the relationship between input and output states is by a timing diagram. A timing diagram is just a voltage-versus-time comparison of inputs and outputs. Figure 3-24a shows a basic AND gate with pulse inputs. The timing diagram for these inputs and the corresponding output is shown in Figure 3-24b. Note that at each instant the output follows the truth table for an AND gate. The output is high only when both clock and control inputs are high. If one input is low, the output is low regardless of the state of the other input. As the timing diagram shows, if the control line is high, then the pulse or clock signal will appear at point Y. If the control line is low, then no clock will appear at Y and the output of the AND gate will be a constant low, regardless of the state of the clock input. The AND gate functions as a logic-controlled switch. A NAND gate can be used as a switch also, but the output will be inverted. When an OR or a NOR gate is used as a switch, the output will follow the clock input when the control input is low.

Figure 3-24c shows a circuit for routing a clock signal to one of two points. A high on the control input enables the upper gate and sends the signal to Y. A low on the control input enables the lower NAND gate and sends the signal to X. Among many other uses, this circuit can route a clock to the count-up or count-down input of a counter, as discussed in Chap. 4. The switch equivalent circuit is shown in Figure 3-24d.

The next switch version, shown in Figure 3-24e, is used to select data from one of two inputs and route them to Y. Figure 3-24f shows the switch equivalent circuit. If the control line is high, data is selected from the A input; if the control is low, data is selected from the B input. This circuit is available ready-made in TTL, as the 74157 quad two-input data selector.

MULTIPLEXERS AND DEMULTIPLEXERS
The circuits of Figure 3-24e and c can be combined to make a two-input time division multiplexer-demulti-

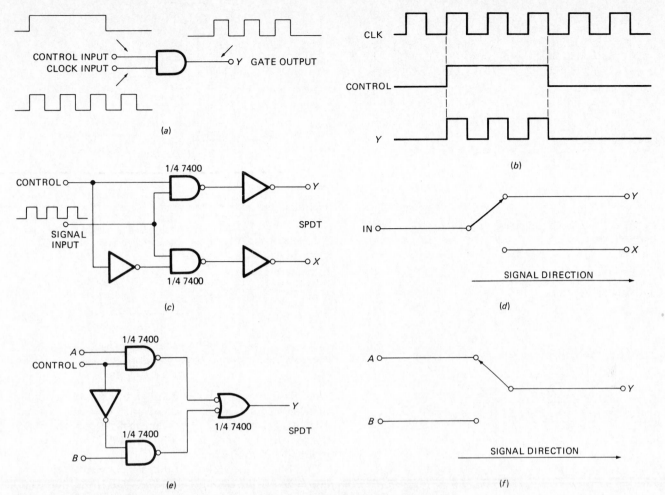

FIGURE 3-24 *(a)* AND gate used as logic-controlled switch, SPST schematic, and signals. *(b)* Timing diagram for AND gate logic-controlled switch shown in part *a.* *(c)* NAND gates used as logic-controlled switches, SPDT output schematic. *(d)* Mechanical-switch equivalent for circuit shown in part *c. (e)* NAND gates used as logic-controlled switches, SPDT input schematic. *(f)* Mechanical-switch equivalent for circuit shown in part *e.*

plexer circuit, as shown in Figure 3-25. *Multiplexing* in this case means transmitting several signals on the same wire by interleaving samples of data from each input. First a sample from *A* is sent, then one from *B*, then another from *A* and another from *B*, and so on. A

high-frequency square-wave pulse on the control input switches the multiplexer back and forth between the two inputs. A synchronized demultiplexer on the other end routes the data coming down the line to the right output. The frequency of the pulse on the control input

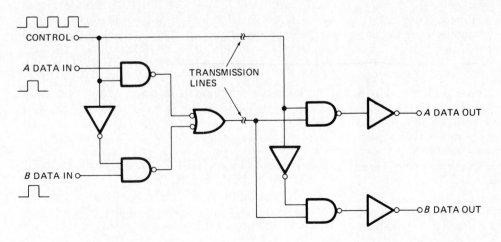

FIGURE 3-25 NAND gate switch circuits are used to transmit two data signals on one wire.

COMBINATIONAL LOGIC, MULTIPLEXERS, CODES, AND ROMS

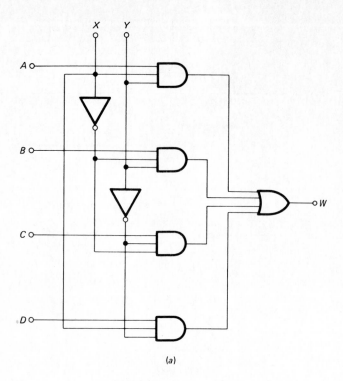

(a)

To Select	X	Y
A	1	1
B	0	1
C	0	0
D	1	0

(b)

FIGURE 3-26 Four-input multiplexer. (a) Schematic. (b) Select table.

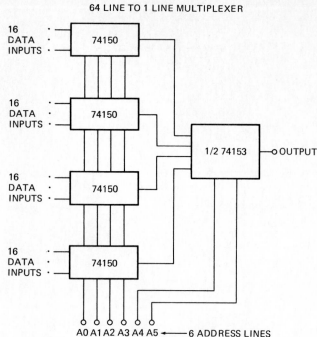

FIGURE 3-27 Sixty-four-line-to-one-line multiplexer.

must be much higher than that of any data being sent. The purpose of this time division multiplexing is to save on the number of wires needed to transmit data from multiple sources. For the two-input example, there is no saving in lines; but as the number of lines multiplexed goes up, the saving increases.

Figure 3-26 shows a circuit for a four-input multiplexer. Notice that it requires two data select lines to determine which of four inputs is connected to the output. The table in Figure 3-26b shows the select inputs to select each data input. Sometimes it is convenient to think of these data select inputs as address lines. Then you can say the address of the A input is 11, the address of the B input is 01, etc. Two of these circuits are available in the 74153 dual four-input multiplexer. If you use this circuit for a data multiplexer as in Figure 3-26, you can send four lines of data with two address or data select lines and one signal line. This is a saving of one wire, which is not very impressive, but let's go on to the next case.

The 74151 is an eight-input multiplexer. To select one of eight inputs, you need three data select or address

lines. This means you can send eight channels of data with only four lines (one signal and three addresses). Analog signals such as phone conversations can be multiplexed in the same manner with the CD4051 eight-input CMOS multiplexer.

The 74150 is a 16-input multiplexer circuit, for it requires four address lines to select one input line at a time out of the 16. Multiplexers can be stacked in multiplexer "trees," as shown in Figure 3-27, to compress the data from many lines onto a single line. A demultiplexer tree at the receiving end restores the data to parallel lines. A 74154 is a 4- to 16-line demultiplexer.

OSCILLOSCOPE EIGHT-INPUT MULTIPLEXER

Switching from one channel of data to another may remind you of the chopped-mode oscilloscope displays discussed in Chap. 1. The principle is the same. Figure 3-28 shows a circuit you can build that will display up to eight channels of digital data on a standard oscilloscope. The circuit uses a 74S151 eight-input digital multiplexer to multiplex the digital data and a CD4051 eight-input analog multiplexer to add voltage steps to separate the traces on the screen. A 555 astable multivibrator serves as the master clock, and a 7493 4-bit binary counter cycles continuously through the eight addresses.

MULTIPLEXERS USED FOR SYNTHESIZING RANDOM LOGIC FUNCTIONS

Multiplexers and demultiplexers have many uses beyond those implied by their names. A multiplexer can be utilized in place of discrete gates to build circuits for

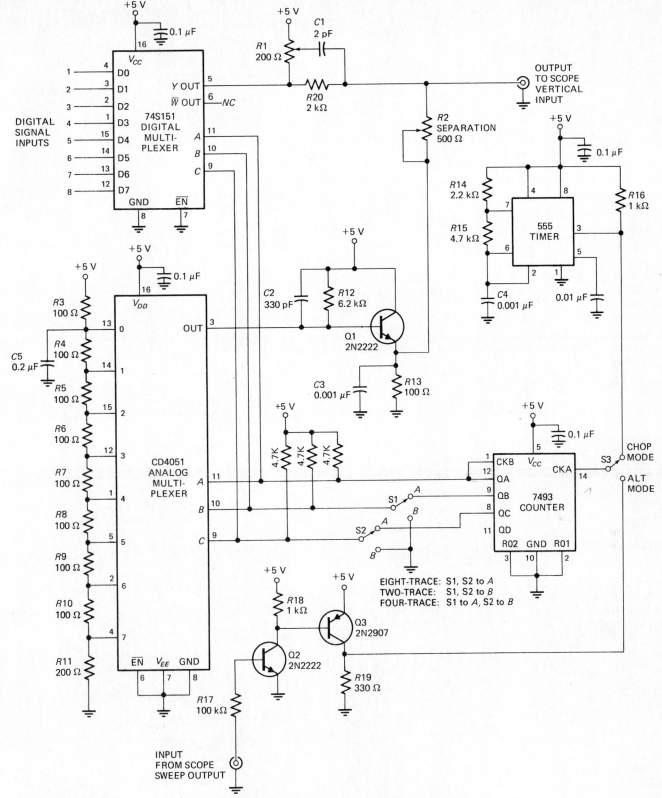

FIGURE 3-28 Circuit to display eight digital input signals on an oscilloscope. This circuit was modified from one found in the October 14, 1976, issue (vol. 49) of *Electronics* magazine, published by McGraw-Hill.

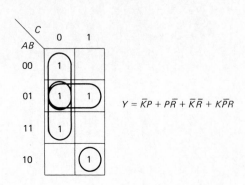

K	P	R	Y
0	0	0	1
0	0	1	0
0	1	0	1
0	1	1	1
1	0	0	0
1	0	1	1
1	1	0	1
1	1	1	0

(a)

$$Y = \overline{K}P + P\overline{R} + \overline{K}\overline{R} + K\overline{P}R$$

(b)

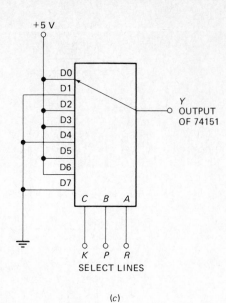

(c)

FIGURE 3-29 A multiplexer is used to implement a logic expression. *(a)* Truth table. *(b)* Karnaugh map. *(c)* Circuit.

so-called random-logic expressions, such as those you previously simplified with Karnaugh maps. Here's how: Figure 3-29*a* shows the truth table for a three-variable logic expression

$$Y = \overline{K}\overline{P}\overline{R} + \overline{K}P\overline{R} + KP\overline{R} + K\overline{P}R + \overline{K}PR$$

With a Karnaugh map as in Figure 3-29*b*, you can simplify this to

$$Y = \overline{K}P + P\overline{R} + \overline{K}\overline{R} + K\overline{P}R$$

but this will still require several packages. Figure 3-29*c*

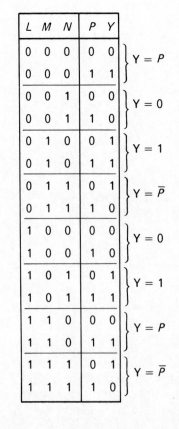

L	M	N	P	Y
0	0	0	0	0
0	0	0	1	1
0	0	1	0	0
0	0	1	1	0
0	1	0	0	1
0	1	0	1	1
0	1	1	0	1
0	1	1	1	0
1	0	0	0	0
1	0	0	1	0
1	0	1	0	1
1	0	1	1	1
1	1	0	0	0
1	1	0	1	1
1	1	1	0	1
1	1	1	1	0

(a)

L	M	N	P	Y	
0	0	0	0	0	$Y = P$
0	0	0	1	1	
0	0	1	0	0	$Y = 0$
0	0	1	1	0	
0	1	0	0	1	$Y = 1$
0	1	0	1	1	
0	1	1	0	1	$Y = \overline{P}$
0	1	1	1	0	
1	0	0	0	0	$Y = 0$
1	0	0	1	0	
1	0	1	0	1	$Y = 1$
1	0	1	1	1	
1	1	0	0	0	$Y = P$
1	1	0	1	1	
1	1	1	0	1	$Y = \overline{P}$
1	1	1	1	0	

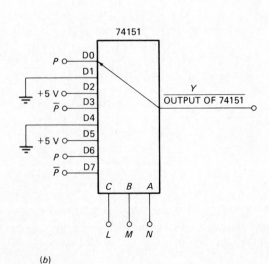

(b)

FIGURE 3-30 Implementation of a logic expression with a multiplexer and a "folding trick." *(a)* Truth table analysis. *(b)* Circuit schematic for "folded" truth table.

shows how this function can be built directly from the truth table with a single IC package, the 74151 eight-input multiplexer. The input variables are connected to the select lines in the same order as the truth table. If input variables K, P, and R are all low or 0, you know the multiplexer is connecting the D0 input to the output. All you have to do is tie the D0 input to +5 V for a 1 or to ground for a 0, whichever value you want Y to have for the input variable combination K, P, and $R = 0$ or $\overline{KPR}$. For the next input combination, $\overline{KP}R$, the D1 input will be addressed. Since you want a 0 or low output for this input, just tie the D1 input to ground. Then continue down the truth table, connecting the D inputs to +5 V or ground as appropriate.

The beauty of this circuit is that you can build it for any expression of three input variables almost as fast as you can write the truth table. Also, if you want to change the output expression, say, to a low output for $\overline{KP}R$, all you have to do is move the D2 input lead from +5 V to ground. With previous methods, you would have to do a new Karnaugh map, simplify the expression, and rebuild the circuit. For this example, it would probably require changing IC packages. With the multiplexer method, a Karnaugh map is not needed; but if your truth table has a lot of 1's in it, you should check with a map to be sure you aren't missing some obvious simplification such as the $F2 = B$ of the furnace controller example. (The 74151, incidentally, has a noninverted Y output and an inverted W output.)

By using the same technique as above, the 74150 16-input multiplexer can synthesize directly any logic function of four input variables. The 74150 comes in a larger, more expensive, 24-pin package, and the output is inverted. A *folding trick* allows you to produce any function of four input variables with only an eight-input multiplexer such as the 74151.

FOLDING TRICK   Figure 3-30a shows the truth table for a four-input-variable expression, and Figure 3-30b shows the construction of the circuit for it by using an eight-input multiplexer. The L, M, and N variables are connected directly to the C, B, and A select inputs. You handle the P variable as follows. Note that in the truth table for the two input conditions where L, M, and N are 0, if P is 0, Y is 0, and if P is 1, then Y is 1. Therefore, for C = 0, B = 0, and A = 0, $Y = P$. On the multiplexer, select inputs of C = 0, B = 0, and A = 0 connect the output to the D0 input. If you just connect the P variable to this input, the output will be correct for either state of the P variable. For the next combination of L, M, and N, $L = 0$, $M = 0$, and $N = 1$. You can see that if $P = 0$, $Y = 0$, and if $P = 1$, $Y = 0$. Therefore, $Y = 0$. You can implement this by tying the D1 input of the multiplexer to ground. As you can see from Figure 3-30a, for $L = 0$, $M = 1$, and $N = 0$, the output is a 1 whether P is 0 or 1. So the corresponding input on the multiplexer can be connected to +5 V. The next combination, $L = 0$, $M = 1$, and $N = 1$, gives a result of $Y = \overline{P}$. A simple inverter will produce $\overline{P}$ to connect to the D3 input of the multiplexer. Work your way through the rest of the truth table to make sure you get the idea. Notice that the only

input possibilities for the data inputs of the multiplexer are 0, 1, P, or $\overline{P}$. If you desire a change in the truth table, all you have to do is move a data input wire to the appropriate one of these.

You can use the 74150 sixteen-input multiplexer with this folding trick to produce any possible output expression for five input variables. For expressions with more than five input variables, usually logic designers use read-only memories or programmable-logic arrays, which are discussed later in this chapter.

DEPENDENCY NOTATION SYMBOLS FOR MULTIPLEXERS AND DEMULTIPLEXERS

As described in App. B, the IEC-IEEE *dependency notation* is an attempt to establish for each logic device a logic symbol that clearly shows the relationships of all inputs and outputs. Previously, MSI (medium-scale integration) and LSI devices have been represented by rectangular blocks with only pin numbers and pin names. It was necessary to study the *truth table* for the device in a data book to determine how the outputs depended on the inputs. As dependency notation becomes more widely understood and used, it should be possible to deduce the operation of MSI and LSI devices *directly* from the system schematic diagrams. Before reading the next section, skim the description of dependency notation in App. B to get an overview of the symbols and labels used.

A MULTIPLEXER USING DEPENDENCY NOTA-
TION   Figure 3-31 shows the dependency notation logic symbol for a 74151 eight-input multiplexer. MUX at the top indicates that the device is a multiplexer. Inputs are shown on the left and outputs on the right. EN inside the device next to the enable input indicates an enable input. The wedge on this input indicates that it is an active low input. If this input is high (disabled), the Y output will be low, and the W output will be high regardless of the signals on the other inputs.

Pins with labels 0 through 7 are the eight data inputs D0 through D7. The binary powers of the select inputs

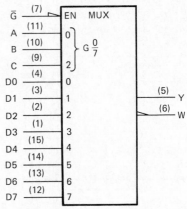

FIGURE 3-31 Dependency notation logic symbol for 74151 eight-input multiplexer.

COMBINATIONAL LOGIC, MULTIPLEXERS, CODES, AND ROMS

A, B, and C are indicated by 0 for the least significant input and 2 for the most significant input. As indicated in App. B, the letter G represents AND dependency. In this case, the bracket before the G and the fraction $\%$ indicate that the output is a function of the binary number on the select inputs AND the data on addressed inputs 0 through 7. In other words, data present on the addressed input will appear on the Y output if the enable input is asserted (L). As indicated by the wedge on it, the W output will have the complement of the logic level on Y.

Here is an example in the language of mixed logic and dependency notation. If the EN input is asserted (L), the binary code for 5 is on the select inputs, AND the input to the pin with 5 next to it in the box is asserted (H), then the output Y will be asserted (H). If the 5 input is not asserted (L), then the Y output will be not asserted (L).

A DEMULTIPLEXER USING DEPENDENCY NOTATION
Figure 3-32 shows the dependency notation symbol for a 74LS138 used as a demultiplexer. DMUX at the top of the symbol is shorthand for demultiplexer. As indicated by the &, the G1, $\overline{G2A}$, and $\overline{G2B}$ inputs are ANDed together. The bracket, G, and $\%$ imply an AND relationship between the binary number on the select inputs and the G1, $\overline{G2A}$, $\overline{G2B}$ function. In other words, if G1, $\overline{G2A}$, and $\overline{G2B}$ are asserted AND a binary number is on the select inputs, then the output labeled with the same number will be asserted. For example, if the G inputs are asserted AND the binary number 5 is on the A, B, and C select inputs, then the output with the number 5 next to it will be asserted (L).

To use the 74LS138 as a demultiplexer, the data is routed to the $\overline{G2A}$ input. The address lines are connected to the select inputs A, B, and C. When an address such as binary 5 is on these three lines, the output labeled 5 will be pointed to. If the data on the $\overline{G2A}$ input is low, the 5 output will be asserted (L). If the data on the $\overline{G2A}$ input is high, the 5 output will be not asserted (H). Note that an output is high except when it is addressed and the data input at that time is low.

FIGURING OUT A NEW DEPENDENCY SYMBOL
Whenever you first see a device dependency notation symbol that you have not seen before, use the truth table for the device and the dependency notation de-

scription in App. B to understand the relationships of inputs and outputs. Then when you see the device again, the dependency notation by itself should indicate the input and output functions to you.

BINARY-BASED CODES

BINARY
In Chap. 2 we introduced the binary number system used for computers and other digital logic. Also shown there is the conversion of decimal numbers to binary and the conversion of binary numbers to decimal.

OCTAL
Binary is not a very compact code. This means that it requires many more digits to express a number than does, for example, decimal. Twelve binary digits can describe only a number up to 4095_{10}. Computers require binary data, but people working with computers have trouble remembering the long binary words produced by the noncompact code. One solution is to use the *octal*, or *base-8*, code. As you can see in Figure 3-33a, the digits in this code represent powers of 8. The symbols, then, are 0 through 7. You can convert a decimal number to the octal equivalent with the same trick that you used to convert decimal to binary. Figure 3-33b shows the technique for decimal-to-octal conversion. Decimal 327 is equal to 507_8. Verification is shown by reconverting the octal to decimal in the second half of Figure 3-33b.

Since 8 is an integral power of 2, conversions from binary to octal and octal to binary are quite simple. Sup-

$$4096 \ 512 \ 64 \ 8 \ 1 \quad \tfrac{1}{8} \quad \tfrac{1}{64} \quad \tfrac{1}{512}$$

$$8^4 \quad 8^3 \quad 8^2 \ 8^1 8^0 . 8^{-1} 8^{-2} 8^{-3}$$

(a)

$$327_{Decimal} = \underline{\quad ? \quad}_{Octal} \qquad 327_D = 507_8$$

LSD

$$8)\overline{327} = 40 \quad R \ \ 7 \ \times 1 \ = 7$$
$$8)\overline{\ 40} = 5 \quad R \ \ 0 \ \times 8 \ = 0$$
$$8)\overline{\ \ 5} = 0 \quad R \ \ 5 \ \times 64 = 320$$

MSD   327

(b)

Binary   101   011   111 .
Octal   5   3   7   Binary Point

(c)

FIGURE 3-33 *(a)* Octal numbers—value of placeholders. *(b)* Conversion of decimal to octal and octal to decimal. *(c)* Binary-to-octal conversion.

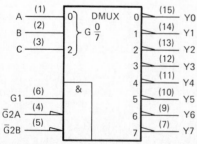

FIGURE 3-32 Dependency notation logic symbol for 74LS138 demultiplexer.

pose you have a binary number such as 101011111. Starting from the binary point and moving to the left, mark off the binary digits in groups of three. Each group of three binary digits is equal to one octal digit. For the example above, 111 is 7, 011 is a 3, and 101 is a 5. Therefore, 101011111 binary is equal to 537 octal.

You convert from octal to binary by replacing each octal digit with its 3-bit binary equivalent.

HEXADECIMAL

Many popular minicomputers such as the PDP-8 have 12 parallel data lines. Four octal digits are an easy way to represent the binary data word on these 12 parallel lines. For example, 100001010111 binary is easily remembered or written as 4127 octal. Most microprocessors have 4-, 8-, or 16-bit data words. For these microprocessors, it is more logical to use a code that groups the binary digits in groups of four rather than three. *Hexadecimal*, or base-16, code does this. Figure 3-34*a* shows the digit values for hexadecimal, which is often just called *hex*. Since hex is base-16, you have to have 16 possible symbols for each digit. The table of Figure 3-34*b* shows the symbols for hex code. After the decimal symbols 0 through 9 are used up, you use the letters A through F for values 10 through 15.

As mentioned above, each hex digit is equal to four binary digits. To convert the binary number 10110110 to hex, mark off groups of four, moving to the left from the binary point. Then write the hex symbol for the value of each group of four. The 0110 group is equal to 6, and the 1011 group is equal to 11. Since 11 is B in hex, 10110110 binary is equal to B6 hex. Eight binary bits require only two hex digits to represent them (see Figure 3-34*c*).

If you want to convert from decimal to hexadecimal, Figure 3-34*d* shows a familiar trick to do this. The result shows that 227_{10} is equal to $E3_{16}$. As you can see, hex is an even more compact code than decimal. Two hexadecimal digits can indicate a number up to 255. Only four hex digits are needed to equal a 16-bit binary number.

To illustrate how hexadecimal numbers are used in digital logic, a service manual tells you that the 8-bit-wide data bus of an 8080A microprocessor should contain $3F_{16}$ during a certain operation. To determine what pattern of 1's and 0's you would expect to find with your oscilloscope or logic analyzer on the parallel lines, convert $3F_{16}$ to 00111111 binary for the patterns on 1's and 0's. Hex 3F is simply a shorthand that is easier to remember and less prone to errors.

To convert from octal to hex code, the easiest way is to write the binary equivalent of the octal and then convert the binary digits, four at a time, into appropriate hex digits. Reverse the procedure to get from hex to octal.

BCD CODES

In applications such as frequency counters, digital voltmeters, or calculators, where the output is a decimal

$$16^3 \quad 16^2 \quad 16^1 \quad 16^0 \, . \, 16^{-1} \quad 16^{-2} \quad 16^{-3}$$

$$4096 \quad 256 \quad 16 \quad 1 \quad \frac{1}{16} \quad \frac{1}{256} \quad \frac{1}{4096}$$

(a)

Dec		Hex
0	=	0
1	=	1
2	=	2
3	=	3
4	=	4
5	=	5
6	=	6
7	=	7
8	=	8
9	=	9
10	=	A
11	=	B
12	=	C
13	=	D
14	=	E
15	=	F

(b)

$$\underbrace{1011}_{B} \quad \underbrace{0110}_{6} \quad \text{Bin}$$
$$\qquad \qquad \qquad \text{Hex}$$

(c)

$$227_D = \underline{\quad ? \quad}_{Hex}$$

$$16\overline{)227} = 14 \qquad R3 \times 1 = 3$$
$$16\overline{)14} = 0 \qquad RE \times 16 = \underline{224}$$
$$\qquad \qquad \qquad \qquad \qquad \qquad 227$$

$$227_{10} = E3_{16}$$

(d)

FIGURE 3-34 Hexadecimal numbers. *(a)* Values of place-holders. *(b)* Symbols. *(c)* Conversion of binary to hexadecimal. *(d)* Conversion of decimal to hex and hex to decimal.

display, often a *binary-coded decimal (BCD)* code is used. The advantage of BCD for these applications is that information for each decimal digit is contained in a separate 4-bit binary word. As you can see in Table 3-1, the simplest BCD code uses the first 10 numbers of standard binary code for BCD numbers 0 through 9.

COMBINATIONAL LOGIC, MULTIPLEXERS, CODES, AND ROMS

TABLE 3-1
COMMON NUMBER CODES

DECIMAL SYSTEM	BINARY	OCTAL	HEXA-DECIMAL	DECIMAL CODES				REFLECTED GRAY CODE	SEVEN-SEGMENT DISPLAY (1 = ON)							
				8421 BCD	2421	5421	EXCESS-3		A	B	C	D	E	F	G	DISPLAY
0	0000	0	0	0000	0000	0000	0011	0000	1	1	1	1	1	1	0	0
1	0001	1	1	0001	0001	0001	0100	0001	0	1	1	0	0	0	0	1
2	0010	2	2	0010	0010	0010	0101	0011	1	1	0	1	1	0	1	2
3	0011	3	3	0011	0011	0011	0110	0010	1	1	1	1	0	0	1	3
4	0100	4	4	0100	0100	0100	0111	0110	0	1	1	0	0	1	1	4
5	0101	5	5	0101	1011	1000	1000	0111	1	0	1	1	0	1	1	5
6	0110	6	6	0110	1100	1001	1001	0101	1	0	1	1	1	1	1	6
7	0111	7	7	0111	1101	1010	1010	0100	1	1	1	0	0	0	0	7
8	1000	10	8	1000	1110	1011	1011	1100	1	1	1	1	1	1	1	8
9	1001	11	9	1001	1111	1100	1100	1101	1	1	1	1	0	1	1	9
10	1010	12	A	0001 0000	0001 0000	0001 0000	0100 0011	1111	1	1	1	0	1	1	1	A
11	1011	13	B	0001 0001	0001 0001	0001 0001	0100 0100	1110	0	0	1	1	1	1	1	B
12	1100	14	C	0001 0010	0001 0010	0001 0010	0100 0101	1010	0	0	0	1	1	0	1	C
13	1101	15	D	0001 0011	0001 0011	0001 0011	0100 0110	1011	0	1	1	1	1	0	1	D
14	1110	16	E	0001 0100	0001 0100	0001 0100	0100 0111	1001	1	0	0	1	1	1	1	E
15	1111	17	F	0001 0101	0001 1011	0001 1000	0100 1000	1000	1	0	0	0	1	1	1	F

5   2   9   Decimal

0101 0010 1001 BCD

FIGURE 3-35 Decimal-to-BCD conversion.

Each decimal digit, then, is individually represented by its 4-bit binary equivalent. Figure 3-35 illustrates this.

Several other common BCD codes are included in Table 3-1 for reference. An important one is the *excess-3 BCD* which is used in some calculators because it simplifies internal arithmetic operations. Excess-3 BCD code is produced by adding 3 to each digit of the standard 8421 BCD. You can see this by comparing corresponding excess-3 and 8421 BCD numbers in Table 3-1.

GRAY CODE

Gray code, another important binary code, is often used for encoding data from computer-controlled machines such as lathes. This code has the same 16 possible combinations as standard binary, but as you can see in Table 3-1, they are arranged in a different order. Notice that only one binary digit changes at a time as you count up in this code. Revising the order of the code in this manner reduces the size of the possible error when data is pulled in from sources such as the shaft-position encoding disk shown in Figure 3-36. The disk converts the position of the rotating shaft to a binary Gray code equivalent. The number of bits (sectors on the disk) will determine the resolution. The disk shown is 4 bits, which gives 22.5° resolution. Eight bits give a resolution of 1 part in 2^8, or 1.4°.

With Gray code, the largest possible error is one *least* significant digit. With a standard binary encoder disk, you can have an error of as much as the *most* significant digit. If, for example, on a change from 7 to 8, the sensor picks up the first 3 bits changing to 0 but doesn't pick up the fourth bit changing to 1, then the output will be 0000, which is an error of a most significant bit.

To demonstrate the advantage of Gray code, try the same transition for it. Now the worst error you can get is

1 *least* significant bit less, or 7, rather than the 8 that should have come through.

If you need to construct a Gray code table larger than that in Table 3-1, a handy method is to observe the pattern of 1's and 0's and just extend it. The least-significant-bit column starts with one 0 and then has alternating groups of two 1's and two 0's as you go down the column. The second-most-significant-digit column starts with two 0's and then has alternating groups of four 1's and four 0's. The third column starts with four 0's and then has alternating groups of eight 1's and eight 0's. By now, you should see the pattern. Try to figure out the Gray code for the decimal number 16. You should get 11000.

SEVEN-SEGMENT DISPLAY CODE

Since seven-segment displays such as that shown in Figure 3-37 are now so common in everything from calculators to gasoline pumps, the segment code for these has been included in Table 3-1. A single seven-segment display will display the last six numbers (10 through 15) of this code as the hexadecimal digits A through F. In Table 3-1, a 1 indicates that the segment is lit, which is true for displays such as the common-cathode LED display in Figure 3-37b. For some displays, such as the common-anode LED display shown in Figure 3-37c, a low actually lights the segment, so you have to invert all the values.

MULTIBIT CODES

PARITY   When communicating with computers from teletypes, punched cards, or a CRT terminal, you need a binary-based code that can represent letters of the alphabet as well as numbers. Common codes used for this

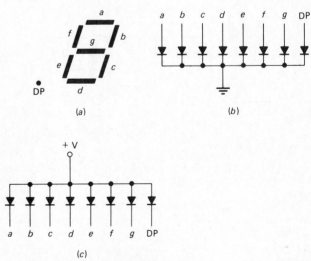

FIGURE 3-37 Seven-segment LED display. *(a)* Segment labels. *(b)* Schematic of common-cathode type. *(c)* Schematic of common-anode type.

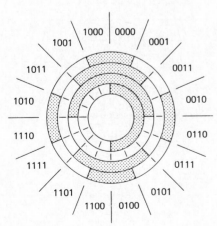

FIGURE 3-36 Gray code shaft position encoder disk.

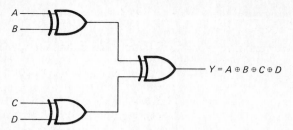

FIGURE 3-38 Exclusive OR "tree" circuit for four variables.

purpose have 5 to 12 binary bits per word. To detect possible errors in multibit codes, an additional bit, called a *parity bit*, is often added as the most significant bit.

Parity is a term used to identify whether a data word has an odd or even number of 1's. If a data word contains an odd number of 1's, the word is said to have *odd* parity. The binary word 0110111 with five 1's has odd parity. The binary word 0110000 has an even number of 1's (two), so it has *even* parity.

The parity of a binary word can be determined with an exclusive OR gate circuit. Figure 3-38 shows a four-input exclusive OR "tree" circuit. The output of this circuit will be high if an odd number of inputs is high and low if an even number of inputs is high.

An eight-input parity "tree" such as that found in the 74180 parity generator checker can tell you whether the parity of up to an 8-bit data word is odd or even. As connected in Figure 3-39, the 74180 at the sending end will detect the odd parity in the data word and generate a high output. This generated parity information, or parity bit as it is sometimes called, is transmitted along with the data on a ninth wire. The overall parity for the nine wires together is now even. Another 74180 then

checks the parity of the nine lines at the receiving end. If the checker finds even parity on the nine lines as expected, it outputs a low, which tells the sending system that the data was received with the same parity as sent. A high output from the receiving-end checker indicates an error in the received data. This high output waves an error flag at the sending system and may be used to tell it to repeat the data word.

Let's summarize this error-detecting scheme. The 74180 at the sending end examines the parity of the data word and generates an output 0 or 1 to always give even parity on the nine lines. The 74180 at the receiving end checks the nine lines and outputs a high as an error flag if the parity of the nine lines is not even, as expected.

A difficulty with this method of detecting errors introduced during transmission is that two errors introduced into a data word may keep the correct parity, and so the parity checker won't indicate an error. A complex method using more lines and "Hamming codes" can be utilized to detect multiple errors in transmitted data and even to correct errors. Hamming-code error detecting/correcting circuits also employ exclusive OR tree-based parity generator/checkers.

ASCII Table 3-2 shows several multibit codes. The first is *ASCII*, the American Standard Code for Information Interchange. This is shown in the table as a 7-bit code. With 7 bits, you can code up to 128 characters, which is enough for full uppercase and lowercase alphabets plus numbers, punctuation marks, and control characters. The code is arranged so that if only uppercase letters, numbers, and a few control characters are needed, the lower 6 bits is all that is required. If a *parity*

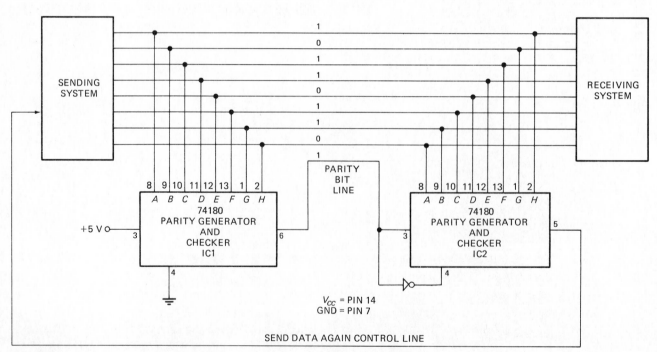

FIGURE 3-39 System using parity generator and checker to check for errors in transmitted data.

check is wanted, a *parity bit* is added to the basic 7-bit code in the most significant bit position. The binary word 11000100, for example, is the ASCII expression for uppercase D with odd parity. You may have seen 1-in [2.5-cm] wide paper tape with holes punched to represent ASCII, as shown in Figure 3-40*a*. Table 3-3 gives the meaning of the control character symbols used in the ASCII table.

BCDIC *BCDIC* stands for the Binary-Coded Decimal Interchange Code used in some computers. It uses 7 bits plus 1 parity bit. The lower 4 bits are referred to as the *numeric bits.* The upper 4 bits contain 1 parity bit and 3 zone bits. The arrangement of these bits is shown at the bottom of Table 3-2. To save space in Table 3-2, the hex equivalent of the binary digits is used for BCDIC expressed with even parity.

EBCDIC Another alphanumeric code commonly encountered in IBM equipment is the Extended Binary-Coded Decimal Interchange Code *(EBCDIC)*. This is an 8-bit code without parity. A ninth bit can be added for parity. To save space in Table 3-2, the eight binary digits of EBCDIC are represented with their two-digit hex equivalents.

SELECTRIC *Selectric* is a 7-bit code used in the familiar IBM spinning-ball typewriters and printers. Table 3-2 shows this code for reference also. Each bit position in the code controls an operation of the spinning ball. From most significant bit to least, the meanings of the 7

bits are ROTATE 5, TILT 1, TILT 2, SHIFT, ROTATE 2A, ROTATE 2, and ROTATE 1. In addition to this 7-bit code, Selectrics have separate machine commands for space, return, backspace, tabs, bell, and index.

Baudot is an obsolete 5-bit code which was used for teletypes. It has been replaced by ASCII and so is not shown in Table 3-2.

HOLLERITH *Hollerith* is a 12-bit code used to encode data from those computer cards that threaten you with a fate worse than death if you "fold, spindle, or mutilate" them. Figure 3-40*b* shows a standard 12-row by 80-column card. The 12 data rows are labeled, starting from the top, 12, 11, 0, 1, 2, 3, 4, 5, 6, 7, 8, 9. The top 3 rows are called *zone punches,* and the bottom 10 rows are called *digit punches.* Note that the zero row is included in both categories. A punched hole represents a 1, and a data word is described by the 12 bits in a vertical column. The card in Figure 3-40*b* shows the Hollerith code for the numbers and letters printed across the top of the card. Table 3-2 shows the entire code and the punched-hole equivalent for each character. Since Hollerith code uses very few of the possible combinations for 12 bits, it is not very efficient. Therefore, it is usually converted to ASCII or EBCDIC for use.

CODE CONVERTERS

What do you do when you have a circuit that puts out a number or character in one code and you need it in an-

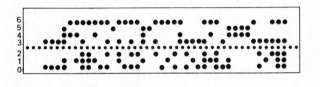

(a)

FIGURE 3-40 Codes used for storage on tapes and cards. (a) ASCII on paper tape. (b) Hollerith on punched cards.

(b)

TABLE 3-2
COMMON ALPHANUMERIC CODES

ASCII SYMBOL	HEX CODE FOR 7-BIT ASCII	BCDIC SYMBOL	HEX CODE FOR EP BCDIC	EBCDIC SYMBOL	HEX CODE FOR EBCDIC	SELECTRIC SYMBOL	HEX CODE FOR SELECTRIC	HOLLERITH SYMBOL	HOLES PUNCHED CODE FOR HOLLERITH
N U L	0 0			N U L	0 0			N U L	12 0 9 8 1
S O H	0 1			S O H	0 1			S C H	12 9 1
S T X	0 2			S T X	0 2			S T X	12 9 2
E T X	0 3			E T X	0 3			E T X	12 9 3
E O T	0 4			E O T	3 7			E C T	9 7
E N Q	0 5			E N Q	2 D			E N Q	0 9 8 5
A C K	0 6			A C K	2 E			A C K	0 9 8 6
B E L	0 7			B E L	2 F			B E L	0 9 8 7
B S	0 8			B S	1 6			B S	11 9 6
H T	0 9			H T	0 5			H T	12 9 5
L F	0 A			L F	2 5			L F	0 9 5
V T	0 B	‡	9 A	V T	0 B			V T	12 9 8 3
F F	0 C			F F	0 C			F F	12 9 8 4
C R	0 D	‡	F F	C R	0 D			C R	12 9 8 5
S O	0 E			S O	0 E			S O	12 9 8 6
S I	0 F			S I	0 F			S I	12 9 8 7
D L E	1 0			D L E	1 0			D L E	12 11 9 8 1
D C 1	1 1			D C 1	1 1			D C 1	11 9 1
D C 2	1 2			D C 2	1 2			D C 2	11 9 2
D C 3	1 3			D C 3	1 3			D C 3	11 9 3
D C 4	1 4			D C 4	3 5			D C 4	9 8 4
N A K	1 5			N A K	3 D			N A K	9 8 5
S Y N	1 6			S Y N	3 2			S Y N	9 2
E T B	1 7			E O B	2 6			E T B	0 9 6
C A N	1 8			C A N	1 8			C A N	11 9 8
E M	1 9			E M	1 9			E M	11 9 8 1
S U B	1 A			S U B	3 F			S U B	9 8 7
E S C	1 B			B Y P	2 4			E S C	0 9 7
F S	1 C			F L S	1 C			F S	11 9 8 4
G S	1 D			G S	1 D			G S	11 9 8 5
R S	1 E			R D S	1 E			R S	11 9 8 6
U S	1 F			U S	1 F			U S	11 9 8 7
S P	2 0	S P	0 0	S P	4 0			S P	NO PNCH
!	2 1	!	6 A	!	5 A	½!	2 7	!	12 8 7
"	2 2	⧻	5 F	"	7 F	"	2 D	"	8 7
#	2 3	#	4 B	#	7 B	#	7 E	#	8 3
$	2 4	$	2 B	$	5 B	$	7 9	$	11 8 3
%	2 5	%	5 C	%	6 C	%	3 D	%	0 8 4
&	2 6	&	3 0	&	5 0	&	7 D		12
'	2 7	V	1 D	'	7 D	'	2 5	'	8 5
(	2 8	Blank	5 0	(	4 D	(	3 8	(	12 8 5
)	2 9	△	6 F	)	5 D	)	3 9	)	11 8 5
*	2 A	*	6 C	*	5 C	*	7 C	*	11 8 4

TABLE 3-2 (Continued)
COMMON ALPHANUMERIC CODES

ASCII SYMBOL	HEX CODE FOR 7-BIT ASCII	BCDIC SYMBOL	HEX CODE FOR EP BCDIC	EBCDIC SYMBOL	HEX CODE FOR EBCDIC	SELECTRIC SYMBOL	HEX CODE FOR SELECTRIC	HOLLERITH SYMBOL	HOLES PUNCHED CODE FOR HOLLERITH
+	2 B			+	4 E	+	0 E	+	12 8 6
,	2 C	,	1 B	,	6 B	,	4 4	,	0 8 3
−	2 D			−	6 0	−	0 0	−	11
.	2 E	.	7 B	.	4 B	.	2 6	.	12 8 3
/	2 F	/	1 1	/	6 1	/	4 1	/	0 1
0	3 0	0	0 A	0	F 0	0	3 1	0	0
1	3 1	1	4 1	1	F 1	1	7 7	1	1
2	3 2	2	4 2	2	F 2	2	3 6	2	2
3	3 3	3	0 3	3	F 3	3	7 6	3	3
4	3 4	4	4 4	4	F·4	4	7 1	4	4
5	3 5	5	0 5	5	F 5	5	3 5	5	5
6	3 6	6	0 6	6	F 6	6	3 4	6	6
7	3 7	7	4 7	7	F 7	7	7 5	7	7
8	3 8	8	4 8	8	F 8	8	7 4	8	8
9	3 9	9	0 9	9	F 9	9	3 0	9	9
:	3 A	:	4 D	:	7 A	:	4 D	:	8 2
;	3 B	;	2 E	;	5 E	;	4 5	;	11 8 6
<	3 C	<	7 E	<	4 C			<	12 8 4
=	3 D	√	0 F	=	7 E	=	0 6	=	8 6
>	3 E	>	4 E	>	6 E			>	0 8 6
?	3 F	?	3 A	?	6 F	?	4 9	?	0 8 7
@	4 0	@	0 C	@	7 C	@	3 E	@	8 4
A	4 1	A	7 1	A	C 1	A	6 C	A	12 1
B	4 2	B	7 2	B	C 2	B	1 8	B	12 2
C	4 3	C	3 3	C	C 3	C	5 C	C	12 3
D	4 4	D	7 4	D	C 4	D	5 D	D	12 4
E	4 5	E	3 5	E	C 5	E	1 D	E	12 5
F	4 6	F	3 6	F	C 6	F	4 E	F	12 6
G	4 7	G	7 7	G	C 7	G	4 F	G	12 7
H	4 8	H	7 8	H	C 8	H	1 9	H	12 8
I	4 9	I	3 9	I	C 9	I	2 C	I	12 9
J	4 A	J	2 1	J	D 1	J	0 7	J	11 1
K	4 B	K	2 2	K	D 2	K	1 C	K	11 2
L	4 C	L	6 3	L	D 3	L	5 9	L	11 3
M	4 D	M	2 4	M	D 4	M	6 F	M	11 4
N	4 E	N	6 5	N	D 5	N	1 E	N	11 5
O	4 F	O	6 6	O	D 6	O	6 9	O	11 6
P	5 0	P	2 7	P	D 7	P	0 D	P	11 7
Q	5 1	Q	2 8	Q	D 8	Q	0 C	Q	11 8
R	5 2	R	6 9	R	D 9	R	6 D	R	11 9
S	5 3	S	1 2	S	E 2	S	2 9	S	0 2
T	5 4	T	5 3	T	E 3	T	1 F	T	0 3

TABLE 3-2 (Continued)
COMMON ALPHANUMERIC CODES

ASCII SYMBOL	HEX CODE FOR 7-BIT ASCII	BCDIC SYMBOL	HEX CODE FOR EP BCDIC	EBCDIC SYMBOL	HEX CODE FOR EBCDIC	SELEC-TRIC SYMBOL	HEX CODE FOR SELEC-TRIC	HOL-LERITH SYMBOL	HOLES PUNCHED CODE FOR HOLLERITH
U	5 5	U	1 4	U	E 4	U	5 E	U	0 4
V	5 6	V	5 5	V	E 5	V	6 E	V	0 5
W	5 7	W	5 6	W	E 6	W	2 8	W	0 6
X	5 8	X	1 7	X	E 7	X	5 F	X	0 7
Y	5 9	Y	1 8	Y	E 8	Y	0 9	Y	0 8
Z	5 A	Z	5 9	Z	E 9	Z	3 F	Z	0 9
[	5 B	[	7 D	[	A D	[	7 F	[	12 8 2
\	5 C	\	1 E	N L	1 5			\	0 8 2
]	5 D	]	2 D	]	D D			]	11 8 2
^	5 E	□	3 C	¬	5 F			^	11 8 7
—	5 F	—	6 0	—	6 D	—	0 8	—	0 8 5
`	6 0			R E S	1 4			`	8 1
a	6 1			a	8 1	a	6 4	a	12 0 1
b	6 2			b	8 2	b	1 0	b	12 0 2
c	6 3			c	8 3	c	5 4	c	12 0 3
d	6 4			d	8 4	d	5 5	d	12 0 4
e	6 5			e	8 5	e	1 5	e	12 0 5
f	6 6			f	8 6	f	4 6	f	12 0 6
g	6 7			g	8 7	g	4 7	g	12 0 7
h	6 8			h	8 8	h	1 1	h	12 0 8
i	6 9			i	8 9	i	2 4	i	12 0 9
j	6 A			j	9 1	j	0 7	j	12 11 1
k	6 B			k	9 2	k	1 4	k	12 11 2
l	6 C			l	9 3	l	5 1	l	12 11 3
m	6 D			m	9 4	m	6 7	m	12 11 4
n	6 E			n	9 5	n	1 6	n	12 11 5
o	6 F			o	9 6	o	6 1	o	12 11 6
p	7 0			p	9 7	p	0 5	p	12 11 7
q	7 1			q	9 8	q	0 4	9	12 11 8
r	7 2			r	9 9	r	6 5	r	12 11 9
s	7 3			s	A 2	s	2 1	s	11 0 2
t	7 4			t	A 3	t	1 7	t	11 0 3
u	7 5			u	A 4	u	5 6	u	11 0 4
v	7 6			v	A 5	v	6 6	v	11 0 5
w	7 7			w	A 6	w	2 0	w	11 0 6
x	7 8			x	A 7	x	5 7	x	11 0 7
y	7 9			y	A 8	y	0 1	y	11 0 8
z	7 A			z	A 9	z	3 7	z	11 0 9
{	7 B			{	8 B			{	12 0
|	7 C			|	4 F			|	12 11
}	7 D			}	9 B			}	11 0
~	7 E			¢	4 A			~	11 0 1
D E L	7 F			D E L	0 7			D E L	12 9 7

BCDIC

$\underbrace{\text{HEX DIGIT}}_{\text{PCBA}} \quad \underbrace{\text{HEX DIGIT}}_{2^3 2^2 2^1 2^0}$

SELECTRIC

$\underbrace{R_5 T_1 T_2}_{\text{HEX DIGIT}} \quad \underbrace{SR_{2A} R_2 R_1}_{\text{HEX DIGIT}}$

TABLE 3-3
DEFINITIONS OF CONTROL CHARACTERS

NUL	NULL	**DC2**	DIRECT CONTROL 2
SOH	START OF HEADING	**DC3**	DIRECT CONTROL 3
STX	START TEXT	**DC4**	DIRECT CONTROL 4
ETX	END TEXT	**NAK**	NEGATIVE
EOT	END OF TRANSMIS-		ACKNOWLEDGE
	SION	**SYN**	SYNCHRONOUS
ENQ	ENQUIRY		IDLE
ACK	ACKNOWLEDGE	**ETB**	END TRANSMIS-
BEL	BELL		SION BLOCK
BS	BACKSPACE	**CAN**	CANCEL
HT	HORIZONTAL TAB	**EM**	END OF MEDIUM
LF	LINE FEED	**SUB**	SUBSTITUTE
VT	VERTICAL TAB	**ESC**	ESCAPE
FF	FORM FEED	**FS**	FORM SEPARATOR
CR	CARRIAGE RETURN	**GS**	GROUP SEPARATOR
SO	SHIFT OUT	**RS**	RECORD
SI	SHIFT IN		SEPARATOR
DLE	DATA LINK ESCAPE	**US**	UNIT SEPARATOR
DC1	DIRECT CONTROL 1		

other code? For example, the output of a decade counter is in BCD, but you need seven-segment code to drive a display. The first method described below is the traditional method of code conversion by using Karnaugh maps and logic gates. This is followed by the MSI and LSI methods of code conversion.

TRADITIONAL METHOD FOR CODE CONVERTERS

Figure 3-41a shows the values for the BCD numbers 0 through 9 and their seven-segment equivalents. Figure 3-41a also shows the other possible combinations of the BCD inputs, followed by the don't-care states in the segment columns. The reason you don't care about the segments for these input combinations is that the BCD code is never supposed to get to these combinations. As you will see, however, including these don't-care states in the table simplifies the final logic expression. To make a converter from BCD to seven-segments, you have to find a simplified logic expression for each segment in terms of the BCD variables A, B, C, and D. This means a Karnaugh map must be made for each segment.

Figure 3-41b shows the Karnaugh map for the a segment expression. As you can see from the circled squares, the don't-care states can be included when you extract the simplified expression. An earlier discussion of Karnaugh maps demonstrated that the more boxes you can circle, the fewer variables you have in the output expression. You *should* include a don't-care state in a loop if the don't-care state occurs *next to a 1*, and if including the don't-care state would make a *larger,*

legal (2^n) *group*. Otherwise, ignore the don't-care state(s). The expression obtained for the a segment, using the K-map and including appropriate don't-care states, is

$$a = BD + \overline{B}\,\overline{D} + A + C$$

Figure 3-41c shows one way to implement the a segment with common TTL logic gates. Note that it takes portions of two different IC packages plus four inverters just for the a segment. You can do the other six Karnaugh maps for practice because in some cases this method is needed. The results for the other six segments are

$$b = \overline{B} + \overline{C}\,\overline{D} + CD$$
$$c = \overline{C} + B + D$$
$$d = \overline{B}\,\overline{D} + A + \overline{B}C + C\overline{D} + B\overline{C}D$$
$$e = \overline{B}\,\overline{D} + C\overline{D}$$
$$f = A + B\overline{D} + B\overline{C} + \overline{C}\,\overline{D}$$
$$g = A + \overline{B}C + B\overline{C} + C\overline{D}$$

MSI APPROACH TO CODE CONVERTERS

A quick look at a TTL or CMOS data book shows that for many code conversions, MSI devices are readily available. Some devices are referred to as decoders, and others are referred to as encoders. A *decoder* usually converts from a more compact code such as BCD to a less compact code such as the seven-segment one. An *encoder* usually converts from a less dense code such as one-of-eight low to a more compact code such as binary.

DECODERS   A good example of a decoder is the 7447 BCD-to-seven-segment decoder. This device does the complete conversion described in the preceding section on traditional code converter methods. Figure 4-15 shows how a 7447 is connected to convert a 4-bit BCD input code to the correct drive signals for a seven-segment, common-anode LED display. A similar device, the 7448, converts a 4-bit BCD input to the proper drive for a common-cathode, seven-segment LED display. The 4511 is a CMOS equivalent of the 7448. These ICs also have additional control inputs that light the entire display for testing, blank the display, or blank the display only if it is a leading zero on, for example, a DVM.

Figure 3-42a and b shows the function table and the dependency notation logic symbol for a 7447 BCD-to-seven-segment decoder. We use this as another example of how dependency notation shows the relationships of inputs and outputs on a more complex device. First, study the function table and notes. Second, try to determine how the dependency notation symbol conveys this information. Refer to App. B for explanation of any notation symbols you have not seen before or do not remember.

Observe the segment outputs a through g in the function table. The wedges on each indicate that they are active low, and the diamonds next to each indicate that these outputs are open-collector. The BCD inputs A, B,

83

	A	B	C	D	a	b	c	d	e	f	g
0	0	0	0	0	1	1	1	1	1	1	0
1	0	0	0	1	0	1	1	0	0	0	0
2	0	0	1	0	1	1	0	1	1	0	1
3	0	0	1	1	1	1	1	1	0	0	1
4	0	1	0	0	0	1	1	0	0	1	1
5	0	1	0	1	1	0	1	1	0	1	1
6	0	1	1	0	1	0	1	1	1	1	1
7	0	1	1	1	1	1	1	0	0	0	0
8	1	0	0	0	1	1	1	1	1	1	1
9	1	0	0	1	1	1	1	1	0	1	1
10	1	0	1	0	X	X	X	X	X	X	X
11	1	0	1	1	X	X	X	X	X	X	X
12	1	1	0	0	X	X	X	X	X	X	X
13	1	1	0	1	X	X	X	X	X	X	X
14	1	1	1	0	X	X	X	X	X	X	X
15	1	1	1	1	X	X	X	X	X	X	X

(a)

FIGURE 3-41 Conversion from BCD to seven-segment code. *(a)* Truth table. *(b)* Karnaugh map and derived expression for *a* segment. *(c)* Circuit implementing expression for *a* segment.

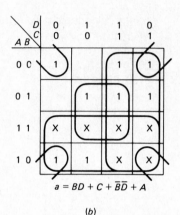

$$a = BD + C + \overline{B}\overline{D} + A$$

(b)

(c)

C, and D are labeled with their binary weights. From the first 16 entries in the function table, you can see that if lamp test ($\overline{LT}$) is high, ripple-blanking input ($\overline{RBI}$) is high, and blanking input/ripple-blanking output ($\overline{BI}/\overline{RBO}$) is high, then the seven-segment outputs will correspond to the binary code on the A, B, C, and D inputs. If the blanking input ($\overline{BI}/\overline{RBO}$) is forced low or if $\overline{RBI}$ is low and $\overline{LT}$ is high, then the outputs will all be off. This blanks the display. If lamp test ($\overline{LT}$) is low and $\overline{BI}/\overline{RBO}$ is not forced low, the outputs will all be on. This mode can be used to see whether all the segments of a display are connected and functioning. The outputs also will all be on if $\overline{LT}$ is high and the A, B, C, and D inputs contain the code for 8.

Now look at the dependency notation symbol in Figure 3-42*b* to see how it summarizes all this. To help you, Figure 3-42*c* shows the internal logic blocks, using more conventional logic symbols. You can see that if $\overline{BI}$

is asserted (L) *or* $\overline{RBI}$ is asserted and $\overline{LT}$ is not asserted (H) while the count on A, B, C, and D is zero, then the signal at X will be asserted. The bubble next to G21 represents a negated internal connection. This means that if the signal at this point is *not asserted* AND a code is present on the A, B, C, and D inputs, then the outputs identified by the number 21 will be asserted (L) according to the input binary code. If the signal at this point is asserted, then all outputs identified by the number 21 will be not asserted (H). This then blanks the display. Note that $\overline{RBO}$ is asserted (L) if $\overline{RBI}$ is asserted, $\overline{LT}$ is not asserted, and the code into A, B, C, and D is 0000.

If the $\overline{LT}$ input is asserted (L), all the outputs identified with the number 20 will be asserted (L). This then lights all segments.

The dependency notation symbols do not eliminate the need for a function table, but they do make it possi-

SEGMENT
IDENTIFICATION

NUMERICAL DESIGNATIONS AND RESULTANT DISPLAYS

'46A, '47A, 'L46, 'L47, 'LS47 FUNCTION TABLE

DECIMAL OR FUNCTION	INPUTS						BI/RBO†	OUTPUTS							NOTE
	LT	RBI	D	C	B	A		a	b	c	d	e	f	g	
0	H	H	L	L	L	L	H	ON	ON	ON	ON	ON	ON	OFF	
1	H	X	L	L	L	H	H	OFF	ON	ON	OFF	OFF	OFF	OFF	
2	H	X	L	L	H	L	H	ON	ON	OFF	ON	ON	OFF	ON	
3	H	X	L	L	H	H	H	ON	ON	ON	ON	OFF	OFF	ON	
4	H	X	L	H	L	L	H	OFF	ON	ON	OFF	OFF	ON	ON	
5	H	X	L	H	L	H	H	ON	OFF	ON	ON	OFF	ON	ON	
6	H	X	L	H	H	L	H	OFF	OFF	ON	ON	ON	ON	ON	
7	H	X	L	H	H	H	H	ON	ON	ON	OFF	OFF	OFF	OFF	1
8	H	X	H	L	L	L	H	ON	ON	ON	ON	ON	ON	ON	
9	H	X	H	L	L	H	H	ON	ON	ON	OFF	OFF	ON	ON	
10	H	X	H	L	H	L	H	OFF	OFF	OFF	ON	ON	OFF	ON	
11	H	X	H	L	H	H	H	OFF	OFF	ON	ON	OFF	OFF	ON	
12	H	X	H	H	L	L	H	OFF	ON	OFF	OFF	OFF	ON	ON	
13	H	X	H	H	L	H	H	ON	OFF	OFF	ON	OFF	ON	ON	
14	H	X	H	H	H	L	H	OFF	OFF	OFF	ON	ON	ON	ON	
15	H	X	H	H	H	H	H	OFF	OFF	OFF	OFF	OFF	OFF	OFF	
BI	X	X	X	X	X	X	L	OFF	OFF	OFF	OFF	OFF	OFF	OFF	2
RBI	H	L	L	L	L	L	L	OFF	OFF	OFF	OFF	OFF	OFF	OFF	3
LT	L	X	X	X	X	X	H	ON	ON	ON	ON	ON	ON	ON	4

H = high level, L = low level, X = irrelevant

NOTES:
1. The blanking input (BI) must be open or held at a high logic level when output functions 0 through 15 are desired. The ripple-blanking input (RBI) must be open or high if blanking of a decimal zero is not desired.
2. When a low logic level is applied directly to the blanking input (BI), all segment outputs are off regardless of the level of any other input.
3. When ripple-blanking input (RBI) and inputs *A, B, C,* and *D* are at a low level with the lamp test input high, all segment outputs go off and the ripple-blanking output (RBO) goes to a low level (response condition).
4. When the blanking input/ripple blanking output (BI/RBO) is open or held high and a low is applied to the lamp-test input, all segment outputs are on.

†BI/RBO is wire-AND logic serving as blanking input (BI) and/or ripple-blanking output (RBO).

(a)

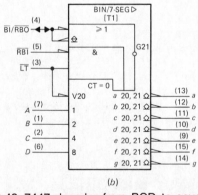

(b)

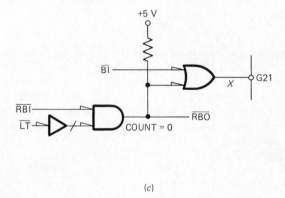

(c)

FIGURE 3-42 7447 decoder from BCD to seven-segment code. (a) Function table. (b) Dependency notation logic symbol. (c) Internal logic for $\overline{BI}$, $\overline{RBI}$, and $\overline{LT}$.

ble to determine most functions of a device directly from the schematic when you have learned how the symbols work.

Another handy group of decoders, the 7442, 7445, and 74145, converts a BCD input code to a 1-line-out-of-10 low code. One of these BCD-to-decimal decoders is used to drive gas-discharge displays that have a common anode tied to a high voltage and separate cathodes shaped as each complete character. When a cathode is grounded, the region around it glows and traces the character.

ENCODERS A good example of an encoder is the 74148, eight-line-to-three-line priority encoder. Figure 3-43 shows the dependency notation symbol and a function table for the 74148. To enable the device, the enable input EI is made low. The device has eight data inputs (0 through 7), three data outputs (A2, A1, and A0), and

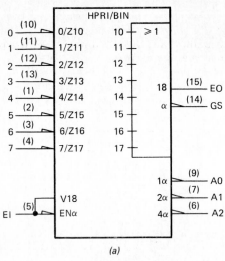

(a)

'148, 'LS148
FUNCTION TABLE

INPUTS									OUTPUTS				
EI	0	1	2	3	4	5	6	7	A2	A1	A0	GS	EO
H	X	X	X	X	X	X	X	X	H	H	H	H	H
L	H	H	H	H	H	H	H	H	H	H	H	H	L
L	X	X	X	X	X	X	X	L	L	L	L	L	H
L	X	X	X	X	X	X	L	H	L	L	H	L	H
L	X	X	X	X	X	L	H	H	L	H	L	L	H
L	X	X	X	X	L	H	H	H	L	H	H	L	H
L	X	X	X	L	H	H	H	H	H	L	L	L	H
L	X	X	L	H	H	H	H	H	H	L	H	L	H
L	X	L	H	H	H	H	H	H	H	H	L	L	H
L	L	H	H	H	H	H	H	H	H	H	H	L	H

(b)

FIGURE 3-43 74148 eight-line-to-three-line priority encoder. (a) Dependency notation logic symbol. (b) Function table.

two control outputs (GS and EO). All inputs and outputs are active low. The EO (enable output) will be low when EI is low and all inputs are high. If a low is present on any of the eight data inputs, the GS (group strobe) output will be low to indicate this. If EI is low and only one data input is low, the data outputs will contain the active low binary code for the number of the input that has the low on it. For example, if only input 3 has a low on it, then A2 will be high, A1 will be low, and A0 will be low. These levels represent the active low or inverted binary code for the number 3. If more than one input has a low, then the code for the highest-number input that has a low on it will appear on the output. This is indicated by the don't-care X's in the function table of Figure 3-43. If, for example, input 3 and input 5 both have low on them, then the outputs will have A2 = L, A1 = H, A0 = L, the inverted code for the number 5. The 5 data input is said to have a higher priority than the 3 data input. To summarize the operation of the 74148 priority encoder, the data outputs will show the inverted binary code for the highest-numbered input that has a low on it.

The dependency notation symbol for the 74148 uses the letter Z to indicate internal logic interconnection. If

the 0 input of the 74148 is asserted, the symbol Z10 indicates that through some internal connections the 10 input of the internal OR block will be asserted. If the $EN\alpha$ input is asserted (L), then all the outputs identified with the Greek letter α will be enabled. In this case, a Greek letter is used to avoid confusion because the outputs already have binary weighted numbers on them. If $EN\alpha$ is asserted (L), then the code for the highest-priority asserted (L) input will be on the A0, A1, and A2 outputs. The EO will be not asserted (H) if EI is not asserted (H) OR one of the data inputs is asserted. GS will be asserted (L) if $EN\alpha$ is asserted (L) and one of the data inputs is asserted (L).

OTHER SIMPLE CODE CONVERTERS

For converting binary code to BCD, the 74185 can be used; for converting BCD to binary, the 74184 is often employed.

If you need to convert one BCD code to another BCD code and you can't find an IC to do it, there is still another MSI way to get there. Earlier in this chapter, we showed how any function of four input variables can be built with a single eight-input multiplexer such as the 74151 or CD4051. With this technique you can, for example, convert 4 bits of standard BCD to the equivalent 4 bits of excess-3 BCD. Figure 3-44a shows the two codes. Figure 3-44b shows the multiplexer inputs to give the proper output for the A digit of the excess-3 code. Three more multiplexers can be used to produce outputs for the remaining excess-3 digits. Four eight-input multiplexers can convert 4 bits of any BCD code to 4 bits of any other BCD code. This is not necessarily a minimum solution, but it is quick and it is an example of a trend in digital electronics. That trend is to use very powerful generalized circuits that can be utilized for many different functions simply by changing their programming.

Conversions between alphanumeric codes, such as from ASCII to EBCDIC, or combinational logic designs with more than four variables require too many gates or multiplexers, so for these we use read-only memories.

READ-ONLY MEMORIES

Read-only memories, or *ROMs*, store binary data as 1's or 0's at the intersections of an *X-Y* grid of lines. By addressing these points in the proper way, you can read out the data stored there.

A HOME-MADE ROM

Figure 3-45 is a circuit for a "home-made" ROM that illustrates how ROMs work and how a ROM handles the standard BCD-to-excess-3-BCD code conversion discussed above. Note that the home-made ROM circuit

BCD				Excess - 3			
J	K	L	M	A	B	C	D
0	0	0	0	0	0	1	1
0	0	0	1	0	1	0	0
0	0	1	0	0	1	0	1
0	0	1	1	0	1	1	0
0	1	0	0	0	1	1	1
0	1	0	1	1	0	0	0
0	1	1	0	1	0	0	1
0	1	1	1	1	0	1	0
1	0	0	0	1	0	1	1
1	0	0	1	1	1	0	0

(a)

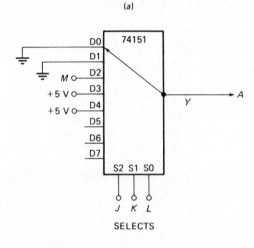

(b)

FIGURE 3-44 Conversion of BCD to excess-3 code by using a multiplexer. *(a)* Truth table. *(b)* Schematic.

consists of a 74LS138 decoder, a 74157 multiplexer, and a matrix containing diodes.

The 74LS138 is referred to by the data book as a *one-line-to-eight-line demultiplexer* or a *three-line-to-eight-line decoder*. Operation in the latter mode is similar to the operation of the 7442 BCD-to-decimal decoder described before. If the G1 input is tied high and the G2A and G2B inputs are grounded, then the 3-bit binary number on the select or address inputs will determine which of the eight output lines is low. Only one line is addressed and pulled low for each input combination. For example, if JKL equals 000, the Y0 output line is low. All the other Y outputs are high. If JKL equals 001, the Y1 line selectively goes low.

Now look at the 74157 quad two-input multiplexer. If the select line M is a 0, the switches are to the right as shown, and the output data is selected from columns

$X0$, $X2$, $X4$, and $X6$. A 1 on the 74157 select input flips the switches to the other direction and selects the output from columns $X1$, $X3$, $X5$, and $X7$. See whether you can figure out the outputs for an input of $JKLM$ equal to 0000. The D output is a 1 or high because it is just tied to +5 V with a 4.7-kΩ resistor. The A output, however, is a little trickier. With JKL equal to 000 on the address lines of the 74LS138, the $Y0$ line will be low. This low will forward-bias the germanium diode $D1$ and give a voltage of about 0.6 V at point X. The 0.6 V is the sum of the voltage across $D1$ and the TTL output low voltage of the $Y0$ line. The multiplexer sees 0.6 V as a 0 or low, and output A is low (0.6 V low has reduced noise margin, but if leads are short, this causes no problem). Since lines Y1 and Y2 are off, diodes $D2$ and $D3$ are off and have no effect. The $X4$ column has a diode to the Y0 line, so the B output also will be low or 0. The $X2$ column has no diode to the Y0 line to pull it low, so the output will be 1. For inputs of JKLM equal to 0000, the output ABCD equals 0011. Checking Figure 3-44 shows that these are corresponding BCD and excess-3 code numbers.

Now try an input combination of JKLM equal to 0001. The $Y0$ line is still low because the JKL address inputs of the 74LS138 are still all 0's. A 1 on the select input of the 74157, however, has flipped the switches over to the left, so that outputs now reflect columns $X1$, $X3$, $X5$, and $X7$. The $X7$ line has a diode to the low Y0 line, so the A output is low. Column $X5$ has no diodes to the Y0 line, so the B output is high. The $X3$ and $X1$ columns have diodes, so the C and D outputs are low. An input of JKLM equal to 0001 gives the expected output of ABCD equal to 0100.

As you can see from this example, you can store *any* truth table you want in this ROM simply by putting in or leaving out diodes. Pin numbers have been included in Figure 3-45, so you can build the circuit and experiment with other codes if you wish. Note that for this example you do not even have to use all the possible lines in this small ROM.

Storage in a ROM is nonvolatile, which means that if you turn off the power and then turn it back on, the information will still be there. This is not the case with some other memories you will meet in Chap. 5.

ROM TYPES

ROMs are available in three basic types: *mask programmed ROMs; programmable ROMs, or PROMs;* and *erasable programmable ROMs, EPROMs and EEPROMs.*

MASK PROGRAMMED ROMs
Mask programmed units are permanently programmed at the time of manufacture by adding or leaving out diodes or transistors, as was done in the home-made ROM. You send the manufacturer the truth table you want, on punched cards or paper tape, and a computer generates a mask for the ROM IC which will give that truth table. Programming is permanent.

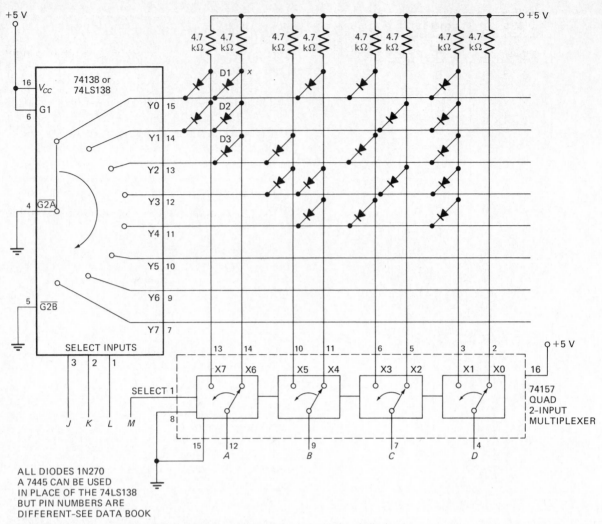

FIGURE 3-45 Home-made ROM using 74138, 74157, and matrix of diodes.

PROMs   Programmable ROMs, or PROMs, such as the 74S287 come with diodes in every bit position; therefore, the output is initially all 0's. Each diode, however, has a fusible link in series with it. If you address a bit and put the proper pulse into the corresponding output, the fuse will blow out and leave a 1 in that bit position. Programming is permanent, so you can't correct errors. Therefore, PROM programming machines, controlled by a paper tape or a correct ROM, are used to minimize errors. A problem with nichrome-fuse PROMs is that the nichrome fuse may regrow and change the programming of one or more bits with aging.

EPROMs   Erasable programmable ROMs use MOS circuitry. They store 1's and 0's as a packet of charge in a buried layer of the IC chip. Programming of EPROMs is discussed later. The important point for now is that you can erase the stored data in the EPROMs by shining a high-intensity, short-wavelength ultraviolet light through the quartz window on top of the IC for about 20 min. This light causes the stored charge to leak off and lets you reprogram if you need to change the output truth table. A disadvantage is that the entire contents will be lost. To change 1 byte, the entire device must be reprogrammed. Figure 3-46 shows a photograph of an Intel 2716 EPROM with its quartz window. An EPROM such as the 2716 may be used to develop a product; then when the design is finalized, a cheaper mask programmed equivalent such as the Intel 2316E may be used.

EEPROMs   Electrically erasable programmable ROMs (EEPROMs) also use MOS circuitry very similar to that of the ultraviolet-light erasable devices described above. Data is stored as charge or no charge on an insulated layer or an insulated floating gate in the device. The insulating layer is made very thin (<200 Å). Therefore, a voltage as low as 20 to 25 V can be used to move charges across the thin barrier in either direction for programming or erasing.

An example of this type of EEPROM is the Intel or Motorola 2817 which stores 2048 bytes of data (16,384 bits). About 10 ms is required to write each byte of data to the device. By using an electric signal of about 21 V, a single byte of data or the entire device can be erased in about 10 ms. This erase time is much less than that for

FIGURE 3-46 Intel 2716 EPROM showing quartz window. *(Intel Corporation.)*

the ultraviolet EPROMs. Also, since the ultraviolet light source is not required, erasing and reprogramming can be done more easily with the device right in the circuit.

ROM STRUCTURE AND ADDRESSING

The home-made ROM contains eight lines and eight columns. This gives 64 possible intersections, so it can store 64 binary bits. Since these bits are decoded and output as 4-bit words, you can store a total of sixteen 4-bit words in this 64-bit ROM. This is referred to as a 16×4 ROM.

Figure 3-47 shows a block diagram of a 74187, 1024-bit ROM. The actual memory matrix is 32×32, because it is easier to make the array square, or nearly so, but it is decoded to give 256 output words of 4 bits. Note that it takes five address lines to select one of 32 rows and three more address lines to select one of eight columns. A total of eight address lines is needed to select one of 256 words. Since 256 is equal to 2^8, you can see that it takes N address lines to select one of 2^N output words.

For another example, the TMS 2532 is a 32,768-bit ROM with 4096×8 organization. How many address lines are needed to select one of the 4096 output words of 8 bits? Since 4096 is 2^{12}, you need 12 address lines.

Most larger ROMs have three-state outputs, so that several ROMs can be connected to a common parallel data bus, as shown in Figure 3-48. The desired $\overline{\text{ROM}}$ is chosen by making its chip select input $\overline{\text{CS}}$ low. If $\overline{\text{CS}}$ is high, the outputs of the ROM float. The chip select signals are decoded from address lines as follows.

The 2732 is a 4096×8 EPROM, so 12 address lines are required to address each of the 4096 words in the ROM. Three additional address lines are used as select inputs for a 74LS138 or 3205 three-line-to-eight-line decoder, as discussed before. Each 3-bit input combination will produce a low on one of eight outputs. If one of the decoder outputs is tied to the chip select input of each ROM, only one ROM at a time will be selected. Some ROMs have more than one chip select or enable input.

ROM DATA SHEET PARAMETERS AND EPROM PROGRAMMING

Appendix C shows the complete data sheets for the Intel 2716 EPROM which stores 2048 words of 8 bits. This device was chosen because it has some interesting features not found on earlier EPROMs. First note that even though it is a high-density, high-speed MOS device, it operates on a single +5-V supply except during programming. Second, the inputs and outputs are solidly TTL-compatible. The outputs can even drive one Schottky TTL input. Third, the device, when it is not being selected, can be switched to a standby mode, which uses only one-quarter as much power as the active mode.

The mode selection chart in table 1 on the data sheet shows that for normal read operation, both the $\overline{\text{CE}}$/PGM input and the $\overline{\text{OE}}$ input must be low. If either input is high, the outputs are in a high-impedance floating state. Since the $\overline{\text{CE}}$/PGM input, when it is brought high, drops the power and also floats the outputs, it can be used in place of the $\overline{\text{OE}}$ input in selecting one of several ROMs with a decoder.

Now look at the ac characteristics table on the data sheets to get some important timing information for the ROM. At first glance the table may seem to be a confusing collection of subscripts, but the waveforms below the table help explain it.

The access time t_{ACC} is the time between valid addresses arriving at the inputs of the ROM and valid data appearing at the outputs. Included in this access time is t_{CE}, the time needed for a low on the chip select input to get the chip out of standby mode. Since the $\overline{\text{CE}}$ signal probably is produced by decoding address lines, the high-to-low transition of $\overline{\text{CE}}$ will come after valid addresses, as shown. If a CMOS decoder is used, its excessively long propagation delay may add to t_{CE} to increase the overall access time. The reason you care about all this is that if the access time is too long, the ROM may not get data out fast enough to work in a particular system.

The second access time t_{OE} is the time required to get the output back from the high-Z mode to the active read mode. If the t_{OE} signal is also produced by decoding address lines as suggested above, then the t_{PD} of the decoder must be included in t_{ACC}. The minimum times of 0 ns for the last two terms in the table tell you that you can't count on the output data being valid anytime after $\overline{\text{OE}}$ goes high, $\overline{\text{CS}}$ goes high, or the addresses change.

The next point to consider is programming this

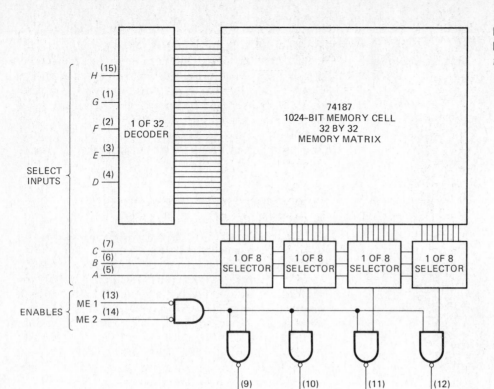

FIGURE 3-47 Logic diagram of DM74187 256 × 4 ROM. (*National Semiconductor.*)

NOTE: A INPUT IS THE LEAST SIGNIFICANT BIT.

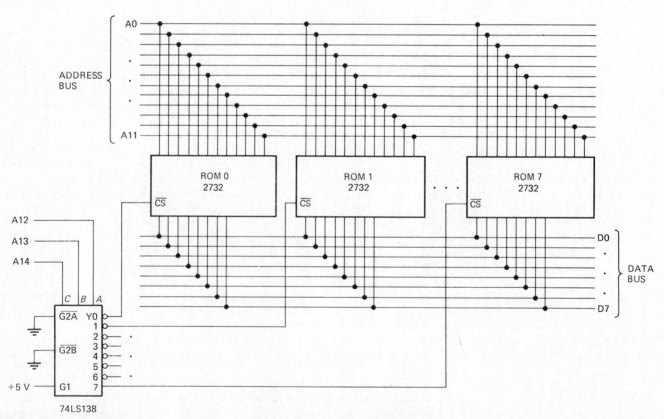

FIGURE 3-48 Memory array using eight Intel 2732 EPROMs to store 32,768 binary words of 8 bits each.

EPROM. Every type of EPROM has its own specific programming procedure, so there is no use memorizing it for one type unless you are going to be programming many of that type. In most cases a PROM programmer machine is going to do the job for you. To give you a rough idea of what is involved, however, here is the programming procedure for the 2716. The chip is powered with a V_{CC} of +5 V and a V_{PP} of +25 V. $\overline{OE}$ is held high. The location in which you wish to store a word is addressed. The data word you want to store is applied to the *outputs* of the 2716. A single 50-ms, positive TTL level pulse is then applied to the $\overline{CE}$/PGM input. You can verify that the programming was successful by removing the data from the outputs and bringing $\overline{OE}$ low. Now the outputs should be reading you back the word you programmed in. The entire memory can be erased with ultraviolet light as described in the data sheet.

ROM APPLICATIONS

AS CODE CONVERTER This chapter introduced a ROM as a code converter. Some further examples of this use are the MM4221RR-ASCII-7-to-EBCDIC converter, the MM4230KP-ASCII-7-to-Selectric code converter, and the MM4230BO-Hollerith-to-ASCII-8 converter. If a standard ROM is not available for a needed conversion, you can always program a PROM or an EPROM to do the job.

AS LOOK-UP TABLE ROMs can be used as look-up tables also. For example, the MM4232 takes as inputs a 9-bit binary equivalent of an angle between 0 and 90°. The outputs are an 8-bit binary equivalent to the sine of the input angle. Several can be cascaded to give up to 16-binary-bit resolution.

CHARACTER GENERATOR ROM character generators, such as the 2513, are used to produce the pattern of dots necessary to display a letter or number on a video terminal screen.

KEYBOARD ENCODER ROM keyboard encoders such as the MM5740 are used to convert the switch-closed code from typewriter-type keyboards to ASCII for computer use.

LOGIC GATE REPLACEMENT Another application for ROMs is replacement of the many gates necessary to build the circuit for a logic expression with a large number of input variables. An eight-input ROM can be programmed to give any output expression for eight input variables.

COMPUTER MEMORY The main use of ROMs, however, is to store the binary-coded instructions and data for a microprocessor or computer. If the instructions are stored in sequential order in the ROM, then the microprocessor can read one instruction, carry it out, and then just step to the next higher address to get the next

instruction. The ROM array in Figure 3-48 will store up to 32,000 words or bytes of 8 bits.

PLAs, FPLAs, FPGAs, AND PALs

PROGRAMMABLE LOGIC ARRAYS AND FIELD PROGRAMMABLE LOGIC ARRAYS

A PLA can best be thought of as a universal sum-of-products (AND-OR) function generator. Figure 3-49 shows the logic diagram of a common mask programmable logic array, the National DM7575. It has 14 data or address inputs, I1 through I14. Inverters produce both the true and complement of these data inputs on row lines. The column lines are connected to the inputs of 96 fourteen-input AND gates. The user specifies to which row lines each AND gate input will be connected. An AND input not tied to any row lines acts as a high. This determines what function of the input variables will produce an output high on each of the AND gates. The outputs of the AND gates form the columns of another matrix. The row lines are the inputs to eight eight-input OR gates. The user specifies to which columns the OR inputs should be connected. Any OR inputs not connected to a column are internally held low. True or inverted outputs are also mask programmable.

The output expression has a sum-of-products form such as

$$F1 = I1\ I2\ I3\ I4\ I5 + \overline{I1}\ \overline{I2}\ I3\ I11\ I14 \cdots$$

An output expression can contain up to eight product terms. Each product term can contain up to 14 variables. However, since there are only 96 AND gates, the total number of different product terms is only 96.

A PLA may be used to replace a board full of logic gates or as a dedicated code converter. The sum-of-products output format makes it easy to determine the internal matrix connections directly from a truth table.

A good application example for this PLA is a 12-bit-Hollerith-to-ASCII-7 code converter. The Hollerith code has 12 bits but uses only about 96 of the possible combinations. The PLA then can be directly programmed to give the ASCII 7-bit output code for each of the 96 used Hollerith input combinations. A much larger ROM is needed to directly do the same job. The MM4230BO-Hollerith-to-ASCII-8 ROM requires additional external logic to reduce the 12 Hollerith lines to 8 before it can do the conversion. Field programmable logic arrays, or FPLAs, such as the Signetics 82S101, are programmed by blowing fuses as is done with bipolar PROMs.

FIELD PROGRAMMABLE GATE ARRAYS AND PROGRAMMABLE ARRAY LOGIC

Another device similar to a PLA or an FPLA is the *field programmable gate array*, or *FPGA*. The FPGA has an

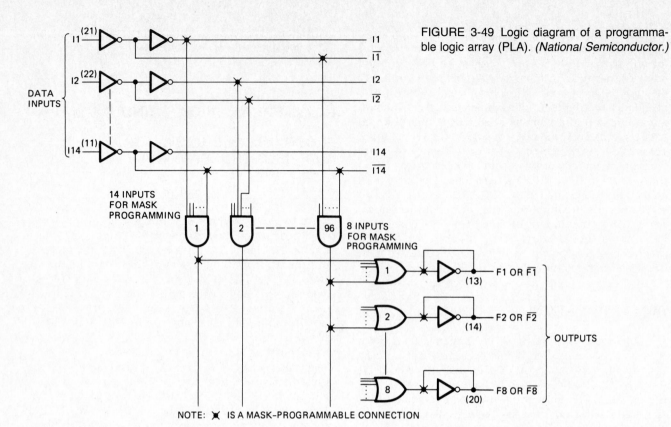

FIGURE 3-49 Logic diagram of a programmable logic array (PLA). *(National Semiconductor.)*

DATA INPUTS

I1 (21)
I2 (22)
I14 (11)

I1
Ī1
I2
Ī2
I14
Ī14

14 INPUTS FOR MASK PROGRAMMING

1   2   96

8 INPUTS FOR MASK PROGRAMMING

1 — F1 OR F̄1 (13)
2 — F2 OR F̄2 (14)
8 — F8 OR F̄8 (20)

OUTPUTS

NOTE: ✖ IS A MASK-PROGRAMMABLE CONNECTION

AND gate matrix as does the PLA in Figure 3-49, but it does not have the output OR gates. Therefore, the output expression is only a programmable AND function of the input variables. These devices are sometimes used for random-logic gate replacement.

Still another version of the AND-OR matrix type of device is known as *programmable array logic*, or PAL. PALs use a programmable AND gate matrix, as does the PLA in Figure 3-49, but the output OR gate connections are fixed. This makes the device easier to program. The output expression for each output is a sum-of-products form, as it was for a PLA.

The main point here is that many times one of these devices can replace an entire printed-circuit board full of discrete gate ICs.

REVIEW QUESTIONS AND PROBLEMS

1. Write the boolean expression and draw the logic gate diagram for the switch circuit in Figure 3-50. Switch closed = 1.

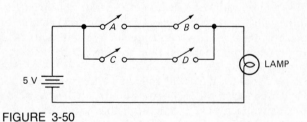

FIGURE 3-50

2. Figure 3-51 shows the schematic diagram for a common household switch circuit. The light can be turned on or off with either switch. Write the boolean expression for the circuit. Switch up is H, switch down is L, and light ON is output H.

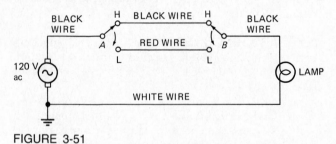

FIGURE 3-51

3. Construct the truth table and determine the output expression for the circuit in Figure 3-52.

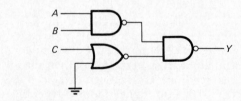

FIGURE 3-52

4. Draw the equivalent circuit for the expression $Y = B \cdot C + D \cdot E$.

5. Draw the equivalent circuit for the expression $Y = (A + B)(C + D)$.

6. The TTL logic block shown in Figure 3-53 is not working correctly. A technician measures the input and output voltage for each gate. The results are shown in Figure 3-53. Where is there a problem in the circuit? Give two possible causes.

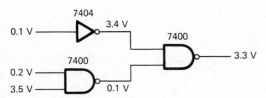

FIGURE 3-53

7. Write the output expression for the circuit in Figure 3-54.

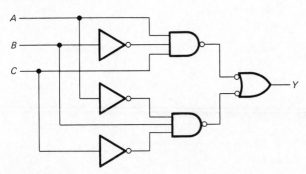

FIGURE 3-54

8. *a.* Show the truth table and output expression for the circuit in Figure 3-55.

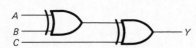

FIGURE 3-55

b. Show with a truth table how an exclusive OR gate can be used to selectively invert or not invert an input signal depending on the state of the other input.

9. Using boolean algebra, simplify the following:
a. $(A + \overline{B})(A\overline{B} + C)C$
b. $(A + \overline{B})C + (A\overline{B} + C)$
c. $AB + ABC + \overline{A}B + A\overline{C}$
d. $\overline{\overline{A \cdot B} + \overline{C \cdot D}}$
e. $A + B + \overline{\overline{CD}}$

10. Describe the relationship between the inputs and the outputs for each device, using dependency notation symbols, shown in Figure 3-56.

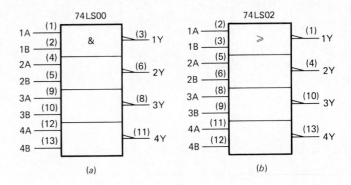

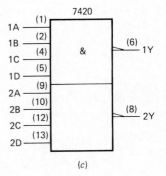

FIGURE 3-56

11. Write the boolean expression for each mixed logic circuit shown in Figure 3-57.

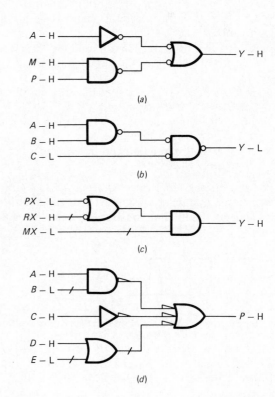

COMBINATIONAL LOGIC, MULTIPLEXERS, CODES, AND ROMS

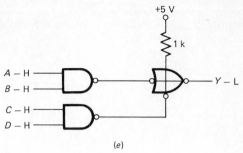

FIGURE 3-57

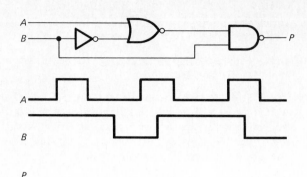

FIGURE 3-58

12. Draw the mixed logic circuits for the circuits in Prob. 9a, b, and c. Assume all input signals are asserted high.

13. Simplify the following equation, using boolean algebra. Also use a Karnaugh map to simplify

$$X = AB + ABC + \overline{A}B + AC$$

Then show how to implement the simplified expression with gates. Remember, to get a minimum expression, try to group as many boxes as possible in groups of 2, 4, 8, or 16.

14. Use a Karnaugh map to simplify

$$X = AB\overline{C}\overline{D} + A\overline{B}CD + ABCD + ABC\overline{D} + \overline{A}B\overline{C}D + \overline{A}BCD$$

and then show the circuit for the simplified expression.

15. Three secretaries and a boss work in an office. The radio will be on if the boss or at least two secretaries want it on.
  a. Write the truth table for the radio ON function. (Let a yes vote equal a 1 in the truth table and ON equal a 1.)
  b. Use a Karnaugh map to produce the simplified expression for the function.
  c. Show how the function can be implemented by using only NAND gates.
  d. Show how the function can be implemented by using only NOR gates.

16. For the circuits in Figure 3-58 show the output waveforms that will be produced by the given input waveforms. Note that the output must obey the

truth table for the circuit at each instant. Using this fact, you can construct the output waveform a section at a time under the input waveforms.

17. Show the circuit for a two-input single-output switch made with NAND gates and with NOR gates.

18. What is the advantage of multiplexing?

19. Show how the function in Prob. 14 can be directly implemented with a 74LS151 eight-input multiplexer.

20. Convert the following numbers to octal:
  a. 1010101_2
  b. 10_{10}
  c. 111001011_2

21. Convert to decimal:
  a. 314_8
  b. 72_8
  c. 40_8

22. Convert to hexadecimal:
  a. 42_{10}
  b. 798_{10}
  c. 01011101_2
  d. 11000011_2

23. Convert to decimal:
  a. $A7_{16}$
  b. $3FF_{16}$
  c. 40_{16}

24. Convert to BCD:
  a. 72_{10}
  b. 59_{10}
  c. 37_{10}

25. The M key is depressed on an ASCII-encoded keyboard. What pattern of 1's and 0's would you expect to find on the seven parallel data lines coming from the keyboard? What pattern would a carriage return, CR, give?

26. Define parity and describe how it is used to detect an error in transmitted data.

27. What are the patterns of 1's and 0's for a G in EBCDIC and Selectric codes?

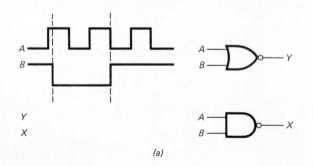

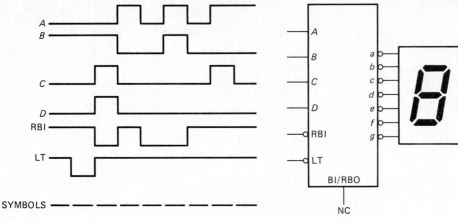

SYMBOLS ▬ ▬ ▬ ▬ ▬ ▬ ▬

FIGURE 3-59 Problem 30.

28. For the 74148 priority encoder shown in Figure 3-43, what code will appear on the A0, A1, and A2 outputs if a low is put on the number 2 data input?

29. Consider the 7447 BCD-to-seven-segment decoder of Figure 3-42.
   a. What effect will making the $\overline{BI}$ input low have on the segment outputs?
   b. Is the lamp test input active low or active high?
   c. What logic level will appear on the seven-segment outputs if the BCD code for 5 is applied to the inputs?

30. Given a 7447 BCD-to-seven-segment decoder/driver, determine the sequence of digits that appears on the display in Figure 3-59.

31. Show where you would put diodes in the homemade ROM matrix of Figure 3-45 to produce an output word of 0011 for input address 1110.

32. The National Semiconductor 1NS8298 is a 65,536-bit ROM organized as 8192 words or bytes of 8 bits. How many address lines are required to address one of the 8192 bytes?

33. Describe the differences between PROMs and EPROMs.

34. How are EPROMs erased?

35. Why do most ROMs have three-state outputs?

36. What logic levels must be present on the A12, A13, and A14 inputs of the 74LS138 decoder in Figure 3-48 to select ROM 7?

37. What is the worst-case access time T_{ACC} for the 2716 EPROM (use data sheet in App. C)?

38. How does a PLA differ from a ROM?

REFERENCES/BIBLIOGRAPHY

Winkel, David, and Franklin Prosser: *The Art of Digital Design*, Prentice-Hall, Englewood Cliffs, NJ, 1980.

Fletcher, William I.: *An Engineering Approach to Digital Design*, Prentice-Hall, Englewood Cliffs, NJ, 1980.

Peatman, John B.: *Digital Hardware Design*, McGraw-Hill, New York, 1980.

Texas Instruments Staff: *TTL Data Book for Design Engineers*, 2d ed., Texas Instruments, Inc., Dallas, 1981.

Texas Instruments Staff: *Bipolar Microcomputer Components Data Book*, 3d ed., Texas Instruments, Inc., Dallas, 1981.

Blakeslee, Thomas R.: *Digital Design with Standard MSI & LSI*, Wiley, New York, 1975.

National Semiconductor Staff: *CMOS Databook*, National Semiconductor Corporation, Santa Clara, CA, 1980.

National Semiconductor Staff: *Logic Databook*, National Semiconductor Corporation, Santa Clara, CA, 1981.

Intel Staff: *Intel Component Data Catalog*, Intel Corporation, Santa Clara, CA, 1982.

National Semiconductor Staff: *MOS/LSI Databook*, National Semiconductor Corporation, Santa Clara, CA, 1980.

CHAPTER 4

FLIP-FLOPS, COUNTERS, AND REGISTERS

For the clocked circuits discussed throughout the rest of this book, timing is very important. To analyze clocked circuits, you use not only truth tables but also timing diagrams. Timing diagrams are just drawings of the input and output waveforms such as you would see if you looked at the circuit with a multichannel oscilloscope. Figure 4-1a and b shows the truth table and its timing diagram equivalent for the familiar NOR gate. Note in Figure 4-1c that a timing diagram can be drawn to show risetimes and the propagation delays of the gate. An example of why you need diagrams to analyze clocked circuits is shown in Figure 4-2. The input waveforms are the same as in Figure 4-1b, but somehow the B waveform has been delayed by 30 ns. The resultant output of $Y = \overline{A + B}$ shows 30-ns-wide unwanted pulses, commonly called *glitches*, at points X and Y. Figure 4-2b shows a circuit that will produce this problem. If inputs C and D are high, gates 1 and 2 simply act as inverters for the B signal. Since two inversions cancel, the propagation delays of gates 1 and 2 just delay the B signal input to NOR gate 3, as shown in Figure 4-2a. One solution is to add equal delay to the A signal path. You can do this by adding two inverters or two NAND gates used as inverters in series with the A signal.

The main point illustrated by this discussion is the importance of using timing diagrams when you analyze pulsed or clocked logic circuits.

Upon completing this chapter, you should be able to:

1. Write the truth tables for the D latch, D flip-flop, JK flip-flop, and T flip-flop.

2. Show the output waveform produced when given input waveforms are applied to D latches, D flip-flops, and JK flip-flops.

3. Draw the output timing diagram for a 4-bit binary or any modulo counter.

4. Draw a circuit to decode any state of a counter.

5. Draw a circuit to convert a 4-bit binary synchronous counter to any modulo between 2 and 15.

6. Use a schematic diagram to explain the operation of a digital clock.

7. Use a schematic diagram to explain the operation of an MSI or LSI frequency counter.

8. Draw the circuit for a D flip-flop register for temporary storage of parallel data.

9. Draw the schematic for a shift register by using D or JK flip-flops.

10. List the differences between static MOS and dynamic MOS shift registers.

11. Draw a circuit showing how UARTs (universal asynchronous receiver-transmitters) can be used to convert parallel data to serial form or serial data to parallel.

12. Describe the operation of MOS bucket brigades and CCDs (charge-coupled devices).

13. Explain the function of flip-flops, counters, and shift registers from their dependency notation symbols.

LATCHES AND FLIP-FLOPS

Latches and flip-flops are memory circuits that store either a 1 or 0 on their outputs. For all but the simplest of these circuits, the output can change only when a special clock or enable input is pulsed.

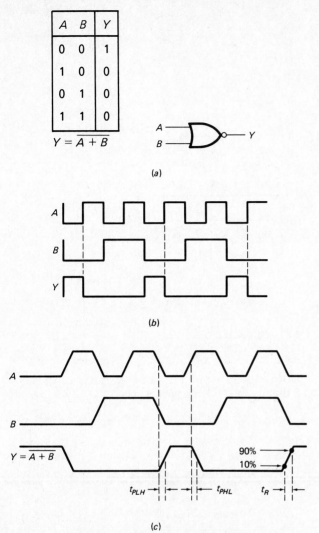

A	B	Y
0	0	1
1	0	0
0	1	0
1	1	0

$Y = \overline{A + B}$

(a)

(b)

(c)

FIGURE 4-1 NOR gate. *(a)* Truth table. *(b)* Ideal timing diagram. *(c)* Timing diagram showing risetimes and propagation delay.

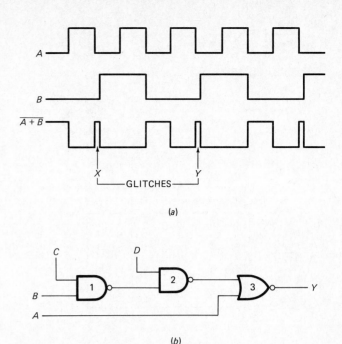

(a)

(b)

FIGURE 4-2 Glitches. *(a)* Timing diagram showing glitches. *(b)* Circuit that can produce glitches.

NAND LATCH

Figure 4-3 shows the schematic, truth table, and timing diagram of a simple NAND latch. First note the input and output labels. Q and $\overline{Q}$ represent complementary outputs. S is the set or preset input, and R is the reset or clear input. Now look at the truth table in Figure 4-3*b*. For S = 0 and R = 1 the Q output is a 1. This 1 on the Q output is described by saying the circuit is *set*. For R = 0 and S = 1, the Q output is a 0, which is referred to as a *reset condition*. Now try S = 0 and R = 0. If you remember the NAND gate trick from Chap. 3 that any low input on a NAND makes the output high, then you can see that both Q and $\overline{Q}$ are high. This violates the definition of Q and $\overline{Q}$ being complementary, so we call this an *indeterminate*, or *prohibited, state* and represent this condition in the table as an asterisk (*). In other words, you don't get any useful information on the outputs if both S and R are low. The Q entry in the table for S = 1 and R = 1 tells you that the Q output is going

to stay fixed at whatever state it is in until either S or R is made low. This is called the *latched* state.

The following example illustrates how the circuit is used. If both inputs are initially high and Q is low, a low pulse applied to the S input as shown in Figure 4-3*c* will drive the Q output high. Beyond a certain minimum, the width of the pulse to S doesn't matter. The high on the Q output is now held, or "latched," there. The circuit is said to be set. The only way to reset the circuit or change the Q output back to a low is to apply a low pulse to the R input after the S input goes back high, as shown in Figure 4-3*c*. Review the switch debounce circuit in Figure 2-31*b*. Note that it is actually just a NAND latch that takes advantage of the fact that once the output is latched, it doesn't matter what that input does.

RS LATCH WITH ENABLE

The NAND latch circuit just discussed has two major problems. First, you have to make sure the inputs don't both go low at the same time. Second, the circuit is asynchronous, which means that it does not have an input to let you control when the S or R pulse changes the output. An enable input can be added as shown in the circuit of Figure 4-4*a*. In this circuit, as long as the enable input is low, the outputs of the input NAND gates are high and the Q and $\overline{Q}$ outputs are latched. Only when the enable input is high can the R and S inputs affect Q and $\overline{Q}$. The states present on the S and R inputs just before the enable goes low will determine the states latched on the Q and $\overline{Q}$ outputs. Note in the truth table of Figure 4-4*b* that the input NAND gates invert the polarity of the pulse required at S or R to set or reset the latch.

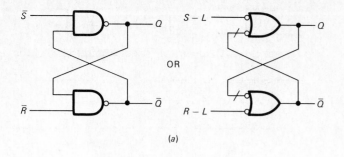

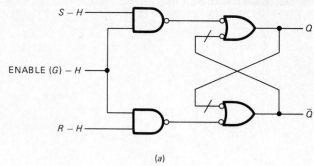

(a)

Function	S	R	Q	$\bar{Q}$
Invalid	0	0	*	*
Set	0	1	1	0
Reset	1	0	0	1
Latch	1	1	Q	$\bar{Q}$

(b)

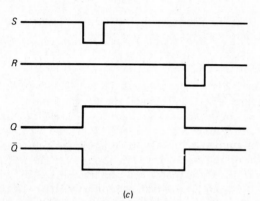

(c)

FIGURE 4-3 Cross-coupled NAND gate latch. *(a)* Schematic. *(b)* Truth table. *(c)* Waveform showing effect of *S* and *R* pulses on outputs.

G	S	R	Q	$\bar{Q}$
0	X	X	Q	$\bar{Q}$
1	0	0	Q	$\bar{Q}$
1	0	1	0	1
1	1	0	1	0
1	1	1	*	*

(b)

(c)

FIGURE 4-4 Enabled RS latch. *(a)* Internal schematic. *(b)* Truth table. *(c)* Schematic symbol.

D LATCH

With the simple RS circuit, there still is the problem of one set of *indeterminate* input conditions when S = 1 and R = 1. For many applications you can employ a circuit that solves this problem by putting an inverter between the S and R inputs, as shown in Figure 4-5a. The resulting circuit is called a *D latch,* and it is available as the 7475 quad D latch or the CMOS 4042. As you can see from the truth table in Figure 4-5b or from analyzing this circuit, *the Q output follows the D input as long as the enable (G) input is high. The logic state present on the D input just before the enable goes low will be latched on the output.* For latches, the enable input is sometimes called the *strobe,* or *gate.*

Figure 4-5c shows the schematic symbol, and Figure 4-5d gives several important timing parameters for a D latch. Three of these are $t_{PW,MIN}$, the minimum width for the enable pulse; t_{SETUP}, the minimum time for which the data must be valid before the enable goes low; and t_{HOLD}, the minimum time for which the data must be

held constant after the enable goes low. For the example shown in Figure 4-5d, you can't be sure that the Q output will catch the low on the D input unless all these times are long enough. For a 7475,

$$t_{PW,MIN} = 20 \text{ ns} \qquad t_{SETUP} = 20 \text{ ns} \qquad t_{HOLD} = 5 \text{ ns}$$

Figure 4-5d also shows T_{PLH} and T_{PHL}. Note that these times are measured from the 50 percent point of an input change to the 50 percent point of the corresponding output change. For a 7475, T_{PLH} is a maximum of 80 ns and T_{PHL} is a maximum of 50 ns.

For a brief example of how you might use a D latch, Figure 4-5e shows the bouncing output of a switch as data input to a D latch. If you wait to enable the latch until after the bouncing stops, as shown, then none of the bouncing will appear at the Q output. This is another way to "debounce" switch signals. The enable pulse can be produced from the switch transition with two 555 timers used as monostables. (Refer to Figure 2-35 to review the circuit.) A negative switch transition triggers one 555 set for an output time of 10 to 15ms. The falling edge of this output triggers a second 555 timer to produce an enable pulse of the desired width. *Strobing* is a term often used in place of *enabling* in referring to latches.

D FLIP-FLOP

A device often confused with the D latch is the D flip-flop. As you remember from the previous section, for a D

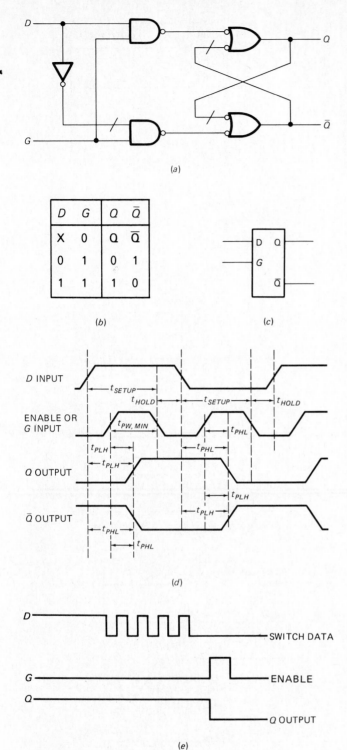

(a)

D	G	Q	Q̄
X	0	Q	Q̄
0	1	0	1
1	1	1	0

(b)

(c)

(d)

(e)

FIGURE 4-5 D latch. (a) Internal schematic. (b) Truth table. (c) Schematic symbol. (d) Timing diagram showing setup and hold times. (e) Timing for debouncing switch contacts.

latch, the Q output follows the D input as long as the enable input is high. A D flip-flop, however, transfers to the Q output only data which is present on the D input at the time of a positive transition on the clock input.

This is shown in Figure 4-6b, which is the truth table for a 74175 quad edge-triggered D flip-flop. The two arrows in the clock column of the table show that Q changes to equal D only at the positive transition of the clock.

The third line in the truth table shows that if the clock is at a high state, the D input is a don't-care, as represented by the X in the D column. This means that a change on D will have no effect on the Q or Q̄ output. The fourth line in the truth table shows that if the clock input is low, a change on D will have no effect on the Q and Q̄ outputs. The last line in the table shows that if the RESET/CLEAR input is made low, the Q output will go to a reset state regardless of the state on the clock and D inputs. Some flip-flops have a direct PRESET input as well as a direct RESET input. These inputs are called *asynchronous* because they directly set or reset the outputs without regard for the state of the clock.

In the connection diagram for the 74175 in Figure 4-6a, note that the clock input of each of the four devices in the IC has a small triangle and a small bubble next to it. The triangle indicates that the actual flip-flop is clocked by an edge of the input clock pulse. The bubble indicates that the flip-flop is clocked by the negative (high-to-low) edge of the clock signal. Since the four clock inputs are all driven by an inverter, the positive (low-to-high) edge of the externally applied clock signal will actually clock the flip-flops. If you think of the bubble on the output of the clock line inverter as canceling the bubbles on the flip-flop clock inputs, you can see directly which edge of the clock is active.

Figure 4-6c shows the dependency notation logic symbol for the 74175. The indented box at the top of the symbol is a common control block. Control signals that enter this block, unless otherwise indicated, affect all the elements below the control block. If the CLR input is asserted, for example, all the flip-flops will be reset. The letter C stands for control dependency. Inputs which have the same identifying number as that after the C will be enabled by the control input. Since in this case the C is in the common control box, it affects all the D inputs. Only one D input needs to be labeled with the 1D label. As indicated by the triangle on the CLK input, data on the D inputs will be transferred to the outputs when the CLK input goes from low to high (positive edge of CLK).

The setup time and hold time for a D flip-flop are measured from the positive edge of the clock, as shown in Figure 4-6d. To give you an idea of typical values, a data book lists $t_{SETUP,MIN}$ as 20 ns, $t_{HOLD,MIN}$ as 0 ns, and $t_{CW,MIN}$ ($t_{PW,MIN}$) as 20 ns. The minimum width for a clear pulse is 20 ns, and the clear pulse must return high at least 30 ns before the positive edge of a clock pulse ($t_{SETUP,CLEAR}$). You certainly don't need to memorize these values, but you do need to understand these parameters when you get them from a data book to help you troubleshoot a flip-flop circuit with timing problems. A flip-flop or latch with marginal timing design often shows the frustrating symptom of sometimes giving the correct data out, and sometimes not. A dual-

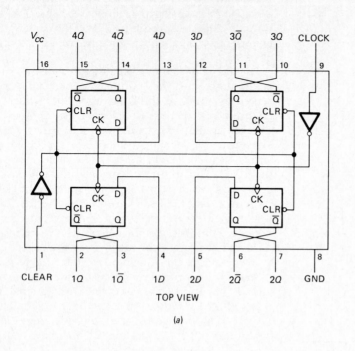

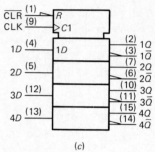

INPUTS			OUTPUTS	
RESET/ CLEAR	CK	D	Q	$\bar{Q}$
H	↑	H	H	L
H	↑	L	L	H
H	H	X	Q	$\bar{Q}$
H	L	X	Q	$\bar{Q}$
L	X	X	L	H

(b)

(c)

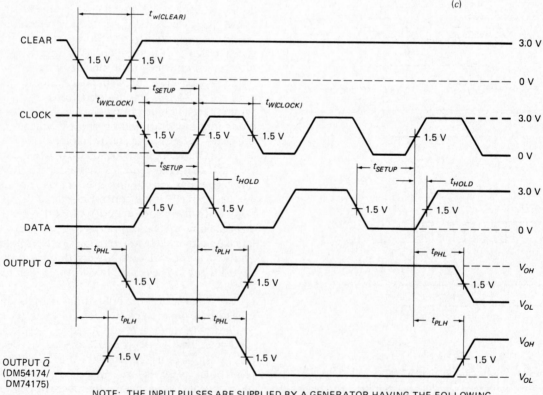

NOTE: THE INPUT PULSES ARE SUPPLIED BY A GENERATOR HAVING THE FOLLOWING CHARACTERISTICS: $t_R \leq 10$ ns, $t_F \leq 10$ ns, PRR ≤ 1.0 MHz, DUTY CYCLE $\leq 50\%$ $Z_{OUT} \approx 50 \ \Omega$. VARY PRR TO MEASURE f_{MAX}.

(d)

FIGURE 4-6 D flip-flop. (a) Connection diagram of 74175 quad D flip-flops. (b) Truth table showing positive-edge triggering. (c) Dependency notation logic symbol. (d) Timing diagram.

trace-triggered oscilloscope usually will expose the timing problems.

The 4013 is an example of a CMOS quad D flip-flop.

D LATCH AND D FLIP-FLOP OUTPUT WAVEFORMS

Figure 4-7 shows how the D latch and D flip-flop output waveforms differ for the same input D and clock waveforms. On the rising edge of each clock pulse, the D flip-flop updates the Q output to show the state present on the D input at that time. Then this state is held on the Q output until it is updated on the next positive clock edge. The device essentially takes a snapshot of D at each rising clock edge and displays it at Q.

When the clock or enable of a D latch is high, the Q output follows the level on the D input as if D and Q were connected by a piece of wire. When the clock goes low, the state left on Q will be whatever state was on D just before the clock went low.

JK FLIP-FLOP

Before we examine the many applications of flip-flops, there is one more very important type to discuss. This is the JK flip-flop, which eliminates the problem of a prohibited input combination found in RS circuits but still provides two inputs, J and K. There is no need to fill this book or your mind with the many internal circuit variations used to accomplish this. The main points to understand are the truth table for the JK flip-flop and the three types of clocking that it may have.

POSITIVE-EDGE-TRIGGERED JK FLIP-FLOP   Figure 4-8a shows the truth table for a CMOS 4027 JK flip-flop. The arrows indicate that the output changes to the state determined by the J and K inputs on the positive transition of the clock. Note that for the condition of both inputs high, which is indeterminate in the simple RS, the output of the JK flip-flop simply changes state or toggles on each positive transition of the clock. The other input combinations produce the same result as they did in the simple RS latch. If both J and K are low, the output will not change when a clock pulse occurs. If J is high and K low, Q will be high after a clock pulse; and if J is low and K high, Q will be low after a clock pulse. The direct set and reset inputs are independent of

the clock, J, and K inputs. Using these, the Q output can be set to a high or reset to a low at any time. As you can see from Figure 4-8a, the set and reset inputs are active high. Both of these cannot be high at the same time. For this type of flip-flop, setup and hold times are measured from the positive edge of the clock.

NEGATIVE-EDGE-TRIGGERED JK FLIP-FLOP   Figure 4-8d shows the truth table for a 74LS76 dual JK flip-flop. The arrows indicate that this device responds to the J and K inputs on the *negative edge* of the clock. However, the J and K inputs still have the same effect on the output as did those of the CMOS 4027. For example, the output will toggle on each clock pulse if J and K are both high. Note that the direct preset and clear (set and reset) inputs for the 74LS76 are active low. Both should not be made low at the same time. Setup and hold times are measured from the negative edge of the clock. Other examples of negative-edge-triggered JK flip-flops are the 74103, 74113, and 74114.

MASTER-SLAVE JK FLIP-FLOP   Still another type of JK clocking is shown for a 7476 dual JK flip-flop in Figure 4-8g. As with many other JK devices, each flip-flop in the 7476 has two parts, a *master* and a *slave*. The circuitry of the 7476 is designed such that when the clock goes to a high level, the J and K data is gated into the master, or input, portion. When the clock drops low again, the J and K data is passed on to the slave, or output, section and appears at the output. A problem with this level-clocked type is that the J and K inputs must not change while the clock is at a high level or else the output will be unpredictable. To minimize the problem, the clock pulse should be kept as narrow as possible.

Note how the dependency notation differentiates between the negative-edge-triggered 74LS76 in Figure 4-8f and the level-triggered 7476 in Figure 4-8i.

JK FLIP-FLOP TIMING DIAGRAMS

For JK flip-flops it is important to predict the output Q waveforms for given J and K input waveforms. Figure 4-9 shows the output waveforms produced by each of the three types of JK flip-flop for some J and K input waveforms. At first glance, it might seem difficult to predict these output waveforms, but if they are drawn out one section at a time, they are quite easy.

For the positive-edge-triggered JK flip-flop output waveform in Figure 4-9, assume that Q is low initially. At the time of the first positive clock transition, the J input is high and the K input is low. As indicated by the JK truth table, this causes the Q output to go high. If we assume no direct PRESET or CLEAR inputs, the Q output cannot change again until the next positive transition of the clock input at 3. Therefore, the waveform can be drawn as a high from clock edge 1 to clock edge 3. At clock edge 3, J is low and K is high. As indicated in the truth table, this causes the Q output to go low, and it stays at this level at least until clock edge 5. At clock

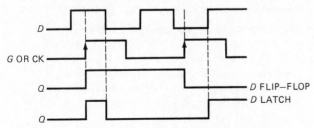

FIGURE 4-7 Diagram comparing D latch and D flip-flop output waveforms for same clock and D input signals.

4027

INPUTS					OUTPUTS	
S	R	CK	J	K	Q	Q̄
L	H	X	X	X	0	1
H	L	X	X	X	1	0
H	H	X	X	X	H*	H*
L	L	↑	L	L	Q0	Q̄0
L	L	↑	H	L	H	L
L	L	↑	L	H	L	H
L	L	↑	H	H	TOGGLE	

(a)

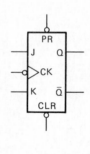

(b)

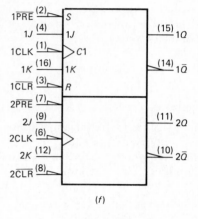

(c)

74LS76

INPUTS					OUTPUTS	
PR	CLR	CK	J	K	Q	Q̄
L	H	X	X	X	H	L
H	L	X	X	X	L	H
L	L	X	X	X	H*	H*
H	H	↓	L	L	Q0	Q̄0
H	H	↓	H	L	H	L
H	H	↓	L	H	L	H
H	H	↓	H	H	TOGGLE	
H	H	H	X	X	Q0	Q̄0

(d)

(e)

(f)

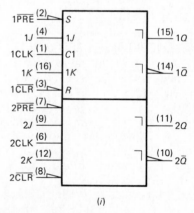

7476, H76

INPUTS					OUTPUTS	
PR	CLR	CK	J	K	Q	Q̄
L	H	X	X	X	H	L
H	L	X	X	X	L	H
L	L	X	X	X	H*	H*
H	H	⊓	L	L	Q0	Q̄0
H	H	⊓	H	L	H	L
H	H	⊓	L	H	L	H
H	H	⊓	H	H	TOGGLE	

(g)

(h)

(i)

NOTES: ⊓ = high-level pulse; data inputs should be held constant while clock is high; data are transferred to output on the falling edge of the pulse.

Q0 = the level of Q before the indicated input conditions were established.

TOGGLE: Each output changes to the complement of its previous level on each active transition (pulse) of the clock.

*This configuration is nonstable; that is, it will not persist when preset and clear inputs return to their inactive (high) level.

FIGURE 4-8 JK flip-flops. *(a)* Positive-edge-triggered 4027 truth table. *(b)* 4027 schematic symbol. *(c)* 4027 dependency notation symbol. *(d)* Negative-edge-triggered 74LS76 truth table. *(e)* 74LS76 schematic symbol. *(f)* 74LS76 dependency notation symbol. *(g)* Level-triggered 7476 truth table. *(h)* 7476 schematic symbol. *(i)* 7476 dependency notation symbol.

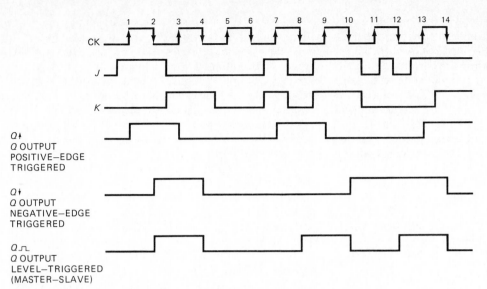

FIGURE 4-9 Output waveforms produced by the three JK flip-flop types for given input clock, *J*, and *K* waveforms.

Q OUTPUT
POSITIVE—EDGE
TRIGGERED

Q OUTPUT
NEGATIVE—EDGE
TRIGGERED

Q OUTPUT
LEVEL—TRIGGERED
(MASTER—SLAVE)

edge 5 both J and K are low; therefore, the output will not change. At clock edge 7 both J and K are high, so the Q output will toggle, which in this case means change from a low to a high. Again, at clock edge 9 both J and K are high, so Q will toggle again, this time from high to low. At clock edge 11 both J and K are low, so Q will stay the same. Finally, at transition 13 the J input is high and the K input is low, so the Q output will go to a high. The device can be thought of as sampling the J and K inputs on each positive transition of the clock and updating the Q output according to the JK truth table rules. This Q state is then held until the next positive clock transition, when the output is updated again. The waveform then is constructed in segments one clock pulse long.

The waveform for the negative-edge-triggered flip-flop in Figure 4-9 is constructed in the same way except that the updating "snapshot" of the J and K inputs is taken on the falling edge (transition) of the input clock.

To draw the waveform for a JK master-slave flip-flop, you must remember not only the truth table, but also how the master-slave operation works. When the clock is high, data on the J and K inputs updates the output of the master section. When the clock goes low, the state on the output of the master section is transferred to the output of the slave section. Therefore, changes of the output Q will coincide with the falling edge of the clock. The bottom waveform in Figure 4-9 shows how this works. At clock edge 1, J is high and K is low, so the master is set. At clock edge 2, the set state is transferred to the output. At clock edge 3, J is low and K is high, so the master is reset. At clock edge 4, the reset state is transferred to the output. For the pulse from edge 5 to edge 6, both J and K remain low, so the output is not changed. However, at clock edge 7, both J and K are high, so the master toggles high. While the clock is high, both J and K go low, latching the high on the master. At clock edge 8, the output reflects the toggle to a high. Again, at clock edge 9, both J and K are high, so the master toggles to a low. The output changes to a low at

edge 10. At clock edge 11, both J and K are low, which should retain the reset state. Between clock edges 11 and 12, however, K is low and J pulses high. This combination sets the master section of the device. When the clock goes low at 12, this set state is transferred to the Q output as shown. It does not matter that J went low again before the clock went low because for J and K low no change can take place. The unpredictable output caused by J or K changing while the clock is high is sometimes called *1's catching*. Between edges 13 and 14 in Figure 4-9, J is a high and K changes to a high. This causes the output Q to toggle on the negative edge of the clock as shown.

FLIP-FLOP PROBLEMS

Flip-flops and latches have three common types of problems:

1. The Q output does not change to the state predicted by the D or by the J and K inputs when clocked. Usually this is caused by a defective IC or by the output being short-circuited to V_{CC}, to ground, or to another trace on the PC board.

2. The output sets or resets at random times. Often this is caused by the set or reset input of the IC being left open. When left open, these inputs may pick up pulses from other lines. Unused set or reset inputs should be tied to ground or to V_{CC}, whichever voltage holds them in the inactive state.

3. The output is sometimes correct and sometimes incorrect. This may be caused by a defective IC or by violations of the setup and/or hold time requirements of the device. If an input data change occurs too close to a clock edge, the output may or may not get the level you want. Since these time parameters vary with temperature and from IC to IC, marginal timing problems can be hard to trace. For example, an instrument works fine with its cover off, but

103

FLIP-FLOPS, COUNTERS, AND REGISTERS

5 min after the cover is on, it shows wrong data on its display. What you do, then, is connect an oscilloscope to look at the input and the clock of the suspected IC to find out whether the timing is marginal. Next bring a heat source near the IC to see whether this causes a malfunction. If it produces a malfunction, you can determine from the oscilloscope display whether the chip is faulty (setup or hold time is longer than minimum value on data sheet) or whether the circuit design is marginal. It is bad to just replace ICs in a case like this because you may find an IC that will make the instrument work fine until it gets to Phoenix.

Other problems are clock pulses or reset pulses that are too narrow. Also, if a risetime of a reset or set pulse is not fast enough, the output of the flip-flop may oscillate and give several pulses out.

ASYNCHRONOUS OR RIPPLE COUNTERS

A JK flip-flop with J and K inputs tied high, or a D flip-flop with the $\overline{Q}$ output connected to the D input as shown in Figure 4-10a, forms what is sometimes called a T, or *toggle, flip-flop*. This name refers to the fact that for either configuration the output state will change or toggle for each clock pulse. Since it takes two input clock pulses to produce one output cycle (Figure 4-10b), the output frequency is half that of the input clock. Two T flip-flops can be connected in series or cascaded to give an additional output frequency equal to the input frequency divided by 4. Adding a third T flip-flop in series gives an $f_{IN}/8$ output, and adding a fourth flip-flop gives an $f_{IN}/16$ output.

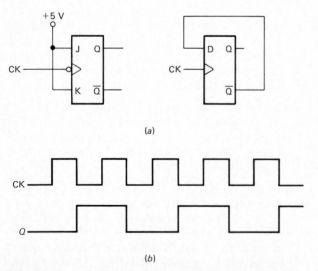

(a)

(b)

FIGURE 4-10 T flip-flops made from JK and D flip-flops. (a) Circuits. (b) Waveforms for T flip-flops made from negative-edge-triggered JK.

Figure 4-11a shows the 7493 four-stage binary divider or 4-bit binary ripple counter, which is effectively four T flip-flops in series. As a divider, one or more of the outputs is used to get a frequency that is some fraction of the input frequency. It is called a *counter* because, as shown by the ideal waveforms in Figure 4-11c, all four outputs together represent a binary number between 0 and 15. The negative edge of each input clock pulse increments the counter to the next-higher binary number. When the counter reaches count 15, or binary 1111, the outputs all roll back to 0 on the next negative clock edge. The count sequence then starts over.

The dependency notation logic symbol for the 7493 in Figure 4-11b shows that the device is made up of two sections, a divide-by-2 section and a divide-by-8 section. The plus symbols near the clock inputs indicate that the counter sections increment (count up) by 1 on the falling edge of each clock pulse. If R0(1) and R0(2) are asserted (H), the output count will be forced to all 0's.

Counters can be cascaded (connected in series) to get a higher count or more bits. The CMOS 4060, for example, has 14 binary divider stages in a single IC package.

The term *ripple* indicates that the effect of a clock pulse propagates sequentially or ripples down through the chain of flip-flops. For the circuit in Figure 4-11a, the output of the last flip-flop will not change until four flip-flop propagation delays after a negative edge of the input clock. This means the counter is *asynchronous* because the outputs are not directly in "sync" with the clock. In a later section we discuss the problems caused by this ripple delay in some circuits.

USING RIPPLE COUNTERS AS MODULO-*N* DIVIDERS

Flip-flops may be cascaded to produce a counter that gives output frequencies equal to $f_{IN}/2^N$, where N is the number of flip-flops. For example, the CD4044 CMOS 14-stage binary divider will have a final stage output frequency of $f_{IN}/2^{14}$, or $f_{IN}/16,384$.

The ratio of the input frequency to the output frequency of a counter is called its *modulo*. A single flip-flop is a modulo-2 divider or counter because the input frequency is 2 times the output frequency. The 7493 counter in Figure 4-11a has a maximum modulo of 16 since the input frequency is 16 times the lowest output frequency. For unmodified ripple counters, the modulo always will be some integral power of 2, such as 2, 4, 8, 16, 32, etc. Many applications of counters as frequency dividers (digital clocks, for example) require some other modulo such as 6 or 10. To produce a modulo less than the maximum for a given counter, you use digital feedback to reset the counter to zero before it can reach its maximum count.

Figure 4-12a shows how a 7493 can be converted to a modulo-10 counter or frequency divider. Starting with 0's on all the outputs, the 7493 counts up in a binary sequence similar to that of Figure 4-11c until the outputs reach the binary state 1010. The high on the Q8 output NANDed with the high on the Q2 output by the

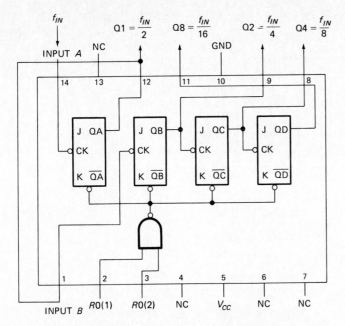

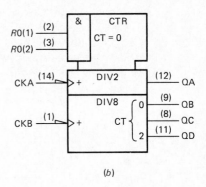

(b)

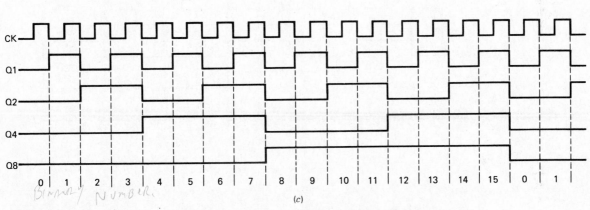

NOTE: ALL J AND K INPUTS TIED HIGH INTERNALLY. R01 AND R02
TIED LOW TO COUNT. A HIGH ON BOTH RESETS ALL Q OUTPUTS.

(a)

(c)

FIGURE 4-11 (a) Diagram for 7493 connected as a four-stage binary divider or counter. (b) Dependency notation symbol for 7493. (c) Clock and output waveforms.

internal NAND gate quickly resets all the outputs to 0's. Then the count sequence starts over. Figure 4-12b shows the clock waveform and the resulting output waveforms. As you can see from the waveforms, 10 clock pulses are required to produce one Q8 output pulse. Using the clock as input and Q8 as the output, then, makes this a modulo-10 frequency divider. Using the clock as input and all four Q outputs together for your output makes this a decade, or modulo-10, counter. The outputs count through the 10 states—0 through 9— much like the cylinders in a car odometer. Note that an eleventh state is present but only for the 30 or 40 ns necessary to reset the outputs to zero. For many applications, such as driving displays for a digital clock, this extra state is too brief to be visible and so doesn't matter. In any application in which the unwanted pulse could cause a problem, synchronous counters are

used. Synchronous counters are discussed later in this chapter.

Actually you usually don't have to synthesize a decade counter because it is available ready-made in TTL as the 7490. The 7490 has a divide-by-5 section and a T flip-flop divide-by-2 section. If the input signal is routed through the divide-by-2 section and then through the divide-by-5 section, the outputs will count up in a BCD sequence (same as binary 0 through 9). If an input signal is routed first through the divide-by-5 section and then through the flip-flop, the final output will be a symmetric square wave with a frequency of one-tenth the input frequency. A divide-by-12 circuit is available in TTL as the 7492 ripple counter.

A scheme similar to that in Figure 4-12a can be used with a 7493 to divide an input frequency by any integer between 2 and 16. The trick is to decode the next-higher

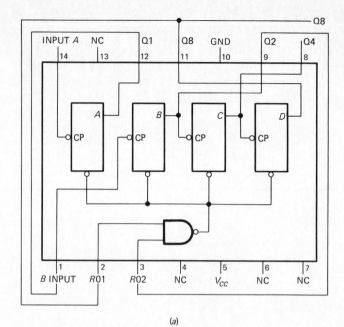

FIGURE 4-12 *(a)* Circuit showing 7493 converted to modulo-10 counter. *(b)* Clock and output waveforms.

(a)

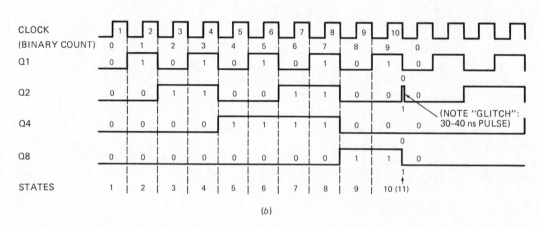

(b)

state and use this to reset the outputs to 0's. This trick may not work with a counter made from discrete flip-flops because a *race* condition is created. The decoded output state tries to reset the counter outputs to 0's. As soon as one output resets, the decoded output state may disappear. If the decoded output state is not present at the reset input long enough, then some flip-flops may not get reset. This, then, is the race. Can all the flip-flops get reset before the decoded output reset pulse goes away? Integrated-circuit counters are designed so that race problems are eliminated for the most part.

Figure 4-13 shows the circuit for a divide-by-11, or modulo-11, counter using a 7493. Note that the modulo-11 circuit requires an external gate because 1's on QA, QB, and QD must be detected.

For modulos above 16 you can cascade two or more counters. You can perform the same trick of decoding the state equal to the modulo you want and using this to reset all the outputs to 0's. Two 7493s can be used to divide an input frequency by any number up to 256.

SOME APPLICATIONS OF RIPPLE COUNTERS

ELECTRONIC ORGAN   Western music is based on an equally tempered chromatic scale of 12 notes per octave. Notes within an octave have a precise mathematical relationship to one another, and the frequency of a note in one octave is exactly half the frequency of the same note in the octave above. For example, a standard "A" is 440 Hz, and the "A" in the next-lower octave is 220 Hz. The next-lower "A" is 110 Hz, and the "A" below that is 55 Hz. For an instrument such as an electronic organ, all the required frequencies can be produced by dividing down from a single reference frequency, as shown in Figure 4-14a. The national MM5833 and MM5832 chromatic frequency generators produce the 12 notes for the top octave with 12 parallel divider chains driven by a master clock of 2.00024 MHz. One of these internal divider chains, for example, divides the 2.00024-MHz master frequency by 284 to give a fre-

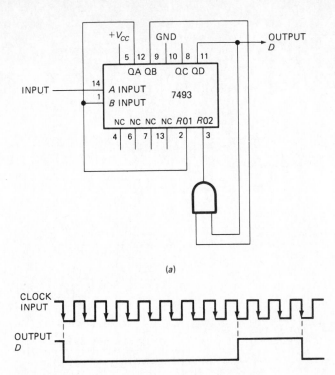

(a)

(b)

FIGURE 4-13 (a) Circuit for 7493 as modulo-11 counter. (b) Clock and output D waveforms.

quency of 7043 Hz, which is very close to the high "A" of 7040 Hz. Figure 4-14b shows the divider modulos and the output frequencies of the MM5832 and MM5833. For comparison, the standard frequencies of the notes in the equally tempered scale (ETS) are given. The frequencies for lower octaves can be derived by dividing each of the top octave frequencies down further with simple binary divider chains. The National MM5824 contains a six-stage binary divider chain. Therefore, the circuit shown in Figure 4-14a produces the frequencies for seven complete octaves plus an extra "C," C 9. These fourteen ICs give you all 85 frequencies—all you need to build an electronic organ or a synthesizer. Note the non-standard voltages—typical for PMOS devices—in the circuit shown.

The outputs are all square waves which, if amplified and played through a speaker, sound somewhat like a clarinet. To produce other sounds, these notes are mixed with their harmonics and passed through formant filters. To avoid a monotonous tone, a slight vibrato or frequency modulation is applied to the master 2.00024-MHz oscillator.

COUNTING UNITS AND DIGITAL CLOCKS Figure 4-15 shows how 7490 decade counters can be joined with 7447 BCD-to-seven segment decoders and common-anode, seven-segment LED displays to form decimal-counting units. Each display will show the decimal number between 0 and 9 corresponding to the binary-coded decimal (BCD) output of the counter. The counter

output will show the number of clock pulses input. As the least significant counter rolls over from 9 to 0, the falling edge of the QD output increments the next counter by one count. More counters can be added to count to as high a number as you want. You might use a light source and a phototransistor as a source of pulses on the input of a circuit such as this to count the number of cans passing on a conveyor belt or the number of people entering a baseball game. *Note:* The order of the display in Figure 4-15 is reversed from the way numbers usually are written.

If you apply a clock with a frequency of one pulse per second, or 1 Hz, to the input of the circuit of Figure 4-15, the display will show elapsed time in seconds. This is a digital readout clock, but not in the familiar format of hours, minutes, and seconds. Figure 4-16 shows how, by changing the modulo of some of the counters to 6 or 12, you can build a clock with digital readouts of hours, minutes, and seconds. The only difficulties with this clock are that it requires a large number of IC packages and dissipates considerable power. Both problems are solved by using one of the many MOS clock ICs such as the National MM53108. As shown by the MM53108 block diagram in Figure 4-17a, this IC contains not only the counters and decoders necessary for a basic clock, but also circuitry for AM and PM, an alarm, a snooze alarm, and a sleep timer to turn off a radio after you go to sleep. Figure 4-17b shows the circuit for a complete digital alarm clock using the MM53108. The only additional input needed is a 50- or 60-Hz pulse. This pulse is supplied to the chip through a 100-kΩ resistor from a tap on the power transformer. It can be produced also by dividing down the output from a crystal oscillator. The clock IC of Figure 4-17a does not display seconds unless the seconds button is pushed, but other clock ICs such as the MM5309 do.

PROBLEMS OF RIPPLE OR ASYNCHRONOUS COUNTERS

Most problems with ripple counters are caused by the delay between the time at which the first flip-flop in a chain changes its output state and the time at which a later stage changes its output state. We use the familiar 7493 to illustrate two of these possible problems. The first problem is errors in the output count. Figure 4-18 shows a timing diagram for a 7493 with a 12.5-MHz input clock. Compare this timing diagram, which includes flip-flop propagation delays, with the ideal 7493 timing diagram in Figure 4-11c. Note the binary states that the outputs show for the 12.5-MHz input. These erroneous states are always produced by a ripple counter, but for low-frequency applications such as driving a clock display they are not visible because they represent such a small part of a cycle. At 12.5 MHz the time for one cycle of the input clock is 80 ns, as shown. This is about the same order of magnitude as the propagation delay of the flip-flops in the counter, so the error states become overwhelming. The output count has

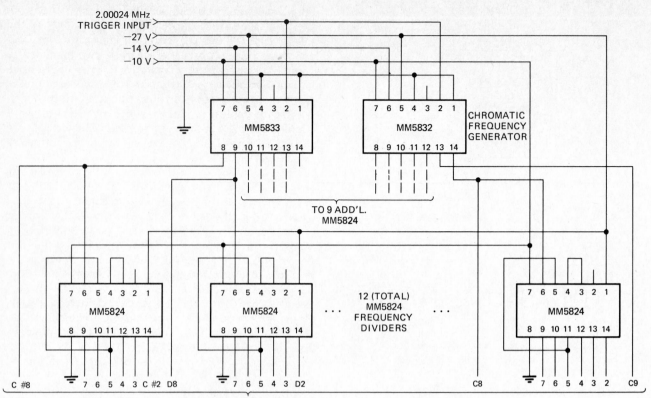

TYPICAL ORGAN TONE GENERATOR

(a)

MM5832

NOTE	DIVISOR	OUTPUT FREQUENCY	E.T.S FREQUENCY	% ERROR
C8	478	4184.61	4186.01	−0.565
C9	239	8369.21	8372.02	−0.565
B8	253	7906.09	7902.13	0.842
A #8	268	7463.58	7458.62	1.119
A8	284	7043.10	7040.00	0.740
G #8	301	6645.32	6644.88	0.112
G8	319	6270.34	6271.93	−0.424

MM5833

NOTE	DIVISOR	OUTPUT FREQUENCY	E.T.S FREQUENCY	% ERROR
F #8	338	5917.87	5919.91	−0.580
F8	358	5587.26	5587.65	−0.117
E8	379	5277.68	5274.04	1.160
D #8	402	4975.72	4978.03	−0.780
D8	426	4695.40	4698.64	−1.159
C #8	451	4435.12	4434.92	0.076

(b)

FIGURE 4-14 *(a)* Schematic for electronic organ tone generator using National MM5833 chromatic frequency generator and MM5824 six-stage binary dividers. *(b)* Comparison of equally tempered scale (ETS) frequencies with those produced by chromatic generator.

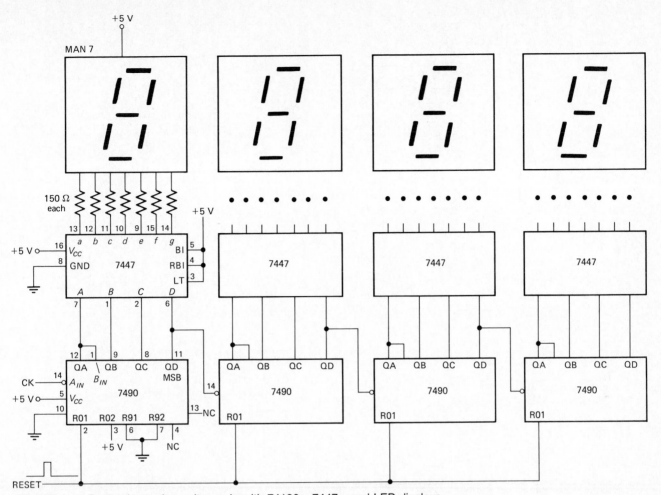

FIGURE 4-15 Decimal counting units made with 74190s, 7447s, and LED displays.

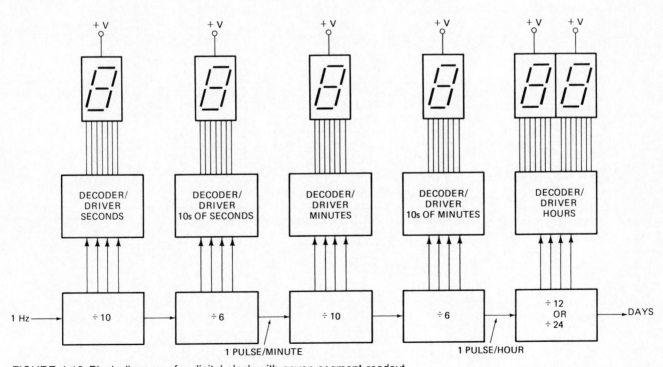

FIGURE 4-16 Block diagram of a digital clock with seven-segment readout.

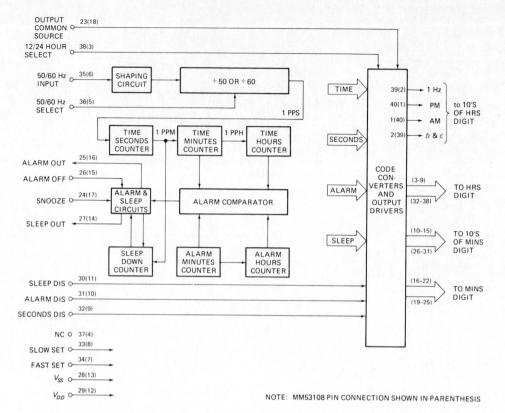

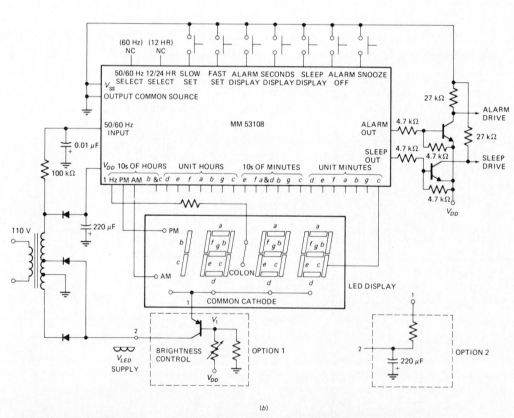

FIGURE 4-17 Large-scale integration digital clock. *(a)* Block diagram of MM53108 clock IC. *(b)* Complete schematic of digital alarm clock using MM53108. *(National Semiconductor.)*

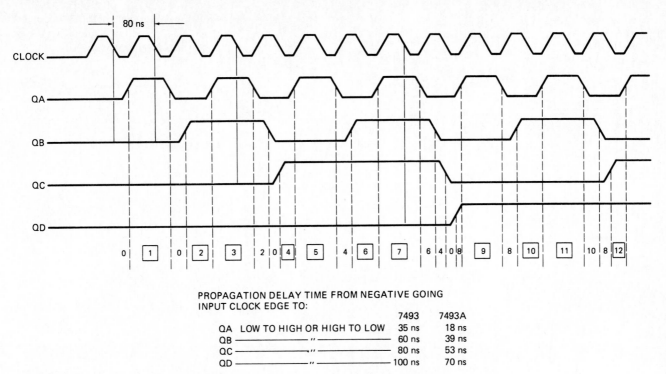

			7493	7493A
QA	LOW TO HIGH OR HIGH TO LOW		35 ns	18 ns
QB	———————— " ————————		60 ns	39 ns
QC	———————— " ————————		80 ns	53 ns
QD	———————— " ————————		100 ns	70 ns

PROPAGATION DELAY TIME FROM NEGATIVE GOING INPUT CLOCK EDGE TO:

FIGURE 4-18 Timing diagram for 7493 counter showing unwanted count states.

little relation to the number of pulses. The device, therefore, is of no use in higher-frequency counter applications, but it is still useful as a frequency divider. You can see on the timing diagram that the frequency of QA is exactly half the clock frequency, QB is $\frac{1}{4}f_C$, QC is $\frac{1}{8}f_C$, and QD is $\frac{1}{16}f_C$.

The second problem with ripple counters is produced by these error states or counts when you try to decode. *To decode a counter is to detect a given state of the counter and produce an output pulse when that state is detected.* The four-input NOR gate in Figure 4-19 will give a pulse out each time all the outputs of the counter are 0's. Ideally, this should happen only once for each 16 input clock pulses. The timing diagram in Figure 4-18 shows that this counter goes through a state of all 0 outputs four times for each 16 clock pulses! This will be true at any frequency of operation. The NOR gate will give out one pulse representing the true zero and three 20-ns-wide unwanted pulses or glitches. The glitches occur during the output 1-to-2 transition, the 3-to-4 transition, and the 7-to-8 transition. The point to remember here is that any asynchronous or ripple circuit

almost always produces glitches when it is decoded. The extra pulses or glitches may be only 20 or 30 ns wide, but they often cause problems. Watch for them. As discussed in Chap. 1, a delayed sweep-time base on an oscilloscope can help you spot them in the actual circuit.

Both these problems would be eliminated if the circuitry could be changed so that the outputs of all the flip-flops changed at the same time. This is exactly what is done in the synchronous counters discussed next.

SYNCHRONOUS COUNTERS

A *synchronous* counter is one in which all the outputs change at the same time because all the flip-flops are directly clocked at the same time by the input clock. Figure 4-20a shows the schematic of a 4-bit synchronous counter made with four JK master-slave flip-flops. Remember, when the J and K inputs are low on a JK flip-flop, the output does not change when the flip-flop is clocked; and if J and K are high, the flip-flop toggles when it is clocked. Now you can work your way through

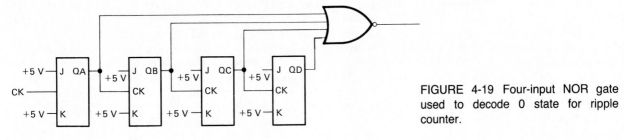

FIGURE 4-19 Four-input NOR gate used to decode 0 state for ripple counter.

FLIP-FLOPS, COUNTERS, AND REGISTERS

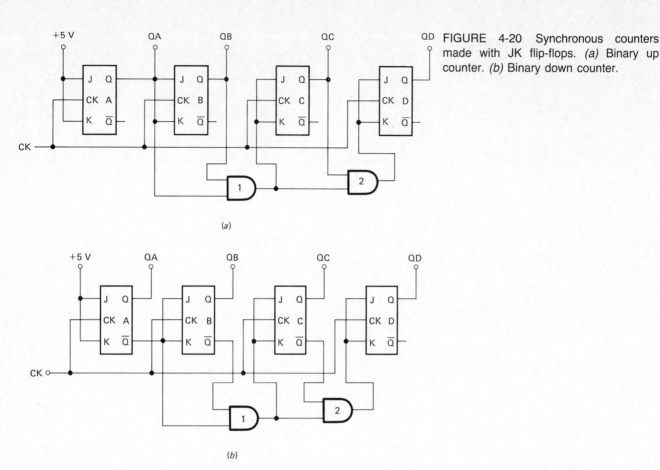

FIGURE 4-20 Synchronous counters made with JK flip-flops. (a) Binary up counter. (b) Binary down counter.

(a)

(b)

a count sequence to see how this circuit works. Starting with a reset condition of all 0's on the outputs, you can see that the J and K inputs of flip-flops B, C, and D are all low; therefore, a clock pulse at this time will not change their outputs. Flip-flop A, however, has its J and K inputs tied to a permanent high, so the first clock pulse will toggle its output from a 0 to a 1 state. Since the J and K inputs of flip-flop B are now high, the next clock pulse will change QB to a 1 and QA back to a 0 to give an output count of 2. With QA now low, only flip-flop A can change on the next clock pulse. This gives an output count of 0011, or 3. Since QA and QB are both high, AND gate 1 will put a high on the J and K inputs of flip-flop C so that it will change to 1 on the next clock pulse. The rest of the count sequence follows a similar pattern. At count 7 when QA, QB, and QC are all 1's, the AND gates enable flip-flop D to toggle to a 1 for count 8. At count 15 all the flip-flops are enabled, so their outputs will all toggle back to 0 on the next clock pulse.

MODULO-N DIVIDERS AND COUNTERS

The trick of synchronous counters is that the output states are decoded and used to set up the J and K inputs. Then the next clock pulse will toggle or not toggle each flip-flop to the desired output state. This very powerful technique can be used to produce counters of any modulo or output count sequence.

To analyze a JK flip-flop circuit with feedback, start with known output states. Trace these states through the decoding logic (feedback gates) to determine the states that are produced on the J and K inputs of the flip-flops. Knowing the J and K inputs will allow you to predict the output state of each flip-flop after a clock pulse. Then these outputs can be traced through the decoding logic to the J and K inputs to predict what the output states will be after the next clock pulse. Repeat the process until, when the flip-flops are clocked, their outputs return to the starting states. In App. A we discuss logic state diagrams that are often used to show the sequence of states in a circuit such as this one.

Figure 4-20b shows the circuit for another useful synchronous counter. Try to trace its count sequence. To start, observe that when the outputs are all 0's in this circuit, the flip-flops are all enabled to toggle to a 1 on the next clock pulse.

The counter in Figure 4-20b is, as you probably discovered, a 4-bit binary down counter. Fortunately, you do not need to synthesize a synchronous counter every time you need one. Several types are readily available. The TTL 74193 and the CMOS 74C193 and CD40193 are synchronous 4-bit up/down binary counters. The TTL 74192 and the CMOS 74C192 and CD40192 are synchronous 4-bit up/down decade counters, respectively. The 74161 is a synchronous binary up counter, and the 74160 is a synchronous decade up counter. All the standard TTL versions of these counters can give a proper count out with an input frequency of 20 MHz or more. When decoded, they produce only very small glitches that are caused by variations in the propagation delays of the flip-flops and the difference between the time from a low to a high output and the time from a high to a low output.

The pin diagram, dependency notation logic symbol, and typical count-sequence diagram of a common 4-bit binary synchronous counter, the 74193, are shown in Figure 4-21. From these you can determine much about its characteristics. This counter has the advantage that it can count up if the clock is routed to the count-up input and count down if the clock is routed to the count-down input. The counter increments or decrements on the positive edge of the clock. The unused input must be held high. When it is counting up, a terminal count of 15 produces a low-going pulse on the carry or terminal count-up output. This occurs on the falling edge of the input clock. When it is counting down, a terminal count of 0 produces a low-going pulse on the borrow or terminal count-down output. The bubbles on these outputs in Figure 4-21a indicate that the output pulse is negative or active low. The bubble on the load input indicates that a low state on this input loads the counter. The clear input is asynchronous, which means that when the clear input is made high, all the outputs will be cleared or reset to 0 with no regard for the state of the clock.

The counter is presettable, which means that a binary number placed on the four data inputs will be loaded onto the outputs as soon as the load input is made low. The output count sequence will proceed upward or downward from the loaded number. This feature can be used to convert this counter to any modulo between 1 and the maximum number of states for the counter. Figure 4-22 shows how this is done. The counter is connected to count down, and the terminal count-down or borrow output is connected to the parallel load input. If you want a modulo-11 counter, for example, just hardwire a binary eleven (1011) on the data inputs as shown. The counter will count down from eleven to 0. When the output count reaches 0, a pulse will be put out on the BORROW output. This pulse will load the 11 from the data inputs onto the counter outputs. The next clock pulses decrement the counter from 11 to 0 again, at which point another terminal count-down pulse is produced to repeat the cycle. The circuit is a modulo-11 divider because it produces one output pulse on the BORROW output for each 11 input pulses.

Figure 4-22b shows the timing waveforms for a 74193 used as a modulo-11 divider. There are three important points to note in these waveforms. First, since BORROW OUT is connected to LOAD, as soon as BORROW OUT goes low, LOAD will go low. This will load 1011 onto the Q outputs and immediately make BORROW OUT high again. The resultant BORROW LOAD pulse will be only 30 to 40 ns wide.

Second, if you count the number of states that are present on the Q outputs, you will find that there are 12 states, 0 through 11. However, since state 0 and state 11 are present for only one-half clock cycle each, the total number of input clock cycles between BORROW output pulses is 11. Because of the presence of these two half-cycle states the circuit cannot be used as a counter in some applications, but it can be used as a frequency divider.

Third, for this circuit the QC and QD outputs both produce one output pulse for each 11 input clock cycles. Since they have closer to a 50 percent duty cycle, often they are more useful than the BORROW OUT pulse.

You can produce any other modulo by simply hardwiring the binary equivalent of the desired modulo onto the data inputs. A modulo greater than 15 can be produced by cascading two or more counters, as shown in Figure 4-22c. When the terminal count is reached, the BORROW OUT of the most significant counter produces the LOAD pulse, which reloads all the counters.

Now that gates, latches, and counters have been discussed, we will show how these components can be combined in a circuit to build a frequency counter giving a digital readout of the frequency of input pulses.

FREQUENCY COUNTERS: AN MSI COUNTER

Figure 4-23 is a schematic for a frequency counter using readily available TTL parts. Since this is a fairly large, complex schematic with many ICs, the trick is to simplify it by looking for familiar "landmarks," or blocks of circuitry whose function you recognize. In this frequency counter, first you might recognize the 1-MHz crystal oscillator and the chain of decade counters used to divide this reference frequency down to the needed output frequencies. Next you might notice the synchronous divider chain of decade counters, each with a 74175 four-bit D flip-flop, a 7447 BCD-to-seven-segment decoder, and a seven-segment LED display. This big block of circuitry is simply a unit counter, to count and display any number up to 999,999.

The 74160 counters, incidentally, are connected so that all the counters after the first one on the left function as one large synchronous counter. This feature allows the counter section to accurately count higher input frequencies.

Now that you have figured out the large blocks of circuitry, you can start looking at the small pieces that tie these together. Note that the unknown frequency is gated into the counting section by a NAND gate. The unknown frequency pulses can get in to be counted only if the pin 1 input of this NAND gate is high. If this gate input is made high for exactly 1 s, then the display will read out the number of unknown input pulses counted in 1 s. This is, of course, the frequency of the unknown (in hertz). Now look to see how this 1-s *count window* is produced. A square-wave pulse with a time high of 1 s has a 0.5-Hz frequency. The divider chain on the crystal oscillator has a 1-Hz output which goes to the clock input of a JK flip-flop. The 7476 JK flip-flop with J and K tied high will give a symmetric square wave out at half the input frequency. The JK flip-flop then produces the precise 1-s count window.

The next part of the circuitry to approach is the 74121 monostable. It is connected so that the negative edge of the count window pulse triggers it to produce a positive pulse. The leading edge of this positive pulse clocks the output states of the counters onto the outputs of the

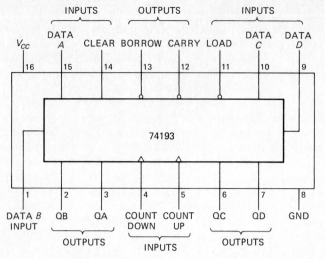

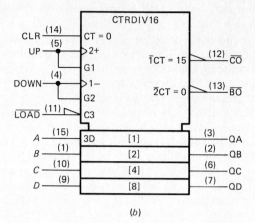

(b)

NOTE: LOW INPUT TO LOAD SETS QA = A, QB = B, QC = C, AND QD = D.

(a)

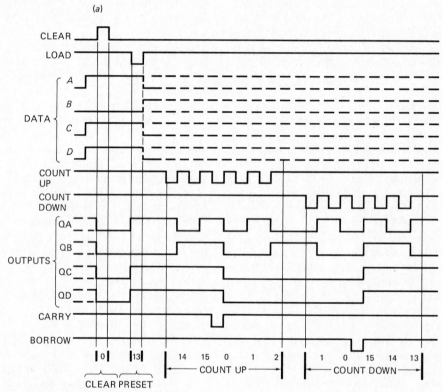

SEQUENCE:

(1) CLEAR OUTPUTS TO ZERO.

(2) LOAD (PRESET) TO BINARY THIRTEEN.

(3) COUNT UP TO FOURTEEN, FIFTEEN, CARRY, ZERO, ONE, AND TWO.

(4) COUNT DOWN TO ONE, ZERO, BORROW, FIFTEEN, FOURTEEN, AND THIRTEEN.

NOTES:

(A) CLEAR OVERRIDES LOAD, DATA, AND COUNT INPUTS.

(B) WHEN COUNTING UP, COUNTDOWN INPUT MUST BE HIGH; WHEN COUNTING DOWN, COUNT–UP INPUT MUST BE HIGH.

(c)

FIGURE 4-21 74193 binary up/down counter. *(a)* Connection diagram. *(b)* Dependency notation logic diagram. *(c)* Typical clear, load, and count sequences.

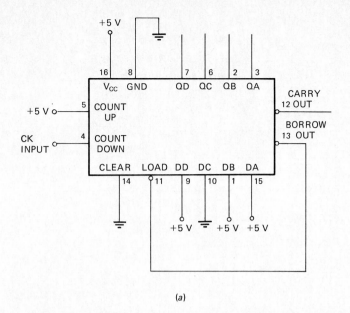

(a)

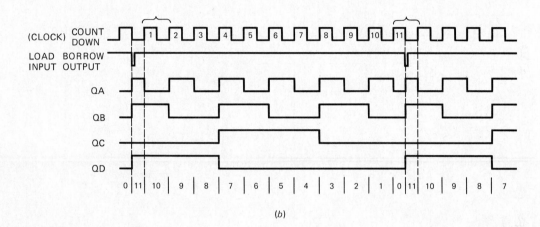

(b)

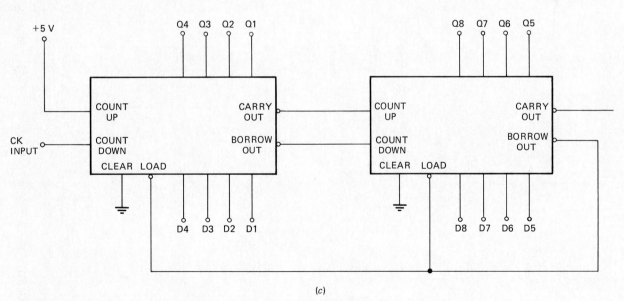

(c)

FIGURE 4-22 Modulo-*N* counter/divider made with the 74193. *(a)* Modulo-11 schematic. *(b)* Modulo-11 waveforms. *(c)* Circuit for modulos greater than 16.

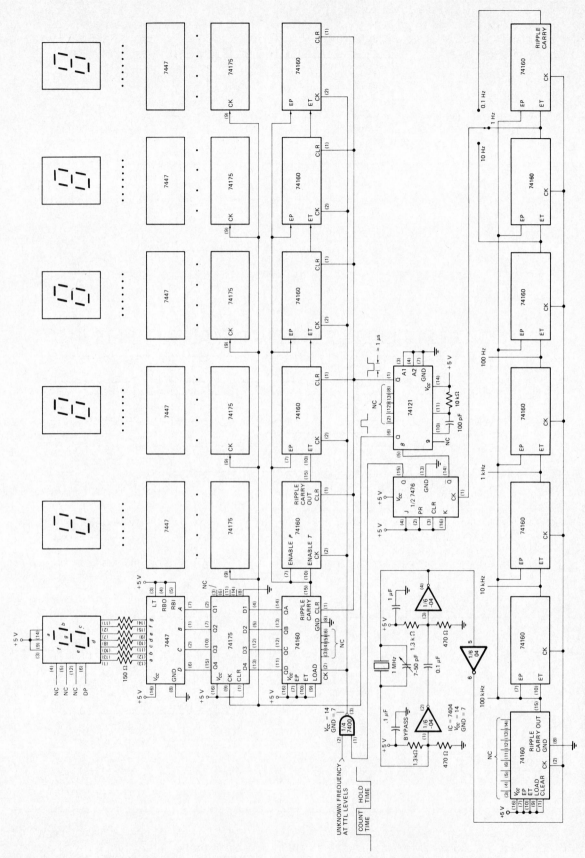

FIGURE 4-23 Complete schematic for an MSI frequency counter.

74175 D flip-flops. Then the flip-flops hold stable data for the entire count cycle. Without the flip-flops the display would alternate between 1 s of garbage during a count and 1 s of reading during the hold time of the window pulse. This would make the display very hard to read. The flip-flops hold the display data constant except for the few nanoseconds every 2 s when the flip-flops are clocked to update the display.

The $\overline{Q}$ pulse from the 74121 is used to reset all the counters to 0 so they are ready for the start of the next count sequence. Figure 4-24 shows the timing of these pulses.

The counter with a 1-s count window and six digits, as shown, can only measure a frequency up to 999,999 Hz before the counter overflows. By adding a decimal point to the display, it can be made to read in kilohertz, with the maximum frequency of 999,999 displayed as 999.999 kHz. The frequency range of the instrument can be extended upward to about 10 MHz by shortening the count-window time to 0.1 s and moving the decimal point so that the display shows up to 9999.99 kHz. Since the count window is only one-tenth as long, 10 times as high a frequency can be put in without overflowing the counters.

Another way to increase the range is to add another decade counter, decoder, and display. To extend the range to still higher input frequencies, you can use one or more prescalers. A *prescaler* is simply a very high speed decade counter, such as the 74S196, which is put in series with the input signal to divide the input signal frequency down to a frequency range that the counter chain can count. A section of a switch used to switch in the prescaler for high-frequency measurement also can move the decimal point so the display reads correctly. The prescalers do not have displays connected directly to them.

For low-frequency measurement, a 1-s window gives poor resolution. At 20 Hz, for example, a 1-s window gives a resolution of only ±1 Hz since the counters cannot count fractions of hertz. This is an error of ±5 percent. If you extend the count window to 10 s and move the decimal point accordingly, the resolution is improved to ±0.1 Hz because it is now counting 200 cycles ±1.

Another improvement you can make to the basic frequency counter is input signal conditioning. The circuit

as shown requires TTL input levels. A high-speed comparator such as the AM 686 can be added in front of the counters to convert small sine waves or other waveforms to TTL levels. The threshold of the comparator can be made adjustable so that the trigger voltage level can be set above noise pulses.

This summarizes the operation of the circuit. Figure 4-23 shows component values and pin numbers, so you can build the circuit and verify its operation if you wish. *Note:* In the schematic, the *least* significant digit is on the left.

MULTIPLEXED DISPLAYS   The circuit in Figure 4-23 works well, but from a practical standpoint it has two problems. It uses both a large number of IC packages and a fairly large amount of current. Both problems can be solved quite easily. For the power problem, the first place to look is the displays, since this is where the most power is dissipated. Assuming 20 mA per segment times a maximum of seven segments per digit gives 140 mA per digit. Six digits give a total maximum current of 840 mA just for the displays. The six 7447s add another 260 mA, for a total of 1.1 A. This total can be cut drastically by multiplexing the display. Multiplexing, as you may remember from Chap. 3, is interleaving segments of several signals onto the same wire. The trick used to produce a multiplexed display is the same one used in Chap. 3 to produce eight oscilloscope traces with a single beam. Only one input is displayed at any given instant; but if you chop or alternate fast enough between inputs, your eyes see the result as a continuous display. With the LED displays only one digit is lighted at a time.

Figure 4-25 shows the counter, flip-flop, and multiplexed display sections of a seven-digit frequency counter. Here, four 74151 eight-input multiplexers multiplex the outputs of the seven 74175 quad flip-flops into a single 7447 BCD-to-seven-segment decoder. Then each output of the 7447 goes to the corresponding segment inputs on all the seven-segment displays. Thus when the seven-segment code for a 5 is on the outputs of the 7447, it is bussed to all the displays. However, since only one of the PNP transistors in series with each display is turned on at a time, only one display will show this 5.

The outputs of a 3-bit binary counter select which decade counter's outputs the four 74151s will route to the 7447. The same 3-bit counter outputs connect to a 74LS138 one-of-eight low decoder, which selects which driver transistor is turned on and therefore which display is lighted. The 74LS138 and the 74151s are synchronized, so that, for example, when the 74151s are sending data from the least significant digit counter to the displays the 74LS138 turns on the digit driver transistor for the least significant digit of the display. Each digit should be refreshed or pulsed on at a rate of 50 to 200 Hz.

The power saving with this multiplexing scheme is about 600 mA. The saving would be greater except that when you multiplex LED displays, you have to at least

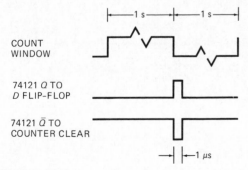

FIGURE 4-24 Timing waveforms for frequency counter control circuitry of Fig. 4-23.

FLIP-FLOPS, COUNTERS, AND REGISTERS

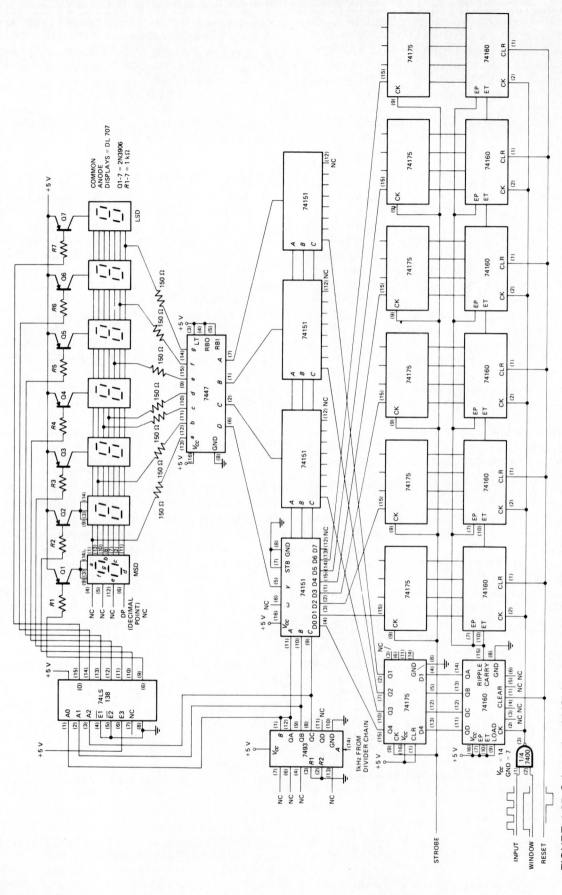

FIGURE 4-25 Schematic of MSI frequency counter with multiplexed displays.

double the current per segment to increase display brightness. The circuit uses as many ICs but saves 36 resistors and much wiring.

TROUBLESHOOTING AN MSI FREQUENCY COUNTER
The time base and control circuitry from Figure 4-23 can be combined with the counter, flip-flop, and display section of Figure 4-25 to make a functioning MSI frequency counter. We refer to this circuit to discuss systematic troubleshooting of a digital system. The details of every system are different, but the basic approach is the same. The major points are:

1. Analyze first.
2. Divide and conquer.

To prototype a circuit such as this one, begin by analyzing the schematic to find which sections can be built separately and tested easily. For example, you might start with the crystal oscillator. Then, when it is working correctly, add and test each counter of the time-base divider chain. The next step would be to build and test the 7476 window generator and the 74121 control-pulse generator.

After this, you can build the display section and test it as a separate module. To determine whether the segments are wired correctly, temporarily hard-wire the BCD code for 2 or 5 on the inputs of the 7447. As the end of each digit driver's base resistor is grounded or made low, that digit should display the correct number. When the 74LS138 and 7493 address counter are added, all the displays should show this number.

Next, build the main counter chain as a module. The count window input can be temporarily tied high instead of to the 7476, and a 1-MHz signal applied to the unknown frequency input. Then the counters will receive a signal continuously. Counters can be added one at a time and checked just as those in the time-base chain were.

Finally, add the flip-flops and multiplexers. Connect the count window, latch strobe, and reset signals to the correct points and see whether the counter functions as a whole. If it does, great! Take a coffee break. If not, this final section can be traced on a static basis as follows. Lift out the 7493, and temporarily hard-wire the QA, QB, and QC output pins to ground. The least significant digit of the display should be enabled now, and the multiplexer inputs all pointed to the least significant 74175 flip-flop on the left. Lift the strobe lead from the 74121 and connect it to ground. Lift the reset lead from the 74121 and connect it to +5 V. Lift the count window lead from the 7476, and apply a 1-MHz signal to the count input to load a count into the counters. Then connect the window input to ground. A count should be present now on the outputs of the counters. This count can be transferred to the outputs of the 74175 flip-flops by briefly lifting the clock input lead from ground. The logic states on the least significant counter now should be present on the outputs of the leftmost 74175 and on the outputs of the four 74151s. The major point here is

that since all these levels are held fixed, they can be checked with a voltmeter or logic probe. This is much easier than trying to decipher the confusing array of pulses that would be seen if the circuit were operating with normal clocks and pulses.

The 74151 multiplexers can be stepped to pick up the outputs of the next 74175 by simply changing the jumpers you put in place of the 7493 to QA = +5 V, QB = ground, and QC = ground. In this way, the connections from each counter to the 7447 inputs can be checked on a dc basis with a logic probe.

TROUBLESHOOTING A PREVIOUSLY FUNCTIONING INSTRUMENT
The procedure for troubleshooting a previously functioning instrument is somewhat different because usually you can assume that the wiring is correct. There is not space here to detail all the possible symptoms and cures, but a few will show you how to go about it.

Start by analyzing the schematic and/or block diagram of the system so that you understand the operation of the major parts. It is not necessary to initially determine the function of every IC. For most instruments or systems, this would take far too much time. Once you isolate a problem to a particular section, you can focus on the details of that section.

If the counter counts but shows erratic counts, one thing to check is that the power-supply voltage is within specifications. TTL devices function erratically with low supply voltages.

If one digit of the display is brightly lit but the others are dark, this is caused by failure of the 74LS138 digit-scanning circuitry. Either the 74LS138 is defective or the 7493 is not supplying sequential addresses to it. An oscilloscope will quickly show you whether the problem is in the 74LS138, the 7493, or the clock pulse coming to the 7493. Since the clock signal to the 7493 comes from the time-base divider chain, no signal here will lead you back to the divider chain and possibly to the nonfunctioning master oscillator. This illustrates that in a digital system, often the cause of a problem is far removed from the section exhibiting it.

If all digits are lighted equally but the numbers do not change when the input frequency is changed, the problem may be no count window or no clock pulse to the 74175 flip-flops. If the digits are all lighted equally but the count constantly changes to random numbers, the counters may not be getting a reset pulse. The point here is to check the control or timing signals. In this system the strobe and reset pulses are very narrow compared with the count window. Using the delayed sweep function of an oscilloscope as described in Chap. 1 will make these pulses clearly visible.

If there is a problem in the counter, flip-flop, and multiplexer section, remove the 7493, 74121, and 7476. Then add jumpers as described earlier so that you can check the levels through this section as static signals.

In summary, the point of systematic troubleshooting is to analyze the symptoms and the schematic and then devise tests you can perform to narrow the problem to a small enough area that it is easily found.

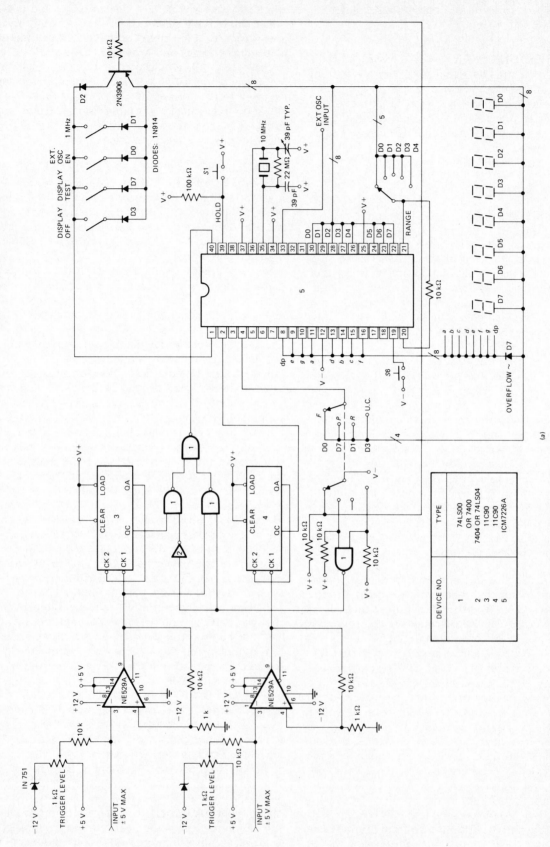

FIGURE 4-26 A 100-MHz multifunction counter using the Intersil ICM7226A. *(a)* Schematic diagram.

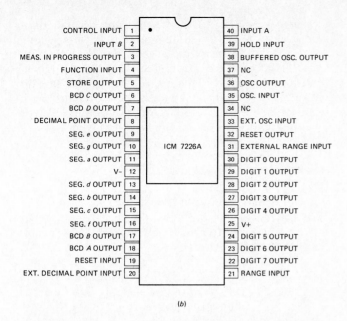

CONTROL INPUT	1		40	INPUT A
INPUT B	2		39	HOLD INPUT
MEAS. IN PROGRESS OUTPUT	3		38	BUFFERED OSC. OUTPUT
FUNCTION INPUT	4		37	NC
STORE OUTPUT	5		36	OSC OUTPUT
BCD C OUTPUT	6		35	OSC. INPUT
BCD D OUTPUT	7		34	NC
DECIMAL POINT OUTPUT	8		33	EXT. OSC INPUT
SEG. e OUTPUT	9		32	RESET OUTPUT
SEG. g OUTPUT	10	ICM 7226A	31	EXTERNAL RANGE INPUT
SEG. a OUTPUT	11		30	DIGIT 0 OUTPUT
V−	12		29	DIGIT 1 OUTPUT
SEG. d OUTPUT	13		28	DIGIT 2 OUTPUT
SEG. b OUTPUT	14		27	DIGIT 3 OUTPUT
SEG. c OUTPUT	15		26	DIGIT 4 OUTPUT
SEG. f OUTPUT	16		25	V+
BCD B OUTPUT	17		24	DIGIT 5 OUTPUT
BCD A OUTPUT	18		23	DIGIT 6 OUTPUT
RESET INPUT	19		22	DIGIT 7 OUTPUT
EXT. DECIMAL POINT INPUT	20		21	RANGE INPUT

(b)

INPUT	FUNCTION	DIGIT
FUNCTION INPUT PIN 4	Frequency	D0
	Period	D7
	Frequency ratio	D1
	Time interval	D4
	Unit Counter	D3
	Oscillator Frequency	D2
RANGE INPUT PIN 21	0.01 s / 1 cycle	D0
	0.1 s / 10 cycles	D1
	1 s / 100 cycles	D2
	10 s / 1k cycles	D3
EXTERNAL RANGE INPUT PIN 31	Enabled	D4
CONTROL INPUT PIN 1	Blank display	D3 & Hold
	Display test	D7
	1 MHz select	D1
	External oscillator enable	D0
	External decimal point enable	D2
	Test	D4
EXTERNAL DECIMAL POINT INPUT, PIN 20	Decimal point is output for same digit that is connected to this input.	

(c)

FIGURE 4-26 *(continued)* *(b)* Pinouts for the ICM7226A. *(c)* Digit strobe connections for range, function, and control inputs of the ICM7226A.

AN LSI FREQUENCY COUNTER   The circuit shown in Figure 4-25 works well, but it is a 1970s-style frequency counter. Since most parts of the circuit operate at relatively low frequency, CMOS and MOS LSI have made it possible to put most of a frequency counter in one or two IC packages. Figure 4-26a shows the schematic for a 100-MHz multifunction counter using the Intersil ICM 7226A. The CMOS 7226 contains a 10-MHz reference oscillator, time-base divider chain, input signal counter chains, latches, decoders, segment drivers, digit drivers, control circuitry, and multiplex circuitry for an eight-digit common-anode display counter system. The NE529A comparators are for input signal conditioning, and the 11C90 high-speed decade counters are used as prescalers. The prescalers extend the frequency range of the unit from the 10 MHz of the 7226A up to 100 MHz.

The ICM7226A is a very versatile IC. Note that it has two signal inputs, A and B. In addition to frequency, the 7226 can measure period, the ratio of two frequencies, time intervals, or just total counts. For frequency, period, or units, the signal is applied to the A input. The A input can count up to 10 MHz. For frequency ratio and time interval measurement, one signal is applied to the A input and the other to the B input. The B input can count up to 2 MHz. For time interval measurement, the 7226 measures the time between a high-to-low transition on the A input and a high-to-low transition on the B input with a resolution of 0.1 μs.

Figure 4-26b shows the complete pinouts for the ICM7226A. Note that the device has multiplexed BCD outputs as well as the multiplexed seven-segment and digit-driver outputs. Each segment driver can sink up to 25 mA, and each digit driver can source up to 170 mA. With the displays off, however, the 7226A itself only uses 5 mA from a +5-V supply.

Observe in the schematic of Figure 4-26a how the eight digit-strobe lines and the eight segment-drive lines are shown as single lines. This is often done when the connections are obvious because it makes the schematic much easier to read.

Another unique feature of the 7226A is the way in which the function, range, and control signals are multiplexed into the device. As shown in Figure 4-26a, digit-strobe lines are connected to the range, function, and control switches. Internal circuitry detects which digit-strobe line is connected to, for example, the range input on pin 21. Then it sets the proper window and decimal point position for that range. Figure 4-26c shows the digit-strobe connections for all the ranges, functions, and controls. This scheme sharply reduces the number of pins needed to get all these signals into the device. Also, if the mechanical range and function switches are replaced with an analog multiplexer such as the CD4051, the selected range and function are determined by the binary word on the address inputs of the multiplexer. In other words, the measurement function and range are programmable with binary words from, for example, a microcomputer or computer.

REGISTERS AND SHIFT REGISTERS

A register is a parallel group of flip-flops, usually D type, that are used for the temporary storage of data bits. Registers may be any number of bits wide. Common register sizes are 4, 8, 12, or 16 bits. In a shift register, the output of each flip-flop is connected to the input of the adjacent flip-flop, and so each successive clock pulse moves the data bits one flip-flop to the left or right depending on the order of the connection. There are four basic types of shift registers which we discuss individually in the following sections.

SERIAL-IN, SERIAL-OUT SHIFT REGISTERS

Figure 4-27a shows four D flip-flops connected as a serial-in, serial-out (SISO) shift register, and Figure 4-27b shows the clock, data, and output waveforms for the circuit. As you can see from these waveforms, each clock pulse moves the data pulse one flip-flop to the right. The output is delayed four clock pulses from the input since it takes four clock pulses for the data pulse to reach the output.

SISO FOR DELAY
One use for a SISO shift register is to delay digital signals. The amount of delay depends on the clock frequency and the number of flip-flops in the register. The time for one clock pulse is 1/frequency (in hertz). This time multiplied by the number of stages gives the total delay for the register. Some MOS SISO shift registers have as many as 4096 flip-flops. Figure 4-28 shows in block diagram form how an analog-to-digital converter, some long shift registers, and a digital-to-analog converter can be utilized to replace the mechanical spring reverb system usually found in gui-

tar amplifiers. The A/D converter converts the input sine wave to binary weighted 8-bit digital words. These digital words are delayed by passing them through eight parallel shift registers. The amount of delay can be controlled by the clock frequency. At the end of the shift registers, a D/A converter changes the digital word back to analog sine waves. A fraction of the output is fed back into the input to give an echo, or reverb, effect.

SERIAL-IN, PARALLEL-OUT SHIFT REGISTERS

A serial-in, parallel-out (SIPO) shift register is connected in the same way as the SISO shift register in Figure 4-27a, but the outputs of each flip-flop are available on pins of the IC package.

SIPO AS A RING COUNTER
An important use of the SIPO shift register is to make a ring counter. If the output of the shift register in Figure 4-27a is connected to the input, then an injected pulse will keep circulating around the ring. The very useful parallel output waveforms of this ring counter are shown in Figure 4-29. As you can see, the outputs are sequential, nonoverlapping pulses. The sequence can be started in the ring by initially setting one output to 1 and resetting all the other outputs to 0, using the direct set and reset inputs. With shift registers such as the TTL 74195 which has parallel-load inputs, you can hard-wire the load inputs as 1000 and load this into the register with a power-on reset circuit to get the count started. The circuit for a power-on reset is shown in Figure 4-30. When power is

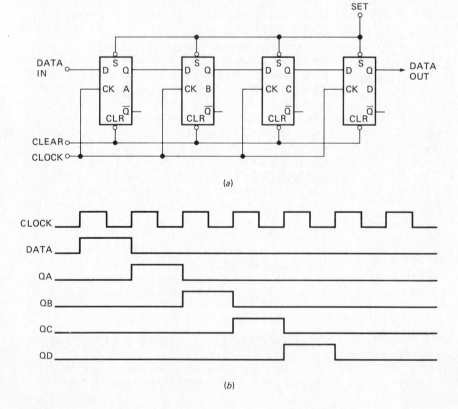

(a)

(b)

FIGURE 4-27 Serial-in, serial-out shift register. (a) Logic diagram. (b) Clock, data, and output waveforms.

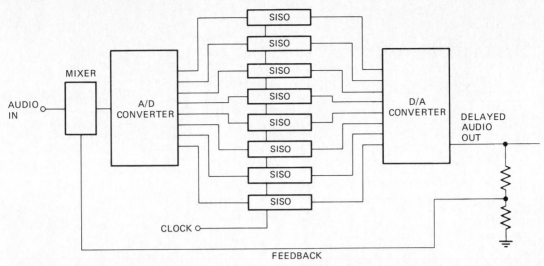

FIGURE 4-28 Serial-in, serial-out shift register used with A/D and D/A converters to make an audio reverb, or echo, unit.

turned on, the output is held low until the large capacitor charges to the threshold voltage of the Schmitt trigger. The Schmitt trigger ensures a sharp transition of the output signal from low to high. You can invert all the pulses by initially loading the register with 0111.

USES OF RING COUNTERS

The stepped output waveforms of ring counters make them suitable for many applications. They can be used as control-state counters, where each output sequentially enables the next state or process in some logic-controlled machine. Also a special type of motor, a *stepping motor,* which rotates in small increments or steps, requires these sequential pulses to rotate its shaft from one step to the next.

Ring counters also can be used in encoding the outputs from electronic keypads such as those in calculators and computers. *Encoding* means converting a keyswitch-closed signal to some standard character code such as hex, ASCII, or EBCDIC. Figure 4-31 shows one way in which a standard 63-key keyboard can be wired as a matrix of eight rows and eight columns and encoded with the help of a ring counter. The 74199, $U1$ in Figure 4-31, is an 8-bit shift register connected as a ring counter. It is initially loaded with 11111110 by a power-on reset circuit. As the register is clocked by an

external 5-kHz scan oscillator, a low is sequentially stepped down the row lines. Only one row will be low at a time. If no keys are pressed, all the columns are held high by the resistors to +5 V. If a key is pressed, the row and column for that key are short-circuited together. When the ring counter cycles a low to that row, the low will also appear on the column in which the key was pressed. The 1N270 germanium diodes prevent two outputs of the ring counter from being short-circuited, if two keys in the same column are pressed simultaneously. The small voltage drop across these diodes does not exceed the $V_{IL.MAX}$ for $U3$ and $U7$.

With no keys pressed, all the inputs of $U7$, an eight-input NAND gate, are high and its output is low. As soon as a low appears on any column, the output of the NAND gate goes high. However, because of contact bounce, this output may go high and low several times, as shown. The first low-to-high transition of this output triggers $U4$, a 74121 monostable with its output pulse width set for about 10 ms. The $\overline{Q}$ output of the 74121 acts through $U9$ to inhibit the clocking of the ring counter until the key stops bouncing and the correct code is output. If the key is held down longer than this, the signal from the 7430 passed through U8C will in-

FIGURE 4-29 Output waveforms for ring counter made with serial-in, parallel-out shift register.

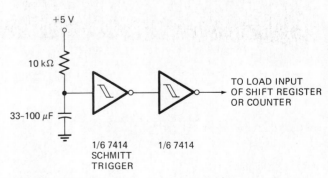

FIGURE 4-30 Circuit for a power-on reset.

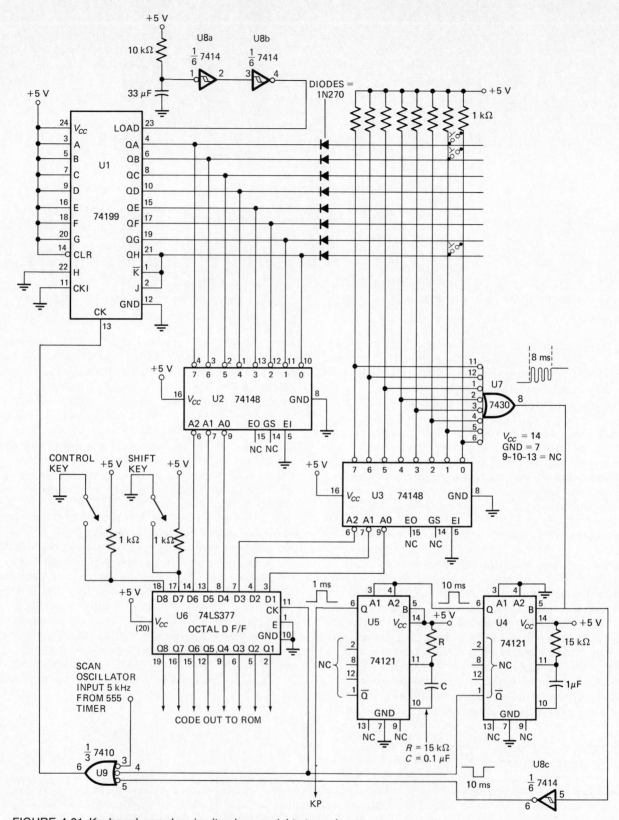

FIGURE 4-31 Keyboard encoder circuit using an eight-stage ring counter.

hibit the clocking of the ring counter until the key is released. This is necessary to prevent the key code from being output several times for each key press.

While the scan oscillator is stopped, the row and column codes are processed by U2 and U3, two 74148s. A 74148 is an eight-line-to-three-line priority encoder.

When enabled by grounding its EI input, this device will output a 3-bit inverted binary code for the highest-number input that has a low on it. For example, if the 74148 has a low on input 5, the outputs will have 010, the inverse of the binary code for 5. If two inputs have lows, only the inverted binary code for the highest-number (priority) input will be output.

When the ring counter is stopped after a key press, one row will contain a low. A 3-bit code for this row number will be output by $U2$. Likewise, when a key is pressed, one column will contain a low. A 3-bit binary code for the number of this column will be output by $U3$.

The 3-bit row code from $U2$, the 3-bit column code from $U3$, and the shift- and control-switch outputs are connected to the D inputs of an octal D flip-flop, $U6$. The rising edge of the clock inhibit pulse from $U4$ clocks the flip-flops and transfers these codes to the flip-flop outputs. The 8-bit code can be sent to a ROM to be converted to ASCII or some other standard code.

A second 74121, $U5$, generates positive and negative key-pressed strobes. One of these can be used to tell a receiving system that the valid code for a new key press is present on the outputs of $U6$.

Complete MOS LSI keyboard encoders that only require wiring to the key switches are available as the General Instruments AY-5-2376 or the National Semiconductor MM5740. These encoders have an additional feature called *two-key rollover*. This means that if two keys are hit very close to each other in time, the code for the second key hit will be stored and output after that for the first key hit.

USE OF A SIPO AS A SERIAL-TO-PARALLEL DATA CONVERTER

To reduce the number of connecting wires, digital data words are often sent from one system to another in serial form. In other words, 1 bit of a word is sent after another on a single wire. Figure 4-32 shows the serial format in which ASCII code is sent from a teletype terminal such as the ASR-33. When no character is being sent, a continuous high is on the line. The start bit which tells the receiver that a character is coming is a low. The data bits for the ASCII-encoded character follow the start bit, as shown. Next are 1 parity bit and then 2 always-high stop bits. The stop bits indicate that sending of the character is complete. As you can see in Figure 4-32, it takes 11 bit times for each character. A standard rate for sending serial data from teletypes is 110 bits/s, or 110 baud (Bd). A bit rate of 110 per second divided by 11 bits per character yields a rate of 10 characters per second.

Microcomputers and other instruments usually require data to be input in parallel form. A SIPO shift register can be used to convert the data format from serial to parallel, as shown in Figure 4-33. The serial data is applied to the serial input of the shift register. Control circuitry detects the start bit 1-to-0 transition and gates a clock which shifts the data into the register. A counter keeps track of the number of shift pulses and for each eight shift pulses generates a data available (DAV) pulse to latch the data in the 8-bit D flip-flops.

PARALLEL-IN, SERIAL-OUT SHIFT REGISTERS

Parallel-in, serial-out (PISO) shift registers, as the name implies, allow you to parallel-load a number into the register and then shift it out the end of the register or serial output. An obvious use of this type of register is to convert parallel-format data to serial format. A start bit, 7 data bits, 1 parity bit, and 2 stop bits can be loaded into an 11-bit PISO shift register and clocked out to produce a serial data character.

UARTS AND USARTS

You seldom need to build a serial-to-parallel or parallel-to-serial data converter from discrete shift registers because MOS LSI devices such as the General Instruments AY-5-1013 or the Signetics 2536 contain in one 40-pin package all circuitry necessary to convert serial data to parallel and parallel to serial. These LSI devices are called *universal asynchronous receiver-transmitters*, or UARTs. Figure 4-34 shows the pin diagram for the AY-5-1013. This device will send or receive data at any rate up to 30,000 baud or bits/s. It requires an input clock of 16 times the desired baud rate. Using a clock frequency higher than the baud rate allows sampling at the center of each bit time to avoid ringing problems. You can select 5 to 8 data bits per character, a parity bit or no-parity bit, odd or even parity, and 1 or 2 stop bits. Output pulses on pins 13 to 15 indicate whether the incoming serial data has a parity error, a *framing error,* or overruns. A framing error indicates the incoming data has a different number of bits than the UART is programmed to accept. The parallel data outputs can be three-stated by making pin 4 high, and so these outputs can be connected directly to a bidirectional data bus and enabled when needed. The send input on pin 23 tells the UART to send the next serial character. The output from pin 19 tells a receiving system that the data on the parallel outputs is valid. The input to pin 18 is a handshake back to the UART from

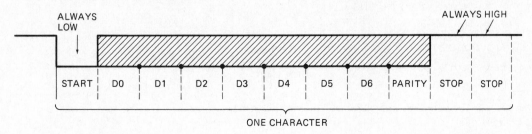

FIGURE 4-32 Serial data format for a teletype.

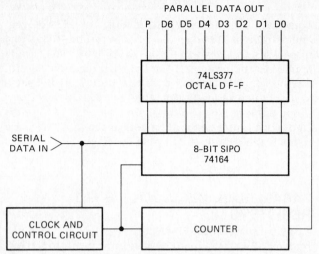

FIGURE 4-33 Serial-to-parallel data converter using SIPO shift register.

the receiving system which says it got the data and that the UART is clear to send the next data character.

A device similar to the UART is the *universal synchronous or asynchronous receiver-transmitter*, or (USART). In synchronous communication, characters are sent at a constant rate. If no data characters are being sent, then the USART sends "null" characters to maintain the constant character rate. UARTs and USARTs are discussed further in Chap. 11.

PARALLEL-IN, PARALLEL-OUT SHIFT REGISTERS

As discussed earlier, a parallel-in shift register allows a binary number, placed on the parallel input pins, to be loaded into the internal registers. With some parallel-load shift registers, this parallel input data appears immediately on the parallel outputs, but with others a clock pulse is required to move the parallel-loaded data to the outputs.

Figure 4-35a shows the schematic for a 74194 universal shift register. This register can be used as any of the four types discussed. It will parallel-load, shift left, shift right, or do nothing, as determined by the programming on the S0 and S1 inputs. For a refresher on tracing gate circuits, you can determine which input states on S0 and S1 produce each operation. Compare your results with the truth table in Figure 4-35b.

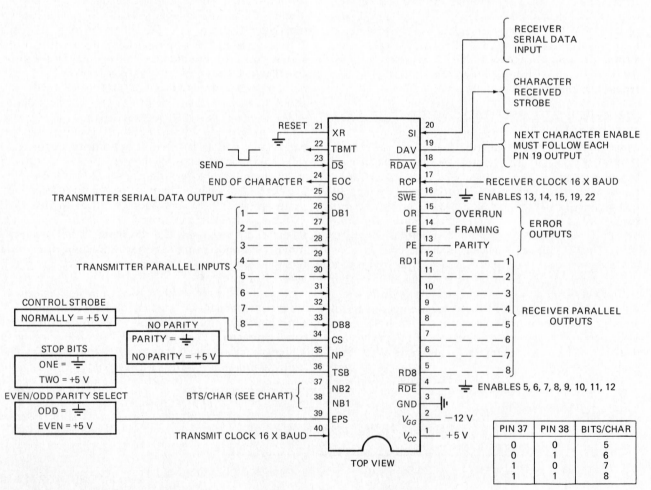

FIGURE 4-34 Pin diagram of AY-5-1013 universal asynchronous receiver-transmitter (UART). *(General Instruments.)*

PIN 37	PIN 38	BITS/CHAR
0	0	5
0	1	6
1	0	7
1	1	8

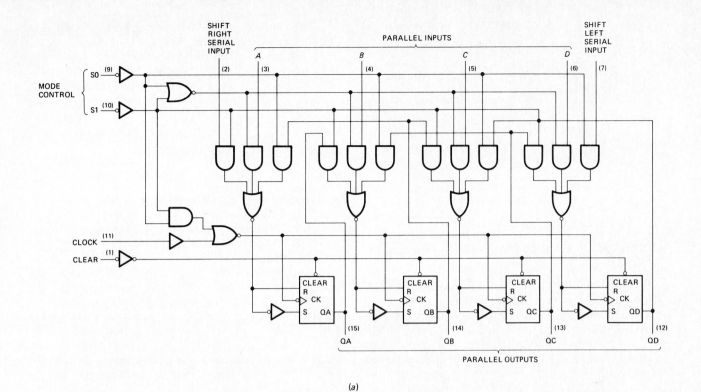

(a)

INPUTS										OUTPUTS			
	MODE			SERIAL		PARALLEL							
CLEAR	S1	S0	CLOCK	LEFT	RIGHT	A	B	C	D	QA	QB	QC	QD
L	X	X	X	X	X	X	X	X	X	L	L	L	L
H	X	X	L	X	X	X	X	X	X	QA0	QB0	QC0	QD0
H	H	H	↑	X	X	a	b	c	d	a	b	c	d
H	L	H	↑	X	H	X	X	X	X	H	QAn	QBn	QCn
H	L	H	↑	X	L	X	X	X	X	L	QAn	QBn	QCn
H	H	L	↑	H	X	X	X	X	X	QBn	QCn	QDn	H
H	H	L	↑	L	X	X	X	X	X	QBn	QCn	QDn	L
H	L	L	X	X	X	X	X	X	X	QA0	QB0	QC0	QD0

H = high level (steady state)
L = low level (steady state)
X = (irrelevant, any input, including transitions)
↑ = transition from low to high level
a, b, c, d = the level of steady-state input at inputs A, B, C, or D, respectively
QA0, QB0, QC0, QD0 = the level of QA, QB, QC, or QD, respectively, before the indicated steady-state input conditions were established
QAn, QBn, QCn, QDn = the level of QA, QB, QC, QD, respectively, before the most recent transition of the clock
Mode controls should be changed only when the clock is high

(b)

FIGURE 4-35 74194 parallel-in, parallel-out shift register. *(a)* Logic diagram. *(b)* Truth table.

Figure 4-36 shows the dependency notation logic diagram for a 74194 shift register. This symbol illustrates mode dependency, as indicated by the letter M next to the S0 and S1 inputs. The mode for this device is selected by the binary number on these S0 and S1 inputs. In mode 3 (S1 = 1 and S0 = 1) the Q outputs are loaded

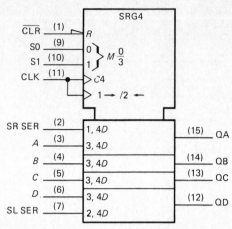

FIGURE 4-36   74194 dependency notation logic diagram.

with the data on the A, B, C, and D inputs. This is indicated by a 3 next to these inputs. In mode 2 (S1 = 1 and S0 = 0) the shift-left serial input is active. This is indicated by a 2 next to this input. In mode 1 the shift-right serial input is active. This is indicated by a 1 next to this input. In mode 0 (S1 = 0, S0 = 0) none of the inputs has any effect on the output, so a 0 is not put next to any input. The symbols next to the lower clock input CLK indicate that in mode 1 the device shifts right on the rising edge of the clock and shifts left in mode 2 on the rising edge of the clock. The control dependency indicated by C4 next to the upper CLK input indicates that the inputs labeled with a 4 will affect the output on the rising edge of the clock signal. The R input indicates that if the CLR signal is asserted, all the outputs will be reset to lows.

USE OF PIPO SHIFT REGISTERS
The most common use of PIPO shift registers is probably for mathematical operations on binary numbers. In Chap. 7 we show how two binary numbers can be multiplied with a series of add and shift-right operations. Binary division can be accomplished with a series of subtract and shift-left operations. For now you can demonstrate for yourself that simply shifting a binary number right one digit position divides it by 2 and shifting a binary number left one digit position multiplies the number by 2. For example, binary 111 (decimal 7) shifted left one digit position becomes 1110, or 14_{10}.

MOS SHIFT REGISTERS
To put a long shift register of many stages into a single IC package, each stage must use minimum chip area or "real estate," and must dissipate minimum power. For lengths over about 16 stages, this usually means that some form of MOS circuitry must be used. A very efficient space and power form is dynamic MOS circuitry.

DYNAMIC MOS   To introduce yourself to dynamic MOS, first do a "memory refresh cycle" back to simple PMOS circuits discussed in Chap. 2. Figure 4-37 shows

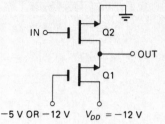

FIGURE 4-37 PMOS inverter. Note negative supply.

a PMOS inverter. Note the negative supply voltage. The transistors in the circuit are E-only MOSFETs, but they are drawn as shown for simplicity. $Q1$ is biased on and serves as an active load resistor for $Q2$. If the input is at a voltage near ground, then $Q2$ is off and the output is pulled to a voltage near V_{DD} by $Q1$. If the input is at a voltage near V_{DD}, then $Q2$ is on. Since the on resistance of $Q2$ is made less than that of $Q1$, the output will be pulled to a voltage near ground which, in the negative logic usually used with PMOS, is a *low*. This circuit works well as an inverter, but it has the problem that there is a continuous current path to ground when $Q2$ is on. This means the inverter is always dissipating power in this state.

A way to reduce the power consumed is shown by the dynamic MOS inverter in Figure 4-38. In this circuit the output state is actually stored on the gate-to-source capacitance of the MOSFET input of the next stage. When the refresh clock is at a voltage level near V_{DD}, both $Q1$ and $Q3$ are on. Since $Q3$ simply acts as a closed switch, the input capacitance C_{GS} of the next stage charges to the output voltage at the junction of $Q1$ and $Q2$. When the refresh clock goes back to ground, $Q1$ and $Q3$ turn off. The output level is held on C_{GS}, and since $Q1$ is now off, it dissipates no power. If the input to the inverter is at a negative logic low, near ground, then $Q2$ is off, and the refresh clock need only keep C_{GS} at the output charged to its high state ($-V_{DD}$), which requires negligible current or power dissipation. If the input is at a logic high, near V_{DD}, then $Q2$ is biased to turn on. But as you can see from the refresh clock waveform, the active load $Q1$ will turn on and conduct current only about 50 percent of the time. This will be adequate to dis-

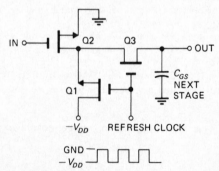

FIGURE 4-38 Dynamic MOS inverter with single-phase clock.

charge C_{GS} and hold the output at the logic low, while dissipating power for only 50 percent of the time. This is a big improvement over the *static* PMOS inverter of Figure 4-37. A clock generator is needed, but this is not a serious drawback since one clock generator can service many gates. This structure is referred to as *dynamic* because logic states must be continuously refreshed by clocking at a rate greater than about 1 kHz. A refresh clock less than about 1 kHz allows time for the charge to leak off the C_{GS}, and the logic states all become invalid.

Another dynamic inverter scheme which saves even more power when used in shift registers is shown in Figure 4-39a. It requires the two-phase clocks shown in the waveforms of Figure 4-39b. Again the output logic level is stored on the input C_{GS} of the following state. A refresh cycle starts when the $\phi 1$ precharge clock turns on $Q1$ and charges the capacitance to $-V_{DD}$. After $Q1$ turns off, the $\phi 2$ refresh clock turns on $Q2$. If the input is high $(-V_{DD})$, then $Q3$ is on and the capacitance will be discharged to ground through $Q2$ and $Q3$ to give a negative logic low output. If the input is low (0 V), then $Q3$ is off. And so when $Q2$ turns on, there will be no conducting path to ground, and the capacitance will remain charged to $-V_{DD}$, which is a negative logic high. This circuit dissipates less power because there is never a direct path from $-V_{DD}$ to ground. A disadvantage is the need for two-phase clocks to continuously refresh the logic state held on C_{IN}.

Two dynamic inverters of either the type in Figure 4-38 or the type in Figure 4-39a can be combined to form a dynamic shift register cell. Figure 4-40a shows the circuit, and Figure 4-40b shows the timing waveforms for a dynamic shift register cell using two single-

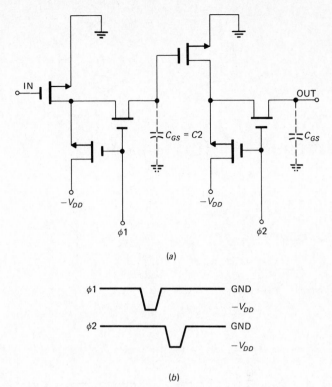

FIGURE 4-40 Dynamic MOS shift register cell. (a) Schematic. (b) Clock waveforms.

phase inverters. The $\phi 1$ clock transfers input data to $C2$, and then the $\phi 2$ clock transfers the data to the output. If only a single-phase clock were tied to all the clock inputs, the input data would race down the entire chain of inverters, rather than step from cell to cell with each clock pulse as it should in a shift register. The two-phase clocks prevent the race condition by enabling only every other inverter at a time.

A typical PMOS dynamic shift register is the National Semiconductor MM4027. It operates on +5- and −12-V supplies and contains 2048 stages or cells. Two-phase −12- to +5-V clocks are required. The minimum frequency of operation varies from 200 Hz at 0°C to 20 kHz at 100°C. Total power dissipation for these 2048 stages and associated circuitry is less than 250 mW when operation is at 1 MHz.

STATIC MOS SHIFT REGISTERS The dynamic shift registers of the previous section have the advantages of low power and many stages per IC package, but they have the disadvantage of a minimum operating frequency. For low-frequency applications, a static or dc-stable shift register must be used. The circuit for a static MOS shift register cell shown in Figure 4-41a uses two static inverters, $Q2$ and $Q4$. On-the-chip clock generators generate the $\phi 1$, $\phi 2$, and $\phi 3$ clocks from a single-phase input clock, as shown in Figure 4-41b. Clock $\phi 1$ shifts data to the output of the first stage, and $\phi 2$ shifts it to the output of the second stage. Between

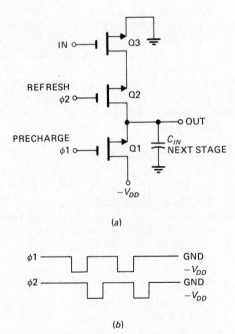

FIGURE 4-39 Dynamic MOS inverter with two-phase clocks. (a) Schematic. (b) Clock waveforms.

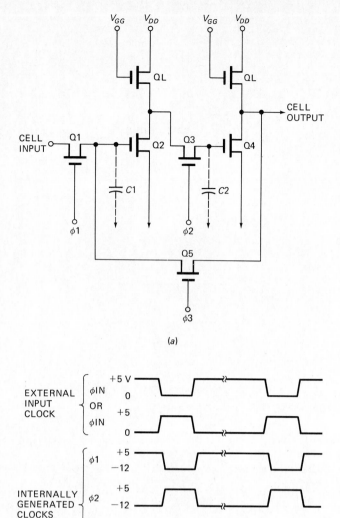

FIGURE 4-41 Static MOS shift register cell. *(a)* Schematic. *(b)* Clock waveforms.

clock pulses $\phi2$ is low and $Q3$ is on, so the output of the first stage is directly coupled to the input of the second stage. However, since $Q1$ is off, $C1$ can discharge and possibly lose the stored logic level if more than a few microseconds elapse between clock pulses. This loss is prevented by $Q5$. If more than 5 μs elapses between clock pulses, an internally generated $\phi3$ turns on $Q5$. Then $Q5$ supplies positive feedback from the output to restore the charge on $C1$. At low frequencies this circuit acts as a stable latch made from cross-coupled inverters. The static circuitry requires more power and uses more chip real estate, and so fewer stages can be put into an IC package. The advantage is that it can operate at frequencies all the way down to dc. A typical MOS static shift register is the 1024-bit Signetics 2533.

BUCKET BRIGADE SHIFT REGISTERS

Another type of serial-in, serial-out MOS shift register that can be used to delay analog as well as digital signals is the *bucket brigade.* It gets its name from its similarity to an old fire-fighting technique. Figure 4-42 shows the circuit and timing diagram for a PMOS bucket brigade. The structure is just a chain of MOSFETs connected drain to source. The substrates of the P-channel E-type MOSFETs are all connected internally to ground. When $\phi1$ goes to V_{DD}, the input FET, $Q1$, samples the input voltage and stores this voltage on C_{IN}. The unlabeled MOSFETs, with their gates tied to V_{BB}, supply biasing. When $\phi1$ goes back to ground, the value of this voltage is held on C_{IN}. When $\phi2$ goes to $-V_{DD}$, $Q2$ is turned on and about 99.9 percent of the charge on C_{IN} is pulled into the gate-to-drain capacitance of $Q2$. The next $\phi1$ negative clock pulse will shift this bucket of charge to the gate-to-drain capacitance of $Q3$ and take a new sample of the input into C_{IN}. Each clock pulse moves the voltage sample or bucket of charge one stage through the brigade. A source follower at the end of the chain gives a low-impedance output. Since the data is stored on the small gate-to-drain capacitance, it will leak off and be lost unless the circuit is clocked at above some minimum frequency.

The voltage that can be shifted through this circuit is not limited to a high or low as it is with digital shift registers. Therefore, analog signals such as audio sine waves can be sampled and delayed by shifting them through a bucket brigade device. The amount of delay is determined by the clock frequency and the number of stages. The output waveform is made up of a number of point values much like a chopped oscilloscope display. A low-pass filter "connects the dots" to give smooth output waves. If you feed part of the output signal back to the input, you will produce an echo, or reverb, effect. A 1024-element bucket brigade such as the Reticon Corporation SAD-1024 can be used to replace the A/D converter, shift registers, and D/A converter in the guitar amplifier reverb unit of Figure 4-28. The SAD-1024 has a minimum clock frequency of 1.5 kHz and a maximum clock frequency of 1 MHz. For audio delay applications, the device gives least distortion with a signal less than 75 mV peak to peak.

CHARGE-COUPLED DEVICES

Charge-coupled devices, or CCDs, are also built as a serial-in, serial-out MOS shift register, but their structure is much simpler and more compact than that of bucket brigades. Figure 4-43 shows the structure for a CCD shift register section. As you can see, the structure consists simply of a P substrate, an insulating oxide layer, and isolated gates. Adjacent shift register sections are built on the same P substrate to form a continuous structure. Data is stored in a CCD as charge. A potential well or charge storage region can be created in the P substrate. At time A in Figure 4-44, a positive voltage on the $\phi2$ gate holds an injected electron charge captive in the region under the $\phi2$ gate. This packet of charge can be shifted to the next cell by the sequence of clock pulses shown in Figure

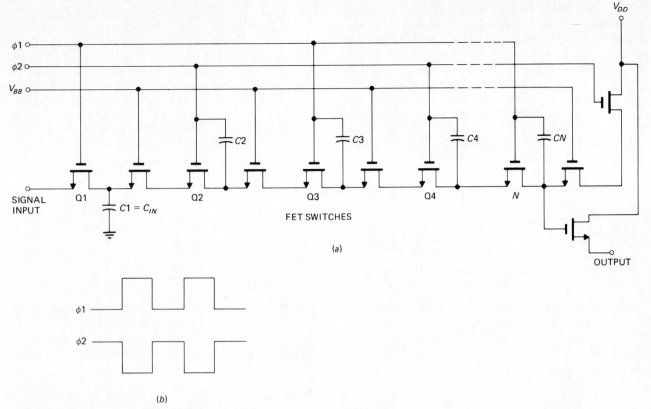

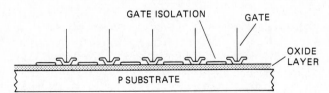

FIGURE 4-42 MOS bucket brigade. *(a)* Schematic. *(b)* Clock waveforms.

4-44. First the $\phi4$ clock creates an empty well under the $\phi4$ gate by making it positive during time B. Then a connecting link between this well and the filled well under the $\phi2$ gate is created by the $\phi3$ clock making the $\phi3$ gate positive during time C. Next the $\phi2$ clock goes low during time D. All the charge is pulled into the $\phi3$ and $\phi4$ gate regions. At the start of time E, the $\phi3$ clock goes low. This forces all the charge to move into the region under the $\phi4$ gate and completes the shift of data from cell 1 to cell 2. This process may seem quite complicated at first, but it is really quite simple and is required to prevent a race condition that would result if fewer steps were used. You can use the rest of the waveforms in Figure 4-44 to show how the packet of charge is moved from the region under $\phi4$ in cell 2 to the region under $\phi2$ in cell 3.

The CCD is dynamic and must be shifted at some minimum rate. Otherwise, the empty potential wells gradually fill up with thermally generated electrons from the P substrate. This is undesirable because it modifies the stored data. An internal amplifier at the input of the CCD injects charge according to the input data. An amplifier on the output of the CCD buffers the output.

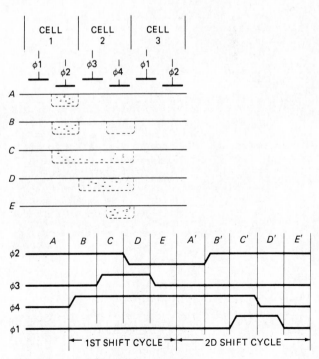

FIGURE 4-44 CCD shifting sequence.

FIGURE 4-43 MOS charge-coupled device structure.

CCDs can be used for digital or analog delay, and as serial data memories, which are discussed in Chap. 5. Another exciting application of CCDs is as the light-sensitive image sensor in television cameras. To make an image sensor, several hundred charge-coupled shift registers are built in parallel on the same chip. A photodiode is doped in under every other gate. When all the gates with photodiodes under them are made positive, potential wells are created. A camera lens system focuses an image on the CCD array. Light shining on the photodiode puts a charge proportional to the intensity of the light in each well. These charges can be shifted out one line at a time to give the dot-by-dot values for a TV picture. Improved resolution is gained by alternating nonlighted charge-coupled shift registers with the lighted registers on the chip. The image data is shifted in parallel from the lighted register to the dark register and then shifted out serially through the dark register. CCD image sensors take up much less space and are much more rugged than traditional vacuum-tube image sensors. Resolution is rapidly approaching that of the vacuum-tube types.

REVIEW QUESTIONS AND PROBLEMS

1. Why are timing diagrams important when you are working with digital circuits?

2. Draw the output waveform for each of the circuits in Figure 4-45.

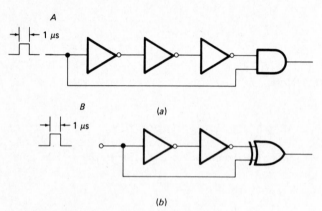

FIGURE 4-45

3. What is the Q output state of a latch when it is "set"?

4. What are two major problems of the simple cross-coupled NAND gate latch of Figure 4-3a?

5. Construct a truth table for the circuit in Figure 4-46. Use the truth table to identify which is the SET input and which is the RESET input.

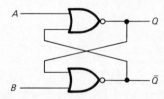

FIGURE 4-46

6. Show the schematic symbol for a D latch. Draw the truth table in timing-diagram form for a D latch that has an active high enable. When is the data actually latched on the output?

7. What is the main difference between a D latch and a D flip-flop?

8. Show the Q output waveforms that would be produced on (a) a D latch and (b) a D flip-flop (positive-edge-clocked) by the clock and D inputs shown in Figure 4-47. Assume Q starts low.

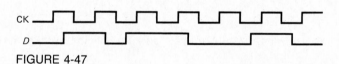

FIGURE 4-47

9. Define setup time and hold time for a flip-flop.

10. Show the output truth table for a JK flip-flop.

11. Given the J, K, and clock input waveforms shown in Figure 4-48, show the Q output waveforms these will produce on a positive-edge-triggered JK flip-flop, a negative-edge-triggered JK flip-flop, and a level-clocked master-slave JK flip-flop. Assume Q starts low.

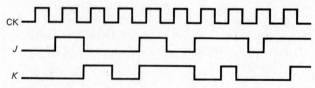

FIGURE 4-48

12. What are two common problems of flip-flop circuits?

13. a. Show how a JK flip-flop is converted to a T flip-flop.
    b. Show how a D flip-flop is converted to a T flip-flop.

14. Show clock and output waveforms for a T flip-flop with a 1-kHz clock.

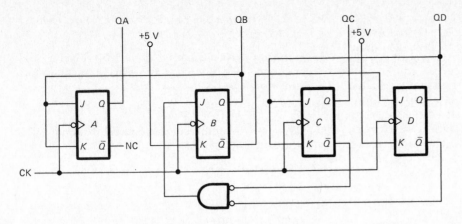

FIGURE 4-49 Problem 22.

15. Draw the clock and output waveforms for a 4-bit binary counter with a 1-kHz input clock. Label each count with its binary value. Calculate the frequency of each output. (Assume flip-flops change state on the negative edge of input clock.)

16. What is the final output frequency of an eight-stage binary divider with an input frequency of 10,240 Hz?

17. What are the maximum counts for an 8-bit, a 12-bit, and a 16-bit binary counter?

18. Show the circuits for a 7493 converted to a modulo-10 (or decade) counter, a modulo-12 counter, and a modulo-13 counter. Show the resulting QD output waveform for each.

19. Show the logic gate circuit you would use to decode the following states from a 4-bit binary counter:
    a. State 4
    b. State 9
    c. State 12

20. Show a block diagram for a simple digital clock.

21. What are two major problems of ripple or asynchronous counters? How do synchronous counters minimize these problems?

22. Determine the Q output count sequence for the circuit in Figure 4-49, using JK flip-flops and digital feedback. Assume that all Qs are low initially.

23. Show the circuit for a 74193 converted to a modulo-11 counter. Show the timing diagram for this modulo-11 counter, including the BORROW OUT pulse.

24. What is meant by the term *asynchronous clear?*

25. Why should the counter used to sequentially address a ROM be synchronous?

26. Consider the frequency counter in Figure 4-23.
    a. What is the purpose of the crystal oscillator and the chain of decade dividers?

b. What is the purpose of the JK flip-flop after the divider chain?
c. What is the highest count that can be displayed?
d. Why are D flip-flops put on each output of the main counter chain?
e. How are the latch pulses for the flip-flops and the main counter reset pulses produced?
f. Why does a frequency counter need a 10-s count window to accurately measure low frequencies?

27. What are the advantages of a multiplexed display such as that in the frequency counter of Figure 4-25? For the multiplexed display frequency counter of Figure 4-25, describe how the data for the most significant digit gets displayed on that digit.

28. In a multiplexed display frequency counter built by a student, the digits read out of order. What is a likely cause?

29. A multiplexed display frequency counter similar to Figure 4-25 worked previously. Describe a possible cause of each of the following symptoms.
    a. The displayed count never changes when a new frequency is input.
    b. One digit of the display is brightly lighted and always shows an 8. The other digits are dimly lighted and show the correct numbers.
    c. The least significant digits are correct for an input frequency, but the three most significant digits are incorrect.

30. What is a prescaler?

31. What is a register?

32. Show the circuit for a 4-bit, serial-out shift register made with D flip-flops. Why can D latches not be used? *Hint:* Remember the difference between a D flip-flop and a D latch.

33. A 200-stage SISO shift register is clocked at 10 kHz. How long will data be delayed in passing through this register?

34. Show the circuit for a 4-bit ring counter made with D flip-flops. Show the timing diagram produced if the counter is initially loaded with 1000. Also show the outputs for an initial loading of 1100 and 0111.

35. The schematic for an interesting shift register counter made with JK flip-flops is shown in Figure 4-50. It is sometimes called a *twisted ring counter*, or *Johnson counter*. Show the output waveforms, starting with 0 on all Q outputs. The advantage of this counter is that all states decode with no glitches since only one flip-flop changes at a time.

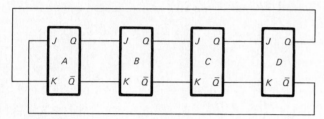

FIGURE 4-50

36. Draw a timing diagram for the keyboard encoder shown in Figure 4-31.

37. Show the serial format in which 7-bit ASCII code is sent by a teletype such as the ASR-33. Include start and stop bits.

38. Define UART, USART, and baud rate.

39. What are the differences between dynamic MOS and static MOS?

40. Describe the operation of a MOS bucket brigade shift register.

41. Explain the operation of a CCD shift register, and compare it with a bucket brigade. What are some uses of CCDs?

REFERENCES/BIBLIOGRAPHY

Texas Instruments Staff: *Bipolar Microcomputer Components Data Book*, 3d ed., Dallas, 1981.

Texas Instruments Staff: *TTL Data Book for Design Engineers*, 2d ed., Dallas, 1981.

Blakeslee, Thomas R.: *Digital Design with Standard MSI & LSI*, Interscience-Wiley, New York, 1975.

MOS Databook, National Semiconductor Corporation, Santa Clara, CA, 1980.

CMOS Databook, National Semiconductor Corporation, Santa Clara, CA, 1978.

Logic Databook, National Semiconductor Corporation, Santa Clara, CA, 1981.

Garen, Eric R.: "Charge-Transfer Devices—Part 1: The Technologies," *Computer Design*, "Around the IC Loop," vol. 16, no. 11, November 1977, pp. 146–152.

"Principles of Analog Discrete-Time Signal Processing Devices," *Application Note*, no. 102, Reticon Corporation, Sunnyvale, CA, 1975.

Intersil ICM 7226 A/B 10 MHz Universal Counter System, Data Sheet, Intersil, Inc., Cupertino, CA, October 1978.

Intersil Timer/Counter Applications Brochure, Intersil, Inc., Cupertino, CA.

General Instrument Corporation-Microelectronics: *AY:5-1013 UAR/T Universal Asynchronous Receiver/Transmitter*, Data Sheet, General Instrument Advance Nitride Technology (GIANT), March 1974.

CHAPTER 5

READ-WRITE MEMORIES

Computers, microprocessors, and many other instruments require memory for storing digital data and binary-coded instructions. Read-only memories, as discussed in Chap. 3, are one type used. ROMs have the advantage that they are nonvolatile. This means that data remains stored in the ROM matrix when power is turned off. Therefore, valid data are still available when the power is turned on again. However, ROMs have the disadvantage that, like a phonograph record, once they are programmed, it is difficult, if not impossible, to alter the stored data. EPROMs and EEPROMs can be erased and rewritten, but not easily and not at normal circuit operating speeds.

In this chapter we discuss memories that can be easily written into, read from, and immediately rewritten with new data as many times as necessary as part of the normal circuit operation.

Upon completion of this chapter, you should be able to:

1. Define register, accumulator, register file, RAM, MBM, and Josephson junction memory.

2. Describe matrix organization and addressing of a RAM.

3. Explain access time and, given a data sheet for a RAM, find the important timing parameters for read and write operations.

4. Describe differences between static and dynamic MOS RAM.

5. Explain how 16 address bits are loaded into a chip on only eight pins.

6. List the advantages and the disadvantages of several common methods of digital data storage.

7. Describe data storage on magnetic tapes and disks.

8. Draw a block diagram of a simple logic analyzer.

9. Describe the four common logic-analyzer display formats and their use.

RANDOM ACCESS MEMORY

A flip-flop is a 1-bit read-write memory. It can be written into by applying a data bit to the input and applying a clock pulse. The data bit can be read from the Q output. Stored data can be changed by simply applying the new data to the input and reclocking. This type of memory is *volatile* because data is lost if power is turned off. The state of the output when power returns is unpredictable.

A group of flip-flops called a *register* can be used to store a binary word of parallel data. To store more than one word, you can use several registers. The TTL 74170 4 × 4 register file contains 16 flip-flops arranged to store 4 words of 4 bits each. The register desired for reading or writing a word is selected by two address inputs.

TTL RAMS

Standard TTL D flip-flops use too much chip real estate and dissipate too much power (heat) to permit more than a few to be built in a single IC package. Therefore, for storage of more bits an *X-Y* matrix of simplified flip-flops is used such as the bipolar version shown in Figure 5-1. The resulting matrix, or array, is called a *random access memory*, or RAM. The name comes from the fact that any location (cell) in the matrix can be accessed (written into or read from) without affecting any other location.

STRUCTURE The RAM cell in Figure 5-1 operates as follows. The X select line is tied to the X select inputs of all the cells in a row of the matrix. The Y select line is tied to the Y select inputs of all the cells in a column of the matrix. Since Q1 and Q2 are cross-coupled inverters, one is always off while the other is on. The X and Y select lines are low normally, causing the current through the conducting transistor to flow through the X and Y select lines to ground. A cell is selected by raising the X and Y select lines that address the cell to the high state. This disables the X and Y emitters and causes the current from the conducting transistor to flow through one of the sense amplifier emitters. A sense amplifier detects whether the current is flowing through the Q1 emitter or the Q2 emitter. For the circuit shown, a 1 is stored in the cell if Q1 is conducting and Q2 is off. The state of the cell is changed to a 0 by pulsing a high on the Q1 sense/write emitter. This turns off Q1. With Q1 off, R1 now supplies base current to turn on Q2. As long as Q2 is on, its collector is low and Q1 is held off. A 1 can be rewritten by pulsing the Q2 sense/write emitter high to turn off Q2.

Figure 5-2a shows the block diagram of a Motorola MCM93425 TTL RAM. The device has a 32 × 32 matrix that can store 1024 bits. The binary address on lines A0 through A4 selects one of the 32 rows in the matrix. The binary address on lines A5 through A9 selects one of the 32 columns. If $\overline{CS}$ is low and $\overline{WE}$ is high, the data stored at the intersection of the selected row and the selected column is sent to D_{OUT}. If $\overline{CS}$ is low and $\overline{WE}$ is low, the data present on D_{IN} is written into (stored in) the addressed cell. During a write operation, the output buffer will be disabled (three-state). The output buffer also will be three-state if the device is deselected by making $\overline{CS}$ high.

Figure 5-2b shows the dependency notation symbol for the MCM93425 RAM. Address dependency is indicated by the letter A followed by 0 over 1023. Other inputs and outputs labeled with an A are used or affected according to the applied address. The letter G stands for

AND dependency. The device is active if $\overline{CS}$ is asserted (L) AND $\overline{WE}$ asserted (H) or $\overline{WE}$ asserted (L). The OR here is implied by the fact that both inputs connected to $\overline{WE}$ are labeled with a 1. If $\overline{WE}$ is asserted high, the three-state output is enabled and data from the addressed cell appears on the output. If $\overline{WE}$ is asserted low, the output is disabled and data on D_{IN} is written to the addressed cell. Use the logic in Figure 5-2a to help you see these relationships.

An important parameter used to compare the speed of memories is T_{AA}, the address access read time. This is simply the time needed for valid data from a memory cell to appear on the outputs after an address is applied to the address inputs. In other words, it is the propagation delay between an applied address and valid output data. For later comparison, the maximum T_{AA} for the MCM93425 is 45 ns.

TTL RAMs have been replaced for the most part by ECL (emitter-coupled logic) or high-speed MOS RAMs which are faster and/or can store more bits. Later we explain the detailed read and write timing parameters for RAMs, using commonly available MOS devices as examples.

ECL RAMS

Emitter-coupled logic has a reputation for using a lot of power. However, by using scaled-down transistors, large ECL RAMs have been produced that do not dissipate excessive power and still have an address access time shorter than that of TTL RAMs. The National Semiconductor DM10470A, for example, is a 4096-bit × 1 device with a maximum address access time of 15 ns. Fairchild has announced a 16K × 1 ECL RAM, the F100480, a larger device with 35-ns maximum address access time.

To store 8- or 16-bit words, several of these devices can be addressed in parallel, as shown with the ROMs in Figure 3-48. The output transistors of these devices are open-emitter, so the outputs of several can be wire-ORed to store more words. A single external pull-down resistor to $-V_{EE}$ or a termination resistor to -2 V is used on each output. Remember that ECL signal transmission lines must be terminated with a resistor if they are more than a few inches long.

STATIC MOS RAMS

STRUCTURE Enhancement-mode MOSFET transistors can be used to make a RAM cell very similar to the TTL cell discussed earlier. Figure 5-3 shows a static MOS RAM cell. Here Q1 and Q2 act simply as load resistors for the cross-coupled inverters Q3 and Q4. Q5 and Q6 are switches that connect the outputs of the inverter latch to the DATA and $\overline{DATA}$ lines. When the row is selected, Q5 and Q6 are turned on. The DATA and $\overline{DATA}$ lines go to the sense amplifier and the write amplifier for each column in an array. Data is written into the cell by selecting it and forcing it to the desired state with the write amplifier.

FIGURE 5-1 TTL RAM cell.

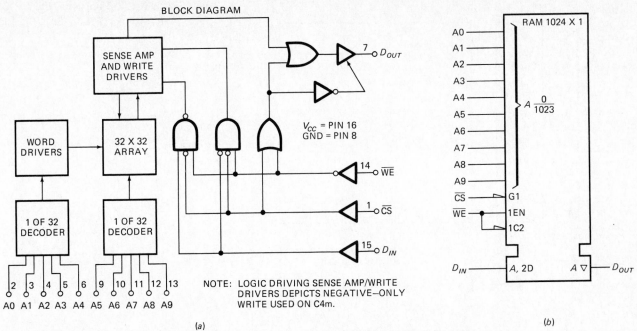

BLOCK DIAGRAM

SENSE AMP
AND WRITE
DRIVERS

WORD
DRIVERS

32 X 32
ARRAY

1 OF 32
DECODER

1 OF 32
DECODER

V_{CC} = PIN 16
GND = PIN 8

7 D_{OUT}

14 $\overline{WE}$

1 $\overline{CS}$

15 D_{IN}

2 3 4 5 6 9 10 11 12 13
A0 A1 A2 A3 A4 A5 A6 A7 A8 A9

NOTE: LOGIC DRIVING SENSE AMP/WRITE
DRIVERS DEPICTS NEGATIVE—ONLY
WRITE USED ON C4m.

(a)

RAM 1024 X 1

A0
A1
A2
A3
A4
A5
A6
A7
A8
A9

$A \dfrac{0}{1023}$

$\overline{CS}$ — G1
$\overline{WE}$ — 1EN
1C2

D_{IN} — A, 2D $A \triangledown$ — D_{OUT}

(b)

FIGURE 5-2 Motorola MCM93425 TTL RAM. (a) Block diagram. (b) Dependency notation symbol.

Figure 5-4 shows a block diagram for the Intel 2167, a 16,384 × 1 static MOS RAM, which requires only a 5-V supply. Note that 14 address lines are needed to select one of the 16,384 words. The device is selected when the chip select ($\overline{CS}$) is low. When $\overline{CS}$ is low and $\overline{WE}$ is low, data on the input will be written into the addressed cell. The output will be in a high-Z state. When $\overline{WE}$ is high, the data from an addressed cell will appear on the output. The input buffer is disabled. Making $\overline{CS}$ high floats both the input and output buffers. It also switches most of the circuitry in the package to a powered-down,

stand-by mode. In stand-by mode, the 2167-55 uses a maximum of only 40 mA as compared with an active current maximum of 125 mA. This drastically cuts the power dissipation when the device is not being selected. The large change in current that occurs when the device is selected or deselected can cause large voltage spikes on V_{CC} unless extra bypassing is used. Intel recommends a 0.1-μF ceramic capacitor between V_{CC} and ground next to every other device, as well as a 22- to 47-μF electrolytic decoupler for every 16 devices.

Also, if many of these high-speed devices are used on a large printed-circuit board, all the address lines, data lines, and control lines must be properly terminated transmission lines.

Many RAMs are decoded to give 1-bit output words. This is done to reduce the number of pins required on the IC package. For storing multiple-bit words, several "by 1" RAMs can be connected in parallel. Eight 2167s, for example, can be connected in parallel to form a 4096 × 8 memory. Using the ×1 RAMs, however, always requires at least eight ICs to give an 8-bit-wide memory. The popularity of 8-bit microprocessors has led some manufacturers to produce RAMs with 8-bit output words. The Mostek MK4802, a 2048 × 8 static MOS RAM, is a good example.

TIMING   RAM data sheets show several important timing parameters. Most manufacturers identify these parameters with subscripts that are the initials of the name for that parameter. You do not need to memorize these values, but you do need to learn the language of "subscriptese" and be able to relate the given times to the waveforms. The reason is that many problems found in RAM circuits (other than obviously bad chips) are re-

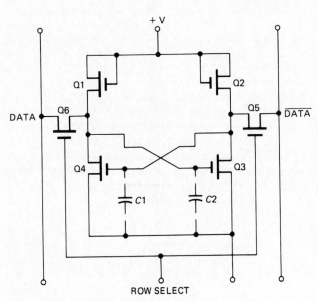

+V

Q1   Q2

DATA Q6   Q5 $\overline{DATA}$

Q4   Q3

$C1$ $C2$

ROW SELECT

FIGURE 5-3 MOS static RAM cell.

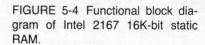

FIGURE 5-4 Functional block diagram of Intel 2167 16K-bit static RAM.

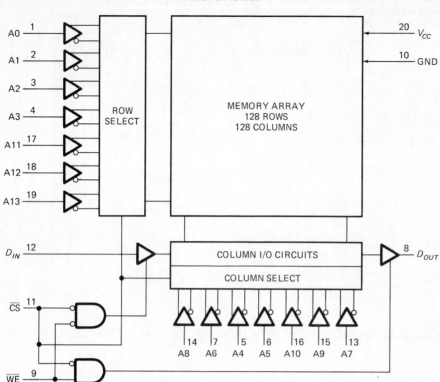

lated to timing. A particular memory IC may work at low frequencies, but its access time may be too long for it to operate consistently in a higher-frequency circuit. A write pulse may be too wide and not allow enough write recovery time before the address lines change. The point is, with an oscilloscope and these timing charts you can usually find the problem.

Figure 5-5 shows the read cycle timing waveforms and values for the Intel 2167. Note that the timing values are different for parts with different suffix numbers. Not all 2167s are created equal. For example, the 2167-55 has a maximum T_{AA} of 55 ns, whereas the 2167-70 and 2167-L70 have a maximum T_{AA} of 70 ns. Accidentally using the wrong suffix part can cause timing problems in a system.

Now look at the read-cycle waveforms and notes. Read cycle 1 assumes that $\overline{CS}$ is low and $\overline{WE}$ is high. The three parameters T_{RC}, T_{OH}, and T_{AA} are measured from the time when all bits on an address are valid on the address inputs. The T_{OH} specification of 5 ns minimum indicates that data from a previously read address will be present on the output for at least 5 ns after a new address is applied. The T_{AA} specification of 55 ns maximum (for the 2167-55) indicates that valid new data will be on the output no later than 55 ns after an address is applied. The cross-hatched region between T_{OH} and T_{AA} indicates that the output may change from the old addressed data to the new addressed data at any point during this time. In other words, the output is unpredictable during this time. The range of uncertainty is caused by differences in propagation delay (skew) through the device and variations from part to part.

Often the region is referred to as the *skew region*, or just *skew*. Read cycle time T_{RC}, is the minimum time for which an address must be held stable. A new address cannot be applied for the next read cycle until after this time has elapsed.

Read cycle 2 waveform shows the timing if $\overline{WE}$ is high and an address is already present or becomes valid at the same time as $\overline{CS}$ is made low. T_{LZ} of 10 ns minimum indicates that the data output will remain three-state for at least 10 ns after $\overline{CS}$ goes low. T_{ACS} of 55 ns minimum (for the 2167-55) indicates that at the latest, valid data will be on the output 55 ns after $\overline{CS}$ goes low. Any time that $\overline{CS}$ is high, the data output will be in a high-impedance state. In the table and waveform diagram T_{HZ} indicates that the output will return to the high-impedance state 0 to 30 ns after $\overline{CS}$ goes high. If $\overline{CS}$ is high, the device is powered down to a stand-by current of 40 mA maximum (for the 2167-55). Within a time T_{PU} after $\overline{CS}$ goes low, the device will be powered up to its normal operating current (125 mA maximum for the 2167-55). Within a time T_{PD} after $\overline{CS}$ goes high, the device will be powered down again. This change in current between selected and deselected modes saves a lot of power in a system, but it does require many bypass capacitors to reduce V_{CC} transients.

Figure 5-6 shows the write cycle timing waveforms and values for the Intel 2167 RAM. Write-cycle waveforms can be best analyzed by thinking about the sequence of events that must occur to write data into an addressed cell or group of cells. Refer to write cycle 1 and the times for the 2167-55 in Figure 5-6 as we explain this sequence. Note that for this waveform most times

$T_A = 0°C$ to $70°C$, $V_{CC} = +5V \pm 10\%$, UNLESS OTHERWISE NOTED.

READ CYCLE

SYMBOL	PARAMETER	2167-55		2167-70, 2167L-70		2167-10		UNIT
		MIN.	MAX.	MIN.	MAX.	MIN.	MAX.	
t_{RC} [1]	READ CYCLE TIME	55		70		100		ns
t_{AA}	ADDRESS ACCESS TIME		55		70		100	ns
t_{ACS}	CHIP SELECT ACCESS TIME		55		70		100	ns
t_{OH}	OUTPUT HOLD FROM ACCESS CHANGE	5		5		5		ns
t_{LZ} [2,3]	CHIP SELECTION TO OUTPUT IN LOW Z	10		10		10		ns
t_{HZ} [2,3]	CHIP DESELECTION TO OUTPUT IN HIGH Z	0	30	0	40	0	40	ns
t_{PU}	CHIP SELECTION TO POWER-UP TIME	40		50		55		ns
t_{PD}	CHIP DESELECTION TO POWER-DOWN TIME		55		70		80	ns

WAVEFORMS

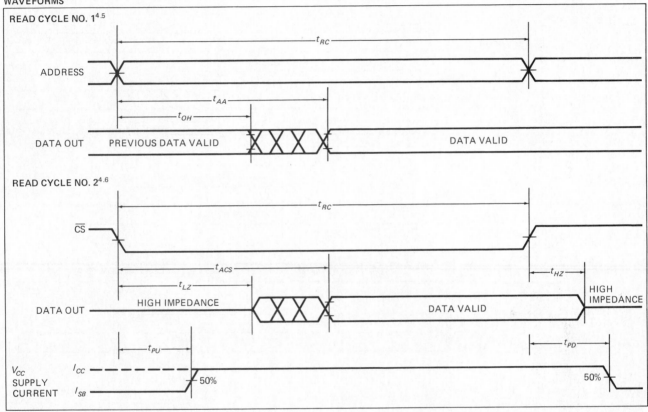

NOTES:
1. ALL READ CYCLE TIMINGS ARE REFERENCED FROM THE LAST VALID ADDRESS TO THE FIRST TRANSITIONING ADDRESS.
2. AT ANY GIVEN TEMPERATURE AND VOLTAGE CONDITION, t_{HZ} max IS LESS THAN t_{LZ} min BOTH FOR A GIVEN DEVICE AND FROM DEVICE TO DEVICE.
3. TRANSITION IS MEASURED ± 500 mV FROM STEADY STATE VOLTAGE WITH SPECIFIED LOADING IN FIGURE 2. THIS PARAMETER IS SAMPLED AND NOT 100% TESTED.
4. $\overline{WE}$ IS HIGH FOR READ CYCLES.
5. DEVICE IS CONTINUOUSLY SELECTED. $\overline{CS} = V_{IL}$.
6. ADDRESSES VALID PRIOR TO OR COINCIDENT WITH $\overline{CS}$ TRANSITION LOW.

FIGURE 5-5 Read cycle timing waveforms and timing values for Intel 2167 static RAM.

are measured from the rising edge of the $\overline{WE}$ pulse. First, an address is applied to the device. This must be done a time T_{AW} of at least 55 ns before $\overline{WE}$ goes high and held for a time T_{WR} of at least 0 ns after $\overline{WE}$ goes high. Second, $\overline{CS}$ is made low to select the device. This must be done at least 55 ns (T_{CW}) before $\overline{WE}$ goes high. The slanted bars on the $\overline{CS}$ waveform indicate that $\overline{CS}$ can be made low before the address is applied and that $\overline{CS}$ can be made high anytime after $\overline{WE}$ goes high. Third, data is applied to the data input of the device. According to the waveforms, this data must be valid at least 25 ns (T_{DW}) before $\overline{WE}$ goes high, and the data must be held valid at least 0 ns (T_{DH}) after $\overline{WE}$ goes high. The write pulse $\overline{WE}$ has a minimum width T_{WP} of

SYMBOL	PARAMETER	2167-55 MIN.	2167-55 MAX.	2167-70, 2167L-70 MIN.	2167-70, 2167L-70 MAX.	2167-10 MIN.	2167-10 MAX.	UNIT
t_{WC}[2]	WRITE CYCLE TIME	55		70		100		ns
t_{CW}	CHIP SELECTION TO END OF WRITE	55		70		90		ns
t_{AW}	ADDRESS VALID TO END OF WRITE	55		70		95		ns
t_{AS}	ADDRESS SETUP TIME	0		0		5		ns
t_{WP}	WRITE PULSE WIDTH	35		40		45		ns
t_{WR}	WRITE RECOVERY TIME	0		0		5		ns
t_{DW}	DATA VALID TO END OF WRITE	25		30		35		ns
t_{DH}	DATA HOLD TIME	0		0		0		ns
t_{WZ}[3]	WRITE ENABLED TO OUTPUT IN HIGH Z	0	25	0	35	0	35	ns
T_{OW}	OUTPUT ACTIVE FROM END OF WRITE	0		0		0		ns

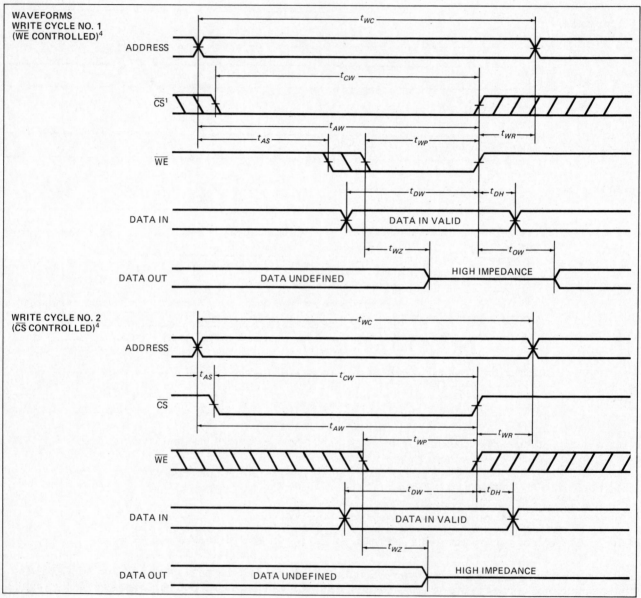

NOTES:
1. IF $\overline{CS}$ GOES HIGH SIMULTANEOUSLY WITH $\overline{WE}$ HIGH, THE OUTPUT REMAINS A HIGH IMPEDANCE STATE.
2. ALL WRITE CYCLE TIMINGS ARE REFERENCED FROM THE LAST VALID ADDRESS TO THE FIRST TRANSITIONING ADDRESS.
3. TRANSITION IS MEASURED ± 500 mV FROM STEADY STATE VOLTAGE WITH SPECIFIED LOADING IN FIGURE 2. THIS PARAMETER IS SAMPLED AND NOT 100% TESTED.
4. $\overline{CS}$ OR $\overline{WE}$ MUST BE HIGH DURING ADDRESS TRANSITIONS.

FIGURE 5-6 Write cycle timing waveforms and timing values for Intel 2167 static RAM.

35 ns. The write pulse can be made low a time T_{AS} after the address is applied. For this device T_{AS} is 0 ns, so $\overline{WE}$ can be made low at the same time as the address is applied. Within a time T_{WZ} of 0 to 25 ns after $\overline{WE}$ goes low, the data output goes to a high-impedance state. If $\overline{CS}$ is still low when $\overline{WE}$ returns high, the output will return to the low-impedance state. The data output will contain the data from the addressed cell.

Write cycle waveform 2 in Figure 5-6 is very similar to waveform 1 except that most times are referenced to the rising edge of the $\overline{CS}$ signal. As indicated by the slanted boxes on the $\overline{WE}$ waveform, the only requirement for $\overline{WE}$ in this mode is that it should go low 35 ns (T_{WP}) (for the 2167-55) before $\overline{CS}$ is made high. For both waveforms the minimum time for a complete write cycle T_{WC} is 55 ns for the 2167-55.

Different manufacturers use various subscripts to identify these timing parameters, but with the help of the waveforms usually you can find the values needed to check out a system.

NONVOLATILE RAMS

By combiniing a static RAM and an electrically erasable PROM in the same chip, Xicor, Inc., created a device that can function as a normal RAM or, in case of a power failure, can save the stored data in the EEPROM. The X2201 contains a 1024-bit RAM and a 1024-bit EEPROM. The entire contents of the RAM can be transferred in parallel to the EEPROM in less than 4 ms by a single signal named $\overline{STORE}$. A signal named $\overline{RECALL}$ can transfer data from the EEPROM back into the RAM. The advantage of these devices is that a battery backup is not required to save working data in the event of a power failure.

CMOS RAMS

In applications where data must be saved during a power failure, CMOS RAMs are often used with a battery backup supply. The static CMOS RAM cell is the same as the static MOS RAM cell in Figure 5-3 except that load transistors $Q1$ and $Q2$ are replaced with P-channel E-only MOSFETs. This converts the structure to cross-coupled CMOS inverters which dissipate power only when switching states, as discussed in Chap. 2. The extremely low current drain of a CMOS RAM when it is not being accessed makes it feasible to include a small nickel-cadmium battery right on the memory board as a backup supply. The Harris HM-6514, for example, is a 1024 × 4 CMOS RAM that requires only a maximum of 50 μA in the stand-by mode. The HM-6514 will retain data with a V_{CC} supply voltage as low as 2.0 V. With V_{CC} between 4.5 and 5.5 V, the device has a maximum address access time of 220 ns.

As of this writing, several manufacturers have announced 2K × 8 CMOS RAMs. Some large static RAM devices use standard NMOS circuits for the storage array to minimize area and CMOS peripheral circuits to minimize stand-by power dissipation.

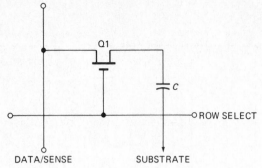

FIGURE 5-7 Single-transistor dynamic RAM storage cell.

DYNAMIC MOS RAMS

STRUCTURE Dynamic MOS RAM cells store data as charge on small capacitors rather than in cross-coupled inverter latches as the static RAM cells do. These dynamic RAM cells, therefore, need *refreshing* every 2 ms or less, as did the dynamic shift register cells discussed in Chap. 4. The advantage of dynamic cells is that they are simpler and smaller, cost less, and need less stand-by power.

Current devices use a single transistor cell such as that shown in Figure 5-7. Here $Q1$ is simply a switch that connects the storage capacitor to the DATA/SENSE line when its row is selected. In a RAM array of these cells, each column has a sense amplifier to read the state of a cell and a write amplifier to write into a cell when it is addressed.

Because of the small area per cell, many dynamic RAM cells can be built on a single chip. The Motorola MCM6664, is a 65,536 × 1 high-speed dynamic RAM. It is packaged in a standard 16-pin DIP and requires only a +5 V supply. Figure 5-8a shows a block diagram and the pin assignment for the MCM6664. As you can see, the device has only eight address inputs. Since 65,536 is 2^{16}, you would expect that 16 address bits would be required to select one of the 65,536 words stored in the device (see the discussion of this idea in Chap. 3). Indeed, 16 address bits are required. However, to reduce the number of pins on the device, the 16 address bits are multiplexed in, 8 at a time, on the eight address inputs A0 through A7.

Using the MCM6664 read cycle timing waveform in Figure 5-8b, you can see how this is done. An 8-bit row address is applied to the device address inputs by an external multiplexer. The row address strobe input $\overline{RAS}$ is asserted low by external circuitry to latch the row address into internal latches in the 6664. After a hold time T_{RAH}, the row address is removed, the write input $\overline{W}$ is made high, and the 8-bit column address is applied to the address inputs. External circuitry then pulses the column address strobe $\overline{CAS}$ low to latch the column addresses into internal latches in the 6664. After a time T_{CAC}, the data from the addressed cell will be present and valid on the Q (data) output. Except for strobing in the row and column addresses, this timing diagram is similar to those for the 2167 MOS static RAM discussed

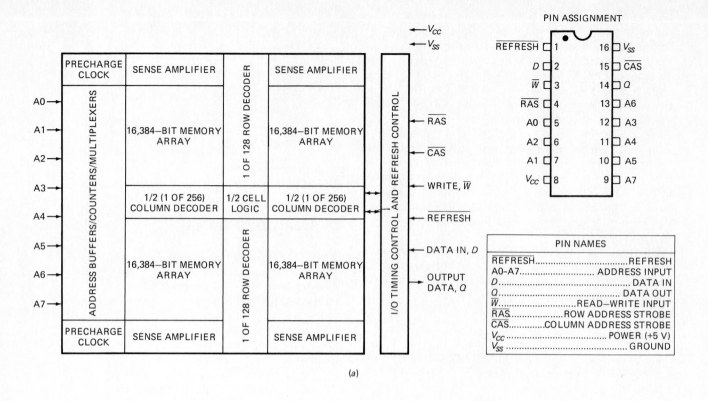

PIN ASSIGNMENT

PIN NAMES	
$\overline{REFRESH}$	REFRESH
A0–A7	ADDRESS INPUT
D	DATA IN
Q	DATA OUT
$\overline{W}$	READ–WRITE INPUT
$\overline{RAS}$	ROW ADDRESS STROBE
$\overline{CAS}$	COLUMN ADDRESS STROBE
V_{CC}	POWER (+5 V)
V_{SS}	GROUND

(a)

READ CYCLE TIMING

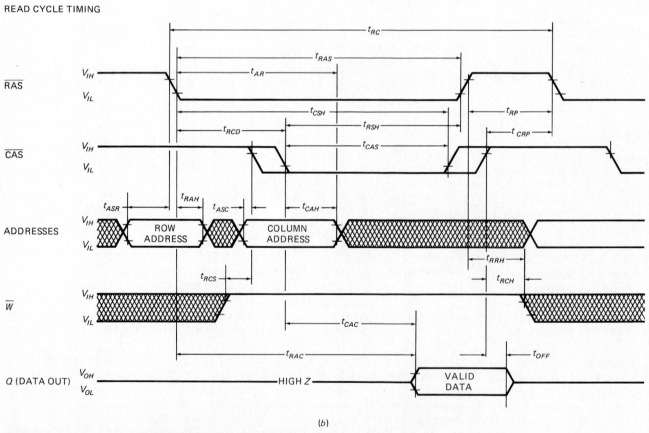

(b)

FIGURE 5-8 Motorola MCM6664 64K-bit dynamic RAM. (a) Block diagram and pin assignment. (b) Read cycle timing waveforms.

AC OPERATING CONDITIONS AND CHARACTERISTICS
(See Notes 2, 3, 6) (Read, Write, and Read-Modify-Write Cycles)
(Full Operating Voltage and Temperature Range Unless Otherwise Noted)

PARAMETER	SYMBOL	MCM6664-15 MIN	MCM6664-15 MAX	MCM6664-20 MIN	MCM6664-20 MAX	UNITS	NOTES
Random Read or Write Cycle Time	t_{RC}	300	—	350	—	ns	8, 9
Read-Write Cycle Time	t_{RWC}	300	—	350	—	ns	8, 9
Access Time from Row Address Strobe	t_{RAC}	—	150	—	200	ns	10, 12
Access Time from Column Address Strobe	t_{CAC}	—	75	—	110	ns	11, 12
Output Buffer and Turn-Off Delay	t_{OFF}	0	30	0	40	ns	18
Row Address Strobe Precharge Time	t_{RP}	120	—	140	—	ns	—
Row Address Strobe Pulse Width	t_{RAS}	150	10,000	200	10,000	ns	—
Column Address Strobe Pulse Width	t_{CAS}	75	10,000	110	10,000	ns	—
Row to Column Strobe Lead Time	t_{RCD}	30	75	35	90	ns	13
Row Address Setup Time	t_{ASR}	0	—	0	—	ns	—
Row Address Hold Time	t_{RAH}	25	—	30	—	ns	—
Column Address Setup Time	t_{ASC}	0	—	0	—	ns	—
Column Address Hold Time	t_{CAH}	45	—	55	—	ns	—
Column Address Hold Time Referenced to $\overline{RAS}$	t_{AR}	120	—	155	—	ns	—
Transition Time (Rise and Fall)	t_T	3	50	3	50	ns	6
Read Command Setup Time	t_{RCS}	0	—	0	—	ns	—
Read Command Hold Time	t_{RCH}	10	—	10	—	ns	14
Read Command Hold Time Referenced to $\overline{RAS}$	t_{RRH}	30	—	35	—	ns	14
Write Command Hold Time	t_{WCH}	45	—	55	—	ns	—
Write Command Hold Time Referenced to $\overline{RAS}$	t_{WCR}	120	—	155	—	ns	—
Write Command Pulse Width	t_{WP}	45	—	55	—	ns	—
Write Command to Row Strobe Lead Time	t_{RWL}	45	—	55	—	ns	—
Write Command to Column Strobe Lead Time	t_{CWL}	45	—	55	—	ns	—
Data in Setup Time	t_{DS}	0	—	0	—	ns	15
Data in Hold Time	t_{DH}	45	—	55	—	ns	15
Data in Hold Time Referenced to $\overline{RAS}$	t_{DHR}	120	—	155	—	ns	—
Column to Row Strobe Precharge Time	t_{CRP}	−10	—	−10	—	ns	—
RAS Hold Time	t_{RSH}	75	—	110	—	ns	—
Refresh Period	t_{RFSH}	—	2.0	—	2.0	ms	—
$\overline{WRITE}$ Command Setup Time	t_{WCS}	−10	—	−10	—	ns	16
$\overline{CAS}$ to $\overline{WRITE}$ Delay	t_{CWD}	45	—	55	—	ns	16
$\overline{RAS}$ to $\overline{WRITE}$ Delay	t_{RWD}	125	—	160	—	ns	16
$\overline{CAS}$ Hold Time	t_{CSH}	150	—	200	—	ns	—
$\overline{RAS}$ to $\overline{REFRESH}$ Delay	t_{RFD}	0	—	0	—	ns	—
$\overline{REFRESH}$ Period (Battery Backup Model)	t_{FBP}	2000	—	2000	—	ns	—
$\overline{REFRESH}$ to $\overline{RAS}$ Precharge Time (Battery Backup Model)	t_{FBR}	390	—	460	—	ns	—
$\overline{REFRESH}$ Cycle Time (Auto Pulse Mode)	t_{FC}	330	—	380	—	ns	—
$\overline{REFRESH}$ Pulse Period (Auto Period Mode)	t_{FP}	60	2000	60	2000	ns	—
$\overline{REFRESH}$ to $\overline{RAS}$ Setup Time (Auto Pulse Mode)	t_{FSR}	30	—	30	—	ns	—
$\overline{REFRESH}$ to $\overline{RAS}$ Delay Time (Auto Pulse Mode)	t_{FRD}	390	—	460	—	ns	—
$\overline{REFRESH}$ Inactive Time	t_{FI}	30	—	30	—	ns	—
$\overline{RAS}$ to $\overline{REFRESH}$ Lead Time	t_{FRL}	390	—	460	—	ns	—

NOTES:
1. All voltages referenced to V_{SS}.
2. V_{IH} min and V_{IL} max are reference levels for measuring timing of input signals. Transition times are measured between V_{IH} and V_{IL}.
3. An initial pause of 100 μs is required after power-up followed by any 8 $\overline{RAS}$ cycles before proper device operation guaranteed.
4. Current is a function of cycle rate and output loading; maximum current is measured at the fastest cycle rate with the output open.
5. Output is disabled (open-circuit) when $\overline{RAS}$ and $\overline{CAS}$ are both at a logic 1.
6. The transition time specification applies for all input signals. In addition to meeting the transition rate specification, all input signals must transmit between V_{IH} and V_{IL} (or between V_{IL} and V_{IH}) in a monotonic manner.
7. Capacitance measured with a Boonton Meter or effective capacitance calculated from the equation: $C = \dfrac{I\Delta_t}{\Delta V}$.
8. The specifications for t_{RC} (min), and t_{RWC} (min) are used only to indicate cycle time at which proper operation over the full temperature range ($0°C \leq T_A \leq 70°C$) is assured.
9. AC measurements assume $t_T = 5.0$ ns.
10. Assumes that $t_{RCD} \leq t_{RCD}$ (max).
11. Assumes that $t_{RCD} \geq t_{RCD}$ (max).
12. Measured with a current load equivalent to 2 TTL (+200 μA, −4 mA) loads and 100 pF ($V_{OH} = 2.0$ V, $V_{OL} = −0.8$ V).

13. Operation within the t_{RCD} (max) limit ensures that t_{RAC} (max) can be met. t_{RCD} (max) is specified as a reference point only; if t_{RCD} is greater than the specified t_{RCD} (max) limit, then access time is controlled exclusively by t_{CAC}.
14. Either t_{RRH} or t_{RCH} must be satisfied for a read cycle.
15. These parameters are referenced to $\overline{CAS}$ leading edge in random write cycles and to $\overline{WRITE}$ leading edge in delayed write or read-modify-write cycles.
16. t_{WCS}, t_{CWD}, and t_{RWD} are not restrictive operating parameters. They are included in the data sheet as electrical characteristics only: if $t_{WCS} \geq t_{WCS}$ (min), the cycle is an early write cycle and the data out pin will remain open circuit (high impedance) throughout the entire cycle; if $t_{CWD} \geq t_{CWD}$ (min) and $t_{RWD} \geq t_{RWD}$ (min), the cycle is a read-write cycle and the data out will contain data read from the selected cell; if neither of the above sets of conditions is satisfied, the condition of the data out (at access time) is indeterminate.
17. Addresses, data-in and $\overline{WRITE}$ are don't care. Data-out depends on the state of $\overline{CAS}$. If $\overline{CAS}$ remains low, the previous output will remain valid. $\overline{CAS}$ is allowed to make an active to inactive transition during the pin #1 refresh cycle. When $\overline{CAS}$ is brought high, the output will assume a high-impedance state.
18. t_{off} (max) defines the time at which the output achieves the open circuit condition and is not referenced to output voltage levels.

FIGURE 5-8 (c) Timing values.

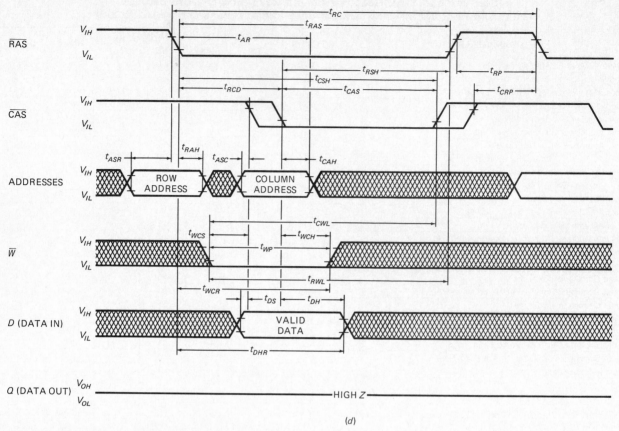

FIGURE 5-8 *(d)* Write cycle timing waveforms.

earlier. Regions of the waveforms shown as cross-hatched represent times when the states of those lines are unknown or are don't-cares. Remember that external hardware must supply the row address $\overline{RAS}$, column address $\overline{CAS}$, and $\overline{W}$ signals with proper timing. Figure 5-8c shows the timing values for the MCM6664-15 and MCM6664-25. Several are worth noting. Using the 6664-15 as a reference, note the overall read access time T_{RAC} of 150 ns maximum and the column address access time T_{CAC} of 75 ns maximum. The difference between these is the price paid for a smaller package with multiplexed address inputs. The minimum time for a complete read cycle T_{RC} is 300 ns. The value of −10 ns for T_{CRP} indicates that $\overline{CAS}$ can actually go high as much as 10 ns after $\overline{RAS}$ starts low for another read cycle.

Figure 5-8d shows the write cycle timing waveforms for the MCM6664. For a write operation, a row address is applied to the address inputs and $\overline{RAS}$ is asserted low. After a hold time T_{RAH}, the row address is removed, $\overline{W}$ is then made low if it is not already low, data is applied to the data input, and a column address is sent to the address inputs of the device. Then $\overline{CAS}$ is asserted low to latch in the column address. The column address must be held stable for a time T_{CAH} after $\overline{CAS}$ goes low, and the data to be written in must be held stable for a time T_{DH} after $\overline{CAS}$ goes low.

REFRESHING A DYNAMIC RAM

As indicated earlier, dynamic RAMs store bits as charge or no charge on a tiny capacitor. Unless the capacitor is refreshed every 2 ms or so, the charge will drain off and the stored data will be lost. To do this, each *row* in the device must be accessed at least once every 2 ms. The MCM6664 allows this to be done in several ways.

Figure 5-8e shows the timing waveforms for three different refresh methods. The $\overline{RAS}$-*only refresh cycle* method is the same as that used for refreshing earlier-generation 16K dynamic RAMs. For this method $\overline{CAS}$ is held high, $\overline{REFRESH}$ is held high, and D_{IN} and $\overline{W}$ are don't-cares. An external counter is used to send a 7-bit row address in on address inputs A0 through A6, and $\overline{RAS}$ is pulsed low. After the required time T_{RAS}, $\overline{RAS}$ is returned high for a time T_{RP}. The external counter is incremented to send out the next row address, and $\overline{RAS}$ is pulsed low again. This process is repeated 128 times every 2 ms. In the 6664 two 8-bit rows are refreshed at a time, so only 128 operations are required. All 128 rows can be refreshed either in a *burst* at the end of each 2-µs period or on a *distributed* basis between read and write operations. The main point is that each row much be addressed and strobed at least once every 2 ms. Since the minimum time for a $\overline{RAS}$-only refresh cycle T_{RC} is 300 ns, it will take 128 rows × 300 ns/row, or 38.4 µs, every 2 ms to refresh the device in the *burst* mode. If the

SELF REFRESH MODE (BATTERY BACKUP)
(SEE NOTE 17)

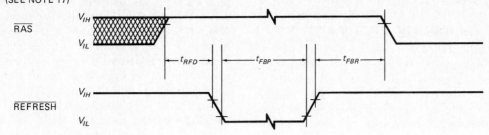

AUTOMATIC PULSE REFRESH CYCLE – SINGLE PULSE
(SEE NOTE 17)

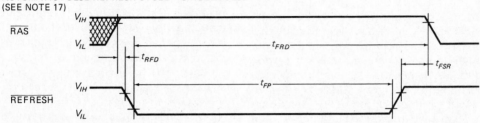

AUTOMATIC PULSE REFRESH CYCLE – MULTIPLE PULSE
(SEE NOTE 17)

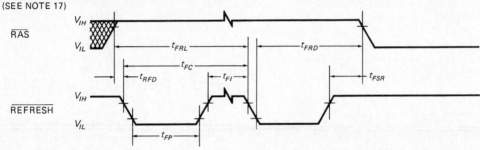

RAS – ONLY REFRESH CYCLE
(DATA–IN AND WRITE ARE DON'T CARE, $\overline{CAS}$ IS HIGH)

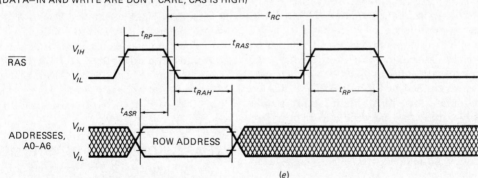

(e)

FIGURE 5-8 (e) Refresh timing waveforms.

distributed mode is spread evenly over the entire 2 ms, then a row will need to be refreshed after every 2000 μs/128 rows, or 15.6 μs. In other words, the distributed refresh will use 300 ns every 15.6 μs to refresh a row.

In the *automatic pulse refresh cycle* mode of the MCM6664, $\overline{RAS}$ is made high and $\overline{REFRESH}$ is pulsed 128 times every 2 ms. An internal row address register is automatically incremented after each pulse, so no ex-

ternal counter is needed. The $\overline{REFRESH}$ pulses can be supplied in either burst or distributed mode. The device cannot be read or written into while $\overline{REFRESH}$ is low.

In the *self-refresh* mode, $\overline{RAS}$ is made high and $\overline{REFRESH}$ is made low. If $\overline{REFRESH}$ is low for more than 2000 ns, an internal oscillator starts up and automatically increments the internal row address register to refresh the rows every 2 ms. The device cannot be read from or written into in this mode. Its main function is to maintain data when a system switches RAM to a battery backup power supply after a power failure.

EXTERNAL HARDWARE REQUIRED FOR DYNAMIC RAMS

Most current dynamic RAMs require considerable external hardware. Those that have $\overline{RAS}$-only refresh cycles need an external counter, a clock for the counter, and a multiplexer to send row addresses, column addresses, or the refresh count to the device. Also, there must be circuitry to arbitrate if the refresh circuitry decides to refresh a row at the same time as a microcomputer is trying to read from or write into the device.

A question that may arise at this point is, Why would anyone use a dynamic RAM, which requires all this complicated refreshing and multiplexing, instead of a static RAM, which doesn't? There are several answers. First, the dynamic RAMs are denser (more bits per IC) and cheaper per bit. Second, dynamic RAMs require considerably less power per bit. The MCM6664, for example, draws only 50 mA when accessed and 5 mA when in stand-by. *Note:* Because of the large difference in these currents, dynamic RAM memory boards require a 0.1-μF capacitor between V_{CC} and ground next to every other device. Third, all the refresh and arbitration tasks for a whole board of dynamic RAMs can be handled by one refresh controller IC. The Intel 8202 does these jobs for 16K devices. No doubt, similar controllers will become available for 64K RAMs as they become more widespread.

FUTURE OF DYNAMIC RAMS

Dynamic RAMs capable of storing 256K bits are available. Pin 1 is used as the required ninth address line input. These devices use *redundancy*. This means that the device is made with extra rows or columns to increase yields. These can be mapped in at final test to replace defective ones by using a laser to "zap" tiny fuses. Some new-generation devices such as the Zilog Z6132 contain on-chip refresh and arbitration circuitry so that they appear essentially static to the user.

Another problem with large dynamic RAMs is "soft," or one-time, errors caused by alpha particle radiation from packaging materials. Most manufacturers now coat the memory chip with a polyimide to soak up the alpha particles. For large memory systems in which these errors cannot be tolerated, error detecting and correcting circuitry is included with refresh control circuitry.

JOSEPHSON JUNCTION MEMORIES

A new technology based on superconducting materials and electron tunneling brings the possibility of having an entire computer and several megabytes of RAM in a single 2-in [5.1-cm] cube. The cycle time of this computer may be 50 times as fast as the fastest 1982 processor, and it will use about 1000 times less power. A difficulty with this computer is that the entire unit must be kept near absolute zero, which is $-273.2°C$.

The storage element is a ring of superconducting metal which has no electric resistance near absolute zero. Once a current is started flowing around this ring, theoretically it will continue to flow forever. A current flowing one way around the ring can represent a stored 0 or low, and a current flowing the other way can represent a stored 1 or high. A sense line detects which way the current is flowing.

One or two Josephson junction switches included in the ring are used to start or change the direction of the circulating current. A Josephson junction switch consists of two strips of superconducting metal separated by an insulator 30 to 50 Å thick. In the absence of a magnetic field, electrons will "tunnel" through this thin insulator in a way similar to the operation of a tunnel diode. A small magnetic field applied to the Josephson junction stops the tunneling. Then the junction acts as an open switch. The advantage of Josephson junction switches is that they can be turned on or off in a few picoseconds.

Since logic gates as well as memory cells can be made by using this technology, it could be used to build an entire computer. Recently substances have been found that superconduct at room temperature. This has raised hopes that it may be possible to build a Josephson junction computer that does not have to be kept near absolute zero.

SHIFT REGISTER MEMORIES

We have discussed how RAMs store digital data by connecting flip-flops in parallel arrays. Flip-flops connected as shift registers can also be used to store digital data. The data can be either loaded in serially or, for short registers, loaded in parallel. The number of bits that can be stored is equal to the number of stages or cells. To get the maximum number of shift register cells in a single IC package, charge-coupled device or magnetic bubble memory technology is used often.

Serial semiconductor memories such as shift registers have the advantage over RAM types that they are easier to make and use less chip real estate per bit. A disadvantage of serial memories is the longer time it may take to shift through to the first bit of a stored data word.

CHARGE-COUPLED DEVICE MEMORIES

Refer to Figure 4-43. The structure of a CCD is simply a substrate, an oxide dielectric layer, and a series of isolated gates. A 1 is stored as a packet of charge and a 0 as no charge in a potential well under a gate. The packet of charge is shifted through a CCD register by four-phase clocks, as shown in Figure 4-44. The CCD is dynamic. It must be shifted at some minimum rate, or else the data becomes invalid. To prevent stored data from being shifted off the end of a register and lost, the output can be tied to the input to make a *recirculating* shift register. Data is then shifted around and around this loop to keep it refreshed.

CCD memories are currently available in two formats: short loop and long loop. Short-loop types typically have 128 or 256 bits per loop, and long-loop types have 1024 to 4096 bits per loop.

An example of the long-loop type is the Texas Instruments TMS3064 which stores a total of 65,536 bits. Internally it has 16 shift registers of 4096 bits each. Four address lines allow one of the 16 registers to be selected and read out or written into in serial form. A separate 4096-bit register is used as a reference for the detector and regenerator circuits. Each stored bit must be refreshed at least once every 4 ms. This is done by clocking each register through the regenerator. Seven clock phases are produced internally from two-phase, 12-V input clocks.

An important parameter for serial memories is *latency time*, which is the time required to get to the first bit of a stored data block. Data sheets usually give an average latency time. For the TMS3064 this is about 410 μs. This time is so long because many bits may have to be shifted through to get to the desired starting bit. Once the first bit of a data block is found, the TMS3064 access time for the next bit is only about 200 ns, assuming the device is using a 5-MHz clock.

Because of the rapid advances made in dynamic RAM memories which don't have the long latency time, CCD memories have not caught on as much as was expected. A new serial memory type that does show signs of expansion is the magnetic bubble memory.

MAGNETIC BUBBLE MEMORIES

The read-write memories already discussed in this chapter are *silicon*-based and store data in latches or as electric charges. A new technology uses *garnet* (a reddish orange mineral often used as an abrasive on sandpaper) and stores data as microscopic magnets. This technology promises storage of more than 1 million bits of data in a single IC package.

STRUCTURE   Figure 5-9 shows the structure of a typical magnetic bubble memory (MBM). The substrate is gadolinium-gallium garnet, or GGG for short. On this substrate is grown a thin layer of pure magnetic garnet.

Strong permanent magnets are placed above and below this chip with their flux lines perpendicular to the chip. These strong magnets cause most of the microscopic magnetic domains in the magnetic garnet to line up parallel to, and with the same polarity as, the flux lines. Many domains are left opposite in polarity, however, and will be shrunk by this external field to tiny, round magnetic "bubbles." These bubbles can be created or removed with a small electromagnet, and they can easily be moved around through the garnet material.

Very thin nickel-iron or Permalloy magnets are built on top of the garnet in a chevron pattern. The bubbles line up under these small magnets. Bubbles can be shifted from one tiny magnet to the next with a rotating magnetic field. This field is produced by currents through the orthogonal coils shown in Figure 5-9. (*Orthogonal* simply means the two coils are wound at right angles to each other.)

For digital data storage a bubble signifies a 1, and no bubble signifies a 0.

Figure 5-10 shows the storage architecture for the Intel 7110 magnetic bubble device which stores 1 million bits in about a 1¾-in [4.4-cm] square package. The device has four sections (quads). Each quad has 80 loops of 4096 bits each. This is 16 more loops than are needed for a total of 1-megabit storage, but the extra loops serve two purposes. First, these extra loops (redundancy) increase production yield. If a quad has a few defective loops, these can be mapped around by programmable logic in the device. (A machine can automatically test each quad and program a loop as a map of the good loops.) Second, the extra storage allows error detection and correction schemes to be easily implemented.

Bubbles are circulated around the loops by the X and Y orthogonal field coils. A bubble passing by a detector creates a pulse of a few millivolts which can be read by external circuitry. A read operation is nondestructive. Data is entered into the loop by a replicating generator. The generators inject a bubble for a 1 and no bubble for a 0. Before data can be written, a loop must be cleared of all bubbles from previously stored data. If you have trouble visualizing this, think of it in terms of magnetic marbles being shifted around on small conveyor belts. The Intel 7110 has an average latency time of a few milli-

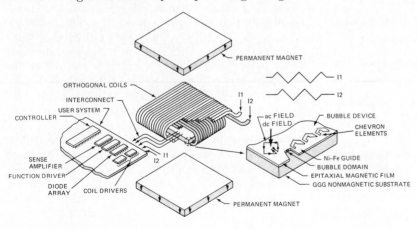

FIGURE 5-9 Construction of a magnetic bubble memory device. (*Texas Instruments.*)

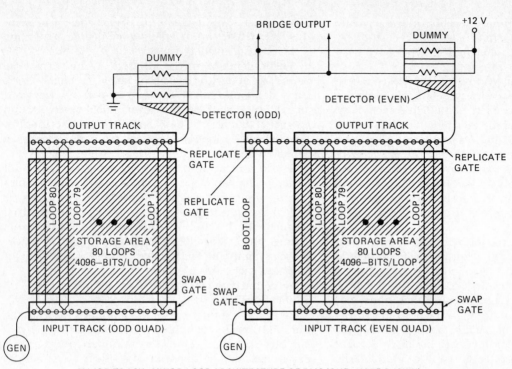

FIGURE 5-10 Intel 7110 MBM storage architecture.

MAJOR TRACK—MINOR LOOP ARCHITECTURE OF 7110 (OHE—HALF SHOWN)

seconds. It transfers data at 100K bits/s once the first bit of a stored block of data is found.

EXTERNAL CONTROL CIRCUITRY

Magnetic bubble memories such as the Intel 7110 require both analog and digital control circuitry. The analog control circuitry include drivers for the X and Y coils. It also contains amplifiers to raise the millivolt signals from the detectors up to logic levels, and buffers to convert logic level signals to the proper internal levels. Intel has available the BPK-70, a printed-circuit board that contains a 1-megabit device and all the analog interface circuitry needed.

The digital control circuitry is quite complex. The first part determines which loops in the device are written into. In addition, circuitry for timing and a UART for serial-to-parallel data conversion are needed. Often a microprocessor is used to coordinate input-output transfers.

ADVANTAGES AND DISADVANTAGES

The advantages of MBMs are:

1. They can store many bits in a single IC package, a million bits or more.

2. The storage is static (does not need refreshing) and nonvolatile. The small chevron magnets hold the bubble in place when the power is off.

3. MBMs are less troublesome than other bulk storage systems such as magnetic tape and disk systems which require motors and many awkward mechanical parts. Fujitsu sells a memory system that uses plug-in cassettes containing magnetic bubble memories.

The disadvantages of MBMs are:

1. MBMs are very slow. With a typical shift rate of 100 Kbits/s, the average latency time to a data bit is about 10 times the average latency time of a CCD such as the TMS3064. The device is obviously not suitable where fast access is needed.

2. The control circuitry is much more complicated than that for ROMs, RAMs, and CCDs.

MAGNETIC CORE MEMORY

Magnetic core memory was used in much computer mainframe memory, but now it is utilized in new designs only if a severe temperature or radiation environment will be encountered. The storage element in core memory is a small magnetic ferrite bead. The bead is doughnut-shaped and typically about 0.050 in [1.27 mm] in diameter with a 0.030-in [0.76-mm] hole. If a wire carrying an electric current is passed through the hole in the bead, a magnetic field is induced in the ferrite bead or core. The direction of the induced magnetic field depends on the direction of current flow. The left-hand rule can be used to determine the direction of the field. If your thumb points in the direction of electron flow, then the other four fingers will point in the direction of the magnetic flux in the core. Once the bead is magnetized in a given direction, it retains the magnetism even when the current is turned off. This is, then, nonvolatile storage. A "1" can be arbitrarily defined as flux pointing in one direction and a "0" as flux pointing in the opposite direction.

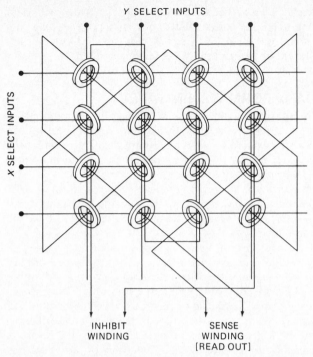

X SELECT INPUTS

INHIBIT
WINDING

SENSE
WINDING
[READ OUT]

FIGURE 5-11 Magnetic core memory plane writing diagram.

To read data from a core, a sense wire is threaded through the hole and connected to a sense amplifier. A 0 is then written into the core by passing a current through the write wire in the proper direction. If the core being read is in the 1 state, then the change in flux from 1 to 0 induces a voltage in the sense line. If the core is already 0, then writing 0 into it induces almost no voltage in the sense wire. This readout is destructive because the core is always in the 0 state after a read. To maintain the data after a read, it must be saved with a latch and rewritten into the core.

RAM-type data storage can be done by wiring cores in an X-Y matrix, as shown in Figure 5-11. In this array a particular core is read by sending a select current equal to half the needed write 0 current to the core on its X line. The same half write 0 current is sent to the core on its Y line. These two currents add to write 0 in the addressed cell. The half currents are not large enough to change the state of any other cores on the X or Y lines. The sense amplifier detects the flux change or no change as a 1 or 0 in the addressed cell. Writing data to a core in a matrix requires two different mechanisms depending on whether 1 or 0 is to be written. To write 1, the core is simply addressed with a positive half write 1 X current and a positive half write 1 Y current. Their effects add to write the 1. A 0 could be written into a core by simply addressing it with a half write 0 X current and a half write 0 Y current, as was done for reading the core. The combined effect of these two would magnetize the core in the opposite direction, to write a 0. However, usually a 0 is written with another wire called the *inhibit wire,* which is threaded through all the cores in a matrix or plane. The core is addressed in the same way

as when a 1 is written, but a current of negative half write 1 in the inhibit wire cancels the effect of one of the select currents and prevents a 1 from being written. This method makes for easier control.

Core planes can be stacked in a three-dimensional array to increase the memory size. Advantages of core memory are its ruggedness, resistance to temperature and radiation, and nonvolatility. Problems with core include its physical volume. An array of core memory planes is much larger than the equivalent modern semiconductor memory. Core memory contains 4096 bits in a 2-in × 2-in [5.1-cm × 5.1-cm] core plane. Eight of these can be stacked to give a 4096 × 8 memory in approximately a 2-in [5.1-cm] cube. Four 16-pin static MOS RAMs such as the Mostek MK4801 could replace this magnetic core memory stack, at a fraction of the cost. Also any readout from core memory is destructive, and circuitry must be included to store and rewrite the data if desired. Another problem is that core memories require high current drivers on the select lines. To get core access times fast enough for many uses, the select line currents may be as much as 1 A.

MAGNETIC TAPE STORAGE

Magnetic tape storage is an important method for storing large amounts of digital data at a low cost per bit. The tape is usually Mylar with a thin coating of magnetic iron oxide. It varies in width from the 0.15-in [3.8-mm] tape used in the small Philips cassettes for portable recorders to the 2- or 3-in [5.1- or 7.6-cm] wide tape commonly used with large mainframe computers.

Figure 5-12 shows a diagram of the record-play "head" used to write data to or read data from the tape. In the write mode, a current through the coil creates a magnetic flux in the iron core of the head. A gap in the head allows the magnetic flux to spill out and magnetize the magnetic material on the tape. Once magnetized, the tape remains so until altered. The polarity of the magnetized region on the tape is determined by the direction of the current flow in the head. One direction of current flow can represent a 1 and the other a 0.

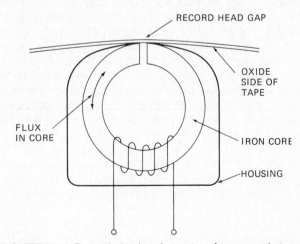

RECORD HEAD GAP

OXIDE
SIDE OF
TAPE

FLUX
IN CORE

IRON CORE

HOUSING

FIGURE 5-12 Record-play head structure for magnetic tape.

Data can be read from the tape with the same head. When a magnetized region in the tape is passed over the gap in the head, a voltage, typically a few millivolts, is induced in the coil. The polarity of the induced voltage corresponds to the polarity of the magnetized region read. An amplifier raises the small signal to logic levels.

RECORDING MODES

In practice, there are many ways in which a 1 or a 0 can be represented by magnetized regions on the tape. Figure 5-13 shows nine of these. (In each waveform, the horizontal time line is drawn through the waveform at the 0 flux level.) The first, *double frequency*, or F2F, represents a 1 with two pulses in the bit time and a 0 with only one pulse in the bit time. This system is referred to as *self-clocking* because speed information can be derived from the data pulses. The speed information can be used to keep a UART synchronized with the data.

The second method is *frequency shift keying*, or FSK. Here a burst of one frequency represents a 1, and a burst of another frequency represents a 0. It is not self-clocking because the two frequencies are not harmonically related.

Non-return-to-zero (NRZ) recording stores a 1 as flux in one direction and a 0 as a flux in the other direction. NRZ also refers to any recording mode that always magnetizes the tape in either one polarity or the other. The F2F recording in Figure 5-13a is NRZ mode, as you can see from the fact that the 0 flux line passes through the center of the waveform. This method is used on many IBM-compatible tapes. A return-to-zero method not shown stores a 1 as flux and a 0 as no flux. NRZ data is not self-clocking. A separate clock track is required just to synchronize a receiving system.

Phase encoding uses a flux transition in the middle of the bit time. A flux change in one direction is a 1 and in the other direction a 0, as shown in Figure 5-13d.

Another type of phase encoding is Manchester. Here the transition is at the start of the bit time. A high-to-low transition at the start of the bit time represents a 0 and no transition a 1.

Pulse width modulation uses a high duty cycle, or wide, pulse to represent a 1 and a low duty cycle pulse to represent a 0, as shown in Figure 5-13f.

Pulse width modulation burst and tone burst methods use gated sine waves to represent 1's and 0's, as shown in Figure 5-13g and h. These and other techniques employing sine waves can be used with standard nonsaturating audio tape recorders. This type of recorder will not record the saturated pulses required by the other techniques.

The Kansas City Standard (KCS) is used in many hobbyist systems. The standard 300-baud version of this method uses eight cycles of 2400 Hz to represent a 1 and four cycles of 1200 Hz to represent a 0. The waveforms usually are created as sine waves or filtered to approximate sine waves so that standard audio recorders can be used. The Kansas City standard also specifies the format for serial data recording. It states that a leader should consist of 30 s of 2400 Hz. Each data character should consist of one 0 stop bit, 8 data bits including parity, and two 1 stop bits. Blocks of data should be separated by 5-s leaders of 2400 Hz.

DATA FORMATS ON TAPE

In addition to the several ways that bits are stored on tape, there are several formats in which the bits are stored as words. On large commercial recorders with wide tape, data words are stored in parallel form on parallel tracks. Multisection read-write heads read or write all the bits of a word at the same time. A standard IBM format uses seven parallel tracks to record each word. Several strips of seven tracks each can be put on the same tape.

Narrow tape systems such as the Philips cassettes have only one or two tracks, so data words are stored in serial form on a single track. Serial words may contain 8 to 19 bits, including start and stop bits. These words are grouped as blocks or records of perhaps 256 or 512 words. Records are separated by IRGs (interrecord gaps) of blank tape. Two or more blocks of related data are called a *file*. End-of-file (EOF) characters signal the end of the file. A beginning-of-file character(s) after an IRG indicates the start of the next file.

ADVANTAGES AND DISADVANTAGES OF TAPE STORAGE

The main advantage of a tape storage system is that a very large number of bits can be stored at a very low cost per bit. Tape storage is nonvolatile, and the tape can be erased and reused many times.

The main disadvantage of tape storage is serial access. A slow system may take as long as several seconds to locate a file. Once a file is found, some parallel-track recorders can spew about 250,000 characters per second, however. An improvement of initial access time has been made by using several cassettes with short tapes rather than one very long tape. Another disadvantage in the large commercial units is the need for a complicated mechanical system to keep tape speed constant and to start and stop the tape without breaking it.

MAGNETIC DISK DATA STORAGE

The serial access or latency time problem of magnetic tape data storage can be partially solved by putting data on a magnetically coated disk similar to a phonograph record. Data is stored on these disks in individual circular tracks rather than in a spiral as is done on phonograph records. The read-write head can be stepped to the desired track to read data without having to read through all the previous tracks.

The magnetic material on the disk and the read-write head construction are the same as those described for magnetic tape. In the following sections we describe two common disk types, floppy disks and rigid disks.

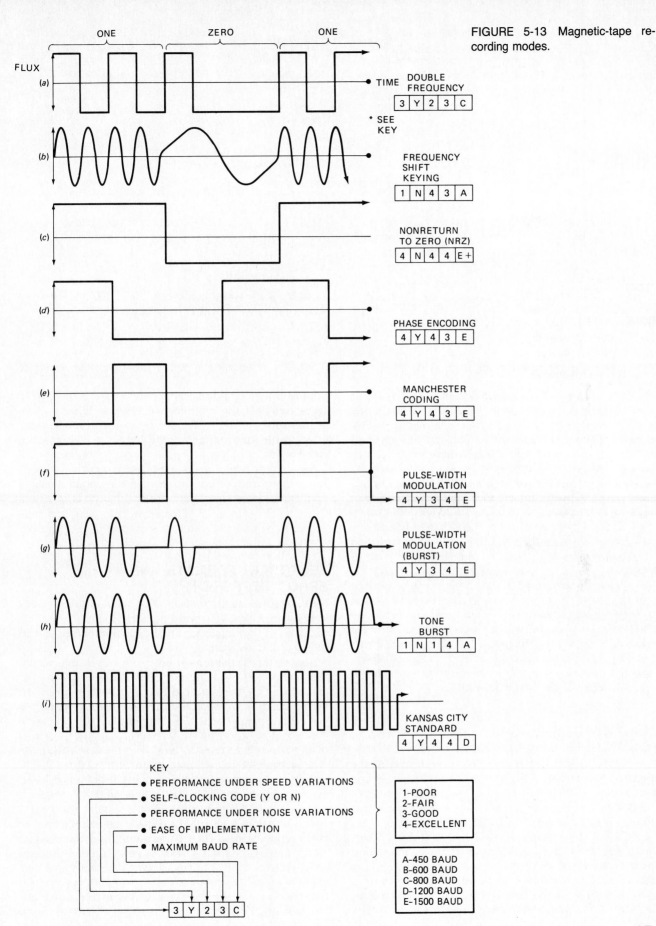

FIGURE 5-13 Magnetic-tape recording modes.

(a) DOUBLE FREQUENCY — 3 Y 2 3 C

* SEE KEY

(b) FREQUENCY SHIFT KEYING — 1 N 4 3 A

(c) NONRETURN TO ZERO (NRZ) — 4 N 4 4 E +

(d) PHASE ENCODING — 4 Y 4 3 E

(e) MANCHESTER CODING — 4 Y 4 3 E

(f) PULSE-WIDTH MODULATION — 4 Y 3 4 E

(g) PULSE-WIDTH MODULATION (BURST) — 4 Y 3 4 E

(h) TONE BURST — 1 N 1 4 A

(i) KANSAS CITY STANDARD — 4 Y 4 4 D

KEY
• PERFORMANCE UNDER SPEED VARIATIONS
• SELF-CLOCKING CODE (Y OR N)
• PERFORMANCE UNDER NOISE VARIATIONS
• EASE OF IMPLEMENTATION
• MAXIMUM BAUD RATE

3 Y 2 3 C

1-POOR
2-FAIR
3-GOOD
4-EXCELLENT

A-450 BAUD
B-600 BAUD
C-800 BAUD
D-1200 BAUD
E-1500 BAUD

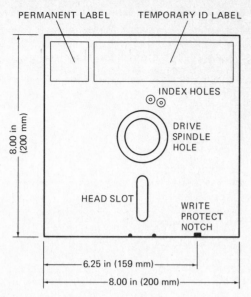

PERMANENT LABEL   TEMPORARY ID LABEL

INDEX HOLES

DRIVE
SPINDLE
HOLE

HEAD SLOT

WRITE
PROTECT
NOTCH

8.00 in (200 mm)

6.25 in (159 mm)

8.00 in (200 mm)

FIGURE 5-14 Floppy disk with protective envelope.

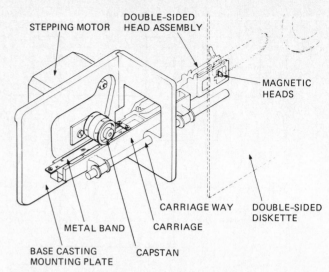

STEPPING MOTOR

DOUBLE-SIDED
HEAD ASSEMBLY

MAGNETIC
HEADS

CARRIAGE WAY

DOUBLE-SIDED
DISKETTE

METAL BAND

CARRIAGE

BASE CASTING
MOUNTING PLATE

CAPSTAN

FIGURE 5-15 Head positioning mechanism for Shugart floppy disk drive unit. *(Shugart Associates.)*

FLOPPY DISK SYSTEM

The floppy disk system uses a 7¾-in-diameter Mylar disk coated with magnetic material and permanently enclosed in a low-friction 8-in envelope, as shown in Figure 5-14. The Mylar disk is only a few thousandths of an inch thick, thus the name *floppy*. When the disk is inserted into a drive unit, a spindle clamps in the large center hole and spins the disk at a constant 360 rpm. The read-write head contacts the disk through the racetrack-shaped opening in the envelope. With floppy disks the heads are permanently *crashed* or in contact with the disk when reading or writing. This increases disk wear but is necessary because of the slower speed of floppies.

The write protect notch in the envelope can be used to protect stored data from being written over, as do the knock-out plastic tabs in tape cassettes. A photodetector (LED and phototransistor combined) can indicate whether the notch is present and disable the write circuit if it is. An index hole indicates the start of the recorded tracks. Another photodetector can be used to detect when the index hole passes.

DISK DRIVE AND HEAD POSITIONING

In addition to the precision feedback-controlled motor required to maintain the disk speed at 360 rpm, some means of precisely positioning the read-write head over the desired track is needed. The most common method uses a stepping motor and either a lead screw or a let-out–take-in steel band, as shown in Figure 5-15. A stepper motor is a special type of motor which, when pulsed, rotates a fixed increment of, for example, 5°. It can be pulsed either forward or reverse. As the stepper motor in Figure 5-15 rotates, the steel band is let out on one side of the motor pulley and pulled in on the other

side. This converts the rotary motion of the motor to linear motion.

To find a given track, usually the motor is stepped to track zero near the outer edge of the disk. When a number of pulses equal to the desired track number is applied to the stepper motor, the head will be moved to this track.

Eight-inch floppy disks are often recorded with one index track and 76 data tracks per side. Each track is divided into sectors. There are two methods of establishing the sectors, or "formatting the disk," as it is called: hard sectoring and soft sectoring.

SECTORING FORMATS AND ERROR DETECTION

Hard-sectored disks typically have 32 additional index holes spaced equally around the disk. Each hole signals the start of a sector. The index hole photodetector can be used to detect them.

Soft-sectored disks have only one index hole which indicates the start of all the tracks. The sector format is established by bytes stored on the track. Figure 5-16 shows the IBM 3740 soft-sectored floppy disk format. As you can see, many bytes are used for identification, synchronization, and buffering between sectors. Soft sectoring allows less data to be stored per track, but the data is returned more reliably than with hard sectoring.

In the soft-sectored IBM format, each sector is divided into four fields: the ID record, ID gap, data field record, and data gap. A 1-byte mark of all 1's starts the ID record field. The rest of this field gives the complete track and sector address. The ID gap field is a buffer zone which gives time for a system to verify an address and switch to write mode to record data. Another mark byte of all 1's signals the start of the data field. The data field consists of the mark byte plus 128 data bytes and two *check-sum*, or *cyclic redundancy*, bytes which are used

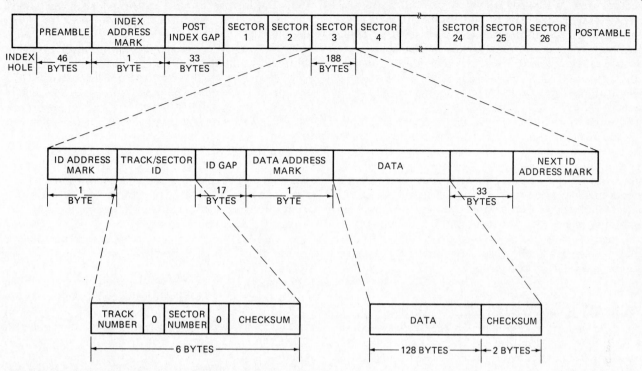

FIGURE 5-16 IBM 3740 floppy disk soft-sectored track format.

to check for errors in data when read. Check-sum bytes are formed by adding all the data bytes and ignoring any carries over 2 bytes. The resultant 2-byte sum is stored in the two byte positions after the data. When the data is read, it is re-added and the sum is compared with the recorded check-sum bytes. If no errors are present in the recorded data, then the two sums will agree. Another method of checking for data errors is the *cyclic redundancy check* (CRC). This method involves dividing the string of data bytes by a constant. Any remainder from this division is written in as 2 CRC bytes, or characters. When the data is read out, these 2 bytes are subtracted from the data string. The result is divided by the original constant. This division should give a zero remainder if the data contains no errors. CRC is the usual method of error checking for disk storage.

Data can be written into disk and then immediately read back and tested for errors. If an error is detected, the data can be written again and rechecked. If 10 write attempts are unsuccessful, then an external circuit can move the attempted write to another sector. A scheme such as this gives floppy disks an error rate about a mil-

lion times better than that typical of Philips tape cassette recording.

The data field is followed by a record gap field which acts as a buffer between the end of one data field and the starting mark of the next sector. This gap also gives time for the system to switch from write mode to read mode when it goes from writing data in one sector to reading the ID field of the next.

As previously discussed, the IBM 3740-compatible soft-sectored floppy disk format has 76 data tracks with 26 sectors each. Since each sector can contain 128 bytes of data, the total storage per IBM-compatible disk is about 250,000 *bytes.* The average access time to a sector is about 500 ms, which is much shorter than that for most tape systems. Once a sector or file is found, data can be read off the disk at about 250,000 *bits*/s. The actual pattern used to store bits on an IBM-compatible disk is an F2F type recorded in NRZ form. This is illustrated in Figure 5-17. Each bit time contains a clock pulse. A 1 is written in the bit time by including another pulse between clock pulses; a 0 is written by leaving out the pulse. It is non-return-to-zero form because each region in the data track has magnetic flux in one direction or the other. The 0 flux line through the middle of the waveform in Figure 5-17 indicates this.

FLOPPY DISK VARIATIONS

Several methods have been used to increase the amount of data that can be stored on a disk. The first, the double-sided method, simply uses both sides of the disk

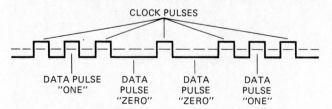

FIGURE 5-17 F2F recording mode for magnetic disks.

rather than only one side as IBM did. Each side can still be formatted for IBM soft-sector compatibility. The second method, double-density, doubles the number of sectors in each track. Either method increases the storage per disk to more than 500,000 bytes, but double-density is not IBM-compatible. These two techniques are combined in double-sided–double-density recording to give over 1 million bytes of storage on a single 8-in [20-cm] disk.

To somewhat reduce the size and cost of a disk storage unit, the mini-floppy system was developed. Mini-floppy systems use the same type of disk and drive mechanisms as the standard 8-in floppies, but the disk size is only 5¼ in. The mini-diskettes have only 35 or 40 tracks including an index track. A 35-track disk stores about 80,000 bytes. Many mini-floppy systems make up for storage space lost with a smaller disk by using double-sided recording, double-density recording, or both. Later we discuss floppy disks and their use with computers and microprocessors.

PRECAUTIONS WITH FLOPPY DISKS

Magnetic disks are subject to damage by dirt, heat, or magnetic fields. Do not touch the exposed areas of the recording surface with your fingers. Keep the disk in its container when you are not using it. Disks should be kept away from other electronic equipment. A disk accidentally laid on top of a power supply can be erased by the alternating magnetic field of a transformer. Also, do not turn off the power to a disk drive unit with the disk in the drive. In some units the formatting circuits or the collapsing field of motors may ruin the data stored on this disk.

HARD DISKS: WINCHESTER DRIVES

Rigid disks also have a magnetic coating on one or both sides. In most cases the disks are permanently fastened in the drive unit rather than being removable as floppy disks are. Some standard hard disk sizes are 5¼, 8, 10½, 14, 20, and 40 in. To increase the amount of storage per drive, several disks may be stacked with spacers between them and an individual read-write head for each disk. Because the rigid disks are more dimensionally stable, they can have more tracks per inch and more bits per track than floppies can. This means that they can store a larger amount of data more easily than an equivalent-size floppy disk. For reference, 5¼-in units can store 3 to 10 million bytes per disk; 8-in units, 5 to 20 million bytes per disk: 10½-in units, 30 to 50 million bytes per disk; and a 14-in unit about 40 to 100 million bytes per disk.

Rigid disks are rotated at 1000 to 3600 rpm. This high speed not only makes it possible to read and write data faster, but also creates a thin cushion of air that floats the read-write head 10 μin or so off the disk. Unless the head "crashes," it never touches the data area of the disk, so wear is minimized. To prevent dust and smoke particles, which usually are larger than 10 μin, from crashing the head, either the disks are enclosed in a sealed dustfree case or cooling air is passed through a fine air filter.

Figure 5-18 is a photograph of an Irwin 5¼-in Winchester hard disk drive. Legend has it that the name *Winchester* given to these drives came from an early IBM dual-drive unit with a planned storage of 30 million bytes per drive. The 30-30 configuration apparently reminded someone of the famous Winchester rifle, and the name stuck.

Heads are positioned over the desired track by a stepper motor and lead screw, a band actuator such as that shown for floppies in Figure 5-15, or a linear "voice coil" actuator. Positioning time for various 5¼-in units ranges from 25 to 150 ms. Once the desired track is located, the data is read out at 500,000 to 900,000 bytes/s. Interface circuitry controls the head position for read and write operations and formats input data for storage.

The advantage of rigid disks is that they can keep a large amount of data quickly available. If rigid disks are used with a word processor, for example, the text for an entire book such as this one could be stored on one disk rather than on a large pile of mini-floppy disks as was done.

To prevent data loss in the event that a hard disk system crashes, backup copies of all files on the disk are kept on some other medium such as magnetic tape or floppy disks. High-speed "streaming" tape units, such

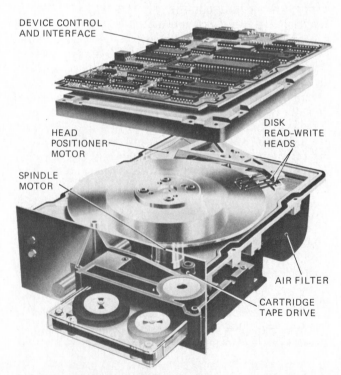

DEVICE CONTROL
AND INTERFACE

HEAD
POSITIONER
MOTOR

SPINDLE
MOTOR

DISK
READ-WRITE
HEADS

AIR FILTER

CARTRIDGE
TAPE DRIVE

FIGURE 5-18 Irwin 5¼-in Winchester hard-disk drive. *(Irwin International, Inc.)*

as that built into the Irwin 510 shown in Figure 5-18, can load 10 million bytes from the disk or dump 10 million bytes to the disk in less than 4 min. Some units solve the backup problem by making the sealed disk enclosure a removable unit. This is more expensive than the streaming tape approach.

ELECTRONIC BEAM ACCESSED MEMORY

Electronic beam accessed memories (EBAMs) are constructed similar to a television picture tube, but the screen is an array of MOS chips. Data is stored on an addressed location on a chip as a packet of charge. The charge is placed there by an electron beam. The beam can be directed to a particular address by the same methods used for horizontal and vertical deflection in an oscilloscope or television set. Data is read out by addressing a cell with the beam and determining the beam current flow. Less beam current will flow if a storage cell already has a packet of charge. Readout is destructive, and storage is volatile.

EBAMs are somewhat bulky because of the various power supplies needed, but they can store 30 to 40 million bits per unit. The access time is only tens of microseconds.

ACCESS TIME VERSUS STORAGE CAPACITY

Figure 5-19 shows a graphical comparison of the storage capacity and access times for the various memory types discussed in this chapter. This graph should help you remember the range of use for each type. For further reference you may wish to write next to each type whether its storage is volatile or nonvolatile and whether it is random access or serial access. Note the human brain and the Josephson junctions at the ex-

tremes of the graph. The factors that determine the type of memory used in a given application are cost, access time, availability, amount of storage, ease of access (serial or random), volatility, and physical size.

TESTING SEMICONDUCTOR MEMORIES

Factory testing of semiconductor memories is of two basic types, parametric and functional. *Parametric* testing includes measurement of maximum and minimum voltage levels, input and output currents. It also includes access times, minimum pulse widths, and other timing relationships. When worst-case time parameters are established, they must be measured for each address in the memory because access times, for example, usually differ for various locations in the matrix.

Functional tests include determining whether data can be written or programmed into each location, read out correctly, and erased or reprogrammed. Obviously each address must be tested. As a first thought, it would seem that a RAM such as the 2147 4K static could be tested by writing a 1 in each address, reading out the data from each address and checking if all 1's were read out—and then doing the same procedure with all 0's. However, in practice it is not enough to test a memory with only the all 1's and all 0's method because many memories are *pattern-sensitive*. This means that certain patterns may not properly program or write into a memory. To completely check a memory for pattern sensitivity, theoretically all the possible combinations should be tried. For a 4K RAM the number of possibilities is 2^{4096}. Since it is impossible to test this many possibilities, various test patterns are used that are likely to show a problem. First fixed patterns such as all 1's, all 0's, or alternating 1's and 0's are used to detect gross failures. Then a "galloping" 1 or 0 pattern may be used to check for pattern sensitivity. A *galpat*, as this is often called, first initializes the memory to all 0's. Then a 1 is written in a location, and the rest of the memory is checked to see whether any other locations were affected. If they were, the IC fails; if not, the written 1 is changed back to a 0 and a 1 written into the next location. The rest of the memory is rechecked for changes. This one is walked or galloped through the entire memory. Then the memory may be written to all 1's and a 0 test galloped through it.

Since the number of tests is so large and since the tests must be done at the maximum operating speed of the memory to be valid, you can't really do it with some switches and LEDs. IC manufacturers use complex computer-controlled test systems such as the Fairchild Sentry shown in Figure 5-20. With the proper options, the Sentry can be programmed to measure all the dc and ac parameters of an IC as well as do high-speed functional tests with galpats. Even with these high-speed testers it takes several seconds to test a large memory.

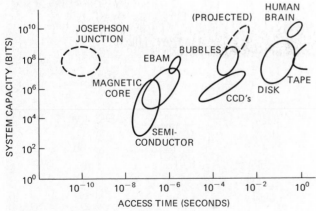

FIGURE 5-19 Comparison of access time and storage capacity of several types of memory.

FIGURE 5-20 Fairchild SENTRY integrated-circuit test system.

LOGIC ANALYZERS: TESTING MEMORIES IN A CIRCUIT

To analyze the functioning of a memory already soldered into a circuit on a PC board, you need an instrument that can simultaneously display the signals on the address bus, the signals on the data bus, and the read, write, and chip select signals. You may need over 40 channels to do this. A big pile of oscilloscopes could be stacked up to get 40 channels, but this is impractical; so logic analyzers and logic state analyzers were developed. Figure 5-21 shows a Hewlett-Packard 1600A and 1607A logic analyzer system which can display up to 32 channels of data. Note the probe "pods" to the left in Figure 5-21. These make for neater and more compact connections than when 32 standard oscilloscope probes are used.

A logic analyzer in its simplest form is basically a multichannel oscilloscope such as the eight-channel multiplexed type discussd in Chap. 3. Figure 5-22 shows a block diagram of a simple logic analyzer. Since logic analyzers are used only to detect 1's or 0's (highs or lows), a comparator is put on each input. The reference input of the comparator is set at a voltage equal to the logic threshold for the family being tested. The signals out of the comparators to the rest of the analyzer are then clear-cut 1's or 0's.

Logic analyzers also have two other features not usually found in standard oscilloscopes. First, they contain a combinational or word-recognizer trigger mode as well as a standard external trigger. The word recognizer circuitry compares the incoming data with a word programmed by the operator with front panel switches. A

trigger is produced when the two words match. Second, logic analyzers contain memory. The memory functions as parallel shift registers of perhaps 256 bits for each channel. Input data is sampled and shifted through the memory by either an internal or external clock. Data samples are continuously being taken and written into the memory until a trigger occurs. When a trigger occurs, the data samples stored in memory are displayed on a CRT screen.

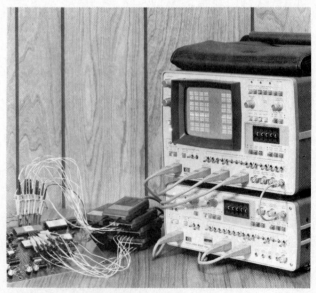

FIGURE 5-21 Hewlett-Packard 1600/1607A logic analyzer system.

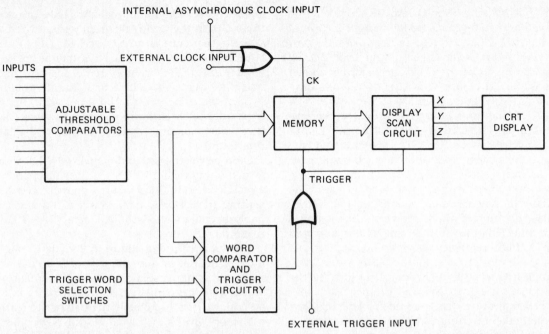

FIGURE 5-22 Block diagram of simple logic analyzer.

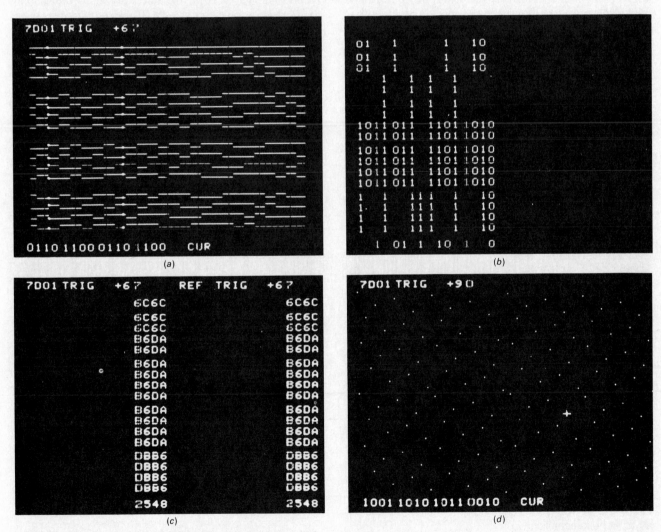

FIGURE 5-23 Logic analyzer display formats. *(Tektronix.)*

DISPLAY FORMATS   The 1's and 0's stored in the internal memory can be displayed for analysis in several formats. Figure 5-23 shows four formats possible with the Tektronix 7D01 logic analyzer and DF2 display formatter. The first is the timing diagram format shown in Figure 5-23a. This format is best used for finding glitches and displaying long data sequences. As shown in Figure 5-24, the timing diagram format can be produced by sampling the inputs with either a synchronous or an asynchronous clock. The synchronous type simply uses the system clock of the system under test. For the waveforms shown in Figure 5-24, the sampling takes place on the negative edge of the system clock. The disadvantage of synchronous clocking is that it may totally miss the induced glitch shown in the incoming channel 2 data. This glitch can be caught and displayed by using a high-frequency asynchronous clock, as shown in the bottom part of Figure 5-24. A disadvantage of high-frequency asynchronous clocking is the requirement of an extra clock and the need for more memory if samples are being stored for later display. Another method of catching glitches used by some analyzers is a *latch-mode* storage. If latch-mode storage is selected, a glitch sets or resets an internal flip-flop. The next clock pulse then updates the display to show that the glitch occurred. Figure 5-25 compares the normal sample-mode with latch-mode storage. A glitch that is a few nanoseconds wide gets stretched to a pulse one clock cycle wide and so is very visible.

The second display format is the 1's and 0's "logic state" type shown in Figure 5-23b. This format is easier to read than the timing diagram format when a data or address sequence is being checked. Digital words are read from left to right. The right half of the screen can display a set of data from a functioning unit, and the left half of the screen data from a malfunctioning unit. The two can be compared to find a problem. Some models have internal circuitry to compare the two halves location by location and intensify in the left display any errors found.

Even easier to read and compare is the hexadecimal format of Figure 5-23c. The four hexadecimal numbers, as you learned in Chap. 3, are a shorthand way of representing the 16-bit binary words of 1's and 0's. Some analyzers offer a choice of hex, octal, decimal, or binary readouts.

The fourth format, Figure 5-23d, is a map display. Each point on the screen represents a data word. The main use of map displays is to identify a malfunctioning system by comparing its map with that of a known good system. Maps are probably the hardest to learn to use.

Memory gives a logic analyzer the powerful capability of displaying words that occurred before the trigger. When a trigger occurs, the data stored in the display memory are, for example, the 1024 samples taken before the trigger, and these will be displayed. This enables you to trigger on an observed error word and observe the conditions that led up to the error. Often this is referred to as *pretrigger recording*. If the trigger is delayed with a counter for 512 clock pulses, then the

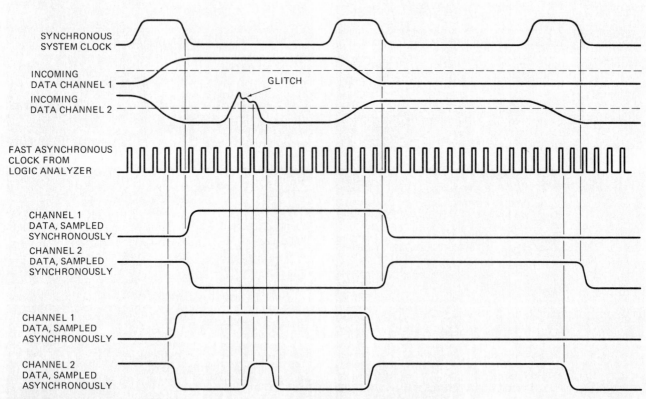

FIGURE 5-24 Waveforms showing how a glitch is captured by a logic analyzer with asynchronous sampling.

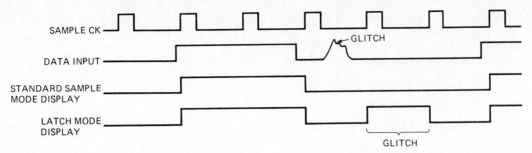

SAMPLE CK

DATA INPUT

STANDARD SAMPLE
MODE DISPLAY

LATCH MODE
DISPLAY

GLITCH

GLITCH

FIGURE 5-25 Waveforms showing how a glitch is captured with the latch mode in a logic analyzer.

display will show 512 samples before the trigger and 512 samples after the trigger. Such a display is called *center trigger recording*. When the trigger is delayed 1024 clock pulses, only samples occurring after the trigger will be displayed. Memory also allows events that occur only once in a long sequence to be stored, displayed, and analyzed.

EXAMPLE OF AN APPLICATION

As an example of how a logic analyzer might be used to troubleshoot a memory system, consider the following:

A memory system puts out an erroneous word at one point in a sequence of data being read. You don't know whether the error is caused by wrong data stored in the memory or by an incorrect address going to the memory at that time. You can find out by connecting a logic analyzer to the address bus and data bus of the memory system. Set the front panel switches so that the word recognizer produces a trigger when the error word occurs on the data bus. Then you can look at the address bus displays to see whether the address is correct. If the address is correct, then a memory IC probably is defective and should be replaced. Further use of logic analyzers is discussed in a later chapter.

REVIEW QUESTIONS AND PROBLEMS

1. Define the following:
   a. Register
   b. Register file
   c. RAM

2. For a RAM, define the following:
   a. Address access time
   b. Chip-select address time

3. How many address lines are required to address 8192 locations in a RAM?

4. A data word is stored in memory at hexadecimal address 04FA. What pattern of 1's and 0's will be found on the 16 address lines when this location is addressed?

5. To what does the expression "1024 × 4" on a RAM data sheet refer?

6. Draw a circuit diagram showing how 2167 RAMs can be connected to form a 16,384 × 8 memory system.

7. Use the timing waveforms and parameters for the 2167-70 in Figure 5-5 to find the following worst-case times:
   a. Address access time
   b. Chip-select access time
   c. Time for which valid data remains on the output after $\overline{CS}$ returns high at the conclusion of a read operation

d. Time required to power down after $\overline{CS}$ returns high

e. Time for which valid data must be present on the input before $\overline{WE}$ goes high during a write operation

f. Minimum write pulse width

8. Why do memories such as the 2167 require bypass capacitors next to every other IC?

9. What is the main advantage of CMOS RAMs?

10. Describe the differences between a static MOS RAM and a dynamic MOS RAM. How is data stored in each?

11. How is a dynamic RAM cell refreshed? How often must most dynamic RAMs be refreshed? List the advantages and disadvantages of dynamic RAMs.

12. Describe the system used with dynamic RAMs to get 16 address bits into the device on only eight pins.

13. Use the timing waveforms and parameters in Figure 5-8 to determine the following worst-case times for an MCM6664-15 dynamic RAM:
    a. Read access time
    b. Read cycle
    c. Minimum $\overline{RAS}$ pulse width

d. Time for which the column address must be held after $\overline{CAS}$ goes low

e. For a write operation, the time for which valid data must be held on the input after $\overline{CAS}$ goes low

f. The minimum time for which $\overline{REFRESH}$ can be held low for self-refresh to occur

14. a. In what form is data stored in a Josephson junction memory?

b. In what environment must these memories operate?

15. a. What are the advantages and disadvantages of serial memories such as CCDs?

b. What is meant by the term *latency time*?

16. a. How are magnetic bubbles formed in garnet?

b. Do magnetic bubble memories store data in serial or random access form?

c. Describe the organization of MBMs.

d. Is MBM storage volatile or nonvolatile?

e. What are the major problems of MBMs?

17. a. How are 1's and 0's stored in core memory?

b. Describe how a particular core in a plane is addressed and read.

c. Explain how a 0 is written into a particular core by using the inhibit wire.

d. Cite the advantages and disadvantages of core memory.

18. a. Describe the operation of a magnetic tape read-write head.

b. Draw a waveform to show how a 1 or a 0 is written on magnetic tape in the F2F format.

c. What is the Kansas City standard for tape-recorded data?

d. In referring to tape-recorded data, what is meant by a *block*, or *record*? A *file*?

e. List the advantages and disadvantages of tape storage of digital data.

19. How is the access for magnetic disks different from that for magnetic tape?

20. How is the read-write head positioned over the desired track in a disk system?

21. a. Define the terms *hard-sectored* and *soft-sectored* in reference to floppy disk data storage.

b. How many bytes of data can be stored in each sector by using the IBM 3740 soft-sectored format?

c. What are check-sum or CRC bytes at the end of a data string used for?

22. List the precautions to be taken in handling floppy disks.

23. Why can hard disks store more data and read it out faster than floppy disks?

24. Define the term *EBAM*.

25. a. Why is it not sufficient to test semiconductor memories only with a write-read of all 0's and write-read of all 1's?

b. What is a galpat?

26. Why is a logic analyzer often needed to troubleshoot a memory system?

27. Draw a simple block diagram of a logic analyzer.

28. a. Which display format is best for finding narrow glitches.

b. What type of clocking will best show the glitch?

29. What is meant by the term *latch-mode storage*?

30. How can a logic analyzer be set to trigger on a particular word or address in a data stream?

31. How is it possible for a logic analyzer to display data that occurred before a trigger?

REFERENCES/BIBLIOGRAPHY

Down, Robert L.: "Technical Note: Understanding Logic Analyzers," *Computer Design*, vol. 16, no. 6, June 1977, pp. 188–191.

Tektronix 1982 Catalogue, Tektronix, Inc., Beaverton, OR, 1982.

Memory Data Manual, Motorola, Inc., Austin, Texas, 1980.

Intel Component Data Catalog, Intel Corporation, Santa Clara, CA, 1982.

Harris Digital Data Book, Harris Semiconductor Products Division, Melbourne, FL, 1981.

Memory Data Book, National Semiconductor, Santa Clara, CA, 1980.

CHAPTER 6

D/A AND A/D CONVERTERS

When you want to store the output data from some sensor such as a temperature or pressure transducer in a memory, usually the data must be converted from an analog form to a digital 1's and 0's form, which the memory can accept. An analog-to-digital (A/D) converter is used to make this conversion. An *analog* quantity can have any value within a range. A *digital* representation of a quantity has only a fixed number of possible values.

In the familiar world a sidewalk ramp is an analog system because there are an infinite number of positions on which you can put your foot between the bottom and top of the ramp. A ladder is a digital system because there are only a fixed number of permitted positions for your feet.

The inverse of an A/D converter is the digital-to-analog (D/A) converter. The D/A converter produces an output current or voltage proportional to the magnitude of a binary word applied to its input. In this chapter we discuss D/A converters first because they are simpler than A/D converters and because several types of A/D converters use D/A converters as part of their circuitry.

At the conclusion of this chapter, you should be able to:

1. Draw the circuit for a weighted resistor D/A converter and describe its operation.

2. Draw the circuit for an R-2R ladder-type D/A converter and describe its operation.

3. Define the terms used to describe D/A and A/D converters, such as resolution, accuracy, linearity, monotonicity, and conversion time.

4. Determine the number of bits required for a given resolution of an A/D or D/A converter.

5. Draw the block diagram and describe the operation of the parallel-type A/D, the single-ramp A/D, the dual-ramp A/D, the counter-type A/D, the tracking A/D, and the successive-approximation A/D converter.

6. Describe the circuits necessary to convert an A/D converter to a digital multimeter.

7. Define CODEC.

DIGITAL-TO-ANALOG CONVERTERS

BINARY WEIGHTED RESISTOR D/A CONVERTER

The simplest type of D/A converter uses binary-weighted resistors and an operational amplifier, or op amp, to convert binary weighted logic signals to a proportional voltage. Figure 6-1 shows the circuit for a 4-bit converter of this type. An op amp has very high gain (usually greater than 100,000), low output impedance, and high input impedance. It has two inputs: an inverting input, signified by the minus sign in the triangular schematic symbol, and a noninverting input, signified by the plus sign in the symbol. Probably the most important thing to remember about op amps is that whenever some of the output is fed back to the inverting input (negative feedback), you will measure the same voltage on the inverting and noninverting inputs. The output of the op amp will source or sink the necessary current to keep these voltages the same. In the circuit of Figure 6-1 the noninverting input is grounded. Therefore, the inverting input will be held at 0 V. Since the inverting

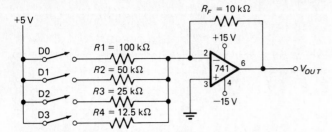

FIGURE 6-1 Binary weighted resistor D/A converter.

input is at 0 V but not actually tied to ground, sometimes it is referred to as a *virtual ground*.

Now consider what happens if switch $D0$ is closed. The 100-kΩ resistor $R1$ has 5 V on one end and 0 V (virtual ground) on the other. Ohm's law tells you that a 100-kΩ resistor with 5 V across it will have a current of 0.05 mA through it. This current cannot go into the input of the op amp because the op amp has a very high input impedance and cannot sink or source much current. Therefore the 0.05 mA must go to the output of the op amp through the 10-kΩ feedback resistor. Ohm's law, applied again, tells you that the voltage on the output of the amplifier must be 10 kΩ × −0.05 mA = −0.5 V to sink the current through switch $D0$ and maintain the virtual ground. If you have trouble visualizing this, draw it as a simple voltage divider with one end at +5 V, the center at 0 V, and the other end at −0.5 V.

Now, mentally open switch $D0$ and close switch $D1$. Since $R2$ is only half the value of $R1$, twice as much current, or 0.1 mA, will flow through R_F, the virtual ground, and $R2$. Twice as much current through R_F will double the output voltage to −1 V. The output voltage is directly proportional to the current through R_F. The next step is to close both $D0$ and $D1$ and determine the output voltage. With both switches closed, 0.05 mA will flow through $R1$ and 0.1 mA through $R2$. The two currents add at the virtual ground point and flow through R_F. Since R_F now has a total of 0.15 mA flowing through it, the voltage on the output of the op amp is −1.5 V. You can try some other combination of switches and predict the output voltages. The resistors produce binary-weighted currents according to the digital word programmed on switches $D0$ through $D3$. These currents are summed at the virtual ground, or *summing point*, and converted to a proportional voltage output by the op amp and R_F. You may wish to build the circuit in Figure 6-1 and connect the outputs of a 4-bit binary TTL or CMOS counter in place of the switches. The output is an interesting staircase waveform with 15 steps of −0.5 V each. The size of the steps can be changed by altering the value of R_F. However, if R_F is too large, the top step will drive the op amp output into saturation at about −14 V.

R/2R LADDER D/A CONVERTER

For a D/A converter of more than about 4 bits, a circuit such as that shown in Figure 6-1 has the problem that

the required range of resistor values is too great. Another circuit for producing binary-weighted currents with only two resistor values is shown in Figure 6-2. As before, the binary weighted currents are converted to a proportional voltage by the op amp and R_F.

At first glance, the $R/2R$ ladder may look like a Kirchhoff's law nightmare, but because of the simple resistor ratios it runs through quite easily. First, assume that switch $D3$, the most significant bit switch, is connected

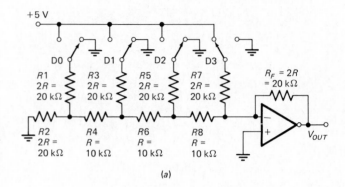

(a)

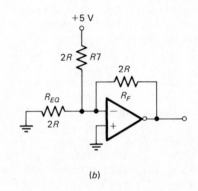

(b)

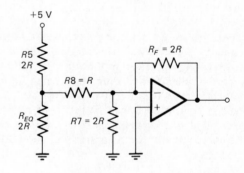

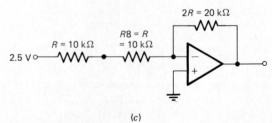

(c)

FIGURE 6-2 *R/2R* ladder D/A converter. (a) Full schematic. (b) Equivalent circuit for MSB switch on. (c) Equivalent circuits for next MSB switch on.

to a precise reference voltage of +5 V and the other switches are closed to ground. Then $R1$ and $R2$ are in parallel to ground. A resistor of $2R$ in parallel with another $2R$ is equivalent to a single R to ground from their junction. This equivalent R adds to $R4$ to form another $2R$ in parallel with $R3$ to ground. The combination of $R3$ and the previous resistors reduces to a single R in series with $R6$. Using the same technique through the rest of the circuit gives the simplified circuit shown in Figure 6-2b. Since the virtual ground of the op amp is at 0 V, there is no current through the equivalent resistor to ground. Therefore, this resistor can be ignored. The 5 V on the end of the 20-kΩ $R7$ produces 0.25 mA through the summing point and the 20-kΩ R_F. The output voltage produced by the most significant bit is -5 V.

To find the voltage produced by the next most significant bit, mentally close switch $D2$ to +5 V and switch $D3$ to ground. All the resistors to the left of $R5$ in Figure 6-2a reduce to a single resistor of $2R$ to ground. The circuit can be simplified further by finding the Thevenin equivalent of the voltage divider formed by $R5$ and the $2R$ to ground, as shown in Figure 6-2c. The Thevenin voltage is just the voltage at the junction, or 2.5 V. The Thevenin resistor is equal to the value of the two resistors in parallel, or R. Everything to the left of $R8$ can be represented by a single resistor of value R connected to 2.5 V. You can ignore $R7$ because both ends are at ground. The total resistance between the summing point and the Thevenin equivalent voltage of 2.5 V is $2R$ or 20 kΩ. The current into the summing point is then 2.5 V/20 kΩ, or 0.125 mA. This current through the 20-kΩ R_F produces an output voltage of -2.5 V for the next to most significant bit. With a similar analysis you can determine that the next lower significant switch produces an output of 1.25 V and the least significant switch an output of 0.625 V. The full-scale output voltage for all switches on is the sum of the four voltages, or 9.375 V.

Even though the $R/2R$ ladder D/A converter is more difficult to analyze than the weighted resistor type, it is easier to build accurately because only one or two values of precision metal-film resistor are needed. The number of bits can be expanded by adding more sections with the same $R/2R$ values. A 4-bit TTL or CMOS binary counter can be connected in place of the switches in Figure 6-2a to give a negative staircase output waveform. Not all the steps may be even because of the variations in logic high voltage levels.

MONOLITHIC AND HYBRID D/A CONVERTERS

Monolithic means "one stone," and when it is used in reference to integrated circuits, it indicates that all the circuitry is contained on a single silicon chip or die. A hybrid circuit contains one or more silicon chips and resistor networks or other microminiature components in a single IC-type package.

A common 8-bit monolithic D/A converter is the MC1408L, whose block diagram is shown in Figure

6-3a. The 1408L comes in a standard 16-pin DIP and requires a V_{CC} of +5 V and a V_{EE} of any voltage from a minimum of -5 to a maximum of -15 V. In the 1408L an $R/2R$ ladder divides a reference current from the current amplifier into eight binary-weighted currents. Bipolar transistor switches then switch the binary-weighted currents to an output line according to the binary word on the TTL-compatible inputs A1 to A8. Note that the labels of the MSB and the LSB are reversed from the way in which counters are normally labeled. Some D/A converters do this and some don't, so you have to read the data sheets carefully. The 1408L has a current output that can be converted to a voltage with an op amp and a resistor as shown in Figure 6-3b. This voltage can be calculated using the following formula:

$$V_{OUT} = \frac{V_{REF}}{R14} \times R0 \left(\frac{A1}{2} + \frac{A2}{4} + \frac{A3}{8} + \frac{A4}{16} + \frac{A5}{32} + \frac{A6}{64} + \frac{A7}{128} + \frac{A8}{256} \right)$$

For the values shown, the full-scale output voltage is

$$2 \text{ V} \times \frac{5 \text{ k}\Omega}{1 \text{ k}\Omega} \times \frac{255}{256} = 9.961 \text{ V}$$

This is referred to as a 10-V full scale converter.

A fun application for inexpensive D/A converters such as the 1408L is to create unique audio sounds and waveforms. The outputs of an 8-bit binary counter can be connected to the D/A inputs. As the counter cycles through its count, the D/A converter puts out sawtooth waveforms made of 255 tiny steps. The output frequency is equal to the input clock frequency of the counter divided by 256. More elaborate waveforms can be created by connecting the D/A converter to the output of an 8-bit-wide ROM or RAM memory. The memory is programmed with the desired binary value on a point-to-point basis. The memory is sequentially read out to the D/A converter over and over by a counter attached to its address inputs. The number of possible sounds is limited by only your imagination, the amount of memory, and your patience in programming the memory.

For applications requiring more bits of resolution, hybrid D/A converters such as the Datel DAC-HZ 12 BCG are available in a 24-pin DIP. A picture of the hybrid with the cover off is shown in Figure 6-4a, and a block diagram is shown in Figure 6-4b. This is a 12-bit converter which requires only a +15- and -15-V supply. Resistors and an op amp are included in the package, so both current and voltage outputs are available. The full-scale output voltage can be changed by using different values of feedback resistor on the op amp. A precision reference voltage for the current-determining network is included. To get the accuracy needed for a 12-bit converter, the resistors in the thin-film resistor network are individually trimmed by a laser to the proper value. Since each resistor has to be trimmed, it is desirable to reduce the number of resistors required in the converter. The $R/2R$ ladder requires two resistors per bit,

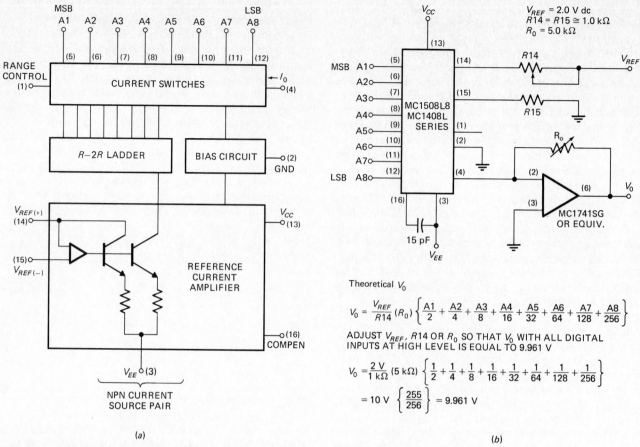

FIGURE 6-3 Motorola MC1408 D/A converter. *(a)* Block diagram. *(b)* Schematic for voltage output.

and the weighted resistor method uses only one resistor per bit. The problem with the weighted resistor method is the extreme range of resistors needed for more than about a 4-bit converter. This problem is solved by using three sections of the 4-bit weighted resistor type as shown in Figure 6-4c. Resistors in series with the less significant sections attenuate the output from these sections, so the same-value current source resistors can be used in all sections. As shown in Figure 6-4c, the precision current sources are all on. Switch circuitry not shown turns off desired current sources by pulling their emitters high. There are several switching schemes, so it is not possible to discuss them all here. The important thing for you as a user of D/A converters is the overall specifications rather than which switching mode is used internally.

D/A CONVERTER CHARACTERISTICS, ERROR SOURCES, AND SPECIFICATIONS

The first characteristic of a D/A converter to consider is *resolution*. This is determined by the number of bits in the input word. An 8-bit converter, for example, has 2^8, or 256, possible output levels; so its resolution is 1 part in 256. A 12-bit converter has a resolution of 1 part in 2^{12}, or 4096. Resolution is sometimes expressed as a

percentage. One part in 4096 is about 0.024 percent.

The next point to consider is the *accuracy* of a D/A converter. By accuracy we mean a comparison of the actual output with the expected output. It is specified as a percentage of full scale, or the maximum output voltage. If a converter has a 10-V full-scale output and ±0.2 percent accuracy, then the maximum error for any output will be 0.002 × 10 V, or 20 mV. Ideally, the accuracy of a D/A converter should be, at worst, ±½ of its LSB. A 10-bit converter has a resolution of 1 part in 1024, or about 0.1 percent. To be consistent, the accuracy should be ±0.05 percent. There are several ways that this error can show up. Figure 6-5 illustrates three.

Linearity errors involve the amount by which the actual output differs from the ideal straight-line output. This error usually is caused by errors in current source resistor values. Another error type is a *gain error*. This is often caused by errors in the feedback resistor on the current-to-voltage converter op amp. *Offset error* means that the output is not 0 when all the inputs are 0's. Offset error adds a constant value to all output values, as shown in Figure 6-5c. This error can be caused by op amp errors or leakage currents in the current switches.

The accuracy of a D/A converter may be less than its resolution. The MC1408L, for example, has a resolution

FIGURE 6-4 Datel hybrid D/A converter.
(a) Photograph. (b) Block diagram. (c) Binary
weighted current sources.

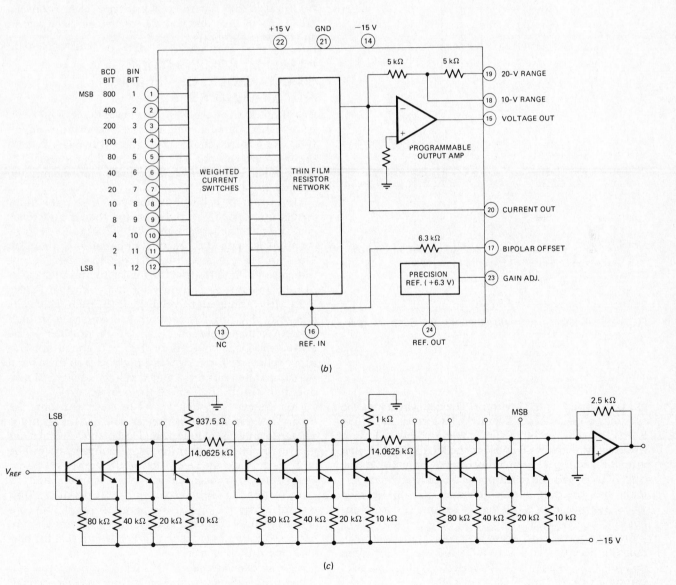

(b)

(c)

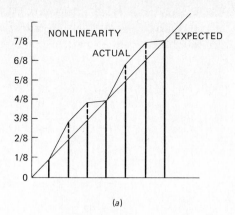

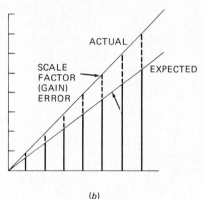

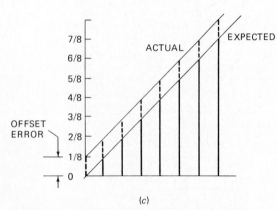

FIGURE 6-5 D/A errors. *(a)* Linearity. *(b)* Scale or gain. *(c)* Offset.

of 1 part in 256, or 0.39 percent. The data sheet lists the accuracy of the MC1408L-8 as ±0.19 percent, or ±½ of a least significant bit which matches the resolution. However, Motorola also markets the MC1408L-6, which is an 8-bit converter with an accuracy of only ±0.78 percent. This equals the resolution of a 6-bit converter. The lower-accuracy part is much cheaper and can be used where the range of 8 bits is needed but the accuracy isn't.

Another term used with D/A converters is *monotonicity*. A converter is said to be monotonic if it does not miss any steps or step backward when stepped through its entire range by a counter.

Output settling time is still another specification given for D/A converters. The settling time is usually defined as the time required for the converter output to settle to within ±½ LSB of the final value after an input word change. If a converter is operated at too high a frequency, it may not have time to settle to the correct value before it is switched to the next.

ANALOG-TO-DIGITAL CONVERSION

In the last section we showed how a digital word can be converted to a proportional current or voltage. In this section we discuss the reverse process of converting an analog voltage to the digital word that best represents the voltage. As with a D/A converter, the resolution of an A/D converter depends on the number of bits. An 8-bit converter, for example, can represent an input voltage with any of 256 possible words. The resolution is said to be 1 part in 256. There are many types of A/D converters. A few of the most common types are discussed in the following sections.

PARALLEL COMPARATOR, SIMULTANEOUS, OR "FLASH" A/D CONVERTERS

The simplest in concept and the fastest type of A/D converter is the parallel comparator type shown in Figure 6-6a. A resistive voltage divider sets threshold voltages on three comparators. The reference voltage at the top of the divider is the full-scale or maximum meaningful input voltage.

The output of each comparator goes high if the input voltage on its positive input is greater than the reference voltage on its negative input. The analog signal to be digitized is applied to all the comparators in parallel. The number of comparators tripped to a high output indicates the amplitude of the analog input voltage. Figure 6-6b shows the outputs for the circuit in Figure 6-6a with various input voltages. If the input voltage is less than 1 V, no comparators are tripped to a high state. A voltage between 1 and 2 V is indicated if the lowest-threshold comparator has a high output. An input voltage of 2 to 3 V gives a high on both the A1 and A2 comparator outputs, and a voltage above 3 V produces an output high on all the comparators.

This system is similar to a 4-in ruler with only the inch marks on it. The converter will resolve an input voltage to only one of four input levels. This is equivalent to 2 binary bits of resolution. More comparators can be used to increase the resolution of the converter. Seven comparators are needed for 3-bit resolution, and 15 comparators for 4-bit resolution, or 16 levels. A N-bit converter requires $2^N - 1$ comparators, so an 8-bit converter requires 255 comparators. This excessive number of comparators is the major disadvantage of the parallel-type A/D converter.

Another disadvantage of this type of converter is that the output code is not binary. It can, however, be con-

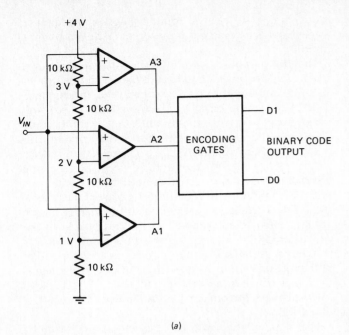

(a)

	Comparator Outputs			Binary Output	
V_{IN} (Volts)	A1	A2	A3	D1	D0
0 to 1	0	0	0	0	0
1 to 2	1	0	0	0	1
2 to 3	1	1	0	1	0
3 to 4	1	1	1	1	1

(b)

FIGURE 6-6 Parallel comparator A/D converter. (a) Schematic. (b) Outputs for various input voltages.

verted to binary with added gates. The conversion of the comparator output code for the circuit in Figure 6-6a is left to the reader as a review of logic gates and Karnaugh maps.

The major advantage of the parallel-type A/D converter is its speed. The entire digital output word is present after just the propagation delay time of the comparators and the encoding gates. Thus the name *flash* is sometimes used for this type. The TRW TCD-1007J can do an 8-bit flash conversion in about 33 ns.

SINGLE-RAMP OR SINGLE-SLOPE A/D CONVERTER

Figure 6-7a shows an A/D converter circuit using a ramp generator, comparator, and BCD, or binary, counters. At the start of a conversion cycle, the ramp and counters have been reset to 0's. An analog voltage is applied to the comparator plus input. Since this input is more positive than the negative input of the comparator, the output of the comparator goes high. This both enables the AND gate to let the clock get to the counters and starts the ramp. The ramp voltage goes positive until it exceeds the input voltage. This causes the comparator output to go low and disables the AND gate. The clock to the counter, therefore, is shut off. Control circuitry such as that in the frequency counter of Chap. 4 can be used to latch the counter data, reset the counters to 0's, and reset the ramp generator.

To further see how this circuit works, assume a clock frequency of 1 MHz, four BCD counters, and a V_{IN} of 2.000 V. Further, assume that the ramp has a slope of 1 V/ms, as shown in Figure 6-7b. From the start of a conversion cycle the ramp will take 2 ms to reach 2 V and shut off the clock pulse to the counters. In this 2 ms, 2000 clock pulses will have reached the counters and been counted. The comparator output going high will strobe the latches and send the count to the dis-

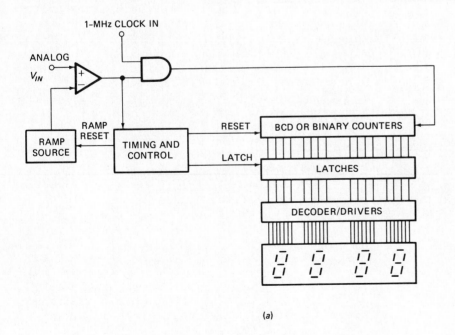

(a)

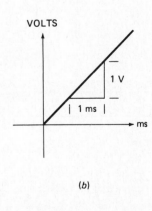

(b)

FIGURE 6-7 Single-slope A/D converter. (a) Block diagram. (b) Ramp slope.

plays. Inserting a decimal point at the proper point in the seven-segment digital display gives a reading of 2.000, which corresponds directly to the analog input voltage of 2.000 V. Any other positive input voltages up to 9.999 V can be converted to a BCD equivalent and displayed in the same manner. This circuit is a simple digital voltmeter. Digital voltmeters are discussed in depth later in this chapter. If binary counters are used instead of the BCD counters, the output will be straight binary-coded.

For applications such as digital voltmeters which may require resolution of 1 part in 20,000 or more, the single-ramp or single-slope A/D converter is not stable enough. Variations in the ramp generator are caused by time, temperature, or input voltage sensitivity. In the dual-slope converter discussed next, most of these effects are canceled out.

DUAL-SLOPE A/D CONVERTERS

The block diagram of a dual-slope A/D converter circuit is shown in Figure 6-8a. The circuit is very similar to the single-slope circuit except for the switch on the

input, which selects the input voltage or a reference voltage, and the reversed connections to the comparator inputs.

The first part of this circuit is the *ramp generator*, or *integrator*. The inverting input of the op amp is held at virtual ground by the op amp. A voltage of, for example, 2 V applied to the input end of the 10 kΩ resistor will cause a constant current of 0.2 mA to flow through the resistor to the summing point. Since this current cannot flow into the high-impedance op amp input, it flows into one plate of the capacitor. To keep the inverting input of the op amp at virtual ground, the op amp must pull this same current from the other plate of the capacitor. As the capacitor charges, the output voltage on the op amp must go more and more negative to keep the constant current flowing. The voltage across a capacitor being charged by a constant current is a *linear ramp*. With a positive input voltage, the output of the integrator ramps negative, as shown in Figure 6-8b. A negative voltage causes the output to ramp linearly in a positive direction.

The slope of the ramp can be determined easily by using the basic relationship $q = CV$ for capacitors and

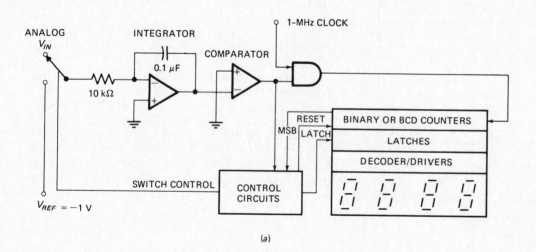

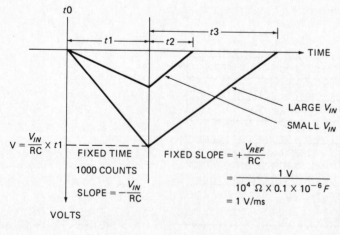

FIGURE 6-8 Dual-slope A/D converter. *(a)* Block diagram. *(b)* Integrator output versus time.

$q = It$. Setting the two equal and shuffling the terms yield $\Delta V/\Delta t = I/C$. Since the current is equal to V_{IN}/R, we know that $\Delta V/\Delta t = V_{IN}/(RC)$. This further demonstrates that for a given V_{IN} the output slope, or $\Delta V/\Delta t$, is a constant. For the values shown and a $+2$-V input, the output slope is -2 V/ms.

Now that the ramp generator has been explained, a conversion cycle can be traced. A cycle starts with the ramp at 0, the counters reset to 0's, and the input switch connected to the input voltage. The positive voltage into the integrator causes its output to ramp negative.

When the integrator output pushes the inverting input of the comparator negative, the comparator output goes positive and enables the AND gate. This lets the clock into the counters. The integrator output is allowed to ramp negative for a fixed number of counts. This is shown for two different input voltages as $t1$ in Figure 6-8b. When the counter reaches the fixed count, control circuits reset the counters to 0's and switch the integrator input to the negative reference voltage. A negative input voltage will cause the integrator output to ramp positive, as shown for $t2$ in Figure 6-8b. When the integrator output reaches just above 0 again, the comparator output goes low. Control circuitry detects this transition and strobes the counter outputs into the latches. It then resets the counters to 0 and switches the integrator input back to the input voltage. This starts another conversion cycle. The number of counts stored in the latches is proportional to the input voltage V_{IN}. Here's why.

The integrator output in fixed time $t1$ ramps down to a voltage V, equal to $(V_{IN}/RC) \cdot t1$. To get back up to 0, the integrator must ramp up the same amount of voltage. For the reference integrate period $t2$, the voltage V is equal to $(V_{REF}/RC) \cdot t2$. The two expressions for V can be set equal:

$$\frac{V_{IN}}{RC} \times t1 = \frac{V_{REF}}{RC} \times t2$$

$$V_{IN} \times t1 = V_{REF} \times t2$$

$$t2 = V_{IN} \times \frac{t1}{V_{REF}}$$

Since RC appears on each side of the equation, it cancels. The practical meaning of this is that, since the same resistor and capacitor are used for the signal integrate period and the reference integrate period, variations in R, and C will have no effect on the accuracy of the output reading. This is a great advantage over the single-ramp converter. The final result of the equation "juggling" above shows that the counter output $t2$ is directly proportional only to V_{IN} since V_{REF} and $t1$ are constants.

For the circuit shown in Figure 6-8a, $t1$ is 1000 counts of the 1 = MHz clock, or 1 ms. And V_{REF} is -1 V. For a 2-V input signal, $t2$ will be $(2 \text{ V}/1 \text{ V}) \times 1000$ counts, or 2000 counts. A decimal point in the right place makes this represent 2.000 V. The inner graph in Figure 6-8b represents the integrator output for a

smaller input voltage, for example, 0.8 V. For this $t2$ will be $(0.8 \text{ V}/1 \text{ V}) \times 1000$ counts, or 800 counts. The readout shows this as 0.800 V.

Since the dual-slope type of A/D converter is used in many digital voltmeters and other instruments, you should understand its operation well. To summarize: The unknown input voltage is applied to an integrator for a fixed number of counts $t1$. The counter is reset to 0 and the integrator input connected to a reference voltage. The number of counts required for the integrator output to get back to 0 is directly proportional to the input voltage.

The advantages of the dual-slope converter are its accuracy, low cost, and immunity to temperature-caused variations in R or C. The disadvantage of the dual-slope type is its slow speed. A converter able to resolve 1 part in 20,000 may take 100 ms or more per conversion cycle. This type is discussed more in a later section on digital voltmeters.

CHARGE BALANCE A/D CONVERTERS

The charge balance type of A/D converter uses essentially the same circuitry as the dual-slope converter in Figure 6-8a. However, instead of the input being switched back and forth between the unknown voltage and a reference voltage, the input is left connected to the unknown voltage. Pulses of a reference current are inserted directly into the summing junction of the integrator for a fixed time. The number of reference current pulses required to keep the summing junction at 0 V for this fixed time is proportional to the unknown input voltage. The advantage of this technique is that the voltage across the integrating capacitor is kept very near 0 V. Therefore, dielectric absorption by the capacitor is not a problem as it is in high-accuracy dual-slope converters.

A/D CONVERTERS USING COUNTERS AND D/A CONVERTERS

SINGLE-COUNTER TYPE   Earlier we discussed how a linear ramp could be produced by connecting the outputs of a counter to the inputs of a D/A converter. The ramp is actually made up of small steps. The size of the steps is determined by the number of bits of resolution of the converter.

Figure 6-9 shows how voltage output D/A converters and counters can be used in place of an integrator to make a single-ramp type of A/D converter. At the start of a conversion cycle, the counters are reset and the output of the D/A converter is 0. When a voltage is applied to the comparator input, its output goes high and gates the clock to the counters. Each clock pulse advances the counters one count and increases the voltage out of the D/A by one step. When the output of the D/A converter becomes greater than the input voltage, the comparator output is tripped low. This shuts off the clock

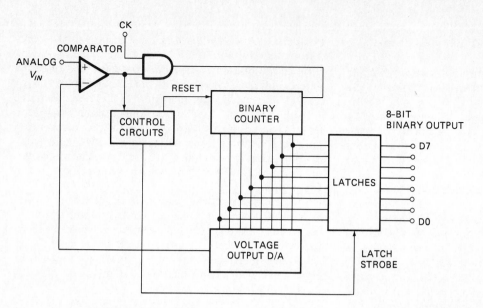

FIGURE 6-9 Binary up counter and D/A converter used to make a single-ramp type of A/D converter.

pulses to the counters. Then control circuits latch the counter outputs and reset the counters to start the cycle over again.

This method is faster than the dual-ramp type, but it requires a precision D/A converter. Another disadvantage is that for each conversion cycle, the count has to start at 0 and count all the way up to where the D/A output matches V_{IN}.

TRACKING A/D CONVERTER   The tracking A/D converter is a method that is somewhat faster for digitizing signals such as audio which change only a small amount from one sample to the next. As shown in Figure 6-10, a tracking A/D converter uses an up/down counter instead of the up-only counter of the previous type. The clock is routed to either the count-up or count-down input by the NAND gate switches on the output of the comparator. At the start of a conversion for some input voltage, the output of the D/A is 0, so the

output of the comparator is high. This routes the clock to the count-up input. The counters count up until the output of the D/A converter is greater than V_{IN}.

The comparator output then changes to a low and routes the clock to the count-down input. However, if V_{IN} has not changed, one down count will drop the output of the D/A converter below V_{IN}. The comparator output will change high and again route the clock to the count-up input. One up count will change the comparator output back low to repeat this loop. The least significant output bit of this type of converter then oscillates for a constant V_{IN}. This is a disadvantage. As the input signal changes, the counter counts up or down to "track" it, thus the name for this type of converter. The advantage is that for digitizing sine waves it only has to count up or down a few counts from one sample to the next instead of counting up from 0 each time. The successive-approximation type described in the next section is faster than either of these and does not oscillate.

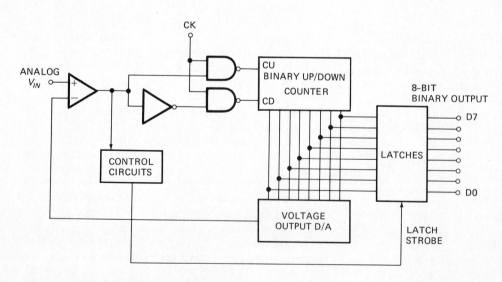

FIGURE 6-10 Binary up/down counter and D/A converter used to make a tracking A/D converter.

SUCCESSIVE-APPROXIMATION A/D CONVERTER

The great advantage of successive-approximation A/D conversion is that N bits of resolution can be produced with only N clock pulses. For example, an 8-bit successive-approximation type requires only eight clock pulses as compared with as many as 256 clock pulses required for a counter type. Figure 6-11 shows a circuit for a successive-approximation (SA) type A/D converter. The heart of this circuit is the successive-approximation register (SAR) such as the MC14549, which functions as follows. At the start of a conversion cycle the SAR, on the first clock pulse, turns on its most significant output bit to the MC1408 D/A converter. Then the SAR waits for a signal from the LM319 comparator indicating whether the D/A output is greater or less than the input voltage. If the comparator output is high, then the D/A output is less than V_{IN} and the SAR will keep the most significant bit set. If the comparator output is low, then the D/A output is greater than V_{IN} and the SAR will reset the MSB. In either case, on the next clock pulse the SAR will set the next most significant bit. It will keep or reset this bit depending on the output from the comparator. The SAR proceeds on down to the least significant bit, trying each bit in turn. It keeps a bit if the D/A output produced is less than V_{IN} and resets a bit if the D/A output is greater than V_{IN}. Only one clock pulse is needed for each bit. When all bits have been tried, the SAR sends out an end-of-conversion (EOC) signal. The EOC signal indicates that the parallel output lines contain a valid binary word representing the size of the analog input signal. If the EOC signal is tied to the start conversion (SC) input, the converter will cycle continuously. The MC14549 also has a serial output which sends the digital word out 1 bit at a time as the SAR determines it.

The successive-approximation method of A/D conversion is similar to finding the height of a table by piling up binary-weighted blocks 128, 64, 32, 16, 8, 4, 2, and 1 cm in height. The most significant block of 128 cm is tried first. If it is too high, it is set aside and a 0 is recorded for that bit position. Then the next most significant block is tried by adding it to the pile. Anytime adding a block makes the pile higher than the table, that block is set aside and a 0 is recorded. If adding a block does not make the pile higher than the table, then a 1 is recorded for that bit. Each block requires only one trial. When all have been tried, the result is an 8-bit binary word which represents the height of the table.

The circuit shown in Figure 6-11 uses readily available parts and is shown with pin numbers, so you can build it if you wish. The maximum input range is ±5 V. The R and C on the clock input are to protect the MC14549 from damage if the power to the A/D converter is accidentally shut off before the pulse generator supplying the clock. Note that the positive input of the current-to-voltage converter op amp is tied to −5 V instead of to ground. This shifts the analog input range to −5 to +5 V, rather than 0 to +10 V. AC sine waves can then be connected directly to the A/D input.

To show how an analog sine wave signal can be reconstructed from its digital values, you can build another D/A converter using an MC1408L-8, an LM741, and the same values as those in the D/A section of Figure 6-11. This D/A converter is connected onto the outputs of the latches. A 5-V *peak*, 1-kHz sine wave input to the A/D converter can be compared to the output of the D/A con-

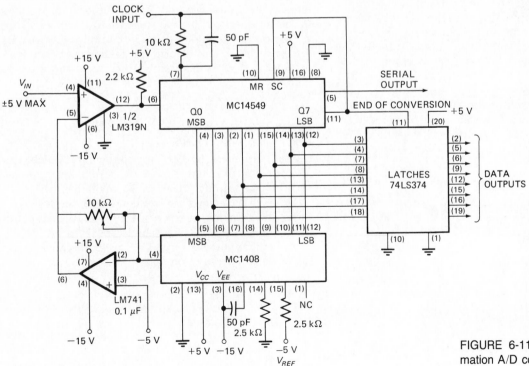

FIGURE 6-11 Successive-approximation A/D converter.

verter. It is instructional to vary the A/D input clock frequency from 1 kHz to about 1 MHz and observe the effect on the output waveform of the D/A converter.

The successive-approximation A/D converter has the disadvantage of requiring a D/A converter, but it has the advantage of high speed with excellent resolution. Analog Devices, Inc. uses I²L to put all the analog and digital circuitry for a 10-bit A/D converter on a single chip in the AD571. The Micro Networks MN5245 does a 12-bit successive approximation conversion in about 900 ns.

A/D SPECIFICATIONS

The specifications discussed earlier for D/A converters such as resolution, accuracy, linearity, etc., are used for A/D converters too. Another term often used with A/D converters is *quantizing error*, or the error between the actual analog value and its digital representation. Ideally, the maximum quantizing error is ±½ of a least significant bit, so this is just another way of specifying resolution.

Another important parameter for A/D converters is *conversion time*, or the time to digitize each sample. This ranges from nanoseconds to milliseconds depending on the type of conversion and the number of bits. High-resolution, slow-speed applications usually use dual-slope types. High-speed, medium-resolution applications use a successive-approximation type of converter, and very high-speed applications are forced to use a parallel comparator scheme.

SAMPLE AND HOLDS

If an attempt is made to digitize a rapidly changing input signal, the input signal will have changed before the conversion is complete. The output of the converter will represent the input at the end of the conversion cycle rather than at the start. Figure 6-12 shows this error. The error voltage ΔV is a function of how fast the input signal is changing. This error is eliminated or greatly reduced by using a *sample and hold* with a buffer amplifier on the input of the A/D converter. An electronic switch which can be turned on and off very rapidly is used to quickly take a sample of the input waveform at time $t1$ and store it on the hold capacitor. Then

the A/D converter digitizes the stored voltage. After the conversion is complete, a new sample is taken and held for the next conversion.

Two important parameters for sample and holds are acquisition time and aperture time. *Acquisition time* is the time required for the switch to close and the capacitor to charge up to the input signal voltage. *Aperture time* is the time needed for the switch to completely open after a hold signal. Ideally both times should be 0 to eliminate any effect of signal change during sampling.

There is another important point to remember when sampling an input signal to digitize it. According to the sampling theorem, an input sine wave must be sampled at a minimum of two nonzero points in the waveform. Actually, a much higher sampling frequency is necessary to get a good digital representation of a sine wave input. Just as with the connect-the-dot pictures children draw, the more dots there are, the better the picture resolution.

If you build the successive-approximation A/D converter of Figure 6-11 and connect a D/A converter on its output, you can vary the clock rate and observe the effect of sampling rate on the reproduction of a sine wave input.

APPLICATIONS

Almost any measurable quantity present as a voltage can be digitized by an A/D converter and displayed with a digital readout. The voltage output of a semiconductor strain gage is amplified, digitized, and displayed in a digital readout scale. Digital readout thermometers use an A/D converter to convert the voltage drop across a temperature-sensitive diode to digital form for display. In the laboratory, A/D converters are the heart of digital voltmeters and digital multimeters. Analog voice signals are converted to digital form for transmission over long distances. At their destination they are reconverted to analog.

DIGITAL READOUT VOLTMETERS AND MULTIMETERS

Digital voltmeters (DVMs) are commonly available as 3½-, 4½-, or 5½-digit units. The ½ digit refers to the fact that the leading digit can display only 0 or 1.

Figure 6-13a shows the schematic for the A/D converter and display sections of a 4½-digit DVM using the Intersil 8052A and 7103A. The full-scale input voltage range of the circuit is ±1.9999 V. The 8052A contains the reference supply, buffer amplifier, integrator, and comparator for a dual-slope converter. The 7103 contains MOS switches, counters, and the control logic for the converter as well as latches and a multiplexer for the 4½-digit display.

The circuit functions as the dual-slope converter discussed previously except for two changes which improve accuracy and make it possible to digitize either positive

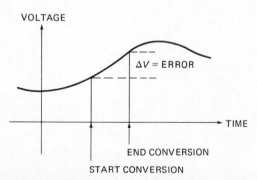

FIGURE 6-12 Error caused by input voltage changing during A/D conversion time.

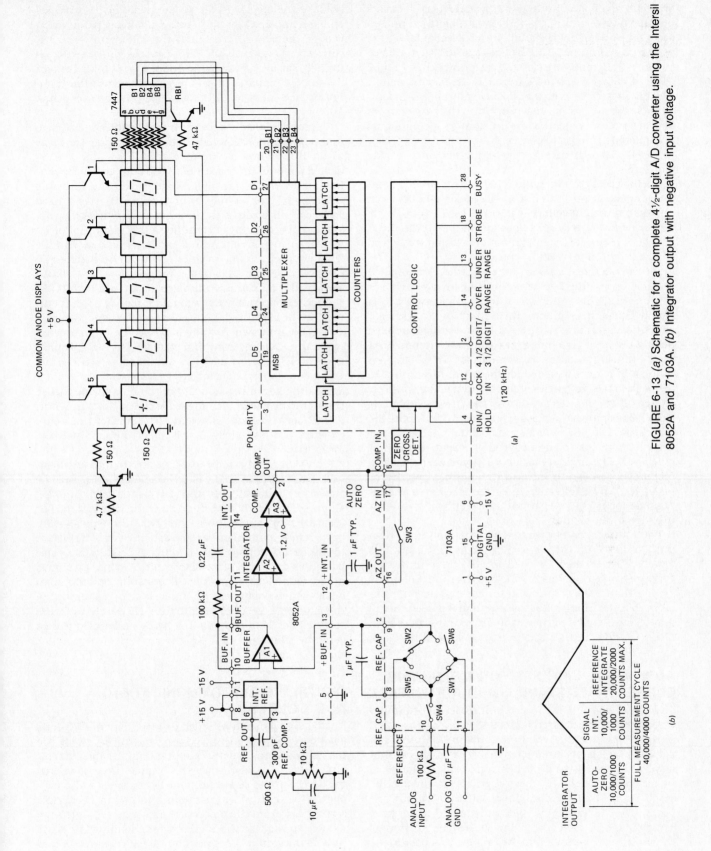

FIGURE 6-13 (a) Schematic for a complete 4½-digit A/D converter using the Intersil 8052A and 7103A. (b) Integrator output with negative input voltage.

173
D/A AND A/D CONVERTERS

or negative voltages. First, the conversion cycle consists of three phases rather than just two. The three phases are auto zero, signal integrate, and reference integrate, as shown in Figure 6-13b. During the auto zero phase, switches 1, 2, and 3 are closed. Closing switches 1 and 2 stores a voltage equal to V_{REF} on the reference capacitor. Closing switch 3 stores the effects of any offsets or leakages on the auto zero capacitor. Since the stored errors become the reference for the integrator during signal and reference integrate phases, the errors are effectively canceled. This greatly increases the accuracy of the converter.

During the signal integrate phase, switches 1, 2, and 3 are opened and switch 4 is closed for 10,000 clock pulses. If the input voltage is positive, the integrator will ramp downward and set a polarity flip-flop. If the input voltage is negative, the integrator will ramp positive and the polarity flip-flop will not be set.

For the reference integrate phase, switch 4 is opened and the polarity flip-flop determines whether switch 5 or 6 is closed. If the input was positive, then switch 6 is closed. Since this connects the buffer input to ground, which is V_{REF} more negative than the voltage applied during the auto zero phase, the integrator ramps back positive until it crosses the comparator threshold. The number of counts required for the integrator output to get back to 0 is proportional to the input voltage, as discussed. If the input voltage was negative, switch 5 is closed. This puts a more positive voltage of V_{REF} plus the V_{REF} stored in the reference capacitor on the buffer input. Then the integrator ramps negative until it reaches 0. The trick of storing V_{REF} on a capacitor and making the integrator able to ramp either positive or negative during reference integrate is what makes it possible for this converter to handle either positive or negative input signals.

This chip set can be operated as a 4½-digit unit when pin 2 of the 7103 is high or a 3½-digit unit if pin 2 is low. A 12-kHz clock is used for 3½-digit units. An overrange output indicates whether the input voltage is greater than 2.0000 V, and an underrange output indicates that the input voltage is less than 9 percent of full scale. These can be connected to driver transistors with LEDs to indicate that another range should be used.

ADDITIONAL CIRCUITS FOR MEASURING CURRENT, RESISTANCE, AND AC VOLTAGE

To make the basic A/D converter into a useful laboratory instrument, several circuits are added to the basic A/D unit of Figure 6-13a. The first addition is a precision metal-film resistor attenuator network such as that shown in Figure 6-14a. This permits higher voltages to be measured. Each step down the network divides the input voltage by 10. For example, with 500-V input, the B junction will have 50 V, the C junction 5 V, and the D junction 0.5 V. Since only the D switch position is within the ±1.9999 V of the A/D range, this position is used for converting the 500 V to a digital readout. Another section on the same switch can move the decimal

point in the display to the proper position. The 100-kΩ resistor and back-to-back zener diodes serve as overvoltage protection. They limit the voltage into the A/D converter to about ±5.7 V in case the range switch is initially set in the wrong position. The 100-kΩ resistor does not attenuate the input signal because the input impedance of the A/D converter is greater than 1000 MΩ.

A second addition is a circuit that enables the DVM to function as an ohmmeter. Figure 6-14b shows one way to do this. When an unknown-value resistor is connected between the output of the op amp and the inverting input, the negative feedback holds the inverting input at 0 V, or virtual ground. Since R1 has a fixed voltage of −2 V across it, the current through R1 is constant. This constant current flows through R unknown to give an output voltage that can be measured with the DVM circuitry. Various values of R1 can be switched in to give different resistance ranges. For the circuit shown, the maximum unknown resistance is 200 kΩ.

Another improvement is to add circuitry so that current can be measured. Figure 6-14c shows one way to do this. The unknown current is passed through a sense resistor to be converted to a voltage. The small voltage across the sense resistor is amplified by 100 before being sent to the A/D converter. Amplifying the voltage across the sense resistor allows a much smaller sense resistor to be used. For a full-scale reading of 2 mA, only a 10-Ω sense resistor is needed. The smaller this sense resistor, the less likely it will affect the circuit being measured.

A DVM or DMM also needs to be able to measure ac as well as dc voltages. To measure ac voltages, they must be converted to dc voltages first. For power applications a simple diode or diode bridge does the conversion. For measurement applications, however, the typical 0.6-V drop across a diode causes too much error. The circuit in Figure 6-14d uses two op amps to produce a precise dc equivalent output for an ac input. The first op amp is an active half-wave rectifier. The second op amp is an adder which combines the input ac with the output of the half-wave rectifier to produce effectively full-wave rectified dc. This dc is filtered by C_F and scaled to rms value by adjusting the 5-kΩ pot.

DIGITAL VOICE COMMUNICATION AND CODECS

Analog signals such as video or voice communications get noisy when transmitted long distances. Since the noise is random, it is difficult (if not impossible) to filter out. Digital signals, however, because they have only two levels, 1 and 0, are not affected by small noise voltages. Therefore, if you are trying to get an intelligible signal back from the moon or Mars, it is well worth the trouble to digitize the signal before sending it and to pass it through a D/A converter at the receiving end to reconstruct the analog signal if necessary.

Figure 6-15 shows a single-channel digital communi-

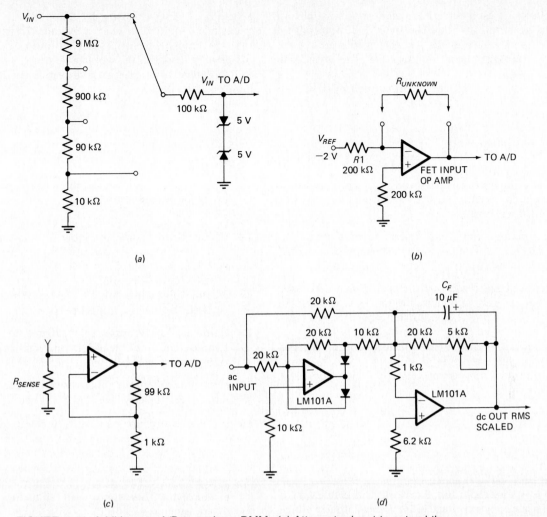

(a)

(b)

(c)

(d)

FIGURE 6-14 Additions to A/D to make a DMM. *(a)* Attenuator input to extend the input voltage range of an A/D converter. *(b)* Resistance-to-voltage converter. *(c)* Current-to-voltage converter. *(d)* AC-to-DC voltage converter.

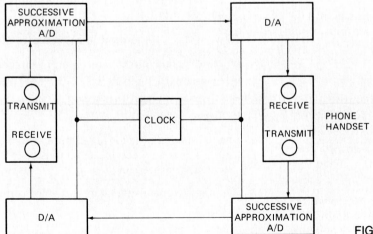

FIGURE 6-15 Digital voice communication link.

cation link. In each end, a successive-approximation A/D converter changes the voice signal to a serial digital form. At the receiver a D/A converter reconstructs the audio waveforms. A common clock keeps the two ends synchronized.

A coder-decoder, or CODEC, integrated circuit such as the Intel 2910 contains the A/D and D/A converters needed for one end of such a system. To save wire or radio channels, several CODEC outputs can be multiplexed on a single pair of wires or channel. Sending data in this digital form is referred to as *pulse code modulation,* or *PCM.*

REVIEW QUESTIONS AND PROBLEMS

1. What voltage will be measured at the negative input of an op amp with negative feedback if the positive input is grounded?

2. What voltage would you expect to find at the inverting input of an op amp with negative feedback if the noninverting input were tied to -5 V?

3. *a.* A 6-bit weighted resistor type of D/A converter has a least significant digit resistor value of 80 kΩ. What values do the other five resistors have?
   b. How many discrete output steps, or levels, does this converter have?
   c. What is a problem of weighted-resistor-type D/A converters?

4. Draw the circuit for a 4-bit $R/2R$ D/A converter. Draw the equivalent circuit for the next-to-MSB switch closed to V_{REF} and the other three closed to ground. What is the advantage of the $R/2R$ type over the weighted resistor type?

5. What is the resolution of a 12-bit D/A converter? If full-scale output voltage is 10 V, how large a voltage step does a least significant bit represent?

6. For a D/A converter define the following: *(a)* linearity error, *(b)* gain error, *(c)* offset error, *(d)* monotonicity. *(e)* Show with a sketch how each of these errors will cause the actual output to differ from the expected output.

7. What accuracy is required of a 12-bit D/A converter to be consistent with its resolution?

8. *a.* How many comparators are required to build a 6-bit parallel A/D converter? Is this number usually practical?
   b. What is the main advantage of the parallel-type A/D converter?

9. Show the gate circuits needed to convert the comparator-tripped output code in Figure 6-6*b* to binary.

10. Draw a block diagram of a single-ramp A/D converter, and briefly describe its operation. What is a disadvantage of the single-ramp types?

11. Draw a block diagram of a dual-slope A/D converter and briefly describe its operation. Include in your explanation the terms *reference integrate* and *sig-nal integrate.* How does the dual-slope A/D converter overcome the main problem of the single-slope type?

12. *a.* For the integrator in Figure 6-8*a*, what slope will the output be for an input voltage of 1.5 V?
   b. What negative voltage will the output reach after 1000 counts of the 1-MHz clock?
   c. How long will it take to digitize the 1.5-V input with the circuit in Figure 6-8*a*?

13. Using a 100-kHz input clock, how long will it take the 8-bit counter type of A/D converter in Figure 6-9 to digitize a full-scale signal?

14. How is a tracking A/D converter different from the single-counter type?

15. *a.* How many clock pulses are required for each conversion cycle of an 8-bit successive-approximation type of A/D converter?
   b. With a 100-kHz clock, how long will a conversion take?
   c. Compare this time with that for the 8-bit counter type.

16. Draw a block diagram of a successive-approximation type of A/D converter, and briefly describe its operation.

17. *a.* Why is it necessary to use a sample and hold when you are digitizing rapidly changing signals?
   b. How often must a sine wave input be sampled?
   c. Define acquisition time and aperture time for a sample and hold.

18. To what does the ½ digit of a 3½-digit DVM refer?

19. What additional circuits are required to change an A/D converter to a digital multimeter?

20. Describe the three conversion phases for the circuit in Figure 6-13*a*.

21. Show the schematic for an attenuator network that might be used on the input of an A/D converter to extend its voltage range.

22. What is the advantage of digital voice communication?

23. Define CODEC.

REFERENCES/BIBLIOGRAPHY

Technical Staff of Datel Systems, Inc.: "DATEL Systems Products Handbook," *Electronic Design's GOLD BOOK*, vol. 3, 1976–77, pp. 3/6–3/18, 3/87–3/90.

Linear Applications Handbook, National Semiconductor Corporation, Santa Clara, CA, 1980.

McMOS Integrated Circuits Data Book, Motorola Semiconductor Products, Inc., Phoenix, AZ, 1973.

Linear Integrated Circuits Data Book, Technical Information Center, Motorola Semiconductor Products, Inc., Phoenix, AZ, 1979.

The Engineering Staff of Analog Devices, Inc. (ed. by Daniel H. Sheingold): *Analog-Digital Conversion Handbook*, Norwood, MA, 1972.

Intersil 4½ Digit Pair 8052A/7103A; 3½ Digit Pair 8052/7103, Data Sheet, Intersil, Inc., Cupertino, CA, 1977.

Data Acquisition Handbook, National Semiconductor Corporation, Santa Clara, CA, 1978.

CHAPTER 7

DIGITAL ARITHMETIC

One of the many uses of A/D converters is to change analog voltages representing weight, pressure, temperature, etc. to digital words. This is done for display purposes and so that mathematical operations such as adding weights or multiplying weight by price per pound are done more easily. D/A converters can change results back to analog voltages or currents if required. Electronic calculators also perform all their operations with binary-coded numbers.

In this chapter we discuss binary arithmetic and the circuits used to perform it.

At the conclusion of this chapter, you should be able to:

1. Correctly add binary, BCD, excess-3 BCD, octal, and hexadecimal numbers.

2. Correctly subtract binary numbers by using the 1's- or 2's-complement method and a sign bit.

3. Correctly subtract BCD, excess-3 BCD, octal, and hexadecimal numbers.

4. Show with a circuit how an ALU such as the 74181 can be used to perform any of 16 logic functions or 16 arithmetic functions on two 4-bit input variables.

5. Perform binary multiplication by the pencil method or add-and-shift-right algorithm.

6. Perform binary division by the pencil method or the subtract-and-shift-left algorithm.

7. Normalize numbers to a binary fractional form.

8. Represent floating-point numbers in a binary, an ANSI FORTRAN hexadecimal, or a BCD format.

ADDITION

BINARY

Figure 7-1a shows the truth table for the addition of two single-bit numbers, A and B. Here S represents the sum of A and B, and C represents a carry into the next digit. Note that since any digit can be only a 0 or 1, $A = 1$ plus $B = 1$ gives a 0 sum and a carry of 1 into the next digit. The logic expression for $S = 1$ is $A \oplus B$, and the expression for C is $A \cdot B$. These functions can be implemented easily with gates as shown in Figure 7-1b.

The circuit of Figure 7-1b is only *half* adder because it

INPUTS		OUTPUTS	
A	B	S	C
0	0	0	0
0	1	1	0
1	0	1	0
1	1	0	1

(a)

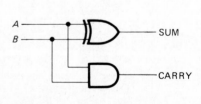

(b)

FIGURE 7-1 Half adder. *(a)* Truth table. *(b)* Circuit.

INPUTS			OUTPUTS	
A	B	C_{IN}	S	C_{OUT}
0	0	0	0	0
0	0	1	1	0
0	1	0	1	0
0	1	1	0	1
1	0	0	1	0
1	0	1	0	1
1	1	0	0	1
1	1	1	1	1

$S = A \oplus B \oplus C_{IN}$

$C_{OUT} = A \cdot B + C_{IN} (A \oplus B)$

(a)

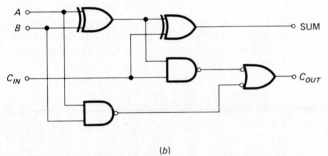

(b)

FIGURE 7-2 Full adder. *(a)* Truth table. *(b)* Circuit.

does not take into account a carry-in from a previous digit. Figure 7-2a shows the truth table for a *full* adder which takes into account a carry C_{IN} from a previous digit. Note that if A, B, and C_{IN} are all 1's, then both sum and carry-out are 1's. The logic expression for S is $A \oplus B \oplus C_{IN}$. Remember that the XOR of three or more variables will be a 1 if an odd number of the variables are 1's. Carry out, C_O, is $A \cdot B + C_{IN} (A \oplus B)$.

The full adder can be built with gates, as shown in Figure 7-2b. Usually it is not necessary to build adders from discrete gates, but the circuits of Figure 7-1b and 7-2b show well the relationship between logic expressions and binary addition. The single-bit full adder is available ready-made in the 7480, and a 4-bit full adder in the 7483. The 7483 adds two 4-bit words to give a 4-bit result plus a carry-out. Several 7483s can be connected in parallel to add two digital words of more than 4 bits each. To cut down on the number of adders needed to add long words, a circuit such as that in Figure 7-3 is used. The words to be added are stored in the two registers. They are shifted to the adder and added 8 bits at a time. The adder may be set up to add 1, 2, 4, or 8 bits at a time. Carry-out is delayed or stored to be used as a carry-in for the next addition. The sum output can be shifted back into the register for storage.

BCD AND EXCESS-3 BCD ADDITION   In systems where the final result of a calculation is to be displayed such as a calculator, it may be easier to work with numbers in a BCD or an excess-3 BCD format. These codes, as shown in Table 3-1, represent each decimal digit, 0 through 9, with a 4-bit binary word. The BCD words are the same as the binary equivalents for 0 through 9, and the corresponding excess-3 BCD words are formed by adding 3 to each standard BCD word.

BCD can have no digit word with a value greater than 9. Therefore, a carry must be generated if the result of a

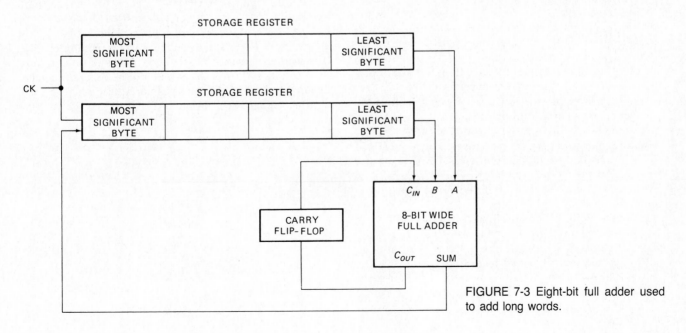

FIGURE 7-3 Eight-bit full adder used to add long words.

```
              BCD
    35      0011  0101
   +23     +0010  0011
   ───      ───────────
    58      0101  1000
```
(a)

```
          BCD
   7      0111
  + 5    + 0101
  ───    ───────
   12     1100      Incorrect  BCD
        +  110      Add 6
        ─────────
         10010      Correct  BCD 12
```
(b)

```
          BCD
   9      1001
  + 8    + 1000
  ───    ───────
   17     10001     Incorrect  BCD
           110      Add 6
        ─────────
         10111      Correct  BCD 17
```
(c)

FIGURE 7-4 Examples of BCD addition.

```
   5      1000      Excess-3
  +2      0101      Excess-3
  ──     ──────
   7      1101      Excess-6
        −  11       Subtract 3
        ────────
          1010      Excess-3 for 7₁₀
```
(a)

```
   7      1010  ⎫
  + 5    + 1000 ⎬  Excess-3
  ───    ────── ⎭
   12     1 0010      Incorrect
        + 11   11     Add 3 because carry
        ──────────
         100 0101     Correct excess-3 for 12₁₀
```
(b)

FIGURE 7-5 Examples of excess-3 addition.

BCD addition is greater than 1001, or 9. Figure 7-4 shows three examples of BCD addition. The first in Figure 7-4a is very straightforward because the sum is less than 9. The result is the same as it would be for standard binary. Adding BCD 7 to BCD 5, as shown in Figure 7-4b, produces 1100. This is a correct binary result of 12, but it is an illegal BCD code. To convert the result to BCD format, a correction factor of 6 is added. The result of adding 6 is 10010, which is the legal BCD code for 12. Figure 7-4c shows another case where a correction factor must be added. The initial addition of 9 and 8 produces 10001. Even though the lower four digits are less than 9, this is an incorrect BCD result because the addition produced a carry. Adding the correction factor of 6 gives the correct BCD of 10111 for 17.

To summarize, a correction factor of 6 must be added to the result if the result in the lower 4 bits is greater than 9 *or* if the initial addition produces a carry. This correction is sometimes called a *decimal adjust operation*.

Excess-3 addition is carried out as shown in Figure 7-5. For the example shown in Figure 7-5a, an excess-3 number plus an excess-3 number gives an excess-6 result. A correction of 3 must be subtracted to give the correct excess-3 result. If a carry into the fifth digit position is produced by an addition as shown in Figure 7-5b, then 3 must be *added* to get the correct result.

OCTAL AND HEXADECIMAL ADDITION

People working with computers or microprocessors often use octal or hexadecimal as a shorthand way of representing long binary numbers such as ROM addresses. Therefore, it is useful to be able to add octal and hexadecimal numbers.

Figure 7-6 shows two ways of adding the octal numbers 47 and 36. The first way is to convert both numbers to their binary equivalents. Remember, each octal digit represents three binary digits. These binary numbers are added by using the rules for binary addition from Figure 7-2a. Then the resultant binary sum is converted to octal.

The second method works directly with the octal form: 7 added to 6 gives 13, which is a carry to the next digit and a remainder of 5. The 5 is written down and the carry added to the next digit column. Then 4 plus 3 plus a carry gives 8, which is a carry with no remainder. The 0 is written down, and the carry is added to the next digit column. This is the same process that you use for decimal addition, but a carry is produced any time the sum is 8 or greater, rather than 10.

```
                                        Carry
                                          ↓
   47₈        100 111               ¹47₈
  +36₈      +  011 110             + 36₈
  ────      ──────────            ────────
             1 000 101          8₁₀ 13₁₀
              1  0   5₈           1 0 5₈
```
(a)   (b)

FIGURE 7-6 Examples of octal addition.

$$
\begin{array}{cc}
7A & 0111\ 1010 \\
+3F & +0011\ 1111 \\
\hline
B9 & 1011\ 1001 \\
\end{array}
$$

Carry
↓

$$
\begin{array}{cc}
^1\,7 & A_{16} \\
+\ 3 & F_{16} \\
\hline
11_{10} & 25_{10} \\
B_{16} & 9_{16} \\
\end{array}
$$

(a) (b)

FIGURE 7-7 Examples of hexadecimal addition.

As shown in Figure 7-7, the same approaches can be used to add two hexadecimal numbers. For converting to binary, remember that each hex digit represents *four* binary digits. The binary numbers are added, and the result is converted to hexadecimal.

The second method works directly with the hex numbers. With hex addition a carry is produced whenever the sum is 16 or greater. An A in hex is a 10 in decimal, and F is 15 in decimal. These add to give 25, which is a carry with a remainder of 9. The 9 is written down and the carry added to the next digit column. So 7 plus 3 plus a carry gives a decimal 11, or B in hex.

You can use whichever method seems easier and gives you consistently right answers. If you are doing a great deal of octal or hexadecimal arithmetic, you might buy a Texas Instruments TI programmer. This is an electronic calculator specifically designed to do decimal, octal, or hexadecimal arithmetic.

SUBTRACTION

STANDARD BINARY

Binary subtraction is complicated by the fact that there are several common methods of performing it. The first method discussed is called *standard* because it is similar to standard binary addition. Figure 7-8a shows the truth table for A minus B. Note that if B is larger than A, a borrow from the next most significant digit is required. The truth table for a full subtraction, which includes the effect of a borrow from the previous digit, is shown in Figure 7-8b. Figure 7-8c shows an example of binary subtraction using these rules.

The logical expression for the difference output for subtraction is $D = A \oplus B \oplus B_{IN}$, and the expression for borrow-out, B_{OUT}, is $\overline{A} \cdot B + \overline{(A \oplus B)}\ B_{IN}$. Since these expressions are very similar to those for the full adder in the last section, it is quite easy to construct a circuit that will either add or subtract two binary digits. Figure 7-9 shows such a circuit. For circuit-tracing practice you can demonstrate that the circuit adds the two inputs with carry-in if the select input is high and subtracts B from A with borrow if the select input is low.

1's-COMPLEMENT SUBTRACTION AND END-AROUND CARRY

Another method of subtraction is the *1's-complement* method. The 1's complement of a binary number is produced by simply inverting each bit of the number. Subtraction of two binary numbers is done by adding the top number (minuend) to the 1's complement of the bottom number (subtrahend). Figure 7-10a shows the operation. Note that the top and bottom numbers must have the same number of digits before complementing. After complementing, the addition is performed. If a carry of 1 is produced from the most significant digit position, it is carried around and added to the least significant digit to give the correct answer. For a positive result this end-around carry is always required. If a 1 does not appear in the box, then you know the bottom number was larger than the top. A 0 in the box also

INPUTS		OUTPUTS	
A	B	D	B
0	0	0	0
0	1	1	1
1	0	1	0
1	1	0	0

Difference = $A \oplus B$

Borrow = $\overline{A} \cdot B$

(a)

INPUTS			OUTPUTS	
A	B	B_{IN}	D	B_{OUT}
0	0	0	0	0
0	0	1	1	1
0	1	0	1	1
0	1	1	0	1
1	0	0	1	0
1	0	1	0	0
1	1	0	0	0
1	1	1	1	1

Difference = $A \oplus B \oplus B_{IN}$

Borrow = $\overline{A} \cdot B + \overline{(A \oplus B)} \cdot B_{IN}$

(b)

$$
\begin{array}{cc}
12 & 1100 \\
-\ 2 & -0010 \\
\hline
10 & 1010 \\
\end{array}
$$

(c)

FIGURE 7-8 Binary subtraction.
(a) Half-subtractor truth table.
(b) Full-subtractor truth table.
(c) Example.

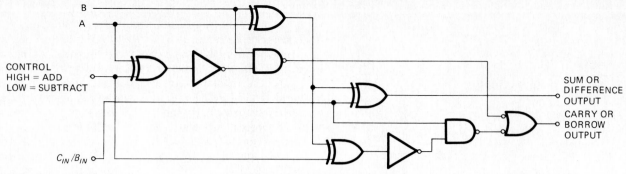

FIGURE 7-9 Circuit to add or subtract two 1-bit binary numbers on command.

indicates the answer is in complemented form. Invert each bit and put a minus sign in front to get a more recognizable answer. Figure 7-10b shows an example of this.

2's-COMPLEMENT SUBTRACTION

The *2's complement* of a binary number is formed by inverting each bit in the number and adding a 1 to the least significant digit. Two binary numbers can be subtracted by adding the top number to the 2's complement of the bottom number. Figure 7-11a shows the operation. You invert each bit of the bottom number and add 1 to form the 2's complement. Then this is added to the top number. No end-around carry is performed because it was essentially done in forming the 2's complement. A carry from the most significant bit, the 1 in the box, indicates that the result is positive. If a 0 appears in the box, it indicates that the result is negative and is shown in 2's-complement form. Figure 7-11b shows an example. To get a number back from 2's-complement form, you just invert each bit and add 1 to the result. This method may seem awkward, but it is easy to implement

in a computer or microprocessor because it requires only the simple operations of inverting, incrementing, and adding.

In a case, for example, where both positive and negative numbers are to be stored in an 8-bit-wide RAM, some way must be set up to identify a number as positive or negative. The usual method is to use the most significant bit as a sign bit. Positive numbers are represented in their true binary form with a 0 sign bit. Negative numbers are represented in their 2's-complement form with the sign bit set, or 1. Figure 7-12 shows examples of numbers expressed with a sign bit as they might be stored in an 8-bit-wide RAM.

To get the 2's-complement sign-and-magnitude form for a negative number:

1. Write the positive binary form of the number, including the sign bit.

2. Invert each bit, including the sign bit.

3. Add 1 to the result.

119_{10} 1110111 Minuend 1110111

$-\ 45_{10}$ -0101101 Subtrahend $+1010010$ (One's complement)

74_{10} $\boxed{1}\ 1001001$

$+\ \ \longrightarrow \boxed{1}$ End around carry

1001010 $= 74_{10}$

(a)

93_{10} 1011101 1011101

-108_{10} -1101100 $+0010011$ Complement

$-\ 15_{10}$ $\boxed{0}\ 1110000$ $-\ 0001111 = -15_{10}$

$\longrightarrow$ No end around carry

(b)

FIGURE 7-10 Examples of 1's-complement subtraction with end-around carry.

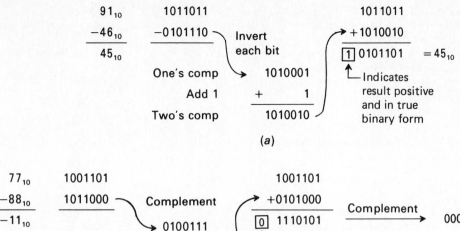

(a)

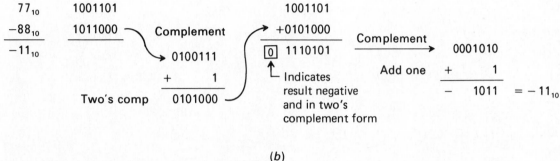

(b)

FIGURE 7-11 Examples of 2's-complement subtraction.

Sign bit
↓

+ 7	0 ¦ 0000111	
+ 46	0 ¦ 0101110	
+105	0 ¦ 1101001	
− 12	1 ¦ 1110100	⎫
− 54	1 ¦ 1001010	⎬ Sign and
−117	1 ¦ 0001011	⎬ two's complement
− 46	1 ¦ 1010010	⎭ of magnitude

FIGURE 7-12 Positive and negative numbers represented with a sign bit.

For example, to produce the 8-bit, 2's-complement, sign-and-magnitude form of −46 decimal, first write the 8-bit binary number for +46, which is 00101110. Second, invert each bit of this, to get 11010001. Third, add 1 to this, to get 11010010, which, as shown in Figure 7-12, is the correct 8-bit, 2's-complement representation of −46.

Note that the 8 bits can still represent 256 values; but instead of representing 0 to 255, they represent −128 to +127. The sign bit for 0_{10} is 0, so 0 counts as a positive number in this scheme. If you like number patterns, you might notice that this scheme shifts the normal codes for 128 to 255 downward to represent −128 to −1.

This sign-and-magnitude form can be used for numbers with any number of bits. Sixteen bits can represent numbers from −32,768 to +32,767.

Figure 7-13 shows some examples of addition of signed binary numbers of this type. Sign bits are added

```
+13    00001101
+ 9    00001001
+22    00010110
         └─Sign bit is 0
           so result is positive
                (a)
```

```
+13    00001101
− 9    11110111 2's complement for −9 with sign bit
+ 4  1│00000100
       └─Sign bit is 0
         so result is positive
     └─Ignore carry
                (b)
```

```
+ 9        00001001
−13        11110011 2's complement for −13 with sign bit
− 4        11111100 Sign bit is 1
           00000011 So invert each bit
      +           1 Add 1
 equals  −00000100 Prefix with minus sign
                (c)
```

```
− 9        11110111⎫ 2's complement,
−13        11110011⎭ sign-and-magnitude form
−22        11101010 Sign bit is 1
           00010101 So invert each bit
      +           1 Add 1
 equals  −00010110 Prefix with minus sign
                (d)
```

FIGURE 7-13 Examples of operations on numbers with a sign bit.

just as the other bits are. Figure 7-13a shows the results of adding two positive numbers. The sign bit of the result is 0, so the result is positive. In the second example (Figure 7-13b), a −9 is added to a +13, or, in effect, 9 is subtracted from 13. As indicated by the 0 sign bit, the result of this, 4, is positive and in true binary form.

Figure 7-13c shows the result of adding a −13 to a smaller positive number, +9. The sign bit of the result is a 1. This indicates that the result is negative, and the magnitude is in 2's-complement form. To reconvert a 2's-complement result to a signed number in true binary form:

1. Invert each bit.

2. Add 1.

3. Put a minus sign in front to indicate that the result is negative.

The final example in Figure 7-13d shows the results of adding two negative numbers. The sign bit of the result is a 1, and the result is negative and in 2's-complement form. Again, inverting each bit, adding 1, and prefixing a minus sign will put the result in a more recognizable form.

SUMMARY   As shown in these examples, subtraction of binary numbers can be done by simply adding the top number to the 2's complement sign-and-magnitude representation of the bottom number. If the sign bit of the result is 0, the result is positive and the magnitude is in true binary form. If the sign bit of the result is 1, the result is negative and the magnitude is in 2's-complement form.

SUBTRACTION IN OTHER NUMBER SYSTEMS

BCD SUBTRACTION   To round out the discussion of subtraction, examples of several other types are given. Figure 7-14a shows a BCD subtraction of 9 from 17. The initial result is an illegal BCD number. Whenever this occurs in BCD subtraction, a correction factor of −6 must be added to produce a legal BCD result. This decimal adjust operation is just the reverse of the correction you make when adding BCD numbers.

EXCESS-3 SUBTRACTION   Excess-3 subtraction is shown in Figure 7-14b and c. Note that since both numbers subtracted are in excess-3 form, the difference comes out in standard BCD form. Three must be added to the initial result to get the result in excess-3 form. Subtraction of excess-3 numbers can also be done by adding the 10's complement of the bottom number to the top, just as you added the 2's complement for binary subtraction. Excess-3 subtraction is used in some calculators rather than standard BCD because the 10's complement is formed more easily. To form the 10's

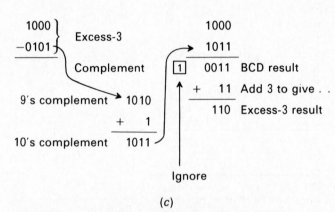

FIGURE 7-14 Examples of BCD and excess-3 BCD subtraction. *(a)* BCD. *(b)* Excess-3 BCD. *(c)* Excess-3 BCD.

complement, an excess-3 number is simply inverted and incremented by 1, as shown in Figure 7-14c. The difference still comes out in standard BCD, so 3 must be added to convert it to excess-3 form.

OCTAL SUBTRACTION   Octal subtraction is shown in Figure 7-15. Since the least significant digit of the top number is smaller than the least significant digit of the bottom number, a borrow must be done. In octal subtraction 8 is borrowed from the next digit position and added to the top number. The bottom number is then subtracted and the remainder written down. The process is continued until all digits are subtracted. If you are uncomfortable "borrowing 8's," you can just convert the number to decimal, subtract, and convert the result back to octal.

$$
\begin{array}{rr}
34_8 & 28_{10} \\
-17_8 & -15_{10} \\
\hline
15_8 & 13_{10}
\end{array}
$$

FIGURE 7-15 Examples of octal subtraction.

$$77_{16} \quad = \quad 119_{10}$$
$$-3B_{16} \quad = \quad -59_{10}$$
$$\overline{3C_{16} \qquad \quad 60_{10}}$$

FIGURE 7-16 Examples of hexadecimal subtraction.

HEXADECIMAL SUBTRACTION

HEXADECIMAL SUBTRACTION Hexadecimal subtraction is similar to octal subtraction except, when a borrow is needed, 16 is borrowed from the next most significant digit. Figure 7-16 shows this. It may help you to follow the example if you do partial conversions to decimal in your head. For example, 7 plus a borrowed 16 is 23. Subtracting B, or 11, leaves 12, or C in hexadecimal. Three from the 6 left after a borrow leaves 3, so the result is $3C_{16}$.

ARITHMETIC LOGIC UNIT

Throughout the rest of this book, binary arithmetic is encountered in several systems. One device for performing addition and subtraction as well as other logic functions is the *arithmetic logic unit,* or ALU.

Figure 7-9 shows a circuit made with simple gates that can add or subtract two single-bit binary numbers with carry or borrow. The operation performed is determined by a high or a low on a select input. An expanded version of this circuit is available in TTL as the 74181 arithmetic logic unit. The logic diagram of the 74181 is shown in Figure 7-17. This device can be programmed by four select inputs and a mode input to perform logic or arithmetic operations on two 4-bit input words. Figure 7-18 shows a table of the 16 logic functions produced when the mode input, M, is high. The A in this table represents the 4-bit binary word A3A2A1A0, and B represents the binary word B3B2B1B0.

Figure 7-19 shows some examples of logic functions performed on two 4-bit words by the 74181. In the first example, the select inputs S3S2S1S0 are connected as HHHL, and the M input is set high. This programs the 74181 to give an output word of $A + B$. Each bit of the A word is ORed with the corresponding bit of the B word. The result of the ORing appears at the corresponding bit output F. The output word will contain a 1 in any bit position where either word A or word B has a 1. Carry-in C_N is ignored whenever M is high.

If the select inputs are changed to HLHH and the mode input M is kept high, then the output word F will be the result of word A ANDed with the corresponding bits of B. Since all 16 possible logic combinations of A and B are present, the 74181 can be used in a positive or negative logic system. All you have to do is select the proper select inputs. For example, LLLH on the select input gives $F = \overline{A + B}$ for positive logic, and select input LHLL gives $F = \overline{A} + B$ in negative logic. For the select inputs, low is ground and high is +5 V.

You can also use Figure 7-18 to refresh your memory about corresponding functions in negative and positive logic. A NOR function in positive logic, for example, is a NAND function in negative logic.

To perform arithmetic functions, the M input of the 74181 is made low. This enables the C_N, and the internal carries. Figure 7-20 shows the arithmetic functions produced by the 16 select input combinations for both positive and negative logic, but here we concentrate on the positive logic functions.

Simple addition of A and B is done when the select inputs are HLLH. The carry input C_N is active low; so if it is low, a carry also will be added to the sum of A and B. If the sum produces a carry-out, the C_{N+4} will go low. The 74181 uses a "look-ahead carry" scheme that does not require the sum to be computed before the C_{N+4} is output. For low-speed addition of words larger than 4 bits, the C_{N+4} output can be connected as a ripple carry to the C_N input of another 74181. For high-speed addition of larger words, the carry propagate and carry generate outputs are used with another part, the 74182 look-ahead carry generator. The 74182 uses the carry generate and carry propagate to anticipate carries for up to four 74181 ALUs. This means that the ALUs adding the more significant bits of a long word do not have to wait for a carry to ripple up through the other ALUs.

The 74181 ALU can give an output of A if the select inputs are all low, M is low, and C_N is high. If C_N is made low, the output will be incremented by 1, to give $A + 1$. When the select inputs are all high, M is low, and C_N is high, then the output will be A decremented by 1 or $A - 1$. An output of $A + A$, or $2A$, can be produced by programming the select inputs with HHLL. As mentioned in a previous chapter, multiplying a binary number by 2 is the same as shifting each digit one place to the left. In a later section you will see how shifting is used in binary multiplication.

Subtraction is another important operation of the 74181. The 74181 does subtraction by internally generating the 1's complement of B and adding this to A. Select inputs of LHHL give an output of $A - B - 1$. To get the correct result, an end-around carry must be performed. This can be done by making C_N low. During a subtraction the C_{N+4} output represents a borrow.

The subtract operation can also be used to determine whether a binary word A is greater than, less than, or equal to a binary word B. The select inputs are made LHHL, M is made low, and C_N is made low. The $A = B$ output on pin 14 will go high if $A = B$. This output is open-collector, so several $A = B$ outputs can be tied together for comparing larger words. The combined output will be high only if all the $A = B$ outputs are high. If the word A is greater than or equal to the B word on a 74181, the C_{N+4} output will be high (indicating no borrow). If word B is greater than word A, then the C_{N+4} output will be low (indicating a borrow).

Pin numbers are included on the 74181 logic diagram in Figure 7-17 so that you can connect one up and experiment with it. The really important points here are that an ALU can perform many logic or arithmetic operations on two binary words, and that the operation performed is determined by a binary-coded instruction programmed on the select and mode inputs. As discussed in the next chapter, an ALU is the "heart" of a microprocessor.

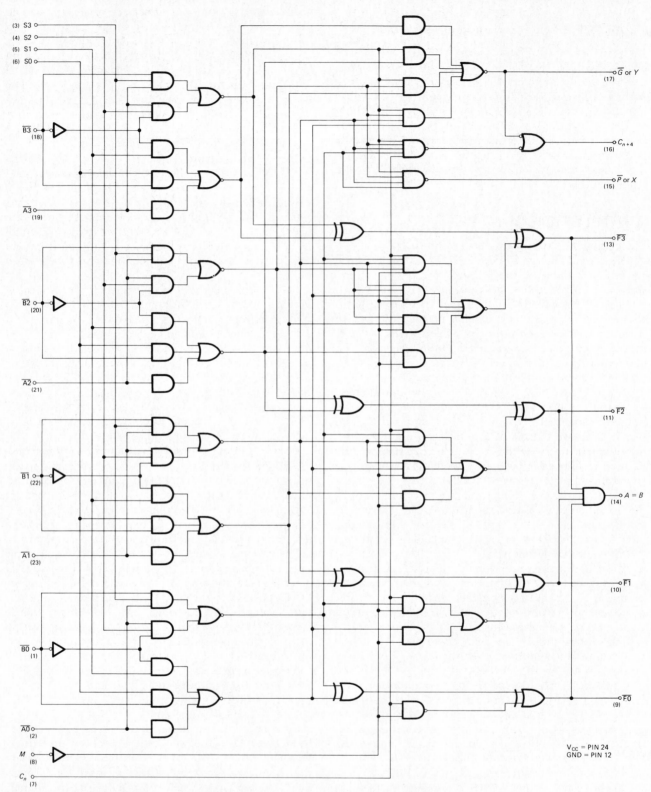

FIGURE 7-17 Logic diagram of DM74181 arithmetic logic unit. *(National Semiconductor.)*

Table of Logic Functions

Function Select				Output Function	
S3	S2	S1	S0	Negative Logic	Positive Logic
L	L	L	L	$F = \overline{A}$	$F = \overline{A}$
L	L	L	H	$F = \overline{AB}$	$F = \overline{A + B}$
L	L	H	L	$F = \overline{A} + B$	$F = \overline{A}B$
L	L	H	H	$F = $ Logical 1	$F = $ Logical 0
L	H	L	L	$F = \overline{A + B}$	$F = \overline{A}B$
L	H	L	H	$F = \overline{B}$	$F = \overline{B}$
L	H	H	L	$F = \overline{A \oplus B}$	$F = A \oplus B$
L	H	H	H	$F = A + \overline{B}$	$F = A\overline{B}$
H	L	L	L	$F = \overline{A}B$	$F = \overline{A} + B$
H	L	L	H	$F = A \oplus B$	$F = \overline{A \oplus B}$
H	L	H	L	$F = B$	$F = B$
H	L	H	H	$F = A + B$	$F = AB$
H	H	L	L	$F = $ Logical 0	$F = $ Logical 1
H	H	L	H	$F = A\overline{B}$	$F = A + \overline{B}$
H	H	H	L	$F = AB$	$F = A + B$
H	H	H	H	$F = A$	$F = A$

With mode control (M) HIGH: Cn irrelevant
For positive logic: logical 1 = HIGH voltage
logical 0 = LOW voltage
For negative logic: logical 1 = LOW voltage
logical 0 = HIGH voltage

FIGURE 7-18 Table of logic functions for 74181 ALU. (National Semiconductor.)

$M = H$ S3 S2 S1 S0 = H H H L			
	BIT3	BIT2	BIT1	BIT0
A	1	1	0	0
B	1	0	1	0
F	1	1	1	0

$$F = A + B$$

(a)

$M = H$ S3 S2 S1 S0 = H L H H			
	BIT3	BIT2	BIT1	BIT0
A	1	1	0	0
B	1	0	1	0
F	1	0	0	0

$$F = A \cdot B$$

(b)

FIGURE 7-19 Examples of 74181 programming. (a) $A + B$. (b) $A \cdot B$. (National Semiconductor.)

BINARY MULTIPLICATION AND DIVISION

MULTIPLICATION

There are several methods of doing binary multiplication. Figure 7-21 shows what might be called the *pencil* method because it is the same as the way you learned to multiply decimal numbers. The top number, or multiplicand, is multiplied by the least significant digit of the bottom number. The partial product is written down. The top number is multiplied by the next digit of the multiplier. The resultant partial product is written down under the last, but shifted one place to the left. Adding all the partial products gives the total product. This method works well when multiplication is done by hand, but it is not practical for a computer because the type of shifts required make it awkward to implement.

One multiplication method used by computers is repeated addition. To do 7×55, for example, the computer can just add seven 55's. For large numbers, however, this method is slow. To do 786×253, for example, requires 252 add operations.

A faster method for computers and microprocessors uses an add-and-shift-right algorithm. An *algorithm* is just a formula or procedure used to solve a problem. The reason why a particular algorithm works to solve a problem may not be immediately obvious. The add-and-shift algorithm for binary multiplication is demonstrated in Figure 7-22a. This method takes advantage of the fact that for binary multiplication the partial product can be either the top number exactly if the multiplier digit is a 1, or a 0 if the multiplier digit is a 0.

The numbers to be multiplied are stored in two registers. The result will be generated in a special register called an *accumulator*, which is initially reset to all 0's. To start, a check is made that the least significant digit

Table of Arithmetic Operations

Function Select				Output Function	
S3	S2	S1	S0	Low Levels Active	High Levels Active
L	L	L	L	$F = A$ minus 1	$F = A$
L	L	L	H	$F = AB$ minus 1	$F = A + B$
L	L	H	L	$F = A\bar{B}$ minus 1	$F = A + \bar{B}$
L	L	H	H	$F =$ minus 1 (2's complement)	$F =$ minus 1 (2's complement)
L	H	L	L	$F = A$ plus $[A + \bar{B}]$	$F = A$ plus $A\bar{B}$
L	H	L	H	$F = AB$ plus $[A + \bar{B}]$	$F = [A + B]$ plus $A\bar{B}$
L	H	H	L	$F = A$ minus B minus 1	$F = A$ minus B minus 1
L	H	H	H	$F = A + \bar{B}$	$F = A\bar{B}$ minus 1
H	L	L	L	$F = A$ plus $[A + B]$	$F = A$ plus AB
H	L	L	H	$F = A$ plus B	$F = A$ plus B
H	L	H	L	$F = A\bar{B}$ plus $[A + B]$	$F = [A + \bar{B}]$ plus AB
H	L	H	H	$F = A + B$	$F = AB$ minus 1
H	H	L	L	$F = A$ plus A †	$F = A$ plus A †
H	H	L	H	$F = AB$ plus A	$F = [A + B]$ plus A
H	H	H	L	$F = A\bar{B}$ plus A	$F = [A + \bar{B}]$ plus A
H	H	H	H	$F = A$	$F = A$ minus 1

With mode control (M) and Cn low for negative logic; (M) low and Cn high for positive logic.
†Each bit is shifted to the next more significant position.

FIGURE 7-20 Table of arithmetic operations for 74181 ALU. (*National Semiconductor.*)

of the multiplier is a 0 or a 1. If the digit is a 1, as shown in the example, the top number, or the multiplicand, is added to the accumulator. The accumulator is then shifted, or "rotated," 1 bit position to the right. The next digit of the multiplier is checked, and either the top number or 0 is added to the most significant bits of the accumulator, as shown. Next another accumulator rotate is done. The operation is continued until all the multiplier digits have been tested and used.

To summarize: If a multiplier bit is a 1, then the multiplicand is added to the accumulator. If the multiplier

```
    11          1011      Multiplicand
  × 9        ×  1001      Multiplier
  ────       ───────
   99          1011  ⎫
             0000   ⎪
            0000     ⎬  Partial products
           1011    ⎭
          ───────
          1100011      Product
```

FIGURE 7-21 Examples of pencil multiplication of binary numbers.

bit is a 0, then 0 is added to the accumulator. The accumulator is then shifted, or rotated, right 1 bit position. The process is continued until all multiplier bits have been used.

A close look at this algorithm will show you that it does the same thing as the pencil method except that the partial products are added as they are produced. Also the sum of the partial products is shifted right instead of each partial product being shifted left, as in the pencil method. The results are the same, but the add-and-shift-right algorithm is easier to do with computers and microprocessors. Figure 7-22*b* shows another example of the method.

Multiplying two 4-bit numbers can give a product with as many as 8 bits, and two 8-bit numbers can yield a 16-bit product. The register used to store the product must be large enough to hold the expected number of bits. If the accumulator is not large enough, then another register can be used to store part of the product.

For fast multiplication outside a computer, one- or two-chip multipliers are available. The SN74284 and SN74285 connected together will give an 8-bit product of two 4-bit binary numbers in about 40 ns. They can be expanded to multiply larger numbers.

```
  1011         Register B
× 1001         Register C

  0000         Reset accumulator register to zero
  1011         Since first multiplier digit is one
  1011           add multiplicand to accumulator
  101|1        Shift result right
  0000         Next multiplier digit is zero
  0101|1         so add zero to accumulator
  0010|11       Shift result right
  0000         Next multiplier digit is zero
  0010|11        so add zero to accumulator
  0001|011      Shift result right
  1011         Next multiplier digit is one so
Product 1100|011      add multiplicand to accumulator
```

(a)

```
     101
  ×  011

     000       Reset accumulator
     101       Add
     101
     010|1     Shift
     101       Add
     111|1
     011|11    Shift
     000       Add
     011|11    Product
```

(b)

FIGURE 7-22 Examples of the add-and-shift-right method of binary multiplication.

BINARY DIVISION

Binary division can be performed in several ways. Figure 7-23 shows two examples of the pencil method. This is the same process as decimal long division. However, it is much simpler than decimal long division because the digits of the result (quotient) can be only 0 or 1. A division is attempted on part of the dividend. If this is not possible because the divisor is larger than that part of the dividend, a 0 is entered into the quotient. Then another attempt is made to divide by using one more digit of the dividend. When a division is possible, a 1 is entered in the quotient. Next the divisor is subtracted from the portion of the dividend used. The process is continued as with standard long division until all the dividend is used. As shown in Figure 7-23b, 0's can be added to the right of the binary point and division con-

```
                    01100   Quotient
Divisor  110) 1001000   Dividend          12
             −110                        6)72
             ────
              110
             −110
             ────
                0
```

(a)

```
                110.01                     6.25
       100) 11001.00                      4)25
          −100
          ────
           100
          −100
          ────
            01 00
```

(b)

FIGURE 7-23 Examples of pencil division of binary numbers.

tinued to convert a remainder to a binary equivalent.

Another method of division which is easier for computers and microprocessors to perform uses successive subtractions. The divisor is subtracted from the dividend and from each successive remainder until a borrow is produced. The desired quotient is 1 less than the number of subtractions needed to produce a borrow. This method is simple, but for large numbers it is slow.

For faster division of large numbers, a subtract-and-shift-left algorithm can be used by computers. Figure 7-24 shows an example of this method. It may look messy, but it is essentially the same process you go through with a pencil long division.

The divisor is first subtracted from the most significant digit of the dividend. If the overflow or carry bit produced by the subtraction is a 0 then the subtraction was possible and a 1 is written into the quotient register, as in line 3 of Figure 7-24. The result of the subtraction is shifted left one place. The divisor is then subtracted from the shifted result, as shown in lines 5 and 6. This time the overflow/carry bit is a 1 because the divisor was too large. A 0 is written in the next digit of the quotient register. The divisor is added back to the result of the subtraction to restore the shifted result, as in lines 7 and 8. This number is shifted left and another subtraction attempted. If the subtraction produces a 0 overflow/carry bit, the subtraction was successful and 1 is written in the quotient register, as in line 11. Whenever a 1 overflow/carry bit is produced, the divisor is added back, the resultant number shifted one place left, and the subtraction tried again. This is the same operation you use in pencil long division. When a division won't work, you write a 0 in the quotient, bring down another digit of the dividend, and try again with the new number. The process is continued until another 0 overflow/carry bit is produced. For each 0 overflow/carry bit

Overflow/carry bit   Quotient Register

	Overflow/carry bit	Register	Quotient Register	Description
1		1010	01000	Dividend
2	↓	−1000		Divisor—subtract
3	0	0010	01000	Overflow/carry bit = 0, so ⟶ 1
4	0	0100	1000	Shift left
5		−1000		Subtract
6	1	1100	1000	Overflow/carry bit = 1, so ⟶ 0
7		+1000		Add divisor back to restore
8		0100	1000	
9		1001	000	Shift left
10		−1000		Try subtraction again
11	0	0001	000	Overflow/carry bit = 0, so ⟶ 1
12		0010	00	Shift left
13		−1000		Subtract
14	1	1010	00	Overflow/carry bit = 1, so ⟶ 0
15		+1000		Add to restore
16		0010	00	
17		0100	0	Shift left
18		−1000		Subtract
19	1	1100	0	Overflow/carry bit = 1, so ⟶ 0
20		+1000		Add to restore
21		0100	0	
22		1000		Shift left
23		−1000		Subtract
24	0	0000		Stop—overflow/carry bit = 0, so ⟶ 1

$$\begin{array}{r} 41 \\ 8\overline{)328} \end{array}$$
Quotient = 101001 = 41_{10}

FIGURE 7-24 Example of subtract-and-shift-left method of binary division.

a 1 is entered in the quotient register. When all of the dividend is used up, the division is complete. If a remainder is present, 0's can be added to the right of the binary point and the division continued to convert the remainder to a binary equivalent.

Later we show how this algorithm for division and the previously described multiplication algorithm are easily implemented with a microprocessor.

FIXED-POINT NUMBERS

INTEGER FIXED-POINT CALCULATIONS

Most simple microprocessor-based systems, such as a scale which multiplies price per pound by weight to display total price, use fixed-point numbers. This means that all numbers are treated as integers with the binary or decimal point to the right of the LSB (least significant bit). For example, 9.6 lb is stored in a register as 10010110. BCD or 1100000. binary. A price of $0.29/lb is stored as 00101001. in BCD or 00011101. in binary.

When the binary representation of the weight is multiplied by the binary price per pound, the result is 101011100000. binary or 2784. decimal. To give the desired display of $2.78, the designer must determine the right position for the decimal point.

To represent very large or very small numbers in a fixed-point system, they must be scaled. For example, 2,260,000 might be represented as 226×10^4. The 226 is stored in a register as 11100010. The scale factor of 10^4 must be remembered by the user. Likewise, 0.000078 must be represented as 78×10^{-6}. The 78 is stored in a register as 1001110 and the scale factor of 10^{-6} remembered.

BINARY FRACTION FIXED-POINT CALCULATIONS

The simple fixed-point scheme discussed above creates a problem with multiplying on some machines. Figure 7-25a shows the multiplication of two 8-bit binary numbers. If you do a multiplication such as this with a pencil, it is no problem to count left four places to position the binary point. However, the machine has no way of counting out to locate the binary point. Furthermore, if the result has to be "truncated," or shortened, to 8 bits so it can be stored in a memory location, most machines will simply take the most significant 8 bits. This leaves no way of determining where the binary point is with respect to these bits.

Figure 7-25b shows one way of overcoming this problem. All numbers are scaled to 8-bit binary fractions

```
      100110.11              38.75
  ×  100001.01            × 33.25
      10011011           1228.4375
    100110110
  100110110000
 101000010000.0111
```
8 most significant digits

(a)

```
     .10011011           ×2⁶
     .10000101           ×2⁶
      10011011
    100110110
  100110110000
 .010100001000 0111      ×2¹²
 └── Point always here
```

(b)

FIGURE 7-25 Multiplication of noninteger binary numbers. (a) Pencil method. (b) Binary fractional method.

with the binary point to the *left* of the MSB. For example, 100110.11 is represented as $.10011011 \times 2^6$. Since this type of machine has no way of storing the 2^6, it must be remembered.

Carrying out the multiplication always give a 16-bit result with the binary point to the left of the MSB (most significant bit). The power of the 2 for the result is determined by adding the exponents of the numbers multiplied. The most significant 8 bits to the right of the binary point and the determined exponent will always give the correct answer. This scheme works in part because it places the binary reference point of the result where it cannot get truncated. Both negative and positive numbers can be handled, but the problem of having to remember and add exponents still remains. The next method solves this problem.

FLOATING-POINT NUMBERS

Fixed-point systems represent all numbers as integers or fractions. The two fixed-point schemes discussed earlier both require the operator to keep track of the proper place for the binary or decimal point.

A floating-point system can represent numbers that contain both an integer part and a fraction part. Such numbers are called *floating-point numbers*, or *real numbers*.

BINARY FLOATING-POINT NUMBER REPRESENTATIONS

There are several common formats for representing floating-point numbers. Figure 7-26a shows one. Four

bytes, or 32 bits, are required to represent each number. The magnitude of the number is represented by a 24-bit binary fraction. This fraction is produced by normalizing a number so that the first digit to the right of the binary point is 1. For example, 1000.11 normalizes to $.100011 \times 2^4$, and .0001011 normalizes to $.1011 \times 2^{-3}$. The 24-bit binary fraction is often referred to as the *mantissa*. Sign-and-exponent information is stored in the first byte of the four. The most significant bit is a sign bit, to indicate whether the number is positive or negative: 0 indicates a positive, and 1 indicates a negative mantissa.

The next bit to the right is a sign bit for the exponent, again 0 for positive and 1 for negative. Six remaining bits in the byte are used for the magnitude of the exponent. The magnitude of the exponent is in 2's-complement form. Some examples of this format are shown in Figure 7-26b. For a positive number, the 2's complement is the same as standard binary. For a negative number, the 2's complement is formed by inverting each bit and incrementing the result by 1.

A format with a 24-bit mantissa gives an accuracy of 1 part in 2^{24}, or about one part in seven decimal places. The 6-bit exponent with sign gives an exponent range of 2^{-63} to 2^{+63}, or about 10^{-19} to 10^{+19} in decimal. This is more than sufficient accuracy and range for many applications.

ANSI FORTRAN REPRESENTATION OF FLOATING-POINT NUMBERS

Another format for representing floating-point numbers is the American National Standards Institute, or ANSI, FORTRAN format shown in Figure 7-27a. This format

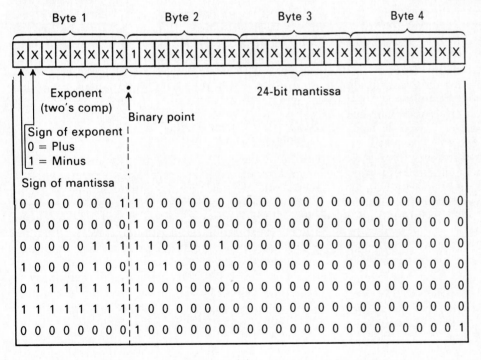

FIGURE 7-26 Binary floating-point numbers. (a) Format. (b) Examples.

$1_{10} = 0.1_2 \times 2^1$

$0.5_{10} = 0.1_2 \times 2^0$

$105_{10} = 0.1101001_2 \times 2^7$

$-10_{10} = 0.1010_2 \times 2^4$

$0.25_{10} = 0.01_2 = 0.1_2 \times 2^{-1}$

$-0.25_{10} = -0.01_2 = -0.1_2 \times 2^{-2}$

0.50000007_{10}

(a)

(b)

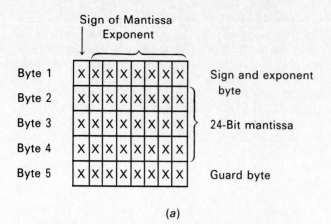

Sign of Mantissa

Exponent

Byte 1 X X X X X X X X — Sign and exponent byte

Byte 2 X X X X X X X X

Byte 3 X X X X X X X X — 24-Bit mantissa

Byte 4 X X X X X X X X

Byte 5 X X X X X X X X — Guard byte

(a)

Decimal	ANSI FORTRAN Hexadecimal Sign & Exponent	Mantissa
1.00	41	10 00 00
100	42	64 00 00
−1.00	C1	10 00 00
.50	40	80 00 00
−.25	C0	40 00 00
0.00	00	00 00 00

(b)

FIGURE 7-27 ANSI FORTRAN hexadecimal floating-point numbers. *(a)* Format. *(b)* Examples.

Byte 1 X X X X X X X X — 8 bit sign 00 = plus
                                      FF = minus
Byte 2 X X X X X X X X — 8 bit exponent

Byte 3 X X X X X X X X

Byte 4 X X X X X X X X — 32 bit BCD mantissa (8 decimal digits)

Byte 5 X X X X X X X X

Byte 6 X X X X X X X X

(a)

Decimal	BCD Sign	Floating Point Format (Hex Digits) Exponent	Mantissa
1.00	00	81	10 00 00 00
5.00	00	81	50 00 00 00
−1.00	FF	81	10 00 00 00
100	00	83	10 00 00 00
−.25	FF	80	25 00 00 00
$2^{16} = 65536$	00	85	65 53 60 00

(b)

FIGURE 7-28 BCD floating-point numbers. *(a)* Format. *(b)* Examples.

also uses a 3-byte, or 24-bit, mantissa with the binary point understood to be at the left of the mantissa. However, in this format a number is considered normalized if any of the four most significant bits is a 1. Thus .1000, .0100, .0010, or .0001 as the most significant digit of the mantissa is considered normalized.

The most significant byte contains the exponent and a sign bit for the mantissa. A 0 sign bit indicates a positive mantissa, and a 1 bit indicates a negative mantissa. The exponent represents a power of 16 rather than a power of 2. It is expressed in excess-64 notation. Excess-64 notation is formed by expressing the desired exponent in a 7-bit 2's-complement form and adding 64, or 1000000 binary. This inverts the most significant bit or sign bit of the exponent. Thus a 1 sign bit indicates a positive or zero exponent, and a 0 sign bit indicates a negative exponent. An exponent of −64 is represented as 0000000, a 0 exponent as 1000000, and a +63 exponent as 1111111. The 7-bit exponent is then essentially binary counted up from −64 to +63. Figure 7-27b shows some numbers expressed in this format. As is often done, hexadecimal digits are employed to represent each four binary digits.

The ANSI FORTRAN floating-point format makes it possible to represent numbers from $.1 \times 16^{-64}$ to $.FFFF \times 16^{+64}$, or about 10^{-79} to 10^{+76}. Although this scheme gives greater range than the previous one, the resolution is still limited to about 1 part in 7 decimal places because the mantissa is still only 24 bits. An extra 8 bits are reserved between numbers as a guard byte to hold some extra bits produced by intermediate calculations. This improves overall accuracy because final results can be rounded rather than just truncated.

BCD FLOATING-POINT REPRESENTATIONS

Electronic calculators that require the result of a calculation to be displayed as decimal digits often use a BCD representation for floating-point numbers. Figure 7-28a shows this format. A 32-bit mantissa allows eight decimal digits to be represented. The 8-bit exponent often is expressed in excess-128 form. Excess-128 form is produced by adding 128_{10}, or 10000000_2, to the 2's complement of the desired exponent. The exponent represents a power of 10. An entire byte is reserved for the sign of the mantissa. FF_{16} represents a negative mantissa, and 00_{16} represents a positive mantissa. Figure 7-28b offers some examples.

The algorithms for performing arithmetic operations on floating-point numbers are beyond the scope of the present discussion.

REVIEW QUESTIONS AND PROBLEMS

1. Add the following:
   - a. 10111 and 1011 in binary
   - b. 37 and 26 in BCD
   - c. 37 and 26 in excess-3 BCD
   - d. 37_8 and 26_8
   - e. $4A_{16}$ and 78_{16}

2. Show the subtraction in binary of the following decimal numbers:
   - a. 7 − 4
   - b. 37 − 26
   - c. 125 − 93

3. Express the following decimal numbers in 8-bit binary sign and 2's-complement magnitude form:
   - a. +26
   - b. −7
   - c. −26
   - d. −125
   - e. 0
   - f. −1
   - g. −97
   - h. +85

4. Convert the following 8-bit binary signed numbers to decimal:
   - a. 00110011
   - b. 10001100
   - c. 11110110
   - d. 01011101

5. Add the following decimal numbers, using their 8-bit binary, 2's-complement, and sign-and-magnitude representations:
   - a. +3, −7
   - b. −3, −7
   - c. −37, +26

6. Do the following subtractions:
   - a. 17 − 8 in BCD
   - b. 17 − 5 in excess-3 BCD
   - c. $22_8 − 13_8$
   - d. $72_{16} − 3A_{16}$

7. Using Figure 7-18, show the programming of the select and mode inputs of the 74181 required to perform the following positive logic functions:
   - a. $\overline{A + B}$
   - b. $A \oplus B$
   - c. $\overline{A \cdot B}$

8. Show the output word produced when the following binary words are ANDed and when they are ORed.
   - a. 1010 and 0111
   - b. 1011 and 1100
   - c. 11010111 and 111000
   - d. ANDing an 8-bit binary number with 11110000 is sometimes called *masking* the lower 4 bits. Why?

9. Using Figure 7-20 show the programming of the select and mode inputs that the 74181 requires to perform the following arithmetic functions:
   - a. $A + B$
   - b. $A − B − 1$
   - c. $A \cdot B + A$

10. Show the multiplication of 1001 and 011 by the pencil method and by the add-and-shift-right algorithm. Do the same for 11010 and 101.

11. Show the division of 1100100 by 1010, using both the pencil method and the subtract-and-shift-left algorithm.

12. Normalize the following numbers to binary fractional format and show the exponent produced:
   - a. .01101
   - b. 110001.0
   - c. .000101

13. Show how each of the following numbers is expressed in binary, ANSI FORTRAN, and BCD floating-point representation:
   - a. 2.0
   - b. 76
   - c. −22
   - d. .125
   - e. −.75

REFERENCES/BIBLIOGRAPHY

Boney, Joel: "Math in the Real World," *BYTE: The Small Systems Journal*, vol. 3, no. 9, September 1978, pp. 114–119.

Digital Integrated Circuits, Digital Data Handbook, National Semiconductor Corporation, Santa Clara, CA, 1974.

Hashizume, Burt: "Floating Point Arithmetic," *BYTE: The Small Systems Journal*, vol. 2, no. 11, November 1977, pp. 76–78, 180–188.

Nashelsky, Louis: *Digital Computer Theory*, chap. 1010_B, Wiley, New York, 1966, pp. 216–235.

CHAPTER 8

MICROPROCESSOR STRUCTURE AND PROGRAMMING

In the preceding chapters all the elements used in a computer or microprocessor system were discussed. It remains to show how these parts are connected to form a computer or microprocessor system. In concept, it is a small step from understanding the ALU of Chap. 7 to understanding a microprocessor. Recall that the 74181 ALU performs logic or arithmetic functions on two 4-bit binary words. The function performed is determined by the mode input and the binary instruction applied to the four select inputs. If RAM to store data, ROM to store sequential binary-coded instructions, and control circuitry are added to an ALU, the ALU can be made to step through a sequence of operations one after another. This is essentially how a computer or microcomputer is structured. The add-and-shift-right multiplication procedure in Chap. 7 is an example of sequential operations that can be performed by such a microcomputer. The sequence of instructions stored in memory is called a program.

At the conclusion of this chapter, you should be able to:

1. Draw a block diagram of a simple computer, showing buses.

2. Trace signal paths on buses for an instruction fetch and execute.

3. Using the chapter as a reference, describe the operations performed by 8080A/8085A instructions.

4. Draw a flowchart for a simple program.

5. Write a simple assembly language program for an 8080A/8085A microprocessor.

6. Hand-assemble an 8080A/8085A program.

7. Explain the use of a text editor program and assembler program to produce a machine language program.

8. Describe the techniques used to debug assembly language programs.

GENERAL ORGANIZATION AND PROGRAM FLOW IN A DIGITAL COMPUTER

COMPUTER SYSTEM BLOCK DIAGRAM
Figure 8-1 shows a block diagram of a simple computer or microprocessor system. The three major parts are the memory, the CPU, and the I/O ports.

MEMORY SECTION   The memory section usually consists of ROM, RAM, and magnetic discs or tape. Memory is used to store instruction sequences or programs which the computer will execute. Memory is also used to store data to be processed by the computer, or data resulting from processing.

CPU SECTION   As will be discussed in greater detail later, the *central processing unit* or CPU contains the control circuitry, an ALU, some registers, and an address/program counter. To execute a program, the CPU sends out to memory the address of the location of the code for the first instruction to be executed. The CPU also sends out a memory enable signal. The instruction code comes to the CPU from memory and gets decoded and executed. After each operation, the program counter increments to address the location of the next instruction or data stored in memory.

INPUT-OUTPUT INTERFACE SECTION   The third part of the computer is the interface of the computer with the outside world. This interface is often called *input-output* (I/O) ports. An input port allows data from

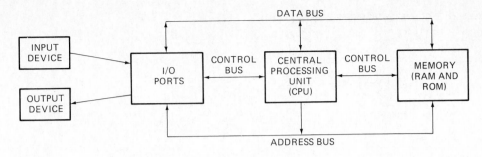

FIGURE 8-1 Block diagram of a simple computer or microcomputer system.

a keyboard, digitized output data from a pressure sensor, or some other data to be taken into the computer under CPU control. An output port is used to send data to some output device such as a teletype, video screen display, or motor-speed control unit. Physically, an output or input port is often just parallel D flip-flops which pass data when strobed by the CPU.

BUS STRUCTURE

Three major buses link the memory, CPU, and I/O ports: the *address bus*, *data bus*, and *control bus*.

ADDRESS BUS   The CPU sends out on the *address bus* the address of a memory location that is to be read from or written into. The address bus is used also by the CPU to select a particular input or output port. It may consist of 8, 12, or more parallel lines.

DATA BUS   As indicated by the double-ended arrows in Figure 8-1, the *data bus* is *bidirectional*. Depending on the operation, data may be coming from or going to the memory, the CPU, or the I/O ports. Any device outputs connected onto the data bus must be three-state, so that they can be floated except when that device is being addressed and read from. A data bus may have 4, 8, 12, or more parallel lines.

CONTROL BUS   The *control bus* carries signals such as *memory read*, *memory write*, *input read*, and *output write*. These signals enable memory or I/O ports for proper operation. For example, to read a memory location, the CPU first sends out the address on the address bus and a memory read signal on the control bus. The memory then outputs the data from the addressed location to the data bus.

EXECUTION OF A THREE-STEP PROGRAM

Before we get into a detailed discussion of a CPU, we will describe the running of a simple program on the generalized computer. This should help you to see how CPU, memory, and I/O ports function as a system.

As illustrated in Figure 8-2, there are three steps in the sample program:

1. Input a number from input port 4.

2. Add 7 to this number.

3. Output the result to port 2.

When the power is first turned on, the CPU automatically does a memory read, or *instruction fetch*, to address 0000H. (The H after the address tells you the address is represented in hexadecimal form.) It is up to the programmer to make sure that this address in memory contains the binary code for the first instruction to be performed. For the sample program the first instruction is an IN (input from port), which might be coded in an 8-bit-wide machine as 11011011 in binary, or DB in hexadecimal.

The arrow labeled 1A in Figure 8-2*a* shows the first instruction fetch address being sent to memory. Memory then returns the contents of memory location 0000H to the CPU (1B). The CPU decodes this word as an IN instruction. Then it does a fetch to memory location 0001H to get the address number of the desired input port (2A). Memory returns this number to the CPU on the data bus (2B). The CPU then carries out the IN instruction by sending out address 04H on the address bus and an input read signal on the control bus (2C). The binary number on input port 4 is returned to the CPU (2D) and stored in its accumulator.

Since this instruction is completed, the CPU must do another fetch to get its next instruction (3A). The ADI (add immediate) instruction coded as 11000110 binary, or C6 hex, comes back to the CPU on the data bus (3B). This add instruction requires the CPU to do another fetch to find out what number to add to the number it has stored (4A). Memory sends 00000111 binary, or 07H, back to the CPU (4B) as the number to be added to that taken in port 4. The CPU then carries out the addition internally and stores the result in its accumulator.

Another fetch (5A) is performed to get the next instruction. The OUT (output to port) instruction coded as 11010011 binary, or D3H, is returned on the data bus (5B). The OUT instruction requires a second fetch to get the address of the output port (6A), which is returned from memory on the data bus (6B). The CPU sends the output port address 02H out on the address lines (6C) and the sum data out on the data bus (6D). It then sends an output write pulse on the control bus to strobe the data to an output device. Figure 8-2*b* shows the required memory contents for this program. Both the addresses and memory contents are expressed in hexadecimal to save space and reduce errors. The memory, of course, can store only the binary 1's and 0's.

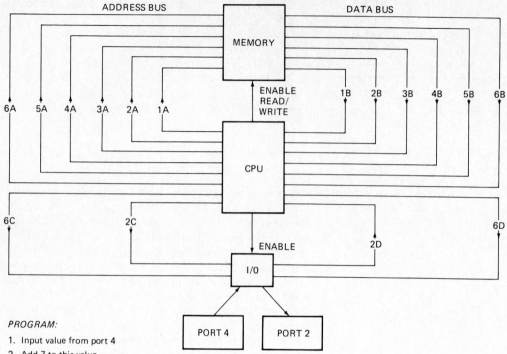

EXECUTION SEQUENCE FOR MICROPROCESSOR PROGRAM

PROGRAM:

1. Input value from port 4
2. Add 7 to this value
3. Output result to port 2

SEQUENCE:

1A Fetch first instruction

1B Instruction comes back (address port whose address is stored in next memory location)

2A Address next memory location (for port address)

2B Port address comes back (#4)

2C Address port 4

2D Data input from port 4 comes back

3A Address next memory location for next instruction

3B Instruction comes back (add number stored in next memory location)

4A Address next memory location (for data)

4B Data comes back (#7) and gets added in CPU

5A Address memory for next instruction

5B Instruction comes back (address port whose address is stored in next memory location)

6A Address next memory location (for port address)

6B Port address comes back (#2)

6C Address port 2

6D Data output to port 2

(a)

Memory Address (Hex)	Memory Contents (Binary)	Memory Contents (Hex)	Op Code	Operand
0000	11011011	DB	IN	04
0001	00000100	04		
0002	11000110	C6	ADI	07
0003	00000111	07		
0004	11010011	D3	OUT	02
0005	00000010	02		
0006				

(b)

FIGURE 8-2 A three-step microprocessor program. (a) Execution sequence. (b) Memory contents and instruction mnemonics.

This may seem like many steps just to input a value, add 7, and output the result. A microcomputer, however, can run through all these steps in a few microseconds. While a computer or microcomputer is very fast, it is somewhat stupid because it must be told by a program the exact sequence of operations to perform. In fact, the only way the CPU knows the difference between a data byte and an instruction is by their order in the program. Therefore, care must be taken in programming to make sure the computer gets an instruction when it needs an instruction, and data when it needs data. Next we examine the internal parts of a CPU so you can see how it processes instructions and data.

MICROCOMPUTER CPUs: THE 8080A AND 8085A MICROPROCESSORS

Since it is easier to learn about microprocessors by using specific examples, we use the 8080A and 8085A throughout this chapter and the next. The 8085A is an improved version of the 8080A. Although the 8085A has replaced the 8080A in new designs, for many years the 8080A was the number one selling microprocessor. Therefore, the 8080A is found in a great deal of existing equipment and is included here for reference.

Figure 8-3a shows a functional block diagram of the 8080A CPU, and Figure 8-3b shows a functional block diagram of the 8085A CPU. The hearts of the CPUs are the arithmetic logic units, or ALUs.

ALU AND ASSOCIATED CIRCUITRY: THE ARITHMETIC LOGIC GROUP

The 8080A and the 8085A are *8-bit processors* because their data buses are 8 bits wide, and their *ALUs* are set up to perform operations on 8-bit words. The ALU in each processor can increment a binary quantity, decrement a binary quantity, and perform AND, OR, XOR, addition, subtraction, or comparison operations on two 8-bit quantities. The ALU functions as part of the *arithmetic logic group* of circuits, which includes the *accumulator, accumulator latch, temporary register, flag flip-flops,* and *decimal adjust circuitry.*

ACCUMULATOR   The *accumulator* is a special 8-bit register that holds one of the quantities to be operated on by the ALU and receives the result of an operation performed by the ALU. For example, an instruction, when executed, might add the contents of the B register to the contents of the accumulator and place the result of the addition in the accumulator.

TEMPORARY REGISTER   The *temporary register* receives data from the internal data bus and holds it for the ALU.

DECIMAL ADJUST CIRCUITRY   If BCD addition or subtraction is being performed, the *decimal adjust accumulator,* or *DAA, circuitry* is used to maintain the correct BCD format in the result.

FLAG FLIP-FLOPS   After each logical or arithmetic instruction execution, five *flag flip-flops* are set or reset to indicate conditions, such as a carry produced by the operation. These flag flip-flops make up the flag register. Figure 8-4 shows the positions of the five flag flip-flops in the flag register. Register bits $D1$, $D3$, and $D5$ are not used, so they can be ignored.

The *carry flag bit,* CY, is set to 1 if a carry is produced by an operation and reset to 0 for no carry. For example, if an instruction adds two binary numbers and the result is greater than FFH, the carry flag will be set to indicate this. For subtraction the carry flag indicates whether a borrow was required to perform the subtraction. In other words, if a number subtracted from the accumulator is larger than that in the accumulator, then the carry/borrow flag will be set to indicate that a borrow was required to perform the subtraction.

The *parity flag bit,* P, is set if the number in the accumulator has an even number of 1's and reset if the accumulator has an odd number of 1's.

The *auxiliary carry flag bit,* AC, is set by a carry-out of bit 3 of the accumulator. When BCD addition or subtraction is being performed, this flag is used to indicate that a decimal adjust must be done. The decimal adjust operation is discussed in Chap. 7.

The *zero flag bit,* Z, will be set if the result of an instruction execution leaves all 0's in the accumulator or in a register. For example, if the accumulator contains 01H and then is decremented by 1, the result will be 00H. Therefore, the zero flag will be set.

The *sign bit,* S, is a copy of bit 7 in the accumulator. As discussed in Chap. 7, in working with 8-bit signed (positive and negative) numbers, the most significant bit, $D7$, is used as a sign bit. In this case, if the sign bit is a 1 (set), then the number in the accumulator is negative and in 8-bit two's-complement form. If the sign bit is 0 (reset), then the number in the accumulator is positive and in normal binary form.

Any of these flags may be used to tell the processor to branch or jump to another program or section of the program. For example, if the result of an operation gives odd parity, the reset parity flag can be used to tell the processor to jump to a subprogram that sends out an error message.

REGISTERS, PROGRAM COUNTER, AND ADDRESS LATCH

REGISTER PAIRS   In addition to the accumulator and flag flip-flop registers, the 8080A and 8085A each contain six *8-bit registers* arranged in pairs. The pairs are B and C, D and E, and H and L. *Temporary register pair W and Z is not available to the user. The BC, DE, and HL registers are sometimes called "scratch pad" reg-

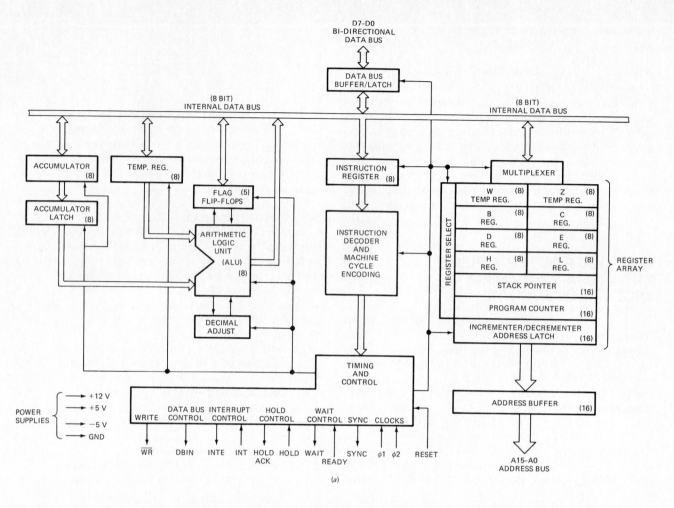

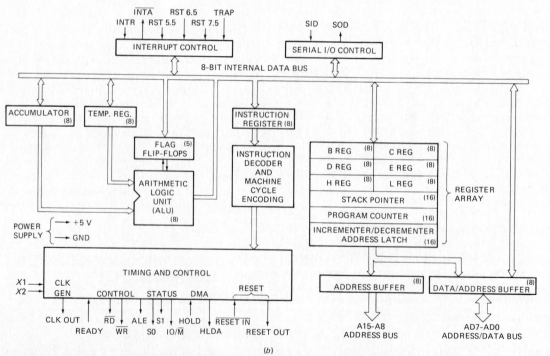

FIGURE 8-3 (a) 8080A CPU functional block diagram. (Intel Corporation.) (b) 8085A
CPU functional block diagram. (Intel Corporation.)

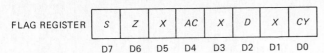

FLAG REGISTER

	S	Z	X	AC	X	D	X	CY
	D7	D6	D5	D4	D3	D2	D1	D0

FIGURE 8-4 8080A/8085A flag register format.

isters. The B register alone, for example, can be used to store a byte of data. With the C register it can be used to store a 16-bit data word or 16-bit address. A multiplexer allows a byte of data on the internal data bus to be routed to either the B column registers or the C column registers.

STACK POINTER The *stack pointer* is a 16-bit register usually employed to store an address. We discuss its use in detail later.

PROGRAM COUNTER The program counter can be thought of as a register or parallel-load counter. It stores the memory address currently being read from or written into by the CPU. After each instruction, the program counter is incremented automatically so that it points to the location of the next instruction or data in memory.

ADDRESS LATCH The *address latch* section serves two functions. First, it selects an address to be sent out from the program counter, from the stack pointer register, or from one of the 16-bit register pairs. Second, it latches this address onto the address lines for the required time. The 16-bit addresses from the 8080A or the 8085A allow each microprocessor to address up to 2^{16}, or 65,536, memory locations. An *incrementer-decrementer* allows the contents of any of the 16-bit registers to be incremented or decremented.

INSTRUCTION REGISTER, INSTRUCTION DECODER, AND CONTROL CIRCUITS

The *instruction register* simply holds instructions coming in on the data bus for the *instruction decoder*. The decoder interprets the instruction and produces the proper signals to carry it out. For example, if the ADD instruction from the three-step program above comes into the decoder, the decoder will recognize that another fetch to memory is needed. A second fetch is needed to get the number that is to be added to the number in the accumulator. The decoder will direct the *control circuitry* to send out another memory read pulse and to steer the data coming back on the data bus into the temporary register, so it can be added to the accumulator. When the addition is completed, the control circuits direct the result back to the accumulator. They then increment the program counter to the next memory address and send out another memory read pulse to get the next instruction from memory.

Note that the 8080A requires supply voltages of +12, +5, and −5 V. The 8085A requires only a +5-V supply. The 8080A requires two clock signals, $\phi1$ and $\phi2$, from an external clock generator. The 8085A requires only an external crystal connected to the X1 and X2 inputs. The rest of the input and output signals for these microprocessors are discussed in Chap. 9.

ASSEMBLY LANGUAGE PROGRAMS FOR THE 8080A AND 8085A

To run a program, a microprocessor must have the program stored in binary form in successive memory locations. This binary form of the program is referred to as *machine language* because it is the form required by the machine. However, for human programmers it is difficult to remember the 1's and 0's form of, for example, all the instructions of the 8080A. Using hexadecimal representation for the binary numbers helps some, but nearly 256 individual instruction codes would still have to be memorized.

Assembly language programming is easier to do and yields much more understandable programs. When completed, an assembly language program is *assembled*. This means that it is translated into the 1's and 0's of machine language for running the microprocessor.

Assembly language uses two- to four-letter *mnemonics* to represent each instruction. A mnemonic is just a shorthand, or device, to help you remember something. The letters in an assembly language mnemonic are usually initials or a shorthand form of the English word for the instruction. For example, the instruction "move the contents of the B register into the accumulator" is represented as MOV A,B. This form keeps the essential information for the instruction but uses minimum space. In the next section we will take a first look at the 8080A/8085A instruction set in assembly language or mnemonic form.

THE 8080A/8085A INSTRUCTION SET

Table 8-1 shows the mnemonic and the hexadecimal code for each of the 8080A/8085A instructions. Except for the SIM and RIM instructions, which have been added to the 8085A, the instructions are the same for the two processors. Each instruction consists of two parts—the *operation code*, or *op code*, and the *operand*—which are separated by a single space. The op code is the mnemonic for the operation performed, and the operand is the register or data acted upon.

For reference, all these instructions are explained together in the following sections. However, do not try to absorb all these instructions at once. The first time you

TABLE 8-1
THE 8080A/8085A INSTRUCTION SET (INTEL CORPORATION.)

DATA TRANSFER GROUP

Move

MOV	A,A	7F	MOV	E,A	5F
	A,B	78		E,B	58
	A,C	79		E,C	59
	A,D	7A		E,D	5A
	A,E	7B		E,E	5B
	A,H	7C		E,H	5C
	A,L	7D		E,L	5D
	A,M	7E		E,M	5E
MOV	B,A	47	MOV	H,A	67
	B,B	40		H,B	60
	B,C	41		H,C	61
	B,D	42		H,D	62
	B,E	43		H,E	63
	B,H	44		H,H	64
	B,L	45		H,L	65
	B,M	46		H,M	66
MOV	C,A	4F	MOV	L,A	6F
	C,B	48		L,B	68
	C,C	49		L,C	69
	C,D	4A		L,D	6A
	C,E	4B		L,E	6B
	C,H	4C		L,H	6C
	C,L	4D		L,L	6D
	C,M	4E		L,M	6E
MOV	D,A	57	MOV	M,A	77
	D,B	50		M,B	70
	D,C	51		M,C	71
	D,D	52		M,D	72
	D,E	53		M,E	73
	D,H	54		M,H	74
	D,L	55		M,L	75
	D,M	56	XCHG		EB

Move Immediate

MVI	A, byte	3E
	B, byte	06
	C, byte	0E
	D, byte	16
	E, byte	1E
	H, byte	26
	L, byte	2E
	M, byte	36

Load Immediate

LXI	B, dble	01
	D, dble	11
	H, dble	21
	SP, dble	31

Load/Store

LDAX B	0A
LDAX D	1A
LHLD adr	2A
LDA adr	3A
STAX B	02
STAX D	12
SHLD adr	22
STA adr	32

ARITHMETIC AND LOGICAL GROUP

Increment*

INR	A	3C	INX	B	03
	B	04		D	13
	C	0C		H	23
	D	14		SP	33
	E	1C			
	H	24			
	L	2C			
	M	34			

Decrement*

DCR	A	3D	DCX	B	0B
	B	05		D	1B
	C	0D		H	2B
	D	15		SP	3B
	E	1D			
	H	25			
	L	2D			
	M	35			

Add*

ADD	A	87	ADC	A	8F
	B	80		B	88
	C	81		C	89
	D	82		D	8A
	E	83		E	8B
	H	84		H	8C
	L	85		L	8D
	M	86		M	8E

Subtract*

SUB	A	97	SBB	A	9F
	B	90		B	98
	C	91		C	99
	D	92		D	9A
	E	93		E	9B
	H	94		H	9C
	L	95		L	9D
	M	96		M	9E

Double Add†

DAD	B	09
	D	19
	H	29
	SP	39

Specials

DAA*	27
CMA	2F
STC†	37
CMC†	3F

Rotate†

RLC	07
RRC	0F
RAL	17
RAR	1F

Logical*

ANA	A	A7	ORA	A	B7
	B	A0		B	B0
	C	A1		C	B1
	D	A2		D	B2
	E	A3		E	B3
	H	A4		H	B4
	L	A5		L	B5
	M	A6		M	B6
XRA	A	AF	CMP	A	BF
	B	A8		B	B8
	C	A9		C	B9
	D	AA		D	BA
	E	AB		E	BB
	H	AC		H	BC
	L	AD		L	BD
	M	AE		M	BE

Arith. & Logical Immediate

ADI byte	C6
ACI byte	CE
SUI byte	D6
SBI byte	DE
ANI byte	E6
XRI byte	EE
ORI byte	F6
CPI byte	FE

BRANCH CONTROL GROUP

Jump

JMP adr	C3
JNZ adr	C2
JZ adr	CA
JNC adr	D2
JC adr	DA
JPO adr	E2
JPE adr	EA
JP adr	F2
JM adr	FA
PCHL	E9

Call

CALL adr	CD
CNZ adr	C4
CZ adr	CC
CNC adr	D4
CC adr	DC
CPO adr	E4
CPE adr	EC
CP adr	F4
CM adr	FC

Return

RET	C9
RNZ	C0
RZ	C8
RNC	D0
RC	D8
RPO	E0
RPE	E8
RP	F0
RM	F8

Restart

RST	0	C7
	1	CF
	2	D7
	3	DF
	4	E7
	5	EF
	6	F7
	7	FF

I/O AND MACHINE CONTROL

Stack Ops

PUSH	B	C5
	D	D5
	H	E5
	PSW	F5
POP	B	C1
	D	D1
	H	E1
	PSW*	F1
XTHL		E3
SPHL		F9

Input-Output

OUT byte	D3
IN byte	DB

Control

DI	F3
EI	FB
NOP	00
HLT	76

New Instructions (8085 only)

RIM	20
SIM	30

ASSEMBLER REFERENCE

Operators

( )
NUL
LOW,HIGH
*,/,MOD,SHL,SHR
+,−
NOT
AND
OR,XOR

ASSEMBLER REFERENCE (Cont.)

Pseudo Instruction

General:
ORG
END
EQU
SET
DS
DB
DW

Macros:
MACRO
ENDM
LOCAL
REPT
IRP
IRPC
EXITM

Relocation:
NAME
ASEG   STKLN
DSEG   STACK
CSEG
PUBLIC MEMORY
EXTRN

Conditional Assembly:
IF
ELSE
ENDIF

Constant Definition

0BDH	Hex
1AH	
105D	Decimal
105	
72O	Octal
72Q	
11011B	Binary
00110B	
'TEST'	ASCII
'A' 'B'	

---

byte = constant, or logical/arithmetic expression that evaluates to an 8-bit data quantity (second byte of 2-byte instructions).

dble = constant, or logical/arithmetic expression that evaluates to a 16-bit data quantity (second and third bytes of 3-byte instructions).

adr = 16-bit address (second and third bytes of 3-byte instructions).

* = all flags (C, Z, S, P, AC) affected.

** = all flags except CARRY affected (except INX and DCX affect no flags).

† = only CARRY affected.

All mnemonics copyright © 1976 by Intel Corporation.

read through these sections, focus on understanding the following types of instructions:

Data transfer group: Move, move immediate, and load immediate

Arithmetic and logic group: Add, subtract, increment, decrement, logical, arithmetic, and logical immediate

Branch control group: Jump

I/O and machine control group: Input-output and HLT

Because of the number of different combinations these look like a lot of instructions, but in reality they are only about 10 different types of operation. The mnemonics should help you remember what each instruction is doing.

After you have read through the descriptions of these instructions, go ahead to programs 1 and 2 in Figures 8-20 and 8-21 to see how some of these instructions are used in simple programs.

On your second pass through the instruction set descriptions, work on understanding the specials group, the rotate group, the call group, the return group, and the stack op codes group. Then look at the program of Figure 8-22 to see how some of these instructions are used.

A third pass through the instruction set should pick up the remaining groups such as load/store and double-add. As you write programs, you will quickly become familiar with most of the available instructions. Knowing most of the available instructions allows you to choose the one that most easily performs the desired operation at a particular point in your program.

DATA TRANSFER GROUP INSTRUCTIONS

MOVE INSTRUCTIONS   The largest group of instructions is the *move* group. This group contains all the possible transfers of a byte of data from one register

to another or between a memory location and a register. The destination register always is listed before the comma, and the source register is listed after the comma. For example, MOV C,D means move the byte in the D register into the C register.

Moving might be thought of as *copying* because it does not change the data in the source register or memory location. Assume, for example, that the D register contains 78H and the C register contains 8AH. After a MOV C,D instruction is executed, the C register will contain 78H and the D register will still contain 78H. Often a *symbolic form* is used to describe the effect of instructions. For the MOV C,D instruction, the symbolic form is (C) ← (D). The parentheses are shorthand for *the contents of*. The arrow is shorthand for *copied to*, or *becomes*. So the symbolic form reads, "The contents of the D register are copied to the C register."

If a move to or from memory such as MOV M,A is used, the desired memory address must first be loaded into the H and L register pair. In other words, the MOV M,A instruction duplicates the contents of the accumulator in the memory location at the address stored in HL. In this case, the HL register pair is being used as a *pointer* to a memory location.

Figure 8-5 shows the use of the pointer for a memory move in diagram form. Since in this case the HL register contains 073AH, the MOV M,A instruction copies the contents of the accumulator, 0B3H, into memory location 073AH. If the instruction were MOV A,M and the HL register contained 073AH, the contents of memory location 073AH would be copied into the accumulator. The contents of memory location 073AH would be unchanged for this second case.

MOVE IMMEDIATE INSTRUCTIONS   The MVI (register, byte), or *move immediate*, instructions load the byte of data written after the comma into the specified register. These instructions require 2 memory bytes: one to store the op code and the next to store the *immediate* number. For the instruction MVI A,07H, the first memory location contains the binary equivalent of 3EH, and the next location contains 07. When run, this instruction will load 07H, or 00000111, into the accumu-

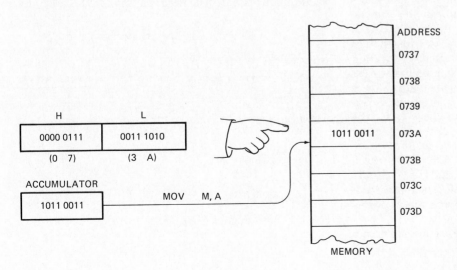

FIGURE 8-5 H and L registers used as a memory pointer for the MOV M,A instruction.

lator. For the MVI M, (byte) instruction, the immediate byte of data will be moved to a memory location pointed to by the contents of HL. Any immediate instruction requires the immediate byte or bytes to be given right after the op code.

LOAD EXTENDED IMMEDIATE

The LXI mnemonic of the next group stands for *load extended immediate.* Here *extended* means that a 16-bit immediate quantity is going to be loaded into the specified register pair. For example, LXI H,234AH will load 23 into the H register and 4A into the L register. However, when this instruction is actually loaded into memory to be run, the low-order byte must be loaded before the high-order byte. For the LXI H,234AH instruction, three successive memory locations will contain: 21H, the code for the LXI H instruction from Table 8-1; 4AH, the low-order byte that will go into the L register; and 23H, the high-order byte that will go into the H register when the instruction executes.

LOAD AND STORE INSTRUCTIONS

By general definition, a *load* instruction means to bring data *into* one of the CPU group registers or accumulator from some place such as memory. The *store* instruction, on the other hand, generally means to *send out* data from the CPU group to some place such as memory.

LDAX B copies data into the accumulator from the memory address contained in the 16-bit extended BC register. In other words, the extended BC register is used as a pointer to a memory address, just as the HL register is used for MOV A,M and other instructions that reference memory with an M.

STAX B copies the contents of the accumulator into memory at the address contained in (pointed to by) the extended BC register.

LDAX D and STAX D use the extended DE register as a pointer to memory. LDAX D copies data from the address pointed to by DE into the accumulator. STAX D copies the contents of the accumulator into the memory address pointed to by the contents of DE.

LHLD adr (address) and SHLD adr instructions provide a way to save or restore the contents of the extended HL register with a single instruction. LHLD adr loads the L register with data from the address given in the instruction and loads the H register with the contents of the following address. For example, the instruction LHLD 43A7H copies the contents of memory location 43A7H into the L register and the contents of the next address, 43A8H, into the H register. This instruction is coded into three successive memory locations as follows: 2A, the instruction code from Table 8-1; A7, the low-order byte of the address; and 43, the high-order byte of the address. And SHLD adr performs the reverse operation. The instruction SHLD 43A7H copies the L register into memory location 43A7H and the H register into the next memory location, 43A8H. The instruction would be coded in three successive memory locations as 22, A7, 43.

LDA adr copies a byte of data to the accumulator from the memory address given in the instruction. For example, LDA 3A78H copies the contents of address 3A78H into the accumulator. The contents of address 3A78H remain unchanged.

STA adr is the reverse of the LDA operation. For example, STA 3A78H copies the contents of the accumulator into memory location 3A78H. The contents of the accumulator remain unchanged. The advantage of these instructions is that they allow a byte of data to be copied from the accumulator to memory, or from memory to the accumulator, without having to use the HL register as a pointer.

Don't worry if all these instructions are not sinking in the first time. They are being explained here for future reference and so that you can see the kinds of operations that the 8080A or 8085A can perform. Some simple program examples in a later section will clarify the more commonly used instructions. Before reading on, glance at the rest of Table 8-1 and see how many of the instructions you can figure out from the mnemonics alone.

ARITHMETIC AND LOGIC GROUP INSTRUCTIONS

ADD INSTRUCTIONS

The ADD register instructions add the contents of the specified register to the contents of the accumulator. The ADD M instruction adds the contents of the memory location pointed to by HL to the contents of the accumulator. The results of the addition (sum) are left in the accumulator. The contents of the source register or memory location are unchanged. All flag flip-flops are affected by add and subtract instructions.

Figure 8-6 shows an example of the effects of an ADD B instruction. Initially the B register contains 01111000, and the accumulator contains 00101001. After an ADD B instruction executes, the accumulator contains the sum of the addition, 10100001. The B register still contains 01111000, since it is unchanged by the ADD B instruction. The flags are left as shown in Figure 8-6. The sign flag bit is 1 because bit 7 in the accumulator is a 1. The zero flag bit is 0 because the

		D7	D6	D5	D4	D3	D2	D1	D0
B REGISTER		0	1	1	1	1	0	0	0
ACCUMULATOR	+	0	0	1	0	1	0	0	1
ACCUMULATOR (AFTER ADD B INSTRUCTION EXECUTES.)		1	0	1	0	0	0	0	1
B REGISTER (AFTER ADD B)		0	1	1	1	1	0	0	0

	S	Z		AC		P		CY
FLAG REGISTER (AFTER ADD B)	1	0	X	1	X	0	X	0

FIGURE 8-6 Effect of ADD B instruction on registers and flags.

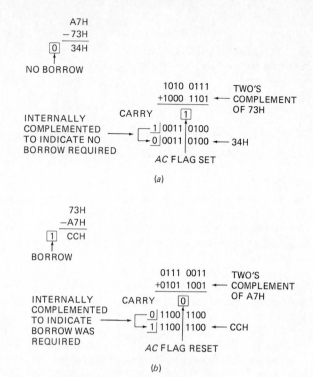

FIGURE 8-7 Comparison of ADD B and ADC B instructions.

FIGURE 8-8 Unsigned binary subtraction.

result in the accumulator is not 0. The AC flag bit is a 1 because there was a carry-out of bit 3 during the addition. The parity flag bit is a 0 because the number of 1's in the accumulator is odd. The CY flag bit is 0 because the result was not greater than FFH.

If the result of an addition is greater than FF_{16} (255_{10}), the carry flag flip-flop is set. When multiple-byte numbers are added, this carry can be included in the addition of the next-higher bytes. The ADC register, or add-with-carry instructions are used for this purpose.

ADC register adds the contents of the indicated register plus the contents of the carry flip-flop to the accumulator. If the carry flag is set (1), 1 is added to the sum of the register and the accumulator. If the carry is reset (0), a 1 is not added. If the result of an add-with-carry operation is greater than FFH, the carry flag is set. For a result equal to or less than FFH, the carry flag is reset. Figure 8-7 shows, for comparison, the effects of the ADD register and the ADC register instructions. Note that both instructions reset the carry flag because the result of the addition is less than FFH.

SUBTRACT INSTRUCTIONS

The SUB register instruction subtracts the contents of the indicated register or memory location from the contents of the accumulator. The result is left in the accumulator. If the result is negative, the carry flag is set. For subtraction, the carry flag being set indicates that a borrow was required to perform the subtraction. In other words, for subtraction the carry flag indicates a borrow.

Figure 8-8 shows examples of how subtraction works. The ALU does subtraction by adding the 2's complement of the bottom number to the top number, as shown. For the case of A7H − 73H, the pencil-computed result shows a difference of 34H and carry set. However, for a subtraction operation, the carry flag is complemented internally. Therefore, the actual result will show the carry flag reset to indicate that no borrow was required to perform the subtraction.

For the case of 73H − A7H in Figure 8-8b, the carry flag is set to indicate that a borrow was required to perform the subtraction. For single-byte subtraction, the carry being set also indicates that the result is negative and in 2's complement form. For 73H − A7H, the result, then, is −34H. For subtracting multiple-byte numbers, the status of the carry (borrow) flag can be subtracted from the next higher byte.

For examples of how the add and subtract instructions work with 8-bit sign-and-magnitude numbers, refer to the examples in Chap. 7.

SBB register (subtract with borrow) instructions sub-

tract the contents of the indicated register and the status of the borrow (carry) flag from the contents of the accumulator. Again, the result is left in the accumulator. If the result of the subtraction is negative, the carry (borrow) flag is set to reflect this.

DOUBLE-ADD INSTRUCTIONS

The DAD register pair, or double-add, instructions add the contents of the specified register pair to the HL extended register. The result is left in the HL register pair. A result greater than FFFFH sets the carry flag. This instruction allows two 16-bit binary numbers to be added with one instruction.

DAD H doubles the contents of the HL register. This is the same as shifting each bit one position to the left and putting a 0 in the LSB position.

INCREMENT AND DECREMENT INSTRUCTIONS

The increment and decrement instructions are quite straightforward. INR register increments the contents of the specified register by 1, and DCR register decrements the contents of the specified register by 1. The results of an INR or a DCR are left in the register specified. As shown by the double-asterisk footnote in Table 8-1, the increment and decrement instructions affect all flags except carry. For example, if a DCR C instruction leaves all 0's in the C register, the zero flag will be set (1). If the accumulator contains FFH, then executing an INR A instruction rolls the accumulator contents over to all 0's. The zero flag will be set to reflect this.

INX register pair increments the specified 16-bit extended register by 1, and DCX register pair decrements

```
CY   AC
     0
 0  1001 1001   ACCUMULATOR – BCD99
 +  0010 1000   B REGISTER – BCD 28
     1
 0  1100 0001   ACCUMULATOR AFTER ADD B
 +  0110 0110   DAA
     0
 1  0010 0111   ACCUMULATOR AND CARRY AFTER DAA – BCD 127
```

FIGURE 8-9 Effects of the DAA instruction for BCD addition.

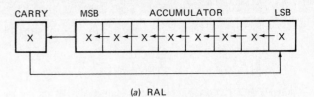

(a) RAL

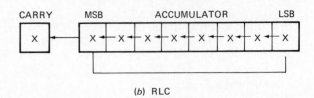

(b) RLC

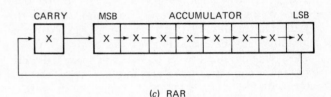

(c) RAR

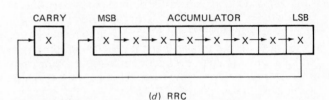

(d) RRC

FIGURE 8-10 Rotate instructions. *(a)* RAL. *(b)* RLC. *(c)* RAR. *(d)* RRC.

the specified register pair by 1. (SP stands for *stack pointer.*) Note, as shown in Table 8-1, that INX and DCX instructions do not affect any flags. It is important to remember this when you use these extended registers in delay loops.

SPECIAL INSTRUCTION GROUP   DAA is the decimal adjust accumulator instruction used to produce a correct result when BCD addition is performed. This correction is discussed in Chap. 7. Figure 8-9 shows the effect of this instruction. Initially the accumulator contains 10011001, which is the BCD code for 99, and register B contains 00101000, which is the BCD code for 28. Execution of the ADD B instruction leaves 11000001, not the correct BCD result, in the accumulator. The DAA instruction solves the problem. It works on each nibble (4 bits) of the sum in the accumulator. If the result in the lower 4 bits is an illegal BCD code, or if the addition produced a carry from bit 3 and set the auxiliary carry flag, then the DAA instruction will add the 0110 correction to the lower nibble in the accumulator. Likewise, the upper nibble will be corrected if a carry was produced by the original addition or if the upper nibble of the result contains an illegal BCD code. For the example shown, both the upper and lower nibbles required correction to give the correct result of 0001 0010 0111 BCD, or 127_{10}. Usually you don't have to think too much about this, because the DAA instruction takes care of it automatically. Note, however, the DAA instruction will function properly only after an instruction that affects the auxiliary carry (AC) flag.

CMA complements or inverts all the bits in the accumulator.

STC sets the carry flag flip-flop to 1.

CMC complements the state of the carry flag.

ROTATE INSTRUCTIONS   The rotate, or shift, instructions have two variations which differ in the way that the carry flag flip-flop is used. Figure 8-10 illustrates these four instructions.

RAL means "rotate left *through* carry." The MSB of the accumulator is shifted left into the carry flip-flop, and each bit of the accumulator is shifted left 1 bit position. The original content of the carry flip-flop is shifted into the LSB position of the accumulator. The carry flip-flop is essentially treated as an extension of the accumulator.

RLC, which means "rotate left *to* carry," shifts the MSB of the accumulator left into the carry flip-flop and

also around to the LSB position of the accumulator. Each of the outer bits in the accumulator is shifted left 1 bit position.

RAR means "rotate (or shift) the accumulator right through carry," as shown in Figure 8-10c.

RRC rotates the LSB of the accumulator to the MSB position and to the carry flip-flop, as shown in Figure 8-10d.

The shift instructions are useful for add-and-shift-right multiplication and subtract-and-shift-left division with a microprocessor.

LOGICAL INSTRUCTIONS: ANA, XRA, ORA, CMP
Again these instructions are quite straightforward.

The ANA register instruction ANDs the contents of the specified register with the contents of the accumulator on a bit-by-bit basis. The result is left in the accumulator. CY is reset.

XRA register produces the *exclusive* OR of the specified register contents and the accumulator contents. CY is reset.

The ORA register instruction produces the simple OR of the specified register contents and the accumulator. CY is reset. For review, Figure 8-11 shows the results produced when two 8-bit numbers are ANDed, ORed, and XORed.

204

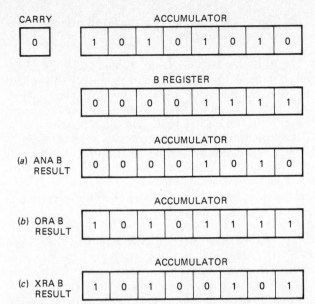

FIGURE 8-11 Result of logical operations on two 8-bit numbers. *(a)* ANA B result. *(b)* ORA B result. *(c)* XRA B result.

The ANA instruction often is used to *mask* bits of a word. As shown in Figure 8-11, for any bit of a word that is ANDed with a 0, the result will be a 0. These bits are said to be masked because after the ANDing they are always 0, regardless of their original state. Bits ANDed with a 1, however, remain unchanged. Suppose, as shown in Figure 8-11, the accumulator contains AAH (10101010) and the B register contains 0FH (00001111). After the ANA B instruction executes, the upper 4 bits of the accumulator are masked and the lower 4 bits are unchanged. The result in the accumulator is 0AH (00001010), as in Figure 8-11*a*. Masking can be done on any number of bits in a word.

ORing a bit with 0 leaves it unchanged. ORing a bit with 1 sets it to a 1. As Figure 8-11*b* shows, ORing a word with 0FH sets the lower 4 bits of the result to 1's but leaves the upper 4 bits unchanged.

Note: ORA A is often used to reset the carry flag without affecting the contents of the accumulator. XORing a bit with a 0 leaves the bit unchanged. XORing a bit with a 1 inverts that bit. As shown in Figure 8-11, XORing AAH with 0FH leaves the upper 4 bits unchanged and inverts the lower 4 bits, to give a result of A5H. This instruction often is employed to complement bits in a word. The XRA A instruction clears the contents of the accumulator to all 0's and resets the carry flag to 0.

The CMP register instruction compares the contents of the specified register with the contents of the accumulator. A compare is done by subtracting the contents of the specified register or memory location from those of the accumulator. Flags are set according to the result, but neither the contents of the accumulator nor the contents of the register are changed. For a CMP B instruction the flags are affected as shown in Figure 8-12. This instruction, then, can be used to determine whether a number is less than, equal to, or greater than some number in the accumulator.

| IF: | THEN: | |
	CARRY	ZERO
Accumulator = register or immediate data	0	1
Register or immediate data > accumulator	1	0
Accumulator > register or immediate data	0	0

FIGURE 8-12 CMP/CPI instructions and flags affected.

ARITHMETIC AND LOGIC IMMEDIATE INSTRUCTIONS: ADI, ACI, SUI, SBI, ANI, ORI, XRI, CPI

These instructions perform the indicated operation on the accumulator and the immediate data contained in the instruction. The instruction ADI 3AH, for example, adds the immediate number 3AH to the accumulator. For all these instructions, the results are left in the accumulator. When coded, each of these instructions requires 2 bytes, the first for the instruction code and the second for the immediate data byte.

The operations of this group of instructions are as follows:

ADI byte adds the immediate byte to the accumulator.

ACI byte adds the immediate byte plus the carry status to the accumulator.

SUI byte subtracts the immediate byte from the accumulator.

SBI byte subtracts the immediate byte and the status of the carry (borrow) flag from the accumulator.

ANI byte logically ANDs the immediate byte with the contents of the accumulator. CY is reset.

ORI byte logically ORs the immediate byte with the contents of the accumulator. CY is reset.

XRI byte logically XORs the immediate byte with the contents of the accumulator. CY is reset.

CPI byte compares the immediate byte to the contents of the accumulator by subtraction. Flags are set as shown in Figure 8-12, but the contents of the accumulator are not changed.

BRANCH AND CONTROL GROUP INSTRUCTIONS

The instructions in this group change the normal sequence program execution.

JUMPS The *unconditional jump* instruction (JMP) is a 3-byte instruction with byte 1 containing the op code for the instruction and bytes 2 and 3 containing an address, as shown in Figure 8-13*a*. As with other instructions mentioned earlier, the low-order address byte is loaded first, followed by the high-order byte. When the JMP instruction is executed, the address in bytes 2 and 3 is loaded into the program counter. The CPU will therefore fetch its next instruction from the memory at this new address. The program then continues to execute sequentially from this "jumped to" address. Since 2 bytes are used for the jump address, a jump to anywhere in 65,536 bytes of memory is possible. A jump may be

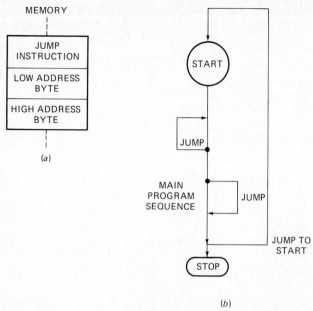

(a)

(b)

FIGURE 8-13 Jump instruction. *(a)* Stored in memory with low address byte first. *(b)* Changes in program flow that can be produced by jump instructions.

forward or backward in the program sequence, as shown in Figure 8-13*b*.

One application of the JMP instruction is to keep a program running over and over in a continuous loop, as is done for a microprocessor-based scale. To do this, the last instruction in the program is JMP. The address bytes of the instruction are loaded with the starting address of the program. Each time the program reaches this instruction, it sets the program counter back to the starting address. The program then runs again from the start.

Conditional jump instructions such as JZ (jump if zero) cause the program counter to go to the jump address only if the specified condition is true. The jump address is contained in the second 2 bytes of the instruction, as before. When the JZ instruction is executed, a jump takes place if the accumulator or register byte is all 0's, as indicated by the zero flag. If the accumulator or register is not all 0's, then the program continues in its normal sequence.

Other conditional jump instructions are as follows:

JNZ (jump if zero flag not set)

JNC (jump if carry flag is 0)

JC (jump if carry flag is 1)

JPO (jump if the parity of the byte in the accumulator is odd)

JPE (jump if the parity of the byte in the accumulator is even)

JP (jump if sign bit is 0, indicating contents of accumulator are positive)

JM (jump if sign bit is 1, to indicate a negative number in the accumulator)

The PCHL instruction, which is also listed with jump instructions in Table 8-1, exchanges the contents of the program counter with those of the HL register. When executed, this instruction produces an unconditional jump to the address stored in the HL register.

The conditional jump instructions can be used to keep running around a program loop until the result satisfies some conditions. For example, the JNZ instruction can be employed to keep looping back to the start of an operation until the result of the operation is 0. Later we give examples of this.

CALL, RETURN, AND SUBROUTINES A CALL address instruction is used to jump the program counter to the starting address of a subprogram, or *routine*, in memory, as shown in Figure 8-14*a*. A *return* instruction (RET) at the end of the subroutine sends the program execution back to the main sequence. The execution returns to the next instruction after the CALL instruction in the main program sequence.

A *routine* is a short program section that performs a specific task, for example, multiplying two numbers. If a routine is to be used many times in a program, it can be written once in memory as a *subroutine* and then just

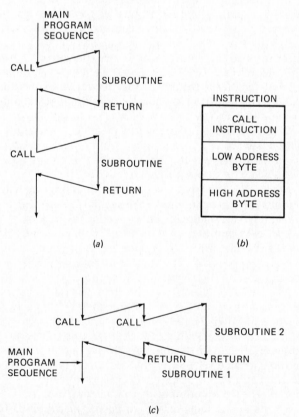

(a)

(b)

(c)

FIGURE 8-14 Subroutines. *(a)* Program flow with subroutines. *(b)* Call instruction stored with low address byte before high address byte. *(c)* Nested subroutines.

called each time it is needed. This method uses much less program memory space than writing the entire routine in the main program each time it is used.

Call instructions contain the starting address of the subroutine in bytes 2 and 3 of the instruction, as shown in Figure 8-14b. A call can be unconditional if the CALL instruction is used or conditional if CNZ, CZ, CNC, CC, CPO, CPE, CP, or CM is used. The meaning of the *conditions* of these instructions is the same as for the conditional jump instructions.

We said previously that the return instruction at the end of a subroutine transfers the program execution back to the next instruction after the CALL. Since a subroutine can be called from anywhere in the main sequence but the return instruction contains no address, a question arises as to how the return instruction determines where to return.

The CALL instruction stores the return address in a special section of RAM called the *stack*. The programmer decides on the starting address for the stack in RAM. This address is loaded into the stack pointer register with an LXI SP, address (2 bytes) instruction at the start of the program. When a CALL instruction is executed, the address of the next main program sequence instruction after the call is stored on, or "pushed onto," the stack.

Figure 8-15a is a small sample program that calls a subroutine, and Figure 8-15b illustrates how the contents of the stack pointer are affected by this program. As shown in Figure 8-15a, the stack pointer is initialized at the start of the program to 20C2H by an LXI SP,20C2H instruction. The address of the next main sequence instruction (return address) after the CALL instruction is 2054H. When the CALL 20A0H instruction executes, several steps occur, as shown in Figure 8-15b.

First, the stack pointer is decremented by 1, to 20C1H, and the high byte of the return address is stored (pushed) at that memory address. Second, the stack pointer is decremented by 1 again, to 20C0H, and the low byte of the return address is stored (pushed) at that memory location. Third, the starting address of the subroutine, 20A0H, is loaded into the program counter. The next instruction will then be fetched from this address. The stack pointer holds the *address* of the last data or address byte pushed on the stack.

When a RET instruction is executed at the end of the subroutine, the microprocessor reads the low byte of the return address from the stack at 20C0H and loads (copies) it into the program counter. The microprocessor then increments the stack pointer to 20C1H and copies the high byte of the return address into the program counter. Finally, it increments the stack pointer to its original value, 20C2H. The program counter now contains the address of the next main program sequence instruction, 2054H, so execution will go on from there.

Returns can be unconditional or conditional. RET is an *unconditional return*. The *conditional returns* are:

RNZ (return if not 0)

RZ (return if 0)

RNC (return if not carry)

RC (return if carry)

RPO (return if parity odd)

RPE (return if parity even)

RP (return if plus)

RM (return if minus)

Subroutines can be *nested*. This means that a CALL instruction in a subroutine can call another subroutine. Figure 8-14c shows a diagram of this. A machine executes the program until it encounters a CALL instruction; then it goes to subroutine 1. It executes subroutine 1 until it encounters another CALL instruction; then it goes to subroutine 2. When subroutine 2 is completed, the machine returns to subroutine 1. When subroutine 1 is completed, the machine returns to the main program. The return addresses for both subroutines are automatically pushed onto the stack in order by the CALL instructions.

RESTARTS *Restart* instructions (RST) are special 1-byte, unconditional call instructions. These instructions store the return address on the stack and jump the program counter to a low address in memory. This instruction is used mostly with *interrupts* (discussed in Chap. 9). The addresses jumped to by the restart instructions are RST 0, 0000H; RST 1, 0008H; RST 2, 0010H; RST 3, 0018H; RST 4, 0020H; RST 5, 0028H; RST 6, 0030H; RST 7, 0038H.

STACK OPERATIONS The *stack* is a location in RAM set aside to store return addresses for subroutine calls. It can also be used to store the contents of registers so they will not be changed or destroyed by a subroutine. For example, data stored in the B register in the main program will be lost if the B register is used in the subroutine. A *push* operation is performed at the beginning of a subroutine to save register contents, as shown in Figure 8-16. For example, the PUSH B instruction stores the contents of the extended register BC on the stack. PUSH D stores the extended DE register, PUSH H stores the extended register HL, and PUSH PSW stores the accumulator and the flag byte. PSW stands for *program status word*.

At the end of the subroutine, the data is restored to the proper register by a POP instruction. A POP register instruction copies the stored data from the stack back into the indicated extended register. For example, POP B restores data from the stack to the B and C registers. Note in Figure 8-16 that data must be popped off the stack in the reverse order from which it was pushed. This type of stack is sometimes called *last-in, first-out memory*, or LIFO. It is similar to the spring-loaded plate holders seen in some restaurants. The last plate pushed onto the stack is the first one popped off.

For the PUSH instruction, the stack pointer keeps

ADDRESS	CODE	INSTRUCTION	OPERAND
2000	21	LXI	SP,20C2H
2001	C2		
2002	20		
.			
.		MAIN PROGRAM CONTINUES	
.			
2051	CD	CALL	20A0H
2052	A0		
2053	20		
2054	D3	OUT	01H
2055	01		
.			
.		MAIN PROGRAM CONTINUES	
.			

MAIN PROGRAM

20A0	E5	PUSH	PSW
20A1	C5	PUSH	B
.			
.			
.			
20B0	C1	POP	B
20B1	F1	POP	PSW
20B2	C9	RET	

SUBROUTINE

(a)

	STACK POINTER		MEMORY ADDRESS	STACK MEMORY CONTENTS
AFTER LXI SP,20C2H	20C2	AFTER RETURN	20C2	XX
CALL	20C1	RET	20C1	20
AFTER CALL 20A0H	20C0	AFTER POP PSW	20C0	54
PUSH PSW	20BF	POP PSW	20BF	(A)
AFTER PUSH PSW	20BE	AFTER POP B	20BE	(FLAGS)
PUSH B	20BD	POP B	20BD	(B)
AFTER PUSH B	20BC	BEFORE POP B	20BC	(C)

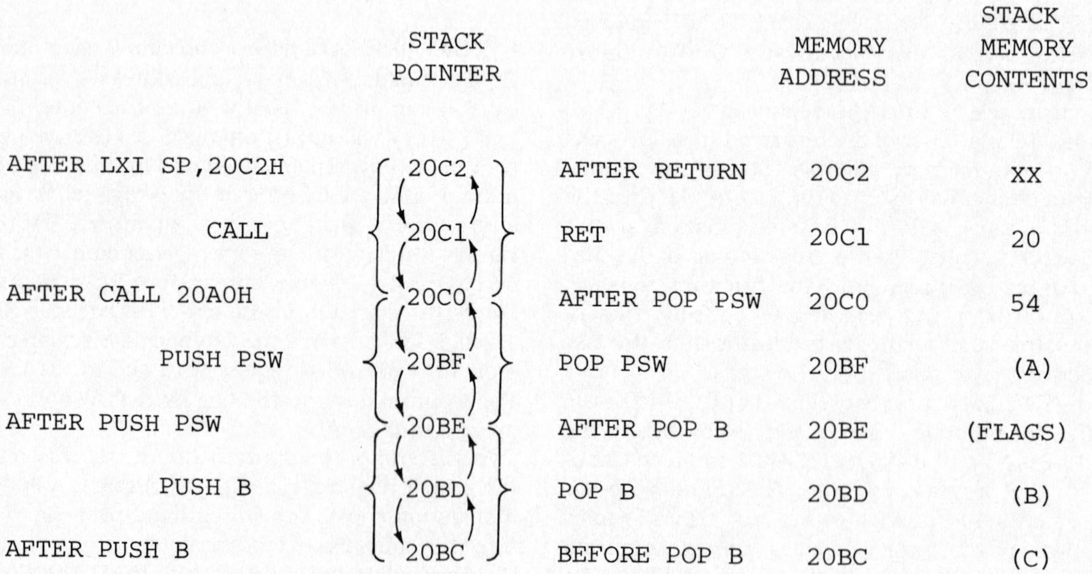

NOTE: ( ) indicates "the contents of" the indicated
register before the push instruction executed

(b)

FIGURE 8-15 (a) Program and subroutine using CALL, PUSH, POP, and RET instructions. (b) Contents of stack during subroutine CALL and return.

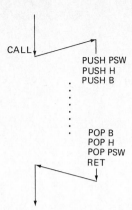

FIGURE 8-16 Diagram of subroutine CALL showing order of PUSH and POP instructions.

track of (holds) the address of the last byte of data pushed onto the stack. A PUSH B instruction, for example, affects the stack and stack pointer as follows. The stack pointer is decremented by 1, and the contents of the B register are copied onto the stack at that address. The stack pointer is decremented again, and the contents of the C register are copied to that address.

A POP B instruction copies the previously saved byte of data from the stack back into the C register, increments the stack pointer, copies the next byte of saved data from the stack to the B register, and increments the stack pointer. After a PUSH B, POP B sequence the stack pointer is left pointing to the same address that it was before the operation.

Figure 8-15b shows an example of how the stack and stack pointer are affected by CALL, PUSH, POP, and RET instructions during a typical subroutine call.

CALL decrements the stack pointer to 20C1H and pushes the high byte of the return address onto the stack. It then decrements the stack pointer to 20C0H, and pushes the low byte of the return address onto the stack. PUSH PSW decrements the stack pointer to 20BFH, stores the accumulator contents on the stack, decrements the stack pointer again (to 20BEH), and stores the flag byte on the stack. PUSH B decrements the stack pointer to 20BDH, stores the contents of the B register, decrements the stack pointer again (to 20BCH), and stores the contents of the C register on the stack.

POP and RET instructions reverse the procedure. POP B copies the contents of the stack at 20BCH onto the C register and increments the stack pointer to 20BDH. It then copies the contents of the stack at 20BDH into the B register and increments the stack pointer to 20BEH. POP PSW restores the flags, increments the stack pointer to 20BFH, restores the contents of the accumulator, and increments the stack pointer to 20C0H. RET loads the low byte of the return address into the program counter, increments the stack pointer to 20C1H, loads the high byte of the return address into the program counter, and increments the stack pointer

to 20C2H. The stack pointer now contains the same address that it did before the subroutine call.

If the number of POP instructions is not kept the same as the number of PUSH instructions in a program, the stack gets "out of balance." A RET instruction may treat what was originally register contents as if they were a return address.

Because the stack grows downward in memory from the starting address, it is usually located at a high address in RAM, so that it does not interfere with other storage or program.

XTHL exchanges 2 bytes of data from the top of the stack with the 2 bytes in the H and L registers. SPHL copies the contents of the H and L registers onto the stack pointer register.

INPUT/OUTPUT   The OUT byte instruction is a 2-byte instruction that moves a byte of data from the accumulator to the output port whose address is contained in the second byte of the instruction. Since the port address is 8 bits, the 8080A/8085A can address up to 256 output ports.

IN byte is a 2-byte instruction that moves a byte of data from an input port to the accumulator. One of 256 possible input port addresses is contained in the second byte of the instruction.

CONTROL   The control instructions DI and EI are *disable interrupt* and *enable interrupt,* respectively. These are discussed in Chap. 9 along with interrupts.

During execution of the NOP (no operation) instruction the processor does nothing but increment the program counter. NOP instructions are often used in programs for timing purposes. Also they are put in so that if an instruction is forgotten, NOPs can be removed and the required instruction inserted without shifting the entire program.

HLT is a *halt* instruction which stops the processor. We examine the full implications of this instruction in Chap. 9.

RIM and SIM instructions are discussed in Chap. 9 with the 8085A interrupts.

You have been introduced to the vocabulary of 8080A/8085A assembly language. Next we show you how to write simple programs by using this vocabulary.

WRITING ASSEMBLY LANGUAGE PROGRAMS

FLOWCHARTS

Before writing the actual assembly language program for a particular operation, many programmers represent the program in a schematic form called a *flowchart.* Figure 8-17 shows some standard flowchart symbols. Templates are available that contain them all.

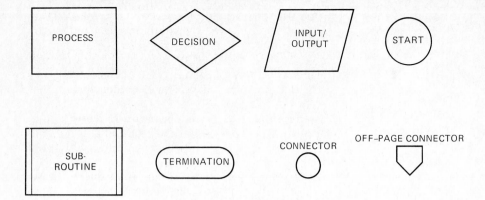

FIGURE 8-17 Flowchart symbols.

Figure 8-18 shows a flowchart of a program to take in 10 samples of data from an input port and store them in memory. *Operations* are shown in rectangular boxes. An operation block may represent several actual instructions. *Decision points* are represented as diamond-shaped boxes. From the flowchart, you can see that this program causes the machine to take in a byte of data, store it in memory, and check whether the operation has been done 10 times. If it has not, the machine loops back and takes in another byte of data and stores it in memory. It does this until 10 samples have been taken in and stored in memory.

A properly written flowchart makes no reference to the assembly language of a specific machine. Therefore, it can be used to write a program for *any* microprocessor. Flowcharts help to break large programs into small pieces, or modules, which can be worked with more easily.

Some experienced programmers draw a flowchart for explanation purposes only after a program is completed, rather than before it is written.

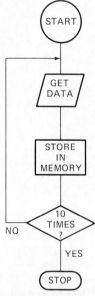

FIGURE 8-18 Flowchart for program to take in 10 samples of data and store in memory.

GRAMMAR RULES FOR ASSEMBLY LANGUAGE

Assembly language programs are usually written in a standard form so that they can be translated to *machine* language by an *assembler* program. In this standard form, assembly language statements have four *fields.* These are shown in Figure 8-19.

The first field is the *label field.* A *label* is a symbol or group of symbols used to represent an address that is not specifically known. Some assembler programs require that labels have no more than five or six characters and that the first character be a letter. The label field is usually ended with a colon. We discuss the use of labels later.

The *op code field* contains the mnemonic for the instruction to be performed. Mnemonics are sometimes referred to as *operation codes,* or *op codes.* They must be written down exactly as shown in Table 8-1 for the 8080A/8085A.

Following the op code field is the *operand field.* This field contains the registers, data, or address on which the instruction is to be performed. A comma is required between register initials or between a register initial and a data byte. Again, note that the instruction MOV B,A copies the accumulator into the B register.

A very important part of any assembly language program is the *comment field.* For most assemblers, the comment field is started with a semicolon or an asterisk. Comments do not become part of the machine program. They are written for the reference of you, the programmer, and those who come after you. If you write a program without comments and set it aside for 6 months, it may be very difficult for you to figure the program out again when you come back to it. It may even seem that someone else must have written it.

Label Field	OP Code Field	Operand Field	Comment Field
LOOP:	IN	00H	;Input data from port 0

FIGURE 8-19 Four fields of correct assembly language statement.

Comments should be written to explain in detail what each instruction or group of instructions is doing. Comments should not just describe the mnemonic, but also the function of the instruction in this particular routine. For best results, write comments as if you were trying to explain the program to someone who initially knows nothing about the program's purpose.

In addition to the comments following each instruction or group of instructions, a program or subroutine should start with a series of comments describing what the program is supposed to do. The starting comments should also include a list of parameters, registers, and memory locations used. It is almost impossible to write too many comments for a program. Figure 8-20 shows a reasonably well-documented, simple assembly language program.

EXAMPLES OF ASSEMBLY LANGUAGE PROGRAMMING

In this section we show, by example, how assembly language programs are written. The first example is the program for the flowchart in Figure 8-18.

EXAMPLE 1: PROGRAM TO TAKE IN DATA AND STORE Figure 8-20 shows an assembly language program for taking in 10 data samples from port 00H and storing them in consecutive memory locations. Note the heading with title, registers used, and subroutines called.

The program is best explained from the middle out because the first instruction that seems to relate to the flowchart is the IN instruction. This instruction simply takes a sample of data from port 00H and stores it in the accumulator. The next instruction, MOV M,A, copies the contents of the accumulator into a memory location pointed to by the HL register. The HL register pair, therefore, must have been previously *initialized* to the memory address at which you desire to start storing the data samples. This is done at the beginning of the program with the LXI H,2050H instruction, which loads 20H into the H register and 50H into the L register. After the first sample is stored in memory, INX H increments the extended HL register pair so that it points to the next storage location in memory.

The next problem to approach is determining when 10 samples have been taken. One method could be to set

PROGRAMMER				SHEET	OF
PROGRAM TITLE				JOB NUMBER	
ADR	**INSTR.**	**LABEL**	**MNEM.**	**OPERAND**	**COMMENTS**
200 0	21		LXI	H,2050H	;INITIALIZE H+L REGISTERS TO 1ST
200 1	50				;STORAGE ADDRESS IN RAM
200 2	20				
200 3	0E		MVI	C,0AH	;INITIALIZE COUNTER REGISTER
200 4	0A				;TO 10 (TEN)
200 5	DB	LOOP:	IN	00H	;TAKE IN DATA SAMPLE FROM
200 6	00				;PORT 0
200 7	77		MOV	M,A	;STORE THE DATA SAMPLE IN RAM
200 8	23		INX	H	;INCREMENT H+L TO POINT TO NEXT STORAGE ADDRESS
200 9	0D		DCR	C	;DECREMENT COUNTER BY ONE
200 A	C2		JNZ	LOOP	;JUMP TO LOOP IF COUNTER REGISTER
200 B	05				;IS NOT ZERO
200 C	20				
200 D	76		HLT		;STOP

COMMENTS: PROGRAM TO TAKE IN TEN SAMPLES OF DATA FROM PORT 0 and STORE THEM IN RAM STARTING AT ADDRESS 2050H. THIS PROGRAM IS INTENDED TO RUN ON AN 8080A, 8085A, OR Z-80 PROCESSOR SYSTEM.

REGISTER USED: A, C, H, L

ROUTINES CALLED: NONE

FIGURE 8-20 Assembly language program for flowchart of Fig. 8-18.

a register initially to 0 and then increment the register after each sample. By comparing the number in the register to 10 (0AH) and detecting a 0 result, you could tell when 10 samples had been taken. An easier method is to set a register initially to 10 (0AH) and decrement the register after each sample is stored. The DCR, or decrement, instruction sets the zero flag directly if the result of the decrement leaves all 0's in a register, so no compare instruction is needed.

This program uses the second method. The C register is set up as a counter by the MVI C,0AH instruction at the start of the program, which moves the immediately following number (0AH) into the C register. The DCR C instruction decrements the number in the C register after each sample is stored. If the result of the decrement leaves a 0 result in the C register, then the zero flag flip-flop will be set to a 1.

The JNZ instruction examines the zero flag flip-flop. If the flag is not set, then this instruction will jump the program counter to the address contained in the second 2 bytes of the instruction. Since you may not know this address when you write the program, you represent the address with a symbolic address called a *label*. You get to choose the name for a label, but you should avoid using letter combinations that are similar to op codes. Most assemblers also require that a label begin with a letter and that the label contain no more than five or six characters. Labels are replaced by addresses when a program is *assembled*, or converted to machine language.

The location of the label in the label field is used to indicate the jump location. For the program in Figure 8-20, program execution jumps to the IN 00H instruction if the result of decrementing the C register is not 0. When the program is run, it circles around this loop until decrementing the C register gives a 0 result. When a 0 result is detected, the program flows on through to the HLT instruction and stops.

To find out whether you understand this program, try to determine what would happen if the loop label accidentally were put next to the MVI C,0AH instruction. In other words, what would happen if the JNZ instruction jumped the program counter back to the address of the MVI C,0AH instruction?

The answer is that the program would never halt because the C register would never reach 0. Each time around the loop, the C register would get 10 (0AH) loaded into it again by the MVI C,0AH instruction. The program would keep incrementing HL after each data sample and fill the entire memory with data samples from port 00H. This type of error is quite common.

Figure 8-20 also shows how the codes for each instruction are entered in successive memory locations. In this case, the starting address was chosen as 2000H. Note the order of the immediate byte for the LXI, H,2050H instruction. Also, note that when the program is coded out, the label "LOOP" in the JNZ LOOP instruction is replaced by the desired jump address, 2005H. The bytes of jump address are, of course, loaded low byte first and high byte second. The special programming paper shown is used to help organize all the parts of the program and reduce errors.

EXAMPLE 2: PROGRAM TO TAKE IN AND STORE DATA WHEN DATA READY PULSE

When the program in Figure 8-20 is run, the machine immediately takes in 10 data samples as fast as it can from port 0, stores them in memory, and then sits in a halt state until the operator tells it what to do next. A simple addition to the above program will make the machine wait for a pulse on the LSB of port 1 before taking in each of the 10 data samples from port 0.

Figure 8-21 shows the flowchart and assembly language program for this version. After the memory pointer in the HL register and the sample counter in register C are initialized, data is taken in from port 1. The ANI 01H instruction ANDs this data with 01H. If the data from port 2 contains a 0 in its LSB, then the ANDing will leave all 0's in the accumulator. The JZ instruction will jump the program back to LOOP. If the data from port 1 contains a 1 in the LSB, then the ANI 01H instruction leaves 01H in the accumulator. Since this is not 0, the program goes on to the IN 00H instruction. The byte of data taken in from port 0 is stored in memory, the memory pointer HL is incremented, the counter register C is decremented, and a check is made to see whether 10 samples have been stored. If they have not, the program jumps back to LOOP to see whether a positive pulse is present on the LSB of port 1. When this pulse is present to indicate the data is ready on port 0, another sample is taken in.

Another way to check for the data ready pulse on port 1 data is to use a RRC or RAR instruction to rotate the LSB of the accumulator into the carry flip-flop. If a 1 is present in the LSB of port 1, then the carry flag flip-flop is set after the rotate right. A JNC instruction can be used to send the program around a loop until the carry flag is set. This illustrates in a small way that there are just about as many ways to write a program as there are programmers.

USE OF SUBROUTINES

EXAMPLE 3: PROGRAM WITH A 1-ms DELAY SUBROUTINE

Another modification of the basic take-in-data-and-store program in Figure 8-20 illustrates the use of subroutines. Suppose that instead of looking for a data ready pulse on the LSB of port 1, as was done in the last example, you want to just take in 10 samples at 1-ms intervals. Figure 8-22a shows the flowchart for a program to do this. Note the double ends on the box for the delay 1-ms step. This indicates that the delay part of the program is done as a subroutine.

Figure 8-22b shows the assembly language program for this example. The C register is loaded with 0AH, the HL memory pointer is set to the starting address for data storage, and the stack pointer register is set to the starting address for the stack. Whenever a program is going to use a subroutine, the stack pointer must be initialized because the call instructions store the return addresses on the stack. The stack is usually assigned an address near the top of available RAM because the stack grows downward from this address as addresses or data are pushed onto it.

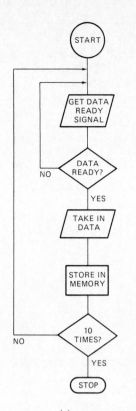

START

GET DATA READY SIGNAL

DATA READY? — NO / YES

TAKE IN DATA

STORE IN MEMORY

10 TIMES? — NO / YES

STOP

(a)

ADR		INSTR.	LABEL	MNEM.	OPERAND	COMMENTS
2000	0	21		LXI	H2050H	;INITIALIZE H+L TO POINT TO 1ST
2001	1	50				; STORAGE ADDRESS
2002	2	20				
2003	3	0E		MVI	C, 0AH	;INITIALIZE COUNTER REGISTER TO 10
2004	4	0A				
2005	5	DB	LOOP:	IN	01	;TAKE IN "DATA READY" SIGNAL
2006	6	01				
2007	7	E6		ANI	01H	;CHECK FOR A 'ONE' IN LSB
2008	8	01				
2009	9	CA		JZ	LOOP	; WAIT FOR A 'ONE' IN LSB
200A	A	05				
200B	B	20				
200C	C	DB		IN	00	;TAKE IN DATA FROM PORT 0
200D	D	00				
200E	E	77		MOV	M, A	;STORE IN MEMORY LOCATION POINTED ;TO BY H+L REGISTERS
200F	F	23		INX	H	;INCREMENT H+L TO POINT TO NEXT LOCATION
2010	0	0D		DCR	C	;DECREMENT COUNTER BY ONE
2011	1	C2		JNZ	LOOP	;JUMP TO LOOP IF 10 SAMPLES WERE
2012	2	05				; NOT TAKEN
2013	3	20				
2014	4	76		HLT		;STOP, IF DONE

COMMENTS: THIS PROGRAM TAKES IN A DATA SAMPLE FROM PORT 0 WHEN A "DATA READY" PULSE IS PRESENT ON THE LSB OF PORT 1. A TOTAL OF TEN SAMPLES IS TAKEN. THE SAMPLES ARE STORED IN MEMORY STARTING AT ADDRESS 2050H.

REGISTERS USED: A, C, H, L
ROUTINES CALLED: NONE

(b)

FIGURE 8-21 Program to take in 10 samples of data when data ready pulse is present and store them in memory. (a) Flowchart. (b) Assembly language program.

After the initializing steps, a data sample is taken in from port 0 and stored in memory. The HL memory pointer register is incremented to point to the next storage location. The C register used as a counter is decremented, and the 1-ms DELAY subroutine is called. The CALL instruction pushes the return address on the stack, decrements the stack pointer by 2, and loads the program counter with the subroutine address represented by the label DELAY. Program execution now continues with the first instruction at the subroutine address.

A first step in any subroutine should be to save the flags and the contents of any registers that might be lost or written over by the subroutine. PUSH PSW stores the accumulator contents and the flags on the stack and decrements the stack pointer register by 2. PUSH B stores the B and C registers on the stack and decrements the stack pointer register by 2. It is a good idea to push all registers used in the subroutine.

POP PSW at the end of the subroutine restores the accumulator and flags for the main program, and POP B restores the B and C registers before returning. Registers are popped off the stack in the reverse order from which they are pushed on. For example, PUSH PSW, PUSH B, and PUSH D at the start of a subroutine require POP D, POP B, and POP PSW at the end of the routine.

The heart of the delay routine is the MVI B,81H; DCR B; and JNZ CNTDN instructions. MVI B,81H loads the number 81H into the B register. DCR B decrements the number, and JNZ checks to see whether the number has reached 0 yet. If not, then the program jumps

around the small loop to CNTDN and decrements the B register again. The delay is created by the time it takes the machine to execute the program loop 81H, or 129, times. In the next section we show how delay constants are calculated for an 8080A or 8085A.

When the delay loop counter is decremented to 0, the program POPs B and increments the stack pointer by 2. It POPs PSW and again increments the stack pointer register by 2. The stack pointer register is now pointing at the memory location where the return address is stored. RET loads this address into the program counter and increments the stack pointer by 2, back to its initial value. It is important to always do as many POP as PUSH operations, or else the stack gets unbalanced.

The main sequence program resumes with the JNZ LOOP instruction. This instruction is supposed to check whether the DCR C instruction set the zero flag. If the flags had not been saved by PUSH PSW and POP PSW instructions in the subroutine, this instruction would not have the proper flags to examine. JNZ LOOP now causes the program to loop around and take the next of its total of 10 samples at 1-ms intervals.

An experienced programmer probably would make at least two observations about this program. First, the DCR C instruction could be placed after the CALL DELAY instruction to avoid having to save the flags. Second, the three active instructions of the delay loop—MVI B,81H; DCR B; and JNZ CNTDN—could be written as part of the main sequence to avoid using a subroutine altogether. For this small program, these comments are quite true. However, for long application

FIGURE 8-22 Program to take in 10 data samples at 1-ms intervals and store them in memory. *(a)* Flowchart. *(b)* Assembly language program.

PROGRAMMER				SHEET	OF	
PROGRAM TITLE				JOB NUMBER		

ADR	INSTR.	LABEL	MNEM.	OPERAND	COMMENTS
200 0	21		LXI	H,2050H	LOAD H+L REGISTERS WITH 1ST
200 1	50				STORAGE ADDRESS
200 2	20				
200 3	0E		MVI	C,0AH	INITIALIZE SAMPLE COUNTER
200 4	0A				TO 10 (TEN)
200 5	31		LXI	SP,20C2H	INITIALIZE STACK POINTER TO
200 6	C2				20C2
200 7	20				
200 8	DB	LOOP:	IN	00	TAKE IN DATA SAMPLE FROM
200 9	00				PORT 0
200 A	77		MOV	M,A	STORE SAMPLE IN MEMORY
200 B	23		INX	H	INCREMENT H+L TO POINT TO NEXT STORAGE ADDRESS
200 C	0D		DCR	C	DECREMENT SAMPLE COUNTER BY ONE
200 D	CD		CALL	DELAY	DELAY 1mSec
200 E	25				
200 F	20				
201 0	C2		JNZ	LOOP	IF SAMPLE COUNTER IS NOT ZERO
201 1	08				THEN TAKE ANOTHER SAMPLE
201 2	20				
201 3	76		HLT		HALT
*** **		DELAY ROUTINE * THIS ROUTINE PRODUCES A 1mSec DELAY			INPUTS: NONE OUTPUTS: NONE REGISTERS USED: A,B,SP
202 5	F5	DELAY:	PUSH	PSW	SAVE ACCUMULATOR CONTENTS + FLAGS
202 6	C5		PUSH	B	SAVE B+C REGISTERS CONTENTS
202 7	06		MVI	B,81H	LOAD DELAY COUNTER REGISTER
202 8	81				
202 9	05	CNTDN:	DCR	B	DECREMENT DELAY COUNTER
202 A	C2		JNZ	CNTDN	CONTINUE COUNTING DOWN UNTIL
202 B	29				B=0
202 C	20				
202 D	C1		POP	B	RESTORE B+C REGISTERS CONTENTS
202 E	F1		POP	PSW	RESTORE ACCUMULATOR + FLAGS
202 F	C9		RET		RETURN TO MAIN PROGRAM

COMMENTS: THIS PROGRAM TAKES 10 SAMPLES OF DATA FROM PORT 0 AT APPROXIMATELY 1mSec (ONE MILLISECOND) INTERVALS AND STORES EACH SAMPLE IN SEQUENTIAL MEMORY LOCATIONS STARTING AT ADDRESS 2050H

REGISTERS USED: A,C,H,L,SP
ROUTINES CALLED: DELAY

(b)

programs, writing the 1-ms delay as a subroutine means that this routine can be called from anywhere in the program. It has to be written in the program only once. For longer routines such as multiplication or a code conversion which may be called many times in a program, this is a big advantage. To make sure a subroutine can be used anywhere in a program, the safe thing to do is to have the subroutine PUSH PSW and any registers used in the subroutine.

Many programmers keep a file of useful subroutines they have found or written. They use these to write application programs which are essentially modules or subroutines joined together. Each subroutine should be documented with a title describing the functions performed, the registers used, and any other subroutines called.

Since delay subroutines are an important part of many programs, next we show how the constant is calculated for a desired amount of delay.

PROGRAM EXECUTION TIME AND DELAY CONSTANTS

CALCULATING DELAY CONSTANTS The heartbeat of a microcomputer is its clock. Each instruction requires a certain number of heartbeats (clock cycles) to execute. If you know the number of clock cycles for each instruction and the time for one clock cycle, you can calculate the time needed to execute each instruction. A clock cycle also is called a *state*. Table 8-2 shows the number of clock cycles or states for each 8080A and 8085A instruction.

A common clock frequency used for the 8080A is 2 MHz. For this frequency, the time of one cycle is 1/(2 MHz), or 500 ns. The number of states multiplied by the time of a state gives the total time for the execution of an instruction. For example, Table 8-2 indicates that the CALL instruction requires 17 states for the

TABLE 8-2
THE 8080A/8085A INSTRUCTION SET

Instruction		Code	Bytes	T States 8085A	T States 8080A	Machine Cycles
ACI	DATA	CE data	2	7	7	F R
ADC	REG	1000 1SSS	1	4	4	F
ADC	M	8E	1	7	7	F R
ADD	REG	1000 0SSS	1	4	4	F
ADD	M	86	1	7	7	F R
ADI	DATA	C6 data	2	7	7	F R
ANA	REG	1010 0SSS	1	4	4	F
ANA	M	A6	1	7	7	F R
ANI	DATA	E6 data	2	7	7	F R
CALL	LABEL	CD addr	3	18	17	S R R W W*
CC	LABEL	DC addr	3	9/18	11/17	S R · /S R R W W*
CM	LABEL	FC addr	3	9/18	11/17	S R · /S R R W W*
CMA		2F	1	4	4	F
CMC		3F	1	4	4	F
CMP	REG	1011 1SSS	1	4	4	F
CMP	M	BE	1	7	7	F R
CNC	LABEL	D4 addr	3	9/18	11/17	S R · /S R R W W*
CNZ	LABEL	C4 addr	3	9/18	11/17	S R · /S R R W W*
CP	LABEL	F4 addr	3	9/18	11/17	S R · /S R R W W*
CPE	LABEL	EC addr	3	9/18	11/17	S R · /S R R W W*
CPI	DATA	FE data	2	7	7	F R
CPO	LABEL	E4 addr	3	9/18	11/17	S R · /S R R W W*
CZ	LABEL	CC addr	3	9/18	11/17	S R · /S R R W W*
DAA		27	1	4	4	F
DAD	RP	00RP 1001	1	10	10	F B B
DCR	REG	00SS S101	1	4	5	F*
DCR	M	35	1	10	10	F R W
DCX	RP	00RP 1011	1	6	5	S*
DI		F3	1	4	4	F
EI		FB	1	4	4	F
HLT		76	1	5	7	F B
IN	PORT	DB data	2	10	10	F R I
INR	REG	00SS S100	1	4	5	F*
INR	M	34	1	10	10	F R W
INX	RP	00RP 0011	1	6	5	S*
JC	LABEL	DA addr	3	7/10	10	F R/F R R†
JM	LABEL	FA addr	3	7/10	10	F R/F R R†
JMP	LABEL	C3 addr	3	10	10	F R R
JNC	LABEL	D2 addr	3	7/10	10	F R/F R R†
JNZ	LABEL	C2 addr	3	7/10	10	F R/F R R†
JP	LABEL	F2 addr	3	7/10	10	F R/F R R†
JPE	LABEL	EA addr	3	7/10	10	F R/F R R†
JPO	LABEL	E2 addr	3	7/10	10	F R/F R R†
JZ	LABEL	CA addr	3	7/10	10	F R/F R R†
LDA	ADDR	3A addr	3	13	13	F R R R
LDAX	RP	000X 1010	1	7	7	F R
LHLD	ADDR	2A addr	3	16	16	F R R R R

Instruction		Code	Bytes	T States 8085A	T States 8080A	Machine Cycles
LXI	RP,DATA16	00RP 0001 data16	3	10	10	F R R
MOV	REG,REG	01DD DSSS	1	4	5	F*
MOV	M,REG	0111 0SSS	1	7	7	F W
MOV	REG,M	01DD D110	1	7	7	F R
MVI	REG,DATA	00DD D110 data	2	7	7	F R
MVI	M,DATA	36 data	2	10	10	F R W
NOP		00	1	4	4	F
ORA	REG	1011 0SSS	1	4	4	F
ORA	M	B6	1	7	7	F R
ORI	DATA	F6 data	2	7	7	F R
OUT	PORT	D3 data	2	10	10	F R O
PCHL		E9	1	6	5	S*
POP	RP	11RP 0001	1	10	10	F R R
PUSH	RP	11RP 0101	1	12	11	S W W*
RAL		17	1	4	4	F
RAR		1F	1	4	4	F
RC		D8	1	6/12	5/11	S/S R R*
RET		C9	1	10	10	F R R
RIM (8085A only)		20	1	4	—	F
RLC		07	1	4	4	F
RM		F8	1	6/12	5/11	S/S R R*
RNC		D0	1	6/12	5/11	S/S R R*
RNZ		C0	1	6/12	5/11	S/S R R*
RP		F0	1	6/12	5/11	S/S R R*
RPE		E8	1	6/12	5/11	S/S R R*
RPO		E0	1	6/12	5/11	S/S R R*
RRC		0F	1	4	4	F
RST	N	11XX X111	1	12	11	S W W*
RZ		C8	1	6/12	5/11	S/S R R*
SBB	REG	1001 1SSS	1	4	4	F
SBB	M	9E	1	7	7	F R
SBI	DATA	DE data	2	7	7	F R
SHLD	ADDR	22 addr	3	16	16	F R R W W
SIM (8085A only)		30	1	4	—	F
SPHL		F9	1	6	5	S*
STA	ADDR	32 addr	3	13	13	F R R W
STAX	RP	000X 0010	1	7	7	F W
STC		37	1	4	4	F
SUB	REG	1001 0SSS	1	4	4	F
SUB	M	96	1	7	7	F R
SUI	DATA	D6 data	2	7	7	F R
XCHG		EB	1	4	4	F
XRA	REG	1010 1SSS	1	4	4	F
XRA	M	AE	1	7	7	F R
XRI	DATA	EE data	2	7	7	F R
XTHL		E3	1	16	18	F R R R R

Machine cycle types:

F Four-clock period instruction fetch
S Six-clock period instruction fetch
R Memory read
I I/O read
W Memory write
O I/O write
B Bus idle
X Variable or optional binary digit
DDD Binary digits identifying a destination register
     B = 000, C = 001, D = 010, Memory = 110

SSS Binary digits identifying a source register
     E = 011, H = 100, L = 101, A = 111
RP   Register pair
     BC = 00, HL = 10
     DE = 01, SP = 11

*Five-clock period instruction fetch with 80
†The longer machine cycle sequence app of condition evaluation with 8080A.
· An extra READ cycle (R) will occur for with 8080A.
All mnemonics copyright © 1976 by Int

```
PROGRAM TITLE                    8080A          8085A
                                 STATES         STATES

DELAY:  PUSH  PSW                 11 ⎫           12 ⎫
        PUSH  B                   11 ⎬ 29        12 ⎬ 31
        MVI   B, _____H            7 ⎭            7 ⎭

CNTDN:  DCR   B                    5 ⎫            4 ⎫
        JNZ   CNTDN               10 ⎬ 15        10 ⎬ 14


        POP   B                   10 ⎫           10 ⎫
        POP   PSW                 10 ⎬ 30        10 ⎬ 30
        RET                       10 ⎭           10 ⎭
                                  ──             ──
                                  59             61    TOTAL EXCLUDING
                                                       CNTDN LOOP
```

FIGURE 8-23 Delay subroutine showing number of states for each instruction.

8080A. Seventeen states times 0.5 μs per state gives 8.5 μs for the CALL instruction.

Now that the calculation of instruction execution time has been explained, the timing of the 1-ms delay loop can be discussed. Figure 8-23 shows the relevant instructions and the number of states for each. If we exclude the DCR B and JNZ instructions, which are repeated many times, the total number of states is 59. At 0.5 μs per state, this is 29.5 μs. The rest of the 1000 μs (1 ms) is made up by running through the DCR B and JNZ loop over and over. The CNTDN loop must generate 1000 μs − 29.5 μs, or 970.5 μs. Each time through the CNTDN loop is 15 states, or 7.5 μs. Dividing 970.5 μs by 7.5 μs gives 129 as the number of times that the loop must be repeated. This is 81 hexadecimal, so 81H is loaded into the B register to start.

The actual time between samples will be slightly longer than 1 ms because of the execution time of the other instructions between samples. For more precise timing these times could be included in the calculations and a smaller number loaded into the delay loop counter. If a different clock frequency is used, then the period of that clock must be used.

Note that the number of states for many instructions is not the same for an 8080A and an 8085A. Also, two numbers are given, for example, 7/10, for instructions such as JNZ in the 8085A. This means that only 7 clock pulses are required if the jump is not taken (fall through), but that 10 clock cycles are required if the jump is taken.

DELAY LOOPS USING 16-BIT REGISTER   Longer delays can be created by loading and counting down a 16-bit register, as shown in Figure 8-24. The DCX instruction, however, does not affect the zero flag. Therefore, another way must be found to determine when the extended register is counted down to 0. As shown in Figure 8-24, this is usually done by moving 1 byte of the 16-bit register into the accumulator and ORing it with the other byte. The zero flag is set by the ORA instruction only if both bytes are 0's.

```
            LXI   D,(16-BIT NUMBER)
LOOP        DCX   D
            MOV   A,D
            ORA   E
            JNZ   LOOP
```

FIGURE 8-24 16-bit delay loop.

NESTED DELAY LOOPS FOR LONGER DELAYS
Another method to produce longer delays uses nested delay loops. To produce a 10-ms delay, for example, the 1-ms loop routine of Figure 8-22 could be placed in another loop that counts the 1-ms loop down 10 times before exiting. Figure 8-25 shows a program to do this. The B register is loaded and counted down 0AH times.

HAND-ASSEMBLING A PROGRAM
Hand-assembling a program consists of assigning memory addresses and machine codes for each of the labels, op codes, and operands. We referred to this earlier as *coding*. The starting address for the program in memory is chosen by the programmer. Figure 8-22 shows the last example program hand-assembled to run in mem-

```
DELAY       PUSH      PSW
            PUSH      B
            MVI       C,0AH
LOAD1       MVI       B,81H
CNTDN       DCR       B
            JNZ       CNTDN
            DCR       C
            JNZ       LOAD1
            POP       B
            POP       PSW
            RET
```

FIGURE 8-25 Nested delay loop for longer delays.

ory starting at 2000H. Program assembly paper is used to organize all the parts of the program during hand-assembly.

The first two columns are in hexadecimal. They represent the actual addresses and the binary code that will be loaded into these addresses to run this program. A common convention is to divide a 65,536-byte memory into 256 pages of 256 bytes each. The page address is then the higher byte of a 16-bit address, and the line address is the lower byte.

It is important to note that for all the 8080A/8085A instructions that have a 2-byte operand, the least significant byte is loaded into memory first. For example, LXI H,2050H is loaded into memory in the order 21, 50, 20; JNZ LOOP is loaded in the order C2, 08, 20; and CALL DELAY is loaded in the order CD, 25, 20.

Hand-assembling long programs is time-consuming and tedious, at best. If you have to do it, here are some hints to minimize your work. First, check and recheck (and then check again!) to make sure you haven't left any instructions out of the assembly language version before you code anything. Otherwise, if you forget an instruction or have to insert one, you will have to change the addresses through all the rest of the program. All the addresses contained in jump and call instructions will have to be changed also. The chance of an error in all this shuffling is very high.

Second, to minimize the problem of inserting a forgotten instruction, a few NOP instructions can be inserted after every 16 program instructions. The NOP instructions do not affect the program flow. The extra memory required is a small price to pay for the ease of adding instructions. If memory space is severely limited, the NOP instructions can be removed from each program section as it is debugged. NOP instructions should also be left between the main program and subroutines, so that the main program can be expanded without having to change the addresses contained in subroutine calls. Often a slashed zero (∅) is used to differentiate 0 from the letter O.

Since it is nearly impossible to hand-assemble large programs without introducing many errors, computer- and microprocessor-based development systems have been designed to do the job for you. These are the topic of the next section.

MICROPROCESSOR DEVELOPMENT SYSTEMS

Physically, a development system usually consists of a microprocessor system with 32K or more of RAM, a typewriter-type keyboard, a video (CRT) display, a printer, and one or more floppy disk drives. A paper-tape reader and a PROM programmer may also be present. Figure 8-26 shows an Intel development system. A good development system is capable of many functions. For now, we examine only the text editor and assembler functions. These functions take much of the drudgery out of writing assembly language programs.

TEXT EDITORS   The first point to fix firmly in your mind is that both a *text editor* and an *assembler* are programs of several thousand bytes each. They are usually stored on floppy disks until needed. The main function of a text editor program is to help you construct an assembly language program in just the right format for the assembler program to translate it correctly to machine language.

When you want to create an assembly language program with a development system, first you load the text editor program into RAM. Then the editor program is run. This program, when run, allows you to type in an assembly language program using the keyboard. The program is stored in the editor workspace in RAM. From 24 to 40 lines of the program are displayed on the CRT screen, so you can see what you are typing. The assembly language program is typed in with comments in the same format (minus addresses and codes), as shown in Figure 8-22. Errors can be corrected or lines can be added or deleted in the middle of a program at the touch of a key.

ASSEMBLER DIRECTIVES, OR PSEUDOINSTRUC-TIONS   To get the program fully ready for the assembler, the program must contain directions to the assembler. For example, the assembler must be told at what address to start assembling the program. These *assembler directives* do not become part of the final program, because they are not op codes. Therefore, they are sometimes known as *pseudoinstructions*, or *false* instructions.

The assembler reference column in Table 8-1 shows the pseudoinstructions used for an 8080A/8085A assembler. These instructions differ for every assembler program, but the seven "general" instructions discussed below will have equivalents for most assemblers.

The *origin* instruction ORG is followed by an address. This instruction tells the assembler at what address to start assembling the program. ORG is entered at the beginning of a program, or at any location in a program at which you wish to start at an address different from the next address in sequence. For example, ORG 0100H tells the assembler to start assembling the immediately following program at 0100H in memory.

The END statement indicates the *end* of the program to the assembler. If no END statement is given, then the assembler just keeps on running through all the memory.

The *equate* instruction EQU defines symbols used in the program. For example, if you are going to use 736AH many times in a program, you can give the number a symbolic name, such as JOE. Then you can write LXI, D,JOE instead of LXI D,736AH. This reduces errors caused by switching digits in numbers, etc. The assembler must be told that it is supposed to plug in 736AH every time it sees JOE. An EQU statement does this. The correct format is JOE EQU 736AH. Equate statements are usually put at the start of a program, but they *must* appear before the first use of the symbolic value.

SET is similar to EQU except that for SET the defini-

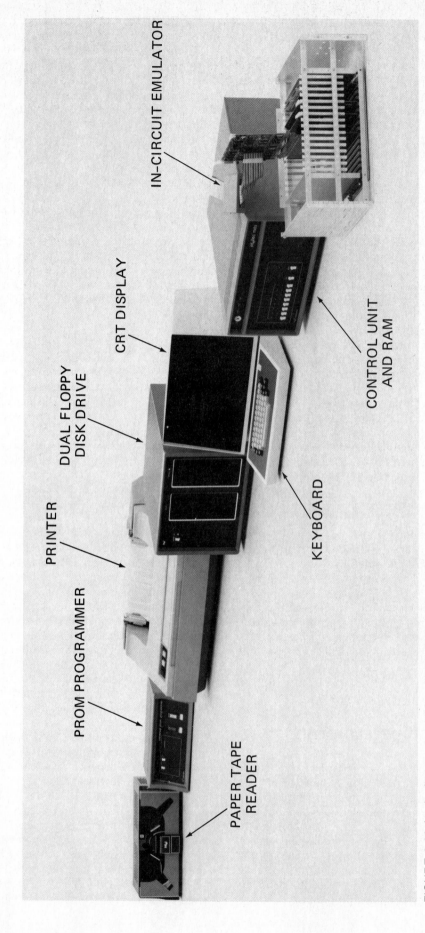

FIGURE 8-26 Intel microprocessor development system. (*Intel Corporation.*)

IN-CIRCUIT EMULATOR

CRT DISPLAY

DUAL FLOPPY
DISK DRIVE

PRINTER

PROM PROGRAMMER

PAPER TAPE
READER

KEYBOARD

CONTROL UNIT
AND RAM

tion of a symbol can be changed by another SET statement later in the program. An EQU value is fixed for the entire program. For example, you might put JOE SET 736AH at the start of a program and 500 lines later change the definition of JOE with JOE SET 4253H.

The next three pseudoinstructions in Table 8-1 are DS, *define storage;* DB, *define byte;* and DW, *define word.* The DS directive tells the assembler to reserve a block of memory for data storage. For example, BUFFER: DS 96 tells the assembler to leave 96 bytes in memory for storage when it assembles the program.

A DB statement stores the data contained in its operand field in successive memory spaces. For example, DB 07AH stores 7AH in memory right after the preceding instruction, and DB JOHN stores 4A, 4F, 48, and 4E in four successive memory locations to represent the string of ASCII characters JOHN.

A DW instruction stores the 16-bit data contained in its operand field in two successive memory locations. The least significant byte is stored first, followed by the most significant byte.

MACROS

Often in long programs a group of instructions may need to be repeated many times. One way to do this is to write the instructions as a subroutine and call it each time it is needed. If the desired group of instructions is too short and/or the group of instructions must be coded in line with the main program for timing reasons, a MACRO is used.

A MACRO is simply a group of instructions that are written and given a name at the start of a program. Each time the assembler reads this name in the program, it codes in the instructions listed in that MACRO. In other words, being able to define a MACRO allows you to avoid typing in a group of instructions over and over. Figure 8-27a shows how a MACRO called LOAD1 would be written to load the accumulator and B register from two memory locations. Here *A1* represents the parameter (variable) that will be loaded into HL each time the

MACRO is used. This parameter can be different each time the MACRO is used. Figure 8-27b shows how this MACRO would be called in a program. In an actual program the specific address desired in HL would be substituted for *A1*.

USING THE ASSEMBLER

When the assembly language program is completed in the editor workspace, it is checked thoroughly and then written into a file on a floppy disk or tape. Next the assembler program is loaded into RAM. When this program is run, it reads through the assembly language program from the disk or tape and translates it to machine language. Usually the assembler has to read the assembly or *source* program twice to do an assembly. On the *first pass,* the assembler program reads through and creates a *symbol table.* This means that it assigns addresses to labels by counting from the starting address.

A *second pass* of the assembler actually does the assembly or translation of the program to machine or *object* code. The resultant machine code can be loaded into a file on a floppy disk or stored in RAM to be run. The object code may also be programmed into a PROM. In addition, the complete assembled program can be printed out on a printer in a format such as that in Figure 8-22.

Most assemblers will indicate errors, such as illegal op codes, on a printout. If an error shows up, it can be corrected easily. The editor program and the assembly language source file are loaded back into RAM. The portion of the program containing the error is scrolled into view on the CRT, and the error is corrected. The corrected program is written back into a source file on the disk. After the assembler program is reloaded, the source file is read by the assembler and assembled. This cycle is repeated until the assembled program shows no errors. An error-free printout does *not* mean the program will work, incidentally. It just means that legal op codes, addresses, etc., were used.

Some development systems also contain compiler and interpreter programs that translate higher-level languages such as BASIC, FORTRAN, or Pascal to machine language. These languages use more English-like statements. Each BASIC, FORTRAN, or Pascal statement may represent many assembly language instruction statements. In Chap. 14 we discuss programming with these languages.

TROUBLESHOOTING (DEBUGGING) ASSEMBLY LANGUAGE PROGRAMS

Quite often when a program is entered into RAM and run the first time, it does not execute correctly. A systematic approach will find the problem more quickly than random poking. Here is a series of steps to follow:

1. Check your algorithm. Mentally step through the program to make sure you have included the instructions to make it do what you want. Did you remember to save the result of a calculation? Did you remember to initialize a memory pointer or stack pointer?

```
LOAD1      MACRO      A1
           LXI        H,A1
           MOV        A,M
           INX        H
           MOV        B,M
           INX        H
           ENDM
```

(a)

```
LOAD1      (actual address
            desired for this
            use of MACRO)
```

(b)

FIGURE 8-27 MACRO Example.

2. Check that you have the right instructions. A common mistake is to switch the operands in an instruction. For example, a MOV A,M instruction may be written when MOV M,A was intended.

3. Check to make sure the codes for each instruction are correct. If the instruction is MOV B,C, make sure you did not write the code for MOV C,B. For instructions such as MVI A,7AH, be sure the immediate data, 7A, is coded into the next byte after the code for the MVI instruction. For 3-byte instructions such as JNZ 2050H, make sure that the 2 bytes of the jump address are entered and in the correct order, low byte first.

4. If everything seems OK up to this point, a possible next thing to do is *single-step* through the program. Most microprocessor development boards allow you to execute a single instruction, stop, and examine the contents of registers and memory locations. This lets you check whether the results are correct after each instruction. If the results are correct, execute the next instruction and check register and memory contents again. If the results are not correct, examine your code, instruction, and thinking to find out why the desired result was not produced.

    To single-step through a count loop such as a delay loop, temporarily reduce the count to 1 or 2 so that it does not take you all day to get out of the loop.

5. For debugging long programs, a *breakpoint* is a handy tool. Most microcomputer development boards allow you to insert an instruction that will preserve the status of all registers and return control to the monitor program. Monitor commands then allow you to examine registers and memory locations. For the Intel SDK-85 board, for example, the RST 1 instruction does this. The code for RST 1 is CFH. If the first 100 bytes of a program seem to execute correctly but problems occur after that, then the CFH code can be substituted temporarily for an instruction code at the start of the problem section. When run, the program will run to this inserted code and return control to the monitor. You can then examine the contents of registers and any relevant memory locations. If they are correct, replace the CFH with the original instruction and move the CFH breakpoint farther down in the program. If this is done carefully, you may be able to run to a breakpoint and then single-step from there to find a problem.

6. The main points in debugging a program are to be systematic and to use all the tools available on your particular system.

A major portion of this chapter has dealt with what computer people refer to as *software*. Software includes instruction sets, programs, and programming. In Chap. 9 we deal mostly with the *hardware* of microprocessors. Hardware includes the ICs, circuit connections, and timing parameters.

REVIEW QUESTIONS AND PROBLEMS

1. Draw a block diagram of a simple computer, showing buses.

2. Describe the functions of memory, I/O ports, and the CPU in a minicomputer or microcomputer system.

3. Describe the signal flow on the buses for fetching and executing an IN instruction.

4. Given an 8080A/8085A system with register and memory contents as shown in Figure 8-28, show how the instructions listed below will affect these contents. Unless listed with a single letter, the instructions are not intended to be sequential.

a.	INR	C
b.	MOV	A,D
c.	MVI	B,7EH
d.	LXI	H,2022H
	MOV	A,M
e.	MOV	A,L
	ADD	D
	INX	H
	MOV	M,A
f.	RAL	
g.	DAD	B

h.	CMP	C
i.	ANI	0F0H
	RRC	
	RRC	
	RRC	
	RRC	
j.	LDA	2023H
	SUB	D
	DCR	B
	MOV	C,A
	ORA	0F0H

REGISTERS			
A	73		
B	4A	5F	C
D	3C	A7	E
H	20	20	L

MEMORY	
2020	3E
2021	7B
2022	2C
2023	F2

FIGURE 8-28

5. Using the register and memory contents shown in Figure 8-28 and the flag register shown in Figure 8-29, show how the flags will be left after each of the following instructions executes. Assume all flags are reset at the start of the instruction.

a. MOV A,D
b. MVI A,59H
   ADD E
c. SUB E

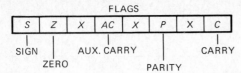

FLAGS

| S | Z | X | AC | X | P | X | C |

SIGN — ZERO — AUX. CARRY — PARITY — CARRY

FIGURE 8-29

6. Show the results of ANDing, ORing, and XORing 11110000 with 10101010.

7. Find and correct any errors in the following instructions and coding:
   a. 3E MVI A,7FH
   b. 21 LXI H,27H
      27
   c. 77 MOV A,M
   d. C6 ADI B
   e. C3 JMP 307AH
      30
      7A

8. Draw a flowchart for a program to take in 100 samples of data from port 01H, to add 7 to each, and to store the result in memory starting at 2050H.

9. Write the assembly language program for the flowchart you drew in Prob. 8.

10. a. Show how the program in Figure 8-21 could be modified to detect a high-to-low pulse on the LSB of port 01.
    b. Show how the program could be modified to detect a high pulse on bit 6 of port 01 instead of bit 0.
    c. Show how the program could be modified to wait for the pulse on the LSB of port 01 to go low before looping back to look for it high again.

11. Show the effect of the following instruction sequence on the stack pointer, stack, and the program counter:

    | 2000 | 21C220 | LXI | SP,20C2H |
    | 2003 | CD5020 | CALL | 2050H |
    | 2006 | 77 | MOV | M,A |

12. How does the RET instruction at the end of a subroutine determine to what address to return the program counter?

13. Why are PUSH and POP instructions used in a subroutine?

14. Use the 1-ms delay routine in Figure 8-22b to write an 8085A assembly language program to produce a 500-Hz square-wave pulse train out of the LSB of port 7. (*Hint:* Use the delay routine twice per loop.)

15. a. Calculate the delay time for the DELAY subroutine in Figure 8-22b if it is run on an 8085A with a 3.072-MHz clock.
    b. Calculate the value of the delay constant needed to give a delay of 1 ms, using an 8085A with a 3.072-MHz clock.

16. Write a routine to show how the DELAY 1-ms subroutine could be used to produce a delay of 100 ms between data samples. (*Hint:* Use the routine 100 times between samples.)

17. Multiplication can be done by successive addition. For example, 3×5 can be performed by zeroing the accumulator and adding 3 to it 5 times. Draw a flowchart for this method of multiplcation. Write an 8085A assembly language program to multiply up to a 4-bit binary number in the B register by another number of up to 4 bits in the C register.

18. Use extended registers H and D and the DAD and DCR instructions to write an assembly language program to multiply two 8-bit binary numbers by the successive-addition method. The multiplicand is in the E register and the multiplier in the C register.

19. Define (a) editor, (b) assembler, and (c) MACRO.

20. Describe a reasonable series of steps you would take to debug a nonfunctioning assembly language program.

REFERENCES/BIBLIOGRAPHY

Intel MCS-80/85 Users' Manual, Intel Corporation, Santa Clara, CA, October 1979.

Intel 8080/8085 Assembly Language Programming Manual, Intel Corporation, Santa Clara, CA, 1977–1978.

Leventhal, Lance A.: *8080A/8085A Assembly Language Programming*, Osborne and Associates, Inc., Berkeley, CA, 1978.

CHAPTER 9

8080A/8085A SYSTEM HARDWARE AND TIMING

Chapter 8 introduced you to writing assembly language programs for 8080A and 8085A microcomputers. This chapter discusses the hardware connections, signals, and timing relationships of 8080A and 8085A microcomputer systems. Although the 8080A is not being used for new designs, it is found in a very large amount of existing equipment. Therefore, from a repair standpoint it is well worth knowing. Also, understanding the 8080A makes it much easier to understand the 8085A.

The approach in this chapter is a spiral. First is an overview of an 8080A-based system showing how the address, data, and control bus signals are generated and how the 8080A is connected with RAM, ROM, and port devices to form a system. Next a similar overview is given for an 8085A-based system. Then comes a discussion of address decoding, with these two systems used as examples. After a discussion of 8085A timing parameters, the specialized wait, hold, and halt states of the 8080A and the 8085A are examined. This is followed by a section explaining how each processor handles interrupt inputs. When the discussion of hardware of the 8080A and 8085A is complete, a systematic method for troubleshooting microcomputers is given. The chapter concludes with an introduction to the Z80 microprocessor and an introduction to the 8048 family of one-chip microcomputers. A major goal of this chapter is to help you learn to read and understand manufacturers' data books to get the hardware information you need for a particular microprocessor.

At the conclusion of this chapter, you should be able to:

1. Describe the functions of the 8080A, 8224, and 8228 CPU group.

2. Define instruction cycle, machine cycle, and state.

3. Describe how the 8085A multiplexes the lower 8 address bits out on the data bus.

4. Initialize programmable port devices such as the 8155 and 8355 for simple input and output operations.

5. Predict the address range enabled by each output of an address decoder, and draw a memory map for a microcomputer system.

6. Define memory-mapped I/O and direct I/O.

7. Describe the 8080A/8085A wait, hold, and halt states.

8. Explain how interrupts are serviced in an 8080A and in an 8085A system.

9. List the major steps for systematically troubleshooting a malfunctioning microcomputer system.

10. Describe the main features of the Z80 and the 8048 microprocessors.

AN 8080A MICROPROCESSOR SYSTEM

CPU GROUP

Figure 9-1 shows how an 8080A *CPU*, an 8224 *clock generator*, and an 8228 *system controller and bus driver* are connected to produce a functional CPU block. The 8080A CPU requires a two-phase clock. This clock can be generated with a crystal oscillator and several discrete ICs, but the usual method is to use a single 8224 clock generator IC. The 8224 requires only an external crystal to produce the clocks and some other control signals required by the 8080A block. The voltage swing of these clocks is +0.6 to +11 V.

The 8228 serves two main functions. First, it acts as a buffer to increase the drive capability of the data lines. The bus driver is bidirectional because data may be either going out or coming in on the data bus. Second,

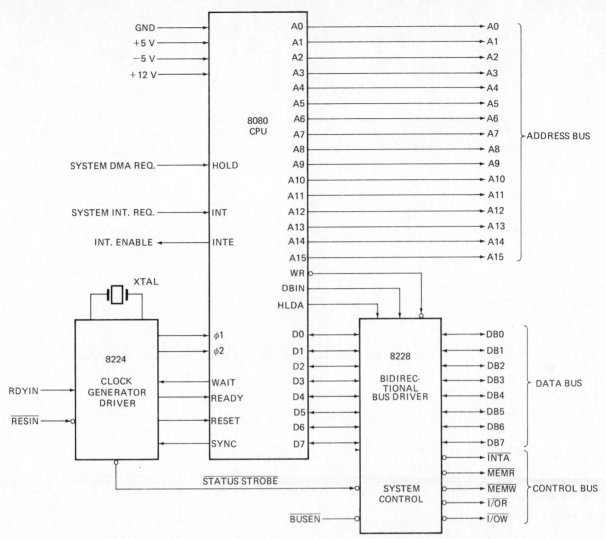

FIGURE 9-1 Schematic for 8080A CPU group with 8080A, 8224, and 8228. *(Intel Corporation.)*

the 8228 produces the proper control bus signals: *memory read* ($\overline{\text{MEMR}}$), *memory write* ($\overline{\text{MEMW}}$), *input-output read* ($\overline{\text{I/OR}}$), *input-output write* ($\overline{\text{I/OW}}$), and *interrupt acknowledge* ($\overline{\text{INTA}}$). The 8080A sends out these signals in a coded form on the data bus during the start of each machine cycle. The *status strobe* ($\overline{\text{STSTB}}$) from the 8224 latches this coded form into the 8228. The 8228 decodes the latched signals to produce the desired control bus signals.

Next we explain how to read a timing diagram which shows the sequence of signals produced by this basic CPU group.

INSTRUCTION CYCLES, MACHINE CYCLES, AND STATES

Figure 9-2 shows a timing diagram for one instruction cycle of an 8080A microcomputer. An *instruction cycle* is the time needed for a computer or microcomputer to fetch all the required parts of an instruction and execute that instruction. For an 8080A each instruction cycle is composed of one to five *machine cycles*. These are labeled $M1$, $M2$, and $M3$ in Figure 9-2. Fetching an instruction byte from memory is one example of a machine cycle operation. Any reference to memory or I/O ports requires a machine cycle. Other examples of machine cycles are memory read, memory write, I/O read, I/O write, and interrupt acknowledge. Machine cycles are further divided into *states*. For the 8080A a machine cycle consists of three, four, or five states. Except for the wait, hold, and halt states discussed later, an 8080A *state* is defined as the time from the positive edge of one $\phi1$ clock pulse to the positive edge of the next. Table 8-2 shows the number of states for each 8080A instruction.

Figure 9-2 shows the machine cycles, states, and bus information for the OUT (port) instruction of the 8080A. The top two lines of Figure 9-2 show the $\phi1$ and $\phi2$ clocks required by the 8080A to produce all the timing cycle waveforms of the instruction cycles, machine cycles, and states. The instruction cycle requires three machine cycles—$M1$, $M2$, and $M3$. Here $M1$ fetches the OUT instruction from memory, $M2$ fetches the port ad-

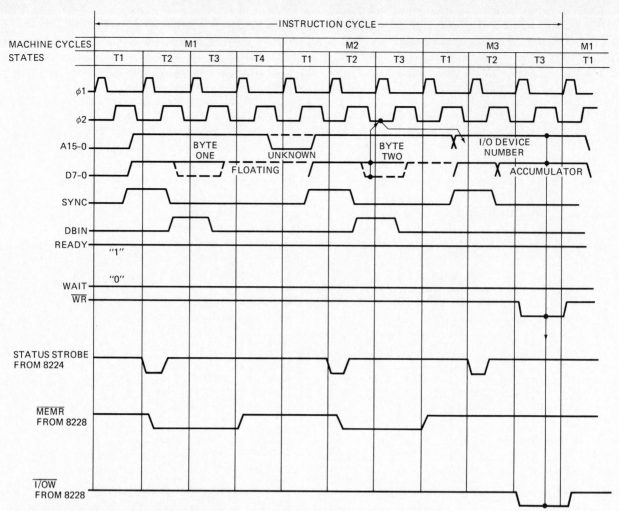

FIGURE 9-2 8080A timing waveforms showing machine cycles and states executing an OUT (port) instruction. *(Intel Corporation.)*

dress from the next memory location, and $M3$ addresses the port and outputs the accumulator contents to it. An *instruction fetch* is always the first machine cycle of an instruction cycle.

During the first state $T1$ of this instruction fetch machine cycle, the address of the instruction is put on the address bus. This is indicated in Figure 9-2 by the waveform labeled A15-0. The point at which this waveform goes high does *not* mean that all 16 address lines go high. It means that a *valid address* will be on the address bus at that time.

During the first state of any machine cycle, the CPU also outputs a *status word* on the data bus. This is shown in Figure 9-2 by the waveform D7-0 going high. The status word indicates the type of activity to be performed during that machine cycle. The 10 possible operations and the hex code for their status bytes are *instruction fetch*, A5H; *memory read*, 82H; *memory write*, 00H; *stack read*, 04H; *input read*, 42H; *output write*, 10H; *interrupt acknowledge*, 23H; *halt acknowledge*, 8AH; and *interrupt acknowledge while halt*, 2BH.

Refer to Figure 9-2, and note that the $M1$ cycle of the

OUT instruction has four states: $T1$, $T2$, $T3$, and $T4$. During the first state, $T1$, a memory address is sent out on the address bus (A15-0) and the status information is sent out on the data bus (D7-0). A SYNC signal from the 8080A indicates that the status word is on the data bus. The 8224 uses the SYNC signal and the $\phi2$ clock to produce the $\overline{\text{STSTB}}$ pulse. The *status strobe* latches the status word into the 8228.

In $T2$ of $M1$, the rising edge of $\phi2$ directs the 8080A to remove the status word from the data bus (dashed line goes low—D7-0). The 8080A then sends out a DBIN signal to indicate that the *data bus* is in the *input mode*. The 8228 decodes the status word to produce a $\overline{\text{MEMR}}$ signal. $\overline{\text{MEMR}}$ is used to enable the addressed memory. After the output enable access time of memory, the byte from memory appears on the data bus (D7-0), at the time labeled "byte 1" in Figure 9-2. For the 8080A the data from memory must be valid for at least 150 ns before the rising edge of $\phi2$ in $T3$ of $M1$. The rising edge of $\phi2$ in $T3$ resets DBIN. Some 30 ns later the 8228 raises $\overline{\text{MEMR}}$ high again. This floats the outputs of the memory again (D7-0, the dashed line labeled "floating" in Figure 9-2).

The $T4$ state of the $M1$ cycle is used for internal operations of the processor such as decoding the instruction. The OUT instruction is decoded during $T4$ to indicate that another memory read operation is required to get the stored port address. This operation is shown as the $M2$ machine cycle in Figure 9-2. The process for the $M2$ cycle is the same as for the first three states of the $M1$ cycle. No $T4$ state is needed for this cycle because the data (the port address) rather than an instruction is read from memory ("byte 2"). This data does not need to be decoded.

At the end of the $M2$ cycle, the CPU has the instruction and the port address. During the $M3$ cycle the instruction is executed. In $T1$ the port address is output on the lower 8 bits of the address bus, and the status word is output on the data bus. The 8080A sends out a SYNC *pulse* to show that the status word is present. The 8224 uses the SYNC pulse to produce $\overline{STSTB}$, which latches the status word into the 8228.

In $T2$ of $M3$ the rising edge of $\phi2$ clears the status word from the data bus. The 8080A then outputs the contents of the accumulator on the data bus. (During a previous instruction cycle the accumulator has been loaded with the data you wish to output to the port.) The rising edge of the $\phi1$ clock in $T3$ triggers the 8080A to put out a $\overline{WR}$ pulse. The 8228 uses this $\overline{WR}$ pulse and the latched status word to produce an $\overline{I/OW}$ pulse which can be used to enable the addressed output port.

Other instructions have a similar sequence of machine cycles and states. All the 8080A instructions start with an $M1$ instruction fetch from memory.

To analyze the sequence of signals in a set of waveforms such as that in Figure 9-2, it usually helps to think of a vertical line moving slowly from left to right across the waveforms. If you work your way carefully through the previously discussed OUT instruction, you should have no difficulty with the rest of the instruction sequences shown in an 8080A data book.

8080A SYSTEM SCHEMATIC

Figure 9-3 shows a schematic for an 8080A-based microcomputer system. In this section we explain how ROM, RAM, I/O ports, and other devices are added to the basic CPU group to form the system.

POWER-ON RESET
In Figure 9-3, note the resistor to V_{CC} and the capacitor to ground connected to the $\overline{RESIN}$ input of the 8224. When power is first turned on, the capacitor has no charge. $\overline{RESIN}$ is held low until the capacitor charges through the resistor. This causes the 8224 to produce a RESET signal to the 8080A and to the system peripherals. When reset, the 8080A loads its program counter with address 0000H and does an instruction fetch machine cycle to that address. Therefore, a monitor program or the first instruction that the programmer wants executed after a reset should be placed at address 0000H.

ADDRESS BUFFERS
Since the address outputs of the 8080A can each drive only one standard TTL input

and about 100 pF of capacitance, address line buffers are often used. The buffers should have a short propagation delay, so that they will not add much to the access time of the memory. Commonly used as address line buffers are the 74LS244 octal noninverting driver and the 8212 eight-bit input-output port. Both have the three-state outputs required for *direct memory access* (DMA) applications discussed later.

The 8212 is an IC of many uses. As shown in Figure 9-4, it contains eight D latches with a three-state buffer on the output of each. In addition, the device has two *device-select* inputs $\overline{DS1}$ and DS2, a *mode* input MD, a *strobe* input STB, and a *clear* input $\overline{CLR}$. Remember, for a latch the output will follow the input as long as the strobe is high. Therefore, for use as a buffer, the *mode* (MD) input and $\overline{DS1}$ are tied low, and STB, DS2, and $\overline{CLR}$ are tied to +5 V. Use the logic diagram in the device to help you understand its truth table. The 8212 can drive 10 standard TTL inputs. Other uses of this device are shown later.

ROMS
The 2716 EPROM shown in the microcomputer system of Figure 9-3 is the 2048 × 8 device discussed in Chapter 3. A $\overline{CS}$ input puts the chip in a low-power stand-by mode when high, and an $\overline{OE}$ input floats the output when high. Both inputs must be low to read the ROM. Eleven address lines are required to address the 2048 bytes stored in the device.

RAM
Some RAM usually is included in a system for temporary data storage and for use as a stack. The system in Figure 9-3 utilizes two Intel 2114s which are 4K (1024 × 4) static RAMs. The advantage of the ×4 format is that only two RAMs are needed for byte storage. "By 1" RAMs require a minimum of eight ICs for parallel byte storage. The two 2114s are connected so that one stores the upper 4 bits (nibble) of a byte and the other stores the lower nibble of a byte. Ten address lines go directly to each 2114 to select one of the 1024 nibbles stored. The *write enable* ($\overline{WE}$) input of each 2114 goes to the $\overline{MEMW}$ line from the 8228. The *chip-select* ($\overline{CS}$) inputs of the two are tied together because, to get a full byte, both must be selected.

PARALLEL I/O PORTS
I/O ports can be implemented in hardware in several ways. A simple input or output port can be set up with an 8212. The 8212 is an octal latch with three-state buffer outputs, each of which can drive up to 10 standard TTL inputs. Figure 9-4 shows the logic diagram for an 8212, and Figure 9-5 shows its connections as an input and as an output port. For an output port, MD is tied high and STB is tied low. $\overline{DS1}$ is connected to the $\overline{I/OW}$ output of the 8228, and DS2 is connected to the A0 address line. Whenever A0 is high and $\overline{I/OW}$ goes low, data on the eight 8212 data inputs is passed to the eight outputs. When $\overline{I/OW}$ goes high, the data is latched on the outputs. The outputs can be floated if desired by pulling MD low.

As shown in Figure 9-5b, the 8212 can be utilized as an input port. $\overline{DS1}$ is tied to $\overline{I/OR}$ from the 8228, and DS2 is connected to an address line. STB is tied

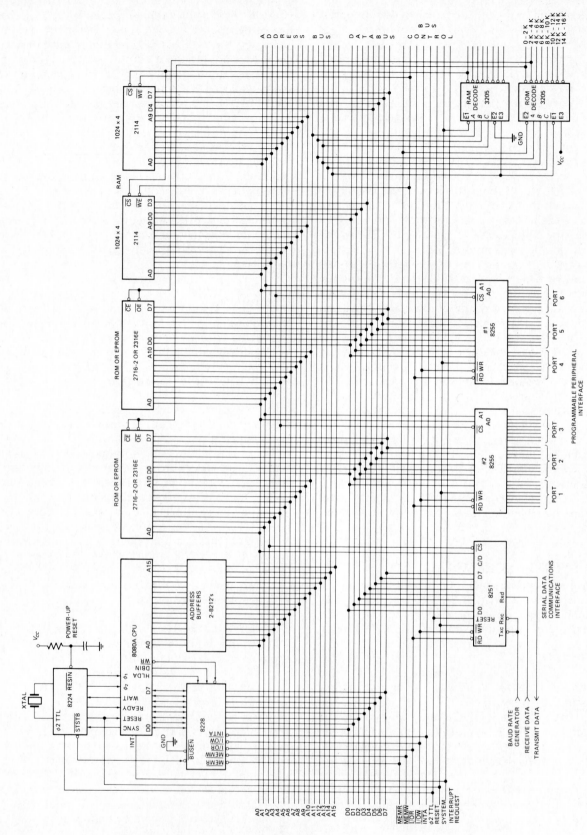

FIGURE 9-3 8080A-based microcomputer system with RAM, ROM, and I/O ports. (*Intel Corporation.*)

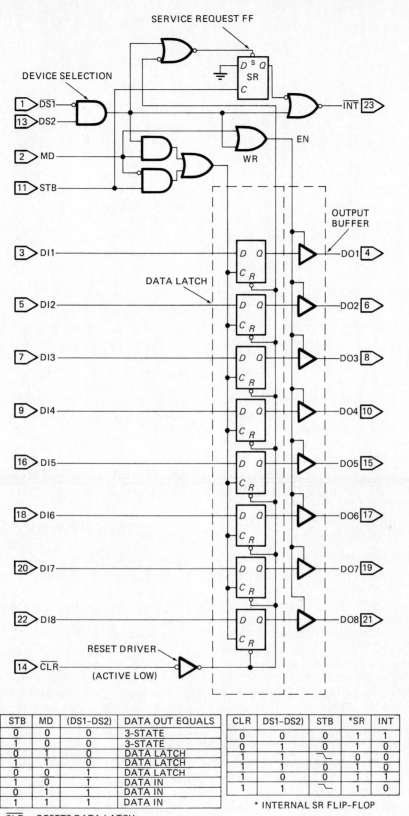

STB	MD	(DS1-DS2)	DATA OUT EQUALS
0	0	0	3-STATE
1	0	0	3-STATE
0	1	0	DATA LATCH
1	1	0	DATA LATCH
0	0	1	DATA LATCH
1	0	1	DATA IN
0	1	1	DATA IN
1	1	1	DATA IN

CLR – RESETS DATA LATCH
SETS SR FLIP-FLOP
(NO EFFECT ON OUTPUT BUFFER)

CLR	DS1-DS2)	STB	*SR	INT
0	0	0	1	1
0	1	0	1	0
1	1	╲_	0	0
1	1	0	1	0
1	0	0	1	1
1	1	╲_	1	0

* INTERNAL SR FLIP-FLOP

FIGURE 9-4 8212 octal latch and I/O port with truth table. *(Intel Corporation.)*

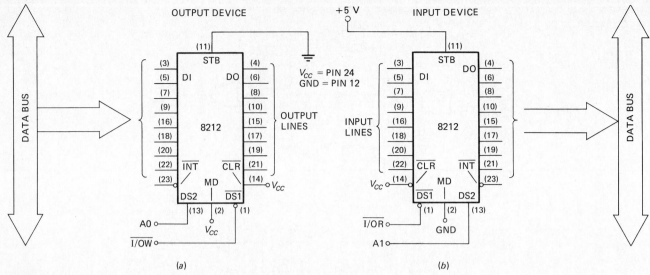

FIGURE 9-5 8212 used as an I/O port. (a) Output port. (b) Input port.

high. MD is tied low, so the output buffers are floated except when the device receives I/OR low and DS2 high signals.

THE 8255 Another method of creating input or output ports is to use a *programmable peripheral interface* IC such as the 8255A. It has three 8-bit ports that can be programmed to be inputs or outputs.

When power is turned on, an 8255A is reset. All the port lines and the data bus lines of the 8255A are in a high-Z state. Before the IC can be used as input or output ports, it must initialized. This is done by a series of instructions in the program that send a control word to the 8255A. Later we explain how programmable parallel ports such as this are initialized.

SERIAL I/O PORTS: THE 8251A USART For communication between a microprocessor system and a teletype, MODEM, or CRT terminal, data must be converted to and from serial form. In Chap. 4 we show how a *universal synchronous-asynchronous receiver-transmitter,* or USART, can be used to perform this conversion. Remember, the UART or USART is essentially a serial-in, parallel-out shift register packaged with a parallel-in, serial-out shift register and some control circuitry.

The 8251A is a programmable USART. The device can receive or transmit data in almost any serial format. Like most programmable ICs, the device must be *initialized* at the start of a program. Initializing the 8251A consists of sending 2 control bytes to it. In Chap. 11 we demonstrate how this device is initialized.

ADDRESS DECODERS The 8080A system in Figure 9-3 uses two 3205s as address decoders. When a microcomputer executes a program, it addresses memory to get instructions and data. It addresses ports to get data from or to send data to the outside world. The function of *address decoder* circuitry is to convert the sent address to a signal that selects the desired IC for that address. This decoding not only must select the desired device, but also must make sure that no other devices are selected at the same time. Later we discuss address decoding techniques in general and show a simple technique to determine the address of each device in a system from the schematic. We then examine the specific address decoding for the 8080A system in Figure 9-3.

AN 8085A MICROPROCESSOR SYSTEM

LEARNING A NEW PROCESSOR

This book cannot teach you about all the microprocessors on the market, but it will try to show you a method for learning a new processor when you encounter it. In the rest of this chapter and the next we give you some practice with a few of the most common microprocessors.

The method can be summarized as follows. First, look in the device manual for the architecture of the CPU. Check the number of registers, the width of the data bus, the number of bits in the address bus, and the power supplies required. Read through the pin descriptions to get a first look at what input and output control signals are available and what interrupt inputs are present.

Second, find the instruction set and read through it to discover what kinds of instructions this CPU can do. Does it have direct I/O instructions or only memory-mapped I/O? Under what conditions can it jump or branch? What arithmetic operations can it perform? How are interrupts handled?

Third, find out whether a schematic is shown for a minimal system using the device. From this you can get an idea of how this CPU connects with other devices and how this system compares with others you have seen.

Fourth, study the detailed description of the control signals, the timing waveforms, and the ac characteristics. Try to understand the function of every pin on the device.

These four steps are actually a loop through which you will cycle many times as you work with a particular processor.

BLOCK DIAGRAM AND PIN DESCRIPTION

Some of the criticisms of the 8080A are that it requires three power supplies, it needs three ICs to form a functional CPU, and it has only one interrupt input. Intel designers answered these criticisms with the 8085A, which solves all these problems and others.

Figure 8-3*b* is a functional block diagram of the 8085A. The 8085A essentially performs all the functions of the 8224, 8228, and 8080A CPU group and more in a single package. It operates on a single 5-V supply. The clocks required by the 8085A are created internally by simply connecting an external crystal between pins X1 and X2.

The internal register structure, ALU, flags, etc., are identical to those in the 8080A, but from there on the 8085A is different. Because of a shortage of pins, the lower 8 address bits are multiplexed out on the data bus, AD0 to AD7. An *address latch enable* output signal, ALE, tells an external device that an address is on the data bus. The ALE signal is used to latch the lower 8 bits in an external latch.

In addition to an *interrupt input* INTR similar to the INT input of the 8080A, the 8085A has four *vectored interrupts*, labeled RST 5.5, RST 6.5, RST 7.5, and TRAP.

The 8085A has a *serial input data* (SID) pin and a *serial output data* (SOD) pin. For simple forms of serial communications such as with a teletype, these two make an external UART unnecessary.

The control signals out of the 8085A are quite straightforward. An $\overline{RD}$ output indicates that the addressed memory or I/O port is to be read. A $\overline{WR}$ indicates that the addressed memory location or I/O port is to be written into. Another output, IO/$\overline{M}$, indicates whether the read or write is to memory or I/O. A high on IO/$\overline{M}$ indicates a read from or write to an I/O port. A low on IO/$\overline{M}$ is sent out during a memory read.

Data bus *status outputs*, S1 and S0, contain coded information as to the type of cycle in progress. An S1S0 of 00 indicates a halt; 01, a write; 10, a read; and 11, an instruction fetch. The other 8085A signal pins are discussed later.

INSTRUCTION CYCLES, MACHINE CYCLES, AND STATES

As with the 8080A, each instruction cycle of the 8085A consists of one to five machine cycles. Each machine cycle consists of three to six clock cycles, or states. Figure 9-6 shows the timing for an 8085A OUT port instruction. As shown, a state is measured from one falling edge of CLK to the next. During T1, status information is sent out on S1 and S0 outputs, the high-order address bits on A8–A15 and the low-order address bits on AD0 through AD7. An address-latch-enable signal ALE is sent out to show that the data bus contains the lower 8 bits of an address. The falling edge of this

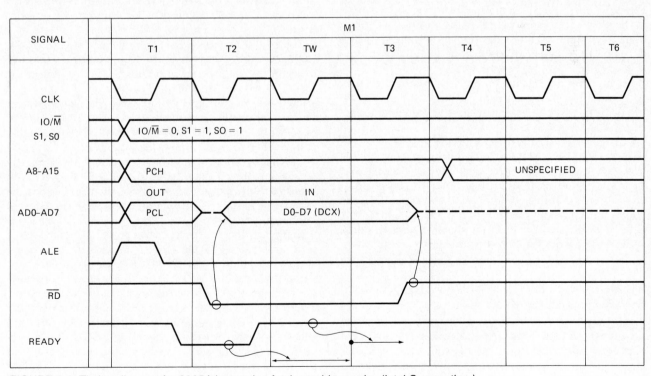

FIGURE 9-6 Timing diagram for 8085A instruction fetch machine cycle. *(Intel Corporation.)*

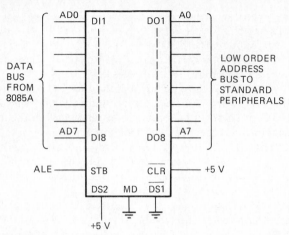

FIGURE 9-7 8212 used as address latch to demultiplex 8085A low-order address from data bus.

signal is used to latch these lower eight addresses into an external latch or peripheral device. Figure 9-7 shows how an 8212 can be used to demultiplex the address bits from the data bus by using ALE to strobe the 8212. Since the 8212 is usually needed as an address line buffer, no extra hardware is added. The resultant address is the same as that in a nonmultiplexed system such as the 8080A in Figure 9-3.

After the start of $T2$ in the 8085A machine cycle in Figure 9-6, $\overline{RD}$ goes low. This signal enables memory to output the op code byte on the data bus. The *curved arrow* from the low on the $\overline{RD}$ waveform to the AD0–AD7 waveform shows that $\overline{RD}$ going to a low level causes a byte to be output from memory. The arrow from $\overline{RD}$ high to the AD0–AD7 waveform in $T3$ indicates that $\overline{RD}$ going high causes the memory to remove the byte from the data bus (memory floats its outputs). The $T4$ state is used for decoding the instruction.

During the second machine cycle, another memory address and $\overline{RD}$ pulse are sent out to get the address of the port that will be output when the instruction executes. IO/$\overline{M}$ remains low.

During the third machine cycle, the OUT instruction is executed. The port address is output on the address bus, AD0 through AD7. IO/$\overline{M}$ goes high, reflecting that an I/O operation is taking place. After the lower 8 address bits are removed from AD0 through AD7, the 8085A outputs a data byte from the accumulator to the data bus and sends out a $\overline{WR}$ signal. The $\overline{WR}$ signal latches the output data in the port device.

MICROCOMPUTER DEVELOPMENT BOARDS: THE SDK-85

When the prototype is developed for a new microprocessor-based product, it is often unfeasible to start from scratch by wiring together the basic CPU, RAM, ROM, and I/O ports. So manufacturers make available *development boards*, which contain these elements already wired and tested, or a kit that can be assembled quickly.

One such kit is the Intel SDK-85, shown in Figure 9-8.

This unit comes with an 8085A CPU, 8155 RAM and I/O, 8355 preprogrammed monitor ROM, 8279 keyboard-display interface IC, and decoders. Also on the board are a 24-key hexadecimal keyboard and seven-segment displays for the hexadecimal displays of address and data bus contents. Designed into the board are sockets for another 8155, another 8355, or 8755A; two 8212 address buffers; and four 8216 data and control bus buffers. These ICs can be simply inserted to expand the system. A large, open area of the board similar to a universal PC board can be used for more memory or interface circuitry.

The monitor program in ROM allows you to enter instructions or data into RAM from a teletype or the hexadecimal keyboard. You can examine the contents of a register or memory location, execute programs, and single-step through a program one instruction at a time. In Chap. 12 we explain how the SDK-85 is used for the prototype of a microprocessor-based scale.

8085A SYSTEM SCHEMATIC FOR THE SDK-85

Since the 8085A contains all the parts of a CPU group as well as a serial port, then all that is needed to complete a system are RAM, ROM, and parallel I/O ports. To minimize the number of ICs needed for a simple system, Intel designed a family of parts that integrate RAM and I/O or ROM and I/O ports on single chips. The 8155/8156 contains 256 bytes of static RAM, two 8-bit parallel I/O ports, a 6-bit I/O port, and a 14-bit counter-timer. The 8355 contains 2048 bytes of ROM and two 8-bit parallel I/O ports. The 8755 has two ports and 2048 bytes of EPROM.

Figure 9-9 shows the schematic for the CPU, RAM, ROM, and ports section of the SDK-85 microcomputer development board shown in Figure 9-8. The schematic for the keyboard and display section of this board is shown and discussed in Chap. 11. The schematic in Figure 9-9 shows how two 8755s, two 8155s, and an 8085A are connected to form a small system roughly comparable to the 8080A system in Figure 9-3. An 8205 decodes chip select signals from address lines. Both the 8155s and the 8755s contain internal 8-bit latches on inputs AD0 through AD7. ALE from the CPU latches the lower 8 address bits into these latches, so no external latches are needed.

Before we discuss these devices in detail, some important features of the system schematic in Figure 9-9 should be pointed out. First, note the large letters A through D and numbers 1 through 8 around the outside of the schematic. These are called *zone coordinates.* They help you find parts or connections on a schematic, just as a similar system on a road map helps you find Bowers Avenue. For example, in Figure 9-9 device A15, an 8755, is in zone C3. Zone coordinates are also used to identify *connecting points* for signal lines. The data bus lines in zone A7 of Figure 9-9, for example, have the notation 1ZD3 next to them. This means that these

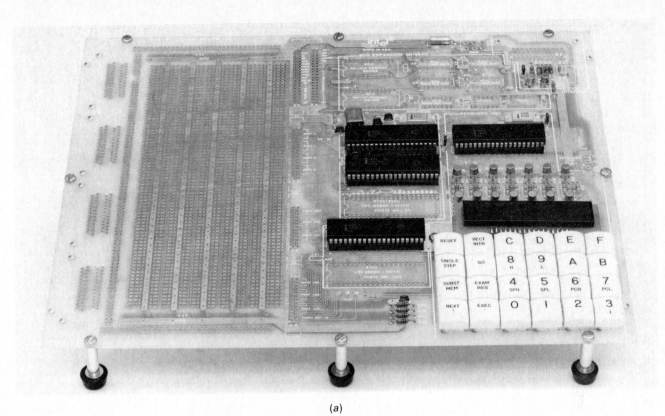

(a)

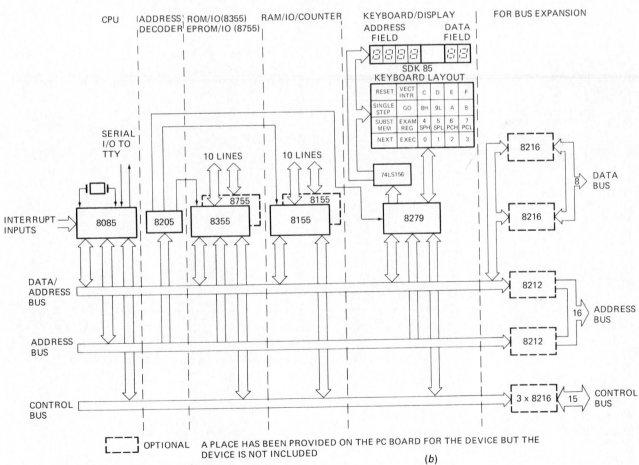

OPTIONAL   A PLACE HAS BEEN PROVIDED ON THE PC BOARD FOR THE DEVICE BUT THE
DEVICE IS NOT INCLUDED

(b)

FIGURE 9-8 SDK-85 microcomputer development board. (a) photograph. (b) Functional block diagram. (Intel Corporation.)

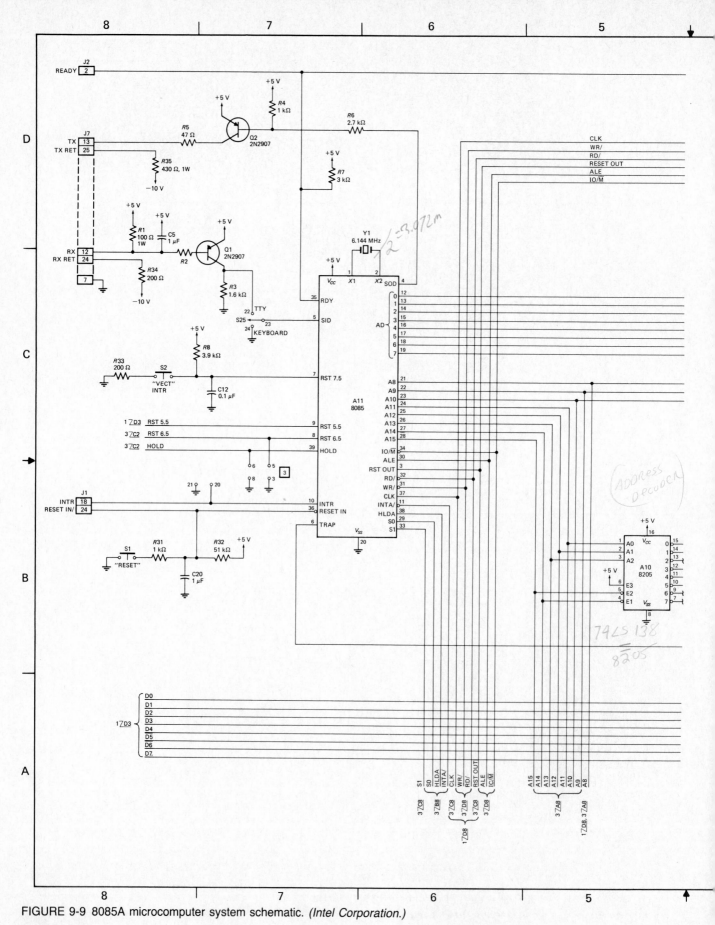

FIGURE 9-9 8085A microcomputer system schematic. *(Intel Corporation.)*

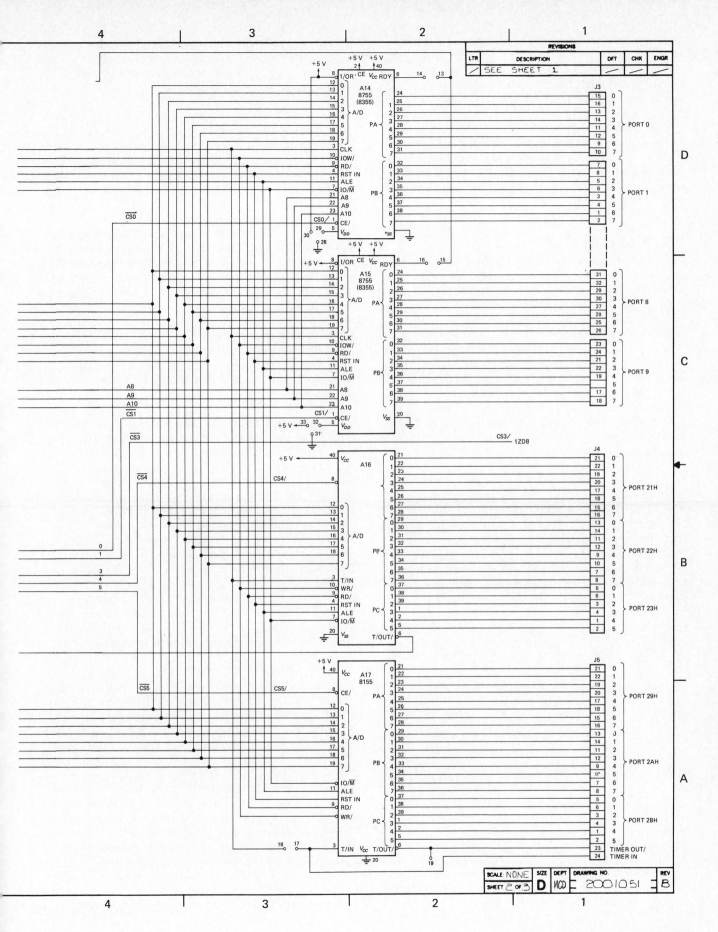

8080A/8085A SYSTEM HARDWARE AND TIMING

lines connect to the corresponding lines on sheet 1, zone D3.

Another feature is the connected boxes labeled J3, J4, and J5 in zones A1, B1, C1, and D1 of Figure 9-9. These are *jacks* that ribbon cables can be plugged into. The numbers in the small boxes are the jack pin numbers.

Still another standard schematic feature in Figure 9-9 is the small connections such as those labeled 18 and 17 in zone A3. These indicate *jumper* or *wire-wrap connections* which can be made by the user. Connecting point 18 to point 17, for example, connects the system clock to the 8155 timer input.

THE 8755A/8355

The 8755A contains 2K bytes of EPROM, two *programmable parallel ports*, and two *data direction registers*. The EPROM is accessed when $\overline{CE}$ is low and the IO/$\overline{M}$ input is low. A0 through A7 are sent in on AD0 through AD7 and latched by the falling edge of ALE. A8 through A10 enter separately.

The I/O ports and data direction registers in the 8755A are addressed by the lower 2 address bits, A1 and A0, when $\overline{CE}$ is low, IO/$\overline{M}$ is high, and either $\overline{RD}$ or $\overline{IOW}$ is low. Figure 9-10 shows the internal addresses for the two ports and two data direction registers in an 8755A.

After a reset, all 16 port lines are programmed as inputs. If any of these lines are required to be outputs in a particular application, they must be initialized as outputs. Each of the 16 lines is individually programmable as an input or output line. To program a port line as an output, a 1 is sent to the corresponding bit in the data direction register. For example, to program the PA0 line as an output, a 1 would be written into bit 0 of the port A data direction register. To establish port A as all outputs, port B7-4 as inputs, and port B3-0 as inputs, the following assembly language instruction sequence would be used for an 8755A with system starting address 00H:

MVI	A,11111111B	;data direction byte for all outputs
OUT	02	;send to DDRA
MVI	A,00001111B	;data direction byte for B7-4 input ;and A3-0 output
OUT	03	;send to DDRB

The advantage of a programmable device such as this

AD₁	AD₀	SELECTION
0	0	PORT *A*
0	1	PORT *B*
1	0	PORT *A* DATA DIRECTION REGISTER (DDR A)
1	1	PORT *B* DATA DIRECTION REGISTER (DDR B)

FIGURE 9-10 8755A internal addresses for ports and data direction registers. *(Intel Corporation.)*

is that it can be used for many different configurations of input and output lines. As some applications require, lines can easily be changed back and forth from input to output in the course of a program.

THE 8155   The 8155 contains 256 bytes of RAM, two 8-bit programmable ports, one 6-bit programmable port, and a 14-bit *programmable timer-counter*. The RAM is accessed by the lower 8 address bits when $\overline{CE}$ is low, $\overline{RD}$ or $\overline{WR}$ is low, and IO/$\overline{M}$ is low. The lower 8 address bits are strobed into internal latches by the falling edge of ALE.

I/O ports and internal registers in the 8155 are accessed by the lower address bits when $\overline{CE}$ is low, $\overline{RD}$ or $\overline{WR}$ is low, and IO/$\overline{M}$ is high. Figure 9-11a shows the internal addresses for each register and port in an 8155. To get the *system address* for each, you just add the given *internal address* to the *system base*, or *starting address*, for the device. The 8155 in Figure 9-9 labeled A16, for example, is decoded to start at address 20H. Therefore, for this device, the *command/status register* has address 20H; *port* A, address 21H; *port* B, address 22H; *port* C, address 23H; the *timer-low register*, address 24H; and the *timer-high and mode register*, address 25H.

After a reset, all port lines are established as inputs. To initialize ports as outputs, a *command word* must be sent to the *command register*. Figure 9-11b shows the format for the 8155 command word. A 1 in bit 0 of this command word defines all eight lines of port A as outputs. A 0 in bit 0 defines port A lines as inputs. Note that the port lines cannot be individually programmed as inputs or outputs as those of the 8755 can. Likewise, bit 1 of the command word defines all eight lines of port B as outputs or inputs. As shown in Figure 9-11c, port C has four *alternatives*, which are selected by bits 2 and 3 of the command word. *Alternative 1* establishes all six lines of port C as inputs, and *alternative 2* establishes all six lines of port C as outputs. *Alternatives 3 and 4* and the use of command word bits 4 and 5 are explained in Chap. 11, in a section on handshake, or strobed, I/O. Bits 6 and 7 of the command word are used to start and stop the timer.

The timer in the 8155 is a 14-bit presettable down counter. The system clock or some other external pulse source is connected to the *timer-in* (TI) of the device. An initial 14-bit binary count is loaded into two internal *count length* registers. The lower-order count length register is at internal address 04H, and the upper-order register is at address 05H. The two most significant bits of the upper register at 05 are not part of the count. They define one of the four output modes shown in Figure 9-11c for *timer-out* (TO). The four modes are as follows: 00, TO goes low during second half of count; 01, TO goes low during second half of count and count is reloaded when counter reaches 0 (square-wave output); 10, a single output pulse on TO when terminal count of 0 is reached; 11, pulse out and count reloaded at each terminal count.

In the *continuous square-wave mode* or *continuous*

I/O ADDRESS†								SELECTION
A7	A6	A5	A4	A3	A2	A1	A0	
X	X	X	X	X	0	0	0	INTERVAL COMMAND/STATUS REGISTER
X	X	X	X	X	0	0	1	GENERAL PURPOSE I/O PORT *A*
X	X	X	X	X	0	1	0	GENERAL PURPOSE I/O PORT *B*
X	X	X	X	X	0	1	1	PORT *C* – GENERAL PURPOSE I/O OR CONTROL
X	X	X	X	X	1	0	0	LOW–ORDER 8 BITS OF TIMER COUNT
X	X	X	X	X	1	0	1	HIGH 6 BITS OF TIMER COUNT AND 2 BITS OF TIMER MODE

X: DON'T CARE.
†: I/O ADDRESS MUST BE QUALIFIED BY CE = 1 (8156H) OR CE = 0 (8155H) AND IO/M = 1 IN ORDER TO SELECT THE APPROPRIATE REGISTER.

(a)

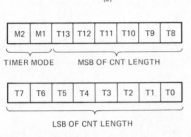

| M2 | M1 | T13 | T12 | T11 | T10 | T9 | T8 |

TIMER MODE   MSB OF CNT LENGTH

| T7 | T6 | T5 | T4 | T3 | T2 | T1 | T0 |

LSB OF CNT LENGTH

M2	M1	
0	0	PUTS OUT LOW DURING SECOND HALF OF COUNT.
0	1	SQUARE WAVE, I. E., THE PERIOD OF THE SQUARE WAVE EQUALS THE COUNT LENGTH PROGRAMMED WITH AUTOMATIC RELOAD AT TERMINAL COUNT.
1	0	SINGLE PULSE UPON TC BEING REACHED.
1	1	AUTOMATIC RELOAD, I.E., SINGLE PULSE EVERYTIME TC IS REACHED.

(c)

Command word format (b):

7	6	5	4	3	2	1	0
TM2	TM1	IEB	IEA	PC2	PC1	PB	PA

DEFINES PA0-7
DEFINES PB0-7   0 = INPUT   1 = OUTPUT

DEFINES PC0-5
00 = ALT 1
11 = ALT 2
01 = ALT 3
10 = ALT 4

ENABLE PORT *A* INTERRUPT
ENABLE PORT *B* INTERRUPT   1 = ENABLE   0 = DISABLE

TIMER COMMAND
00 = NOP – DO NOT AFFECT COUNTER OPERATION
01 = STOP – NOP IF TIMER HAS NOT STARTED; STOP COUNTING IF THE TIMER IS RUNNING
10 = STOP AFTER TC – STOP IMMEDIATELY AFTER PRESENT TC IS REACHED (NOP IF TIMER HAS NOT STARTED)
11 = START – LOAD MODE AND CNT LENGTH AND START IMMEDIATELY AFTER LOADING (IF TIMER IS NOT PRESENTLY RUNNING). IF TIMER IS RUNNING, START THE NEW MODE AND CNT LENGTH IMMEDIATELY AFTER PRESENT TC IS REACHED.

(b)

PIN	ALT 1	ALT 2	ALT 3	ALT 4
PC0	INPUT PORT	OUTPUT PORT	A INTR (PORT *A* INTERRUPT)	A INTR (PORT *A* INTERRUPT)
PC1	INPUT PORT	OUTPUT PORT	A BF (PORT *A* BUFFER FULL)	A BF (PORT *A* BUFFER FULL)
PC2	INPUT PORT	OUTPUT PORT	A S̄T̄B̄ (PORT *A* STROBE)	A S̄T̄B̄ (PORT *A* STROBE)
PC3	INPUT PORT	OUTPUT PORT	OUTPUT PORT	B INTR (PORT *B* INTERRUPT)
PC4	INPUT PORT	OUTPUT PORT	OUTPUT PORT	B BF (PORT *B* BUFFER FULL)
PC5	INPUT PORT	OUTPUT PORT	OUTPUT PORT	B S̄T̄B̄ (PORT *B* STROBE)

(d)

FIGURE 9-11 8155. *(a)* Internal addresses. *(b)* Command word format. *(c)* Timer modes. *(d)* Port *C* alternatives. *(Intel Corporation.)*

pulse mode, the frequency on the timer output is equal to the timer input frequency divided by the programmed count. To produce a 1-kHz square-wave output from a 3.072-MHz input, for example, the timer would be loaded with 3072_{10}, or 0C00H, and the mode bits set to 01 for continuous square-wave mode. Shown below is the assembly language sequence to load the timer-low register, timer-high, and mode registers. It also sends a command word to start the timer and establishes port A as input, port B as output, and port C as output. The base address for the device in this example is 20H.

```
MVI   A,00H         ;low-byte timer count
OUT   24H           ;send to timer-low register
MVI   A,01001100    ;square-wave mode
OUT   25H           ;high 6 bits of timer count
                    ;to timer-high register
```

```
MVI   A,11001110    ;command word to start timer
                    ;disable port interrupts,
                    ;port C = out, port B = out,
                    ;port A = in
OUT   20H           ;send to command register
```

As we demonstrate later, this timer can be used with an interrupt input on the microprocessor to replace delay loops for program timing. If you do not use the timer, you only have to send a command word to initialize the ports in the desired manner.

ADDRESS DECODING

A microcomputer system with 16 address lines can directly address 2^{16}, or 65,536 locations. In hexadecimal this range is 0000 through FFFF. If you think of the address lines as the outputs of a 16-bit counter, then the binary weight of each bit or address line is as shown in the *decoder worksheet* in Figure 9-12. We will use this and the schematic in Figure 3-48 to show how *address decoding* works.

The circuit in Figure 3-48 shows eight 2732 EPROMs connected in a system. Each device contains 4096 bytes of data. Since 4096 is 2^{12}, 12 address lines (A0 through A11) are needed to address one of the 4096 bytes in the device. If you have trouble with this, think of how many bits a counter has to have to count the 4096 states from 0 to 4095_D, or 0000H to 0FFFH.

Each 2732 in Figure 3-48 has a $\overline{CS}$ input that must be low for the addressed byte in a device to be output on the data bus. The trick is to make sure that the $\overline{CS}$ input of only one device at a time is low. This is done by the 74LS138. If the 74LS138 is enabled by making its $\overline{G2A}$ and $\overline{G2B}$ inputs low and its G1 input high, then only one output of the 74LS138 will be low. The output that will be low is determined by the 3-bit address applied to the C, B, and A select inputs. For example, if CBA is 000, then the Y0 output will be low and all the others will be high. If CBA is 010, the Y2 output will be low and the others high. If CBA is 111, only the Y7 output will be low. Now, in Figure 3-48, system address lines A14, A13, and A12 are connected to the CBA inputs of the 74LS138. Knowing this, you can use the worksheet in Figure 9-12 to determine the addresses for each device in the system.

The worksheet in Figure 9-12 shows all the possible CBA input combinations under the address lines connected to them. Address line A15 is shown as a 0. Actually, in this system it is a don't-care because it is not connected to the decoder or to devices. However, we let it be 0 to give the simplest addresses. From the worksheet you can see that ROM 0 in Figure 3-48 is enabled if A15, A14, A13, and A12 are all low. To address the first location in ROM 0, A0 through A11 will also be low. There-

fore, the *starting* address in ROM 0 is 0000H. The *ending* address of ROM 0 is that address where A0 through A11 are all high. Since A12 through A15 are still low, the ending address for ROM 0 is 0FFFH. The *address range* of ROM 0 is said to be 0000H to 0FFFH, a 4K-byte block.

ROM 1 is enabled if A15 is 0, A14 is 0, A13 is 0, and A12 is 1. For the starting address in ROM 1, A0 through A11 are all low. Therefore, the starting address of ROM 1 is 1000H. Its ending address, when A0 to A11 are all high, is 1FFFH. Using the worksheet in Figure 9-12, you can see that the ranges for the other six devices are 2000H to 2FFFH, 3000H to 3FFFH, 4000H to 4FFFH, 5000H to 5FFFH, 6000H to 6FFFH, and 7000H to 7FFFH. Some people like to think of A12, A13, and A14 as counting off 4K blocks of memory.

ADDRESS DECODING FOR AN 8080A SYSTEM

To analyze the addressing in the 8080A system of Figure 9-3, begin by determining the number and size of the words stored in each device. An easy way to do this is to note the number of address and data lines connected to each device. The 2716s, for example, each have eight data lines connected to them. Therefore, they output 8-bit words (bytes). The 2716s each have 11 address lines, A0 through A10, connected to them. Therefore, each contains 2^{11}, or 2048_D, words.

Each RAM in Figure 9-3 has four data lines, so it outputs 4-bit words (nibbles). Each has 10 address lines (A0 to A9), so it contains 2^{10}, or 1024_D, words.

Now make up a worksheet with headings such as those shown in Figure 9-12, and use it to determine which address lines connect to each decoder. We will start with the 3205 ROM decoder.

The 3205 functions identically to the 74LS138 discussed earlier, but it has different labels on its enable inputs. $\overline{E1}$ and $\overline{E2}$ are active low, and E3 is active high. E3 of the ROM decoder is permanently tied high to V_{CC}.

		HEX DIGIT				HEX DIGIT				HEX DIGIT				HEX DIGIT				HEX EQUIVALENT ADDRESS
		A15	A14	A13	A12	A11	A10	A9	A8	A7	A6	A5	A4	A3	A2	A1	A0	
BLOCK 1	START	0	0	0	0	0	0	0	0	0	0	0	0	0	0	0	0	= 0000
	END	0	0	0	0	1	1	1	1	1	1	1	1	1	1	1	1	= 0FFF
BLOCK 2	START	0	0	0	1	0	0	0	0	0	0	0	0	0	0	0	0	= 1000
	END	0	0	0	1	1	1	1	1	1	1	1	1	1	1	1	1	= 1FFF
BLOCK 3	START	0	0	1	0	0	0	0	0	0	0	0	0	0	0	0	0	= 2000
	END	0	0	1	0	1	1	1	1	1	1	1	1	1	1	1	1	= 2FFF
BLOCK 4	START	0	0	1	1	0	0	0	0	0	0	0	0	0	0	0	0	= 3000
	END	0	0	1	1	1	1	1	1	1	1	1	1	1	1	1	1	= 3FFF
BLOCK 5	START	0	1	0	0	0	0	0	0	0	0	0	0	0	0	0	0	= 4000
	END	0	1	0	0	1	1	1	1	1	1	1	1	1	1	1	1	= 4FFF
BLOCK 6	START	0	1	0	1	0	0	0	0	0	0	0	0	0	0	0	0	= 5000
	END	0	1	0	1	1	1	1	1	1	1	1	1	1	1	1	1	= 5FFF
BLOCK 7	START	0	1	1	0	0	0	0	0	0	0	0	0	0	0	0	0	= 6000
	END	0	1	1	0	1	1	1	1	1	1	1	1	1	1	1	1	= 6FFF
BLOCK 8	START	0	1	1	1	0	0	0	0	0	0	0	0	0	0	0	0	= 7000
	END	0	1	1	1	1	1	1	1	1	1	1	1	1	1	1	1	= 7FFF

DECODER ADDRESS INPUTS

FIGURE 9-12 Decoder worksheet demonstrating address decoding for eight 2732 EPROMs in schematic of Fig. 3-48.

E2 is tied to the system $\overline{\text{MEMR}}$ signal. The ROM decoder, then, is enabled only for a memory read operation. This prevents an accidental write to ROM which might damage the device. $\overline{\text{E1}}$ is connected to system address line A14, so mark a 0 under this bit on the worksheet. In this system A15 is not connected to anything, so it is a don't-care. Mark a 0 under this bit in the worksheet. A13, A12, and A11 are connected to the C, B, and A inputs of the decoder, respectively. Mark all possible combinations of these address lines on the worksheet under the appropriate bits. The address range for each decoder output can now be read almost directly off the sheet, as was done for the previous example. The starting address of the first ROM occurs when A11 to A15 are 0 and A0 to A10 are 0. This block ends when A11 to A15 are 0 and A0 to A10 are all 1's. In hex this corresponds to 0000 through 07FF. The second ROM decoder output is enabled when A15 is 0, A14 is 0, A13 is 0, A12 is 0, and A11 is 1. The starting address that will enable this second output is 0800H. The range for this output is 0800H to 0FFFH. The ranges of the other outputs are 1000H to 17FFH, 1800H to 1FFFH, 2000H to 27FFH, 2800H to 2FFFH, 3000H to 37FFH, and 3800H to 3FFFH.

The RAM decoder in Figure 9-3 has its $\overline{\text{E1}}$ enable input tied to the system reset line. This is done so that RAM will be disabled during a reset operation. $\overline{\text{E2}}$ is permanently enabled by being tied to ground. The E3 enable input is tied to A14. Since E3 is an active high enable input, A14 must be high to enable this decoder. Mark a 1 under A14 on a worksheet such as that in Figure 9-12. The C, B, and A inputs of the RAM decoder are connected to system address lines A12, A11, and A10, respectively. Mark all possible combinations of these three, in order, on the worksheet as shown in Figure 9-12. A13 and A15 are don't-cares, so 0's can be marked in those columns on the worksheet. The first block of RAM will start at address 4000H and end at 43FFH. The ranges of the other RAM decoder outputs are 4400H to 47FFH, 4800H to 4BFFH, 4C00H to 4FFFH, 5000H to 53FFH, 5400H to 57FFH, 5800H to 5BFFH, and 5C00H to 5FFFH.

PORT DECODING: MEMORY-MAPPED I/O AND DIRECT I/O

MEMORY-MAPPED I/O   In a *memory-mapped* I/O system, the addresses for port devices are decoded in the same way as the ROM and RAM addresses were earlier. Unused outputs on the RAM decoder might be used to enable port ICs. For example, the 05, 06, and 07 outputs of the 8205 RAM decoder in Figure 9-3 could be connected to the $\overline{\text{CS}}$ inputs of three port ICs. These ports would then have full 16-bit addresses 5400H, 5800H, and 5C00H. The advantage of this approach is that any instruction that references memory can send data to or read data from the port. For example, if HL is initialized to 5400H, the instruction MOV D,M reads data directly from the port to the D register. The CPU treats the port just as it would any other memory location. The disadvantages of memory-mapped I/O are: each read from or write to the port requires a full 16-bit (2-byte) address; and ports use up some of the system address space.

DIRECT I/O   Some microprocessors such as the 8080A and 8085A set aside separate address spaces for input ports and output ports. This is known as *direct I/O*. These addresses are accessed by the IN port and OUT port instructions of the 8080A and 8085A. When an 8080A or 8085A executes an IN port or OUT port instruction, it sends out an 8-bit port address on the lower eight address lines. (*Note:* This 8-bit port address is duplicated on the upper eight address lines. For example, OUT 29H would produce 2929H on the address bus.) Control signals such as $\overline{\text{I/OR}}$ and $\overline{\text{I/OW}}$ indicate that the address on the bus is for a port, and not for a memory location. These control signals are used to turn on the desired port device. Since an 8-bit address is sent out, $\overline{\text{CS}}$ signals for up to 256 input and 256 output ports could be decoded by using devices such as the 3205. Input port decoders would be enabled by $\overline{\text{I/OR}}$, and output port decoders would be enabled by $\overline{\text{I/OW}}$.

The advantages of direct I/O are that each IN or OUT instruction requires only a single-byte address and that none of the memory space is used for ports. The disadvantages are that more control signals are needed and not as many instructions can be used to access a port.

In small systems such as that in Figure 9-3 where only a few ports need to be selected, address lines are used directly to enable port devices. This technique is called *linear select*. Since the direct I/O addresses are 8 bits wide, up to eight input and eight output ports can be selected.

In the system in Figure 9-3, for example, A2 is connected to the $\overline{\text{CS}}$ input of the 8251 serial port device. $\overline{\text{I/OR}}$ is connected to the $\overline{\text{RD}}$ input and $\overline{\text{I/OW}}$ is connected to the $\overline{\text{WR}}$ input of the 8251. The device is selected for a read operation if A2 is low and $\overline{\text{I/OR}}$ is low. It is selected for a write operation if A2 is low and $\overline{\text{I/OW}}$ is low. A0 is connected to the CONTROL/$\overline{\text{DATA}}$ (C/$\overline{\text{D}}$) input of the device. If C/$\overline{\text{D}}$ is high, a word sent to the device goes into its internal control register. If C/$\overline{\text{D}}$ is low, a sent byte goes out the port. For the 8251A in Figure 9-3, the word required on the lower 8 bits of the address bus is XXX110X1 to address the control register and XXX110X0 for the serial input-output port. Bits A3 and A4 must be 1's so that neither of the 8255As is selected at the same time as the 8251A. If the don't-care bits are all made 1's, then the control port address is FBH and the serial output port address is FAH.

The $\overline{\text{CS}}$ of the #1 8255 in Figure 9-3 is connected to system address line A3. The device is enabled if A3 is low and either $\overline{\text{RD}}$ or $\overline{\text{WR}}$ is low. A0 and A1 are used to select one of the three ports or the control register in the device. The first of the four addresses occupied by the device is XXX10100. If the don't-cares are made 1's, then this address is F4H. The other addresses in this device are F5H, F6H, and F7H.

The $\overline{CS}$ of the #2 8255 in Figure 9-3 is connected to system address line A4. The first of the four addresses occupied by this device is XXX01100, or ECH; the other three addresses are EDH, EEH, and EFH.

MEMORY MAPS

The previous sections have discussed methods of address decoding for microcomputer systems and used a worksheet to determine the address range for each device in a particular system. The address ranges in the device in a system often are summarized graphically in a *memory map*. This allows the system memory usage to be seen at a glance. Figure 9-13 shows a memory map for the system in Figure 9-3.

ADDRESS DECODING FOR AN 8085A SYSTEM

MEMORY   For further practice with address decod-

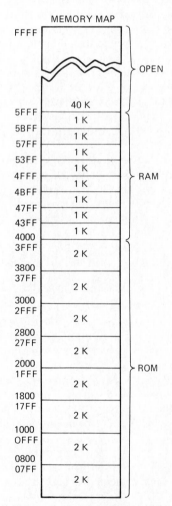

FIGURE 9-13 Memory map for 8080A system in Fig. 9-3.

ing, we now examine the decoding on the SDK-85 system shown in Figure 9-9. To start, draw an address worksheet, as was done in the previous sections. Mark in the value for each bit as you determine it.

For the 8205 address decoder in Figure 9-9, the E3 enable input is permanently enabled by being tied to +5 V. The $\overline{E2}$ enable input is tied to address line A15, and the $\overline{E1}$ enable input is tied to A14. Since both must be low to enable the device, A15 and A14 must be low. Mark these two bits as lows on your address worksheet. System address lines A13, A12, and A11 are connected to 8205 select inputs A2, A1, and A0, respectively. Mark in the binary codes for 000 to 111 under these bits on your address worksheet. Since A10 through A0 are used to select bytes within, for example, the 8355 ROM, mark in 0's under bits A10 through A0 on your worksheet. Compare your worksheet with that in Figure 9-14.

From your worksheet you should be able to see that the 0 output of the 8205 is selected when the address lines are all 0's, 0000H. This 0 output then selects IC number A14, the 8355 which contains the monitor program ROM, so the starting address of the monitor program is 0000H.

The 1 output of the 8205 is selected if A13 = 0, A12 = 0, and A11 = 1. The first address that produces this combination is 0800H. The 1 output of the 8205 selects A15, the second 8355 ROM. The starting address for the ROM in the A15 8355, then, is 0800H. The difference between the starting address for the 8205 0 output and the 8205 1 output is 800H, or 2048_D. Another way of looking at this is to think of A13, A12, and A11 as counting off 2048 (2K) blocks of address space.

Use your worksheet to determine the starting and ending addresses for each of the 8205 outputs in Figure 9-9. You should get 0 = 0000H to 07FFH; 1 = 0800H to 0FFFH; 2 = 1000H to 17FFH; 3 = 1800H to 1FFFH; 4 = 2000H to 17FFH; 5 = 2800H to 2FFFH; 6 = 3000H to 37FFH; and 7 = 3800H to 3FFFH.

The 3 output goes to the 8279 keyboard display interface device, which is discussed in Chap. 11.

The 4 output enables A16, an 8155. The starting address for the 256 bytes of RAM in this device, then, is 2000H. According to the range for the 4 output shown above, any address in the 2048-bit range between 2000H and 27FFH will enable the device. Since the 8155 contains only 256 bytes of RAM, it requires only addresses from 2000H to 20FFH. However, any other address between 2100H and 27FFH also will select the device and "fold back" to some address in the basic 256-byte space. This address space from 2100H to 27FFH is often called *foldback*. If you refer to your address worksheet, another way to describe this foldback is that A13, A12, and A11 select the A16 8155; A7 through A0 select one of the 256 bytes in the RAM; and A10, A9, and A8 are don't-cares because they have no connection to the device.

The 5 output from the 8205 selects A17, the second 8155, at starting address 2800H. Any address from 2800H to 2FFFH will select the device. The region from 2900H to 2FFFH is the foldback for this device.

HEX				HEX				HEX				HEX				HEX EQUIVALENT ADDRESS	START OF BLOCK
A15	A14	A13	A12	A11	A10	A9	A8	A7	A6	A5	A4	A3	A2	A1	A0		
0	0	0	0	0	0	0	0	0	0	0	0	0	0	0	0	= 0000	1
0	0	0	0	1	0	0	0	0	0	0	0	0	0	0	0	= 0800	2
0	0	0	1	0	0	0	0	0	0	0	0	0	0	0	0	= 1000	3
0	0	0	1	1	0	0	0	0	0	0	0	0	0	0	0	= 1800	4
0	0	1	0	0	0	0	0	0	0	0	0	0	0	0	0	= 2000	5
0	0	1	0	1	0	0	0	0	0	0	0	0	0	0	0	= 2800	6
0	0	1	1	0	0	0	0	0	0	0	0	0	0	0	0	= 3000	7
0	0	1	1	1	0	0	0	0	0	0	0	0	0	0	0	= 3800	8

DECODER ADDRESS INPUTS

FIGURE 9-14 Address decoding worksheet for SDK-85.

PORT ADDRESSING FOR THE SDK-85 SYSTEM

When an 8080A or an 8085A executes an OUT (port) or IN (port) instruction, the 8-bit address of the port is sent out on the lower eight address lines. The same address is copied on the upper eight address lines. You must remember this when tracing 8085A port decoding as in Figure 9-9, because the upper eight address lines are decoded to select ports. It is easier to use the upper address lines because they are not multiplexed as A7 through A0 are.

To determine port addresses for a system such as the SDK-85, you can use an address worksheet as before. From the worksheet, you can determine the address that must be present on A15 to A8 to produce a $\overline{CS}$ signal to the port device. As previously stated, this same address must be the port address on A7 to A0. In Figure 9-9, for example, all 0's on A15 to A8 select the top 8355. The starting, or base, address for the ports in this device then is 00H. As shown in Figure 9-10, the 8355 has four internal addresses that will be selected by A1 and A0. The second 8355, in Figure 9-9, A15, is selected if A15 to A8 are equal to 08H. Therefore, the base address of this device is 08H. Adding the internal address selected by A1 and A0 gives addresses of 08H for port A, 09H for port B, 0AH for DDRA, and 0BH for DDRB.

The A16 8155 will be selected if A15 to A8 contain 20H. Therefore, the base address for this device is 20H. As explained in a previous section on the 8155, the six addresses for the device are 20H, command/status register; 21H, port A; 22H, port B; 23H, port C; 24H, timer-low register; and 25H, timer-high and mode register. The A17 8155 is selected when A15 to A8 contain 28H, so the base address for this device is 28H.

A thorough understanding of address decoding is very important when you are troubleshooting microcomputer systems. A memory device or port cannot read in correct data if it is not getting enabled properly by the address decoding circuitry.

MICROCOMPUTER TIMING PARAMETERS

Timing relationships are very critical in most microcomputer systems. Figure 9-15 shows the read cycle waveforms and related timing values for the 8085A.

First note the minimum and maximum times for the *clock cycle period* T_{CYC}. The minimum T_{CYC} of 200 ns for the 8085A-2 indicates that the device cannot operate with an internal clock frequency greater than 5 MHz. The clock frequency produced by connecting a crystal between the 8085A X1 and X2 inputs is divided internally by 2. Therefore, a 10-MHz crystal produces a 5-MHz internal clock. The maximum clock cycle period of 2000 ns indicates that if a clock frequency less than 500 kHz is used, the dynamic circuits within the 8085A will start to lose stored data between clock pulses.

If the 8085A is supplied with a clock or crystal in the legal operating range, most of the timing values such as the width of ALE, etc., are taken care of internally by the processor. The timing values that are of greatest concern to us here are those where there is an interaction between the 8085A and other devices in the system. For example, during a read operation the 8085A sends out an address, an ALE signal, and a $\overline{RD}$ signal. Within a time T_{AD} after the *address* is sent out and within a time T_{RD} after $\overline{RD}$ goes low, the addressed memory device or port must return *valid data* on the data bus to the 8085A. Maximum T_{AD} is 575 ns for a 3-MHz 8085A and 350 ns for a 5-MHz 8085A-2. For successful operation the *address access time* of the addressed device plus the *propagation delay* of the address decoder used to select the device must be less than this T_{AD}. For a system such as that in Figure 9-9, this works out as follows. The maximum propagation delay of the 8205 decoder is 18 ns. The maximum address access time for an 8755A is 450 ns, and for an 8755A-2 it is 330 ns. Adding the decoder delay to these gives a *total access time* of 468 ns for the 8755A and 348 ns for the 8755A-2. Comparing these values with the T_{AD} values above, you can see that the 8755A will work only with the 3-MHz 8085A, and the 8755A-2 will work correctly with either the 8085A or 8085A-2. This again points out that for proper operation of a given system, the correct suffix number part must be used. You cannot just replace a defective 8755A-2, for example, with an 8755A without checking the timing to see whether it will work properly in that particular system.

Another timing parameter to check in a case such as this is T_{RD}. The maximum T_{RD} is 300 ns for an 8085A and 150 ns for the 8085A-2. The corresponding $\overline{RD}$ *access times* are 170 ns for the 8755A and 140 ns for

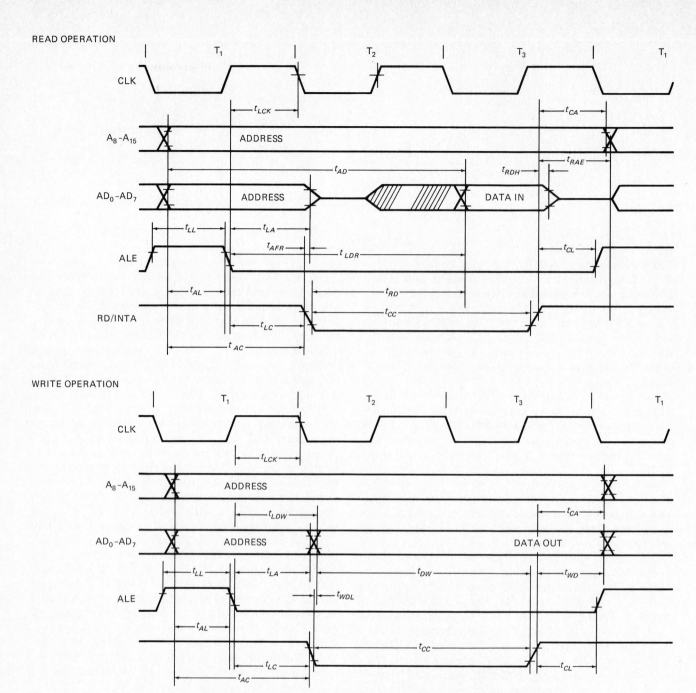

FIGURE 9-15 8085A timing waveforms and chart of related timing values *(p. 241). (Intel Corporation.)*

the 8755A-2. This again illustrates that the faster 8755A-2 must be used with the faster 8085A-2.

In summary, the access time of the addressed memory device or port must be less than the response time (T_{AD}, etc.) required by the processor. Violation of this rule is a very common cause of a system sometimes working correctly and sometimes not.

WAIT, HOLD, AND HALT STATES

Now we will explain the functions of a few more pins on the 8080A and 8085A.

WAIT STATES

If the READY input of an 8080A or 8085A CPU is made low at the right time, the CPU enters a *wait state*, T_W, after $T2$ of the current machine cycle. Figure 9-16 shows the timing waveforms for an 8085A read machine cycle both with and without a wait state. The bottom waveform in Figure 9-16 shows the effect of the READY input. If READY is high when it is checked by the CPU during $T2$, the CPU proceeds directly to state $T3$ of the machine cycle. If READY is low when it is checked by the CPU in $T2$, then a wait state T_W is inserted. If READY goes high during the wait state, then after the wait state the CPU continues with $T3$ of the machine cycle. If

SYMBOL	PARAMETER	8085A* MIN.	8085A* MAX.	8085A-2* (PRELIMINARY) MIN.	8085A-2* (PRELIMINARY) MAX.	UNITS
t_{CYC}	CLK Cycle Period	320	2000	200	2000	ns
t_1	CLK Low Time (Standard CLK Loading)	80		40		ns
t_2	CLK High Time (Standard CLK Loading)	120		70		ns
t_r, t_f	CLK Rise and Fall Time		30		30	ns
t_{XKR}	X1 Rising to CLK Rising	30	120	30	100	ns
t_{XKF}	X1 Rising to CLK Falling	30	150	30	110	ns
t_{AC}	A8–15 Valid to Leading Edge of Control†	270		115		ns
t_{ACL}	A0–7 Valid to Leading Edge of Control	240		115		ns
t_{AD}	A0–15 Valid to Valid Data In		575		350	ns
t_{AFR}	Address Float After Leading Edge of READ (INTA)		0		0	ns
t_{AL}	A8–15 Valid Before Trailing Edge of ALE†	115		50		ns
t_{ALL}	A0–7 Valid Before Trailing Edge of ALE	90		50		ns
t_{ARY}	READY Valid from Address Valid		220		100	ns
t_{CA}	Address (A8–15) Valid After Control	120		60		ns
t_{CC}	Width of Control Low (RD, WR, INTA) Edge of ALE	400		230		ns
t_{CL}	Trailing Edge of Control to Leading Edge of ALE	50		25		ns
t_{DW}	Data Valid to Trailing Edge of WRITE	420		230		ns
t_{HABE}	HLDA to Bus Enable		210		150	ns
t_{HABF}	Bus Float After HLDA		210		150	ns
t_{HACK}	HLDA Valid to Trailing Edge of CLK	110		40		ns
t_{HDH}	HOLD Hold Time	0		0		ns
t_{HDS}	HOLD Setup Time to Trailing Edge of CLK	170		120		ns
t_{INH}	INTR Hold Time	0		0		ns
t_{INS}	INTR, RST, and TRAP Setup Time to Falling Edge of CLK	160		150		ns
t_{LA}	Address Hold Time After ALE	100		50		ns
t_{LC}	Trailing Edge of ALE to Leading Edge of Control	130		60		ns
t_{LCK}	ALE Low During CLK High	100		50		ns
t_{LDR}	ALE to Valid Data During Read		460		270	ns
t_{LDW}	ALE to Valid Data During Write		200		120	ns
t_{LL}	ALE Width	140		80		ns
t_{LRY}	ALE to READY Stable		110		30	ns
t_{RAE}	Trailing Edge of READ to Re-Enabling of Address	150		90		ns
t_{RD}	READ (or INTA) to Valid Data		300		150	ns
t_{RV}	Control Trailing Edge to Leading Edge of Next Control	400		220		ns
t_{RDH}	Data Hold Time After READ INTA‡	0		0		ns
t_{RYH}	READY Hold Time	0		0		ns
t_{RYS}	READY Setup Time to Leading Edge of CLK	110		100		ns
t_{WD}	Data Valid After Trailing Edge of WRITE	100		60		ns
t_{WDL}	LEADING Edge of WRITE to Data Valid		40		20	ns

Notes: * *Test conditions:* t_{CYC} = 320 ns (8085A)/200 ns (8085A-2): C_L = 150pF.

†A8–A15 address Specs apply to IO/M̄, S0, and S1 except A8–A15 are undefined during T4–T6 of OF cycle whereas IO/M̄, S0, and S1 are stable.

‡Data hold time is guaranteed under all loading conditions.

FIGURE 9-15 *(continued)*

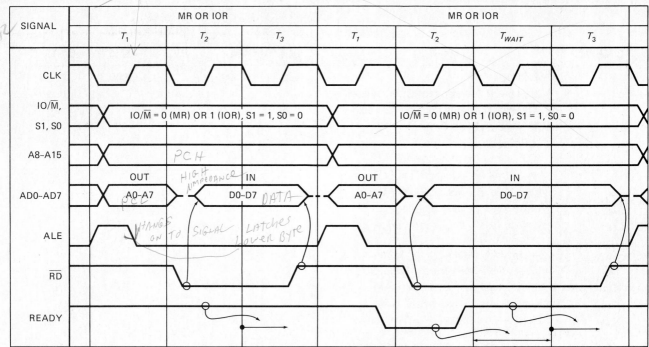

FIGURE 9-16 8085A timing waveform for read machine cycle with and without wait state. *(Intel Corporation.)*

READY is still low, then another wait state is inserted. Wait states continue to be inserted as long as READY is low. To indicate that it is in a wait state, the 8080A will output a high on its WAIT output. The 8085A does not have a WAIT output.

Figure 9-16 shows that during a wait state the contents of the address bus, the data bus, and the control bus are all held constant. The wait state then gives an addressed memory or port an extra clock cycle time to output valid data on the data bus. This feature permits the use of cheaper memory devices that have longer access times. A monostable multivibrator such as the 74LS121 can be triggered by the 8080A $\overline{MEMR}$ or $\overline{MEMW}$, or by the 8085A $\overline{RD}$ or $\overline{WR}$, to pulse READY low each time the slower device is addressed. The monostable can be enabled by the same signal that is sent to the $\overline{CS}$ input of the addressed device. This prevents a wait state from being inserted during each read or write operation. Some devices, such as the 8755A, contain internal circuitry to request a wait state. Note in Figure 9-9, zones D2 and C2, that the open-drain RDY outputs of the 8755As can be jumpered to the READY input of the 8085A. If this connection is made, the 8755A will cause the 8085A to insert one wait state each time the 8755 is read.

Refer to the 8085A timing parameters in Figure 9-15*b* and the waveforms in Figure 9-15*a*. To be recognized, the READY input must be made low a time T_{RYS} before the *rising edge* of CLK in T2. To exit the wait state, READY must be made high T_{RYS} before the rising edge of CLK in T_W.

HOLD STATES AND DIRECT MEMORY ACCESS

Raising the HOLD input of an 8080A or 8085A to a high requests a *hold state.* In a hold state the address, data, and control buses are floated so they can be used by an external device or system. The HOLD *request signal* is internally synchronized so that when the processor finishes using the buses for its current *machine cycle,* it sends out a *hold acknowledge signal* on its HLDA output. The processor then floats the buses. This is near the end of T3 in the machine cycle. If the processor is doing a machine cycle that requires four, five, or six states, these states are internally completed after the buses float. The processor then sits and idles. Internal registers are kept refreshed. When HOLD is made low, the program starts again with T1 of the next machine cycle.

An application of the hold state is for *direct memory access,* or DMA. DMA is used where some external system puts out bursts of data bytes faster than an 8080A or 8085A can input them and store them in memory. A device such as the 8257 *programmable DMA controller* sends a HOLD request signal to the 8080A or 8085A when a block of data is ready. When finished with the buses for the current machine cycle, the 8080A or 8085A sends out a HLDA signal and floats the address and data buses. Note in the 8080A system of Figure 9-3 that HLDA is connected to the 8228, so HLDA floats the data bus. A signal from the DMA controller to the $\overline{BUSEN}$ input of the 8228 floats the 8228 control bus

242
CHAPTER NINE

outputs. In an 8085A all this is done internally. The DMA controller then has possession of all three buses. It supplies the proper addresses and control signals to write the block of data into memory. When the data transfer is complete, the DMA controller drops the HOLD input low, and the 8080A or 8085A continues with the next machine cycle.

HALT STATE

When the *halt instruction* HLT is executed, an 8080A or 8085A will enter a *halt state* after T2 of the next machine cycle. In the halt state, the processor is stopped; the address and data buses are floating. There are only three ways to get out of the halt state. First, a low on the RESET IN input of the 8085A resets the entire system and loads the program counter with all 0's. For the 8080A, a low on the RESIN of the 8224 will output a high to the RESET input of the 8080A (and other system components), to reset the system and load the program counter with all 0's. After a reset, 8080A/8085A program execution begins immediately from address 0000H.

The second way to get out of the halt state is to make the HOLD input high. The processor then enters the hold state, but when the HOLD input goes low again, the CPU returns to the halt state.

The third method of exiting a halt state is with an *interrupt* signal on an interrupt input of the 8080A or 8085A. This method works only if interrupts were enabled with an *enable interrupt* (EI) instruction in the program. The next section discusses interrupts and their uses.

For now, the important thing for you to remember is that when you use a HLT instruction at the end of a program, the CPU executes the program and then halts. The only way to get it to do anything after this is with a RESET, a HOLD, or an INT input.

INTERRUPTS

An *interrupt* is one way in a microcomputer-based system of dealing with input signals that occur seldom and randomly in time. For example, suppose a microcomputer system is being used to control a large papermaking machine. Part of the microcomputer's job is to make sure the steam boiler does not exceed its safety limit pressure. If the boiler pressure exceeds the safety limit, then the microcomputer should immediately cut off the fuel and open a pressure relieve valve. Ideally, the microcomputer should interrupt what it is doing and take care of this problem before the boiler blows up.

8080A INTERRUPTS

The quickest way of servicing such a problem is with an interrupt. The output signal from the boiler pressure transducer could, for example, be connected to the interrupt input (INT) of the 8080A. The INT input is enabled previously by an EI instruction in the program. When the 8080A receives an INT signal, it completes the instruction in progress and then does an M1 memory fetch machine cycle. The status word put out during T1 indicates that an interrupt operation is taking place. DBIN is put out during T2 to indicate that the data bus is in the input mode. The 8228 uses the status word to produce an *interrupt acknowledge* signal INTA. The CPU is essentially saying at this point, "Now that you have interrupted me, tell me what to do." The data bus is waiting for an instruction to be loaded on it. This must be done by external hardware.

The 8080A designers have made this task relatively easy by including the *restart* (RST) instructions in the instruction set. The RST instructions are 1-byte, unconditional call instructions. A 3-bit address is contained as bits $D3$, $D4$, and $D5$ of the instruction. There are eight possible restart addresses: RST 0 to RST 7. When the RST instruction is loaded onto the data bus by external hardware, the 8080A decodes the instruction. Since RST is essentially a call instruction, the CPU stores the program return address on the stack. It then uses the RST number, 0 to 7, as the address bits $A3$, $A4$, and $A5$ of the subroutine starting address. The other address bits are assumed all 0's. The program counter is loaded with this address, and program execution starts from there. For example, the RST 7 instruction stores the return address on the stack and jumps the program counter to address 0038H. The eight RST addresses are, in hex, 00, 08, 10, 18, 20, 28, 30, and 38.

Starting at the chosen RST address, the programmer writes a sequence of instructions to shut off the fuel and open the safety valve. Since it is not known where the interrupt will occur in the program, the subroutine to service the interrupt will probably start by pushing all registers. When the boiler is saved, the registers can be popped off the stack and execution returned to the main program. When the 8080A recognizes an interrupt, it disables the interrupt input so that a signal on this input can't keep interrupting the processor over and over. When the interrupt has been serviced, an EI instruction must be used to enable the interrupt input again. The processor can then respond to a new interrupt signal.

An interrupt instruction can be loaded onto the data bus by a simple 8212 circuit. The RST instruction is hard-wired on the inputs of the 8212, and the INTA signal from the 8228 is used to strobe the RST instruction onto the data bus.

If only one interrupt is needed in a system using an 8228, this can be accomplished by tying the INTA output to +12 V with a 1-kΩ resistor. Connected this way, the 8228 automatically gates a RST 7 instruction onto the data bus when the 8080A receives an INT signal.

8085A INTERRUPTS

The designers of the 8085A recognized the need for multiple-interrupt capability. Therefore, the 8085A has five interrupt inputs: INTR, RST 5.5, RST 6.5, RST 7.5, and TRAP.

The INTR input functions very similarly to the INT input of the 8080A. After a reset the INTR input is disabled. An EI instruction must be executed to enable the INTR input. At the end of each instruction cycle, the 8085A checks to see whether interrupts are enabled and whether an INTR interrupt is being requested. If these two conditions are met, then the 8085A disables interrupts, sends out an interrupt acknowledge signal, INTA, and puts the data bus in the input mode. External hardware must put a RST or CALL instruction on the data bus, as described earlier for the 8080A. In response to the RST or CALL instruction, the 8085A pushes the return address on the stack, vectors to the RST or CALL address, and begins execution of the interrupt service subroutine at that address. At the end of the interrupt service subroutine, an EI instruction is used to reenable interrupts so that the next interrupt will be recognized. A RET instruction then pops the stored return address off the stack and returns execution to the main program.

The 8085A interrupt inputs RST 5.5, RST 6.5, RST 7.5, and TRAP eliminate the need for external hardware to insert a RST instruction. If an interrupt signal is asserted on one of these four inputs, and they are enabled and unmasked, the 8085A proceeds as follows. Interrupts are disabled to prevent the same interrupt signal from continuing to interrupt. The return address is pushed onto the stack. The program counter is loaded with a specific RST address for that interrupt input, and execution continues from the loaded address. The addresses for these four are TRAP, 24H; RST 5.5, 2CH; RST 6.5, 34H; and RST 7.5, 3CH. The start of the subroutine to service each interrupt is placed at the corresponding address in memory. In the service routines, interrupts must be reenabled with an EI instruction before any further interrupts except TRAP can be recognized. This is usually done near the end of the routine. A RET instruction at the end of routine pops the return address off the stack and returns execution to the main program.

The five interrupt inputs of the 8085A have a *fixed priority* of service. This means that if two interrupt signals occur at the same time, the one with the highest priority is serviced. Priority of the inputs from highest to lowest is TRAP, RST 7.5, RST 6.5, RST 5.5, and INTR.

MASKING AND ENABLING

Not only does the TRAP interrupt input have the highest priority, but also it cannot be disabled or masked in any way. The 8085A always recognizes an interrupt request on the TRAP. Therefore, this input is usually used to quickly save data in case of a power failure or to respond to a catastrophic situation such as an atomic reactor starting to melt down.

The INTR input is disabled after a reset, after a *disable interrupts* instruction (DI), or after an interrupt request. It is enabled with the EI instruction.

After a reset the RST 7.5, RST 6.5, and RST 5.5 interrupt inputs are both *disabled* and *masked*. They are *enabled* by the EI instruction, and they can be *selectively unmasked* with the *set interrupt mask* instruction (SIM). Figure 9-17 shows a simple logic diagram to help you visualize what is required to permit each of the five 8085A interrupt inputs to cause a valid interrupt to the processor. For example, *masking* allows RST 6.5 to be *selectively disabled* without disabling RST 7.5 and RST 5.5.

Figure 9-18a shows the bit format for the word used with the SIM instruction. If bit 3 of this word is a 1, then bits 0 through 2 will selectively mask or unmask the RST 5.5, RST 6.5, and RST 7.5 inputs, respectively. A 0 in a bit unmasks that interrupt. If bit 3 is a 0, bits 0 through 2 are ignored. For example, to unmask RST 7.5 and RST 5.5 but leave RST 6.5 masked, bit 3 would be a 1, bit 2 a 0, bit 1 a 1, and bit 0 a 0. The upper 4 bits of the SIM word can all be a 0 for now. The SIM word is then 0AH. To unmask interrupts, first this word is loaded into the accumulator with a MVI A,0AH instruction. Then the SIM instruction is executed. Next the EI instruction can be executed to enable the inputs if this was not done previously.

If bit 6 of the SIM word in the accumulator is a 1, then the data present in bit 7 of the word is sent out the

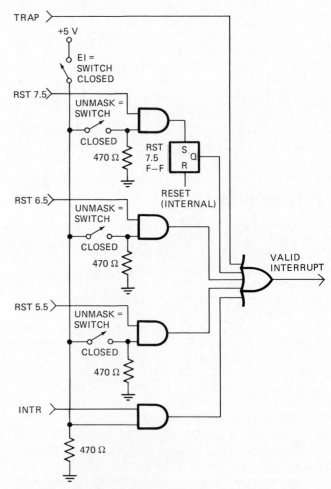

FIGURE 9-17 8085A interrupt enabling and unmasking.

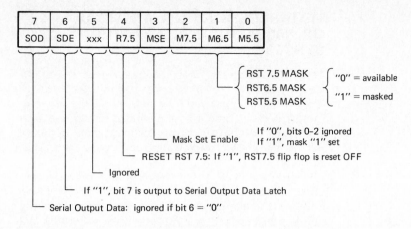

7	6	5	4	3	2	1	0
SOD	SDE	xxx	R7.5	MSE	M7.5	M6.5	M5.5

RST 7.5 MASK
RST6.5 MASK
RST5.5 MASK

"0" = available
"1" = masked

Mask Set Enable — If "0", bits 0–2 ignored / If "1", mask "1" set

RESET RST 7.5: If "1", RST7.5 flip flop is reset OFF

Ignored

If "1", bit 7 is output to Serial Output Data Latch

Serial Output Data: ignored if bit 6 = "0"

(a)

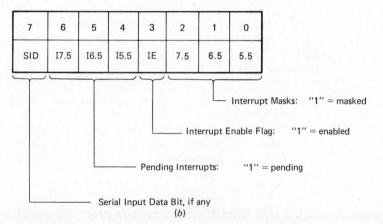

7	6	5	4	3	2	1	0
SID	I7.5	I6.5	I5.5	IE	7.5	6.5	5.5

Interrupt Masks: "1" = masked

Interrupt Enable Flag: "1" = enabled

Pending Interrupts: "1" = pending

Serial Input Data Bit, if any

(b)

8085A SOD (serial output data) line when the SIM instruction is executed. This feature allows the 8085A to easily send serial data to a teletype or printer by using a program routine rather than a hardware UART.

Figure 9-18b shows the 8085A *read-interrupt-mask* (RIM) instruction word format. After the RIM instruction executes, this word is in the accumulator. Bits 0, 1, and 2 indicate whether RST 5.5, RST 6.5, and RST 7.5 are currently masked or unmasked. Bit 3 tells whether 8085A interrupts are enabled or disabled. A 1 in this bit means that interrupts are enabled. Bits 4, 5, and 6 indicate whether an interrupt is being requested on RST 5.5, RST 6.5, or RST 7.5. This allows a program to disable interrupts during a critical process but still be able to check bits 4, 5, and 6 with a RIM instruction to find out whether a particular interrupt is being requested. Bit 7 of the accumulator, after a RIM instruction, contains the data present on the 8085A SID (serial input data) pin. This feature allows the 8085A to read in serial data from a teletype or terminal with a program routine rather than a hardware UART.

8085A TRANSITION STATE DIAGRAM

Figure 9-19 shows the 8085A *transition state diagram.* This is a compact way of showing when during an instruction cycle the 8085A will enter a halt state, insert a wait state, respond to a HOLD input, or respond to an interrupt input.

If $\overline{\text{RESET IN}}$ is asserted, the 8085A stays in a reset state with the address bus floating. When $\overline{\text{RESET IN}}$ is not asserted or the previous machine cycle is finished, the CPU enters T1 of a new machine cycle. If the previous instruction was a HLT instruction, the CPU goes directly to a halt state. The three ways to exit the halt state are by a reset (not shown), a valid interrupt, and a HOLD request. Note that the hold exit is only temporary. As soon as HOLD is not asserted, the CPU returns to the halt state.

If a halt state was not entered, then the CPU proceeds to T2 of the machine cycle. Here it checks the READY input. If the READY line is not asserted (L), the CPU inserts wait states until READY goes high.

The HOLD input is checked at several points. If a hold request is present on it, the hold acknowledge flip-flop is set. However, the CPU will not enter the hold state until the end of bus use for the machine cycle.

Note that the 8085A does not check whether a valid interrupt request is present until the end of the instruction cycle. This is necessary so that the address of the next instruction can be pushed onto the stack.

To summarize: A halt state is entered during T1, a wait state is entered after T2, a hold state is entered after a machine cycle is completed, and an interrupt is responded to after an instruction cycle is completed.

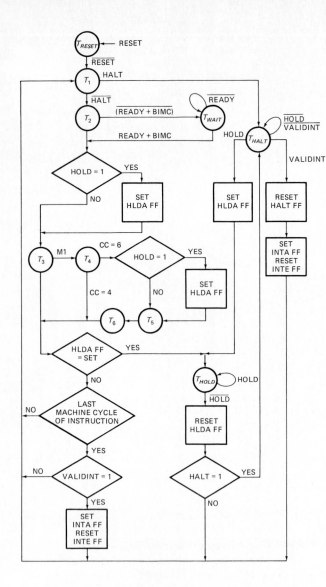

NOTE: SYMBOL DEFINITION

 = CPU STATE T_x. ALL CPU STATE TRANSITIONS OCCUR ON THE FALLING EDGE OF CLK.

$\diamondsuit$ = A DECISION (X) THAT DETERMINES WHICH OF SEVERAL ALTERNATIVE PATHS TO FOLLOW.

$\boxed{X}$ = PERFORM THE ACTION X.

$\longrightarrow$ = FLOWLINE THAT INDICATES THE SEQUENCE OF EVENTS.

$\xrightarrow{X}$ = FLOWLINE THAT INDICATES THE SEQUENCE OF EVENTS IF CONDITION X IS TRUE.

CC = NUMBER OF CLOCK CYCLES IN THE CURRENT MACHINE CYCLE.

BIMC = "BUS IDLE MACHINE CYCLE" = MACHINE CYCLE WHICH DOESN'T USE THE SYSTEM BUS.

VALIDINT = "VALID INTERRUPT" — AN INTERRUPT IS PENDING THAT IS BOTH ENABLED AND UNMASKED (MASKING ONLY APPLIES FOR RST 5.5, 6.5, AND 7.5 INPUTS).

HLDA FF = INTERNAL HOLD ACKNOWLEDGE FLIP—FLOP. NOTE THAT THE 8085A SYSTEM BUSES ARE 3—STATED ONE CLOCK CYCLE AFTER THE HLDA FLIP—FLOP IS SET.

FIGURE 9-19 8085A transition state diagram.

TROUBLESHOOTING AN 8080A OR 8085A MICROCOMPUTER SYSTEM

Chap. 13 discusses the instruments and techniques used to develop and debug microcomputer-based instruments. We show you here how to use the programming knowledge from Chap. 8 and the hardware knowledge from this chapter to troubleshoot a malfunctioning 8080A or 8085A microcomputer system that once worked.

In troubleshooting a microcomputer or microcomputer-based instrument, a sequential approach is more effective in the long run than random poking, probing, and hoping. For either a gross failure or a subtle malfunction, it is best to work your way down from the start of the debug procedure described below. For practice, work through this procedure on the microcomputer you are using.

INITIAL CHECKS AND SIGNAL "ROLL CALL"

1. Identify the symptom. List the symptoms that you find or a customer describes. If someone else describes the symptoms to you, check these yourself or have that person demonstrate the symptoms to make sure the problem is not just an operator error.

2. Do a careful visual and tactile inspection of your microcomputer, including the following suggestions.

   a. Check for signs of excessive heat.

   b. Check that all ICs are firmly seated in their sockets and that the ICs have no bent pins. Sometimes a bent pin will make contact for a while, but after heating, cooling, and vibration it no longer does. Also, inexpensive IC sockets may oxidize with age and fail to make contact.

   c. Check that PCB edge connectors are clean and seated fully. A film that may be present on edge-connector fingers can be removed with a clean pencil eraser.

   d. Check the ribbon-cable connectors. If they have been moved around a lot and do not have stress relief, they may no longer make dependable contact. If moving the cable around alters the operation of the machine, try replacing the cable.

3. Check the power supplies. Power supplies are a very common source of problems. Low voltage or excessive ripple may make you think an IC is bad.

   a. From the operation manual determine the power-supply specifications for your microcomputer. Then measure the power-supply voltages, and check that they are within specifications.

   b. Also check with an oscilloscope to make sure the power supplies do not have excessive noise on them.

4. Verify the clock signals. Using an oscilloscope, check that the clock signals are present and correct.

5. Verify the CPU *input* control signals.

a. If everything thus far seems correct but the unit still has problems, check that the CPU input control signals, such as RESET, READY, HOLD, INT, $\overline{NMI}$, etc., are at the proper level for normal operation. (Refer to the CPU data book or manual for proper levels.) A common problem is that the processor will get stuck in a wait, hold, reset, or interrupt condition because of some other hardware problem.

b. If one of these signals is at the wrong level, find out from the schematic what is connected to that input and track down the problem.

6. Verify the CPU *output* signals.

a. If the input signals are correct, check the active CPU or CPU-group output signals such as SYNC, ALE, DBIN, STATUS STROBE, $\overline{MEMR}$, $\overline{RD}$, etc., to see whether they are present. The absence of these signals probably indicates a bad CPU or CPU-group IC. Try replacing the IC that is the source of the missing signal.

b. Also quickly check whether pulses are present on lower address lines and data lines. With a normal oscilloscope these pulses will appear random, but at least it is a quick way to find out whether the CPU is sending out addresses and whether the data bus is active.

c. Also use an oscilloscope to determine whether the memory decoders are producing chip-select signals ($\overline{CS}$). If a decoder is not producing $\overline{CS}$ signals, the decoder may be bad or may not be getting enabled.

VERIFYING THE STEP-BY-STEP CPU OPERATION DURING INSTRUCTION EXECUTION

If the CPU or one of the CPU-group ICs is not bad, then you should find out whether the correct instructions are coming into the CPU in proper order from memory.

1. A logic analyzer is a good means to check detailed CPU operation.

a. Connect eight of the signal inputs of the analyzer to the data bus, and connect any remaining signal inputs to the address bus.

b. The analyzer can be clocked on $\overline{MEMR}$ for an 8080A or $\overline{RD}$ for an 8085A.

c. The analyzer should be set to trigger on the first instruction word to be read in from memory and the address of that word.

d. The analyzer trace should show the first 256 or so address and data words after the trigger word.

e. If the words read back are incorrect, there are several possible sources for the problem.

(1) An address or data line that should change but never does may be short-circuited to V_{CC} or ground.

(2) An address bus or data bus buffer IC may be bad or may not be getting enabled.

(3) Two address lines or two data lines might be shorted together. Watch for this pattern in the logic-analyzer trace.

2. *Single-stepping* is another way to check detailed CPU operation. If a logic analyzer is not available, then for 8080A, 8085A, and Z80 processors a *hardware single-step* method can be used to let you see this same information.

a. *Discussion:* From a previous section, remember that putting a low on the READY input of the 8224 or the READY input of an 8085A causes the 8080A or 8085A to enter a wait state at the end of the $T2$ state. As long as READY is held low, the processor keeps inserting wait states. During a wait state, the address, data, and control signals for the machine cycle in progress are held on the buses. The processor can be held in a wait condition as long as necessary. The levels on all the bus lines can be checked with an oscilloscope or even a logic probe to see whether they are correct.

When the signals have been checked for one machine cycle, the READY line is sent high again, just long enough for the 8080A or 8085A to finish that machine cycle and start the next machine cycle. Since READY is low again, the processor enters another wait condition at the end of $T2$ for this machine cycle. The bus signals can be checked again for this machine cycle. If they are correct, another short pulse applied to the READY line will step the processor to $T2$ of its next machine cycle.

This technique uses wait states to single-step through and check the operation of a program, machine cycle by machine cycle. It usually does not reveal timing problems. However, it is an excellent way of determining why a program stops at a particular point or doesn't read a port because no $\overline{IOR}$ signal is sent. Refer to Figure 9-2 to see the signals you would expect, for example, on the buses for the three machine cycles of the 8080A OUT instruction.

b. Hardware implementation. Figure 9-20 shows a simple circuit that can be used to single-step an 8080A system such as that in Figure 9-3 or an 8085A system such as that in Figure 9-9. Two 7400 NAND gates debounce a spring-loaded pushbutton switch. When triggered by a signal from the switch, the 74121 monostable produces a 500-ns pulse.

(1) For an 8080A, the 74121 output pulse can be applied to the READY input of the 8224 to single-step the 8080A from a wait state in one machine cycle to a wait state in the next.

(2) For an 8085A system, the 74121 output of

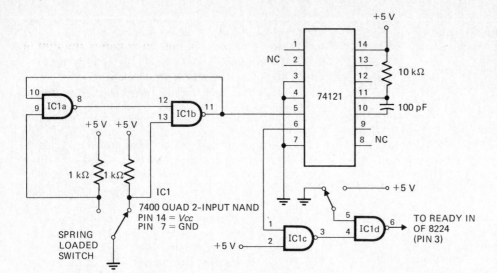

FIGURE 9-20 Circuit for single-stepping an 8080A or 8085A system by inserting wait states.

the single stepper is connected directly to the READY input. The two leftover NAND gates are used as a switch for the single-step pulses. If the control input for the switch is low, READY will be high and the processor will run normally.

c. An important point about the single-stepper approach is that it allows you to trace signals on a dc basis. LEDs with driver transistors can be connected to the address and data lines, so you can see their levels more easily as you step through a program from machine cycle to machine cycle.

TESTING ADDRESS AND DATA BUS BUFFERS

In microcomputer systems that use address bus and data bus buffers, these buffers are common cause of failure. Two common methods of finding a problem with a buffer follow.

1. Using a logic analyzer, or the hardware single-step method and an oscilloscope or a voltmeter, you can check the inputs and outputs of each buffer for proper levels.

   a. If a signal is getting to the input pin of a buffer but not appearing on the output pin:

   (1) First check that the buffer is getting enabled. If not, find the problem with the enable.

   (2) If the buffer is enabled, then it may be defective; so try replacing it and recheck the output.

   (3) If the output level of the suspected buffer is still incorrect after it is replaced probably some other device on that bus line has an internally shorted input or output. To find the shorted device, use the schematic to locate all the devices which have inputs or outputs connected to the malfunctioning line. Remove these ICs one at a time, or lift pins until the line goes to the correct logic level. The last one lifted, then, is defective and should be replaced.

(4) Unfortunately, it is not uncommon for more than one input or output on a bus line to be shorted. So after you replace the malfunctioning IC, reconnect the other ICs one at a time to make sure none of them is also defective.

b. If the input signal to a buffer is not correct:

(1) Remove the buffer. If the signal input line now has the correct value, the buffer input probably was shorted and so the buffer should be replaced.

(2) If the input signal line is still incorrect after the buffer is removed, probably some other device on that line is shorted. Remove other devices (or pins) on this line, one at a time, as described in (3) above until the signal level is correct. Replace the defective IC. Also use a procedure similar to that described in (4) above to reconnect the other ICs and verify their proper operation.

(3) If removing all devices on the line does not correct the problem, the signal source may be at fault.

c. Remember that the cost of simply replacing a few inexpensive buffer or decoder ICs may be much less than the cost of labor involved in unsoldering and lifting pins to find a defective one.

2. If you have a *pulser probe* such as the Hewlett-Packard 546A and a *current tracing probe* such as the Hewlett-Packard 547A (see Figure 1-5), a shorted IC usually can be found quite quickly. The 546A is used to pulse current into the shorted line. Then the current probe is moved along the short-circuited line, as shown in Figure 1-5. When the probe passes the IC with a shorted input, the light in the probe tip goes out.

TESTING RAM

If the microcomputer reads in instructions and data properly from ROM but still does not perform all its functions correctly, the remaining sections to test are RAM and I/O ports. To test RAM, use either step 1 or step 2 below depending on the capabilities of your specific microcomputer.

1. *Using the monitor program.* If the system has a monitor program, then this can be used to write to RAM and read back to see whether the write was successful. Use a routine that tests a block of RAM by first writing all 1's (FF) to each address and reading them back, then writing and reading all 0's and finally writing and reading alternating 1's and 0's (55 or AA).

2. *Using a test PROM.* If the system does not have a monitor program, then the system ROM can be replaced with a test PROM containing a routine, similar to that described in item 1 above, to test all the RAM in the system and give an error indication if the test fails.

TESTING I/O PORTS

If RAM works correctly, then program routines can be used to exercise I/O ports and help you track down a problem with them.

1. To check an input port, for example, a routine is written to initialize and read over and over from that port.

2. A standard oscilloscope can then be used to check whether the port is getting addressed (enabled) and whether the data on the inputs of the port device is getting onto the system data bus when the device is enabled.

THE ZILOG Z80

While Intel designers were working on the 8085A as an improvement of the 8080A, some of the designers who had worked on the 8080 at Intel moved to Zilog, Inc., and developed the Z80. The Z80/Z80A is considered an enhancement of the 8080A because it uses a similar architecture and because machine code programs made for an 8080A will run on a Z80. The Z80 is the CPU used in the Radio-Shack TRS-80 home computers.

Z80/Z80A BLOCK DIAGRAM, REGISTERS, AND PIN DEFINITIONS

The complete data sheet and product specifications for the Zilog Z80/Z80A are found in App. C. These are used for the following discussion of the Z80. Incidentally, the Z80A is the same as the Z80 except that it operates with up to a 4-MHz input clock whereas the Z80 operates only up to 2.5 MHz.

First, on page 1 of the data sheet, observe that the Z80 requires only a single 5-V supply and a single-phase TTL

level clock. The CPU block diagram shows the same functions as for the 8080A. The Z80A has an 8-bit data bus and a 16-bit address bus. In registers the Z80 is much richer. The A, PSW, B, C, D, E, H, and L registers of the 8080A have all been duplicated. In addition to the 16-bit program counter and stack pointer, the Z80A has two 16-bit index registers, IX and IY. An 8-bit interrupt vector register and an 8-bit memory refresh register are included, too.

Page 2 of the data sheet shows the Z80 pin-out descriptions. A 16-bit three-state address bus and an 8-bit three-state data bus are shown, as with the 8080A. The control bus signals are internally decoded so that an external device such as the 8228 is not needed. $\overline{M1}$, $\overline{MREQ}$, $\overline{IORQ}$, $\overline{RD}$, and $\overline{WR}$ functions are quite clear from their descriptions, but $\overline{RFSH}$ may require some explanation. Because dynamic MOS memories are cheaper than static MOS memories, the designers included in the Z80 circuitry to refresh dynamic memories. This removes the need for an external refresh controller. Remember from Chap. 5 that dynamic RAMs are refreshed by addressing each row once every 2 ms or so. While a fetched instruction is being decoded, the contents of the refresh register are sent out on the address bus. A $\overline{RFSH}$ signal and an $\overline{MREQ}$ signal are used to gate the lower 7 bits of this address to all the dynamic memories in a system. This refreshes the addressed rows. The register contents are then incremented to refresh the next row in the next cycle. All this is transparent to the user. In other words, the refresh is accomplished while the processor is not using the address and data buses for other purposes.

$\overline{HALT}$ indicates the CPU is in a halt state. The $\overline{WAIT}$ input serves the same function as the READY input of the 8080A. As long as $\overline{WAIT}$ is low, wait states are inserted. The $\overline{INT}$ interrupt input functions the same as the INT input on the 8080A. The $\overline{NMI}$ nonmaskable interrupt input functions similar to the TRAP input of the 8085A. It is nonmaskable, and a low on this input vectors the program execution to address 0066H.

$\overline{BUSRQ}$ corresponds to the 8080A HOLD input. A low on this input floats the address, data, and control buses at the end of the current machine cycle. The buses can then be used by another processor or DMA controller. $\overline{BUSAK}$ (bus acknowledge) indicates the buses are floating.

Z80A INSTRUCTION SET

The Z80A has 158 instructions. The 78 instructions of the 8080A are included, but the Z80A uses different mnemonics for them (mnemonics are usually copyrighted). Therefore, the 8080A instructions are not immediately recognizable. Many of the new instructions require 2 bytes in memory for their op codes. For example, the CPIR instruction has the 2-byte op code EDB1 hex.

In addition to the 8080A instructions, the Z80A has single instructions that are equivalent to many 8080A instructions. For example, a single LDIR instruction moves a block of data bytes from one location in memory to another. The data sheet in App. C shows the entire Z80A instruction set in symbolic form. This is a good opportunity for you to learn to read the symbolic form of instructions often used on data sheets. For a complete description of all the details of the Z80A instructions, consult the *Z80 Assembly Language Programming Manual* from Zilog. The top of page 4 in the data sheet gives a key for the symbols used. Two that may need some explanation are "( )" and "e." The parentheses around a register pair name refer to the contents of the memory location or I/O port pointed to by the enclosed register. For example, (HL) is read as "the contents of the memory location pointed to by the 16-bit number in the H and L registers." And "e" represents a displacement of between $+128$ and -128 which can be added to the program counter or to one of the index registers. The value of "e" is put in as the second or third byte of an instruction. It is used for program relative jump instructions and indexed addressing. These are explained later.

The first instruction on page 4 of the data sheet is LD r,s. It actually represents many MOV type of instructions. It says to load the contents of some source register or memory location, s, into the indicated register, r. The source may be another register, an immediate number, or a memory location pointed to by HL. The source may also be a memory location pointed to by the sum of the 16-bit index register IX and a displacement contained in the third byte of the instruction. This is an example of *indexed addressing.* If the IX register contains 25AFH, then the instruction LD C,(IX + 18H) adds 25AFH to 18H and copies the contents of memory location 25C7H into the C register. The IY index register can also be used for this indexed addressing.

There is not space here to explain all the Z80A instructions, but we will discuss a few more to show the sophistication of this processor. The block move instruction LDIR is a good example. This instruction moves a block of data from one location in memory to another. Before this instruction, the starting address of the source is loaded into the HL register, the starting address of the destination is loaded into DE register, and the number of bytes to be moved is loaded into BC. When the LDIR instruction is executed, the contents of the memory location pointed to by HL are moved to the memory location pointed to by DE. DE and HL are incremented to point to the next source location and destination location. The byte counter is decremented by 1. The one LDIR instruction continues the operation until all the bytes have been moved and BC = 0.

The CPIR instruction searches a block of memory to find a byte equal to a byte contained in the accumulator. The starting address is loaded into HL, and the number of memory locations to be searched is loaded into BC. The CPIR instruction, when executed, continues the search until a match is found or BC = 0.

A significant feature of the Z80A instruction set is that any register or memory location can be rotated right or left, not just the accumulator. RLD and RRD even rotate nibbles.

Bit set and reset instructions allow individual bits to be set or reset without changing the rest of the bits in a register or memory byte.

IN and OUT instructions of the Z80A have more varieties than those of the 8080A. An IN instruction can bring a byte to a register or memory location as well as to the accumulator. INIR will input a block of data to successive memory locations pointed to by HL.

The jump instructions are expanded also. The Z80A has all the flags that the 8080A has and, in addition, the parity flag is used to indicate overflow during arithmetic operations. An overflow flag is produced if the result in the accumulator is greater than $+127$ or less than -128. Here JP nn and JP cc,nn represent all the 8080A unconditional and conditional jump instructions. These are 3-byte instructions that contain the jump address in bytes 2 and 3. Since most program jumps are to a location within ±128 bytes, a 2-byte program relative jump would save 1 byte of memory. The JR e instruction unconditionally jumps program execution a displacement "e" from the program counter address after the instruction and displacement. For example, assume the JR instruction is at address 0250H and there is a displacement of 05H in 0251H. The CPU fetches the JR instruction, increments the program counter, and fetches the displacement. It then automatically increments the program counter to 0252H, and the program jumps to 0257H and continues. Some assembler programs allow you to enter an "e" based on the address of the JR instruction.

Another interesting instruction useful for timing loops is DJNZ e. The B register is decremented; if it is not 0, program execution is jumped a displacement "e." This concludes an introduction to the Z80A instruction set. Next consider briefly the Z80A timing.

Z80A TIMING AND SYSTEM CONNECTIONS

As with the 8080A and 8085A, each Z80A instruction cycle consists of several machine cycles. Each machine cycle consists of three to six states, or clock cycles. For a Z80A with a 4-MHz input clock, the time for each state is $0.25\ \mu$s.

The data sheet in App. C shows the detailed ac characteristics and timing diagrams for the Z80 and Z80A.

A Z80A is easily built into a microprocessor system with standard RAM and ROM parts. Parallel I/O ports can be added with the Z80-PIO, which contains two 8-bit programmable ports. Serial I/O ports can be added with devices such as the Z80A-SIO programmable USART.

ONE-CHIP MICROCOMPUTERS

A strong trend in microprocessor evolution has been to put the CPU and enough RAM, ROM, and I/O ports for a small system in a single IC package. A good example is the Intel 8048 family.

INTEL 8048 FAMILY

ARCHITECTURE Figure 9-21 is a block diagram of the internal structure of the 8048/8049. The 8048 and 8049 are identical except that the 8048 has 1K of ROM and 64 bytes of RAM whereas the 8049 has 2K of ROM and 128 bytes of RAM. The ROM program memory is mapped into addresses 0000 to 3FFH, so that a reset can vector the CPU to the start of a program in ROM.

Portions of the data memory RAM are used as scratchpad registers. There are two 8-byte "banks" of registers, register bank 0 (RB0) and an optional second register bank (RB1). An 8-byte stack space is also present in the data RAM.

The 8048 has 27 lines that can be used for input or output. There are three 8-bit ports and three test inputs. Each pin of ports 1 and 2 can be programmed to be an input or output. BUS is a bidirectional port which is an extension of the internal bus. Note the single internal bus that transfers addresses, data, and instructions. The TEST 0, TEST 1, and $\overline{\text{INT}}$ inputs can be tested by conditional jump instructions.

Internal clock logic requires only an external crystal to produce the 8048 clock. An 8-bit programmable timer also is included on the chip.

INSTRUCTION SET The 8048 family instruction set is not in any way compatible with that of an 8080A. Different mnemonics and op codes are used. The 90 instructions for the 8048 are shown in Table 9-1. Read through these instructions and compare them with the 8080A and Z80A instructions.

The SWAP A instruction swaps the nibbles of the accumulator. A wide variety of jump and branch instructions are present. Note the DJNZ R, addr instruction which means decrement the register and, if it is not 0, jump to the indicated address. This is an example of two instructions combined in one. Another example is the return and restore status (RETR), which does the work of the POP, PSW, and RET instructions in the 8080A.

Also note the timer instructions. Used as a counter, a MOV T,A instruction can set the counter to 0. The STRT CNT instruction connects the timer input to the TEST 1 input. The counter then increments for each pulse that enters TEST 1. The count can be read at any time with a

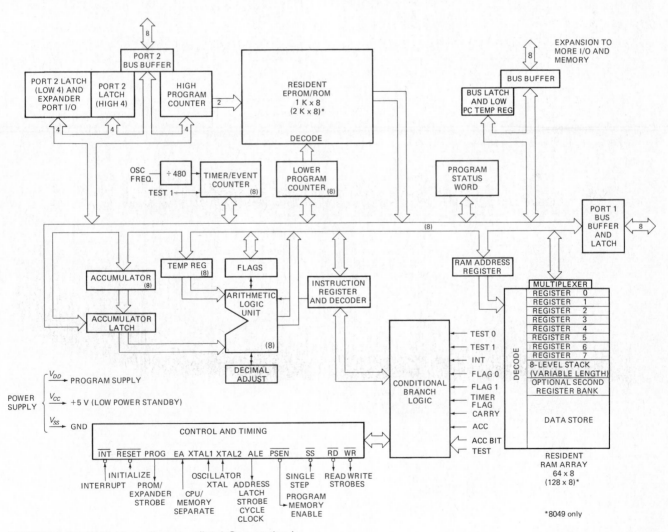

FIGURE 9-21 8048 block diagram. *(Intel Corporation.)*

251

TABLE 9-1

MNEMONIC	DESCRIPTION	BYTES	CYCLES
	Accumulator		
ADD A, R	Add register to A	1	1
ADD A, @R	Add data memory to A	1	1
ADD A, # data	Add immediate to A	2	2
ADDC A, R	Add register with carry	1	1
ADDC A, @R	Add data memory with carry	1	1
ADDC A, # data	Add immediate with carry	2	2
ANL A, R	AND register to A	1	1
ANL A, @R	AND data memory to A	1	1
ANL A, # data	AND immediate to A	2	2
ORL A, R	OR register to A	1	1
ORL A @R	OR data memory to A	1	1
ORL A, # data	OR immediate to A	2	2
XRL A, R	XOR register to A	1	1
XRL A, @R	XOR data memory to A	1	1
XRL, A, # data	XOR immediate to A	2	2
INC A	Increment A	1	1
DEC A	Decrement A	1	1
CLR A	Clear A	1	1
CPL A	Complement A	1	1
DA A	Decimal adjust A	1	1
SWAP A	Swap nibbles of A	1	1
RL A	Rotate A left	1	1
RLC A	Rotate A left through carry	1	1
RR A	Rotate A right	1	1
RRC A	Rotate A right through carry	1	1
	Input/Output		
IN A, P	Input port to A	1	2
OUTL P, A	Output A to port	1	2
ANL P, # data	AND immediate to port	2	2
ORL P, # data	OR immediate to port	2	2
INS A, BUS	Input BUS to A	1	2
OUTL BUS, A	Output A to BUS	1	2
ANL BUS, # data	AND immediate to BUS	2	2
ORL BUS, # data	OR immediate to BUS	2	2
MOVD A, P	Input expander port to A	1	2
MOVD P, A	Output A to expander port	1	2
ANLD P, A	AND A to expander port	1	2
ORLD P, A	OR A to expander port	1	2
	Registers		
INC R	Increment register	1	1
INC @R	Increment data memory	1	1
DEC R	Decrement register	1	1
	Branch		
JMP addr	Jump unconditional	2	2
JMPP @A	Jump indirect	1	2
DJNZ R, addr	Decrement register and skip	2	2
JC addr	Jump on carry = 1	2	2
JNC addr	Jump on carry = 0	2	2
JZ addr	Jump on A zero	2	2
JNZ addr	Jump on A not zero	2	2
JTO addr	Jump on TO = 1	2	2
JNTO addr	Jump on TO = 0	2	2
JT1 addr	Jump on T1 = 1	2	2
JNT1 addr	Jump on T1 = 0	2	2
JF0 addr	Jump on F0 = 1	2	2
JF1 addr	Jump on F1 = 1	2	2
JTF addr	Jump on timer flag	2	2
JN1 addr	Jump on INT = 0	2	2
JBb addr	Jump on accumulator bit	2	2

TABLE 9-1 (cont.)

MNEMONIC	DESCRIPTION	BYTES	CYCLES
Subroutine			
CALL addr	Jump to subroutine	2	2
RETR	Return	1	2
RETR	Return and restore status	1	2
Flags			
CLR C	Clear carry	1	1
CPL C	Complement carry	1	1
CLR F0	Clear flag 0	1	1
CPL F0	Complement flag 0	1	1
CLR F1	Clear flag 1	1	1
CPL F1	Complement flag 1	1	1
Data Moves			
MOV A, R	Move register to A	1	1
MOV A, @R	Move data memory to A	1	1
MOV A, # data	Move immediate to A	2	2
MOV R, A	Move A to register	1	1
MOV @R, A	Move A to data memory	1	1
MOV R, # data	Move immediate to register	2	2
MOV @R, #data	Move immediate to data memory	2	2
MOV A, PSW	Move PSW to A	1	1
MOV PSW, A	Move A to PSW	1	1
XCH A, R	Exchange A and register	1	1
XCH A, @R	Exchange A and data memory	1	1
XCHD A, @R	Exchange nibble of A and register	1	
MOVX A, @R	Move external data memory to A	1	2
MOVX @R, A	Move A to external data memory	1	2
MOVP A, @A	Move to A from current page	1	2
MOVP3 A, @	Move to A from page 3	1	2
Timer-Counter			
MOV A, T	Read timer-counter	1	1
MOV T, A	Load timer-counter	1	1
STRT T	Start timer	1	1
STRT CNT	Start counter	1	1
STOP TCNT	Stop timer-counter	1	1
EN TCNT1	Enable timer-counter interrupt	1	1
DIS TCNT1	Disable timer-counter interrupt	1	1
Control			
EN 1	Enable external interrupt	1	1
DIS 1	Disable external interrupt	1	1
SEL RB0	Select register bank 0	1	1
SEL RB1	Select register bank 1	1	1
SEL MB0	Select memory bank 0	1	1
SEL MB1	Select memory bank 1	1	1
ENT 0 CLK	Enable clock output on T0	1	1
NOP	No operation	1	1

MOV A,T instruction. As a timer the STRT T instruction connects the timer to a divided-down version of the system clock. For a 6-MHz crystal, the timer increments every 80 μs.

A single instruction SEL RB0 or SEL RB1 can switch the working register bank from RB0 to RB1. The 8048 can address up to 4K of program memory. This is divided into two banks of 2K each. The SEL MB0 and SEL MB1 instructions select the desired memory bank.

8048 FAMILY SYSTEMS For very small systems an 8048 or 8049 may contain all the required functions. Several other versions of the 8048 are available with different options. The 8748, for example, has 1K of EPROM instead of the mask programmed ROM of the 8048. The 8035 has RAM but no internal program memory. The 8021 is an inexpensive, simplified version in a 28-pin package which performs a subset of the 8048 instructions. The 8022 version contains an 8-bit A/D converter.

The 8031/8051/8751 is an improved subfamily of the 8048 family. The 8051 contains 4K bytes of ROM, 128 bytes of RAM, four 8-bit ports, two 16-bit timer-event counters, and a complete UART. The instruction set has been increased to include multiply, divide, subtract, and compare instructions.

The 8048 can be expanded to address 4K of external program memory. Eight lower bits of an address are sent out on BUS and latched into an 8212 by the ALE signal. Four upper address bits are sent out in the lower half of port 2. To use external program memory, the PSEN output from the processor is tied to the CS input of the memory. 8048 family devices interface easily with 8085A peripherals such as the 8155 and 8755.

REVIEW QUESTIONS AND PROBLEMS

1. Describe the functions of the 8224 and 8228 in the 8080A CPU group.

2. Why should the main program ROM in a system start at address 0000H?

3. Define (a) instruction cycle, (b) machine cycle, and (c) state.

4. Describe the sequence of events during an instruction fetch (M1) machine cycle for the 8080A. Use Figure 9-2 for reference.

5. Why are address buffers usually required in a microcomputer system?

6. The 8251A is a programmable USART. What is the function of a USART in a microprocessor system?

7. What is the purpose of the address decoder in a microcomputer system?

8. Describe the sequence of events during an instruction fetch (M1) machine cycle for an 8085A.

9. a. What is the purpose of the 8085A ALE signal? b. Which edge of this signal is the active edge?

10. In what state is the 8085A IO/M̄ output during an instruction fetch?

11. Briefly describe the functions found in an Intel 8755A and an Intel 8155.

12. Show the assembly language instructions required to initialize an 8755 at base address 08H as follows: ports A7 to A4 as outputs, ports A3 to A0 as inputs, port B as outputs.

13. List the assembly language instructions required to initialize an 8155 with base address 40H as follows: timer loaded with count of 309AH; timer mode set for continuous square wave; port A as output, port B as input, port C as output, and start timer.

14. A 3205 decoder has its three address inputs connected to A12, A13, and A14 of the system address bus. It has Ē1 connected to A15, Ē2 connected to MEMR, and E3 connected to +5 V. What eight ROM address blocks will the decoder outputs select? Why is MEMR used as one of the enables on a ROM decoder?

15. Show a memory map for the ROM in Prob. 14.

16. Draw a circuit to show how another 3205 can be connected to select one of eight 1K byte RAMs starting at address 8000H.

17. Define memory-mapped I/O and direct I/O.

18. In the system of Figure 9-3, how does the 8080A select an I/O port at address F4H?

19. Explain how the 8205 in Figure 9-9 selects the A15 8755 when port 08H is addressed.

20. Why do the 8080A and 8085A have minimum operating frequencies of 500 kHz?

21. By referring to the 8085A timing diagram and parameters in Figure 9-15, determine:
a. Time between ALE going low and RD going low
b. Time for which memory must hold data stable after RD goes high
c. Minimum time the clock must be low
d. Time that the lower 8 address bits remain on the data bus after ALE goes low

22. The 2732 is a standard 4K × 8 EPROM with an address access time of 450 ns and a chip-select access time of 450 ns. Will this device work correctly in a 5-MHz 8085A system that uses an 8205 decoder with a T_{PD} of 18 ns maximum? Will it work correctly in a 3-MHz 8085A system?

23. a. How is an 8080A or an 8085A entered into a wait state?
b. At what point in a machine cycle does the 8080A or 8085A enter a wait state?
c. What information is on the buses during a wait state?
d. How long is a wait state?
e. How many wait states can be inserted?

24. Describe conditions present on the 8080A and 8085A buses during a hold state. Name one use for the hold state.

25. What operations is the 8080A and 8085A performing during a halt state? List three ways the 8080A or 8085A can be exited from a halt state.

26. a. For what are interrupts used?
b. Describe how a RST instruction is used to service an interrupt.
c. Why must an interrupt subroutine PUSH and POP all registers?

27. Describe the major operations that an 8085A performs after it receives a RST 6.5 interrupt signal. Assume interrupts are enabled and unmasked.

28. List the five 8085A interrupts in order of decreasing priority. For what might the TRAP interrupt input be used?

29. Show the assembly language instructions required to enable 8085A interrupts and unmask only RST 6.5.

30. At what time during an instruction cycle does an 8085A peform the following? (Refer to the transition state diagram in Figure 9-19.)
   a. Enter a halt state
   b. Enter a wait state
   c. Enter a hold state
   d. Respond to a valid interrupt request

31. Describe a logical sequence of steps you would take to troubleshoot a microcomputer system that once worked.

32. Explain how wait states are used to single-step an 8080A or 8085A microcomputer system for troubleshooting.

33. Describe how the Z80A microprocessor structure and instruction set represent improvements over the 8080A.

34. The LD HL,0082H instruction of the Z80A corresponds to what 8080A instruction?

35. Explain the function of the following Z80A instructions: JR e; LDDR; LD B,IX+e; EXX.

36. In what applications would a single-chip microcomputer such as the 8048 most likely be found?

REFERENCES/BIBLIOGRAPHY

Intel MCS-80/85 User's Manual, Intel Corporation, Santa Clara, CA, October 1979.

Intel SDK-85 System Design Kit User's Manual, Intel Corporation, Santa Clara, CA, 1978.

Intel 8080A/8085A Assembly Language Programming Manual, Intel Corporation, Santa Clara, CA, 1977–1978.

Intel Components Data Catalog, Intel Corporation, Santa Clara, CA, 1982.

Intel MCS-48 Family of Single Chip Microcomputers User's Manual, Intel Corporation, Santa Clara, CA, 1978.

Osborne, Adam: *An Introduction to Microcomputers*, vol. 2: *Some Real Products*, Adam Osborne and Associates, Inc., Berkeley, CA, 1976.

Z80/Z80A CPU Technical Manual, Zilog, Inc., Cupertino, CA, 1977.

Z80-Assembly Language Programming Manual, Zilog, Inc., Cupertino, CA, 1977.

CHAPTER 10

THE MC6800 MICROPROCESSOR AND MICROPROCESSOR EVOLUTION

Before we discuss another group of processors, a few historical notes are in order. In December 1971, Intel introduced the first 8-bit parallel processor, the 8008. This first-generation processor was made with PMOS and required many external components to make a functioning system. In December 1973, Intel released a second-generation microprocessor, the 8080, made with NMOS and requiring only two additional chips for a CPU group. At about the same time Motorola introduced the MC6800, also a second-generation processor. The MC6800 uses a different architecture and instruction set, so it is not in any way compatible with the 8080. For several years the 8080A and the MC6800 were, respectively, the number one and two best-selling processors. Each produced a series of evolutionary offspring.

In Chap. 9 we showed the evolution along the 8080A branch. It started with the second-generation 8080A, continued with the 8085A and Z80A, and finished with the 8048.

In this chapter we explain the MC6800 thoroughly as a base and then discuss the 6502. The MC6801 and MC6809 are included. Third-generation 16-bit processors are briefly mentioned.

At the conclusion of this chapter, you should be able to:

1. Describe the architecture of the MC6800 microprocessor.

2. List and describe the operation of the seven addressing modes of the 6800.

3. Write simple assembly language programs for a 6800 microcomputer.

4. Calculate the execution time for 6800 programs or routines.

5. Determine the 6800 clock and control bus timing parameters from timing waveforms and tables.

6. Describe the reset process for a 6800.

7. Explain the response of a 6800 to a signal on its NMI input or IRQ input.

8. Determine the address decoding for a 6800-based microcomputer system.

9. Initialize the MC6821 PIA as a simple input or output port.

10. List a logical sequence of steps for troubleshooting a malfunctioning 6800 microcomputer system.

11. Describe the architecture of the MCS6502 microprocessor and the operation of its indirect addressing modes.

MC6800 MICROPROCESSOR

MC6800 CPU GROUP

Figure 10-1 is a block diagram of the MC6800 CPU. The MC6800 has an *8-bit three-state data bus* and a *16-bit three-state address bus* with each line buffered to drive one standard TTL input. It also has a *16-bit program counter register, 16-bit stack pointer register,* and *16-bit index register*. The index register is used in the same manner as the IX or IY index registers of the Z80A.

Although it has *two accumulators*, the MC6800 does not have the rich array of scratchpad registers that the 8080A and Z80A do. This is one of the major differences

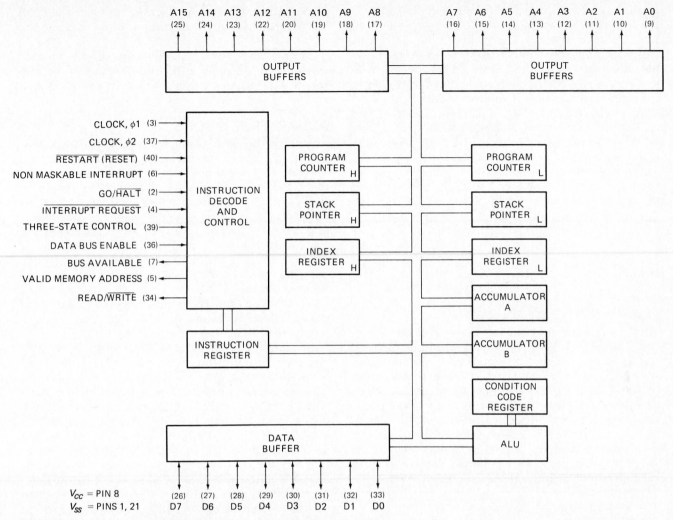

FIGURE 10-1 Motorola MC6800 microprocessor block diagram.

between the two groups. The 8080A group processors are register- and I/O-oriented. The 6800 group processors are *memory-oriented.* They use RAM locations as a scratchpad, and all I/O ports are *memory-mapped.* As you will see, the MC6800 has no instruction with the same meaning as the 8080A IN or OUT direct I/O instruction. The *condition code register* attached to the ALU contains the flags for the MC6800. These are discussed in the section on the 6800 instruction set. The condition code register corresponds to the flag register of the 8080A.

The MC6800 operates on a single +5-V supply. Since it requires a two-phase clock, an external clock generator such as the MC6875 two-phase clock generator driver is usually used. The MC6875 uses an external crystal of 4 times the $\phi 1$ frequency to produce the two clocks. In addition, the 6875 contains circuitry to synchronize other signals such as power-on RESET, with the clocks so that the signals can be input to the 6800.

Because the MC6800 does not have direct I/O, its control signals are simpler. Therefore, an equivalent to the 8228 of the 8080A CPU group is not needed.

MC6800 SIGNAL DESCRIPTIONS

A high on the *read-write* ($R/\overline{W}$) output indicates the CPU is doing a read of a memory location or an I/O port. A low on this output indicates a write. The *valid memory address* (VMA) output, as the name implies, indicates that the address on the bus is valid. *Data bus enable* (DBE) is normally held high. When it is brought low, the data bus and $R/\overline{W}$ output are floated. The *three-state control* (TSC) is normally held low. When it is made high, the address bus is floated. Making DBE low and TSC high is one way to float the buses for a DMA operation. When not used for DMA operations, DBE is often tied to the $\phi 2$ clock pulse. This enables the data bus only when a read or write is being done.

Another way to float the buses for a DMA operation is with a low input on the $\overline{HALT}$ input. The 6800 halt state is equivalent to the 8080A hold state. When the $\overline{HALT}$ input is brought low, the CPU finishes its current instruction and floats the buses and $R/\overline{W}$ output. A *bus available* (BA) signal is sent out on the BA pin to indicate the buses are floating.

The 6800 has four *direct interrupt* methods (three in

hardware and one in software): *reset, nonmaskable interrupt, interrupt request,* and *software interrupt.* When the RESET input is made low, the CPU automatically jumps the program counter to FFFE. The CPU reads the high-order byte of a stored address from this location. It then increments the program counter to FFFF and reads the low-order byte of the stored address. The CPU loads this address into the program counter and fetches its next instruction from this address. In other words, addresses FFFE and FFFF are always used to store the beginning address for the program desired after a reset. Since a 16-bit address is stored, the startup program can be anywhere in memory. Unlike the 8080A, which resets to 0000, the 6800 allows 0000 and following addresses to be used for RAM. This has some programming advantages, which we explain later.

The nonmaskable interrupt input, the maskable interrupt request input, and the software interrupt instruction are also discussed later.

MC6800 INSTRUCTION SET

Table 10-1 shows the complete instruction set for the 6800. In Table 10-1 find the legend explaining the symbols used in the table. OP stands for *operation code,* which is given in hexadecimal for each instruction. The column under the "~" contains the number of clock cycles for each instruction. The column under the # shows the number of memory bytes required by each instruction. Table 10-1 also shows the flags affected by each 6800 instruction, the addressing modes available for each instruction, and a symbolic representation of the function of each instruction.

FLAGS

The six condition flags for the 6800 are *half carry* (H), *interrupt mask* (I), *negative* (N), *zero* (Z), *overflow* (V), and *carry* (C).

The half-carry *flag,* H, is set if a carry is produced from bit 3. It corresponds to the 8080A auxiliary carry flag and is used during BCD arithmetic operations.

The I flag, indicates that the interrupts are disabled when it is 1 and enabled when it is 0.

N is the *sign* or *negative, bit flag.* A 1 in this bit position indicates the result of an operation is negative. This flag copies bit 7 of the result.

The *zero flag,* Z, is set if the result of an operation leaves all 0's in an accumulator or memory location.

The *overflow flag,* V, indicates when the result of a signed addition or subtraction exceeds the legal range of -128 to $+127$. For example, if bit 7 is used as a sign bit and 96_{10}, or 01100000_2, is added to 48_{10}, or 00110000_2, then the result is 10010000_2, or 142_{10}, which is greater than the permitted maximum of 127. The 1 in bit 7 incorrectly indicates the result is negative. This is called an *overflow error,* and the 6800 sets the V flag to indicate that such an error has been made.

The *carry flag* is set when a 1 is carried out from bit 7. Register or memory contents can also be rotated

through the carry flag flip-flop. In addition, the 6800 has instructions to clear carry (CLC) and set carry (SEC).

The TAP instruction loads the entire command code register with a byte from the *A* accumulator, and TPA copies the command register byte into accumulator *A.*

ADDRESSING MODES

An *addressing mode* is simply a way of referring to how or from where an instruction gets the operand to act on. The 6800 has seven addressing modes: immediate, direct, indexed, extended, implied, accumulator, and relative. As these modes are discussed, you will see that you have met most of them in the discussions of the 8080A and Z80A. Because of the way in which the 8080A and Z80A instruction sets are organized, it was not necessary to emphasize addressing modes much in their discussions. However, if you look at the first 6800 instruction, ADDA, in Table 10-1, you see that this one instruction has four possible addressing modes. For clarity, all seven modes are discussed before we approach the instruction set as a whole.

Immediate addressing indicates that the operand is contained in the second or in the second and third bytes of the instruction. For example, ADDA #65 means to add the immediate number 65_{10} to the contents of accumulator *A.* The # sign tells the assembler that the addressing mode is immediate. For an MC6800 assembler, a $ after the # indicates the number is hexadecimal, an @ specifies an octal number, a % specifies a binary number, and an apostrophe specifies an ASCII character. For example, ADDB #$3A adds the immediate number 3AH to the *B* accumulator. The ADDA # instruction corresponds to the 8080A ADI instruction.

Direct addressing indicates that the 8-bit *address* of the operand is contained in the second byte of the instruction. For this mode the operand must be stored in the lowest 256 memory locations from 0000 to 00FF. Direct addressing is sometimes called *base page* addressing because the operand must be in the first 256-byte page of memory. The 6800 uses address locations FFF8 to FFFF to store restart and interrupt addresses to keep the lowest 256 addresses free for direct addressing. As an example of this mode, the ADDA $09 instruction adds the contents of memory location 09H to the contents of the *A* accumulator. The result is left in the *A* accumulator. An advantage of this mode is that only 2 memory bytes are needed for the instruction. IN and OUT are the only 8080A instructions that use essentially the direct addressing mode.

Indexed addressing mode gets the operand from an *effective* address. The address is computed by adding the offset number in the second byte of the instruction to the contents of the 16-bit index register. An X in the operand field signifies that the indexed mode is being used. If, for example, the index register contains a *base address* of 4321H, the ADDA $30,X instruction, when it executes, first computes the *effective address* where it will get its operand. It does this by adding the indicated

TABLE 10-1
MC6800 INSTRUCTION SET (MOTOROLA). ACCUMULATOR AND MEMORY INSTRUCTIONS

OPERATION	MNEMONIC	Immed OP	Immed ~	Immed #	Direct OP	Direct ~	Direct #	Index OP	Index ~	Index #	Extnd OP	Extnd ~	Extnd #	Implied OP	Implied ~	Implied #	BOOLEAN/ARITHMETIC OPERATION (All register labels refer to contents)	5 H	4 I	3 N	2 Z	1 V	0 C
Add	ADDA	8B	2	2	9B	3	2	AB	5	2	BB	4	3				$A + M \rightarrow A$	↕	·	↕	↕	↕	↕
	ADDB	CB	2	2	DB	3	2	EB	5	2	FB	4	3				$B + M \rightarrow B$	↕	·	↕	↕	↕	↕
Add accumulators	ABA													1B	2	1	$A + B \rightarrow A$	↕	·	↕	↕	↕	↕
Add with carry	ADCA	89	2	2	99	3	2	A9	5	2	B9	4	3				$A + M + C \rightarrow A$	↕	·	↕	↕	↕	↕
	ADCB	C9	2	2	D9	3	2	E9	5	2	F9	4	3				$B + M + C \rightarrow B$	↕	·	↕	↕	↕	↕
AND	ANDA	84	2	2	94	3	2	A4	5	2	B4	4	3				$A \cdot M \rightarrow A$	·	·	↕	↕	R	·
	ANDB	C4	2	2	D4	3	2	E4	5	2	F4	4	3				$B \cdot M \rightarrow B$	·	·	↕	↕	R	·
Bit test	BITA	85	2	2	95	3	2	A5	5	2	B5	4	3				$A \cdot M$	·	·	↕	↕	R	·
	BITB	C5	2	2	D5	3	2	E5	5	2	F5	4	3				$B \cdot M$	·	·	↕	↕	R	·
Clear	CLR							6F	7	2	7F	6	3				$00 \rightarrow M$	·	·	R	S	R	R
	CLRA													4F	2	1	$00 \rightarrow A$	·	·	R	S	R	R
	CLRB													5F	2	1	$00 \rightarrow B$	·	·	R	S	R	R
Compare	CMPA	81	2	2	91	3	2	A1	5	2	B1	4	3				$A - M$	·	·	↕	↕	↕	↕
	CMPB	C1	2	2	D1	3	2	E1	5	2	F1	4	3				$B - M$	·	·	↕	↕	↕	↕
Compare accumulators	CBA													11	2	1	$A - B$	·	·	↕	↕	↕	↕
Complement, 1's	COM							63	7	2	73	6	3				$\overline{M} \rightarrow M$	·	·	↕	↕	R	S
	COMA													43	2	1	$\overline{A} \rightarrow A$	·	·	↕	↕	R	S
	COMB													53	2	1	$\overline{B} \rightarrow B$	·	·	↕	↕	R	S
Complement, 2's (Negate)	NEG							60	7	2	70	6	3				$00 - M \rightarrow M$	·	·	↕	↕	(1)	(2)
	NEGA													40	2	1	$00 - A \rightarrow A$	·	·	↕	↕	(1)	(2)
	NEGB													50	2	1	$00 - B \rightarrow B$	·	·	↕	↕	(1)	(2)
Decimal adjust, A	DAA													19	2	1	Converts binary add. of BCD characters to BCD format	·	·	↕	↕	↕	(3)
Decrement	DEC							6A	7	2	7A	6	3				$M - 1 \rightarrow M$	·	·	↕	↕	(4)	·
	DECA													4A	2	1	$A - 1 \rightarrow A$	·	·	↕	↕	(4)	·
	DECB													5A	2	1	$B - 1 \rightarrow B$	·	·	↕	↕	(4)	·
OR, exclusive	EORA	88	2	2	98	3	2	A8	5	2	B8	4	3				$A \oplus M \rightarrow A$	·	·	↕	↕	R	·
	EORB	C8	2	2	D8	3	2	E8	5	2	F8	4	3				$B \oplus M \rightarrow B$	·	·	↕	↕	R	·
Increment	INC							6C	7	2	7C	6	3				$M + 1 \rightarrow M$	·	·	↕	↕	(5)	·
	INCA													4C	2	1	$A + 1 \rightarrow A$	·	·	↕	↕	(5)	·
	INCB													5C	2	1	$B + 1 \rightarrow B$	·	·	↕	↕	(5)	·
Load accumulator	LDAA	86	2	2	96	3	2	A6	5	2	B6	4	3				$M \rightarrow A$	·	·	↕	↕	R	·
	LDAB	C6	2	2	D6	3	2	E6	5	2	F6	4	3				$M \rightarrow B$	·	·	↕	↕	R	·

259

TABLE 10-1 (continued)

Operations	Mnemonic	IMMED OP	IMMED ~	IMMED #	DIRECT OP	DIRECT ~	DIRECT #	INDEX OP	INDEX ~	INDEX #	EXTND OP	EXTND ~	EXTND #	IMPLIED OP	IMPLIED ~	IMPLIED #	Boolean/Arithmetic Operation	H	I	N	Z	V	C
OR, inclusive	ORAA	8A	2	2	9A	3	2	AA	5	2	BA	4	3				$A + M \rightarrow A$	•	•	↔	↔	R	•
	ORAB	CA	2	2	DA	3	2	EA	5	2	FA	4	3				$B + M \rightarrow B$	•	•	↔	↔	R	•
Push data	PSHA													36	4	1	$A \rightarrow M_{SP},\ SP - 1 \rightarrow SP$	•	•	•	•	•	•
	PSHB													37	4	1	$B \rightarrow M_{SP},\ SP - 1 \rightarrow SP$	•	•	•	•	•	•
Pull data	PULA													32	4	1	$SP + 1 \rightarrow SP,\ M_{SP} \rightarrow A$	•	•	•	•	•	•
	PULB													33	4	1	$SP + 1 \rightarrow SP,\ M_{SP} \rightarrow B$	•	•	•	•	•	•
Rotate left	ROL							69	7	2	79	6	3				M) rotate left, C ↔ b7…b0	•	•	↔	↔	⑥	↔
	ROLA													49	2	1	A) rotate left, C ↔ b7…b0	•	•	↔	↔	⑥	↔
	ROLB													59	2	1	B) rotate left, C ↔ b7…b0	•	•	↔	↔	⑥	↔
Rotate right	ROR							66	7	2	76	6	3				M) rotate right, C → b7…b0	•	•	↔	↔	⑥	↔
	RORA													46	2	1	A) rotate right, C → b7…b0	•	•	↔	↔	⑥	↔
	RORB													56	2	1	B) rotate right, C → b7…b0	•	•	↔	↔	⑥	↔
Shift left, arithmetic	ASL							68	7	2	78	6	3				M) C ← b7…b0 ← 0	•	•	↔	↔	⑥	↔
	ASLA													48	2	1	A) C ← b7…b0 ← 0	•	•	↔	↔	⑥	↔
	ASLB													58	2	1	B) C ← b7…b0 ← 0	•	•	↔	↔	⑥	↔
Shift right, arithmetic	ASR							67	7	2	77	6	3				M) b7…b0 → C	•	•	↔	↔	⑥	↔
	ASRA													47	2	1	A) b7…b0 → C	•	•	↔	↔	⑥	↔
	ASRB													57	2	1	B) b7…b0 → C	•	•	↔	↔	⑥	↔
Shift right, logic	LSR							64	7	2	74	6	3				M) 0 → b7…b0 → C	•	•	R	↔	⑥	↔
	LSRA													44	2	1	A) 0 → b7…b0 → C	•	•	R	↔	⑥	↔
	LSRB													54	2	1	B) 0 → b7…b0 → C	•	•	R	↔	⑥	↔
Store accumulator	STAA				97	4	2	A7	6	2	B7	5	3				$A \rightarrow M$	•	•	↔	↔	R	•
	STAB				D7	4	2	E7	6	2	F7	5	3				$B \rightarrow M$	•	•	↔	↔	R	•
Subtract	SUBA	80	2	2	90	3	2	A0	5	2	B0	4	3				$A - M \rightarrow A$	•	•	↔	↔	↔	↔
	SUBB	C0	2	2	D0	3	2	E0	5	2	F0	4	3				$B - M \rightarrow B$	•	•	↔	↔	↔	↔
Subtract accumulators	SBA													10	2	1	$A - B \rightarrow A$	•	•	↔	↔	↔	↔
Subtract with carry	SBCA	82	2	2	92	3	2	A2	5	2	B2	4	3				$A - M - C \rightarrow A$	•	•	↔	↔	↔	↔
	SBCB	C2	2	2	D2	3	2	E2	5	2	F2	4	3				$B - M - C \rightarrow B$	•	•	↔	↔	↔	↔
Transfer accumulators	TAB													16	2	1	$A \rightarrow B$	•	•	↔	↔	R	•
	TBA													17	2	1	$B \rightarrow A$	•	•	↔	↔	R	•
Test, zero or minus	TST							6D	7	2	7D	6	3				$M - 00$	•	•	↔	↔	R	R
	TSTA													4D	2	1	$A - 00$	•	•	↔	↔	R	R
	TSTB													5D	2	1	$B - 00$	•	•	↔	↔	R	R

TABLE 10-1 (Continued)

LEGEND		CONDITION CODE SYMBOL:	
OP	Operation code (hexadecimal)	+	Boolean inclusive OR
~	Number of MPU cycles	⊕	Boolean exclusive OR
#	Number of program bytes	$\overline{M}$	Complement of M
+	Arithmetic plus	→	Transfer into
−	Arithmetic minus	0	Bit = zero
	Boolean AND	00	Byte = zero
M_{SP}	Contents of memory location pointed to by stack pointer		

Condition Code Symbol:

H	Half-carry from bit 3
I	Interrupt mask
N	Negative (sign bit)
Z	Zero (byte)
V	Overflow, 2's complement
C	Carry from bit 7

R	Reset always
S	Set always
↕	Test; set if true, cleared otherwise
.	Not affected

Note: Accumulator addressing mode instructions are included in the column for implied addressing.

Index Register and Stack Manipulation Instructions

POINTER OPERATION	MNEMONIC	Immed OP	Immed ~	Immed #	Direct OP	Direct ~	Direct #	Index OP	Index ~	Index #	Extnd OP	Extnd ~	Extnd #	Implied OP	Implied ~	Implied #	BOOLEAN/ARITHMETIC OPERATION	H (5)	I (4)	N (3)	Z (2)	V (1)	C (0)
Compare index reg.	CPX	8C	3	3	9C	4	2	AC	6	2	BC	5	3				$X_H - M, X_L - (M+1)$	.	.	⑦	↕	⑧	.
Decrement index reg.	DEX													09	4	1	$X - 1 \to X$	.	.	.	↕	.	.
Decrement stack pntr.	DES													34	4	1	$SP - 1 \to SP$	.	.	.	.	.	.
Increment index reg.	INX													08	4	1	$X + 1 \to X$	.	.	.	↕	.	.
Increment stack pntr.	INS													31	4	1	$SP + 1 \to SP$	.	.	.	.	.	.
Load index reg.	LDX	CE	3	3	DE	4	2	EE	6	2	FE	5	3				$M \to X_H, (M+1) \to X_L$	.	.	⑨	↕	R	.
Load stack pntr.	LDS	8E	3	3	9E	4	2	AE	6	2	BE	5	3				$M \to SP_H, (M+1) \to SP_L$	.	.	⑨	↕	R	.
Store index reg.	STX				DF	5	2	EF	7	2	FF	6	3				$X_H \to M, X_L \to (M+1)$	.	.	⑨	↕	R	.
Store stack pntr.	STS				9F	5	2	AF	7	2	BF	6	3				$SP_H \to M, SP_L \to (M+1)$	.	.	⑨	↕	R	.
Index reg. → Stack pntr.	TXS													35	4	1	$X - 1 \to SP$	.	.	.	.	.	.
Stack pntr. → Index reg.	TSX													30	4	1	$SP + 1 \to X$	.	.	.	.	.	.

TABLE 10-1 (Continued)
Jump and Branch Instructions

OPERATION	MNEMONIC	Relative OP	Relative ~	Relative #	Index OP	Index ~	Index #	Extnd OP	Extnd ~	Extnd #	Implied OP	Implied ~	Implied #	BRANCH TEST	5 H	4 I	3 N	2 Z	1 V	0 C
Branch always	BRA	20	4	2										None	•	•	•	•	•	•
Branch if carry clear	BCC	24	4	2										C = 0	•	•	•	•	•	•
Branch if carry set	BCS	25	4	2										C = 1	•	•	•	•	•	•
Branch if = zero	BEQ	27	4	2										Z = 1	•	•	•	•	•	•
Branch if ≥ zero	BGE	2C	4	2										$N \oplus V = 0$	•	•	•	•	•	•
Branch if > zero	BGT	2E	4	2										$Z + (N \oplus V) = 0$	•	•	•	•	•	•
Branch if higher	BHI	22	4	2										$C + Z = 0$	•	•	•	•	•	•
Branch if ≤ zero	BLE	2F	4	2										$Z + (M \oplus V) = 1$	•	•	•	•	•	•
Branch if lower or same	BLS	23	4	2										$C + Z = 1$	•	•	•	•	•	•
Branch if < zero	BLT	2D	4	2										$N \oplus V = 1$	•	•	•	•	•	•
Branch if minus	BMI	2B	4	2										N = 1	•	•	•	•	•	•
Branch if not equal zero	BNE	26	4	2										Z = 0	•	•	•	•	•	•
Branch if overflow clear	BVC	28	4	2										V = 0	•	•	•	•	•	•
Branch if overflow set	BVS	29	4	2										V = 1	•	•	•	•	•	•
Branch if plus	BPL	2A	4	2										N = 0	•	•	•	•	•	•
Branch to subroutine	BSR	8D	8	2											•	•	•	•	•	•
Jump	JMP				6E	4	2	7E	3	3				None	•	•	•	•	•	•
Jump to subroutine	JSR				AD	8	2	BD	9	3					•	•	•	•	•	•
No operation	NOP										01	2	1	See Special Operations	•	•	•	•	•	•
Return from interrupt	RTI										3B	10	1		(10)					
Return from subroutine	RTS										39	5	1	Advances Prog. Cntr. Only	•	•	•	•	•	•
Software interrupt	SWI										3F	12	1	See Special Operations	•	•	•	•	•	•
Wait for interrupt*	WAI										3E	9	1		•	(11)	•	•	•	•

*WAI puts address bus, R/W, and data bus in the three-state mode while VMA is held low.

| | | Implied | | | | Condition Code Register | | | | | |
| | | | | | | 5 | 4 | 3 | 2 | 1 | 0 |
OPERATION	MNEMONIC	OP	~	#	BOOLEAN OPERATION	H	I	N	Z	V	C
Clear carry	CLC	0C	2	1	$0 \rightarrow C$	•	•	•	•	•	R
Clear interrupt mask	CLI	0E	2	1	$0 \rightarrow I$	•	R	•	•	•	•
Clear overflow	CLV	0A	2	1	$0 \rightarrow V$	•	•	•	•	R	•
Set carry	SEC	0D	2	1	$1 \rightarrow C$	•	•	•	•	•	S
Set interrupt mask	SEI	0F	2	1	$1 \rightarrow I$	•	S	•	•	•	•
Set overflow	SEV	0B	2	1	$1 \rightarrow V$	•	•	•	•	S	•
Accumulator A → CCR	TAP	06	2	1	$A \rightarrow CCR$			⑫			
CCR → accumulator A	TPA	07	2	1	$CCR \rightarrow A$	•	•	•	•	•	•

Condition Code Register Notes (Bit set if test is true and cleared otherwise):

1 (Bit V) Test: Result = 10000000?

2 (Bit C) Test: Result ≠ 00000000?

3 (Bit C) Test: Decimal value of most significant BCD character greater than 9? (Not cleared if previously set.)

4 (Bit V) Test: Operand = 10000000 prior to execution?

5 (Bit V) Test: Operand = 01111111 prior to execution?

6 (Bit V) Test: Set equal to result of N ⊕ C after shift has occurred.

7 (Bit N) Test: Sign bit of most significant (MS) byte = 1?

8 (Bit V) Test: 2's complement overflow from subtraction of MS bytes?

9 (Bit N) Test: Result less than 0? (Bit 15 = 1)

10 (All) Load condition code register from stack (see Special Operations)

11 (Bit 1) Set when interrupt occurs. If previously set, a nonmaskable interrupt is required to exit the wait state.

12 (All) Set according to contents of accumulator A.

SOURCE: Motorola.

displacement of 30H to the base address of 4321H, to get an effective address of 4351H. *Note:* The contents of the index register are *not* changed. The index register still contains the base address of 4321H. The 6800 then fetches the byte from memory at the effective address of 4351H and adds this byte to the contents of the *A* accumulator. The result of the addition is left in the *A* accumulator. The byte in memory at the effective address is not changed. Since the offset or displacement is an 8-bit quantity, it can be anywhere between 0 and 255. For the 6800, however, it can only be positive. The 8080A has no indexed addressing mode, but the Z80A does with both its IX and IY registers.

Extended addressing uses the second and third bytes of the instruction to store the address of the operand. Unlike with 8080A and Z80A processors, the second byte contains the high-order address byte. The third byte of the instruction contains the low-order address byte. For example, ADDA $073A is written in three successive memory bytes as BB, 07, 3A. This instruction adds the contents of memory location 073A to the *A* accumulator. Several 8080A and Z80A instructions use this mode. Examples are the 8080A JMP, CALL, LDA, and STA instructions.

Implied addressing mode contains the operand address within the instruction op code. For example, the decrement index register (DEX) instruction needs only 1 byte because the reference to the index register is coded into the 09 op code. The 8080A MOV instructions are examples of this addressing mode.

Accumulator addressing is a special case of implied addressing. The operand is one of the accumulators, and this is indicated in the op code for the instruction. The CLRA instruction to clear accumulator *A* is an example. Instructions using this mode also require only 1 byte. The DAA, CMA, and rotate instructions are examples of this mode for the 8080A.

Relative addressing is the last of the 6800 addressing modes. In relative addressing a displacement contained in the second byte of the instruction is added to the program counter. The next instruction is fetched from the resultant address. This mode is used for jump and branch instructions. For the 6800, the displacement is stored in the second byte as a signed 2's-complement number. A number from -128 to $+127$ can be represented. Since the program counter automatically increments after each fetch, the displacement is added to the address after that in which the displacement byte was stored. Therefore, when measured from the instruction address, the relative addressing range is -126 to $+129$ addresses.

When the displacement is added to the program counter, a carry-out of the lower 8 bits is automatically added to the upper address byte. Later we show how the displacements are calculated.

INSTRUCTION SET EXPLANATIONS

Many manufacturers use a symbolic shorthand to describe the function of microprocessor instructions. The ADDA instruction in Table 10-1, for example, is represented by $A + M \rightarrow A$. This means that an immediate byte or the contents of an addressed memory location are added to the contents of the *A* accumulator, and the sum is left in the *A* accumulator. As you can see, the symbolic form is much shorter and perhaps clearer than the verbal description. The symbols under the flag initials indicate whether a flag is affected by that particular instruction. A vertical arrow indicates that that flag is affected, a dot indicates that that flag is not affected. R means that a flag is always reset, S means that a flag is always set, and a circled number refers you to a note at the end of the table. It is very important to become comfortable with these symbolic representations because they allow you to see the effect of an instruction at a glance.

Most of the 6800 instructions in Table 10-1 are quite clear from the name or from the symbolic expression. We explain and illustrate some of the less obvious instructions in the following sections.

ACCUMULATOR AND MEMORY INSTRUCTIONS

The ANDA and BITA are similar in that they both AND the contents of the indicated memory location with the *A* accumulator. However, the ANDA instructions put the result of the ANDing into the *A* accumulator. The BITA instructions affect the flags, but do not affect the contents of the memory location or the *A* accumulator. The BITA instruction, therefore, permits a test without changing the numbers tested.

The COM instructions form the 1's complement of the operand by inverting all bits. The NEG instructions form the 2's complement.

Two of the most commonly used 6800 instructions are LDAA and LDAB. These instructions copy an immediate byte or the contents of a memory location into the indicated accumulator. Four addressing modes are available with these instructions. LDAA #$7A, for example, copies the *immediate* hex number 7A into the *A* accumulator. LDAB $37 copies the contents of the direct memory address 37H into accumulator *B*. LDAA $4FC0 copies the contents of memory at the extended address 4FC0H into the *A* accumulator. LDAB $53,X copies into *B* the contents of the memory at the effective address computed by adding 53H to the index register. Since all ports in a 6800 system are addressed as memory locations, the LDAA and LDAB instructions also function as input from port instructions.

The 6800 also uses a stack in RAM to store return addresses and registers during subroutines. The PSHA instruction copies the contents of the *A* accumulator into the memory location pointed to by the stack pointer. The stack pointer is then automatically decremented by 1. Note that the 6800 pushes a *byte* and decrements the stack pointer, whereas the 8080 decrements the stack pointer and then pushes a byte. This is important to remember when you are debugging programs. To restore data to the accumulators at the end of a subroutine, the PUL instructions are used. PULA increments the stack pointer by 1 and then copies into *A* the contents of the memory location pointed to. The data stored in the stack is not affected.

The 6800 rotate and shift instructions can shift the contents of the *A* accumulator, the *B* accumulator, or a memory location. ROL is rotate left through carry, and ROR is rotate right through carry. ASL shifts a 0 into the least significant bit position and the most significant bit into carry. LSR shifts a 0 into the MSB position and the LSB into carry. ASR shifts the byte right with the LSB going into carry. The MSB, however, stays the same. Refer to the diagrams in Table 10-1 for these instructions.

Another very commonly used pair of 6800 instructions is STAA and STAB. These instructions copy the indicated accumulator to a direct, an extended, or an indexed memory address. The contents of the accumulator are not changed. Since all ports in a 6800 system are addressed as memory locations, these instructions function as output to port instructions.

The TST instructions are used to determine whether the contents of an accumulator or memory location are 0 or negative. The TST instructions affect the flags without changing the byte tested.

INDEX REGISTER AND STACK INSTRUCTIONS

Remember that the 6800 index register and stack pointer register are both 16 bits.

The LDX instruction is used to load a base address into the index register for use with indexed addressing mode instructions. It is also used to initialize the starting value in the index register when it is used as a counter. LDX loads 2 immediate bytes or the contents of two consecutive memory locations into the index register. For example, LDX #$075A loads 075AH into the index register. LDX $3A7B copies the contents of address 3A7BH into the upper 8 bits of the index register and the contents of address 3A7CH into the lower 8 bits of the index register.

DEX decrements the index register by 1, and INX increments the index register by 1. Note that both set the zero flag if the result of the operation leaves 0's in all 16 bits of the index register. This makes it easy to use the index register for long delay loops.

STX copies the index register to two consecutive memory locations. STX $3A7B, for example, copies the upper 8 bits of the index register to address 3A7BH and the lower 8 bits to address 3A7CH. Since the 6800 has no instruction to push the index register during a subroutine, this is accomplished by a STX instruction at the start of the subroutine and an LDX instruction at the end of the routine. For example, STX $3A7B saves the index register at 3A7BH and 3A7CH. LDX $3A7B restores the values to the index register from 3A7BH and 3A7CH.

CPX compares 2 immediate bytes or the contents of two consecutive memory locations with the contents of the index register. The first byte in memory is compared with the high-order byte of X, and the next byte is compared with the low-order byte of X. A compare is done by subtraction, but neither the index register nor the memory contents are affected. If the index register is equal to the compared word, then the zero flag is set. The negative flag is set if bit 15 of the result is a 1. The overflow flag is set if the subtraction produced a result outside the legal 2's-complement range for 16 bits.

LDS is used to initialize the stack pointer. For instance, LDS #$00FF initializes the stack pointer with the immediate number 00FFH. LDS $0A00 loads the stack pointer high byte from 0A00H and the stack pointer low byte from 0A01H.

TSX loads the index register with 1 plus the contents of the stack pointer. The stack pointer is unchanged.

TXS loads the stack pointer with the contents of the index register minus 1. The contents of the index register are not changed.

JUMP AND BRANCH INSTRUCTIONS

The terms *branch* and *jump* are often used interchangeably. However, with the 6800 there is a difference. Branch instructions use only relative addressing, so they can change program execution by only −126 to +129 addresses. Jump instructions can jump the program execution to any location in memory because they use indexed or extended addressing. Unconditional jump (JMP), unconditional branch (BRA), and all the conditional branch instructions simply move program execution to the indicated address.

JSR (jump to subroutine) and BSR (branch to subroutine) are similar to the call instructions of the 8080A/8085A. Both store the return address on the stack and load the program counter with the starting address of the subroutine. Although the 6800 has a wide variety of conditional branch instructions, it does not have the conditional call instructions or parity flag of the 8080A/8085A.

The RTI and RTS pull the return address off the stack and return program execution to that address. RTI also restores the condition code register, accumulators, and index register that were pushed onto the stack when the 6800 recognized an interrupt.

SWI is the software interrupt. When executed, it pushes all the registers and the program counter contents onto the stack. The CPU then goes to addresses FFFA and FFFB to get the starting address of the desired interrupt service routine.

Wait for interrupt (WAI) pushes the register contents and program counter onto the stack. The CPU then floats the address bus, data bus, and R/W control line. VMA output is low. The processor is in a wait-for-interrupt state. It is just idling and waiting for a nonmaskable interrupt (NMI) or a maskable interrupt (IRQ) to occur. These, and RESET, are the only ways the processor can get out of the wait-for-interrupt state. In addition, IRQ cannot cause an exit unless it has been unmasked with a clear interrupt mask (CLI) instruction. Motorola suggests a NOP-CLI-WAI instruction sequence to unmask the IRQ interrupt and enter the wait-for-interrupt state. Since all registers, etc., have already been pushed, the WAI instruction allows quicker response to an interrupt than the normal method. When interrupted, the processor vectors directly to the interrupt service routine. The processor, however, is doing

265

```
;PROGRAM TO TAKE 10 DATA SAMPLES FROM PORT AT ADDRESS 8004H.
;EACH SAMPLE WILL BE TAKEN WHEN A "DATA READY" PULSE IS PRESENT ON THE LSB
;OF THE PORT AT 8006H.  THE SAMPLES WILL BE STORED IN MEMORY STARTING AT
;ADDRESS 00A0H.
;MEMORY USED:  00A0-00A9H; 8004H, 8006H, AND 15H BYTES FOR PROGRAM
;ROUTINES CALLED:  NONE
```

ADR	INSTR.	LABEL	OPCODE	OPERAND	COMMENTS
0000	CE		LDX	#S00A0	;LOAD INDEX REGISTER WITH
01	00				;STARTING ADDRESS FOR STORAGE
02	A0				; (IMMEDIATE)
03	C6		LDAB	#$0A	;INITIALIZE SAMPLE COUNTER IN
04	0A				;B ACCUMULATOR (IMMEDIATE)
05	B6	LOOP:	LDAA	$8006	;TAKE IN DATA READY SIGNAL FROM
06	80				;PORT AT ADDRESS 8006H
07	06				; (EXTENDED)
08	85		BITA	#$01	;TEXT LSD FOR ONE
09	01				; (IMMEDIATE)
0A	27		BEQ	LOOP	;REPEAT UNTIL LSB FROM 8006
0B	F9				;IS EQUAL ONE (RELATIVE)
0C	B6		LDAA	$8004	;DATA READY - TAKE IN DATA
0D	80				;FROM PORT 8004H (EXTENDED)
0E	04				
0F	A7		STAA	$00,X	;STORE DATA IN MEMORY AT ADDRESS
0010	00				;POINTED TO BY INDEX REGISTER (INDEXED)
11	08		INX		;INCREMENT INDEX REGISTER (IMPLIED)
12	5A		DECB		;DECREMENT B ACC COUNTER (ACCUMULATOR)
13	26		BNE	LOOP	;REPEAT LOOP UNTIL 10 SAMPLES
14	F0				;ARE TAKEN (RELATIVE)

FIGURE 10-2 6800 program for taking in 10 samples of data when a data signal is present.

nothing while waiting. A later section of this chapter summarizes the effects of the 6800 hardware and software interrupts.

CONDITION CODE REGISTER MANIPULATION INSTRUCTIONS Most of these instructions are clearly defined by their names. Note that SEI (set-interrupt-mask) disables the $\overline{IRQ}$ interrupt, and CLI enables it. TAP and TPA interchange the *A* accumulator and the condition code register contents.

This concludes the discussion of the 6800 instruction set. Some sample programs are shown now to give you more feeling for how these instructions are used.

PROGRAM EXAMPLES

EXAMPLE 1   The first program example shows how the flowchart of Figure 8-21*a* is programmed for the 6800. The program takes in a sample of data when a data ready signal is present. The data sample is stored in memory. The process is repeated until 10 samples have been taken and stored.

Figure 10-2 shows the 6800 program for this flowchart. This specific example is written to check the data ready pulse in the LSB of the port at address 8006H. It takes in the data from the port at address 8004H. Data samples are stored in RAM starting at address 00A0H.

In the program of Figure 10-2, a $ in front of a number indicates to the assembler that the number is hexadecimal. A # indicates the immediate mode of addressing is being used. This program, incidentally, illustrates six of the seven possible addressing modes.

LDX #$00A0 loads the index register with the starting address of the storage location for the 10 samples. LDAB #$0A loads 10 into the *B* accumulator to set it up as a sample counter. LDAA $8006 takes in a byte from the data ready signal port at address 8006H to the *A* accumulator. BITA #$01 ANDs the number 01H with the number in the *A* accumulator to set the flags. ANDA #$01 also could be used, but BITA #$01 often has the advantage that it sets the flags without altering the contents of the accumulator. For either instruction the zero flag will remain high unless the LSB of the byte from the port at 8006H is a 1. The BEQ LOOP instruction keeps branching the program back until the data ready signal appears in this LSB.

When the data ready pulse appears, the LDAA $8004 instruction takes the sample from the data port at address 8004H into accumulator A. The STAA $00,X instruction uses indexed addressing to move the data bytes to memory. Since a displacement or an offset of $00 is indicated, the first sample is stored at 00A0H, the number initially loaded into the index register. The X says indexed addressing is being used.

INX increments the index register to point to the next storage address. DECB decrements the number-of-samples counter in the B accumulator. BNE sends the program back to wait for another data ready pulse if the counter has not been counted down to 0.

Figure 10-2 also shows the assembled version, or machine code, for the program. You can use the machine code and Table 10-1 to verify the addressing mode for each instruction. Note that addresses are written into memory with the high-order byte first. This is opposite to the way it is done for the 8080A/8085A and Z80A.

For the BEQ and BNE instructions which use relative addressing, a displacement must be calculated. If the program is coded with an assembler, this is done automatically by the assembler. For hand assembly, the displacement is calculated as follows. The program counter automatically increments after fetching the displacement, so the displacement is measured from address 010C for the BEQ instruction. A branch to address 0105 is desired. The difference is a displacement of −7. For the BEQ instruction, this displacement is represented in 2's complement as 11111001 binary , or $F9_{16}$. The 1 in the MSB position indicates a negative displacement. BNE LOOP requires a displacement of -16_{10}, or -10_{16}. This is coded as 2's complement 11110000 binary, or $F0_{16}$.

EXAMPLE 2   This program takes in 10 samples of data at timed intervals, as shown in Figure 8-22. For the program in Figure 8-22 the interval was 1 ms, but here we show how to create longer delays as well.

Figure 10-3 shows the 6800 assembly language program to take in 10 samples from the port at address 8004H and store them in memory starting at address 00A0H. Also shown are three subroutines that can space the samples at intervals of 1 to 65,535 ms.

The program starts by initializing the stack pointer to an address in RAM. This is necessary when BSR, JSR, or interrupts are used in a program because all these use the stack to store return addresses, etc. Since the stack grows downward in memory as bytes are pushed onto it, the stack pointer can be initialized to the highest address in available RAM.

Next the index register and sample counter are initialized. A sample of data is then taken in from the port at address 8004H and stored in memory at the address pointed to by the index register. When this is completed, one of the delay routines is called. If the routine is located within 129 bytes, a BSR can be used. If not, then a JSR must be used. If the delay routine is going to be used only once in a program, then it can be written

right into the main program. However, writing the delay loop as a subroutine allows it to be written once and used anywhere in a program.

The WAITMS routine simply loads the number 165_{10} into the A accumulator and decrements it to 0. The number of cycles for each instruction is given in Table 10-1. If the 6800 uses a 1-MHz clock, then the time for each instruction in microseconds is equal to the number of cycles. Knowing this, you can calculate the time for the WAITMS routine by adding together the instruction times.

The wait more (WAITMR) delay routine uses nested loops to get a delay up to 255 ms. At the start of the routine it is necessary to push the B accumulator contents onto the stack because the B accumulator is used in both the main program and the subroutine. The routine loads and counts down the 1-ms routine as many times as needed. The desired number of milliseconds is loaded into the B accumulator and counted down each time the 1-ms loop is run.

A third routine, WAITLT (wait a long time), which allows delays up to 65,535 ms, uses a similar method. The 16-bit index register is used as the millisecond counter. STX $00B0 saves the contents of the index register by storing them in memory. The desired delay (in milliseconds) is loaded into the index register with the LDX instruction. LOOP2 is the 1-ms routine, and LOOP1 counts down the number in the index register. Before returning to the main program, the index register is restored by LDX $00B0.

These routines are quite simple, but they have the disadvantage that the processor is tied up while they are running. In many applications, this is a waste of processor time. Therefore, external timers such as that discussed in the 8155 may be used instead of delay routines.

After a return from the delay subroutine, the program increments the index register to point to the next storage location in memory. It then decrements the sample counter. The program keeps looping to GETDAT until 10 samples have been taken and stored.

EXAMPLE 3: MULTIPLE-PRECISION ADDITION AND SUBTRACTION   Much more 6800 programming is shown in Chap. 12, but here two more short examples illustrate direct and indexed addressing.

The 6800 accumulators and data bus are 8 bits wide and the 6800 arithmetic instructions operate only on 8-bit quantities. This limits the range of numbers that can be added or subtracted to −128 to +127. Larger numbers can be added or subtracted 1 byte at a time, starting with the least significant byte. A borrow or carry-out of 1-byte operation is included in the next-higher byte operation. Operations with 2-byte numbers are called *double precision*, and operations with more than 2-byte numbers are called *multiple precision*.

Figure 10-4a shows a short routine to add 2-byte numbers by using direct addressing. One number to be added is stored in memory at 0040H and 0041H; the other number to be added is stored at 0050H and 0051H, and the sum is stored at 0060H and 0061H. The

```
;MC6800 PROGRAM TO TAKE IN TEN SAMPLES OF DATA FROM THE PORT AT
;ADDRESS 4003H AND STORE THEM IN MEMORY STARTING AT ADDRESS 0A00H.
;USING DIFFERENT DELAY ROUTINES, THE SAMPLES CAN BE TAKEN AT INTERVALS
;OF 1MSEC TO 65,535MSEC.
;MEMORY USED:  00F0-00FFH (STACK); 00A0-00A9H (STORAGE); 8004H (INPUT PORT)
;ROUTINES CALLED: WAIT MS, WAIT MR OR WAIT LT.
```

LABEL	OPCODE	OPERAND	COMMENTS
	LDS	#$00FF	;INITIALIZE STACK POINTER
	LDX	#$00A0	;INITIALIZE INDEX REGISTER
	LDAB	#$0A	;SET SAMPLE COUNTER
GET DAT:	LDAA	$8004	;GET DATA FROM PORT AT ADDRESS 8004H
	STAA	$00,X	;STORE DATA SAMPLE IN MEMORY
	BSR	WAIT--	;BRANCH TO DELAY SUBROUTINE
	INX		;INCREMENT INDEX REGISTER
	DECB		;DECREMENT COUNTER AND
	BNE	GET DAT	;LOOP UNTIL TEN SAMPLES TAKEN
WAITMS:	LDAA	#165 (DECIMAL)	;LOAD 1MSEC DELAY CONSTANT
LOOP:	DECA		;DECREMENT A AND
	BNE	LOOP	;LOOP UNTIL ZERO
	RTS		;RETURN FROM SUBROUTINE
WAITMR:	PSHB		;SAVE B ACCUMULATOR ON STACK
	LDAB	# NUMBER	;DELAY (IN MSEC) UP TO 255 MSEC
LOOP1:	LDAA	#165	;LOAD 1MSEC DELAY CONSTANT (165 DECIMAL)
LOOP2:	DECA		;DECREMENT A AND
	BNE	LOOP2	;LOOP UNTIL ZERO
	DECB		;DECREMENT B AND
	BNE	LOOP1	;LOOP UNTIL ZERO
	PULB		;RESTORE PREVIOUS B CONTENTS
	RTS		;RETURN FROM SUBROUTINE
WAITLT:	STX	$0AF0	;STORE INDEX REGISTER
	LDX	# NUMBER	;DELAY (IN MSEC) UP TO 65,535
LOOP1:	LDAA	#165	;LOAD 1 MSEC DELAY CONSTANT
LOOP2:	DECA		;DECREMENT A AND LOOP
	BNE	LOOP2	;UNTIL ZERO
	DEX		;DECREMENT INDEX REGISTER
	BNE	LOOP1	;AND LOOP UNTIL ZERO
	LDX	$00B0	;RESTORE INDEX REGISTER
	RTS		;RETURN FROM SUBROUTINE

FIGURE 10-3 6800 program to take in 10 samples of data at intervals of 1 to 65,535 ms.

least significant byte is in the lower address.

The least significant byte of one number is moved to the A accumulator and the most significant byte into the B accumulator. ADDA $50 adds the contents of memory location 0050 to the A accumulator and leaves the result in the A accumulator. This instruction uses direct addressing, so only a 1-byte address is given. Direct addressing automatically assumes the upper address byte is all 0's. ADCB $51 adds the most significant byte stored at address 0051, plus carry, to the B accumulator. Again direct addressing is used, so only 2 bytes

are needed in memory for the instruction. After execution of the ADCB instruction, the A accumulator contains the less significant byte of the result, and the B accumulator contains the more significant byte. STAA $60 and STAB $61 use direct addressing to move these results to the proper memory locations. The advantage of direct addressing is that only 2 bytes are required for a memory reference. Extended addressing requires 3 bytes for each instruction. The disadvantage of direct addressing is that it can be used only for locations 00H to FFH. This routine can be used for subtraction by sim-

```
LABEL    OPCODE    OPERAND      COMMENTS


;ROUTINE TO ADD TWO TWO-BYTE NUMBERS.  ONE OF THE NUMBERS TO BE
;ADDED IS STORED IN LOCATION 0040H AND 0041H, AND THE OTHER IS
;STORED IN ADDRESSES 0050H AND 0051H.  THE RESULT IS STORED IN
;ADDRESSES 0060H AND 0061H.  IN ALL CASES THE LEAST SIGNIFICANT
;BYTE IS STORED FIRST.


         LDAA      $40          ;LOAD 16 BIT NUMBER FROM MEMORY
         LDAB      $41          ;TO A AND B ACCUMULATORS
         ADDA      $50          ;ADD LEAST SIGNIFICANT BYTES
         ADCB      $51          ;ADD MOST SIGNIFICANT BYTES
         STAA      $60          ;STORE RESULT IN MEMORY
         STAB      $61
         ROLA                   ;CARRY INTO LSB OF A
         ANDA      #$01         ;MASK ALL BUT LSB
         STAA      $62          ;SAVE FINAL CARRY


                              (a)



LABEL    OPCODE    OPERAND      COMMENTS


;ROUTINE TO ADD TWO MULTIPLE BYTE NUMBERS.  ONE NUMBER
;IS STORED IN MEMORY STARTING AT 00A0H.  THE NEXT TO BE
;STORED STARTING AT 00B0H.  THE RESULT STORED STARTING
;AT 00C0H.


         CLC                    ;CLEAR CARRY
         LDAB      #N           ;LOAD NUMBER OF BYTES IN B AS A COUNTER
         LDX       #$00A0       ;LOAD BASE ADDRESS INTO INDEX REGISTER
LOOP:    LDAA      $00,X        ;LOAD FIRST BYTE INTO ACCUMULATOR A
         ADCA      $10,X        ;ADD FIRST BYTES
         STAA      $20,X        ;STORE RESULT
         INX                    ;INCREMENT INDEX REGISTER
         DECB                   ;DECREMENT BYTE COUNTER
         BNE       LOOP         ;LOOP UNTIL ALL BYTES ADDED
         ROLA                   ;CARRY INTO LSB OF A
         ANDA      #$01         ;MASK ALL BUT LSB
         STAA      $20,X        ;SAVE FINAL CARRY


                              (b)
```

FIGURE 10-4 6800 routines to add multiple-byte numbers. *(a)* Double-precision addition using direct addressing. *(b)* Multiple-precision addition.

ply replacing ADDA with SUBA and ADCB with SUBB. Packed BCD numbers, where each byte represents two BCD digits, can be added correctly by inserting a DAA instruction after ADDA and after ADCB.

Indexed addressing also uses only 2 bytes for a memory reference. Figure 10-4*b* shows a good example of how indexed addressing is used in a routine to add multiple-byte numbers.

Carry is cleared, and the number of bytes to be added is loaded into the *B* accumulator. One of the numbers to

be added is stored in memory starting with its least significant byte at 00A0H. The other number to be added is stored in memory with its least significant byte starting at 00B0H. The result is stored in memory starting at 00C0H.

LDX #$00A0 loads the base address into the index register. LDAA $00,X loads a byte into the A accumulator from location 00A0H. ADCA $10,X adds the number from address 00B0H to the *A* accumulator with carry. STAA $20,X loads the result into memory location

269

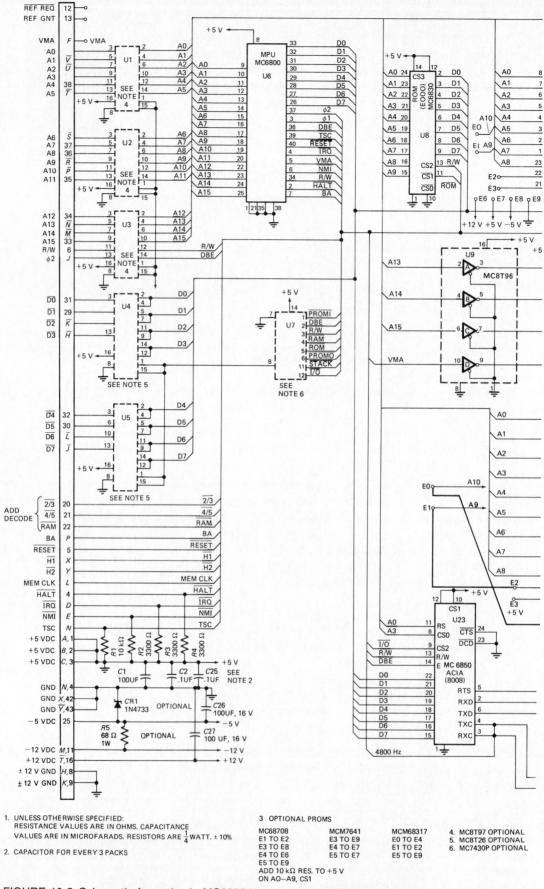

1. UNLESS OTHERWISE SPECIFIED:
   RESISTANCE VALUES ARE IN OHMS. CAPACITANCE
   VALUES ARE IN MICROFARADS. RESISTORS ARE $\frac{1}{4}$ WATT. ± 10%

2. CAPACITOR FOR EVERY 3 PACKS

3 OPTIONAL PROMS

MC68708	MCM7641	MCM68317
E1 TO E2	E3 TO E9	E0 TO E4
E3 TO E8	E4 TO E7	E1 TO E2
E4 TO E6	E5 TO E7	E5 TO E9
E5 TO E9		

ADD 10 kΩ RES. TO +5 V
ON A0–A9, CS1

4. MC8T97 OPTIONAL
5. MC8T26 OPTIONAL
6. MC7430P OPTIONAL

FIGURE 10-5 Schematic for a simple MC6800-based system. *(Motorola.)*

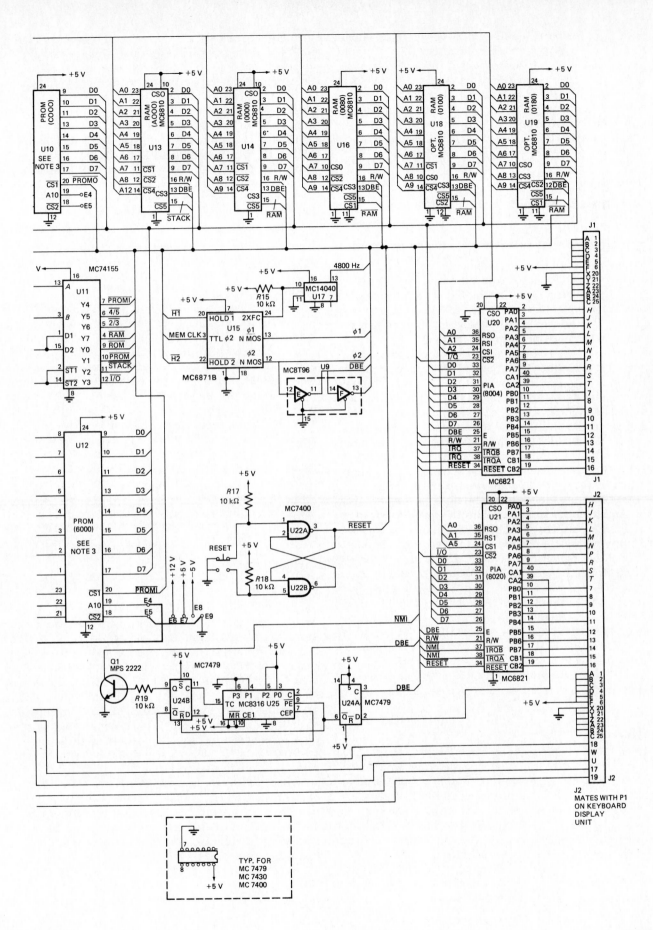

THE MC6800 MICROPROCESSOR AND MICROPROCESSOR EVOLUTION

00C0H. INX increments the index register by 1. Therefore, LDAA, ADCA, and STAA each get their operand from the next-higher address when the program loops. One simple INX instruction then changes the operand source for all three at once. DECB decrements the byte counter, and if it is not down to 0, BNE loops back to add another byte. The second bytes to be added come from addresses 00A1H and 00B1H, and the result is stored in 00C1H, for example.

A SIMPLE MC6800 SYSTEM

OVERVIEW

Figure 10-5 shows the schematic for a simple MC6800 system with ROM, RAM, and I/O ports. This circuit is the microprocessor part of the Motorola MEK6800D2 development kit shown in Figure 12-12. The two boards of this kit are used as the basis of a microprocessor-controlled EPROM programmer in Chap. 12.

Note in Figure 10-5 that the buses are compressed to a single line after they leave the ICs. This reduces congestion in schematics, but makes it a little harder to trace some signal paths. When approaching a schematic such as this, start with a part you know, such as the CPU, and work your way around from there.

The MC6800 CPU is in the upper left of the schematic. Address and data buses run from here to ROM, RAM, and I/O ports. They also run to some optional buffers which increase the drive for system expansion.

Next examine the source of the clock for the CPU. The two-phase clock required by the MC6800 is generated by U15, an MC6871B. The MC6871B is a hybrid package containing both a crystal and the oscillator circuitry. For this particular board, the clock frequency chosen was 614.4 kHz. Versions of the MC6800 operate with up to a 2-MHz clock. However, the 614.4-kHz clock was chosen so that a 4800-Hz clock for the MC6850 UART could be derived easily. In addition to the $\phi 1$ and $\phi 2$ outputs at 614.4 kHz, the MC6871B has an output of 2 F, or 1,228,800 Hz. The MC14040 binary divider divides this by 256 to give the desired 4800 Hz. HOLD 1 and HOLD 2 are inputs to the 6871B. They are used for clock cycle stretching and are discussed later.

ROM

ROM and its mapping in the memory space are good to examine next. Set up an address decoding worksheet, as in Chap. 9 to help you follow this. ROM U8 is an MC6830 1 K × 8 mask-programmed ROM. It contains the system monitor routines that enable the user to write programs into RAM, to run programs, etc. As indicated on the schematic, the starting address for the ROM is E000H. This ROM is selected when CS2 is high and $\overline{CS1}$ is low. The $\overline{CS1}$ signal is produced by the 74155, which is used as a three-line-to-eight-line decoder. In other words, if the 74155 is enabled by a valid

memory address signal to its strobe inputs, then one output addressed by the A, B, and D inputs will be low. This is similar to the 8205 discussed previously. Address bits A13, A14, and A15 are inverted by the MC8T96 buffers and connected to the A, B, and D inputs of the 74155. Therefore, when A15, A14, and A13 are all 1's and VMA is high, the Y0 output on pin 9 of the 74155 will go low and select ROM U8. A15, A14, and A13 high corresponds to address E000. Actually, since A12 is not decoded, FXXX also selects this ROM. This is important because the CPU sends out addresses FFF8 to FFFF to get interrupt and restart routine addresses. The other two ROMs are selected by $\overline{PROM}$ 0 and $\overline{PROM}$ 1 from the 74155 to start at C000H and 6000H, respectively.

RAM

RAM ICs are also selected by outputs from the 74155. U13 is an MC6810 128 × 8 RAM used as a stack and scratchpad by the monitor program. The Y2 output of the 74155 selects this RAM, starting at address A000H. Four more MC6810s (U14, U16, U18, and U19) give 512 bytes of RAM, starting at address 0000H. The Y7 output of the 74155 decoder sends a $\overline{CS}$ signal to these four when A15 = 0, A14 = 0, and A13 = 0. Since A12, A11, and A10 are not connected to the RAMs, they are don't-cares. Address lines A7, A8, and A9 determine which of the four is selected. RAM is put in the address space from 00 to FF, so the direct addressing mode can be used. Two more outputs from the 74155 can be used to select external ROM or RAM or I/O ports. The $\overline{2/3}$ line selects the 8000-byte block from 2000H to 3FFFH, and $\overline{4/5}$ selects the 8K byte block from 4000H to 5FFFH. The I/O output selects I/O ports, starting at address 8000H.

PARALLEL I/O PORTS

Parallel I/O ports are implemented in this circuit with MC6821 peripheral interface adapters. This device is similar to the 8255 and 8155 described in Chap. 9. It must be initialized in the program before use. A block diagram of the 6821 is shown in Figure 10-6. Each half of the device contains an 8-bit input-output port and two control lines. Each port line is individually programmable as an input or an output. The direction for each line is programmed by a word loaded into the data direction register. A 1 in a bit position of the data direction register sets the corresponding line as an output, and a 0 sets a line as an input.

In addition to the two data direction registers, the MC6821 has two control registers, CRA and CRB. These must also be programmed. The control word format is shown in Figure 10-7a. Before we discuss the initialization procedure, we need to explain the rest of the pin functions of the 6821.

CA1 and CB1 are programmable interrupt inputs. CA2 and CB2 are programmable as input or output control signals. When enabled, an interrupt into the CA1 or

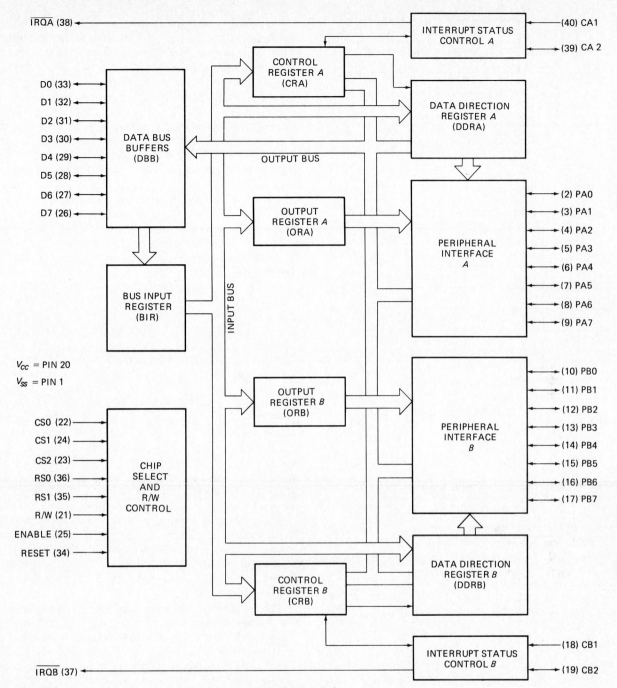

FIGURE 10-6 Block diagram of the MC6821 PIA. *(Motorola.)*

CB1 input sends an $\overline{\text{IRQA}}$ or $\overline{\text{IRQB}}$ to the $\overline{\text{IRQ}}$ or $\overline{\text{NMI}}$ input of the CPU.

When made low the $\overline{\text{RESET}}$ input resets all internal register bits to 0's, so all port lines are programmed as inputs. This input is usually connected to a power-on or manual reset. Note the cross-coupled NAND gate contact debouncer on the manual reset in Figure 10-5.

The enable input, E, makes sure the device is active only at the proper part of a cycle. It is connected to a delayed $\phi 2$ clock signal called DBE.

CS0, CS1, and CS2 are chip-select inputs that are connected to address lines or decoder outputs to select the device. RS0 and RS1 inputs are connected to address lines A0 and A1, respectively. Each MC6821 uses four addresses. For example, *U20* in Figure 10-5 uses addresses 8004H to 8007H. These two RS inputs together with a bit loaded into the control register determine which internal register is addressed. Figure 10-7*b* shows the internal addressing determined by bit 2 of the control register and the RS0 and RS1 inputs.

CRA	7	6	5	4	3	2	1	0
	IRQA1	IRQA2	CA2 Control			DDRA Access	CA1 Control	

CRB	7	6	5	4	3	2	1	0
	IRQB1	IRQB2	CB2 Control			DDRB Access	CB1 Control	

(a)

		Control Register Bit		
RS1	RS0	CRA2	CRB2	Location Selected
0	0	1	X	Peripheral Register A
0	0	0	X	Data Direction Register A
0	1	X	X	Control Register A
1	0	X	1	Peripheral Register B
1	0	X	0	Data Direction Register B
1	1	X	X	Control Register B

X = Don't Care

(b)

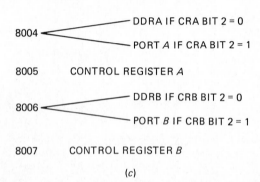

8004 — DDRA IF CRA BIT 2 = 0
     — PORT A IF CRA BIT 2 = 1

8005   CONTROL REGISTER A

8006 — DDRB IF CRB BIT 2 = 0
     — PORT B IF CRB BIT 2 = 1

8007   CONTROL REGISTER B

(c)

FIGURE 10-7 MC6821 PIA control word and internal addressing format. *(a)* Control word format. *(b)* Internal address format. *(c)* System addresses for *U*20 6821 in Fig. 10-5.

Now we examine the initialization of the 6821 as simple input and output ports. In Chap. 11 we explain how the 6821 is initialized for handshake input and output by using the CA and CB pins.

To initialize any programmable port device, first determine the system address for each data direction register, control register, and port register in the device, so you know to what address to send control words, etc. Do this by adding the internal addresses, such as those given in Figure 10-7*b*, to the base address of the device determined by the system address decoding. The base address of the *U*20 PIA in Figure 10-5 is 8004H. Figure 10-7*c* shows the system addresses of each register of this 6821. Note that if bit 2 of control register A is a 0, then a byte sent to address 8004H goes to data direction register A (DDRA). If bit 2 of control register A is a 1, then a byte sent to address 8004H goes to the port A register and is output from port A. Likewise, if bit 2 of control register B is a 0, then a byte sent to address 8006H goes to data direction register B (DDRB). If bit 2 of control register B is a 1, then a byte sent to 8006H goes to the port B register and is output on port B.

After a reset, all bits of the data direction registers and control registers are 0's. Therefore, all ports are inputs. To initialize a port line as an output, a 1 must be sent to the data direction register bit corresponding to that line. After the data direction word is sent to the data direction register, a control word with bit 2 a 1 must be sent to the control register. After the control word is sent to gain access to the port register, data can be output to the port. Figure 10-8 shows an assembly language routine to initialize port A of a 6821 as all outputs, the upper four lines of port B as inputs, and the lower 4 bits of port B as outputs. The base address for the device is 8004H.

The two control registers are first loaded with all 0's by the CLR instructions to gain access to the data direction registers. This is only required if a reset is not done. A byte with all 1's (FFH) is loaded into the *A* accumulator and sent to the A data direction register of the 6821. This configures port A as all output lines. Next, a byte with bit 2 a 1 (04H) is loaded into the *A* accumulator and sent to control register A of the 6821. This switches access from DDRA to port A, so data can be written to the port. The procedure is repeated for port B, except that 0's are sent to the bits in the data direction register corresponding to input lines. LDAA $50 copies a byte of data from address 0050H. Then STAA $8004 sends this byte of data out port A. LDAB $8006 reads a byte of data from port B into the *B* accumulator. A port line can still be read, even if it is set as an output.

The process is summarized as follows. First reset, or send control words with bit 2 equal to 0. Second, send data direction words with a 1 for each output line to the data direction registers. Third, send new control words with bit 2 equal to 1, so that port registers can be read from or written to.

The last part of the circuit in Figure 10-5 to discuss is the MC6850 asynchronous communication interface adapter, or ACIA. This device is a programmable UART that functions similarly to the asynchronous mode of the 8251 discussed in Chap. 9. It must be initialized before use.

Figure 11-30 shows a block diagram of the MC6850 ACIA. The device requires an external clock of 1, 16, or 64 times the desired baud rate. For the system in Figure 10-5, a 4800-Hz clock is used. This is 16 times the desired 300-Bd rate.

Three chip-select inputs (CS0, CS1, and $\overline{CS2}$) select the device. The enable input is connected to a delayed version of the $\phi2$ clock. A register select input is used to select data registers or control/status registers. A high on this input selects the data transmit/receive registers. This input is usually connected to the A0 address line. The addresses of the MC6850 in the system of Figure 10-5 are 8008H for the control/status register and 8009H for the data port register.

```
CLR        $8005        ;SET CONTROL REGISTERS

CLR        $8007        ;BIT 2 TO ZERO TO ACCESS DDR'S

LDAA       #$0FF        ;DATA DIRECTION WORD FOR ALL OUTPUTS

STAA       $8004        ;SEND TO PORT A DATA DIRECTION REGISTER

LDAA       #$04         ;CONTROL WORD FOR PORT ACCESS

STAA       $8005        ;SEND TO CONTROL A

LDAA       #$0F         ;DDR WORD FOR UPPER 4 IN, LOWER 4 OUT

STAA       $8006        ;SEND TO PORT B DDR

LDAA       #$04         ;CONTROL WORD FOR PORT ACCESS

STAA       $8007        ;SEND TO CONTROL B

LDAA       $50          ;GET DATA BYTE TO SEND

STAA       $8004        ;SEND TO PORT A

LDAB       $8006        ;READ PORT B TO ACC B
```

FIGURE 10-8 Assembly language routine to initialize MC6821 PIA at 8004H.

6800 SYSTEM TIMING

INSTRUCTION CYCLES AND MACHINE CYCLES

As indicated earlier, the MC6800 requires a two-phase nonoverlapping clock. For the 6800 a machine cycle is always equal to one clock period measured from the rising edge of $\phi1$. An instruction cycle consists of two to eight machine cycles, as shown in Table 10-1. Since the number of machine cycles is equal to the number of clock cycles, it is quite easy to calculate the execution time for any instruction by multiplying the number of machine cycles by the clock period.

READ CYCLE

The 6800 has only three basic types of machine cycle: a read cycle, a write cycle, and an internal operation cycle. During an internal operation cycle, the buses are not being used. So from a signal standpoint, only the read and write cycles are of interest. Figure 10-9 shows the simple timing waveforms and values for an MC6800 read cycle. Within time t_{AD} from the rising edge of $\phi1$, the CPU sends out a read signal, an address, and a valid memory address signal. Memory has a time t_{ACC} to respond with valid data and must hold the data stable for a time t_{DSR}. The falling edge of the $\phi2$ clock signals the end of the operation, and the rising edge of the next $\phi1$ starts the next cycle.

If the access time of the memory is longer than the t_{ACC} allowed by the CPU, the clocks can be "stretched" to increase the available t_{ACC}. This is done by making the $\overline{HOLD2}$ input of the MC6871B low. The $\phi1$ clock is held low and the $\phi2$ clock held high until $\overline{HOLD2}$ is made high. The clocks can only be stretched for a maximum of 5 μs, because internal dynamic registers are not being refreshed during this time. The $\overline{HOLD2}$ input signal must be synchronized to go low while $\phi1$ is high. The $\overline{HOLD1}$ input of the MC6871 is used to stretch the clocks with $\phi1$ high and $\phi2$ low.

WRITE CYCLE

A write cycle for the MC6800 is very similar to a read cycle. Within time t_{AD} after $\phi1$ goes high, the CPU puts out a write signal, addresses, and a VMA signal. When the CPU receives a data bus enable (DBE) signal, it outputs data on the data bus. DBE is usually just a delayed $\phi2$ clock, so the data will be valid during the later part of $\phi2$ high.

6800 INTERRUPTS

Figure 10-10 summarizes the effects of the 6800 $\overline{RESET}$ ($\overline{RESTART}$), $\overline{IRQ}$, and $\overline{NMI}$ hardware inputs as well as the WAI and SWI instructions. The point is not to memorize all this, but to learn how to use Figure 10-10 to help you determine the response of the 6800 when you need to.

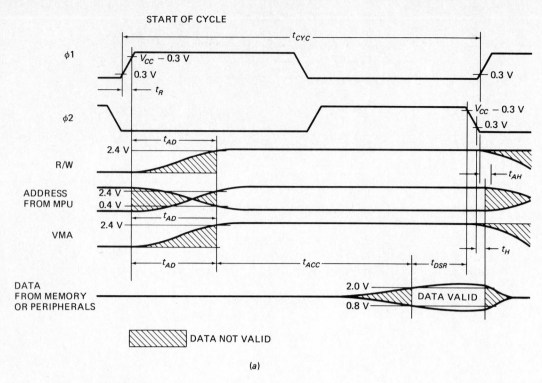

START OF CYCLE

DATA NOT VALID

(a)

Read–Write Timing

Characteristic	Symbol	Min	Typ	Max	Unit
Address Delay	t_{AD}	—	220	300	ns
Peripheral Read Access Time $t_{ACC} = t_{UT} - (t_{AD} + t_{DSR})$	t_{ACC}	—	—	540	ns
Data Setup Time (Read)	t_{DSR}	100	—	—	ns
Input Data Hold Time	t_H	10	—	—	ns
Output Data Hold Time	t_H	10	25	—	ns
Address Hold Time (Address, R/W, VMA)	t_{AH}	50	75	—	ns

(b)

FIGURE 10-9 MC6800 read cycle timing. *(Motorola.)*

RESTART

As shown on the right side of Figure 10-10, a low on the RESET (RESTART) input of the 6800 causes it to go to address FFFEH and fetch the most significant byte of the restart address stored there. The 6800 then increments its program counter to FFFFH and fetches the least significant byte of the restart address stored there. After setting the interrupt mask (bit 4 of the condition code register), the 6800 loads the restart address into its program counter and fetches its next instruction from that address. Since the 6800 always goes to FFFEH and FFFFH to get its restart vector (address), the programmer must make sure to have the starting address of the RESTART routine stored there.

NMI

A low on the hardware nonmaskable interrupt (NMI) input of the 6800 causes it to finish its current instruction and push the return address, index register, accumulators, and condition code register on the stack. As shown by the vertical arrow next to RESTART in Figure 10-10, the 6800 then goes to address FFFCH to get the most significant byte of the service routine starting address. Then it goes to address FFFDH to get the least significant byte of the service routine starting address. After setting the interrupt mask to disable any other hardware interrupts, the 6800 loads the interrupt vector (address) that it got from FFFCH and FFFDH into its program counter. It fetches its next instruction from this address. The programmer, therefore, must load the

INTERRUPTS:

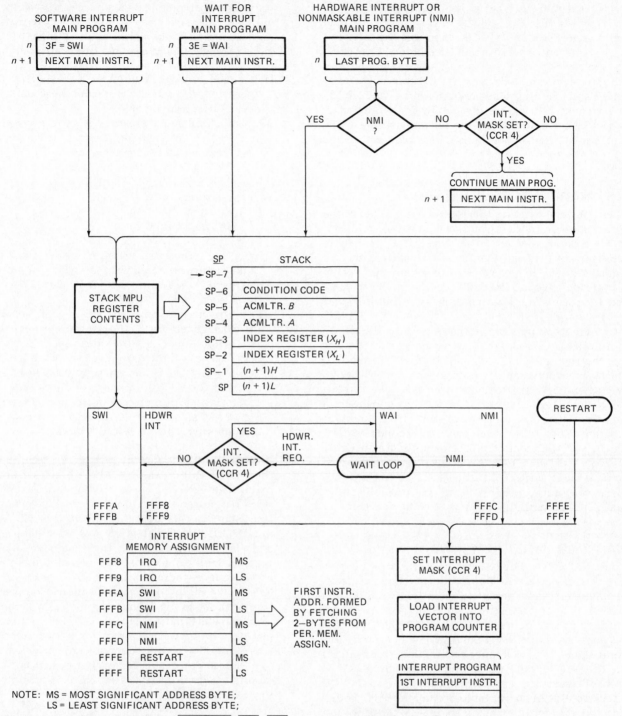

NOTE: MS = MOST SIGNIFICANT ADDRESS BYTE;
LS = LEAST SIGNIFICANT ADDRESS BYTE;

FIGURE 10-10 Effects of MC6800 $\overline{\text{RESTART}}$, $\overline{\text{NMI}}$, $\overline{\text{IRQ}}$ inputs and the SWI and WAI instructions.

starting address for the $\overline{\text{NMI}}$ interrupt service routine into FFFCH and FFFDH. A return-from-interrupt instruction (RTI) at the end of the service routine pulls all the registers off the stack and returns execution to the next instruction of the main program. If other hardware interrupts need to be recognized in the main program, a CLI instruction must be used before the RTI to unmask this interrupt.

INTERRUPT REQUEST

A low on the $\overline{\text{INTERRUPT REQUEST}}$ ($\overline{\text{IRQ}}$) input of the 6800 will have no effect unless the interrupt mask (condition code register bit 4) has been cleared by a CLI instruction. If the interrupt mask has been cleared, then an $\overline{\text{IRQ}}$ input causes the 6800 processor to push the return address, index register, accumulators, and condition code registers on the stack after it finishes its current instruction. Then it goes to address FFF8H to get the most significant byte of the interrupt vector (address) and to FFF9H to get the least significant byte of the interrupt vector. The 6800 then sets the interrupt mask to prevent the IRQ signal from immediately causing another interrupt. It loads the starting address of the service routine (interrupt vector) that it got from FFF8H and FFF9H into its program counter and fetches its next instruction from that address. A CLI instruction at the end of the service routine unmasks the interrupt. An RTI instruction pulls all the registers off the stack and returns execution to the main program.

SOFTWARE INTERRUPT (SWI)

When the 6800 executes this instruction, it pushes the return address, index register, accumulators, and condition code register onto the stack. Then it goes to FFFAH to get the most significant byte of the interrupt vector and to FFFBH to get the least significant byte of the interrupt vector. The 6800 sets the interrupt mask and then loads the starting address of the interrupt service routine that it got from FFFAH and FFFBH into the program counter. The processor executes the service routine. Some 6800 system monitor programs allow this instruction to be used as a program breakpoint. The monitor program is written so that when the 6800 executes an SWI instruction in your program, it interrupts your program, stores all the registers and program counter on the stack, and returns control to the monitor program. Thus you can employ the examine register function of the monitor program to discover whether the registers, etc., had the correct contents at that point. If the registers, etc., are correct, you can move the SWI breakpoint to a later point in your program, run it again, and recheck the registers, etc., at that point.

WAIT FOR INTERRUPT (WAI)

When the 6800 executes the wait-for-interrupt instruction WAI, it pushes the return address, index register, accumulators, and condition code registers on the stack. It then enters a wait-for-interrupt loop, as shown in the middle of Figure 10-10. The only exits from this loop are a hardware NMI input signal or a hardware IRQ input signal. If an NMI input occurs, the processor gets the NMI service routine starting address from FFFCH and FFFDH, masks the interrupt, and goes to the NMI service routine. If an IRQ input occurs, the 6800 cannot exit the wait-for-interrupt loop unless the interrupt mask bit was cleared previously by a CLI instruction.

TROUBLESHOOTING A 6800-BASED MICROCOMPUTER SYSTEM

For a malfunctioning 6800-based microcomputer system that once worked correctly, the following sequence of steps should help you troubleshoot it.

1. With an oscilloscope, check that the power-supply voltages are correct and do not have excessive ripple or noise. Excessive power-supply ripple can cause a microcomputer to do many strange things. It may work correctly sometimes and other times not.

2. Do a signal "roll call" around the CPU.
   a. Are the CPU's dual heartbeats, the $\phi 1$ and $\phi 2$ clocks, present and correct?
   b. Is the DBE signal present? Since this signal is produced by delaying $\phi 2$, it should look like $\phi 2$ when seen on an oscilloscope.
   c. RESET should be high.
   d. NMI and IRQ should be high.
   e. HALT should be high.
   f. TSC should be low.
   g. VMA should show pulses.
   h. With an oscilloscope, check that there is activity on the address bus and data bus. You will not be able to tell what exactly is going on, but pulses on the address lines at least show that the processor is sending out addresses.
   i. If the clock signals are not present, suspect the clock generator IC. If clocks are present but DBE is not, then check the gates used to produce DBE from $\phi 2$. If the gates used to debounce the RESET input fail, RESET may get stuck low. This will hold the 6800 in a reset condition, and no processing will take place. Failure of gates supplying an NMI signal to the 6800 could cause a continuous interrupt if it forced NMI low.

3. If steps 1 and 2 do not uncover the problem, hold the RESET button down with tape or a rubber band. This puts a continuous low on the 6800 RESET input and holds the CPU in the start of a reset state. In this state the address bus will contain FFFEH, BA should be low, VMA should be low, and R/W should be high. These signals can be checked with an oscilloscope. Signals on inputs and outputs of address buffers also can be checked. Release the RESET button, and connect your oscilloscope probe to the CS input of the ROM containing the restart routine. When the RESET button is pressed over and over, pulses should be seen on this CS input. If no pulses are present here, suspect the address decoder. If pulses are present but the monitor program doesn't execute, suspect the monitor ROM or CPU.

4. If you get the monitor program executing correctly, you can use short routines to write data to RAM and verify that the data can be read back correctly. Test routines can also be used to write to a port or read from a port over and over, so that you can look at its signals with a standard oscilloscope.

5. For more difficult problems, a logic analyzer can be used to follow the signal flow on the address, data, and control buses as the processor runs.

DESCENDANTS OF THE MC6800

In this section we briefly show Motorola's evolution of the MC6800 comparable with Intel's evolution of the 8080A to the 8085A and 8048 family.

THE MC6802

A first step in the MC6800 evolution is the MC6802. The device uses the same instruction set as the MC6800, but the IC contains a clock and 128 bytes of RAM. The RAM is in addresses 0000 to 007F, so that direct addressing can be used on data stored there. Since this RAM can easily be used as scratchpad registers, the register shortage problem of the 6800 is solved.

A companion IC for the MC6802 is the MC6846. It contains 2K ROM, 10 I/O lines, and a 16-bit timer. The two devices together form a small system.

THE MC6801

The MC6801 compresses even more functions in a single IC. It contains 2K of mask programmable ROM, 128 bytes of RAM, 31 I/O lines, a 16-bit timer, and a full UART. Figure 10-11 shows a functional block diagram of the MC6801.

The device has three modes of operation to choose from in regard to how the I/O lines are used. In applications where the internal RAM and ROM are enough, all lines can be used for I/O. This is called the *single-chip mode*. Another mode, the *expanded nonmultiplexed mode*, uses port 3 as the data bus and port 4 as the lower 8 address bits. This mode allows the MC6801 to communicate with peripheral ICs or 256 more memory locations.

The third mode, the *expanded multiplexed* mode, uses port 4 for the upper address bits. The lower address bits and the data bus are multiplexed on port 3. Port 1 and port 2 are still available for I/O. This mode allows the MC6801 to use a full 65K memory and peripherals.

In addition to the 6800 instruction set, the MC6801 has instructions to do 16-bit addition, 16-bit subtraction, and an 8-bit × 8-bit multiply to give a 16-bit result.

THE MC6809

The MC6809 is a semi- 16-bit processor. It executes all the instructions of the 6800 and an additional group of instructions that treat the 8-bit A and B accumulators as one 16-bit D register. This allows 16-bit arithmetic operations to be performed more easily. The MC6809 has two 16-bit index registers, two 16-bit stack pointers, a 16-bit program counter, an 8-bit condition code register, and a direct page register.

One main improvement in the MC6809 is the increased number of addressing modes. The direct page register allows direct addressing to be used within the page whose address is contained in the register.

Also the MC6809 has added several types of indirect addressing. With indirect addressing the CPU sends to memory to get the *address* of the next instruction, rather than the next instruction. An example of indirect addressing is the interrupt vector addresses of the 6800. In the case of an $\overline{\text{NMI}}$ signal, the CPU goes to ad-

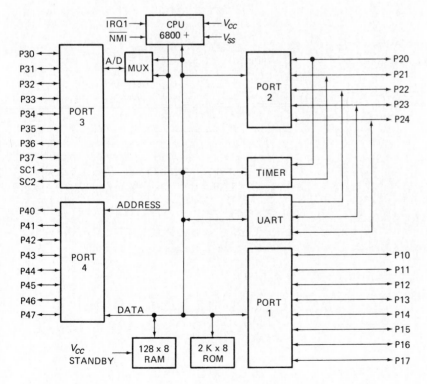

FIGURE 10-11 MC6801 functional block diagram. *(Motorola.)*

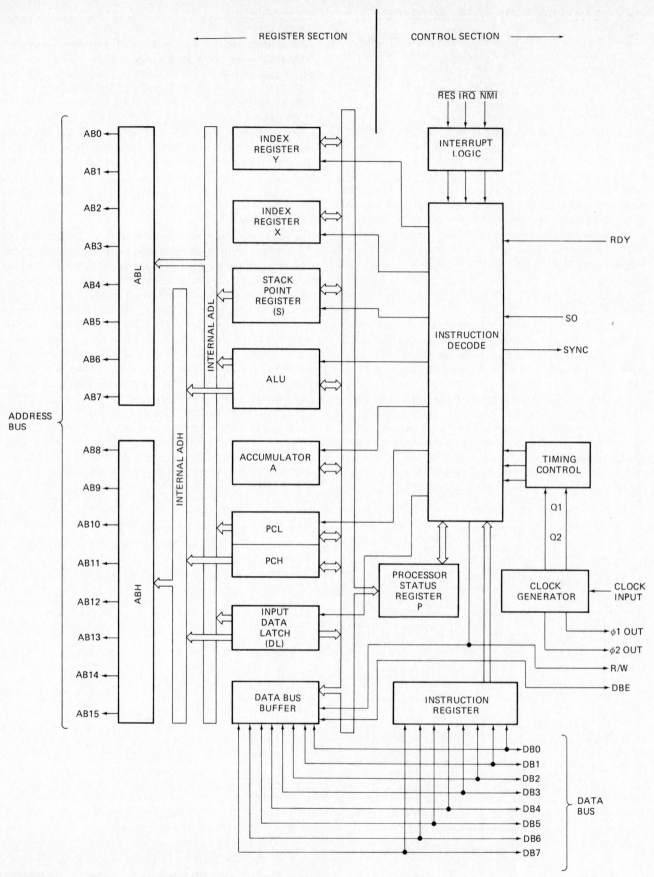

FIGURE 10-12 MOS technology MCS6502 internal architecture.

dresses FFFC and FFFD to get the *address* of the interrupt service routine start. In other words, the *effective address* for the next instruction to be executed is contained in FFFC and FFFD. This is called *absolute indirect addressing* because the CPU always goes to the absolute addresses to get the $\overline{\text{NMI}}$ routine address.

Extended indirect addressing uses the second and third bytes of the instruction to point to the memory location where the effective address is stored. The CPU fetches the effective address and then puts it out on the address bus to fetch the next instruction.

Indexed indirect addressing adds a displacement contained in the second byte of the instruction to the index register. This result points to the location in memory where the effective address is stored.

Relative indirect addressing adds the displacement contained in the second and third bytes of the instruction to the program counter. The result points to the memory location where the effective address is stored. Standard relative addressing of the MC6809 can have a 2-byte displacement rather than just the 1-byte displacement of the 6800.

These addressing modes are mainly useful for manipulating large tables of data. This book has no space to illustrate their applications. However, it is useful to understand their concept.

THE MOS TECHNOLOGY 6500 MICROPROCESSOR FAMILY

Soon after the introduction of the MC6800 by Motorola, MOS Technology, Inc., introduced the 6500 family as enhancements of the MC6800. The 6500 family was later second-sourced by Synertek and Rockwell International. The family at present has 10 versions that differ in package size, addressing capability, and the presence or absence of internal clock logic. A member of this family, the 6502, which has grown to a very prominent sales position, is discussed here. While the 6502 was designed as conceptually similar to the MC6800, it is not compatible with the MC6800 as the Z80A is with the 8080A/8085A. The 6502 is the CPU used in the Apple II computer.

ARCHITECTURE AND CONTROL SIGNALS

ARCHITECTURE   Figure 10-12 shows the architecture of the 6502. An internal clock generator produces the required two-phase clocks from a single TTL-level clock input. Therefore, no external clock generator such as the 6871B is needed.

The 6502 has only one 8-bit accumulator. It has an 8-bit stack pointer register; two 8-bit index registers, X and Y; and an 8-bit status register. Address and data buses are buffered to drive one TTL load.

CONTROL SIGNALS   One reason for the popularity of the 6502 is the simplicity of its control signals. For simple systems these control signals are quite adequate.

A read-write output indicates whether the processor is doing a memory read or write. Like the 6800, the 6502 has no direct I/O instructions, so all I/O is memory-mapped.

The $\overline{\text{RESET}}$, $\overline{\text{IRQ}}$, and $\overline{\text{NMI}}$ inputs are similar in operation to those with the same names on the 6800. If $\overline{\text{RESET}}$ is held low for eight or more clock cycles, the CPU goes to addresses FFFC and FFFD to get the beginning address of the desired restart routine in memory. $\overline{\text{NMI}}$ is the nonmaskable interrupt input. When $\overline{\text{NMI}}$ is made low, the processor completes its current instruction and then pushes the program counter and the status register contents on the stack. When this is done, it fetches the beginning address of the service routine from addresses FFFA and FFFB. The $\overline{\text{IRQ}}$ input is a maskable interrupt. If this interrupt is enabled and the $\overline{\text{IRQ}}$ input held low, the CPU pushes the program counter and the status register onto the stack. Then it vectors to addresses FFFE and FFFF to get the beginning address of the service routine.

The 6502 has no capability for clock stretching and no halt state. In fact, it has no condition where the buses are floated. External three-state buffers must be used if DMA operations are required. However, the 6502 can insert wait states in a maneuver similar to that of the 8080A. When the ready input (RDY) of the 6502 is pulled low during a read cycle, the CPU enters a wait state or cycle. If a write cycle is in progress, it is completed and a read cycle is entered before the processor will enter a wait state. As many wait cycles as necessary can be inserted. Wait states can increase the time available for slow memories or for single-stepping the processor, as described in Chap. 9.

The SYNC output sends out a positive pulse when the CPU is doing an instruction fetch. The set-overflow input (SO) sets the overflow flag in the status register when made low.

6502 FLAGS

Figure 10-13 shows the format for the 6502 status register. The carry, zero, overflow, and negative (sign) flags are identical to those discussed for the MC6800. The other three bear some discussion.

If the interrupt-disable flag, I, is a 1 then the $\overline{\text{IRQ}}$ interrupt input is masked. This flag can be set or reset by program instructions. It is also set by the CPU during a reset or an interrupt response and must be cleared before the processor can respond to an $\overline{\text{IRQ}}$ input.

The decimal-mode flag, D, if set, does an automatic decimal adjust on the results of any add or subtract operations. This puts the results in correct BCD form. The set-decimal-mode instruction (SED) sets this flag and CLD (clear decimal mode) resets this flag.

The break-command flag, B, is set by execution of the 6502 software interrupt instruction BRK. This flag is

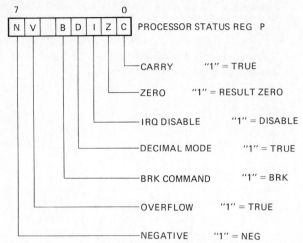

FIGURE 10-13 MCS-6502 status register format.

used by the processor to determine whether an interrupt came from software or hardware. There are no instructions which directly set or reset the break flag.

INSTRUCTION SET AND ADDRESSING MODES

Table 10-2 shows the 56 instructions of the 6502 with the addressing modes, number of bytes, number of clock cycles, and op codes for each. The function of most instructions is quite clear from the symbolic representation. In the key at the bottom of the table, note the symbols used for logical AND, OR, and XOR. Memory per effective address means the contents of the memory location at the effective address, and MS means the contents of the memory location pointed to by the stack pointer.

A striking feature of Table 10-2 is the 13 addressing modes available in the 6502. Some of these are used for only one or two instructions, however. Most modes you have already met in the discussion of the MC6800 and MC6809, but they are briefly discussed here as they apply to the 6502.

IMMEDIATE ADDRESSING   This contains the operand in the second byte of the instruction. For example, AND $7E logically ANDs the accumulator with 7EH and puts the result in the accumulator.

ABSOLUTE ADDRESSING   This corresponds to extended addressing for the 6800. In this mode bytes 2 and 3 of the instruction contain the 16-bit memory address where the operand is stored. For example, AND 3F08 logically ANDs the contents of the accumulator with the contents of memory location 3F08H. The instruction is actually stored in memory with the lower-order address byte first as 2D, 08, 3F.

ZERO-PAGE ADDRESSING   This corresponds to the direct addressing mode of the 6800. Sometimes it is called *base-page* addressing. The 6502 divides memory into pages of 256 bytes each. The upper 8 bits of an address can be thought of as giving the page number and the lower 8 bits as giving the address within a page. *Zero page* refers to the first 256 memory locations from 00H to FFH. With zero-page addressing, the 8-bit address of the operand is contained in the second byte of the instruction. For example, AND 3A logically ANDs the contents of the accumulator with the contents of memory location 003A.

ACCUMULATOR ADDRESSING   This is a special case of implied addressing where the accumulator already contains the operand. For example, ROL rotates the accumulator left through carry.

IMPLIED ADDRESSING   This has the source of the operand included in the instruction op code. For example, INX increments the X index register by 1.

RELATIVE ADDRESSING   For the 6502, this is the same as discussed for the 6800. A 2's-complement displacement contained in the second byte of the instruction is added to the number in the program counter after the instruction is executed. Only the branch instructions use this mode. A branch can be from −126 to +129 from the address where the branch instruction is stored. For example, BEQ 09H stored at 0072, when executed, will cause a branch to 007D.

ABSOLUTE X AND ABSOLUTE Y   These are indexed addressing modes. The operand is fetched from an address calculated by adding the contents of the X or Y index register to the number contained in the second and third bytes. The added value can be from 0 to 255 (00 to FFH). If the X index register contains 93H, the instruction AND 00F2,X will fetch the operand from location 0185H. If the addition of the index register to the contained address produces a carry-out of the lower 8 bits, it will be added to the upper 8 bits to give the correct 16-bit address in the program counter. This is called a *page boundary crossing.*

ZERO-PAGE X AND ZERO-PAGE Y   These are also indexed addressing modes. An 8-bit zero-page address in the second byte of the instruction is added to the index register to get the effective address. For example, if X contains 33H, then AND 0A,X will AND the contents of memory location 3D with the accumulator. The result will be in the accumulator. Care must be taken in using this mode because no carry is produced if the sum of the zero-page address plus the index is greater than FFH. If FFH is exceeded, the program will wrap around into the zero page. This is called a *page boundary error.*

INDIRECT ADDRESSING   This was discussed for the MC6809. In this mode the second 2 bytes of the instruction contain the address of the effective address. For example, JMP FF08 sends the program counter to FF08 and FF09 to fetch the address where the next instruction is stored.

Clock Input	Processor	OP Code	Operand	Cycles	Time
2 MHz	R6502A	ADC	Immediate	2	1 μs
2 MHz	68B00	ADCA	Immediate	2	1 μs
5 MHz	8085A	ACI	Immediate	7	1.4 μs
4 MHz	Z80A	ADC	A, Immediate	7	1.75 μs

FIGURE 10-14 Execution time comparison for four processors.

INDEXED INDIRECT ADDRESSING (IND,X)
A zero-page address in the second byte of the instruction is added to the X index register with no carry. The sum is a zero-page address. The CPU goes to this resultant zero-page address and the one after it to get the effective address. For example, if X contains 33H, the instruction AND (IND 0A,X) will go to addresses 3DH and 3EH to get the *address* of the byte to be ANDed with the accumulator.

INDIRECT INDEXED ADDRESSING (IND), Y
In this mode the second byte of the instruction contains a zero-page address. From this zero-page address and the one after it, the CPU fetches a stored address. The contents of the Y index register are added to this fetched address to get the effective address. This is sometimes called *post-indexed addressing* because the index Y is added after the indirect address is fetched. The process described for IND,X is pre-indexed because the index is added before fetching the indirect address.

For application examples of all these addressing modes, consult the *MOS Technology MCS6500 Programming Manual*.

A SIMPLE 6502 SYSTEM
As with the other microprocessors discussed in this book, the 6502 has a family of peripheral ICs that can be connected with it to form a system.

The 6522 is a peripheral interface adapter very similar to the MC6821 PIA discussed earlier. The device contains two 8-bit I/O ports and four control lines as well as two programmable timers.

Another useful peripheral IC is the 6530. It contains 1K bytes of ROM, 64 bytes of RAM, a programmable interval timer, and two I/O ports. Standard RAM or ROM also can be used with the 6502 to build a system.

6502 CYCLES, TIMING, AND PIPELINING
As with the 6800, a machine cycle for the 6502 is equal to one period of the $\phi 1$ clock. The 6502A operates with up to a 2-MHz input clock. This gives a machine cycle time of 500 ns. Using the number of machine cycles given for each instruction in Table 10-2, you can compute their execution times.

An important point here is that you cannot judge the speed of a processor just by the input clock frequency. Figure 10-14 compares the execution times of equivalent instructions on the fastest currently available versions of four processors discussed in this book. The R6502A and MC68B00 each execute the instruction with a 2-MHz input clock in less time than the Z80A does with a 4-MHz clock or the 8085A does with a 5-MHz clock.

The R6502A and MC68B00 can have a shorter instruction cycle time with a lower-frequency clock because they make extensive use of a technique called *pipelining*. Pipelining means fetching the next byte from memory while interpreting the byte fetched last. An example will show how this works. According to Table 10-2, the ADC instruction with zero-page addressing takes three clock cycles to execute. Figure 10-15 shows the activities and timing of this execution. Taken by itself, the instruction requires four steps. However, since step 1 and step 4 each overlaps with the adjacent instruction, effectively only three cycles are needed to execute the instruction.

To be fair, we should note at this point that the relative speed of a processor in executing a program cannot be judged accurately from the execution time of a single instruction. Test programs called *benchmark programs* compare the overall speeds of microprocessors.

Performance depends greatly on the benchmark used. For example, a benchmark using a parity check would probably run faster on an 8085A or a Z80A because they

Cycle	Processor Activity
1	Finish previous instruction, fetch OP code
2	Interpret OP code, fetch zero page effective address
3	Fetch number from zero page effective address
4	Add fetched number to accumulator, fetch next OP code

FIGURE 10-15 Processor operations for the ADC instruction with zero-page addressing.

TABLE 10-2
R6502 INSTRUCTION SET

Each addressing-mode cell lists: OP (hex) / n (cycles) / # (bytes). Processor Status Code columns: 7=N, 6=V, 5=—, 4=—, 3=D, 2=I, 1=Z, 0=C.

MNEMONIC	OPERATION	Immediate	Absolute	Zero Page	Accum.	Implied	(Ind,X)	(Ind),Y	Z.Page,X	ABS,X	ABS,Y	Relative	Indirect	Z.Page,Y	Processor Status Code
ADC	A+M+C→A (1)	69 2 2	6D 4 3	65 3 2			61 6 2	71 5 2	75 4 2	7D 4 3	79 4 3				N V · · · · Z C
AND	A∧M→A (1)	29 2 2	2D 4 3	25 3 2			21 6 2	31 5 2	35 4 2	3D 4 3	39 4 3				N · · · · · Z ·
ASL	C←[7 0]←0		0E 6 3	06 5 2	0A 2 1				16 6 2	1E 7 3					N · · · · · Z C
BCC	BRANCH ON C=0 (2)											90 2 2			· · · · · · · ·
BCS	BRANCH ON C=1 (2)											B0 2 2			· · · · · · · ·
BEQ	BRANCH ON Z=1 (2)											F0 2 2			· · · · · · · ·
BIT	A∧M		2C 4 3	24 3 2											M7 M6 · · · · Z ·
BMI	BRANCH ON N=1 (2)											30 2 2			· · · · · · · ·
BNE	BRANCH ON Z=0 (2)											D0 2 2			· · · · · · · ·
BPL	BRANCH ON N=0 (2)											10 2 2			· · · · · · · ·
BRK	BREAK					00 7 1									· · · 1 · 1 · ·
BVC	BRANCH ON V=0 (2)											50 2 2			· · · · · · · ·
BVS	BRANCH ON V=1 (2)											70 2 2			· · · · · · · ·
CLC	0→C					18 2 1									· · · · · · · 0
CLD	0→D					D8 2 1									· · · · 0 · · ·
CLI	0→I					58 2 1									· · · · · 0 · ·
CLV	0→V					B8 2 1									· 0 · · · · · ·
CMP	A-M	C9 2 2	CD 4 3	C5 3 2			C1 6 2	D1 5 2	D5 4 2	DD 4 3	D9 4 3				N · · · · · Z C
CPX	X-M	E0 2 2	EC 4 3	E4 3 2											N · · · · · Z C
CPY	Y-M	C0 2 2	CC 4 3	C4 3 2											N · · · · · Z C
DEC	M-1→M		CE 6 3	C6 5 2					D6 6 2	DE 7 3					N · · · · · Z ·
DEX	X-1→X					CA 2 1									N · · · · · Z ·
DEY	Y-1→Y					88 2 1									N · · · · · Z ·
EOR	A∀M→A	49 2 2	4D 4 3	45 3 2			41 6 2	51 5 2	55 4 2	5D 4 3	59 4 3				N · · · · · Z ·
INC	M+1→M (1)		EE 6 3	E6 5 2					F6 6 2	FE 7 3					N · · · · · Z ·
INX	X+1→X					E8 2 1									N · · · · · Z ·
INY	Y+1→Y					C8 2 1									N · · · · · Z ·
JMP	JUMP TO NEW LOC		4C 3 3										6C 5 3		· · · · · · · ·
JSR	JUMP SUB		20 6 3												· · · · · · · ·
LDA	M→A (1)	A9 2 2	AD 4 3	A5 3 2			A1 6 2	B1 5 2	B5 4 2	BD 4 3	B9 4 3				N · · · · · Z ·

Instruction set reference table (source: Rockwell). Addressing-mode data given as OP/n/# (hexadecimal op code / number of cycles / number of bytes).

Mnemonic	Operation	Immediate	Zero Page	Zero Page,X	Zero Page,Y	Absolute	Absolute,X	Absolute,Y	(IND,X)	(IND),Y	Accum	Implied	Cond. Codes
LDX	M → X (1)	A2/2/2	A6/3/2		B6/4/2	AE/4/3		BE/4/3					N Z
LDY	M → Y (1)	A0/2/2	A4/3/2	B4/4/2		AC/4/3	BC/4/3						N Z
LSR	0→[7 0]→C		46/5/2	56/6/2		4E/6/3	5E/7/3				4A/2/1		0 Z C
NOP	NO OPERATION											EA/2/1	
ORA	A∨M → A	09/2/2	05/3/2	15/4/2		0D/4/3	1D/4/3	19/4/3	01/6/2	11/5/2			N Z
PHA	A → Ms, S-1 → S											48/3/1	
PHP	P → Ms, S-1 → S											08/3/1	
PLA	S+1 → S, Ms → A											68/4/1	N Z
PLP	S+1 → S, Ms → P											28/4/1	(RESTORED)
ROL	[7 0]←C		26/5/2	36/6/2		2E/6/3	3E/7/3				2A/2/1		N Z C
ROR	C→[7 0]		66/5/2	76/6/2		6E/6/3	7E/7/3				6A/2/1		N Z C
RTI	RTRN INT											40/6/1	(RESTORED)
RTS	RTRN SUB											60/6/1	
SBC	A-M-C → A (1)	E9/2/2	E5/3/2	F5/4/2		ED/4/3	FD/4/3	F9/4/3	E1/6/2	F1/5/2			N V(3) Z C(1)
SEC	1 → C											38/2/1	C=1
SED	1 → D											F8/2/1	D=1
SEI	1 → I											78/2/1	I=1
STA	A → M		85/3/2	95/4/2		8D/4/3	9D/5/3	99/5/3	81/6/2	91/6/2			
STX	X → M		86/3/2		96/4/2	8E/4/3							
STY	Y → M		84/3/2	94/4/2		8C/4/3							
TAX	A → X											AA/2/1	N Z
TAY	A → Y											A8/2/1	N Z
TSX	S → X											BA/2/1	N Z
TXA	X → A											8A/2/1	N Z
TXS	X → S											9A/2/1	
TYA	Y → A											98/2/1	N Z

(1) ADD 1 TO n IF PAGE BOUNDARY IS CROSSED
(2) ADD 1 TO n IF BRANCH OCCURS TO SAME PAGE
    ADD 2 TO n IF BRANCH OCCURS TO DIFFERENT PAGE
(3) CARRY NOT = BORROW
(4) IF IN DECIMAL MODE Z FLAG IS INVALID, ACCUMULATOR MUST BE CHECKED FOR ZERO RESULT

SOURCE: Rockwell.

X   INDEX X
Y   INDEX Y
A   ACCUMULATOR
M   MEMORY PER EFFECTIVE ADDRESS
Ms MEMORY PER STACK POINTER

+   ADD
−   SUBTRACT
∧   AND
∨   OR
∀   EXCLUSIVE OR

M7 MEMORY BIT 7
M6 MEMORY BIT 6
n   NO CYCLES
NO BYTES

have a parity flag that the 6800 and 6502 do not. However, a benchmark that involves operating on data in a memory buffer and restoring it to a buffer might be faster on the 6800 and 6502 because they have direct and indexed addressing.

SUMMARY AND COMPARISON OF CURRENTLY AVAILABLE MICROPROCESSORS

Currently available microprocessors, for the most part, fall into four major categories: general-purpose CPUs, single-chip microcomputers, super processors, and bit-slice processors.

GENERAL-PURPOSE MICROPROCESSORS

Examples of general-purpose microprocessors are the Intel 8085A, the Zilog Z80A, the Motorola MC6800, and the MOS Technology 6502. These devices use external ROM, RAM, and I/O port ICs to build a microcomputer system of any size.

ONE-CHIP MICROCOMPUTERS

Devices such as those of the Intel 8048 family and the Motorola MC6801 contain ROM, RAM, clock logic, and I/O ports all in a *single* IC. External ROM and RAM can be added to these, but for many applications, such as replacing random logic in "smart" instruments, the on-board ROM and RAM is sufficient.

SUPER PROCESSORS

One direction of microprocessor evolution has been toward devices that perform all the CPU functions of a *minicomputer*. Minicomputers typically have 16- or 32-bit-wide data buses. Their address buses are also very wide, so they can address megabytes of memory. Minicomputers have the ability to work with words, bytes, nibbles, bits, and strings of characters. Instruction cycle times are typically about 1 μs.

Examples of minicomputer-type CPUs are the Intel iAPX 86/10 and iAPX 88/10, the Motorola MC68000, and the Texas Instruments TMS 9900. In the following subsections we give a brief overview of each.

THE iAPX 86/10 and iAPX 88/10   The iAPX 86/10, commonly called the 8086, is a logical extension of the 8080A to a 10-MHz, 16-bit processor. It contains only a CPU in its 40-pin package. An *external 8284 clock generator*, an *external 8288 bus controller*, and *external address demultiplexers* are needed to form a CPU group. Sixteen data and 20 address lines are multiplexed out on a 20-bit bus. The 20 address lines permit up to 1,048,576 bytes of memory to be accessed.

As shown in Figure 10-16, the iAPX 86/10 is divided into two independent functional parts, the *bus interface unit* (BIU) and the *execution unit* (EU). The BIU handles all transfers of data and addresses on the buses for the EU. To speed up execution, the BIU fetches as many as 6 instruction bytes ahead of time from memory and holds them in the *QUEUE registers* for the EU. The BIU can be fetching instructions while the EU is decoding an instruction or executing an internal instruction. Also, the EU can get its next instruction from the QUEUE faster than it could from memory. Except for the case of jump instructions, this *prefetch and QUEUE scheme* greatly speeds up processing.

The BIU sends out a 20-bit address so it can address 1,048,576 addresses. However, at any one time the BIU works with only four 64K-byte *segments* within this address space. The upper 16 bits of the base address for each of these four segments are contained in the *segment registers* of the BIU. The lower 4 bits are assumed all 0's. Segments may overlap. The *code segment* is used for instructions, the *stack segment* for stacks, and the *data* and *extra segments* for data. The advantage of this segmented-memory approach is that several programs can be present in different areas of the memory space. The processor can switch from one program to another by simply changing the contents of the four segment registers.

Also contained in the BIU is the *instruction pointer register* (IP) which functions as a program counter. To compute the address from which to fetch an instruction, the BIU shifts the contents of the code segment register four bit positions left and adds the contents of the instruction pointer register. As shown in Figure 10-16b, the result is a 20-bit address. The instruction pointer then specifies an offset within the code segment.

The *execution unit* (EU) of the iAPX 86/10 contains an *arithmetic logic unit, nine flag flip-flops,* and *eight 16-bit registers*. As shown in Figure 10-16, the A, B, C, and D registers can be used as 8- or 16-bit registers. The AL register corresponds to the 8085A accumulator; the BL and BH to the 8085A HL; the CH and CL to the 8085A BC; and the DH and DL to the 8085A DE. This similar register structure makes it easy to translate 8085A programs to run on an iAPX 86/10.

The other four registers in the EU are the *stack pointer* (SP), *base pointer* (BP), *source index* (SI), and *destination index* (DI). These index registers give the device a rich array of addressing modes. Figure 10-16c shows the ways in which an address can be computed. The EU first computes an *effective address*, or EA. One way it does this is by adding the contents of BX, BP, SI, or DI to a displacement contained in an instruction. Another way is to add the contents of BX or BP, plus the contents of SI or DI, plus a displacement contained in an instruction. The computed effective address (EA) is passed to the BIU. There it is added to the *shifted* contents of a segment register to generate a 20-bit address.

In addition to the usual microprocessor instructions, the iAPX 86/10 instructions include several forms of

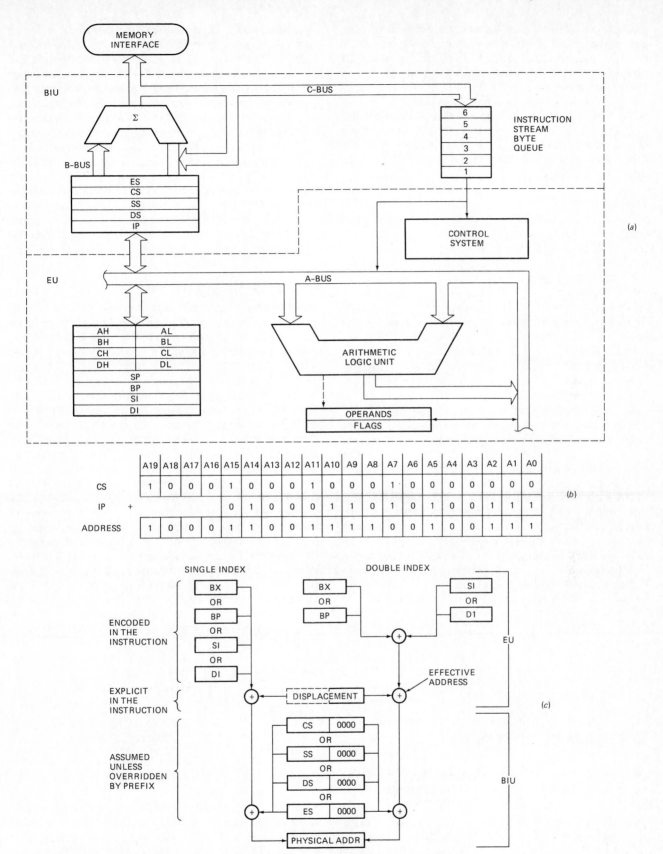

FIGURE 10-16 *(a)* Intel 8086 functional block diagram. *(b)* 8086 address computation using code segment register and instruction pointer. *(c)* Memory address computation.

multiply and divide. *Indexed* and *indirect addressing* modes are available, and the iAPX can work with bits, bytes, strings of ASCII characters, words, or blocks. It can do arithmetic operations directly on numbers in ASCII character form.

The iAPX 88/10 has the same instruction set as the iAPX 86/10, and it has similar internal architecture. However, it has only an 8-bit multiplexed data bus. This makes it easier to interface with 8-bit programmable peripheral devices. The iAPX 88/10 is the CPU used in the *IBM Personal Computer*.

THE MC68000   The Motorola MC68000 is also a 16-bit processor. It is housed in a 64-pin package so that the 23-bit *address bus* and 16-bit *data bus* are not multiplexed. These 23 address lines plus 1 that is multiplexed let the 68000 address up to 16 megabytes.

Borrowing a page from the register-oriented processors, Motorola has included eight 32-bit *general-purpose* registers, seven 32-bit *address* registers, and two 32-bit *stack pointers* in the 68000. Although the 68000 is considered a 16-bit processor, it has a 32-bit-wide internal structure. Also included are a 32-bit *program counter* and a 16-bit *status register*. The 68000 can use a single-phase clock of any frequency from dc to 12 MHz.

The 68000 has only 56 basic instruction types, but it has *14 addressing modes* which are available for most basic instructions. Therefore, the total of possible operations is large. The MC68000 also includes hardware and software features to aid debugging.

TEXAS INSTRUMENTS TMS 9900 FAMILY   The TMS 9900 family of microprocessors contains both NMOS and I²L versions. The CPUs contain a *program counter, workspace register,* and *status register*. The most notable feature of this family is that most members do not have an internal accumulator or other data registers. A 256-byte section of RAM is set aside for use as *sixteen 16-bit general-purpose registers. General*

purpose means that any one can be used as an accumulator, as a counter, or for data storage. The starting memory address for this *register bank* is kept in the workspace register in the CPU. This system of using memory for all registers is called *memory-to-memory* architecture. An advantage of this architecture is that an interrupt service routine can access a new bank of registers just by putting a new value in the workspace register. This technique is called *context switching*. It is faster than pushing a set of registers on a stack. However, many instruction times are longer because each operation requires a read from memory and a write to memory. This takes longer than accessing registers on the same chip. The TMS 9995, a 12-MHz member of this family, contains 256 bytes of on-board RAM to solve this problem.

BIT-SLICE PROCESSORS

Another type of processor not previously discussed is the *bit-slice* type. For applications where the general-purpose microprocessors are not fast enough or do not have a desired instruction, bit slices are used to custom build a CPU.

The starting block, or slice, is a 4- or 8-bit ALU with perhaps some registers and multiplex circuitry. The basic bit slices can be connected in parallel to handle wider words. To perform useful operations, the bit slice must be told every little step it must perform. These simple steps are called *microinstructions*. The designer actually creates each instruction by specifying a series of microinstructions for it. A sequence of microinstructions can be stored in a ROM. An instruction decoder selects the desired sequence for each instruction. A series of microinstructions corresponds to a machine language instruction for the processors such as the 8080A. Fortunately, most applications do not require a CPU to be built from bit slices. Examples of bit-slice processors are the Motorola 10800 ECL series and the Advanced Micro Devices 2900 Schottky TTL series.

REVIEW QUESTIONS AND PROBLEMS

1. Draw a block diagram of the MC6800 architecture. What does the 6800 use in place of internal scratchpad registers?

2. Describe the 6800 CPU response to a low on the $\overline{\text{RESET}}$ input. How is this different from the 8080A response?

3. Describe the 6800 response to a low on the $\overline{\text{NMI}}$ input.

4. What condition is indicated by the 6800 overflow flag?

5. Define the following 6800 addressing modes:
 a. Immediate *e.* Implied
 b. Direct   *f.* Accumulator
 c. Indexed *g.* Relative
 d. Extended

6. If the index register contains 227AH, where will the CPU go to get the operand for the instruction ADDB $3B,X?

7. If the instruction BRA $32 is stored at memory locations 07A5H and 07A6H, to what address will the

CPU go to get its next instruction? *Note:* Assume hand assembly, so displacement is measured from the address after that where displacement is stored.

8. Using Table 10-1, describe the operation performed by the following 6800 instructions:

   a. ADDA #$2A   *f.* STAB $0076
   b. BITA $09   *g.* WAI
   c. LDX $345E   *h.* RTI
   d. SWI   *i.* TST $31,X
   e. ROR $27,X   *j.* BGE $10

9. Describe the 6800 wait state. How is it different from the 8080A wait state?

10. Rewrite the WAITMS routine in Figure 10-3 to give a 0.3-ms delay.

11. Write a short assembly language routine for the 6800 to subtract two multibyte numbers.

12. Write an assembly language routine for the 6800 to multiply two 8-bit numbers by successive additions.

13. Write an assembly language routine for the 6800 to convert an ASCII character for 0–9 and A–F to the equivalent 4-bit hexadecimal. Consult Table 3-2 for the ASCII code. Assume the ASCII is in the *A* accumulator and the result is to be stored in the lower 4 bits of accumulator *B*.

14. Describe the condition present on the buses during a 6800 halt state.

15. For the system in Figure 10-5, what address block in memory does the $\overline{2/3}$ output of the 74155 enable? Draw a memory map for the system in Figure 10-5.

16. Define ACIA and PIA.

17. Calculate the execution times for the 6800 programs in Figure 10-4*a* and *b*. Assume a 1-MHz clock. For Figure 10-4*b* let $N = 5$.

18. *a.* Show the assembly language instructions required to initialize an MC6821 with base address 8010H as follows: Ports A7 to A2 as outputs, ports A1 to A0 as inputs, and ports B7 to B0 as outputs. Assume that a $\overline{\text{RESET}}$ has occurred.
   b. Write an assembly language program that turns an LED connected to bit 7 of port A on for 1 s and off for 1 s.

19. *a.* Show how you would add another 6821 PIA to the system in Figure 10-5 to make its starting address 8040H.
   b. How you would connect a 2732 EPROM in the system of Figure 10-5 so that its starting address is 4000H?

20. For a 6800-based microcomputer system such as that in Figure 10-5, describe the steps you would take to determine the source of the following problems:
   a. Power supply OK, clock present, no processor activity, address bus shows FFFEH.
   b. Power supply OK, clock present, activity on address bus after reset, system does not execute monitor program.
   c. Monitor program executes, but you cannot write data to RAM.

21. Describe the improvements that the MC6801 has over the MC6800.

22. Define *(a)* indirect addressing, *(b)* extended indirect addressing, *(c)* indexed indirect addressing, and *(d)* relative indirect addressing.

23. Rewrite the program of Figure 10-2 to operate on the 6502A with a 2-MHz clock.

24. Define zero-page addressing and pipelining.

25. For the 6502, what is the difference between indexed indirect addressing and indirect indexed addressing?

26. How do the 8086, Z8000, and MC68000 resemble minicomputers?

REFERENCES/BIBLIOGRAPHY

MEK6800D2 Manual, Motorola Semiconductor Products, Inc., Austin, TX, 1976.

Osborne, Adam: *An Introduction to Microcomputers,* vol. 2: *Some Real Products,* Adam Osborne and Associates, Inc., Berkeley, CA, 1976.

M6800 Microcomputer System Design Data, Motorola Semiconductor Products, Inc., Phoenix, AZ, 1976.

MC6800 Programming Reference Manual, M68PRM(D), Motorola Semiconductor Products, Inc., Phoenix, AZ, 1976.

The 8086 Family User's Manual, Intel Corporation, Santa Clara, CA, October 1979.

Microprocessor Application Manual, Motorola Semiconductor Products, Inc. (Contributors: Alvin W. Moore,

Vinay Khanna, Gary G. Sawyer, Thomas C. Daly, Karl E. Fronheiser, John M. DeLaune, Mark E. Eidson, James F. Vittera), McGraw-Hill, New York, 1975.

Zilog Z8000 Advance Specification, Zilog, Inc., Cupertino, CA, April 1978.

Motorola Semiconductor Data Sheets: MC6800, MC68A00, MC68B00, MC6801, MC6802, MC6809(E), MC68A09(E), MC68B09(E), MC68000(L4), (L6), & (L), Motorola Semiconductor Products, Inc., Austin, TX, 1978, 1980.

Rockwell R6500 Microcomputer System Data Sheet, document no. 29000 D39, revision 2, August 1978, parts R650X and R651X, Rockwell International Corporation, Anaheim, CA, 1978.

Simpson, William D., et al.: *9900 Family Systems Design and Data Book*, Texas Instruments, Inc., Houston, 1978.

CHAPTER 11

INPUT AND OUTPUT INTERFACING

Most of the work involved in developing a microprocessor-based instrument involves programming and I/O interfacing. In this chapter we collect and describe some very important techniques and devices used to interface microprocessors to the real world.

At the conclusion of this chapter, you should be able to:

1. Define polled and interrupt types of I/O.

2. Write interrupt service subroutines for an 8080A/8085A or a 6800.

3. Describe the sequence of signals during a handshake input or output data transfer.

4. Initialize a programmable parallel port IC for handshake input or output data transfer.

5. Describe the detect, debounce, and encode operations of a keyboard encoder.

6. Draw a flowchart for a program to scan and decode a keyboard.

7. Draw a flowchart for multiplexing an eight-digit, seven-segment display.

8. Draw a circuit showing how a D/A or an A/D converter is connected to a microprocessor.

9. Describe how to input data from a nonmultiplexed or multiplexed output A/D converter.

10. Define S-100, multibus, and IEEE-488.

11. Define RS-232-C, space, mark, 20-mA current loop, RS-422, RS-423, MODEM, and frequency shift keying.

12. Initialize a programmable serial port IC such as the 8251A or MC6850 for asynchronous serial data communication.

13. Describe how digital data is sent in serial form using BISYNC or SDLC format.

14. Explain how alphanumeric characters are displayed on a CRT.

SIMPLE, POLLED, AND INTERRUPT I/O

SIMPLE INPUT OR OUTPUT

For many applications a microcomputer can just read data from a parallel port when it needs the data or send data to a parallel port when it has data ready to send. An example of simple output would be turning a light connected to a port line on and off every 30 s.

POLLED I/O

In the *polled I/O* mode, a ready signal line from some external device is read over and over by the microcomputer through a port line. When this ready signal is found asserted, the microcomputer reads a byte of data sent to a port by the external device, or sends a byte of data to the external device. The programs in Figures 8-21 and 10-2 are examples of the polled input of data. The ready signal might be the keypressed strobe from an ASCII-encoded keyboard. When the strobe is asserted, valid ASCII code for the key pressed is on the parallel outputs from the encoder and can be read in by the microcomputer.

The disadvantage of polled I/O is that while the microcomputer is polling the ready signal, it cannot easily be doing other tasks. In systems where the microcomputer must be doing many tasks, polling is a waste of time. Interrupt-driven I/O is therefore used in these systems.

INTERRUPT I/O

Almost all microcomputers have one or more interrupt inputs. An interrupt input, if enabled, allows an external signal to tell the microcomputer to finish the instruction that it is executing and then jump to a subroutine that responds to the need indicated by the interrupt. At the end of the interrupt service subroutine, a return instruction can send the CPU back to the program it was running before it was interrupted.

Interrupts have several important functions and advantages. They can relieve the microcomputer of the burden of polling an input over and over to see if a signal such as a keypressed strobe is present. They allow a microcomputer to respond very quickly to a catastrophe or a situation requiring immediate attention. They can also relieve the microcomputer of the burden of timing and delay loops.

In the following sections we discuss the hardware and software considerations when interrupts are used. Following that, examples of interrupt use are shown for both the 8085A and the 6800.

HARDWARE CONSIDERATIONS Important hardware points to consider when you use interrupts with a particular microcomputer are as follows:

1. How many interrupt inputs does the microcomputer have?

2. Do these inputs have priorities?

3. Do these inputs require active high, active low, or edge-active signals to assert them?

4. For how long do the input signals have to be asserted?

5. Is external hardware required to insert a restart instruction, or is this done automatically by the CPU when it responds to the interrupt request?

SOFTWARE CONSIDERATIONS

1. What instructions are required to enable and/or unmask the interrupts?

2. How is the stack pointer initialized?

3. Does the CPU automatically save main program register contents and flags by pushing them on the stack when it responds to the interrupt? Or does the interrupt service routine have to use push instructions at its start to do this?

4. How can data required by the service routine be accessed no matter where an interrupt occurs in the main program?

5. What is required to restore main program registers, enable interrupts, and return to the main program when the interrupt service routine is completed?

AN 8085A INTERRUPT EXAMPLE As we discussed in Chap. 9, the 8085A has five interrupt inputs: TRAP,

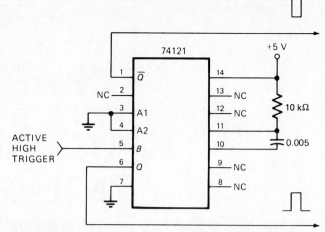

FIGURE 11-1 Pulse narrower for interrupt input.

RST 7.5, RST 6.5, RST 5.5, and INTR. When two interrupts occur at the same time, TRAP has the highest priority and INTR the lowest priority. The TRAP input requires a signal that has a low-to-high edge and remains high until recognized. RST 7.5 requires only a rising edge because this edge sets an internal RST 7.5 flip-flop. RST 6.5, RST 5.5, and INTR all require a high signal that stays high until recognized. For this example we use RST 6.5.

When an 8085A receives a high input on its RST 6.5 input, it automatically disables interrupts, pushes the program counter contents (return address) on the stack, and vectors to address 0034H to get its next instruction. Therefore, no external hardware is required to insert a restart instruction. However, the fact is that whenever the RST 6.5 input is high, an interrupt occurs. If the interrupt request signal from an external device remains high too long, the 8085A may service the interrupt request when it first detects the high on RST 6.5, return to its main program, and immediately be reinterrupted by the same request signal from the external device. This problem can be solved with the simple pulse narrower circuit shown in Figure 11-1. When an active high signal from an external device is applied to the 74121 B input, its Q output pulses high for a time determined by the R and C values. This is about 20 μs for the values shown. The Q output of the 74121 is connected to the RST 6.5 input of the 8085A. This ensures that RST 6.5 will be high long enough to be recognized, but not so long that multiple interrupts will occur.

After a reset, the 8085A RST 6.5 is both masked and disabled. Before it will respond to an interrupt request, it must be unmasked with the SIM instruction, as shown in Chap. 9, and then enabled with the EI instruction.

Figure 11-2 shows an 8085A assembly language program that initializes pointers and simulates the main program of a system. Also shown is the service routine to read in a byte of data from port 29H, store the data in a table in memory, and return to the main program to await the next interrupt. The programs are intended to run on an SDK-85 microcomputer board, but can be

```
PROGRAM TITLE - 8085A - READ DATA ON RST 6.5 INTERRUPT

ADDR     INSTR.    LABEL    OPCODE    OPERAND      COMMENTS

2000     31C220             LXI       SP,20C2H     ;INITIALIZE STACK POINTER
2003     218020             LXI       H,2080H      ;LOAD START OF TABLE ADDRESS
2006     22B020             SHLD      20B0H        ;SAVE TABLE ADDRESS
2009     3E0A               MVI       A,0DH        ;CODE TO UNMASK RST 6.5
200B     30                 SIM                    ;SET MASK
200C     FB                 EI                     ;ENABLE INTERRUPTS
200D     C30D20    WAIT     JMP       WAIT         ;WAIT FOR INTERRUPT (simulates
                                                            mainline program)


20C8     C32020             JMP       RDATA        ;JUMP TO MAIN PART OF INTERRUPT
                                                   ;SERVICE ROUTINE
```

(a)

```
PROGRAM TITLE - 8085A - READ DATA ON RST 6.5 INTERRUPT

ADDR     INSTR.    LABEL    OPCODE    OPERAND      COMMENTS

2020     F5        RDATA    PUSH      PSW          ;SAVE A AND PSW
2021     E5                 PUSH      H            ;SAVE H AND L
2022     2AB020             LHLD      20B0H        ;GET POINTER FROM POINTER STORE
2025     DB29               IN        29H          ;READ DATA
2027     77                 MOV       M,A          ;STORE IN TABLE
2028     23                 INX       H            ;POINT TO NEXT TABLE LOCATION
2029     22B020             SHLD      20B0H        ;SAVE POINTER
202C     E1                 POP       H            ;RESTORE H AND L
202D     F1                 POP       PSW          ;RESTORE A AND FLAGS
202E     FB                 EI                     ;ENABLE INTERRUPTS
202F     C9                 RET                    ;BACK TO MAIN PROGRAM
```

(b)

FIGURE 11-2 (a) 8085A assembly language program to initialize and wait for interrupt. (b) Interrupt service routine to read a byte of data on interrupt.

easily adapted to run on any 8085A-based microcomputer.

After initializing the stack pointer to 20C2H, the simulated mainline program in Figure 11-2a initializes the
HL register to point to the starting address of the table
in memory where data bytes will be stored.

Because a normal mainline program would use HL for
many things, this pointer must be saved in memory,
where it won't get destroyed and the subroutine can access it as needed. The SHLD 20B0H instruction stores
the contents of L into memory address 20B0H and the
contents of H into address 20B1H. Addresses 20B0H
and 20B1H are then set up as a *pointer store*. An LHLD
20B0H instruction returns the pointer value to HL
whenever it is needed. MVI A, 0DH loads the accumulator with the SIM word required to unmask only RST 6.5.

SIM then unmasks RST 6.5. Interrupts are enabled by
the EI instruction. The mainline program is simulated
by the JMP WAIT instruction. This instruction is executed over and over until an interrupt occurs.

When a RST 6.5 interrupt occurs, the 8085A disables
interrupts, pushes the return address (200DH) on the
stack, and vectors to address 0034H for its next instruction. At this address the SDK-85 monitor program ROM
has a JMP 20C8H instruction. Since the address 20C8H
is in RAM, you can write an instruction there that will
jump to the actual start of the interrupt service routine.
For the program in Figure 11-2, the jumps are from
0034H to 20C8H and then to 2020H, where the main
part of the interrupt service routine begins.

The service routine in Figure 11-2b must first push
on the stack any registers used in the routine, so that

293

they can be restored before return to the main program. Then LHLD 20B0H reloads the memory pointer from 20B0H and 20B1H back into HL. Data is read in from port 29H and stored in the table. The HL pointer is incremented to point to the next location in the table. The new value is saved in the pointer store by SHLD 20B0H. POP H and POP PSW restore the values that HL, A, and flags had before the interrupt. Since the 8085A disabled interrupts when it responded to the RST 6.5 input, interrupts must be enabled by EI if you want the 8085A to respond to any further interrupts. The return instruction will pop the return address off the stack and return execution to the next mainline instruction. In this case, since the JMP WAIT instruction loaded 200DH before the interrupt was serviced, the return will be to 200DH. The 8085A will then execute this instruction over and over again until another interrupt occurs.

A 6800 INTERRUPT EXAMPLE

As discussed in Chap. 10, the 6800 has two interrupt inputs, NMI and IRQ. NMI requires a high-to-low edge, and IRQ requires a low level. A pulse narrower such as that in Figure 11-1 is needed to make sure the 6800 does not get interrupted several times by the same requesting signal. When the 6800 receives an IRQ input, it pushes the program counter, index register, accumulator, and condition code register on the stack. The CPU then goes to FFF8H and FFF9H to get the address of the start of the interrupt service routine. After setting the interrupt mask to disable interrupts, the CPU goes to the address it got from FFF8H and FFF9H and executes from there. After a reset, the 6800 IRQ input is disabled. It must be enabled by a clear-interrupt-mask instruction (CLI) before the 6800 will respond to an IRQ input.

Figure 11-3 shows 6800 assembly language programs to prepare for an interrupt, simulate a mainline program, and respond to an IRQ input. The service routine takes in a byte of data from port 8004H and stores it in a memory table, starting at address 0080H.

After initializing the stack pointer, the program in Figure 11-3 initializes the index register as a pointer to the first address in the table where the data will be stored. Since a normal mainline program would write over this pointer in the index register, it must be saved in memory where it can be picked up by the service routine when needed. The STX $00E0 instruction stores the contents of the index register at addresses 00E0H and 00E1H. An LDX $00E0 instruction copies the pointer back into the index register any time it is needed. CLI resets bit 4 of the condition code register. This enables the interrupt. The JMP WAIT instruction executes over and over until an interrupt occurs. This simulates the mainline program.

As discussed earlier, an IRQ input causes the 6800 to go to addresses FFF8H and FFF9H to get the starting address of the service routine. For the MEK6800D2 microcomputer development board that this program was written to run on, these addresses are in the monitor ROM. The JBUG monitor program has the address E014H in these locations, so the 6800 goes to E014H. At

E014H the monitor program contains a jump to address A000H, which is in RAM. At address A000H you must put in a jump to the start of the actual interrupt service routine. This may sound confusing, but all you really must remember to do is put the jump instruction at A000H.

The main part of the service routine in Figure 11-3b first gets the pointer back into the index register from the pointer store with LDX $00E0. Data is next read in from port 8004H and stored in memory at the location pointed to by the index register. The index register is then incremented to point to the next storage location in the table. The incremented pointer is returned to the pointer store by STX $00E0. CLI reenables the interrupt. The return-from-interrupt instruction (RTI) pulls the condition code register, accumulators, and index register values off the stack. This restores to those registers the values they had before the interrupt. RTI also pulls the return address off the stack and returns execution to the mainline program. For the programs in Figure 11-3, execution returns to the simulated mainline, JMP WAIT, at address 000AH.

INTERRUPTS FOR TIMING

We showed how an interrupt could be used to replace the polled method of reading in or sending out data. An interrupt input also can be used to replace delay loops for timing. A programmable timer, such as that in the 8155, discussed in Chap. 9, can be programmed to pulse an interrupt input every 1 ms. This interrupt can tell the processor to input or output a byte of data. The service routine can be written to count 10 interrupts or 100 interrupts before inputting or outputting a byte. This gives precise but programmable timing. A service routine can be written that increments the number in a memory location after every 1000 interrupts. The number in the memory location will be a count of seconds. The service routine could be extended to increment a minute count in another memory location after each 60 s and increment an hours count after each 60 min. This interrupt and service routine then produces a real-time clock which can be used to time any processes in the mainline program. The mainline program can simply read the count of seconds, minutes, and hours from their memory locations as needed.

MULTIPLE INTERRUPTS

Some applications require that the processor can be interrupted by any one of several signals. They may require also that an interrupt signal with the highest priority be serviced first. For example, if the atomic reactor in the basement is "going critical," this had better be serviced before an overflowing soap machine.

An IC such as the 8259 *programmable interrupt controller unit* (PICU) allows up to eight interrupts to be serviced in order of their priority. The priority of the interrupt signal from a device is established by the interrupt request input of the 8259 to which it is connected. IR0 has the highest priority and IR7 the lowest in normal operation.

PROGRAM TITLE - 6800 READ DATA ON IRQ

ADDR	INSTR.	LABEL	OPCODE	OPERAND	COMMENTS
0000	8E00FF		LDS	#$00FF	;INITIALIZE STACK POINTER
0003	CE0080		LDX	#$0080	;INITIALIZE TABLE POINTER
0006	FF00E0		STX	$00E0	;SAVE IN POINTER STORE
0009	0E		CLI		;ENABLE INTERRUPT
000A	7E000A	WAIT	JMP	WAIT	;WAIT FOR INTERRUPT (simulates mainline program)
A000	7E0010		JMP	RDATA	;LOAD JMP TO STARTING ADDRESS OF ;INTERRUPT SERVICE ROUTINE

(a)

PROGRAM TITLE - 6800 READ DATA ON IRQ

ADDR	INSTR.	LABEL	OPCODE	OPERAND	COMMENTS
0010	CE00E0	RDATA	LDX	$00E0	;GET TABLE POINTER
0013	B68004		LDAA	$8004	;READ DATA FROM PORT
0016	A700		STAA	$00,X	;STORE DATA IN TABLE
0018	08		INX		;POINT TO NEXT LOCATION IN TABLE
0019	FF00E0		STX	$00E0	;SAVE POINTER FOR NEXT READ
001C	0E		CLI		;ENABLE INTERRUPT
001D	3B		RTI		;BACK TO MAIN PROGRAM

(b)

FIGURE 11-3 MC6800 interrupt service routine to read in data. (a) Initialize and wait for interrupt. (b) Interrupt service routine.

When the 8259 receives an interrupt request on one of the eight IR inputs, it sends an INT signal to the 8080A or 8085A. The 8080A or 8085A responds with an INTA signal. The 8259 uses this to load a standard call (CD hex) instruction onto the system data bus. The 8080A or 8085A decodes the call instruction and sends out two more INTA pulses. The second INTA stimulates the 8259 to put out the lower byte of the desired subroutine address on the data bus. A third INTA pulse from the 8080A or 8085A gates the high-order address byte onto the data bus. (The eight subroutine addresses are stored in the 8259 during its initialization.) After the 8259 has output the address, the 8080A or 8085A then executes the call instruction. It stores the return address on the stack and loads the program counter with the address given to it by the 8259. Since the 8259 supplies a standard call instruction and two address bytes, it can vector the program to anywhere in memory. Remember, the RST instructions can vector the program to only one of eight addresses between 00H and 38H.

If one interrupt is being serviced and a lower-priority interrupt signal comes in, it is ignored until the higher-priority interrupt servicing is completed. An EI instruction in the interrupt service subroutines reenables the 8080A or 8085A interrupt input, so that the next interrupt signal can interrupt the processor and be serviced when the first service is finished. The 8259 vectors the program to another address for the second service routine.

If an interrupt is being serviced and a higher-priority interrupt signal comes in, two things may happen. If the operating routine did not do an EI instruction at its start, then the new interrupt is ignored until an EI is done. If an EI was done, then the higher-priority interrupt interrupts the lower-priority routine. The 8259 sends out a call instruction and the vector address for the higher-priority routine. When the higher-priority servicing is completed, program execution returns to the lower-priority routine. When this is finished, execution returns to the main program. It is essentially just a nested subroutine process.

HANDSHAKE INPUT AND OUTPUT WITH A MICROCOMPUTER

The orderly transfer of data from one system to another is done in one of three ways: simple strobe I/O, single-layer handshake I/O, and double handshake I/O.

SIMPLE STROBE I/O

Figure 11-4a illustrates the *simple strobe I/O* technique for data from a peripheral to a microcomputer. A peripheral such as an encoded keyboard outputs a byte of data to a parallel port. Then it sends a strobe signal to indicate that valid data is on its output lines. The strobe signal can be connected to an interrupt input on the CPU, so that the data is read in on an interrupt basis. The strobe signal could also be connected to a port line, and the line polled by the CPU until it was found asserted. Programs in Figures 8-21 and 10-2 show how this is done. The simple strobe method works well if the peripheral outputs data at a low rate. In all but the simplest systems, however, data transfer between systems is done on a "handshake" basis. This means that the peripheral will not output the next byte of data until it receives an acknowledge "handshake" signal back from the receiver to indicate that the first byte was read.

SINGLE-HANDSHAKE I/O

Figure 11-4b shows the signal sequence for a *single handshake data transfer* from a peripheral to a micro-computer. The peripheral outputs a byte of data and sends a strobe to indicate that valid data is present. The CPU detects the strobe on an interrupt or on a polled basis and reads in the data. It then sends an acknowledge signal back to the peripheral, telling the peripheral to send the next byte of data. The peripheral will not send data faster than the receiving system can accept it.

DOUBLE-HANDSHAKE I/O

In systems where even more communication is required during data transfer, a *double handshake* is used. Figure 11-4c shows a double handshake signal sequence for data from a peripheral to a microcomputer. The peripheral sends its strobe output low to ask the microcomputer, "Are you ready?" The CPU replies "yes" by sending its acknowledge output high. The peripheral then outputs a byte of data and raises its strobe signal back high to tell the microcomputer, "Here is some valid data for you." The CPU reads in the data and drops its acknowledge signal low. The acknowledge signal going low says to the peripheral, "I have the data, thank you, and I await your request to send the next byte."

GENERATING HANDSHAKE SIGNALS: PROGRAMMABLE PORT ICs

The examples given above are all for data being sent from a peripheral to a microcomputer. However, similar sequences of signals could be drawn for data output from a microcomputer to a peripheral such as a printer. The handshake signals from the microcomputer can be generated by software, hardware, or combinations of the two. To relieve the CPU of the burden of generating handshake signals, programmable parallel port ICs such as the Intel 8155 and Motorola MC6821 can be programmed to automatically produce these signals. The following subsections show how these devices are initialized and used for handshake I/O operations.

INITIALIZATION OF THE 8155   As discussed in Chap. 9, the 8155 contains 256 bytes of RAM, two 8-bit programmable ports, a 6-bit programmable port, and a 14-bit programmable timer. The discussion in Chap. 9 shows how the three ports are initialized for simple input and output operations. (If this process is not current in your mind, do a memory refresh cycle to Chap. 9 before going on.) Now we show how port C can be programmed to generate handshake signals for handshake (strobed) input or output on ports A and B. As shown in Figure 9-11d, port C of the 8155 can be initialized as one of four alternatives. Alternative 1 programs all six lines of port C as inputs. Alternative 2 programs all six lines of port C as outputs. Alternative 3 programs port C lines 0 to 2 as handshake signals for port A, and port C lines 3 to 5 as outputs. Alternative 4 programs port C lines 0 to 2 as handshake signals for port A and port C lines 3 to 5 as handshake signals for port B.

8155 HANDSHAKE INPUT EXAMPLE: TAPE READER   Figure 11-5a shows how an 8155 is connected to do a handshake input from a peripheral such

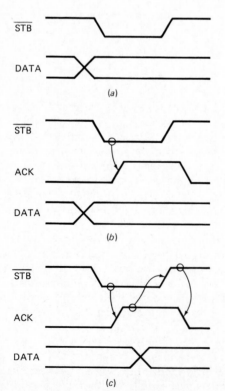

FIGURE 11-4 Handshake data transfer. *(a)* Simple strobe. *(b)* Single level. *(c)* Double level.

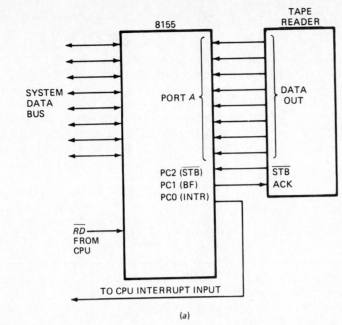

8155

SYSTEM
DATA
BUS

PORT A

PC2 (STB)
PC1 (BF)
PC0 (INTR)

RD
FROM
CPU

TO CPU INTERRUPT INPUT

TAPE
READER

DATA
OUT

STB
ACK

(a)

FIGURE 11-5 Handshake data transfer with 8155. (a) Connection to high-speed tape reader. (b) Signal sequence for handshake data input.

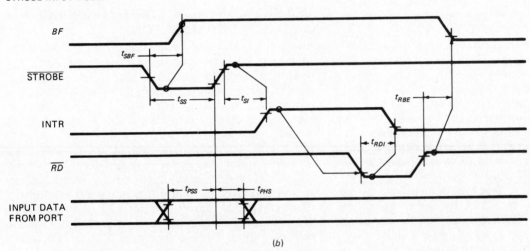

STROBE INPUT MODE

BF

t_{SBF}

STROBE

t_{SS} t_{SI} t_{RBE}

INTR

t_{RDI}

RD

t_{PSS} t_{PHS}

INPUT DATA
FROM PORT

(b)

as a high-speed tape reader. Port A of the 8155 is initialized as an input port, port C is initialized as alternative 3 or alternative 4, and the port A interrupt is enabled. A command word of 15H or 19H will do this. Figure 11-5b shows the sequence of signals that occurs as a handshake input data transfer takes place.

First the tape reader sends a *strobe* signal (STB) low to tell the 8155 that it has a byte of data to send. When it detects a low on its PC2 input (STB), the 8155 automatically raises its PC1, *buffer full* (BF), output high. This indicates that the 8155 is ready to receive data. If it has not already done so, the tape reader sends a byte of data and raises its STB output high again. The STB going high latches the byte of data into the port A buffers. STB going high also causes the 8155 to send out an interrupt signal on PC0. This signal is connected to an interrupt input on the CPU. If interrupts are enabled, the CPU vectors to an interrupt service routine which reads the data from the 8155. To read port A, the CPU addresses the device and sends a RD signal. RD going

low causes the 8155 to reset its INTR output signal on port C line 0. When the CPU raises its RD signal high again, the 8155 automatically drops its BF signal from port C line 1 low again. BF going low indicates to the tape reader that it can now repeat the sequence to send the next character.

The CPU can determine the state of the 8155 control signals at any time by reading the 8155 status register. The status register has the same address as the control register. Figure 11-6 shows the bit format for the 8155 status word. Instead of servicing the peripheral on an interrupt basis, the CPU can poll bit 0 of the status register to determine when the peripheral has sent, or is ready to receive, a character.

8155 HANDSHAKE OUTPUT EXAMPLE: SPEECH SYNTHESIZER   The 8155 can also be used for handshake or strobed output data transfers. Figure 11-7 shows how an 8155 can be connected to a Votrax SC-01 speech synthesizer IC to give your programs a voice.

297

INPUT AND OUTPUT INTERFACING

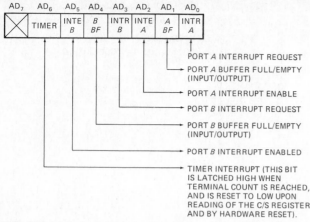

FIGURE 11-6 Bit format for the 8155 status register.

The SC-01 uses phonemes to produce speech. *Phonemes* are parts of words. By linking phonemes, you can create any word. A 6-bit binary code sent to the SC-01 determines which of its 64 phonemes will be output. An additional 2 bits determine the pitch of the sounded phoneme. To produce words, phrases, or sentences, then, the microcomputer simply has to output a series of bytes to the SC-01 with the proper timing. Phonemes require between 47 and 250 ms to sound. The SC-01 A/R output goes high when the device is ready to receive its next phoneme. This request from the SC-01 could be serviced on a polled or on an interrupt basis. However, because of the relatively long time between requests, the interrupt basis is more reasonable.

For this operation the 8155 is initialized as follows. Port A is initialized as an output by a 1 in bit 0 of the command word, port C is initialized in alternative 3 or 4 by 10 or 01 in bits 2 and 3 of the command word, and the port A interrupt output is enabled by a 1 in bit 4 of the command word. The other bits in the command word are determined by the use of port B and the timer.

Figure 11-7b shows the timing diagram for a handshake output data transfer from an 8155 to the SC-01. When power is first turned on, the acknowledge/request output (A/R) of the SC-01 will be high. After shifting to TTL levels, this signal is connected to the STB input of the 8155. A high on this input causes the 8155 to send an interrupt signal to the CPU. When the CPU services the interrupt, it writes a byte of data to port A. Writing to the 8155 port A will reset the interrupt output from port C bit 0. When the WR signal from the CPU returns high, the 8155 automatically raises its buffer-full line (BF) high to tell a peripheral that valid data is available on port A.

However, the SC-01 requires that data be present on its data inputs for at least 500 ns before it receives an STB signal. Two 74C906 buffers are used to delay the strobe as required. When the SC-01 receives a STB input from the second 74C906 buffer, it drops its A/R output line low to acknowledge receipt of the phoneme code. A/R going low causes the 8155 to drop its BF signal low again.

The BF signal going low normally would be the signal to a peripheral that it can request another byte of data from the computer any time. For the example here, however, the output of the second 74C906 buffer going low, rather than BF going low, tells the SC-01 that it can request the next phoneme code. After 47 to 250 ms the SC-01 raises its A/R output high again to request another phoneme. This signal into the 8155 STB input causes the 8155 to send an interrupt request to the CPU. An interrupt service routine then writes the next phoneme into port A of the 8155, and the process continues as described.

The SC-01, incidentally, requires only a few phonemes per word. For example, sending the hexadecimal sequence 9F, 42, 63, 58, 5D, 3E, 3E, AA, 7B, 5F, 6A, 3E, 3E, 59, 63, 4C, E5, 58, 6C, 6A, 3E, 3E, 3E produces the statement, "Self-test complete." In case you want to build up one of these to keep you company on a cold night, a table of hex codes for all 64 phonemes is given in App. D.

USING THE MC6821 FOR HANDSHAKE DATA I/O: INITIALIZATION

Another very common programmable port that can be used for handshake input or output is the MC6821. The 6821 has two 8-bit ports and four interrupt input or control lines. Recall from Chap. 10 that initializing a port on the 6821 for input or output is a three-step process. First, reset the device or send control words with bit 2 equal to 0 to each control register. This gains access to the data direction registers. Second, send a data direction word to each *data direction register* (DDR). Sending a 1 to a particular bit in the DDR makes the corresponding port line an output. Sending a 0 to a bit in a DDR makes the corresponding port line an input. Third, control words with bit 2 equal to 1 are sent to each control register, so that the CPU can now access the two ports.

The four interrupt inputs and control lines on the 6821 are labeled CA1, CA2, CB1, and CB2. CA1 and CB1 can be programmed as interrupt inputs. CA2 and CB2 can be programmed as interrupt inputs or as control signal outputs. Figure 11-8 shows the format for the 6821 control register. Sending a control word to control register A determines the function of the CA lines. Sending a control word to control register B determines the function of the CB lines. For simplicity, we use the CA lines and the *A control register* (CRA) as an example. The B side of the device is nearly identical.

If CRA bit 0 is a 1, the CA line is enabled as an interrupt input. CRA bit 1 determines whether the rising or falling edge of a signal input to CA1 produces an interrupt. When the CA1 input receives an active signal edge, the 6821 does two things. First, it sets bit 7 in control register A as an interrupt flag. Second, the 6821 sends out a low on its IRQA output. Usually this is connected to an interrupt input on the CPU.

Also CA2 can be programmed as an interrupt input. If

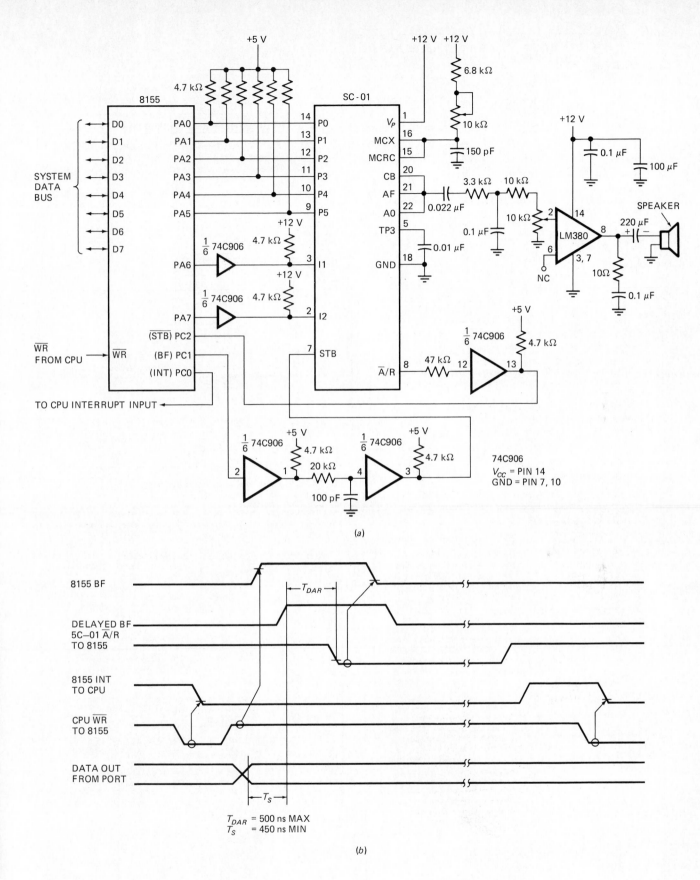

FIGURE 11-7 Votrax SC-01 speech synthesizer used with 8155. (a) Circuit connections. (b) Timing diagrams.

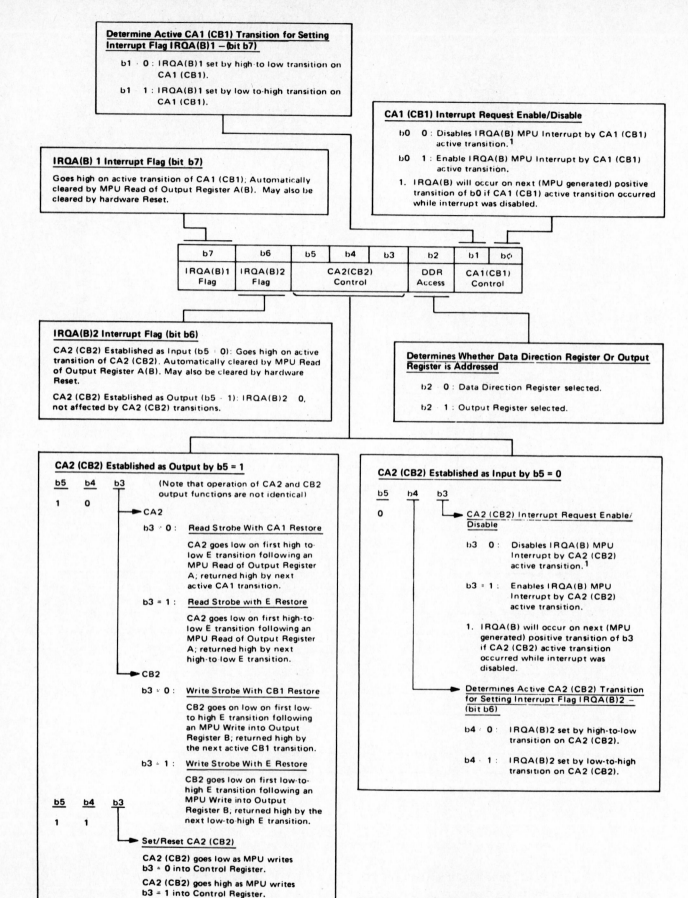

FIGURE 11-8 Format for 6821 control register.

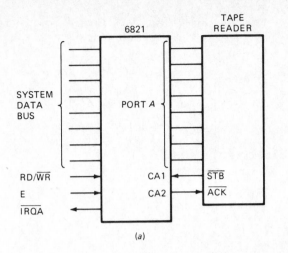

(a)

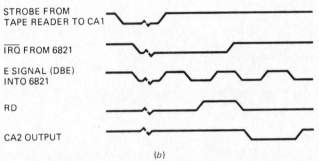

(b)

FIGURE 11-9 Handshake data transfer with a 6821. *(a)* Tape reader connected to a 6821 for input data transfer. *(b)* Signal sequence: CA1 as interrupt input, CA2 as read strobe with E restore.

bit 5 of the control word is a 0, then a 1 in bit 3 enables CA2 as an interrupt input, and bit 4 determines which edge of the CA2 input signal causes an interrupt. When the CA2 input receives an active signal, the 6821 does two things. First, it sets bit 6 of the control register as an interrupt flag. Second, it sends out a low on the IRQA output. If both CA1 and CA2 lines are being used as interrupt inputs, the question arises as to how the CPU knows whether CA1 or CA2 is causing the interrupt. The answer is that the interrupt service routine must read the 6821 control/status register and check bits 6 and 7. A 1 in bit 7 indicates that CA1 produced the interrupt; a 1 in bit 6 indicates that CA2 produced the

interrupt. In this case, the 6821 "funnels" two interrupt signals into one interrupt input on the CPU. CB1 and CB2 can also be programmed as interrupt inputs. Since the 6821 IRQA and IRQB outputs are open-drain, they can be connected to the same CPU interrupt input. The 6821 then acts as a four-input interrupt funnel.

The CA2 line can be programmed to function as an output by making bit 5 of the control word a 1. If bit 4 is also a 1, then whatever logic level is present in bit 3 is output from the CA2 line when the control word is sent to the control register. In this case, then, the CA2 line functions as a single-bit output port. If bit 5 of the control word is a 1 and bit 4 a 0, then CA2 functions as an automatic handshake signal such as those described in a previous section for the 8155. Note that CA2 will function only as a read strobe (input) and CB2 will function only as a write strobe (output).

6821 HANDSHAKE INPUT EXAMPLE: TAPE READER

Figure 11-9a shows how a 6821 can be connected to a high-speed tape reader for a handshake (strobed) input operation. Figure 11-9b shows the sequence of signals that occur during a data transfer. The 6821 CRA is programmed with CA1 enabled and active on the falling edge and CA2 as an output of read strobe with E restore. The tape reader sends a STB signal to the CA1 input and a byte of data to port A. The 6821 then sends an interrupt request (IRQA) to the CPU. The CPU performs an interrupt service routine that reads the data from port A of the 6821. The RD signal going high causes the IRQA signal from the 6821 to return high. The CA2 output goes low following the falling edge of the E signal after the read. As programmed, the CA2 signal returns high just after the next high-to-low edge of the E signal. CA2 returning high is a signal to the tape reader that it can send the next byte of data.

6821 HANDSHAKE OUTPUT

Control register B can be programmed to produce a similar sequence of signals for an output data transfer from port B. Figure 11-10 shows the sequence of instructions required to initialize a 6821 at addresses 8004H through 8007H as follows: port A as strobed input, CA1 active on rising edge, and CA2 read strobe with E restore; port B as strobed output, CB1 active on falling edge, and CB2 write strobe with CB1 restore.

```
LDAA   #$00          ;CONTROL WORD FOR DDRA ACCESS
STAA   $8005         ;ACCESS DDRA
STAA   $8007         ;ACCESS DDRB
STAA   $8004         ;00H TO DDRA - ALL INPUTS
LDAA   #00101111     ;CONTROL WORD FOR PORT A
STAA   $8005         ;SEND TO CONTROL REGISTER A
LDAA   #$FF          ;DDR WORD FOR PORT B - ALL OUTPUTS
STAA   $8006         ;SEND TO DDRB
LDAA   #00100101     ;CONTROL WORD FOR PORT B
STAA   $8007         ;SEND TO CONTROL REGISTER B
```

FIGURE 11-10 MC6800 routine to initialize 6821 port A as strobed input, port B as strobed output.

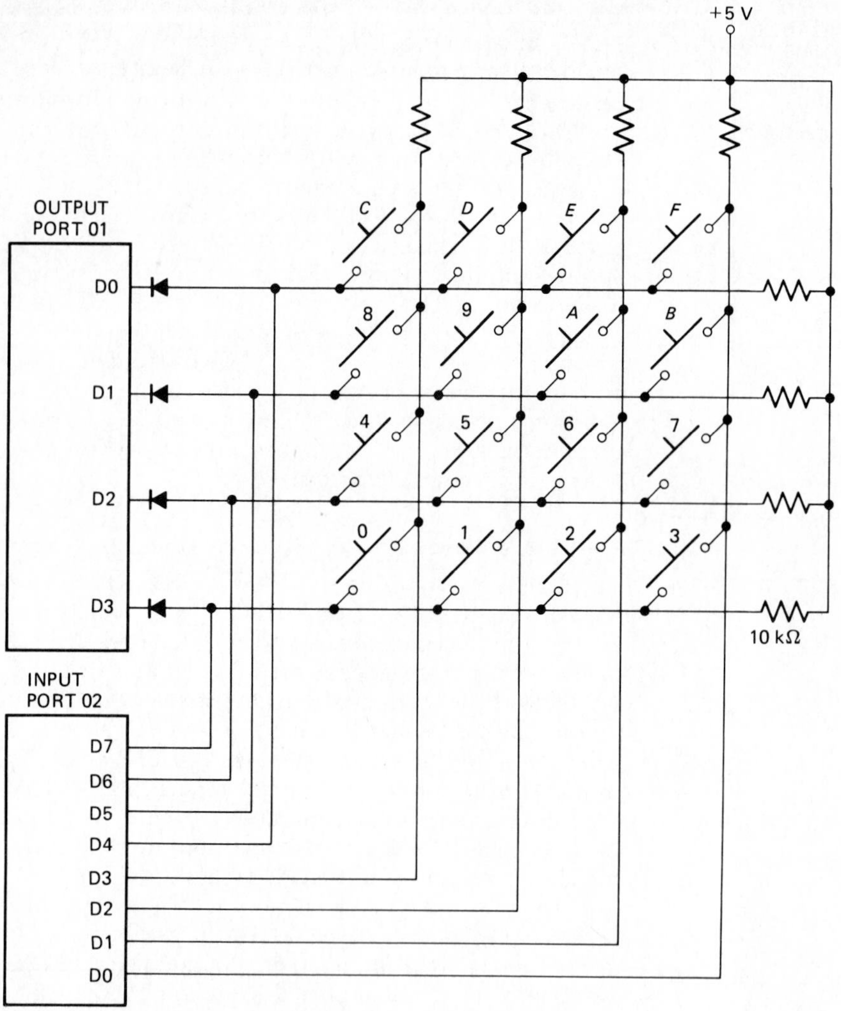

FIGURE 11-11 Connection of a keyboard matrix to I/O ports.

INTERFACING KEYBOARDS TO MICROCOMPUTERS

A common method of entering programs, data, or commands into a microcomputer is with a keyboard. The keyboard may be a simple hex type such as that shown on the SDK-85 in Figure 9-8a, or it may be a typewriter-type keyboard. In either case, the keyboard is a matrix of switches arranged in columns and rows, as discussed in Chap. 4. Getting meaningful data from a keyboard requires performing three tasks:

1. Detect a key closed.

2. Debounce it.

3. Encode it.

As you may remember, depressing a key shorts a row to a column. The general procedure for detecting a key closed is to put a low on one row at a time and scan the columns to see whether any of the normally high have been pulled low by being shorted to the low row. A somewhat faster method of detecting a key closed is to put lows on all the rows at once and scan the columns for a

low. If a closed key is found, then the rows are scanned one at a time to find the row in which the key is located. When the row is found, row scanning is stopped so that debouncing can take place.

Debouncing is done by waiting 10 or 20 ms and then rechecking to see whether the key is still pressed. If it is, the keypress is considered valid.

Encoding of the row and column code to a more standard code such as hex or ASCII can be done with a ROM or a program routine.

When a keyboard is interfaced to a microprocessor, the three tasks can be done with external hardware or internal program routines.

HARDWARE APPROACH

With the hardware approach, all the scanning of the keyboard matrix, debouncing, and encoding are done by an external device. The microprocessor merely takes in the encoded character when the external keyboard encoder sends it a data ready signal or valid keypressed signal. An example of a complete MOS LSI keyboard encoder is the General Instruments AY-5-2376. This device contains all the circuitry needed to interface to a type-

302

CHAPTER ELEVEN

writer-type keyboard. It contains a scan oscillator, debounce timer, and encoding ROM. Characters are output in ASCII with bit 8, a selectable odd or even parity bit. When a valid character is on the eight parallel output lines, the AY-5-2376 sends out a positive keypressed pulse. A CPU can use either polled or interrupt mode to detect this keypressed pulse and take in the character.

Another feature of the AY-5-2376 is *two-key rollover*. If two keys are pressed at nearly the same time, the code for the first pressed is sent out. When that key is released, the code for the second is sent out. Without this feature, pressing two keys would give a wrong or illegal output code.

The advantage of the hardware approach is that the processor is not tied up with the three tasks. For many applications, however, the processor has plenty of time available. In that case, a software approach is used to eliminate the cost and PC board space of the hardware encoder.

SOFTWARE APPROACH

The software approach consists basically of doing the three tasks with program routines. The row lines of the keyboard matrix are connected to an output port, and the column lines to an input port, as shown in Figure 11-11. Figure 11-12 shows a simple flowchart for a program to scan this keyboard.

Zeros are output on all four row lines. The four column lines on the input port are read and checked for 0's. A 0 indicates a key pressed. The program loops until it finds all keys open before accepting a key pressed. In other words, a new key cannot be pressed until the last one is released. This is used in place of two-key rollover. When the keyboard is found all open, the program loops, looking for a keypress. If a keypress is found, the program calls a 20-ms delay routine to give the key time to stop bouncing.

The next problem is to find out which key was pressed. This is done by outputting a 0 on the first row and 1's on all the others. Then the columns are read and checked for a 0. If no 0 is found, the pressed key must not be in that row, so the 0 is rotated into the next row and the columns are rechecked. If a 0 is found in a column, then the 8-bit word read from the input port will contain the 4-bit row code and the 4-bit column code for the key pressed. This 8-bit code is converted to the desired 4-bit hexadecimal code by using a lookup table.

Figure 11-13 shows an 8080A/8085A assembly language program for this keyboard scan routine. For the most part, the program follows the flowchart quite closely, but a few points bear explanation. The XRA A instruction is a handy 1-byte way to clear both the accumulator and the carry flag flip-flop. It does this by XORing the accumulator with itself ($A \oplus A = 0$). The 8080A/8085A has no specific instruction for clearing carry, but ORA A will clear carry without changing the accumulator contents. Note how the RAL instruction is

used in a loop to rotate a 0 from carry to successive rows.

The 20-ms delay routine uses the extended BC register and decrements a loaded count down to 0. Since the DCX instruction affects no flags, a test for 0 is made by ORing B with C. The result is 0 if B and C are both all 0's, and the zero flag will be set.

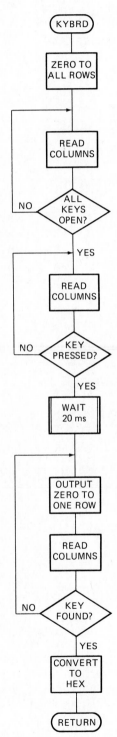

FIGURE 11-12 Flowchart for keyboard scan routine.

```
;KYBRD
;THIS ROUTINE SCANS AND DECODES A 16 SWITCH KEYPAD.
;ROW LINES ARE CONNECTED TO OUTPUT PORT 01, D0 - D3.  ROW AND COLUMN
;INPUTS ARE INPUT ON PORT 02, D0 - D7.
;REGISTERS USED:  HL, BC, A
;ROUTINES CALLED: WT20MS


KYBRD:      PUSH    H           ;SAVE CONTENTS OF REGISTERS
            PUSH    B           ;USED
            XRA     A           ;CLEAR ACCUMULATOR AND CARRY
            OUT     01          ;SEND ZEROS TO ALL ROWS OF
                                ;KEYBOARD

KYOPN:      IN      02          ;READ COLUMNS
            CPI     0FH         ;LOOK FOR ZEROS ON COLUMN
            JNZ     KYOPN       ;IF ANY KEYS CLOSED, LOOP UNTIL
                                ;ALL OPEN

KYCLOS:     IN      02          ;WHEN ALL KEYS OPEN CHECK FOR
            CPI     0FH         ;KEY CLOSED
            JZ      KYCLOS      ;LOOP UNTIL KEY PRESSED
            CALL    WT20MS      ;DELAY 20 MS
            MVI     A,FF        ;SET ACC TO ALL ONES
            ORA     A           ;CLEAR CARRY

NXTROW:     RAL                 ;ROTATE ZERO INTO FIRST ROW POSITION
            MOV     B,A         ;SAVE ROW MASK
            OUT     01          ;OUTPUT TO ROWS
            IN      02          ;READ ROWS AND COLUMNS
            MOV     C,A         ;SAVE ROW AND COLUMN CODE
            ANI     0FH         ;MASK ROW CODE
            CPI     0FH         ;CHECK COLUMNS FOR A LOW
            JNZ     DECODE      ;IF LOW PRESENT, ROW FOUND.
                                ;C HAS ROW AND COLUMN CODE
            STC                 ;
            MOV     A,B         ;RESTORE ROW MASK
            JMP     NXTROW      ;TRY NEXT ROW

DECODE:     LXI     H,TBLTOP    ;LOAD STARTING ADDRESS OF
                                ;CONVERSION TABLE
            MVI     B,0FH       ;SET CHARACTER COUNTER
            MOV     A,C         ;MOVE ROW AND COLUMN CODE TO ACC

NXTBYT:     CMP     M           ;CHECK FOR MATCH WITH TABLE ENTRY
            JZ      DONE        ;DONE IF MATCH FOUND
            INX     H           ;INCREMENT HL TO POINT TO NEXT
                                ;TABLE ENTRY
            DCR     B           ;DECREMENT CHARACTER COUNTER
            JP      NXTBYT      ;CHARACTER COUNTER STILL POSITIVE
                                ;TRY NEXT BYTE FOR MATCH

ERROR:      STC                 ;SET CARRY TO INDICATE CHARACTER
            JMP     EXIT        ;NOT FOUND AND EXIT
```

FIGURE 11-13 8085A assembly language program for keyboard scan routine.

```
DONE:       MOV    A,B        ;MOVE HEX CODE TO ACC
            ORA    A          ;CLEAR CARRY
            POP    B          ;RESTORE REGISTERS
            POP    H          ;
            RET               ;RETURN

WT20MS:     LXI    B,09C4H    ;LOAD 16 BIT COUNTER

LOOP:       DCX    B          ;DECREMENT
            MOV    A,B        ;   CHECK FOR ZERO
            ORA    C          ;
            JNZ    LOOP       ;LOOP UNTIL ZERO IN B AND C
            RET               ;RETURN

TBLTOP:     F = 1110 1110       7 = 1011 1110
            E = 1110 1101       6 = 1011 1101
            D = 1110 1011       5 = 1011 1011
            C = 1110 0111       4 = 1011 0111
            B = 1101 1110       3 = 0111 1110
            A = 1101 1101       2 = 0111 1101
            9 = 1101 1011       1 = 0111 1011
            8 = 1101 0111       0 = 0111 0111
```

FIGURE 11-13 (continued)

Probably the most interesting part of this program is the routine for converting the 8-bit column and row code to 4-bit hexadecimal code. The 16 possible combinations of column and row codes are stored as a table in 16 successive memory locations, as shown in Figure 11-13. The 8-bit code for key F is stored at the first address, the code for key E at the next address, and so on for the other 14. The B register is initialized to 0F, which corresponds to the hex code for the first table entry in memory. The 8-bit column-row code is moved from the C register where it ended up to the accumulator. Then this is compared with the value at the first table address in memory pointed to by HL. If the two codes match, then the zero flag is set and the routine exits.

If the column-row code in the accumulator did not match the table value, then HL is incremented, B is decremented, and the next value in the table is compared with that in the accumulator. The program steps through the table until a match is found. When a match is found, B will contain the correct hex code for the key pressed. If no match is found in the table, then carry is set to indicate an error and the key scan routine exits.

Note that the table values are put in memory with the value corresponding to 0F first. This is done so that B can be decremented to 0 and the sign flag used to detect an error. If B decrements to a negative number and no match is found, this indicates an error. As long as B is positive, the JP instruction keeps the program looping through the table, looking for a match.

The software technique shown for interfacing a keyboard to a microprocessor can be expanded easily to work with typewriter-type keyboards, but the hex type was shown here for simplicity. The disadvantage of this scheme is that it ties up processor time.

INTERFACING SEVEN-SEGMENT DISPLAYS TO MICROCOMPUTERS

Microprocessor-based instruments often use seven-segment displays to display data, instructions, or addresses. For efficiency, the displays are usually multiplexed. The multiplexing can be done primarily with software routines or primarily with hardware components.

SOFTWARE MULTIPLEXING OF SEVEN-SEGMENT DISPLAYS

Figure 11-14 shows the schematic of the keyboard-display module for the Motorola MEK6800D2 microprocessor prototyping kit. The kit consists of two printed-circuit board modules; an MC6800-based CPU board, whose schematic is shown in Figure 10-5, and the keyboard-display module, whose schematic is shown here. This kit is used as the basis of an EPROM programmer prototype in Chap. 12.

The entire lower right portion of the schematic in Figure 11-14 shows the circuit for a Kansas City standard cassette-tape interface. This allows programs to be stored on inexpensive tape cassettes. (See Chap. 5 for a description of the Kansas City format.)

The rest of the schematic in Figure 11-14 shows the keyboard and display circuitry. Temporarily ignore the keyboard circuitry.

$U1$ through $U6$ are seven-segment, common-cathode LED displays. $Q1$ through $Q7$ are the segment driver transistors. A low on the base of one of these turns on the transistor and sends current to that segment. Each segment driver is connected to an output line of an MC6821 peripheral interface adapter (PIA).

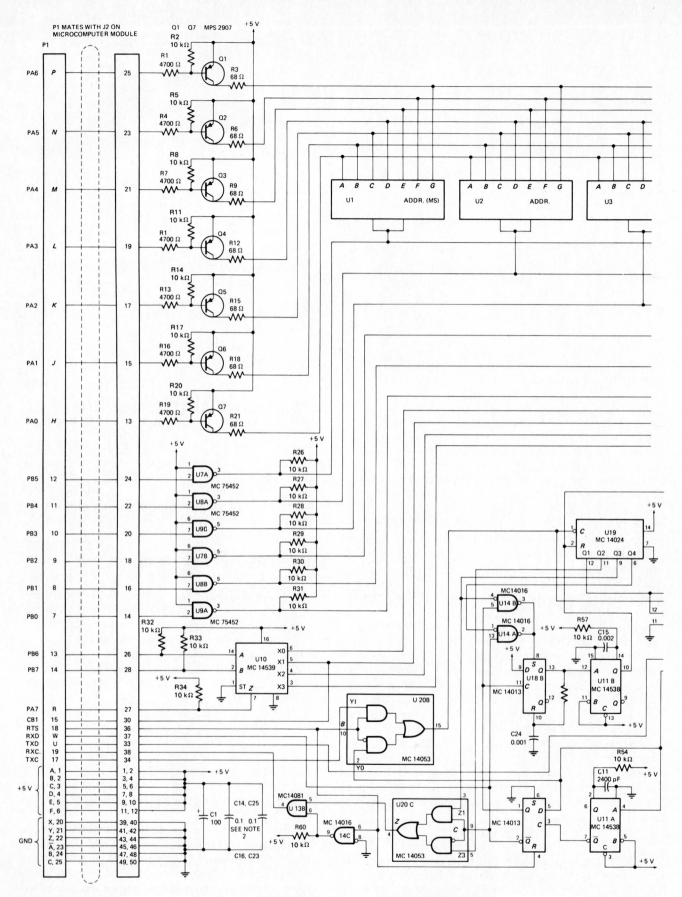

FIGURE 11-14 MEK6800D2 keyboard-display module schematic. *(Motorola.)*

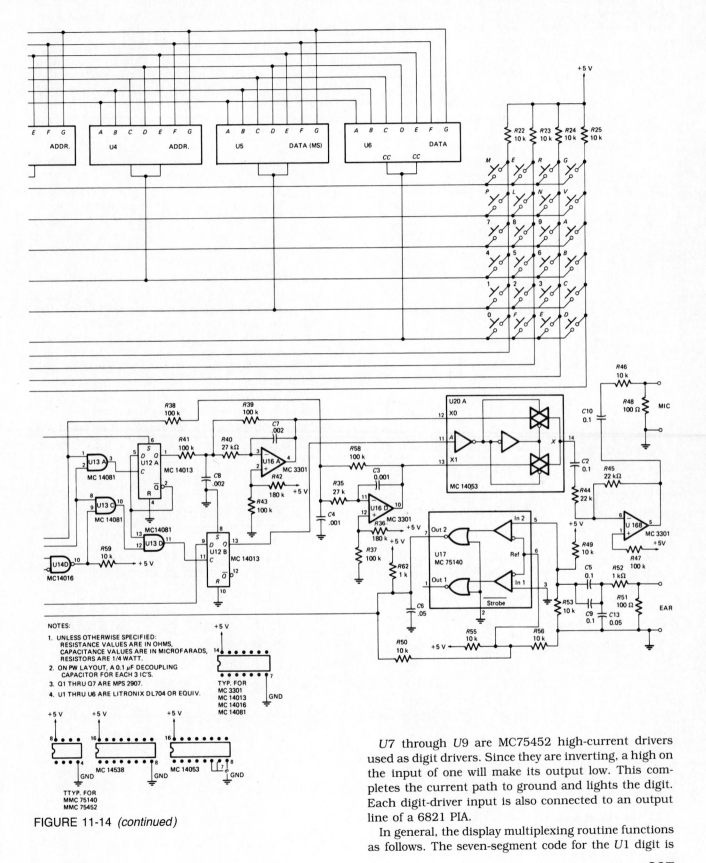

NOTES:
1. UNLESS OTHERWISE SPECIFIED:
   RESISTANCE VALUES ARE IN OHMS,
   CAPACITANCE VALUES ARE IN MICROFARADS,
   RESISTORS ARE 1/4 WATT.
2. ON PW LAYOUT, A 0.1 μF DECOUPLING
   CAPACITOR FOR EACH 3 IC'S.
3. Q1 THRU Q7 ARE MPS 2907.
4. U1 THRU U6 ARE LITRONIX DL704 OR EQUIV.

TYP. FOR
MC 3301
MC 14013
MC 14016
MC 14081

TYP. FOR
MC 14538

MC 14053

TTYP. FOR
MMC 75140
MMC 75452

FIGURE 11-14 (continued)

U7 through *U9* are MC75452 high-current drivers used as digit drivers. Since they are inverting, a high on the input of one will make its output low. This completes the current path to ground and lights the digit. Each digit-driver input is also connected to an output line of a 6821 PIA.

In general, the display multiplexing routine functions as follows. The seven-segment code for the *U1* digit is sent out on lines PA0 to PA6. Then a high is sent out on line PB5 while PB0 to PB4 are held low. This turns on *U7A* and pulls the *U1* LED cathodes low to light that digit. After about 1 ms, the segment code for the *U2* digit is put out and the *U2* digit cathodes are strobed low for 1 ms. The pattern is repeated until all the digits have been refreshed. When all digits have been refreshed, the program can cycle back to refresh digit 1 again.

However, if a keyboard is also being used, as in Figure 11-14, then the digit-strobe lines may be used to scan the keyboard before a return to refresh the display again. This is the technique used in the MEK6800D2.

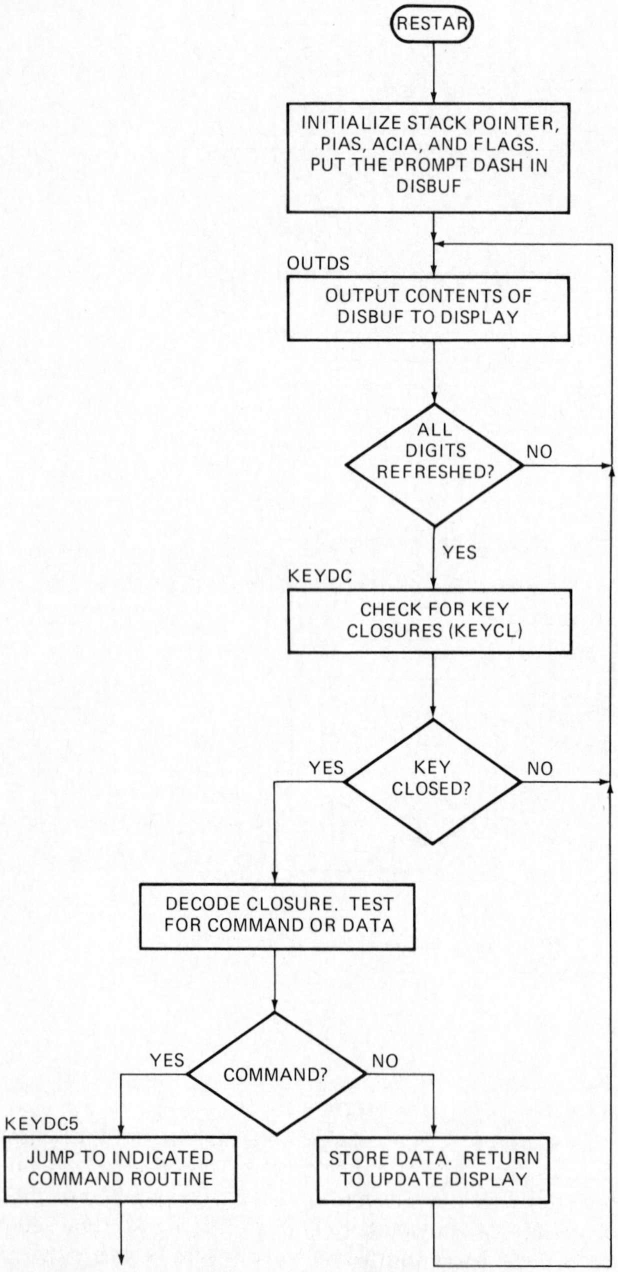

Figure 11-15 shows a general flowchart for the monitor program it uses.

DISBUF is a buffer storage location in memory where the characters to be displayed are stored. The program gets a character from DISBUF and displays it. Characters are fetched and displayed until all six display digits are refreshed. The program then blanks the segments and puts all lows on the digit-strobe/keyboard row scan lines to check for a key closed. If no closed key is detected, the program returns to refresh the display. If a closed key is found, the display is left blanked and the key is debounced and encoded. To reduce the number of inputs required for the column data, the four column lines are multiplexed in on one line with an MC14539 four-line-to-one-line multiplexer.

Since the MEK6800D2 is used in an EPROM programmer prototype in Chap. 12, it is useful to get a head start by discussing here the details of the display routine OUTDS. Figure 11-16 shows a flowchart, and Figure 11-17 is the 6800 assembly language program for it.

It is very common to have to figure out a program written by someone else. To complicate this, many programs use tricks that are not documented and not immediately obvious. Also most professionally written programs such as the one in Figure 11-17 contain many labels or symbolic addresses.

The first step in approaching such a program is to find the definitions of these symbolic addresses. Usually they are at the beginning as *equates*. Most programmers try to make the symbolic addresses mnemonic. For example, DISBUF refers to the starting address of the buffer where the six display digits are stored. If the la-

FIGURE 11-15 Flowchart for MEK6800D2 keyboard-display scan routine. *(Motorola.)*

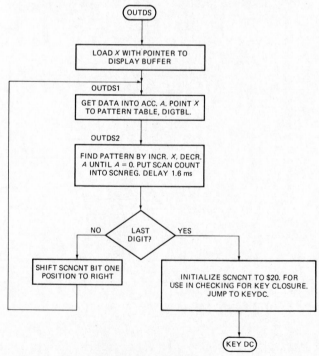

FIGURE 11-16 Flowchart for the MEK6800D2 display routine OUTDS. *(Motorola.)*

```
;SCNREG - SYMBOLIC NAME FOR ADDRESS OF OUTPUT PORT TO DISPLAY CATHODES.
;SCNCNT - ADDRESS WHERE MASK FOR DIGIT DRIVERS IS STORED.  THIS LOCATION
;         IS INITIALLY LOADED WITH 20H WHICH LIGHTS THE LEFTMOST DISPLAY
;         WHEN OUTPUT.
;DISREG - SYMBOLIC NAME FOR ADDRESS OF OUTPUT PORT TO DISPLAY SEGMENTS.
;XDSBUF - STARTING ADDRESS FOR LOCATION IN MEMORY WHERE INDEX REGISTER
;         CONTENTS ARE TEMPORARILY STORED.
;DIGTBL - ADDRESS IN MEMORY WHERE SEVEN SEGMENT CODES FOR ALL HEX
;         CHARACTERS ARE STORED.
;DISBUF - ADDRESS IN MEMORY WHERE CHARACTERS TO BE DISPLAYED ARE STORED.


OUTDS:    LDX     #DISBUF       ;LOAD STARTING ADDRESS OF CHAR
                                ;BUFFER IN INDEX
OUTDS1:   LDAA    0,X           ;LOAD FIRST CHARACTER TO BE
                                ;DISPLAYED INTO A
          INCA                  ;INCREMENT CHARACTER VALUE
          INX                   ;INCREMENT INDEX REGISTER POINTER
          STX     XDSBUF        ;STORE POINTER IN MEMORY
          LDX     #DIGTBL-1     ;LOAD STARTING ADDRESS OF SEVEN
                                ;SEGMENT CODE TABLE - 1
OUTDS2:   INX                   ;INCREMENT INDEX REGISTER POINTER
          DECA                  ;DECREMENT A
          BNE     OUTDS2        ;BRANCH IF NOT EQUAL TO ZERO TO OUT
          CLR     SCNREG        ;BLANK DISPLAY (SEND ONES TO ALL
                                ;DIGIT CATHODES)
          LDAA    0,X           ;GET SEVEN SEGMENT CODE FROM TABLE
                                ;ADDRESS POINTED TO BY X
          STAA    DISREG        ;SEND SEGMENT CODE TO SEGMENT
                                ;DRIVERS
          LDAA    SCNCNT        ;LOAD DIGIT STROBE MASK
          STAA    SCNREG        ;SEND MASK TO DIGIT DRIVERS
          LDX     #$4D          ;LOAD DELAY CONSTANT FOR 1MS DELAY
          BSR     DLY1          ;DELAY 1MS
          LDX     XDSBUF        ;RELOAD POINTER TO CHARACTER IN
                                ;DISBUF
          CPX     #DISBUF+6     ;SEE IF ALL 6 CHARACTERS IN DISBUF
                                ;HAVE BEEN DISPLAYED
          BEQ     OUTDS3        ;IF YES, EXIT TO CHECK FOR KEY
                                ;PRESSED
          LSR     SCNCNT        ;NO, ROTATE DIGIT MASK TO DISPLAY
                                ;NEXT DIGIT
          BRA     OUTDS1        ;LOOP BACK TO GET CODE FOR NEXT
                                ;DIGIT AND DISPLAY IT
```

FIGURE 11-17 MC6800 assembly language program for OUTDS display routine. *(Motorola.)*

bels are only equated at the end of the program to their absolute addresses without definitions, then try to make a list of definitions for yourself. An example is shown at the top of the program in Figure 11-17.

The actual OUTDS routine is shown just as found in the JBUG monitor program of the MEK6800D2 except that the comments have been expanded for clarity. The routine functions as follows. The starting address of DISBUF is loaded into the index register, and the first character to be displayed is loaded from the first buffer location to the *A* accumulator. The character value and the index register are then incremented. (The reason is explained later.) STX XDSBUF stores the index register contents in memory, starting at a memory location labeled XDSBUF. Then 1 less than the starting address of the table which contains the seven-segment codes is loaded into the index register. The OUTDS2 loop decrements the character to be displayed to 0 and increments

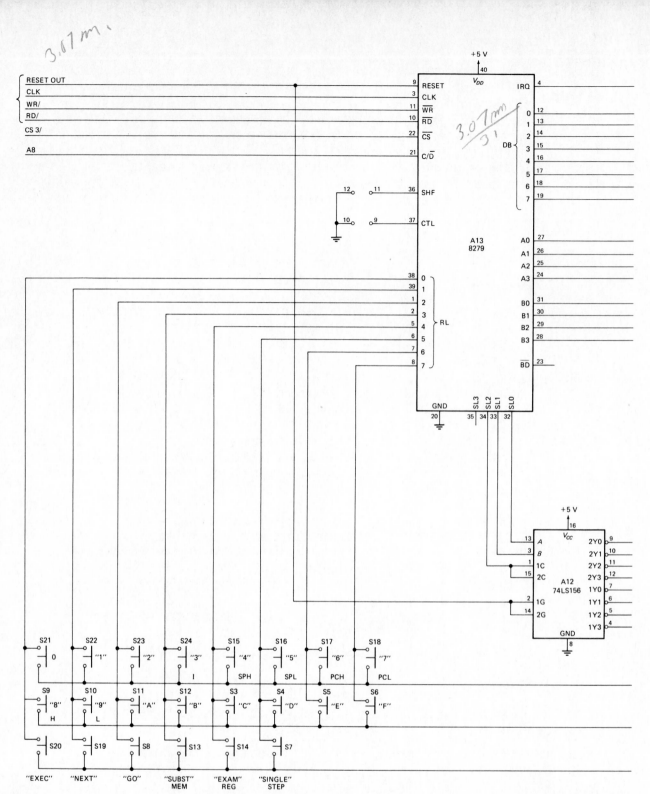

FIGURE 11-18 8279 keyboard-display section of Intel SDK-85 microcomputer development board. *(Intel Corporation.)*

the index register. This leaves the index register pointing to the correct seven-segment code for the character in the table.

CLR SCNREG blanks the display by sending 0's to all the digit drivers. The OUTDS2 loop in the routine produces the correct seven-segment code for a character as follows. The seven-segment codes are stored in a table in memory. The code for 0 is stored in the first location, the code for 1 in the next, the code for 2 in the next, etc. LDX #DIGTBL − 1 loads the starting address minus 1 into the index register, and INX increments this to point to the first entry in the table, the code for the character 0. If the character in *A* is a 0, however, the DEC A instruction in the loop decrements it to FF instead of to 0.

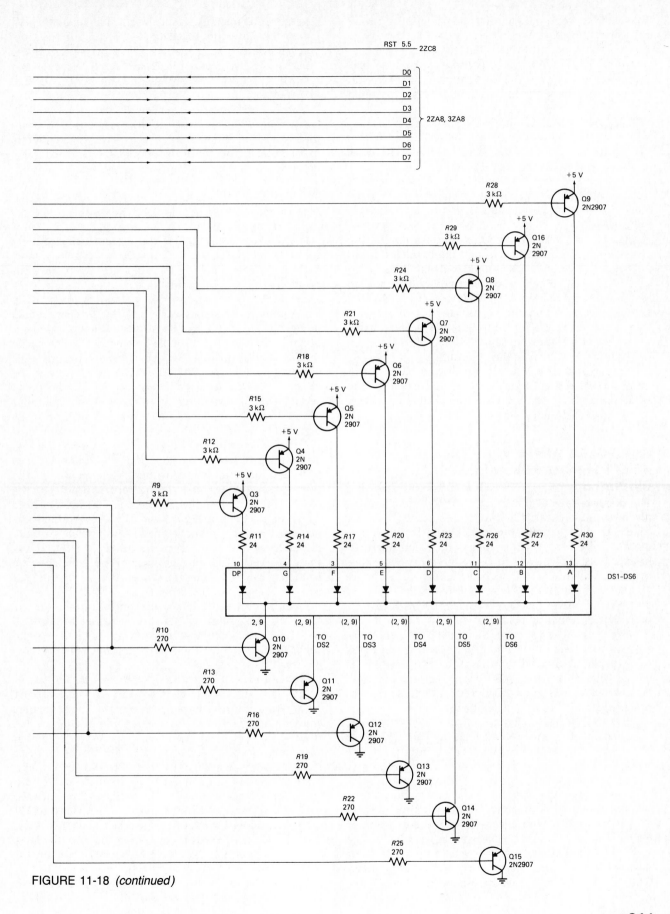

FIGURE 11-18 *(continued)*

To have the seven-segment table line up with the proper character, the character in the *A* accumulator must be incremented before this loop. This is done by the INC A instruction in the OUTDS1 section of the loop. The seven-segment code is then brought from the table address pointed to by the index register to the accumulator.

STAA DISREG sends out the code to the segment drivers. No digits will be lighted yet, however, because no digit drivers are on. LDAA SCNCNT loads the digit-strobe mask into the accumulator from a memory location labeled SCNCNT. The value in SCNCNT is set initially to 20_{16}, which lights the leftmost of the six displays. STAA DISREG sends the digit-strobe mask to the digit drivers. This lights the strobed digit.

After a 1-ms delay, the pointer address for the displayed character in DISBUF is loaded into the index register again. The value in X is compared with the starting address of DISBUF plus 6 to see whether all six characters have been displayed. If so, BEQ exits the program to the key-closed routine through OUTDS3. If not, the 1 in the digit-strobe mask stored in SCNCNT is moved right to the next digit position. Then BRA OUTDS1 loops the program back to display the next digit.

It is good practice to work your way through a program such as this until you understand each step. This is often necessary when you are troubleshooting microprocessor-based instruments.

HARDWARE APPROACH TO MULTIPLEXING SEVEN-SEGMENT DISPLAYS

The disadvantage of the software method of scanning displays and keyboards is that it uses much processor time. The display may have to be blanked while a program is running. Another approach, which does not have these problems, is to use a programmable keyboard-display interface IC such as the Intel 8279. Figure 11-18 shows how an 8279 is connected to the keyboard and seven-segment displays on the Intel SDK-85 microprocessor prototyping board.

Basically the device functions as follows. After the 8279 is initialized, the segment codes for the characters to be displayed are loaded into a 16-byte register in the device. The 8279 then sends out the segment code and digit strobe to each digit in turn. Once it is loaded, the device needs no attention from the processor to keep the displays refreshed. In Figure 11-18, the 74LS156 is used as 3-line-to-one-of-8-lines-low-decoder. It selects the digit addressed by the digit-select outputs of the 8279.

The stepping pattern of lows on the digit-strobe driver inputs is also used to scan the keyboard. Columns of the keyboard are read by the return inputs R0 through R7 of the 8279. If a key is found closed, an internal debounce circuit waits 10 ms and checks again to see whether the switch is still closed. If it is, the code for the pressed key is stored in an internal first in, first out (FIFO) memory. A high level is sent out on the IRQ output to indicate

that the FIFO memory has a character. Usually the IRQ is tied to an interrupt input such as RST 5.5 on the processor. If interrupts are enabled, and a key is pressed, the processor will vector to a keyboard service routine that reads in the character from the FIFO memory. The 8279 FIFO memory, incidentally, can hold up to eight characters to be read before it overflows.

When an 8279 is used, the microprocessor program merely has to initialize the device, send the characters to be displayed to the internal register, and read keyboard characters when signaled by an IRQ.

INITIALIZING THE 8279 Since the SDK-85 is used in a prototype example in Chap. 12, we discuss the 8279 initialization here.

First, take a closer look at the inputs and outputs of the 8279 in Figure 11-18. DB0–DB7 is a bidirectional data bus. Reset, clock, write, and $\overline{RD}$ inputs are connected to the corresponding system lines. $\overline{CS}$ is a chip-select input. $C/\overline{D}$, if high, indicates a control word is being written or a status word read. If $C/\overline{D}$ is low, data is being written or read. In the SDK-85, the 8279 is memory-mapped at address 1900H for a control write-status read and at 1800H for data read-write. The two 4-bit outputs, A0 to A3 and B0 to B3, can be used individually to send out BCD for segments or together to send out seven-segment code directly as shown. The four digit scan lines SL0 to SL3 can be decoded to scan up to 16 digits.

The decoded scan lines also are used to scan the keyboard rows. Typewriter-type keyboards can be serviced by simply using more scan lines. The columns of the keyboard are connected to the return inputs of the 8279. These inputs have internal pull-up resistors which hold a high until pulled low by a key closed. If the keyboard has shift and control keys, these are connected to the shift and control inputs of the 8279 so that their effect is included in the output character. The keys in Figure 11-18, incidentally, are wired so that output code for a key is in hex from 00_{16} to 15_{16}.

To initialize the 8279, several control words must be sent. For the circuit in Figure 11-18, they are sent to address 1900H. The first control word to send sets the keyboard-display mode. Figure 11-19*a* shows the format for this word.

Right entry is the form used by many electronic calculators. The first character is entered in the rightmost digit. When another character is entered, the first character is shifted left one place. As each new character is entered, the previous characters are all shifted left to make room for it. *Left entry* just enters characters one after the other from the left as you would write them.

For the keyboard modes, *encoded* and *decoded* refer to the four scan line outputs SL0 to SL3. In the encoded mode, the count on these outputs cycles from 0 to 15. An external decoder is used to give 1 of 16 or 1 of 8 low signals. In the decoded mode, one of the four SL lines will be low at a time. For small keyboards and four displays or less, this saves the external decoder.

Two-key lockout in this case means that if two keys

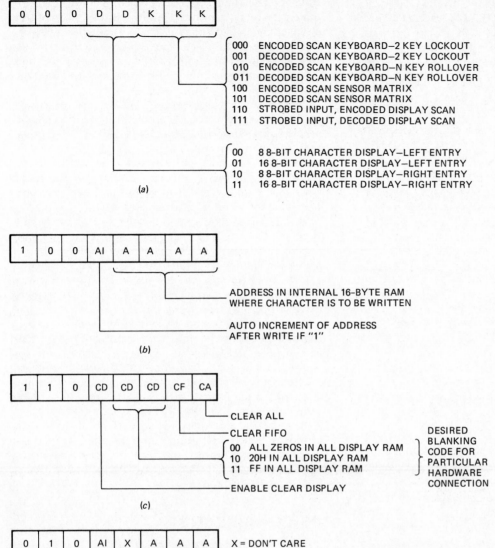

are pressed at nearly the same time, the last key released will be entered. With *N-key rollover*, two or more keys pressed at nearly the same time are all entered in the order in which they were pressed.

Not all the possible command words can be discussed here, but another important one to consider is the word used to point to a write location in the internal 16-byte display RAM. The format for this word is 100AIAAAA, where AI signifies automatic increment (see Figure 11-19b). The four As represent the 4-bit address of one of the 16 locations in the internal display RAM. To write in a character for display, first a control word and then the character to be displayed are sent.

Another useful control word sets the character used for blanking the display. The control word is shown in

Figure 11-19c. Since an output of all 1's, or FF hex, will blank the display of the SDK-85, for example, CC hex is the control word sent to do this.

An example may clarify the initializing process for the 8279. Figure 11-20 shows a short program section to initialize the 8279 for its use in the SDK-85 board. The keyboard-display command word is sent to the control port address 1900H. Then the command word is sent to move the internal pointer to the first address of the internal RAM. Next the command word for the character used to blank the display is sent. The SDK-85 requires all 1's for blanking, and the command code for this is CC hex. The required seven-segment code for the number 4, 99_{16}, is written into this RAM location by sending it to the read-write port address of 1800H. This will dis-

```
*INITIALIZATION
MVI     A,00H       ;SEND KEYBOARD/DISPLAY
                    ;MODE
STA     1900H       ;FOR LEFT ENTRY, 8
                    ;DIGITS, ENCODED 2 KEY
                    ;LOCKOUT
MVI     A,90H       ;SEND COMMAND WORD TO
                    ;SET
STA     1900H       ;INTERNAL POINTER TO
                    ;FIRST LOCATION IN
                    ;8279 CHARACTER RAM
                    ;AND AUTO INCREMENT
MVI     A,0CCH      ;SEND COMMAND WORD FOR
STA     1900H       ;CHARACTER USED TO
                    ;BLANK DISPLAY (ALL
                    ;ONES FOR SDK-85)
*SEND FIRST CHAR TO BE DISPLAYED
MVI     A,99H       ;SEND 7 SEGMENT CODE
                    ;FOR
STA     1800H       ;NUMBER 4 TO BE
                    ;DISPLAYED AS LEFTMOST
                    ;DIGIT


READ FIFO RAM

MVI     A,40H       ;CONTROL WORD FOR FIFO
                        ACCESS

STA     1900H       ;TO CONTROL ADDRESS


LDA     1800H       ;READ FIFO
```

FIGURE 11-20 Assembly language program for initializing the 8279.

play a 4 in the leftmost digit. After a write, the internal pointer increments to the next location in the internal RAM.

To read the FIFO RAM after an interrupt caused by a key being pressed, the FIFO RAM must be accessed. To do this, a control word is sent that points an internal pointer to the FIFO RAM. Figure 11-19d shows the format for this control word. A control word of 40H will do this. After this word is sent to the control address, a read from the data address will read the character from the FIFO memory, as shown in Figure 11-20. To write to the display RAM again, another write-to-display-RAM control word must be sent to the control address.

The main point of the 8279 approach is that it relieves the CPU of the housekeeping chores of refreshing a display and scanning a keyboard. Consult the 8279 data sheet for further command words and operating modes.

INTERFACING TO LIQUID CRYSTAL DISPLAYS

Liquid crystal displays (LCDs) are created by sandwiching a liquid crystal fluid between two glass plates. A transparent electrically conductive film or back plane is put on the rear glass sheet. Transparent sections of conductive film in the shape of the desired characters are coated on the front glass plate. When an electric field is created between the front film character and the back plane, the transmission of light through the liquid crystal is altered.

There are two general types of LCD: dynamic scattering and field effect. The dynamic scattering type scrambles the molecules where the field is present. This scatters the light hitting the region and produces an etched-glass-looking light character on a dark background. Field-effect types use polarization to absorb light where the electric field is present. This produces dark characters on a light background.

Most LCDs turn on and off too slowly to be easily multiplexed. Therefore, each segment must be driven individually. However, LCDs deteriorate rapidly when a steady dc voltage is applied to them, and so a 30- to 125-Hz ac square-wave drive is usually used. The square-wave drive can be either added to each segment drive or applied to the common back plane.

The RCA CD4054 does BCD-to-seven-segment decoding and provides the ac drive for one LCD digit. The Intersil ICM 7211 provides binary-to-hexadecimal decoding and segment drive for four LCD digits. Internal latches make it possible to load the digits in one at a time from a microprocessor. The 7211 also has the required ac output for the LCD back plane.

Although it is intended for use with multiplexed BCD output A/D converters and counters, the 7211 can be connected to an output port. A short program routine can send the digit strobe and BCD digit code for each digit to the 7211.

POWER CIRCUITS

NONDESTRUCTIVE TURN-ON

Microprocessors are often used to control high-power devices such as incandescent lights, heaters, and motors. Interfacing to these devices must be done carefully to avoid catastrophe. Probably the most important factor is setting up each device so that it is in a nondestructive state when power is turned on.

For example, suppose the input to a solid-state relay such as those described in Chap. 2 is connected to an output line on an 8255 PIA. The output of the relay is connected to a siren, and a logic high on the input of the relay turns it on. When the processor is first turned on, all the port lines on the 8255 are in the high-impedance floating state. Therefore, the input to the solid-state relay is unpredictable. It may float to a high state and turn on the siren. The problem is solved by using a resistor to hold the solid-state relay input low while the 8255 is being initialized. The safe state must be determined for each device.

POWER DRIVE

Actual interface of these devices is mostly a matter of greatly increasing the current drive of the output port line and, if the device operates on ac, isolating it. These functions can be done with Darlington transistors, mechanical relays, or solid-state relays. With mechanical relays, remember to include the reverse-biased protective diodes.

STEPPER MOTORS

A unique type of motor useful for making robots and such is the *stepper motor*. Instead of rotating smoothly around and around as most motors do, stepper motors rotate or "step" from one fixed position to the next when pulsed. Common step sizes range from 0.9 to 30°. Steppers might be thought of as "digital" motors.

A stepper motor is stepped from one position to the next by the proper combination of fields in the motor. Figure 11-21 shows the schematic and stepping sequences for a Superior Electric SLO-SYN motor. If the coils are energized with SW1 and SW2 on, then changing to SW1 and SW4 on causes the motor shaft to move one step of 1.8° clockwise. A change to SW3 and SW4 on will cause your robot's hand to rotate another 1.8° clockwise. Changing to switches 2 and 3 gives another 1.8°, and switching back to 1 and 2 on gives another 1.8° step clockwise. The sequence can be repeated to produce as many steps as desired. If the switch step sequence is reversed, the motor steps in a counterclockwise direction, as indicated by the arrows in Figure 11-21b. If all switches are turned off, the motor shaft is held in its last position by the permanent magnet in the motor. However, it is not held in place as firmly then as it is with the field currents on. VFET or bipolar power transistors can be used as the switches in Figure 11-21a.

A close look at the step sequence table in Figure 11-21b reveals an interesting pattern. To move from position 1 to position 2, the pattern of 1's and two 0's is rotated right one place. The 1 from switch 4 is rotated around into switch 1. For each step the pattern is rotated right one more place. To reverse the direction of rotation, the pattern is shifted left.

This drive sequence can be produced by connecting the switches to the outputs of a 4-bit ring counter initially loaded with 0011. The ring counter is clocked once for each step. If a shift-left–shift-right register is used for the ring counter, the motor can be stepped in either direction. Another method is to connect the four switches to four output port lines of a microprocessor. The step sequence is then created with a simple program.

If the switches are connected to the lower four port lines, the pattern 00110011 is loaded into the accumulator. Rotating this right and outputting it produces a clockwise step. Rotating it left and outputting it produces a counterclockwise step. Since the motor can step only 500 steps per second, a short delay must be introduced between steps.

In addition to their use in robots, stepper motors are used to position the heads in a floppy disk drive unit. They are used also in *X-Y* plotters which draw graphs

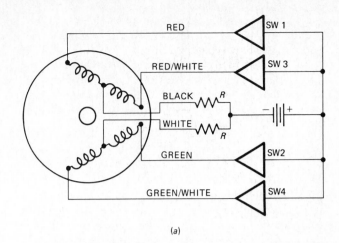

(a)

	Switch				
Step	SW4	SW3	SW2	SW1	CW
1	0	0	1	1	
2	1	0	0	1	
3	1	1	0	0	
4	0	1	1	0	
1	0	0	1	1	

CCW

1 = Switch On

(b)

FIGURE 11-21 Superior Electric SLO-SYN stepper motor. *(a)* Schematic of drive. *(b)* Step sequence.

under computer control. One stepper motor and a belt control the position of the pen in the *X* direction, and another stepper motor and belt control the position of the pen in the *Y* direction. In short, stepper motors are useful almost anywhere where movement in increments is needed.

D/A CONVERTER INTERFACING AND APPLICATIONS

INTERFACING A D/A CONVERTER TO A MICROPROCESSOR

In many cases the parallel inputs of a D/A converter can simply be connected to the parallel lines of a microprocessor output port. A word written to the port is changed to a proportional voltage or current by the D/A converter.

A problem arises, however, if the D/A converter has more input lines (bits) than the port has output lines. For example, if a 12-bit D/A converter is connected to the output of an 8-bit processor, the D/A converter must be connected to two output ports. Since the processor cannot output to two ports at the same time, data gets to some bits of the D/A converter sooner than to others. This means the D/A output will go to an incorrect value before it gets the second part of the data. The problem is solved by using 12 external latches. Data is written into the two ports; then the latches are strobed to transfer all 12 bits to the D/A converter at the same time. Some more expensive D/A converters have latches built into their inputs for this purpose.

SOME MICROPROCESSOR AND D/A APPLICATIONS

The number of applications in which a D/A converter is used on the output of a microprocessor is very large. This is done in almost every case where a digitally programmable voltage or current output is desired.

PROGRAMMABLE POWER SUPPLY One application is for a *programmable power supply*. The output of the D/A converter is connected to the reference input of the power supply. As the output of the D/A converter is changed, the power-supply output changes proportionately.

MOTOR-SPEED CONTROL The speed of a small dc motor is proportional to the current flowing through it. The output of a D/A converter can be converted to an amplified current and used to control the speed of the motor.

SPEECH SYNTHESIS Another exciting application of D/A converters with microprocessors is for *speech synthesis*. Talking calculators and digital voltmeters using speech synthesis are a great aid to the blind and others. One method of producing speech stores the digital data for each word in ROM. A processor reads out the data for the desired word and passes it through some complicated digital filtering. The filtering actually simulates the human vocal tract.

The result of the digital filtering is passed through a D/A converter to produce the analog audio waveforms. These are amplified and used to drive a speaker. The number of words in the vocabulary is determined by the amount of ROM used. A talking calculator from Telesensory System, for example, has a 64-word vocabulary. One IC from Texas Instruments, the TMC 0280, contains signal generators, digital filters, and a D/A converter. ROM and a microprocessor are all that must be added to make a talking spelling game.

A/D CONVERTER INTERFACING AND APPLICATIONS

INTERFACING A/D CONVERTERS TO MICROPROCESSORS

NONMULTIPLEXED OUTPUT A/D CONVERTERS Many A/D converters have parallel outputs with one line for each bit of resolution. An *end-of-conversion* (EOC) signal is also sent out by the converter when the data on the output lines is valid. This type of converter can be easily interfaced by connecting the parallel outputs to one or more input ports of a microprocessor.

The EOC signal, indicating when data is valid, can be connected to an interrupt input of the processor, and an interrupt service routine can be used to take in the data. An alternative is to connect the EOC signal line to an input port line and poll the line with a program loop. When the EOC signal is detected, data is read in. The parallel output code for this type of converter is usually binary or 2's complement binary.

MULTIPLEXED OUTPUT A/D CONVERTERS A/D converters intended for use in digital voltmeters and digital panel meters usually have a multiplexed BCD output. The multiplexed BCD format is used because it is easily interfaced to displays. Since this type of converter is produced in large quantities for DVMs, it is often quite inexpensive and is therefore used in many other applications.

To interface this multiplexed BCD type of converter to a microprocessor, the four BCD lines and each of the digit-strobe lines are connected to input port lines. The first digit-strobe line is polled; when it is found true, the BCD data for that digit is pulled in and stored in memory. The next digit-strobe line is polled; when it is found true, data for that digit is pulled in and stored. An 8080A/8085A assembly language example of this input process is shown in Chap. 12. It is handy to know how to pull in this type of data because many frequency counters, digital voltmeters, and other test instruments have a multiplexed BCD output available on their rear panel. The scale in Chap. 12 shows an example of interfacing this type of A/D converter.

MICROPROCESSORS AND SAR A/D CONVERTERS

For medium-speed applications, a microprocessor and a program routine can replace the successive-approximation register (SAR) in a successive-approximation type of A/D converter. (Review Chap. 6 if necessary.) A digital-to-analog converter is connected to an output port. The output of the D/A converter is connected to the reference input of a comparator. The comparator output is connected to an input port line. A program routine sends out each bit in turn to the D/A converter. If the comparator response is high, the bit is kept and the next is tried. If the comparator response is low, the bit is reset and the next bit is tried. This technique has the disadvantage of tying up the processor.

Since so many microprocessor applications require an A/D converter, Intel responded to the need by producing the 8022. The 8022 contains an 8048 family microprocessor and a complete 8-bit successive-approximation A/D converter in one IC package. The device also contains 2 kilobytes of ROM, 64 bytes of RAM, 28 I/O lines, and an 8-bit timer.

The A/D converter uses an on-board SAR, so that the processor time is required only to start a conversion and to store the result in memory. Conversion time is around 50 μs. One of two inputs can be program-selected for digitizing.

DATA ACQUISITION SYSTEMS

Some microprocessor applications require analog data from many sources to be digitized and stored. A separate A/D converter can be used on each channel, but this is expensive, and it is difficult to keep all the A/D converters calibrated with one another. Another approach is to use one A/D converter and select one analog signal at a time with an analog multiplexer such as the MC14051 or CD4051. The 4051 produces a low-resistance path between the output and one of eight binary addressed inputs. You may remember that this device was used in the oscilloscope multiplexer circuit in Chap. 3.

An A/D converter with an analog multiplexer and the necessary control circuity is called a *data acquisition system*. The industrial controller in Chap. 12 shows the interfacing and application of a data acquisition system.

MICROPROCESSOR AND A/D CONVERTER APPLICATIONS

The number of applications of A/D converters with microprocessors is almost unlimited. Most temperature sensors, pressure transducers, liquid flowmeters, and photodetectors have a small current or voltage output. This output must be amplified and converted to digital form for use in a microprocessor. Chapter 12 shows an application of an A/D converter with a load-cell pressure transducer to make a microprocessor-based digital scale.

SPEECH RECOGNITION

Another exciting application of an A/D converter with microprocessors is as part of a *speech recognition* unit. These units allow direct voice communication of data or instructions to a microprocessor. Several techniques are used for recognizing spoken words. None is 100 percent accurate, but the day of listening and talking computers is not far distant.

The Speech Lab Model 50 from Heuristics, Inc., is a speech-to-digital conversion module contained on an S-100 bus-compatible PC board. It is intended to be used with a microprocessor. With the programs supplied, it permits a microprocessor to be taught to recognize up to 64 words with better than 90 percent accuracy. Figure 11-22 shows a block diagram of the Model 50.

The microphone is used as both a small speaker for signaling and a microphone for receiving speech. The microphone output is amplified by a preamplifier and applied to the inputs of three bandpass filters with 80-dB/decade skirts. The three filters are centered on three resonant, or "formant," frequencies of an average human voice. These filters are set for 150 to 900, 900 to 2200, and 2200 to 5000 Hz. Averagers convert the output of the filters to a voltage proportional to the amplitude of the signal in each band.

An unfiltered portion of the signal is amplified again and applied to a zero crossing detector. The pulse train

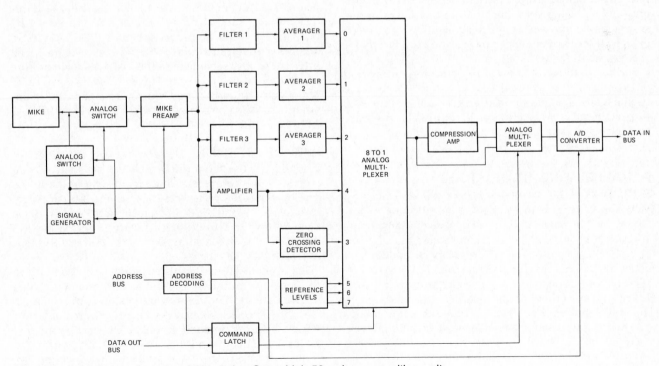

FIGURE 11-22 Block diagram of Heuristics Speechlab-50 voice recognition unit.

output of the zero crossing detector is converted to a voltage and used to determine the dominant frequency in the input. As shown in Figure 11-22, the averager outputs, amplified signal, zero crossing detector output, and three reference voltages are all sent to an eight-input analog multiplexer. A 3-bit address input to the 4051 multiplexer will select the voltage to be sent to the 6-bit A/D converter.

The device works as follows. A microprocessor program is used to first train the unit. To do this, the operator says the first desired word. The program samples the word every 10 ms for 1.5 s. Sixteen samples evenly spaced in the spoken word are analyzed. The digital information for the three frequency band outputs and the zero crossing output for these 16 samples are stored in memory. This creates a stored template, or pattern, for the desired word. In a similar manner, a template is stored in memory for each desired vocabulary word. For each vocabulary word 64 bytes of memory are required.

When the unit is asked to "perform," a spoken word is likewise sampled and the four parameters are stored. The values for the spoken word are compared with the templates stored in memory, and the best match is assumed to be the correct word. Each word can direct a processor to perform a desired operation such as turning on lights, opening a door, or mixing a drink.

COMMON BUS AND DATA COMMUNICATION STANDARDS

The standard for ac power in the United States, 120 V ac at 60 Hz, makes it possible to just plug in appliances from most manufacturers without concern for compatibility. When communication is from one microprocessor system to another or to a peripheral unit such as a printer, it is also desirable to have a standard format for the connecting wires or buses.

For communication within a microcomputer system or to another microcomputer system, parallel buses are used usually. For communication over long distances or to peripheral units such as CRT terminals, printers, and teletypes, data is often sent in serial form.

PARALLEL BUS STANDARDS

S-100 BUS   The S-100 bus format was developed by MITS, Inc., for use in their 8080-based Altair microcomputer. The basic Altair unit consists of a large chassis with a power supply, a CPU printed-circuit board, and slots where other PC boards can be added. The added PC boards may be RAM boards, ROM boards, a speech recognition board, a floppy disk controller board, or one of many other possible boards. Each board plugs into a 100-pin edge connector—thus the name S-100. The connector is actually double-sided with 50 pins on each side of the board.

This standard has been used in many hobbyist computer boards and systems. Table 11-1 shows the pin definitions for the S-100 bus. For a more detailed description of each signal, consult the MITS Altair 8800B manual.

Note that power-supply voltages are sent on the bus in unregulated form. Each board or system connected to the bus is assumed to have its own voltage regulator ICs. Also note that data-out lines are separated from the data-in lines, as required by some systems. On boards such as RAM boards which require the DO and DI lines to be combined, this can be done with some three-state buffers on the board. The S-100 format has an abundance of control and interrupt signal pins, so it can be used with processors other than the 8080.

Some criticisms of the layout of this bus include cross-talk from clock lines to control lines on adjacent pins and potential shorting between the adjacent +8- and −16-V pins.

INTEL MULTIBUS   An industrial bus standard that has gained wide acceptance is the Intel multibus. This bus, used for connecting RAM, ROM, or I/O boards within a microcomputer system, has an 86-pin edge connector and an optional 60-pin edge connector. As shown in Table 11-2, the 86-pin connector has power sections, a 20-bit address bus, a 16-bit bidirectional data bus, and an extensive control section. A slash (/) after a mnemonic in Table 11-2 indicates that the signal on that pin is active low.

A major feature of the multibus is that several CPU boards can be connected to the bus at the same time to make a multiprocessor system. Each CPU board can be performing dedicated tasks, using its on-board ROM, RAM, and I/O ports until it needs to access a system memory board, another CPU board, or an I/O board connected to a system peripheral such as a printer or hard-disk storage unit. When a CPU board needs to use the bus, it asserts its *bus request signal* (BREQ/) and its *bus priority output signal* (BPRO/). A parallel or serial priority encoder scheme uses these signals to determine which CPU board has control of the bus at a given time. The orderly transfer of bus control from one CPU to another is handled by a bus controller IC such as the Intel 8218 or 8288.

HPIB OR IEEE-488   The Hewlett-Packard interface bus (HPIB) is used to connect instruments such as computers, digital voltmeters, and signal generators. It was developed mostly by Hewlett-Packard and later accepted as standard by the IEEE. In fact, Hewlett-Packard holds a patent on the complex handshaking scheme used in this standard. It is also known as the *general-purpose interface bus* (GPIB).

The standard describes three types of devices that can be connected on the HPIB. First is a *listener*, which can receive data from other instruments. Examples of a listener are a printer, display device, programmable power supply, or programmable signal generator. Second is a *talker*, which can send data to other instruments. Examples of talkers are tape readers, digital voltmeters, and frequency counters. A device can be both a talker

TABLE 11-1
S-100 Parallel Bus Pin Definitions

PIN NUMBER	SYMBOL	NAME	FUNCTION
1	+8V	+8 Volts	Unregulated voltage on bus, supplied to PC boards and regulated to 5 V.
2	+18V	+18 Volts	Positive preregulated voltage.
3	XRDY	External ready	External ready input to CPU board's ready circuitry.
4	VI0	Vectored interrupt line 0	
5	VI1	Vectored interrupt line 1	
6	VI2	Vectored interrupt line 2	
7	VI3	Vectored interrupt line 3	
8	VI4	Vectored interrupt line 4	
9	VI5	Vectored interrupt line 5	
10	VI6	Vectored interrupt line 6	
11	VI7	Vectored interrupt line 7	
12	XRDY2	External ready 2	A second external ready line similar to XRDY.
13 to 17	To be defined		
18	STAT DSB	Status disable	Allows the buffers for the eight status lines to be three-stated.
19	C/C DSB	Command/control disable	Allows the buffers for the six output command/control lines to be three state.
20	UNPROT	Unprotect	Input to the memory protect flip-flop on a given memory board.
21	SS	Single-step	Indicates that the machine is in the process of performing a single step (i.e., that SS flip-flop on display/control is set).
22	ADD DSB	Address disable	Allows the buffers for the sixteen address lines to be three-stated.
23	DO DSB	Data out disable	Allows the buffers for the eight data output lines to be three-stated.
24	02	Phase 2 clock	
25	01	Phase 1 clock	
26	PHLDA	Hold acknowledge	Processor command/control output signal that appears in response to the HOLD signal; indicates that the data and address bus will go to the high-impedance state and processor will enter HOLD state after completion of the current machine cycle.
27	PWAIT	Wait	Processor command/control signal that appears in response to the READY signal going low; indicates processor will enter a series of a 0.5-μs wait states until READY goes high again.
28	PINTE	Interrupt enable	Processor command/control output signal; indicates interrupts are enabled, as determined by contents of CPU internal interrupt flip-flop. When the flip-flop is set (enable interrupt instruction), interrupts are accepted by the CPU; when it is reset (disable interrupt instruction), interrupts are inhibited.
29	A5	Address line 5	
30	A4	Address line 4	
31	A3	Address line 3	
32	A15	Address line 15	
33	A12	Address line 12	
34	A9	Address line 9	
35	DO1	Data out line 1	
36	DO0	Data out line 0	(LSB)
37	A10	Address line 10	
38	DO4	Data out line 4	
39	DO5	Data out line 5	
40	DO6	Data out line 6	

TABLE 11-1 (continued)

PIN NUMBER	SYMBOL	NAME	FUNCTION
41	DI2	Data in line 2	
42	DI3	Data in line 3	
43	DI7	Data in line 7	(MSB)
44	SM1	Machine cycle 1	Status output signal that indicates that the processor is in the fetch cycle for the first byte of an instruction.
45	SOUT	Output	Status output signal that indicates the address bus contains the address of an output device and the data bus will contain the output data when PWR is active.
46	SINP	Input	Status output signal that indicates the address bus contains the address of an input device and the input data should be placed on the data bus when PDBIN is active.
47	SMEMR	Memory read	Status output signal that indicates the data bus will be used to read memory data.
48	SHLTA	Halt	Status output signal that acknowledges a HALT instruction.
49	CLOCK	Clock	Inverted output of the $\phi2$ Clock.
50	GND	Ground	
51	+8V	+8 Volts	Unregulated input to 5-V regulators.
52	−18V	−18 Volts	Negative preregulated voltage.
53	SSWI	Sense switch input	Indicates that an input data transfer from the sense switches is to take place. This signal is used by the display/control logic to enable sense switch drivers, enable the display/control board driver's data input (FD10–FD17), and disable the CPU board data input drivers (D10–D17).
54	EXT CLR	External clear	Clear signal for I/O devices (front-panel switch closure to ground).
55	*RTC	Real-time clock	60-Hz signal is used as timing reference by the real-time clock/ vectored interrupt board.
56	*STSTB	Status strobe	Output strobe signal supplied by the 8224 clock generator. Primary purpose is to strobe the 8212 status latch so that status is set up as soon in the machine cycle as possible. This signal is also used by display/control logic.
57	*DIGI	Data input gate 1	Output signal from the display/control logic that determines which set of data input drivers has control of the CPU board's bidirectional data bus. If DIGI is high, the CPU drivers have control; if it is low, the display/control logic drivers have control.
58	*FRDY	Front panel ready	Output signal from display/control logic that allows the front panel to control the READY lines to the CPU.
59 to 67	To be defined		
68	MWRITE	Memory write	Indicates that the data present on the data out bus is to be written into the memory location currently on the address bus.
69	PS	Protect status	Indicates the status of the memory protect flip-flop on the memory board currently addressed.
70	PROT	Protect	Input to the memory protect flip-flop on the board currently addressed.
71	RUN	Run	Indicates that the 64/RUN flip-flop is reset; i.e., machine is in RUN mode.
72	PRDY	Processor ready	Memory and I/O input to the CPU board WAIT circuitry.
73	PINT	Interrupt request	The processor recognizes an interrupt request on this line at the end of the current instruction or while halted. If the processor is in the HOLD state or the interrupt enable flip-flop is reset, it will not honor the request.

TABLE 11-1 (continued)

PIN NUMBER	SYMBOL	NAME	FUNCTION
74	PHOLD	Hold	Processor command/control input signal that requests the processor enter the HOLD state; allows an external device to gain control of address and data buses as soon as the processor has completed its uses of these buses for the current machine cycle.
75	PRESET	Reset	Processor command/control input; while activated, the content of the program counter is cleared and the instruction register is set to zero.
76	PSYNC	Sync	Processor command/control output provides a signal to indicate the beginning of each machine cycle.
77	PWR	Write	Processor command/control output is used for memory write or I/O output control. Data on the data bus is stable while the PWR is active.
78	PDBIN	Data bus in	Processor command/control output indicates to external circuits that the data bus is in the input mode.
79	A0	Address line 0	(LSB)
80	A1	Address line 1	
81	A2	Address line 2	
82	A6	Address line 6	
83	A7	Address line 7	
84	A8	Address line 8	
85	A13	Address line 13	
86	A14	Address line 14	
87	A11	Address line 11	
88	DO2	Data out line 2	
89	DO3	Data out line 3	
90	DO7	Data out line 7	
91	DI4	Data in line 4	
92	DI5	Data in line 5	
93	DI6	Data in line 6	
94	DI1	Data in line 1	
95	DI0	Data in line 0	(LSB)
96	SINTA	Interrupt acknowledge	Status output signal acknowledges signal for interrupt request.
97	SWO	Write out	Status output signal indicates that the operation in the current machine cycle will be a WRITE memory or output function.
98	SSTACK	Stack	Status output signal indicates that the address bus holds the pushdown stack address from the stack pointer.
99	POC	Power-on-clear	
100	GND	Ground	

*New bus signal for 8800b.
SOURCE: MITS-Altair.

and a listener. A programmable DVM whose function can be selected under program control is, for example, a talker and a listener. Third is a *controller*, which determines who talks and who listens on the bus.

As shown in Figure 11-23a, the GPIB has eight *bidirectional data lines*. These lines transfer data, addresses, commands, and status bytes among as many as 8 to 15 instruments. Figure 11-23b shows the formats for the combination command-address codes that a controller can send to talkers and listeners. Bit 8 of these words is a don't-care, bits 7 and 6 specify which command is being sent, and bits 5 through 1 give the address of the talker or listener to which the command is being sent. For example, to establish a device at address 04 as a talker, the controller simply asserts the *attention line* (ATN) low and sends out a com-

TABLE 11-2
INTEL MULTIBUS PIN DEFINITIONS

		Component Side				Circuit Side	
	PIN	**MNEMONIC**	**DESCRIPTION**	**PIN**	**MNEMONIC**	**DESCRIPTION**	
Power supplies	1	GND	Signal GND	2	GND	Signal GND	
	3	+5V	+5 V dc	4	+5V	+5 V dc	
	5	+5V	+5 V dc	6	+5V	+5 V dc	
	7	+12V	+12 V dc	8	+12V	+12 V dc	
	9	−5V	−5 V dc	10	−5V	−5 V dc	
	11	GND	Signal GND	12	GND	Signal GND	
Bus controls	13	BCLK/	Bus clock	14	INIT/	Initialize	
	15	BPRN/	Bus pri. in	16	BPRO/	Bus pri. out	
	17	BUSY/	Bus busy	18	BREQ/	Bus request	
	19	MRDC/	Mem. read cmd.	20	MWTC/	Mem. write cmd.	
	21	IORC/	I/O read cmd.	22	IOWC/	I/O write cmd.	
	23	XACK/	XFER acknowledge	24	INH1/	Inhibit 1 disable RAM	
Bus controls and address	25		Reserved	26	INH2/	Inhibit 2 disable PROM or ROM	
	27	BHEN/	Byte high enable	28	AD10/		
	29	CBRQ/	Common bus request	30	AD11/		
	31	CCLK/	Constant clock	32	AD12/	Bus	
	33	INTA/	Intr. acknowledge	34	AD13/		
Interrupts	35	INT6/	Parallel	36	INT7/	Parallel interrupt requests	
	37	INT4/	Interrupt	38	INT5/		
	39	INT2/	Requests	40	INT3/		
	41	INT0/		42	INT1/		
Address	43	ADRE/		44	ADRF/		
	45	ADRC/		46	ADRD/		
	47	ADRA/		48	ADRB/		
	49	ADR8/	Address bus	50	ADR9/	Address bus	
	51	ADR6/		52	ADR7/		
	53	ADR4/		54	ADR5/		
	55	ADR2/		56	ADR3/		
	57	ADR0/		58	ADR1/		
Data	59	DATE/		60	DATF/		
	61	DATC		62	DATD/		
	63	DATA/		64	DATB/		
	65	DAT8/	Data bus	66	DAT9/	Data bus	
	67	DAT6/		68	DAT7/		
	69	DAT4/		70	DAT5/		
	71	DAT2/		72	DAT3/		
	73	DAT0/		74	DAT1/		
Power supplies	75	GND	Signal GND	76	GND	Signal GND	
	77		Reserved	78		Reserved	
	79	−12V	−12 V dc	80	−12V	−12 V dc	
	81	+5V	+5 V dc	82	+5V	+5 V dc	
	83	+5V	+5 V dc	84	+5V	+5 V dc	
	85	GND	Signal GND	86	GND	Signal GND	

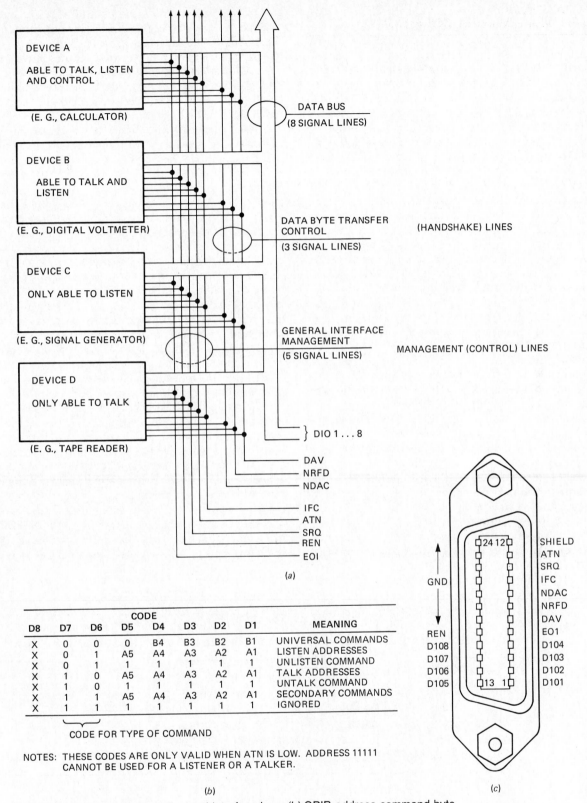

				CODE				
D8	D7	D6	D5	D4	D3	D2	D1	MEANING
X	0	0	0	B4	B3	B2	B1	UNIVERSAL COMMANDS
X	0	1	A5	A4	A3	A2	A1	LISTEN ADDRESSES
X	0	1	1	1	1	1	1	UNLISTEN COMMAND
X	1	0	A5	A4	A3	A2	A1	TALK ADDRESSES
X	1	0	1	1	1	1	1	UNTALK COMMAND
X	1	1	A5	A4	A3	A2	A1	SECONDARY COMMANDS
X	1	1	1	1	1	1	1	IGNORED

CODE FOR TYPE OF COMMAND

NOTES: THESE CODES ARE ONLY VALID WHEN ATN IS LOW. ADDRESS 11111 CANNOT BE USED FOR A LISTENER OR A TALKER.

(b)

(c)

FIGURE 11-23 *(a)* Hewlett-Packard interface bus. *(b)* GPIB address command byte formats. *(c)* GPIB connector.

mand address byte of X1000100. Each listener is enabled when the controller sends it a byte containing $X01A_5A_4A_3A_2A_1$. When a data transfer is complete, lis- teners are turned off by the controller sending an *unlisten command*, X0111111, and the talker is turned off by the controller sending an *untalk command*,

X1011111. *Universal commands* sent by the controller with bits 7, 6, and 5 all 0's will go to all instruments. The lower 4 bits of this word specify one of 16 universal commands.

The GPIB also has five *bus management lines* which function basically as follows. The *interface clear* line (IFC), when asserted by the controller, puts all devices in a known or starting state. It is similar to a system reset. *Attention* (ATN), when asserted (low), indicates that the controller is putting either a universal command or an address-command such as listen on the data bus. When ATN is not asserted, the data lines contain a data or status byte from a talker. *Service request* (SRQ) is similar to an interrupt. It is sent to the controller by a device to indicate that it needs attention. An instrument, for example, might send an SRQ when it had completed a series of measurements and its buffer was full of data. When asserted by the controller, *remote enable* (REN) enables an instrument to be controlled by the system controller (remote) rather than by the front-panel controls (local) of the instrument. *End or identify* (EOI) usually is sent by a talker to the system controller to indicate that transfer of a block of data is complete.

In addition to the data bus and the management bus, the GPIB has three *handshake lines* that coordinate the transfer of data bytes on the data bus: *data valid* (DAV), *not ready for data* (NRFD), and *not data accepted* (NDAC). A major feature of the GPIB handshake data transfer scheme is that it allows devices with very different response rates to be connected together in a system.

Figure 11-24 shows the sequence of signals on the handshake lines for a transfer of data from a talker to several listeners. DAV, NRFD, and NDAC are all open-collector active low signals. Since NRFD and NDAC are open-collector, any listener can pull NRFD low to indicate that it is not ready for data or NDAC low to indicate

that it has not accepted the data. The NRFD line, for example, will not go high to indicate ready until all addressed listeners have released it. When all listeners have indicated that they are ready (5 in Figure 11-24), the talker asserts its DAV signal to indicate that valid data is on the bus. The addressed listeners all assert the NRFD line low and start accepting the data. As each listener accepts the data, it releases its hold on the NDAC line. When the slowest listener has accepted the data, the NDAC line goes high. The talker senses this and prepares to send the next byte of data by raising the DAV line high. When the DAV goes high, all the addressed listeners assert NDAC low again. The process is repeated. The slowest listener, then, determines the rate at which data is sent. The specified maximum data rate for the GPIB is 1 megabyte/s, but a typical data rate is 250K bytes/s or less.

An example of one possible sequence will show how all these lines work together. Upon power-up, the controller takes control of the bus and sends out an IFC signal to set all instruments on the bus to a known state. It may also send an unlisten command and an untalk command. Then the controller proceeds to use the bus and instruments as needed. Periodically the controller checks the SRQ line for a service request. If the SRQ line is low, the controller polls each device on the bus one after another (serial) or all at once (parallel) until it finds the device(s) requesting service. A talker such as a DVM, for example, might be indicating that it has completed a conversion and has some data to send to a listener such as a chart recorder. When the controller determines the source of the SRQ, it asserts ATN low and sends listener address commands to each listener to receive the data. The controller then sends a talk address command to the desired talker and raises ATN high again. The talker now sends data to the listener(s) using the NRFD-DAV-NDAC handshake sequence described previously. When the data transfer is complete, the talker sends out an

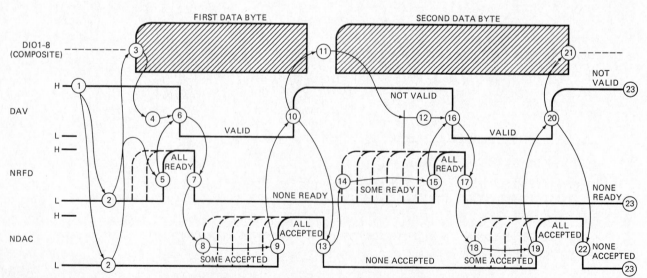

FIGURE 11-24 GPIB signal sequence for transfer of a byte of data from a talker to several listeners.

EOI signal to the controller. The controller then takes control again. It sends an untalk address command and an unlisten address command. The bus is now available for other data transfers.

A standard 8080A or 8085A data bus can be interfaced to the HPIB with the Intel 8291 GPIB talker-listener. The Intel 8292 is a GPIB controller. For interfacing 6800-type processors to the HPIB, the MC68488 general-purpose interface adapter is available. It can serve as a listener or talker.

The importance of the IEEE-488 bus is that it allows a computer or microcomputer to be connected with several test instruments to form an integrated test system.

SERIAL DATA INPUT AND OUTPUT

To reduce the number of connecting wires required to send digital data over long distances, usually the data is sent in serial form. In other words, one bit of a word is sent after another on a single wire. The data words may be sent *synchronously* or *asynchronously*.

In asynchronous format the *bits* of a data word or character are sent at a constant rate, but characters can come at any rate as long as they do not overlap. When no characters are being sent, the line is kept high or *marking*. Figure 11-25 reviews the format for asynchronous serial data (see Chap. 4 for the initial discussion). The beginning of a character is indicated by a start bit which is always low. After the start bit, the data bits are sent out in order, with the least significant bit first. Different systems may use 5, 6, 7, or 8 data bits. A parity bit may follow the data bits. The end of the character is indicated by one or more always-high stop bits. Different systems use 1, 1½, or 2 stop bits.

The rate at which bits are sent is called the *baud rate*. The baud rate is equal to 1 divided by the time for 1 bit. If, for example, the bit time is 3.33 ms, the baud rate is 1/3.33 ms = 300 Bd. Common rates are 75, 110, 150, 300, 600, 1200, 2400, 4800, 9600, and 19,200 Bd.

Microcomputers work with data in parallel form. Thus, to interface with serial lines, the data must be converted to and from the serial form. A parallel-in, serial-out (PISO) shift register and a serial-in, parallel-out (SIPO) shift register can do this. Packaged together for this purpose, they are called a *universal asynchronous receiver-transmitter* (UART). A good example of a UART is the AY-5-1013 described in Chap. 4. This device is hardware-programmable for number of data bits, parity, and number of stop bits. It requires a transmit-and-receive clock of 16 times the desired baud rate. The

receiver data outputs have three-state capability, so they can be connected directly onto a microcomputer data bus and enabled by a $\overline{CS}$ signal. A device such as this works well in a dedicated system where the transmit and receive parameters are always the same. However, in a general-purpose microcomputer, a UART that is software-programmable for different baud rates, different numbers of data bits, different numbers of stop bits, and different parity settings is needed. Often this UART also needs the capability for handshake data transfer with modulator-demodulators (MODEMs), CRT terminals, and printers.

Most microprocessor manufacturers produce software-programmable UARTs that have these handshake capabilities. Two of the most common are the Intel 8251/8251A and the Motorola MC6850. Later we show how these devices are initialized to send and receive serial data. First, however, it is necessary to describe the commonly used signal and handshake standards. Serial data 1's and 0's may be sent as current or no current, as two voltage levels, or as two audio tones.

20 AND 60 mA CURRENT LOOPS: TELETYPES

Teletypes often send and receive serial 1's and 0's as a fixed current or no current. A 1, or *mark*, is a nominal current of 20 or 60 mA, and a 0, or *space*, is no current. Teletypes send data at a rate of 300 Bd or less. A rate of 110 Bd is very common. Figure 11-26 shows the circuitry used on the SDK-85 microcomputer board to interface the TTL levels of the 8085A SID and SOD ports to 20-mA current loop levels. Also shown is a representation of the internal circuitry of a teletype.

A low on the SOD port turns on Q2 and allows a nominal 20-mA current to flow through an electromagnet in the teletype and back to a negative supply through the TX RET line. A high to the SOD port shuts off Q2, and thus the current to the teletype. *Note:* Data sent to the SOD port must be inverted in order to maintain the convention that a logic high is 20 mA and a logic low is no current. The pattern of current or no current to the teletype sets up internal print bars to print the desired character.

To send data, the teletype represents each data bit simply as either a closed or an open switch, as shown in Figure 11-26. If the teletype switch is closed, current flows from +5 V through R1, through the teletype switch, and back through RX RET and R34 to a negative supply. This current flow turns on Q1. Current flowing through Q1 produces a voltage across R3 and sends a logic high to the 8085A SID port. When no character is

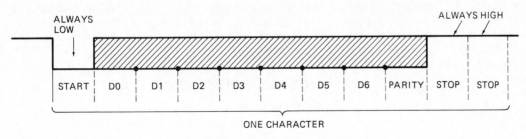

FIGURE 11-25 Format for asynchronous serial data.

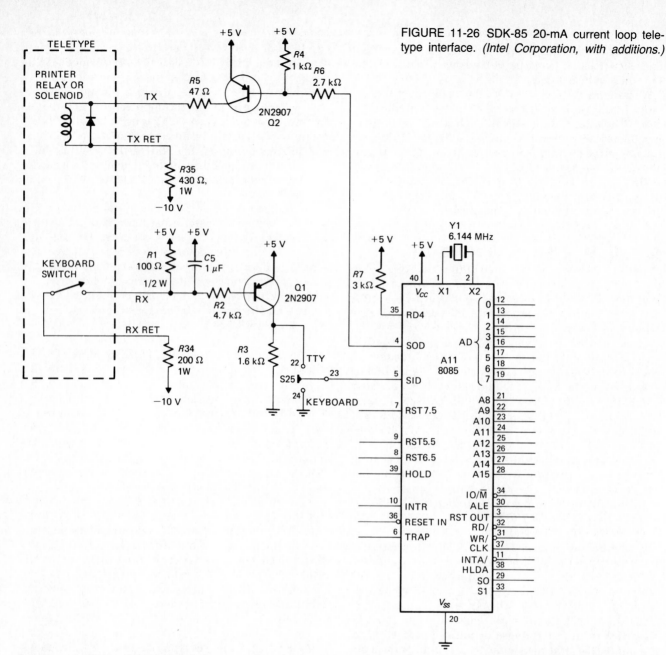

being sent, the switch is closed, so a constant high, or mark, is sent. If the teletype switch is off during a bit time, no current flows, Q1 is off, and SID receives a logic low. The purpose of C5, the 1-μF capacitor next to R1 in Figure 11-26, is to filter out noise generated by the rotating switch in the teletype.

Connecting a teletype to a computer with two independent current loops, as shown in Figure 11-26, is referred to as *full-duplex mode*. The term *full duplex* means separate lines for sending and receiving data. Therefore, data can be sent and received at the same time. When a key is pressed on the teletype keyboard in Figure 11-26, the code for that key is sent to the computer. The computer then "echoes" this character back to the printer. One advantage of this scheme is that if you see the character printed, you know the computer got the character you typed.

The teletype also may be connected in *half-duplex mode*. In this mode the keyboard switch and printer electromagnet are connected in a single loop with one current source. Referring to Figure 11-26, if R35 and R1 are removed and the TX RET line is connected to the RX line, the circuit functions as a half-duplex link. *Note*: The SOD output must be low so that a marking current of 20 mA is flowing. Now when a key is pressed, the switch opening and closing interrupts the current flow through the printer electromagnet. Therefore, a character is printed out directly as it is typed. No echo from the computer is required.

MODEMS AND RS-232C

MODEMS   In the 1960s as the use of timeshare computer terminals became more widespread, it was necessary to find some easy way to connect these terminals to large computers many miles away. Standard phone lines are a logical choice since they are usually already in place. A problem with standard phone lines is that they cannot be used to send voltage- or current-pulse-type signals. The bandwidth of standard phone lines is about 3 kHz. Sending pulses through this low-pass filter distorts them beyond recognition. Therefore, digital data is often sent over phone lines in one of three sine wave forms: amplitude modulation, phase shift modulation, or frequency shift keying (FSK). In *amplitude modulation*, the presence of a sine wave carrier indicates a 1. No carrier indicates a 0. In *phase shift modulation*, the phase of a constant-frequency carrier is shifted by 180° to indicate a change from a 0 to a 1 or from a 1 to a 0. In *frequency shift keying* a digital 1 is represented by a "burst" of sine waves of one frequency, and a digital 0 by a burst of sine waves of another frequency. (Refer to Figure 5-13.) Both frequencies are within the bandpass of the telephone lines.

A *modulator-demodulator,* or MODEM, sends and receives telephone-line-compatible signals. If, for example, FSK is being used, the modulator section of the MODEM converts logic level signals to bursts of the proper sine wave output frequencies. The demodulator portion of the MODEM translates the two received sine wave signals to digital 1's and 0's.

The Motorola MC6860 contains the majority of the circuitry needed for a MODEM that can send and receive FSK data at up to 600 Bd. The device requires external analog buffers, limiters, and filters.

By using four different frequencies, MODEMs can communicate in both directions at once (full-duplex mode). A MODEM can function in either the *originate mode* or the *answer mode.* In the originate mode, it may use frequencies of 1270 Hz for a mark, or 1, and 1070 Hz for a space, or 0. In the answer mode, a MODEM accepts the 1270- and 1070-Hz signals from the originate-mode MODEM and communicates back with a 2225-Hz sine wave for a mark and a 2025-Hz sine wave for a space. Handshake signals are also necessary in this system to make sure, for example, that the terminal does not send data unless the computer is ready to receive it. Handshaking between MODEMs is done by the presence or absence of the frequency for a high. An originate-mode MODEM, for example, will not transmit data until it receives a 2225-Hz signal back from an answer-mode MODEM.

RS-232C

MODEMs and other equipment, such as microwave links used to send serial data over long distances, are often referred to as *data communication equipment,* or DCE. The terminals and computers that are sending or receiving the serial data are called *data terminal equipment,* or DTE. In response to the need for signal and handshake standards between DTE and DCE, the Electronic Industries Association (EIA) published *EIA standard RS-232C.* This standard defines the voltage levels of the signals, the handshake signals, and a 25-pin connector. Figure 11-27 shows the names

and functions for all 25 pins. In most simple systems, only pins 1 through 8 and 20 are used. Figure 11-28 shows how a CRT terminal and a computer are connected to MODEMs using these signals. Note the arrows that indicate the direction of signal flow for each signal.

Basically, a handshake between the terminal and the MODEM functions as follows. After the terminal power is turned on and the terminal runs any self-check routines, it sends $\overline{DTR}$ to the MODEM. The MODEM responds with $\overline{DSR}$ to indicate it is operational. The MODEM at the computer end is then dialed. This answer-mode MODEM responds with a 2225-Hz carrier frequency. When the MODEM connected to the terminal receives this carrier, it asserts *carrier detect* ($\overline{CD}$) to the terminal. The terminal then asserts a *request-to-send* signal ($\overline{RTS}$) to the MODEM. After the proper timing interval, the MODEM responds with a *clear-to-send* signal ($\overline{CTS}$). The terminal then sends serial data on its TxD output. A similar handshake takes place for receiving data.

A problem that often occurs when various types of RS-232 equipment are connected together can be shown by removing the MODEMs in Figure 11-28 and trying to connect the terminal directly to the computer. As you can see, both units are trying to output data on pin 2. If they are connected directly, these outputs will fight each other. The solution is to make an adapter with two connectors which crosses over the signals as shown in Figure 11-29. This connection is called a *null MODEM.*

The voltage levels for all RS-232C signals are as follows. A logic high, or *mark,* is a voltage between −3 and −15 V under load. A logic low, or *space,* is a voltage between +3 and +15 V under load. Typical voltages used are +10 or +12 V. To limit cross-talk between adjacent wires, the risetime rate and falltime rate must be limited to 30 V/μs. RS-232C is specified for a maximum distance of 50 ft [15 m] at a maximum baud rate of 20,000 Bd. For lower baud rates, wires as long as 2000 to 3000 ft [610 to 915 m] are often used.

To interface RS-232C signal levels with standard TTL or CMOS, drivers and receivers must be used. The Motorola MC1488 contains four TTL-to-RS-232C drivers, and the MC1489 contains four RS-232C-to-TTL receivers.

RS-422

An improvement of RS-232C is EIA standard RS-422 which uses low-impedance differential signals. The differential signal is created by *differential line drivers* such as the MC3487 and sent on a twisted pair line, as discussed in Chap. 2. The differential signal is received and translated to TTL logic levels by differential line receivers such as the MC3486. *Differential line receivers* have the advantage that they attenuate any noise induced equally in the twisted pair lines. RS-422 increases the maximum baud rate to 10 MBd for 40-ft [12-m] connecting lines or 100,000 Bd for 4000-ft [1220-m] connecting wires.

RS-423

Another enhancement of the RS-232C is EIA standard RS-423. This standard achieves longer trans-

PIN NUMBER	COMMON NAME	RS-232-C NAME	DESCRIPTION
1		AA	Protective Ground
2	TxD	BA	Transmitted Data
3	RxD	BB	Received Data
4	$\overline{\text{RTS}}$	CA	Request to Send
5	$\overline{\text{CTS}}$	CB	Clear to Send
6	$\overline{\text{DSR}}$	CC	Data Set Ready
7	GND	AB	Signal Ground (Common Return)
8	$\overline{\text{CD}}$	CF	Received Line Signal Detector
9		—	(Reserved for Data Set Testing)
10		—	(Reserved for Data Set Testing)
11			Unassigned
12		SCF	Secondary Rec'd. Line Sig. Detector
13		SCB	Secondary Clear to Send
14		SBA	Secondary Transmitted Data
15		DB	Transmission Signal Element Timing (DCE Source)
16		SBB	Secondary Received Data
17		DD	Receiver Signal Element Timing (DCE Source)
18			Unassigned
19		SCA	Secondary Request to Send
20	$\overline{\text{DTR}}$	CD	Data Terminal Ready
21		CG	Signal Quality Detector
22		CE	Ring Indicator
23		CH/CI	Data Signal Rate Selector (DTE/DCE Source)
24		DA	Transmit Signal Element Timing (DTE Source)
25			Unassigned

FIGURE 11-27 RS-232C signal names and pin numbers.

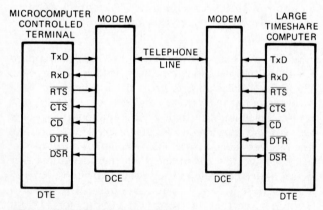

DTE = DATA TERMINAL EQUIPMENT
DCE = DATA COMMUNICATION EQUIPMENT

FIGURE 11-28 Communication of a terminal with a timeshare computer using MODEMs.

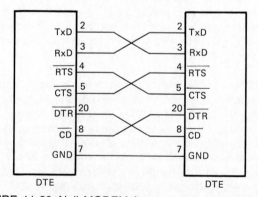

FIGURE 11-29 Null MODEM for connecting two RS-232C data terminal type of devices. *Note:* Many systems do not use all these handshake signals.

mission distances and higher baud rates by using a low-impedance, single-ended driver instead of a differential driver. With the specified 50-Ω impedance of RS-423, coaxial cable can be employed for connections. Line drivers and receivers are required to interface to standard logic families.

PROGRAMMABLE SERIAL PORT ICs: THE MC6850 AND i8251A

MOTOROLA MC6850   Motorola refers to this device as an *ACIA*, which stands for *asynchronous communication interface adapter.* It functions as a programmable UART. Figure 11-30 is a block diagram for the 6850. Figure 10-5 shows how this device is connected to a microcomputer system. D0 through D7 are connected to the system data bus. R/W is connected to the system R/W line, and ENABLE is connected to the DBE signal. Chip-select inputs can be connected to address lines or decoder outputs. The register-select input usually is connected to system address line A0. A high on this input selects the transmit-data register if R/W is low

and the receive-data register if R/W is high. A low on the register-select input will access the control register if R/W is low and the status register if R/W is high. If the base address of the device in a system is, for example, 8008H, then the data registers have address 8009H. The control and status registers have address 8008H.

The 6850 requires transmit and receive clock inputs of 1, 16, or 64 times the desired baud rate. In addition to TRANSMIT DATA and RECEIVE DATA, the 6850 has three signal pins for interfacing with MODEMs: $\overline{CTS}$, $\overline{RTS}$, and $\overline{DCD}$ (data carrier detect). The interrupt request output of the 6850 can be connected to the CPU interrupt request input. This allows the 6850 to interrupt the processor when it has a character ready to be read in its receive-data register or when its transmit-data register is empty.

A 6850 is initialized by sending a control word to its control register. Figure 11-31a shows the 6850 control word format. Bits CR1 and CR0 specify the ratio between the transmit clock input frequency and the output baud rate. If the transmit clock frequency is 4800 Hz and the desired baud rate is 300 Bd, then the

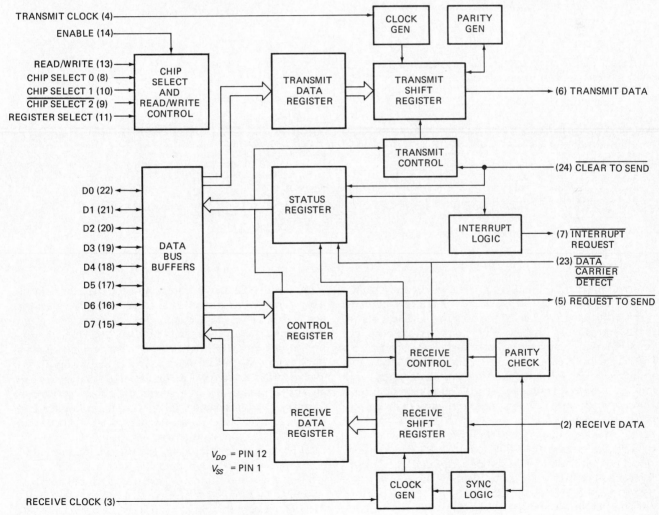

FIGURE 11-30 Block diagram of MC6850. *(Motorola.)*

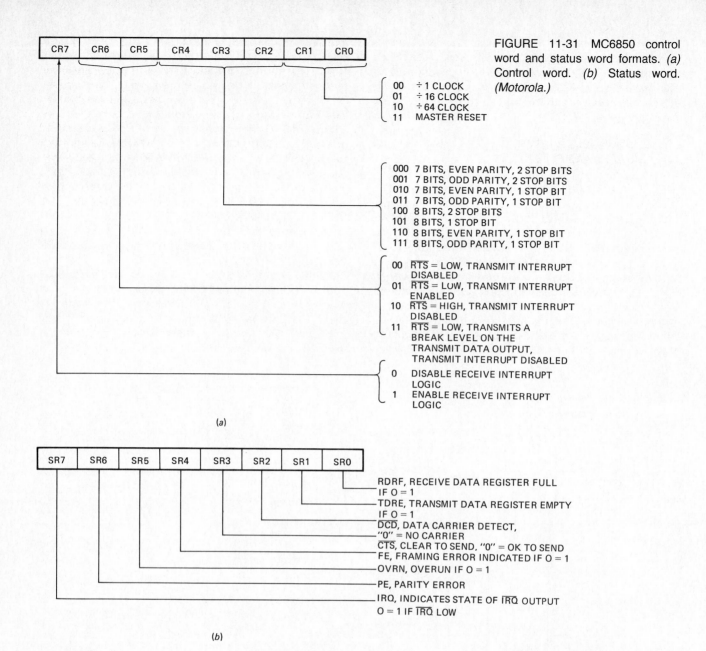

FIGURE 11-31 MC6850 control word and status word formats. *(a)* Control word. *(b)* Status word. *(Motorola.)*

CR7	CR6	CR5	CR4	CR3	CR2	CR1	CR0

00 ÷ 1 CLOCK
01 ÷ 16 CLOCK
10 ÷ 64 CLOCK
11 MASTER RESET

000 7 BITS, EVEN PARITY, 2 STOP BITS
001 7 BITS, ODD PARITY, 2 STOP BITS
010 7 BITS, EVEN PARITY, 1 STOP BIT
011 7 BITS, ODD PARITY, 1 STOP BIT
100 8 BITS, 2 STOP BITS
101 8 BITS, 1 STOP BIT
110 8 BITS, EVEN PARITY, 1 STOP BIT
111 8 BITS, ODD PARITY, 1 STOP BIT

00 $\overline{RTS}$ = LOW, TRANSMIT INTERRUPT DISABLED
01 $\overline{RTS}$ = LOW, TRANSMIT INTERRUPT ENABLED
10 $\overline{RTS}$ = HIGH, TRANSMIT INTERRUPT DISABLED
11 $\overline{RTS}$ = LOW, TRANSMITS A BREAK LEVEL ON THE TRANSMIT DATA OUTPUT, TRANSMIT INTERRUPT DISABLED

0 DISABLE RECEIVE INTERRUPT LOGIC
1 ENABLE RECEIVE INTERRUPT LOGIC

(a)

SR7	SR6	SR5	SR4	SR3	SR2	SR1	SR0

RDRF, RECEIVE DATA REGISTER FULL IF O = 1
TDRE, TRANSMIT DATA REGISTER EMPTY IF O = 1
$\overline{DCD}$, DATA CARRIER DETECT, "0" = NO CARRIER
$\overline{CTS}$, CLEAR TO SEND, "0" = OK TO SEND
FE, FRAMING ERROR INDICATED IF O = 1
OVRN, OVERUN IF O = 1
PE, PARITY ERROR
IRQ, INDICATES STATE OF $\overline{IRQ}$ OUTPUT O = 1 IF $\overline{IRQ}$ LOW

(b)

desired divide ratio is 16. This is selected by programming bits CR1 as 0 and CR0 as 1. Sending a control word with these bits both 1's resets the device. This is necessary when the device is first powered up because the 6850 does not have a hardware reset input. Bits CR4, CR3, and CR2 specify the format in which the data will be sent. Bit CR5 enables the transmitter interrupt circuitry. A 1 in this bit allows the 6850 to send an interrupt request to the CPU when its transmit-data register is empty. An interrupt service routine then writes the next character to be transmitted into this register. Bit CR6, if low, causes the 6850 to assert its $\overline{RTS}$ output signal to a MODEM or some other peripheral. Bit CR7 of the control register enables or disables the receiver interrupt response. If bit CR7 is a 1, then the 6850 sends out an interrupt to the CPU when its receive-data register has a character ready for reading. An interrupt service routine can then read the character into the CPU.

Figure 11-31*b* shows the format for the 6850 status word. The CPU can read this status word to determine when the receive-data register has a character, when the transmit-data register is ready to receive a character, or when carrier-detect and clear-to-send signals have been received from a MODEM. Error conditions such as framing error, overrun error, or parity error can also be determined from reading the status register. A *framing error* indicates that a received character had the wrong number of data bits or stop bits. An *overrun error* indicates that characters were received and loaded into the receive-data register, but they were not read by the CPU. Therefore, some data characters will have been lost. A *parity error* indicates that the parity of a received character is different from the parity it was sent with. Bit SR7 of the status register indicates whether the IRQ output of the 6850 is asserted.

Characters can be read from a 6850 on a polled or an

interrupt basis. For the interrupt basis, the CPU initializes the 6850 and goes on with its mainline program until an interrupt request from the 6850 indicates that it has a character ready to read. An interrupt service routine reads the character in and returns control to the main program until the next interrupt. For a polled operation, the CPU initializes the 6850 and then reads the status register over and over until SR0, the receive-data register full bit, is a 1. The CPU then reads the character in, checks whether it is the last character of a block of data, and, if it is not, returns to waiting for SR0 to go high again. Reading a character from the receive-data register resets bit SR0. Figure 11-32a shows a 6800 assembly language program to initialize a 6850 and read characters in on a polled basis until the end-of-text character, 03H, is read. The characters are stored in a table in memory.

Characters can be sent to the 6850 for transmission on either an interrupt or a polled basis. For a polled basis, the status register is read over and over until the

```
          LDX     #$0100          ;INITIALIZE MEMORY POINTER

          LDAA    #00000011       ;CONTROL WORD FOR MASTER RESET

          STAA    $8008           ;SEND TO CONTROL REGISTER

          LDAA    #00000101       ;CONTROL WORD FOR ÷ 16 CLOCK, 7 BITS

                                  ;ODD PARITY, 2 STOP BITS, RTS LOW,

                                  ;TRANSMIT INTERRUPT DISABLED

                                  ;RECEIVER INTERRUPT DISABLED

          STAA    $8008           ;SEND TO CONTROL REGISTER

WAIT      LDAA    $8008           ;READ STATUS REGISTER

          ANDA    #$01            ;MASK ALL BUT RDRF

          BEQ     WAIT            ;WAIT UNTIL CHARACTER READY

          LDAA    $8009           ;READ CHARACTER

          CMPA    #$03            ;CHECK IF END OF TEXT

          BEQ     EXIT            ;YES, DONE

          STAA    00,X            ;STORE IN MEMORY TABLE

          INX                     ;INCREMENT POINTER

          BRA     WAIT            ;WAIT FOR NEXT CHARACTER

EXIT

          6800 ROUTINE TO INITIALIZE MC6850 AT ADDRESSES 8008H

          AND 8009H AND READ CHARACTERS ON A POLLED BASIS UNTIL

          AN END-OF-TEXT CHARACTER IS DETECTED.
```

(a)

FIGURE 11-32 MC6800 assembly language program to initialize an MC6850 and poll it for receive and transmit. (a) Receive. (b) Transmit (shown on p. 332).

```
WAIT    LDAA    $8008       ;READ STATUS REGISTER

        ANDA    #$0E        ;MASK ALL BUT SR3, SR2, SR1

        CMPA    #$02        ;CHECK FOR TDRE = 1, DCD = 0, CTS = 0

        BNE     WAIT        ;WAIT UNTIL ALL READY

        LDAA    00,X        ;LOAD CHARACTER FROM MEM

        STAA    $8009       ;SEND TO 6850 TRANSMIT DATA

        CPX     #$01FF      ;END OF BLOCK?

        BEQ     EXIT        ;YES, DONE

        INX                 ;NO, INCREMENT POINTER

        BRA     WAIT        ;WAIT UNTIL READY FOR NEXT
```

6800 assembly language routine to transmit characters
on a polled basis with an MC6850 when TDRE = 1, $\overline{DCD}$ =
0, and $\overline{CTS}$ = 0

FIGURE 11-32 *(continued)*   *(b)*

transmit-data register bit SR1 is found to be a 1. A character is then read from memory and sent to the 6850. Writing to the transmit-data register resets the SR1 bit. For sending data to a MODEM, the status word should also be checked to make sure the $\overline{DCD}$ bit, SR2, and the $\overline{CTS}$ bit, SR3, are both low before a character is sent. Figure 11-32*b* shows a 6800 routine to poll status register bits SR3, SR2, and SR1 before sending a character to the 6850 transmit-data register. Note the mask-and-compare sequence that must be used to see whether the 3 bits are in the desired state.

THE 8251A: UNIVERSAL SYNCHRONOUS-ASYNCHRONOUS RECEIVER-TRANSMITTER

PIN AND SIGNAL DESCRIPTIONS   Another very common programmable serial port device is the Intel 8251A. Figure 11-33 shows the block diagram and pin descriptions for the 8251A. RESET is connected to the system reset line. CLK is connected to the CPU clock. $C/\overline{D}$ is connected to system address line A0. If A0 is low, then the transmit-data buffer or the receive-data buffer is accessed. If A0 is high, the control or the status register is accessed. The $\overline{RD}$ and $\overline{WR}$ inputs determine which is selected in each case. For example, with A0 ($C/\overline{D}$) low,

a low on $\overline{RD}$ selects the receive buffer, and a low on $\overline{WR}$ selects the transmit buffer. As with most UARTs, an external clock of 1, 16, or 64 times the desired baud rate must be applied to the $\overline{TxC}$ and $\overline{RxC}$ inputs.

The 8251A is *double-buffered*. This means that one character can be loaded into a holding buffer while another is being shifted out. The TxRDY output from the 8251A indicates that the holding buffer is ready for the next character to be sent by the CPU. The TxEMPTY output signal indicates that the 8251A has sent all the characters that it got from the CPU. RxRDY indicates that the receiver buffer has a character ready for the CPU to read. Incidentally, if this character is not read out before the 8251A gets the next character shifted in, the first character is written over.

The SYNDET/BRKDET pin of the device has several uses. In the asynchronous mode, this pin goes high if the receiver line stays low for more than two character times. It then indicates a *break* in the data stream. In the synchronous mode, this pin goes high to indicate that a sync character(s) has been found and that a block of data characters is coming.

8251A INITIALIZATION   To initialize an 8251A, two control words must be sent to the control register address of the device. Figure 11-34 shows the format for these words and the status word. After a hardware or software reset, the mode word must be sent first. This

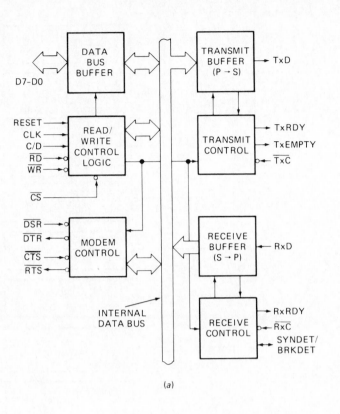

Pin Name	Pin Function
D7–D0	Data bus (8 bits)
C/$\overline{D}$	Control or data is to be written or read
$\overline{RD}$	Read data command
$\overline{WR}$	Write data or control command
$\overline{CS}$	Chip select
CLK	Clock pulse (TTL)
RESET	Reset
$\overline{TxC}$	Transmitter clock
TxD	Transmitter data
$\overline{RxC}$	Receiver clock
RxD	Receiver data
RxRDY	Receiver ready (has character for CPU)
TxRDY	Transmitter ready (ready for char. from CPU)
$\overline{DSR}$	Data set ready
$\overline{DTR}$	Data terminal ready
SYNDET/BD	Sync detect/break detect
$\overline{RTS}$	Request to send data
$\overline{CTS}$	Clear to send data
TxEMPTY	Transmitter empty
V_{CC}	+5-V supply
GND	Ground

(b)

FIGURE 11-33 (a) Block diagram for the 8251. (b) Pin functions of the 8251. *(Intel Corporation.)*

word sets the format for the data to be sent and received. Mode word bits B2 and B1 specify synchronous mode if they are both 0's. In synchronous mode, the baud rate is equal to the $\overline{TxC}$ and $\overline{RxC}$ input frequencies. For asynchronous data transfer, mode word bits B2 and B1 set the ratio between, for example, the $\overline{TxC}$ input frequency and the baud rate of the transmitted data. A $\overline{TxC}$ input of 4800 Hz and a baud rate factor of 16 produce 300-Bd transmission. As shown in Figure 11-34a, other bits of the mode word specify the number of data bits in the character, parity or no parity, odd or even parity, and the number of stop bits.

After the mode word is sent to an 8251A, a *command* word is sent. Figure 11-34b shows the format for this word. A 1 in D0 of this word enables the transmitter section. Before enabling, the TxD output is in a continuous high, or marking, state. Bit D2 of the command word enables the RxRDY output, so that it goes high when the receiver buffer has a character ready. With this bit disabled, characters are still shifted into the receiver buffer. Therefore, the first character read from the receiver buffer after initialization may be garbage and should be discarded. A high in bit D1 asserts the *data-terminal-ready* output signal ($\overline{DTR}$), and a high in bit A5 asserts the request-to-send ($\overline{RTS}$) output sig-

nal. These signals may go to a MODEM to indicate the device is ready and wants to send data. A 1 in bit D6 resets the 8251A, so that the next word sent to its control register address is treated as a mode word. This feature allows the input or output data format to be changed during a program without doing a system reset. Sending a command word with bit D3 a 1 causes the 8251A to output a continuous low (break) on its TxD output. Sometimes this is used as an indication of the end of a message to the receiving system.

INTERRUPT MODE   After the 8251A is initialized by sending it a mode word and a command word, it must be checked to see whether it's ready for the CPU to write a character to its transmitter or read a character from its receiver buffer. This can be done on an interrupt or a polled basis. If the TxRDY output of the 8251A is connected to an interrupt output on the CPU, the mainline program is interrupted only when the transmitter is ready for the next character. Likewise, if the 8251A RxRDY output is connected to a CPU interrupt input, the mainline program is interrupted only when the receiver buffer has a character ready to be read by the CPU. An interrupt service routine can read in the character, store it in memory, and return to the mainline

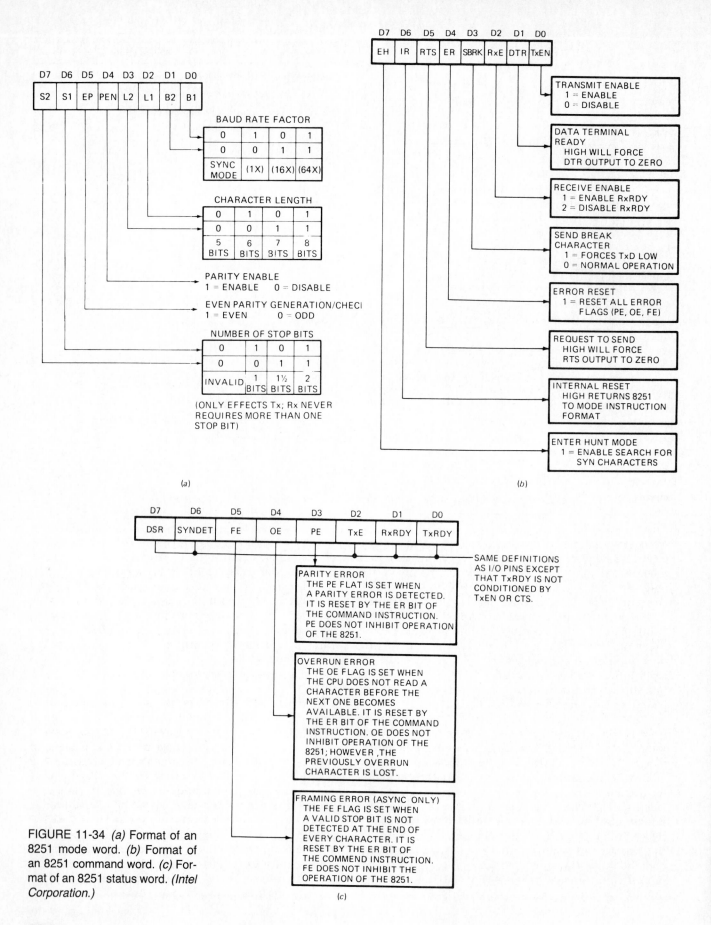

FIGURE 11-34 *(a)* Format of an 8251 mode word. *(b)* Format of an 8251 command word. *(c)* Format of an 8251 status word. *(Intel Corporation.)*

program until the next character is ready in the receiver buffer.

POLLED MODE   For sending or receiving characters on a polled basis, the 8251A status register is read and the appropriate bits checked to see whether the transmitter or receiver is ready. Figure 11-34c shows the 8251A status register format. This status byte is read from the same control address to which the mode and command words were sent. To determine whether the receiver buffer has a character ready to read, a program loop reads the status register over and over until bit D1 is found high. Then the CPU exits the loop and reads in the character from the receiver buffer. Status register bits D5, D4, and D3 can be checked to see whether a framing error, an overrun error, or a parity error has occurred. If it has, a message to retransmit the data can be sent to the transmitting device.

Figure 11-35a shows an 8080A/8085A assembly language program to initialize an 8251A at address 50H (data) and 51H (control/status). Some pre-1982 8251As have a small problem; they do not always respond fully to a hardware system reset, and so they do not always initialize correctly. The program in Figure 11-35a shows a solution. The device is sent a dummy mode word, one or two nulls (00H), and a software reset command (40H). In any case, these extra instructions do no harm. The device then correctly accepts the desired mode and command words.

The mode word (4F) shown in Figure 11-35a sets the data format as 64 times the baud rate factor, 8 data bits, no parity, and 1 stop bit. The command word (27H) enables the transmitter and receiver, asserts $\overline{\text{DTR}}$, and asserts $\overline{\text{RTS}}$. The program in Figure 11-35a then reads and checks the status register over and over until it finds D1, the RxRDY bit, high. When it finds that the receiver is ready with a character, it reads the status register again to find out whether any errors were detected. If not, it reads the character in and checks that it is the last character in the message. If the character read is the end-of-text character (03H), the program exits to the main program. If it is not the end-of-text character, the program increments the memory pointer and returns to polling the status register until the next character is ready.

Figure 11-35b shows an assembly language loop to poll the 8251A status register until a data-set-ready signal is received and the transmitter is ready. Note that when you are checking 2 or more bits in a word, both masking with ANI and comparing with CPI must be used.

SERIAL DATA INPUT AND OUTPUT WITH ALL SOFTWARE

In small systems where the CPU is not expected to do any other tasks while it is sending out or receiving serial data, the whole process can be done with program routines instead of a UART. Data can be input or output from parallel port lines or from single-bit ports, such as SID and SOD on the 8085A. Timing is done with delay loops.

Transmitting a character might be done as follows. Initialize the port line as an output, and send a high for at least 2 bit times. Send a low start bit, and delay 1 bit time. Then rotate the first data bit into position and output it. Delay 1 bit time, and rotate the next data bit into position to output to the port line. Output this bit, and delay 1 bit time. Repeat this process until all the data bits are sent, and then send 1 or 2 stop bits.

Receiving data on a serial port line can be done in a similar manner. The parallel port line is polled over and over until a start bit is detected. After a delay of ½ bit time, the port line is checked again to determine whether it is still low. In other words, it checks to see whether the first low detected was a valid start bit or just a noise pulse. If the start bit was valid, the program delays 1 bit time until the center of the first data bit time. The data bit is then read in. The bit read in is rotated to make room for the next bit to be read in. After another delay of 1 bit time, the next data bit is read in and rotated. The process is repeated until all the data bits of a character are read in. Reading each data bit at the center of its bit time reduces errors that might occur if data bits were read near their leading or trailing edges.

SYNCHRONOUS SERIAL DATA TRANSMISSION

For asynchronous communication, a start bit is required to identify the beginning of each data character, and at least 1 stop bit is needed to identify the end of each data character. This means that a total of 10 bits must be sent for each 8-bit data character. In other words, the start and stop bits waste 20 percent of the time. If the receiver and the transmitter could use the same clock, however, they could be started simultaneously and no start or stop bits would be needed. The receiver would automatically know that every 8 bits received was a data character. Locking the transmitter and receiver together in this way is the basis of *synchronous communication*. There are many formats and protocols for synchronous data transmission. Consult the references at the end of the chapter for detailed information. We describe here two of the most common formats.

BISYNC   A very widely used serial protocol is IBM's *binary synchronous protocol*, or BISYNC. Figure 11-36a shows the message format for BISYNC. If no data is being sent, the line is in an idle condition. During idle, continuous 1's are sent. To signal the start of a message, the transmitter sends two or more previously agreed-upon "sync" characters such as the ASCII 10010110. The receiver uses these to synchronize its clock with that of the transmitter. A USART such as the 8251A uses the sync characters and discards them. A *start-of-header* (SOH) character indicates that a *header* with labels, control codes, or addresses will follow. A *start-*

```
              MVI    A,80H                    ;DUMMY MODE

              OUT    51H                      ;TO CONTROL ADDRESS

              MVI    A,00H                    ;NULL TO

              OUT    51H                      ;CONTROL ADDRESS

              MVI    A,00H                    ;NULL TO

              OUT    51H                      ;CONTROL ADDRESS

              MVI    A,40H                    ;RESET COMMAND

              OUT    51H                      ;TO CONTROL ADDRESS

              MVI    A,4FH                    ;MODE WORD

              OUT    51H                      ;

              MVI    A,27H                    ;COMMAND WORD

              OUT    51H                      ;

LOOP          IN     51H                      ;READ STATUS

              ANI    02H                      ;CHECK FOR RXRDY

              JZ     LOOP                     ;NO, WAIT

              IN     51H                      ;YES, CHECK FOR

              ANI    38H                      ;ERRORS

              JNZ    ERROR                    ;YES, GO TO ERROR ROUTINE

              IN     50H                      ;NO, READ CHARACTER

              CPI    03H                      ;LAST CHARACTER?

              JZ     EXIT                     ;YES, GO ON

              INX    H                        ;NO, INCREMENT POINTER

              JMP    LOOP                     ;WAIT FOR NEXT CHARACTER
```

(a)

FIGURE 11-35 8080/8085A assembly language routine to initialize 8251A, receive, and transmit data. *(a)* Initialize 8251A and receive data on polled basis. *(b)* Transmit data on polled basis.

of-text (STX) character indicates the start of a 256-byte block of data. *End-of-text* (ETX) indicates the end of the data character block. The *block check* characters usually are cyclic redundancy characters sent by the transmitter so that the receiver can check the received block of data for errors. The Intel 8251A can handle BISYNC communication.

BISYNC is usually set up for half-duplex transmission. The transmitter sends a block of data and then waits while the receiver checks for errors. If the receiver

```
LOOP          IN    51H          ;READ STATUS REGISTER

              ANI   81H          ;MASK ALL BUT DSR+TXRDY

              CPI   81H          ;BOTH EQUAL ONE?

              JNZ   LOOP         ;NO, WAIT

              MOV   A,M          ;YES, LOAD CHARACTER

              OUT   50H          ;WRITE TO 8251A
```

FIGURE 11-35 (continued)   (b)

detects no errors, it sends an *acknowledge* (ACK) to the transmitter. The transmitter then sends the next block of data. The transmitter spends about 50 percent of its time waiting for ACK signals. A more efficient protocol for serial transmission is SDLC.

SDLC   BISYNC is a *byte-controlled protocol* (BCP). This means that character codes or bytes such as SOH, STX, or ETX are used to indicate parts of the message. IBM's *synchronous data link control* (SDLC) is a bit-oriented protocol. Meaning is determined by the number and position of bits in the message. Figure 11-36b shows the message format for SDLC. Messages are sent in *frames* which consist of *fields*.

Between messages the line idles with a string of constant 1's. Start of a message is indicated by a specific bit pattern known as the *beginning flag.* The 8-bit address of the receiving station is sent next. Then an 8-bit control word is sent. Each bit in this control word has a specific meaning. The information field which follows can have any number of bits. As much as 10,000 or 20,000 bits may be sent per frame. After the data bits, a 16-bit *cyclic redundancy* character is sent so that the receiver can check for errors. The frame ends with an *ending flag* pattern of 01111110. In order for data bits in the message not to be taken as an ending flag, SDLC requires that the transmitter insert a 0 after each succeeding pattern of five 1's. The receiver automatically removes these added 0's. SDLC systems use full-duplex communication. Only when an error occurs does the

receiver tell the transmitter to resend a data field. The Intel 8273A is a microprocessor-bus-compatible SDLC.

INTERFACING A CRT DISPLAY MONITOR TO A MICROPROCESSOR

CRTs are a convenient way to display microprocessor programs, addresses, or data. In this section we describe how numbers, letters, and graphics displays are produced on a CRT screen. Since a *CRT display* unit or *video monitor* is just a specialized TV, the discussion starts with a basic review of TV operation.

BASIC TV OPERATION

As discussed in Chap. 1, images are produced on a CRT screen by sweeping a beam of electrons across a phosphor coating on the inside of the tube face. When the beam reaches the right-hand side of the screen, it is blanked and retraced rapidly back to the left side of the screen to start over. If the beam is slowly swept from the top of the screen to the bottom while it is sweeping back and forth horizontally, the entire screen appears lighted. When the beam reaches the bottom of the screen, it is blanked and quickly retraced to the top to start over. This display is called a *raster*.

To produce a picture made of light and dark regions, a video amplifier changes the intensity of the beam as it sweeps. Turning the beam on at a particular spot produces a light spot, and turning the beam off produces a dark spot.

The horizontal sweep frequency, vertical sweep frequency, and video information must be synchronized to give a stable display. Black-and-white TVs in the United States use a horizontal sweep frequency of 15,750 Hz and a vertical sweep frequency of 60 Hz. This means the beam sweeps 262.5 times horizontally for each vertical sweep. One sweep of the beam from top to bottom of the screen is called a *field*. Sixty fields per second are swept out. To get better picture resolution, TVs use interlaced scanning. As shown in Figure 11-37a, *interlaced scanning* means the scan lines for one field are offset and interleaved with those of the next field. After every two fields, the scan pattern repeats. Therefore, two fields

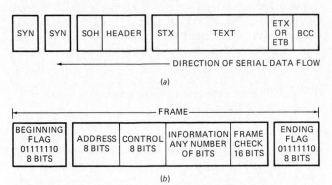

SYN	SYN	SOH	HEADER	STX	TEXT	ETX OR ETB	BCC

◄────── DIRECTION OF SERIAL DATA FLOW

(a)

├──────────────────── FRAME ────────────────────┤

BEGINNING FLAG 01111110 8 BITS	ADDRESS 8 BITS	CONTROL 8 BITS	INFORMATION ANY NUMBER OF BITS	FRAME CHECK 16 BITS	ENDING FLAG 01111110 8 BITS

(b)

FIGURE 11-36 (a) BISYNC message format. (b) SDLC message format.

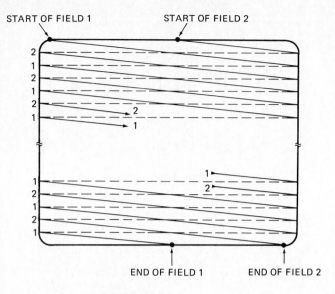

START OF FIELD 1   START OF FIELD 2

END OF FIELD 1   END OF FIELD 2

262½ LINES/FIELD
2 FIELDS/FRAME
525 LINES/FRAME FOR 15,750 Hz
HORIZONTAL AND 60 Hz VERTICAL

(a)

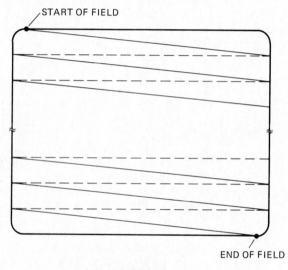

START OF FIELD

END OF FIELD

260 LINES/FIELD
1 FIELD/FRAME
260 LINES/FRAME FOR
15,600 Hz HORIZONTAL AND
60Hz VERTICAL

(b)

FIGURE 11-37 CRT scan patterns. *(a)* Interlaced. *(b)* Noninterlaced.

make a complete picture, or *frame*. The frame rate is 30 per second.

CRT units used for computer readout usually have *noninterlaced scanning*, as shown in Figure 11-37b. A horizontal sweep frequency of, for example, 15,600 Hz and a vertical sweep frequency of 60 Hz give 260 lines per frame. In this case, the frame rate and field rate are each 60 Hz.

A bare-bones television receiver often used as the basis of a CRT computer readout device is called a *video monitor*. A video monitor contains only a vertical oscillator, a horizontal oscillator, a video amplifier, a CRT, and all the necessary deflection circuitry. It requires as input a single composite video signal which contains horizontal sync pulses, vertical sync pulses, blanking pulses to blank the beam during retrace, and video intensity information. Figure 11-38 shows a typical composite video signal. It is hard to show in a figure, but there is one vertical sync pulse for each 260 horizontal sync pulses. Next we describe how the composite video signal is created to generate characters on the screen.

GENERATING A PAGE OF CHARACTERS ON A CRT

CHARACTER GENERATORS Each character on the screen is generated with a pattern of light and dark dots. Patterns for the desired characters are stored in a mask-programmed ROM called a *character generator*. A common character generator is the Motorola MC6571. The MC6571 uses a 7 × 9 dot matrix for each character, as shown in Figure 11-39. The basic 7 × 9 matrix for each character can be shifted so that small letters with descending tails appear in proper position on a line. If both capital and small letters are used, then a minimum of 13 rows must be used (12 for the characters and 1 for spacing). The MC6571 can display uppercase and lowercase alphabets, numbers, the Greek alphabet, and many symbols.

A character generator, the MC6571 for example, works as follows. The ASCII or other code for the character to be displayed is applied to the data inputs of the character generator. A row address is applied to the four row inputs. The 6571 then outputs on its seven parallel lines the dot pattern for the addressed row of the character. An 8-bit parallel-in, serial-out shift register such as the 74165 is used to convert the parallel dot code to the serial form needed to turn the CRT beam on and off. The clock for this shift register is known as the *dot clock* because each pulse shifts out a dot.

DISPLAY FORMAT AND RAM Characters may be displayed on the CRT in several different page formats. Common formats are 16 rows of 32 characters each and 24 rows of 80 characters each. For the following discussion, a format of 16 rows of 32 characters each is used. This displays a total of 512 characters at a time.

The ASCII codes for 512 characters to be displayed are stored in RAM and read as needed. The RAM may be part of a CPU memory or an independent memory dedicated to the CRT display.

TIMING The process to display a row of characters proceeds as follows. The ASCII code for the first character of the row is addressed in memory and read to the character generator data inputs. The first dot row in the character generator is addressed. The parallel dot code put out by the character generator is shifted out to the

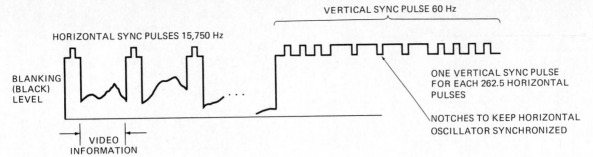

FIGURE 11-38 Composite video signal.

video amplifier a dot at a time. As the CRT beam sweeps across the screen, it produces the dot pattern for the top row of the first character.

The next information that must be given to the video amplifier is for the top row of dots of the second character. This is done by addressing and reading the next character from memory and applying it to the character generator inputs. The shift register sends the dot pattern for the top line of the second character to the video amplifier. This procedure is repeated for each of the 32 characters in the top row. When the beam has traced the top dot row of all 32 characters in the row, the dot row address input to the character generator is incremented to output the dot pattern for the second dot row of each character. Each of the 32 characters must be read again from memory. For the 6571, each character must be read from memory 13 times to display the row of characters across the screen.

When the scan of the first row of characters is completed, the first character for the second row must be read from memory and sent to the character generator for conversion to dot code. The shift register sends the dot code serially to the video amplifier. The process for this row of characters and the remaining 14 rows continues as for the first row of characters. When all 16 rows of 32 characters have been traced out, the beam is blanked and retraced to the top of the screen to start over.

Figure 11-40 shows with a block diagram how a chain of counters can be used to synchronize the timing of all this. The highest frequency is the 6-MHz crystal oscillator used to clock the dots out of the shift register. After every eight dots, A0 through A4 change to address the next character. These five address lines can address the 32 characters needed for a row. The output of the divide-by-384 section is the 15,600-Hz horizontal sweep frequency. It produces one pulse for each sweep of the beam across the screen, so it is used to clock the dot row counter. Each pulse increments the four counter outputs connected to the address inputs of the character generator to point to the next dot row pattern. After all 13 lines of one row of characters are traced, a pulse is output from the divide-by-13 counter. This increments the character row counter to the starting address in memory of the next row of characters. The character row counter is a divide-by-20. Sixteen of the counts are for the 16 rows of characters. The other four row line counts are for vertical retrace time. The final output of the divide-by-20 section is the 60 Hz needed for a vertical sync pulse.

In addition to the chain of counters, other circuitry is needed to produce horizontal and vertical blanking pulses, a cursor, and display scrolling. A *cursor* is a blinking box or underline which shows where the next character is to be written on the screen. A full-function CRT character display requires considerable circuitry. Fortunately, LSI has made it possible to put much of the circuitry in a single IC called a CRT controller.

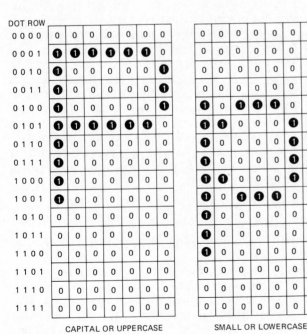

FIGURE 11-39 Character generator formats: Motorola MC6571.

PROGRAMMABLE CRT CONTROLLERS

Several manufacturers offer CRT controllers that contain different amounts of the required circuitry. The two discussed here are the Intel 8275 and the Motorola MC6845.

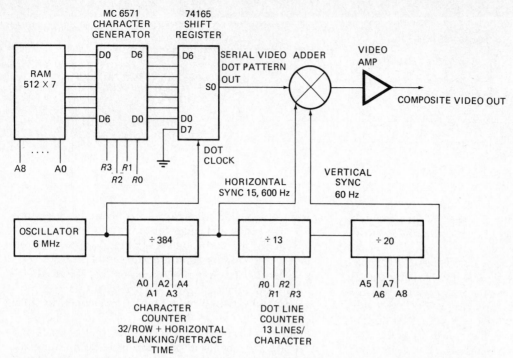

FIGURE 11-40 Block diagram of video character generator.

INTEL 8275 Figure 11-41a shows a block diagram of the Intel 8275 programmable CRT controller, and Figure 11-41b shows one way it can be connected to a microprocessor system. The 8275 requires external RAM to hold the page of characters to be displayed, a character generator, a shift register, and a dot clock. It also requires the analog interface circuits that produce the composite video signal at proper voltage levels for the video monitor. However, the 8275 does contain the character counter, line counter, and row counter. It also contains the timing signals for vertical and horizontal sync and blanking. Two 80-byte row buffers in the 8275 allow one row of characters to be displayed while another row of characters is being fetched from memory.

The 8275 can be programmed to display 1 to 80 characters per row and 1 to 64 rows per frame. The number of dot lines per character is programmable from 1 to 16. The number of dots horizontally in each character is determined by the width of the output word from the character generator and the shift register length. Initializing all the programmable features is quite complex.

As shown in Figure 11-41b, the 8275 is set up to pull characters from a microprocessor main memory on a direct memory access, or DMA, basis. When the 8275 needs a new row of characters, it outputs a DMA request signal to the 8257 DMA controller. The DMA controller sends a hold request to the hold input of, for example, an 8080A. When the 8080A floats its buses and acknowledges the hold, the 8257 addresses memory and reads the next row of characters into an 8275 row buffer. The DMA approach uses only a small percentage of the microprocessor's time. If a character from a keyboard is to be displayed, it can easily be written into the section of memory accessed by the DMA controller.

MOTOROLA MC6845 CRT CONTROLLER As with the 8275, the MC6845 contains the character counter, row counter, line counter, and the circuitry to produce the horizontal and vertical sync and blanking. The external parts needed for a typical application with a microprocessor are shown in Figure 11-42. Note that the 6845 does not use a DMA controller. Characters to be displayed are stored in a refresh RAM. The 6845 outputs a 14-bit refresh memory address to read the character to be displayed. Characters can be written into the refresh RAM when the multiplexer connects the processor address lines to it.

Refresh RAM can actually be a part of the microprocessor main memory if some scheme is used to prevent arguments between the processor and the controller over access to it.

CRT GRAPHICS

In addition to displaying alphanumeric characters, CRTs often are used to display computer-generated graphics. Good examples are the video games now finding their way to your home.

A common way to produce graphics on a CRT is to treat the entire screen as a matrix of dots. These dots are sometimes called *pixels*. Pictures are drawn by specifying whether each pixel is light or dark. Instead of using a character generator ROM, the dot pattern for

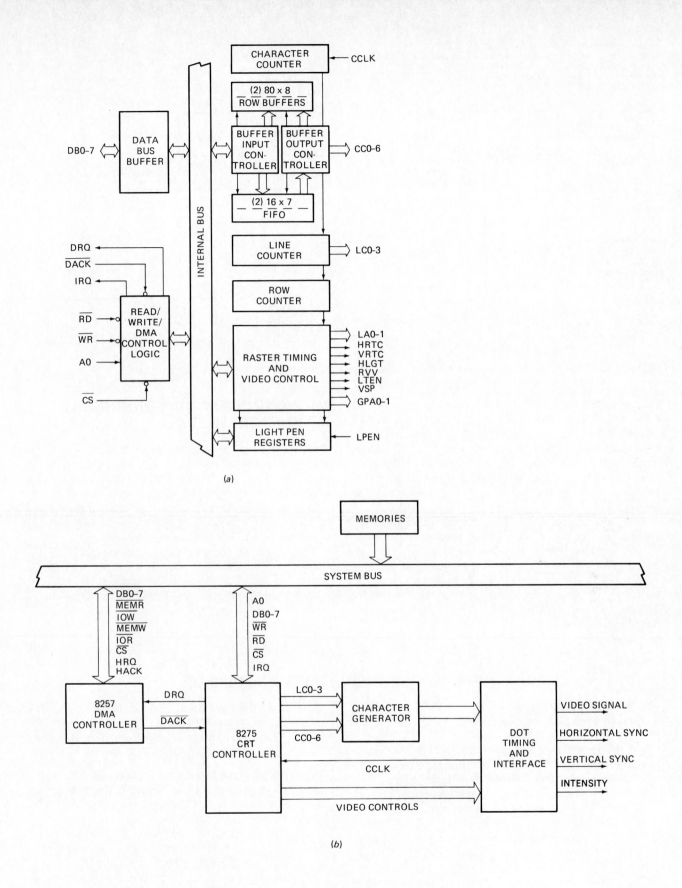

FIGURE 11-41 Intel 8275 programmable CRT controller. (a) Internal block diagram.
(b) System connection to a microprocessor.

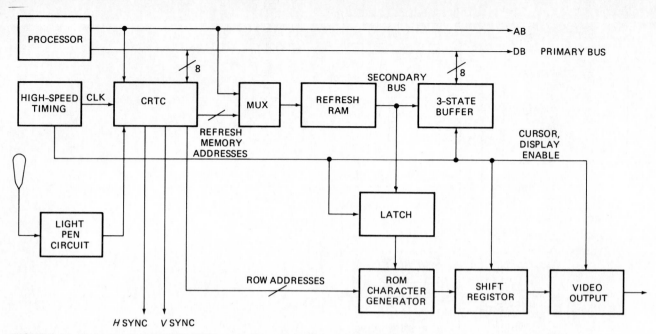

FIGURE 11-42 Motorola MC6845 CRT controller used with a microprocessor.

each successive group of eight dots on a line is stored in RAM. This byte is loaded into a shift register and shifted out to a video amplifier one dot at a time. Since at least 1 byte of memory is required for each eight dots, a large memory is needed to produce graphics with good resolution. Producing color graphics requires even more memory because the color information for each dot must also be stored.

The Motorola MC6847 video display generator provides a large part of the circuitry necessary to produce graphics or characters on a color or black-and-white television receiver. In the alphanumeric mode, it displays 16 rows of 32 characters per row. In the graphics mode, it can do a four-color dot matrix as large as 128 dots by 192 dots. One use of the MC6847 is for microprocessor-based video games.

MICROPROCESSOR PERIPHERAL INTERFACE TRENDS

The trend in peripheral interface to microprocessors seems to be toward more and more complex peripheral controller ICs. A look at the Intel or Motorola data books reveals many more peripheral IC types than it is possible to discuss here. Some peripheral ICs such as the Intel 8041 actually contain preprogrammed, dedicated microprocessors to control their functions.

The advantage of these "smart" peripheral ICs is that they relieve much of the burden from the main microprocessor and simplify its programming. The main microprocessor program simply has to initialize the peripheral devices and respond to them periodically or when signaled.

REVIEW QUESTIONS AND PROBLEMS

1. Describe polled I/O and interrupt I/O. What are the advantages and disadvantages of each type?

2. *a.* Show the instructions required to unmask and enable the 8085A RST 6.5 interrupt input.
   b. Assuming that RST 6.5 is unmasked and enabled, explain the three major responses that an 8085A will make to a RST 6.5 input after it finishes its current instruction.
   c. Why is it often necessary to make sure the interrupt request pulse is not too wide?

3. *a.* Why is it necessary to initialize the stack pointer in programs that use interrupts?
   b. Why should an interrupt service routine push registers on the stack?

4. *a.* Describe the response of a 6800 to an interrupt signal on its NMI input.
   b. What instruction is required to enable interrupts after a reset in a 6800 system?

5. *a.* Define handshake data transfer.
   b. Using Figure 11-5b, describe the sequence of signals during a handshake or strobe input data transfer with an 8155.
   c. For an 8155 at addresses 28H through 2DH, show the instructions necessary to initialize the device as follows: port A strobed output, port B strobed input, interrupt A enabled, interrupt B enabled, timer NOP.

6. Fill in the bits required to program the interrupt control lines as specified in Figure 11-43.

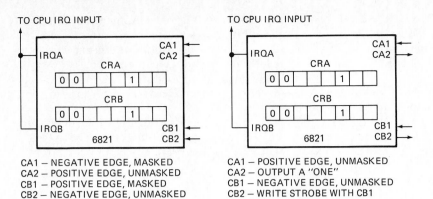

TO CPU IRQ INPUT

IRQA

CRA
| 0 | 0 | | | | 1 | | |

CRB
| 0 | 0 | | | | 1 | | |

IRQB

6821

CA1
CA2

CB1
CB2

CA1 – NEGATIVE EDGE, MASKED
CA2 – POSITIVE EDGE, UNMASKED
CB1 – POSITIVE EDGE, MASKED
CB2 – NEGATIVE EDGE, UNMASKED

(a)

TO CPU IRQ INPUT

IRQA

CRA
| 0 | 0 | | | | 1 | | |

CRB
| 0 | 0 | | | | 1 | | |

IRQB

6821

CA1
CA2

CB1
CB2

CA1 – POSITIVE EDGE, UNMASKED
CA2 – OUTPUT A "ONE"
CB1 – NEGATIVE EDGE, UNMASKED
CB2 – WRITE STROBE WITH CB1
RESTORE

(b)

FIGURE 11-43

7. Show the series of assembly language instructions required to initialize a 6821 at addresses 8020H through 8023H as follows: port A: strobed input, CA1 active on negative edge and unmasked, CA2 as read strobe with E restore; port B: strobed output, CB1 active on positive edge and unmasked, CB2 as write strobe with CB1 restore.

8. A 6800 receives an interrupt request from the 6821 shown in Figure 11-43a. How can the CPU determine which of the four inputs caused the interrupt request?

9. Describe the three major tasks required to get useful data from a matrix keyboard.

10. *a.* Draw a flowchart for a program to scan and encode a 16-key keyboard.
  b. Modify the flowchart so that the program waits for the keypressed strobe to go low before returning to the main program.

11. Show the assembly language instructions you would add to the program in Figure 11-13 to make it jump to the error exit if all four rows are checked and no key is found pressed. (*Hint:* Use a row counter.)

12. List the advantages and disadvantages of software-multiplexed and hardware-multiplexed displays.

13. Draw a flowchart for multiplexing an eight-digit seven-segment display.

14. Write an 8085A or a 6800 assembly language routine to drive the eight-digit display in Prob. 13. Use port 01 to drive segments and port 02 to drive digits. Assume common-anode displays, inverting segment drivers, and inverting digit drivers. Store seven-segment codes in one table in memory and the characters to be displayed in another table in memory. Light each digit for 1 ms before moving to the next.

15. Describe the drive required by liquid crystal displays. Why are LCDs not usually multiplexed?

16. Define stepper motor. Write a short assembly language routine to step a stepper motor an inserted number of steps clockwise or counterclockwise.

17. Draw a circuit showing how a 12-bit D/A converter can be interfaced to 8-bit microcomputer ports so that the D/A converter gets all 12 bits at the same time.

18. Write an assembly language routine to input data from a four-BCD-digit A/D converter. (*Hint:* Test for a digit strobe true and read in the data for that digit.)

19. Explain the operation of a data acquisition system.

20. *a.* Describe S-100 bus, multibus, and IEEE-488 (GPIB).
  b. What is the main use of each?
  c. In a multibus system with several CPU boards, how is it determined which CPU board controls the bus?
  d. Describe the sequence of signals on NRFD, DAV, and NDAC during the transfer of a data byte on the GPIB data bus.

21. Define the following:
  a. RS-232-C
  b. Space
  c. Mark
  d. 20-mA current loop
  e. RS-422
  f. RS-423
  g. MODEM
  h. Frequency shift keying
  i. Full duplex
  j. Half duplex
  k. Baud
  l. Stop bit
  m. Start bit
  n. DTE
  o. DCE

22. *a.* Describe the function of the following RS-232C signals: TxD, RxD, $\overline{RTS}$, $\overline{CTS}$, $\overline{CD}$, $\overline{DTR}$, $\overline{DSR}$.
  b. Give the voltage levels for an RS-232C logic high and logic low.
  c. What type of connector is used for RS-232C signal cables?
  d. For how long a transmission distance is the RS-232C specified?

23. *a.* Show the control word that must be sent to an MC6850 ACIA to initialize it as follows: baud rate factor of 16; 7 bits; odd parity; 1 stop bit; $\overline{RTS}$ asserted (low); transmit interrupt enabled; receiver interrupt enabled.

b. Show the sequence of assembly language instructions needed to initialize a 6850 at addresses 8008H and 8009H, as shown in part (*a*) and to poll the status register to determine when a character can be sent to the transmit buffer.

c. If the MC6850 as initialized above is going to send data at 1200 Bd, what frequency transmit clock must be supplied to the 6850 transmit clock input?

d. In what two ways can the CPU determine that the receiver buffer of the 6850 has a character ready to be read?

24. *a.* Show the mode word and command word that must be sent to an 8251A to initialize the device as follows: baud rate factor of 64; 7 bits per character; even parity; 1 stop bit; transmit interrupt enabled; $\overline{DTR}$ and $\overline{RTS}$ asserted (low); receiver interrupt enabled; reset error flags; no break character; no hunt mode.

b. List the sequence of assembly language instructions required to initialize an 8251A at addresses 80H and 81H, as in part (*a*), and to poll the status register to determine whether the receiver buffer has a character ready to read.

c. How could you determine whether a character read in contained a parity error?

d. What frequency transmit and receive clock will an 8251A as initialized above require to send data at 2400 Bd?

e. In what two ways can the CPU determine whether the receiver buffer of an 8251A has a character ready to read?

25. Draw a flowchart showing how asynchronous serial data can be sent from a port bit by using a software routine.

26. Describe interlaced and noninterlaced scanning used to produce character or graphics displays on a CRT. Why must the horizontal and vertical sweep be synchronized?

27. Explain how a character generator such as the 6571 produces a pattern of dots for a character on a CRT. Draw a block diagram showing RAM, a shift register, and timing chain to aid the explanation.

28. Describe how the 8275 CRT controller uses DMA to get the next row of characters from the main microprocessor memory. Why do you suppose a program routine usually is not used to output the character to be displayed to the 8275? (*Hint:* Time.)

REFERENCES/BIBLIOGRAPHY

SC-01 Speech Synthesizer Data Sheet, Votrax, Division of Federal Screw Works, Troy, MI, 1980.

Intel SDK-85 User's Manual, Intel Corporation, Santa Clara, CA, December 1977.

MEK6800D2 Evaluation Kit II Manual, Motorola Semiconductor Products, Inc., Austin, TX, 1976.

Intel Component Data Catalog 1982, Intel Corporation, Santa Clara, CA, 1982.

Lesea, Austin, and Rodnay Zaks: *Microprocessor Interfacing Techniques*, 2d ed., Sybex, Inc., Berkeley, CA, 1978.

The Complete Motorola Microcomputer Data Library, prepared by Technical Information Center, Motorola, Inc., Phoenix, AZ, 1978.

EIA Standard RS-232-C Interface between Data Terminal Equipment and Data Communication Equipment Employing Serial Binary Data Interchange, Electronic Industries Association Engineering Department, Washington, 1969.

Condensed Description of the Hewlett-Packard Interface Bus, Application Note HP-IB, Hewlett-Packard, Palo Alto, CA, 1977.

Slo-Syn Synchronous/Stepping Motors and Motor Controls, Catalog MD174-1, The Superior Electric Company, Bristol, CT, 1974.

8800 System Bus Structure, Altair 8800 Manual, MITS, Inc., Albuquerque, NM.

Speech Plus the Talking Calculator, Data Sheets, Telesensory Systems, Inc., Palo Alto, CA, 1976.

Speech Synthesizer Module TS1, Data Sheet, Telesensory Systems, Inc., Palo Alto, CA, 1977.

SpeechLab, Data Sheet, Heuristics, Inc., Los Altos, CA, 1977.

"Revised data-interface standards permit faster data rates and longer cables. New chips, and RS-232 adapters, simplify their use," Technology, *Electronic Design*, vol. 18, September 1, 1977, pp. 138–141.

CHAPTER 12

PUTTING IT ALL TOGETHER

In this chapter we show how the microprocessors and other components discussed earlier are used to build microcomputer-based products. Three examples are shown. The first is an 8085A-based "smart" scale such as those used at supermarket checkout stands. The second is a 6800-based EPROM programmer. The third example uses an 8085A-based temperature controller to demonstrate the principles of industrial control with microcomputers. These commonplace examples were chosen and developed to be simple enough to understand yet complex enough to be realistic. The examples were also chosen so that the discussed programs contain many routines you can use in writing your own programs.

The section of the chapter for each instrument is set up in logical steps for learning any new instruments. These instruments were built on microcomputer prototyping boards so that they can be used for examples in Chap. 13 of how a product is debugged during the development stage.

At the conclusion of this chapter, you should be able to:

1. List logical steps for learning the hardware and software of a microprocessor-based instrument.

2. Define monitor program, equates, buffer, and modular programming.

3. Explain how a lookup table can be used to convert one code to another.

4. Describe a process that can be used to convert BCD numbers to their binary equivalent.

5. Describe three methods of multiplying binary numbers with a microprocessor.

6. Describe the operation of a microprocessor-based scale.

7. Describe the operation of a microprocessor-based EPROM programmer.

8. Describe how a microcomputer can monitor and control several industrial processes at the same time.

A MICROPROCESSOR-BASED SMART SCALE

PRODUCT DESCRIPTION

The first step in learning about a new instrument is to read the product description, if available.

This scale is intended for use in applications such as the checkout stand of a supermarket or hardware store. The item to be weighed is placed on a platform. The weight is determined by the scale and displayed on four seven-segment LED displays and two additional LED digit display units. Maximum weight is 10.00 lb, and resolution is 0.01 lb.

The weight display is continuously updated until the operator enters the price with a calculator-type keyboard. Then price per pound is displayed on the four LED displays. Maximum price with full-scale weight is $9.99/lb. If an error is made in entering the price per pound, there are two ways to correct it. One is pressing the zero key 4 times. This clears all digits from the price-per-pound display, and the correct price can then be entered. Another way is to press the INTR key. This returns the scale to weighing mode until a new price is entered.

Pressing the price (F) key indicates that all the digits of the price have been entered. The scale then automatically multiplies the weight by the price per pound and displays the total price. After 3 s the scale returns to determining and displaying weight. A return to this weighing mode can be produced at any time by pressing the INTR key.

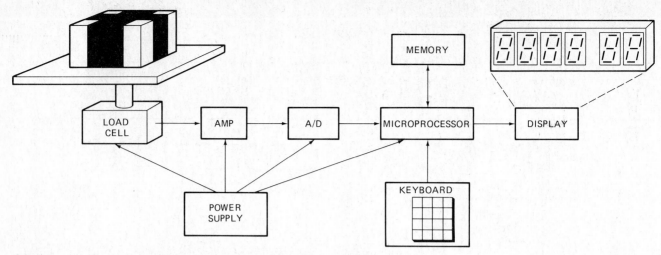

FIGURE 12-1 Block diagram of a microcomputer-based smart scale.

THEORY OF OPERATION: HARDWARE

The next step in learning about a new instrument is to examine the hardware portion of the theory of operation. This includes schematics, memory maps, and block diagrams.

As shown in Figure 12-1, the scale starts with a load-cell transducer which converts weight to a proportional electric signal. An amplifier amplifies the load-cell signal to the proper range for the A/D converter. The A/D converter changes the amplifier output to a digital word which can be operated on by the microprocessor. Memory stores the program for the microprocessor and the digital version of the weight from the A/D converter.

The display shows first the weight, then the price per pound, and finally the total price. The keyboard is used to enter the price per pound and the command to compute the total price.

FIGURE 12-2 Photograph of strain gage load cell with platform for scale. (Transducers, Inc.)

LOAD CELL AND AMPLIFIERS   A load-cell pressure transducer converts weight to a proportional electric signal. Figure 12-2 shows a picture of the Model C462-10#-10P1 load cell of Transducers, Inc. This unit is specified for accuracy over a 0- to 10-lb [0- to 4.5-kg] range, but up to 100 lb [45.4 kg] can be applied without mechanically damaging it. Accuracy of the unit is better than ±0.1 percent, or ±1 part in 1000. This translates to ±0.01 lb [±0.004 kg] for a 10-lb [4.5-kg] unit.

The actual sensing elements in the load cell are equivalent to a 350-Ω resistive bridge, as shown in the left-hand side of Figure 12-3. A nominal 10 V is applied as bias to the bridge. With no load on the cell, the two outputs of the bridge are at about the same voltage, 5 V. An applied load will increase the resistance of one of the resistors. This produces a differential output voltage. The maximum differential voltage out for a 10-lb [4.5-kg] load is 2 mV per volt of bias voltage. With a 10-V bias voltage as shown, this gives a full-scale differential output voltage of 20 mV.

To amplify this small signal to the proper levels for the A/D converter without introducing errors, the instrumentation amplifier circuit shown in Figure 12-3 is used. IC-1 and IC-2 form a high-input-impedance differential-input, differential-output amplifier. The 5-V common-mode signal to IC-1 and IC-2 is not amplified. The 20-mV differential signal is amplified by $(R3 + R4 + 2R1) \div (R3 + R4)$. And $R4$ is adjusted to give a gain of about 5, so the differential output is 100.0 mV for a 10.00-lb [4.54-kg] load on the cell. IC-3 is a standard unity gain differential-input amplifier. It converts the differential signal from IC-1 and IC-2 to a single-ended signal needed for the A/D converter.

A/D CONVERTER   The A/D converter used in this scale is the Motorola MC14433. It is an inexpensive dual-slope converter intended for use in 3½-digit digital voltmeters, etc. As shown in Figure 12-3, only an external voltage reference and a few resistors and capacitors are needed for the complete A/D converter.

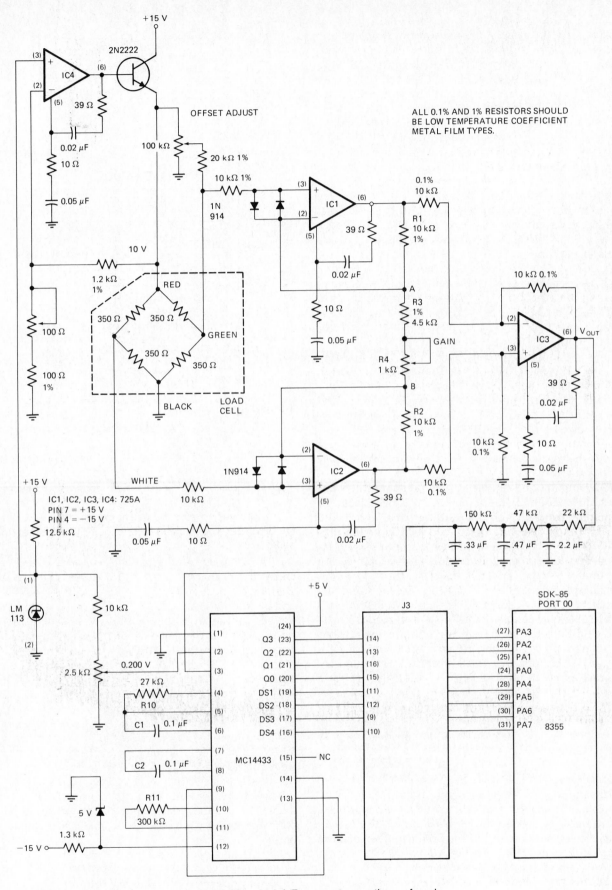

FIGURE 12-3 Schematic of load-cell amplifier and A/D converter sections of scale.

The converter can easily be used for 2.000 V full scale or 200.0 mV full scale. A 2.000-V reference voltage and an *R*10 of 470 kΩ give 2.000 V full scale. A 200.0-mV reference voltage and an *R*10 of 27 kΩ give 200.0 mV full scale as shown in Figure 12-3. The LM113 in Figure 12-3 is a 1.2-V precision, temperature-stable reference diode used to produce the converter reference voltage.

Output of the converter is in multiplexed BCD form. Q0 through Q3 are the data outputs, and DS1 through DS4 are the data strobe outputs. A data strobe output goes high when the data for that digit is on the Q0 to Q3 outputs. Conversion rate and multiplexing frequency are controlled by an internal oscillator. The frequency of the oscillator is determined by *R*11. An *R*11 of 300 kΩ gives an internal clock frequency of 66 kHz, a multiplex frequency of 0.8 kHz, and about four complete conversions per second.

Accuracy of the converter is ±.05 percent of reading and ±1 count, which is comparable to the accuracy of the load cell. Another way of thinking of the accuracy of the 3½-digit converter is ±2 parts of the full count of 1999 counts.

SDK-85 MICROPROCESSOR DEVELOPMENT BOARD

The microprocessor, RAM, ROM, keyboard, and display for the scale are contained on an Intel SDK-85 microprocessor prototyping board. Figure 9-9*a* and *b* shows the schematic of the RAM, ROM, and microprocessor part of the board; Figure 9-8*a* and *b* shows a photograph and block diagram of the board; and Figure 11-18 shows a schematic of the keyboard and display part of the SDK-85.

Referring to these figures, you can see that the board uses an 8085A microprocessor with a 3-MHz clock. An 8355 that comes with the kit contains two I/O ports and 2 kilobytes of ROM preprogrammed with a monitor program. The monitor program allows programs to be entered into RAM and executed during development. In the final prototype version of the scale, the 8355 monitor is replaced with an 8755 EPROM containing the complete scale program and the same two I/O ports.

An 8155 supplies 256 bytes of RAM, three I/O ports, and an 8-bit programmable timer. The scale requires a small amount of RAM for use as a stack and for data storage. In the final version of the scale, a less expensive IC containing only RAM might be used if the I/O ports and timer are not needed.

The 8279 keyboard-display controller chip interfaces the processor to a 24-key keyboard and six seven-segment LED displays. Refer to Chap. 11 for a discussion of the 8279 operation.

MEMORY MAP FOR THE SDK-85 BOARD

An important part to look for when you are studying a new system is the memory map. If one is not given, it is helpful to make one up from the schematic. The memory map allows you to see at a glance which areas of memory are assigned to RAM, ROM, or I/O ports. Figure 12-4 shows the memory map for the SDK-85. The 8205, A10 in Figure 11-18, decodes address lines to produce six

MEMORY ADDRESS

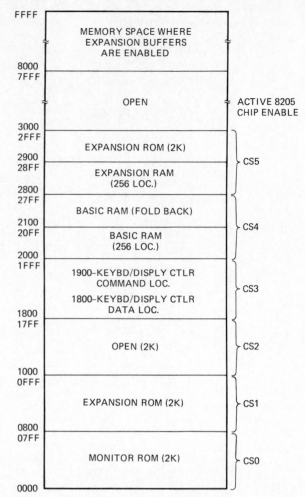

FIGURE 12-4 Memory map of the SDK-85 development board.

chip-select outputs, CS0 to CS5. Two are used to enable 2-kilobyte blocks of ROM in addresses 0000 to 07FFH and 0800 to 0FFFH. The CS2 output is left open. It can be used to enable another block of memory or added peripheral IC. The CS3 enables the 8279 keyboard-display controller at address 1800H. Address line A8 connected to the 8279 C/$\overline{D}$ input selects the command address as 1900H and the data address for the device as 1800H.

CS4 decodes a 2-kilobyte block of memory from 2000H to 27FFH. This output is used to select A16, which is the 8155. Since the 8155 has only 256 bytes of RAM, the 2K space is not filled. The unused area is referred to as *foldback*. What this actually means is that any address from 2000H to 27FFH will "fold back" and select a location in the basic 256-byte RAM. CS5 selects another 2-kilobyte block for RAM from 2800H to 2FFFH. All the memory space above 3000H, or 12228 decimal, is available for memory expansion or memory-mapped I/O ports.

The actual scale program uses an 8755 EPROM at 0000 to 07FFH, an 8279 keyboard-display controller at 1800H, and an 8155 RAM at 2000H.

For processors such as the 8080A and 8085A which separate I/O ports from memory, another important map is that for the I/O port addresses. For the SDK-85 board, I/O port addresses are shown in Figure 9-9b. The scale uses port A of the A14 8755 as an input port. As shown in Figure 9-9b, the address for this port is 00H. Another port is available in the 8755 at address 01H.

THEORY OF OPERATION: SOFTWARE

After you have at least a preliminary understanding of the hardware operation of a new instrument, the next part to approach is the software, or programming. Before tackling the program itself, look for a flowchart to give you an idea what the program is supposed to be doing. Some instrument manuals may have a flowchart for the overall operation and additional flowcharts for the operations in each block of the main flowchart.

FLOWCHART FOR THE SDK-85-BASED SCALE

Figure 12-5 shows the overall flowchart for the SDK-85-based scale. After initializing, the program takes in the weight one digit at a time from the A/D converter. The weight is displayed in a group of four seven-segment LEDs and Lb is displayed on two additional seven-segment LED displays.

A check is made to see whether a key on the keyboard has been pressed. If no key has been pressed, the program loops back to what became known as *dumb-scale* mode during the development of the unit. The dumb-scale loop is repeated until a key is pressed.

When a key is pressed, the weight is removed from the four-digit display, and the character for the pressed key is displayed. The two-digit display is changed to read SP., as an abbreviation for "selling price." As additional digits of the price per pound are entered on the keyboard, they are shown on the four displays. When an 'F' key is pressed, the program exits the keyboard read loop.

The price per pound is converted from BCD to binary, and the weight is converted from BCD to binary, because binary multiplication is easier to program than BCD multiplication. Weight in binary is then multiplied by price per pound. The result of this multiplication is the total price in binary. The binary total price is converted to BCD and rounded to the nearest cent. Total price is displayed on the four LED displays and "Pr." for price, is displayed on the two LED displays.

After 3 s, which is long enough to read the total price and enter it on a cash register, the program returns to dumb-scale mode. Pressing the INTR key returns the program to dumb scale from anywhere in the program. This is handy if an error is made in entering the price per pound.

ASSEMBLY LANGUAGE PROGRAM ANALYSIS

The next step in analyzing a new instrument is to examine the assembly language program to see how the blocks of the flowchart are actually done. A complete program for this scale is shown in Figure 12-6. The pro-

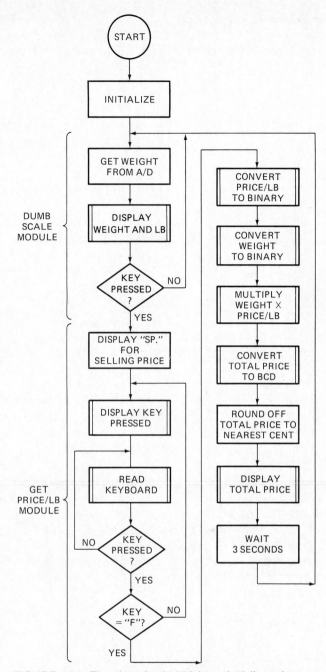

FIGURE 12-5 Flowchart for 8085A-based 10-lb scale.

gram was written to be simple enough to be understandable, yet complex enough to be realistic. Before we get started on a discussion of the program, a few general comments are in order.

First, it is important not to be overwhelmed by the many pages of program printout found with an instrument. In earlier days of microprocessors, when memory costs were quite high, programmers prided themselves on writing programs in the cleverest and sneakiest ways that used minimum memory. As memory cost per byte has gone down, the emphasis in programming has moved to writing programs for clarity rather than minimum memory usage.

349

PUTTING IT ALL TOGETHER

```
LOC   OBJ       SEQ          SOURCE STATEMENT


               1 ;****************************************************************
               2 ;HALL 8085 10 LB. SCALE PROGRAM SCALE *
               3 ;D.V.HALL
               4 ;12/11/78
               5 ;****************************************************************
               6
               7 ;----------------------------------------------------------------
               8 ;EQUATES FOR 8085 BASED SCALE
               9 ;----------------------------------------------------------------
              10
20A0          11 BUFFR1 EQU   20A0H
20A4          12 BUFFR2 EQU   20A4H
20A6          13 BUFFR3 EQU   20A6H
20AA          14 BUFFR4 EQU   20AAH
20AC          15 SAVBP  EQU   20ACH
20AE          16 KYBUF  EQU   20AEH
              17
              18 ;----------------------------------------------------------------
0000 31C220   19 START: LXI   SP,20C2H    ;INITIALIZE SP IN RAM
0003 3E00     20        MVI   A,00        ;INITIALIZE KEYBOARD/DISPLAY MODE
0005 320019   21        STA   1900H
0008 3ECC     22        MVI   A,0CCH      ;SET BLANKING CHARACTER
000A 320019   23        STA   1900H       ;SEND TO CONTROL OF 8279
000D 3E00     24        MVI   A,00        ;INITIALIZE COMMAND/STATUS REGISTER
000F D320     25        OUT   20H         ;      OF 8155
0011 21A020   26 GETWT: LXI   H,BUFFR1    ;LOAD STARTING ADDRESS OF CHARACTER STORE
0014 DB00     27 DS1:   IN    00          ;READ PORT 00
0016 E610     28        ANI   10H         ;CHECK FOR STROBE
0018 CA1400   29        JZ    DS1         ;LOOP UNTIL STROBE FOUND
001B DB00     30        IN    00          ;READ MSD DATA FROM PORT 0
001D E60F     31        ANI   0FH         ;MASK UPPER FOUR BITS
001F FE04     32        CPI   04          ;CHECK IF MSD=1 (BIT 3)
0021 C22F00   33        JNZ   BLANK       ;IF MDS=0 BLANK
0024 3E01     34        MVI   A,01        ;IF MSD=1 LOAD ONE
0026 C33100   35        JMP   STORE
0029 00       36        NOP
002A 00       37        NOP
002B 00       38        NOP
002C C32801   39        JMP   KYBINT      ;RST 5.5 KEYBOARD INTERRUPT ADDRESS
002F 3E15     40 BLANK: MVI   A,15H       ;LOAD CHARACTER FOR BLANK
0031 77       41 STORE: MOV   M,A         ;STORE MS CHARACTER
0032 23       42        INX   H           ;MOVE POINTER TO NEXT LOCATION
0033 C34000   43        JMP   DS2         ;JUMP OVER RST 7.5 ADDRESS
0036 00       44        NOP
0037 00       45        NOP
0038 00       46        NOP
0039 00       47        NOP
```

FIGURE 12-6 Complete assembly language program for 10-lb scale.

LOC	OBJ	SEQ		SOURCE STATEMENT	
003A	00	48		NOP	
003B	00	49		NOP	
003C	C1	50		POP	B	;UNLOAD STACK RST 7.5 INTR
003D	C30000	51		JMP	START	;BACK TO DUMB SCALE FROM ANY POINT
0040	DB00	52 DS2:	IN	00	;IN PORT 0
0042	E620	53		ANI	20H	;CHECK FOR DIGIT 2 STROBE
0044	CA4000	54		JZ	DS2	;LOOP UNTIL STROBE = 1
0047	DB00	55		IN	00	;TAKE IN DATA FOR DIGIT 2
0049	E60F	56		ANI	0FH	;MASK UPPER FOUR BITS
004B	77	57		MOV	M,A	;STORE CHARACTER IN MEMORY
004C	23	58		INX	H	;POINT TO NEXT LOCATION IN MEMORY
004D	DB00	59 DS3:	IN	00	;IN PORT 0
004F	E640	60		ANI	40H	;CHECK FOR DIGIT STROBE 3
0051	CA4D00	61		JZ	DS3	;LOOP UNTIL STROBE = 1
0054	DB00	62		IN	00	;TAKE IN DIGIT 3 DATA
0056	E60F	63		ANI	0FH	;MASK UPPER FOUR BITS
0058	77	64		MOV	M,A	;STORE DIGIT 3
0059	23	65		INX	H	
005A	DB00	66 DS4:	IN	00	
005C	E680	67		ANI	80H	;CHECK FOR LEAST SIGNIFICANT DIGIT STROBE
005E	CA5A00	68		JZ	DS4	;LOOP UNTIL STROBE = 1
0061	DB00	69		IN	00	;TAKE IN LSD DATA
0063	E60F	70		ANI	0FH	;MASK UPPER FOUR BITS
0065	77	71		MOV	M,A	;STORE LSD
0066	3E00	72		MVI	A,00H	;SET TO DISPLAY IN 4 LED DIGITS
0068	21A020	73		LXI	H,BUFFR1	;SET POINTER TO FIRST CHARACTER OF WEIGHT
006B	CD3901	74		CALL	DISPLY	;DISPLAY WEIGHT
006E	21A420	75		LXI	H,BUFFR2	;LOAD STARTING ADDRESS OF UNITS BUFFER
0071	3611	76		MVI	M,11H	;LOAD CODE FOR "L"
0073	23	77		INX	H	;POINT TO NEXT LOCATION
0074	360B	78		MVI	M,0BH	;LOAD CHARACTER FOR "B"
0076	3E01	79		MVI	A,01	;SET TO DISPLAY IN UNITS LED
0078	21A420	80		LXI	H,BUFFR2	;SET POINTER BACK TO FIRST CHAR OF UNITS
007B	CD3901	81		CALL	DISPLY	;DISPLAY LB
007E	3E08	82		MVI	A,08H	;LOAD CHARACTER TO UNMASK INTERRUPTS
0080	30	83		SIM		;UNMASK INTERRUPTS
0081	FB	84		EI		;ENABLE INTERRUPTS
0082	21AE20	85		LXI	H,KYBUF	;LOAD ADDRESS OF KYBINT BUFFER
0085	7E	86		MOV	A,M	;READ KYBINT BUFFER
0086	B7	87		ORA	A	;SET SIGN FLAG WITHOUT CHANGING CHARACTER
0087	F28E00	88		JP	EXIT1	;EXIT DUMB SCALE AND DISPLAY IF CHARACTER
008A	FB	89		EI		;NO CHARACTER - RE-ENABLE INTERRUPTS
008B	C31100	90		JMP	GETWT	;LOOP TO DUMB SCALE
008E	3680	91 EXIT1:	MVI	M,80H	;PUT 1 IN MSB OF KYBUF TO INDICATE
0090	FB	92		EI		; CHARACTER READ
0091	57	93		MOV	D,A	;SAVE CHARACTER
0092	21AA20	94		LXI	H,BUFFR4	
0095	3605	95		MVI	M,05H	;"S"
0097	23	96		INX	H	

FIGURE 12-6 *(continued)*

LOC   OBJ   SEQ   SOURCE STATEMENT

```
0098  3612      97           MVI   M,12H        ;"P"
009A  3E01      98           MVI   A,01
009C  21AA20    99           LXI   H,BUFFR4
009F  CD3901    100          CALL  DISPLY       ;DISPLAY SP. IN UNITS FIELD
00A2  7A        101          MOV   A,D          ;MOVE SAVED CHARACTER BACK TO A
00A3  110000    102          LXI   D,0000       ;CLEAR D AND E
00A6  CD6F01    103  SHOWSP:CALL   INSRT        ;INSERT NEW CHARACTER IN PACKED BCD IN D&E
00A9  CD7801    104          CALL  EXPND        ;EXPAND TO ONE CHAR/BYTE FOR DISPLAY
00AC  21A620    105          LXI   H,BUFFR3     ;LOAD STARTING ADDRESS OF DISPLAY BUFFER
                106                             ;  FOR PRICE PER LB.
00AF  3E00      107          MVI   A,00         ;SET FOR DISPLAY IN 4 DIGITS
00B1  CD3901    108          CALL  DISPLY       ;DISPLAY ENTERED DIGIT OF PRICE/LB.
00B4  CD9701    109          CALL  RDKY         ;READ NEXT CHARACTER FROM KEYBOARD
00B7  FE0A      110          CPI   0AH          ;CHECK IF TERMINATOR
00B9  DAA600    111          JC    SHOWSP       ;NO-DISPLAY CHAR & READ KEYBOARD AGAIN
00BC  CDA701    112          CALL  BCDCVT       ;PRICE/LB IN DE, CONVERT TO BINARY IN HL
00BF  EB        113          XCHG               ;MOVE BINARY PRICE TO DE
00C0  21AC20    114          LXI   H,SAVBP
00C3  73        115          MOV   M,E          ;SAVE BINARY PRICE/LB IN MEMORY
00C4  23        116          INX   H
00C5  72        117          MOV   M,D
00C6  21A020    118          LXI   H,BUFFR1     ;START PACK BCD WT FOR BCDCVT
00C9  7E        119          MOV   A,M          ;LOAD MS CHARACTER
00CA  FE15      120          CPI   15H          ;SEE IF MOST SIG CHARACTER IS A BLANK
00CC  C2D000    121          JNZ   SHIFT        ;NO, GOTO SHIFT
00CF  AF        122          XRA   A            ;YES, CLEAR ACC AND CARRY
00D0  07        123  SHIFT: RLC                 ;MOVE TO UPPER 4 BITS
00D1  07        124          RLC
00D2  07        125          RLC
00D3  07        126          RLC
00D4  23        127          INX   H            ;POINT TO NEXT BCD CHAR OF WT
00D5  86        128          ADD   M            ;ADD IN LOWER 4 BITS OF A
00D6  57        129          MOV   D,A          ;PUT IN D FOR BCDCVT
00D7  23        130          INX   H            ;POINT TO NEXT CHAR OF WT
00D8  7E        131          MOV   A,M          ;LOAD BCD CHARACTER
00D9  07        132          RLC                ;SHIFT TO UPPER FOUR BITS
00DA  07        133          RLC
00DB  07        134          RLC
00DC  07        135          RLC
00DD  23        136          INX   H            ;POINT TO LS BCD CHARACTER
00DE  86        137          ADD   M            ;ADD IN LOWER FOUR BITS
00DF  5F        138          MOV   E,A          ;BCD WT NOW IN DE
00E0  CDA701    139          CALL  BCDCVT       ;CONVERT WT TO BINARY IN HL
00E3  EB        140          XCHG               ;MOVE BIN WT TO DE
00E4  21AC20    141          LXI   H,SAVBP      ;GET BACK BINARY PRICE/LB.
00E7  4E        142          MOV   C,M
00E8  23        143          INX   H
00E9  46        144          MOV   B,M
00EA  CDD001    145          CALL  MULTO        ;MULTIPLY WT X PRICE/LB.
```

FIGURE 12-6 (continued)

LOC   OBJ   SEQ   SOURCE STATEMENT

```
                146                          ;        BC X DE = DEHL
00ED CDEE01     147          CALL  BINCVT     ;CONVERT BIN TOTAL PRICE IN REG E,H,L
                148                          ;        TO BCD PACKED IN HLBC
00F0 3E49       149          MVI   A,49H      ;START ROUND OFF TO NEAREST CENT
00F2 B9         150          CMP   C          ;CARRY SET IF C > 49H
00F3 3E00       151          MVI   A,00       ;CLEAR ACC, LEAVE CARRY
00F5 88         152          ADC   B          ;PROPAGATE CARRY TO NEXT DIGITS
00F6 27         153          DAA              ;MAINTAIN BCD FORMAT
00F7 5F         154          MOV   E,A        ;MOVE TO E FOR EXPND
00F8 3E00       155          MVI   A,00
00FA 8D         156          ADC   L          ;PROPAGATE CARRY TO NEXT DIGITS
00FB 27         157          DAA              ; AND MAINTAIN BCD
00FC 57         158          MOV   D,A        ;MOVE TO D FOR EXPND
00FD CD7801     159          CALL  EXPND      ;EXPAND TO ONE CHAR/BYTE
                160                          ;RESULT IN BUFFR3
0100 21A620     161          LXI   H,BUFFR3   ;LOAD START OF BUFFR3
0103 3E00       162          MVI   A,00       ;SET FOR 4 DIGIT DISPLAY
0105 CD3901     163          CALL  DISPLY     ;DISPLAY TOTAL PRICE
0108 21AA20     164          LXI   H,BUFFR4   ;POINT TO UNITS BUFFER
010B 3612       165          MVI   M,12H      ;"P"
010D 23         166          INX   H          ;POINT TO NEXT LOCATION
010E 3614       167          MVI   M,14H      ;"R"
0110 2B         168          DCX   H          ;POINT TO START OF BUFFER
0111 3E01       169          MVI   A,01       ;SET FOR UNITS DISPLAY
0113 CD3901     170          CALL  DISPLY     ;DISPLAY "PR."
0116 110000     171          LXI   D,0000H    ;LOAD LONG LOOP DELAY CONSTANT
0119 1B         172 LNGLP:   DCX   D          ;DECREMENT LONG LOOP
011A 0608       173          MVI   B,08H      ;LOAD SHORT LOOP DELAY CONSTANT
011C 05         174 SHRTLP:  DCR   B          ;DECREMENT SHORT LOOP CONSTANT
011D C21C01     175          JNZ   SHRTLP     ;LOOP UNTIL B = 0
0120 7A         176          MOV   A,D        ;CHECK LONG LOOP CONSTANT
0121 B3         177          ORA   E          ;        FOR ZERO
0122 C21901     178          JNZ   LNGLP      ;LOOP UNTIL ZERO (3SEC)
0125 C31100     179          JMP   GETWT      ;JUMP BACK TO DUMB SCALE
                180
                181 ;****************************************************************
                182 ;KYBINT - SERVICES INT OUTPUT OF 8279
                183 ;       - CHARACTER IS READ FROM 8279 AND STORED IN BUFFER KYBUF
                184 ;INPUTS:  NONE
                185 ;OUTPUTS: NONE
                186 ;CALLS:   NOTHING
                187 ;DESTROYS:NOTHING
                188 ;****************************************************************
                189
0128 E5         190 KYBINT:PUSH H             ;SAVE HL
0129 F5         191        PUSH PSW           ;SAVE A AND FF'S
012A 210019     192        LXI   H,1900H      ;CONTROL ADDRESS FOR 8279
012D 3640       193        MVI   M,40H        ;OUTPUT CONTROL CHAR FOR 8279 READ KEYBD
012F 25         194        DCR   H            ;POINT TO 8279 READ ADDRESS
```

FIGURE 12-6 *(continued)*

```
LOC   OBJ    SEQ       SOURCE STATEMENT

0130 7E      195       MOV  A,M        ;READ CHARACTER FROM FIFO
0131 E60F    196       ANI  0FH        ;MASK UPPER FOUR BITS
0133 32AE20  197       STA  KYBUF      ;STORE CHARACTER IN KYBUF
0136 F1      198       POP  PSW        ;RESTORE A AND FF'S
0137 E1      199       POP  H          ;RESTORE HL
0138 C9      200       RET
             201
             202 ;*****************************************************************
             203 ;DISPLY - READS CHARACTERS FROM A BUFFER IN MEMORY, CONVERTS THEM TO
             204 ;         - 7 SEGMENT CODE, AND SENDS THEM TO 8279 FOR DISPLAY
             205 ;INPUTS:       ACC - 00 = DISPLAY IN 4 DIGIT FIELD
             206 ;                  - 01 = DISPLAY IN 2 DIGIT FIELD
             207 ;              HL  - ADDRESS OF BUFFER START
             208 ;OUTPUTS:      NOTHING
             209 ;CALLS:        NOTHING
             210 ;DESTROYS:     A,H,L,FF'S
             211 ;*****************************************************************
             212
0139 D5      213 DISPLY:PUSH D         ;SAVE DE
013A C5      214       PUSH B          ;SAVE BC
013B 0E04    215       MVI  C,04H      ;SET CHARACTER COUNTER AT 4
013D 0F      216       RRC             ;CHECK FOR DISPLAY IN 2 DIGIT FIELD
013E DA4601  217       JC   DISP1      ;YES
0141 3E90    218       MVI  A,90H      ;8279 CONTROL WORD FOR 4 DIGIT FIELD
0143 C34801  219       JMP  DISP2      ;
0146 3E94    220 DISP1: MVI A,94H      ;8279 CONTROL WORD FOR 2 DIGIT FIELD
0148 320019  221 DISP2: STA 1900H      ;SEND CONTROL WORD TO 8279
014B 7E      222 DISP3: MOV A,M        ;GET CHARACTER FROM BUFFER
014C EB      223       XCHG            ;SAVE POINTER IN DE
014D 212802  224       LXI  H,DSPTB    ;LOAD 7 SEGMENT TABLE TOP ADDRESS
0150 85      225       ADD  L          ;ADD BCD CHAR TO HL TO
0151 6F      226       MOV  L,A        ;   POINT TO 7 SEG IN TABLE
0152 3E00    227       MVI  A,00       ;CLEAR ACC BUT KEEP CARRY
0154 8C      228       ADC  H          ;PROPAGATE CARRY TO H
0155 67      229       MOV  H,A        ;RESTORE H
0156 6E      230       MOV  L,M        ;MOVE 7 SEG CODE TO L
0157 0D      231       DCR  C          ;DECREMENT CHARACTER COUNT
0158 79      232       MOV  A,C        ;
0159 FE02    233       CPI  02H        ;CHECK IF SECOND CHARACTER FROM LEFT
015B 7D      234       MOV  A,L        ;MOVE 7 SEG TO ACC
015C C26101  235       JNZ  DISP4      ;NOT SECOND CHAR THEN DISPLAY
015F F608    236       ORI  08H        ;MASK DEC POINT AFTER 2ND CHARACTER
0161 2F      237 DISP4: CMA            ;COMPLEMENT CODE
0162 320018  238       STA  1800H      ;SEND CHARACTER TO 8279
0165 EB      239       XCHG            ;BRING BACK BUFFER POINTER
0166 23      240       INX  H          ;FROM DE TO HL AND INCREMENT
0167 79      241       MOV  A,C        ;CHECK IF CHARACTER COUNTER = 0
0168 B7      242       ORA  A          ;SET FLAGS WITHOUT CHANGING
0169 C24B01  243       JNZ  DISP3      ;NOT ZERO GET NEXT CHARACTER
```

FIGURE 12-6 (continued)

LOC OBJ   SEQ   SOURCE STATEMENT

```
016C C1      244            POP    B
016D D1      245            POP    D
016E C9      246            RET
             247
             248  ;******************************************************************
             249  ;INSRT - SHIFTS CONTENTS OF DE LEFT 4 BITS AND INSERTS NEW BCD
             250  ;          CHARACTER IN LOWER 4 BITS OF E
             251  ;INPUTS:         A BCD DIGIT TO BE INSERTED
             252  ;                DE - 4 PACKED BCD DIGITS
             253  ;OUTPUTS:        DE - 4 PACKED BCD DIGITS WITH A IN E LOWER NIBBLE
             254  ;CALLS:          NOTHING
             255  ;DESTROYS:       A, FF'S
             256  ;******************************************************************
             257
016F EB      258  INSRT:  XCHG           ;MOVE DE TO HL
0170 29      259          DAD    H       ;SHIFT H AND L LEFT 4 BITS
0171 29      260          DAD    H
0172 29      261          DAD    H
0173 29      262          DAD    H
0174 85      263          ADD    L       ;INSERT NEW BCD DIGIT
0175 6F      264          MOV    L,A     ;MOVE TO L
0176 EB      265          XCHG           ;MOVE FROM HL BACK TO DE
0177 C9      266          RET
             267
             268  ;******************************************************************
             269  ;EXPND-EXPANDS 4 PACKED BCD CHARACTERS IN DE TO 1 CHARACTER PER BYTE
             270  ;          IN A BUFFER IN MEMORY
             271  ;INPUTS:         DE - 4 PACKED BCD DIGITS
             272  ;OUTPUTS:        NONE
             273  ;CALLS:          NOTHING
             274  ;DESTROYS:       A, HL, FF'S
             275  ;******************************************************************
             276
0178 7A      277  EXPND:  MOV    A,D     ;GET FIRST 2 BCD DIGITS
0179 0F      278          RRC            ;SHIFT MS DIGIT TO LOWER NIBBLE
017A 0F      279          RRC
017B 0F      280          RRC
017C 0F      281          RRC
017D E60F    282          ANI    0FH     ;MASK UPPER FOUR BITS
017F 21A620  283          LXI    H,BUFFR3
0182 77      284          MOV    M,A     ;STORE IN BUFFER
0183 7A      285          MOV    A,D     ;GET FIRST 2 BCD DIGITS AGAIN
0184 E60F    286          ANI    0FH     ;MASK UPPER FOUR BITS
0186 23      287          INX    H       ;POINT TO NEXT LOCATION IN BUFFR3
0187 77      288          MOV    M,A     ;STORE SECOND MS DIGIT
0188 7B      289          MOV    A,E     ;GET SECOND 2 BCD DIGITS
0189 0F      290          RRC            ;SHIFT NEXT BCD DIGIT TO LOWER NIBBLE
018A 0F      291          RRC
018B 0F      292          RRC
```

FIGURE 12-6 *(continued)*

```
LOC   OBJ      SEQ        SOURCE STATEMENT

018C  0F       293        RRC
018D  E60F     294        ANI   0FH        ;MASK UPPER 4 BITS
018F  23       295        INX   H
0190  77       296        MOV   M,A        ;STORE 3RD BCD DIGIT
0191  7B       297        MOV   A,E        ;GET LOWER BYTE AGAIN
0192  E60F     298        ANI   0FH        ;MASK UPPER 4 BITS
0194  23       299        INX   H          ;NEXT BUFFR3 POSITION
0195  77       300        MOV   M,A        ;STORE 4TH BCD DIGIT
0196  C9       301        RET
               302
               303  ;******************************************************************
               304  ;RDKY-CHECKS KYBUF LOADED BY KYBINT, THE KEYBOARD INTERRUPT
               305  ;     ROUTINE, TO SEE IF A CHARACTER HAS BEEN ENTERED.  IF NOT,
               306  ;     RDKY LOOPS UNTIL A CHARACTER IS ENTERED.  AN ENTERED
               307  ;     CHARACTER IS RETURNED IN A
               308  ;INPUTS:     NONE
               309  ;OUTPUTS:    A - CHARACTER READ FROM KEYBOARD
               310  ;CALLS:      NOTHING
               311  ;DESTROYS:   A, H, L, FF'S
               312  ;******************************************************************
               313
0197  21AE20   314  RDKY:  LXI   H,KYBUF    ;POINT TO KYBUF ADDRESS
019A  7E       315         MOV   A,M        ;READ KYBUF
019B  B7       316         ORA   A          ;SET FLAGS, BIT 7 = SIGN IF 0,
019C  F2A301   317         JP    EXIT2      ;CHARACTER PRESENT
019F  FB       318         EI               ;NO CHARACTER - ENABLE INTERRUPTS
01A0  C39701   319         JMP   RDKY       ; AND LOOP TO RDKY
01A3  3680     320  EXIT2: MVI   M,80H      ;SET MSB OF KYBUF TO 1 TO INDICATE
               321                          ;    BUFFER EMPTY
01A5  FB       322         EI               ;ENABLE INTERRUPTS
01A6  C9       323         RET
               324
               325  ;******************************************************************
               326  ;BCDCVT-CONVERTS 4 PACKED BCD DIGITS IN D AND E TO A 16 BIT BINARY
               327  ;          RESULT IN HL
               328  ;INPUTS:     DE - 4 PACKED BCD DIGITS
               329  ;OUTPUTS:    HL - 16 BIT BINARY EQUIVALENT
               330  ;CALLS:      MSBCD, LSBCD
               331  ;DESTROYS:   A, B, C, D, E, H, L, FF'S
               332  ;******************************************************************
               333
01A7  210000   334  BCDCVT:LXI   H,0000H    ;CLEAR HL
01AA  01E803   335         LXI   B,03E8H    ;LOAD BC WITH 03E8H = 1000 DECIMAL
01AD  7A       336         MOV   A,D        ;LOAD BCD THOUSANDS AND HUNDREDS
01AE  CDC701   337         CALL  MSBCD      ;CONVERT MOST SIG BCD DIGIT
01B1  016400   338         LXI   B,0064H    ;LOAD BC WITH 64H = 100 DECIMAL
01B4  CDBF01   339         CALL  LSBCD      ;CONVERT LEAST SIG BCD DIGIT OF D
01B7  7B       340         MOV   A,E        ;LOAD BCD DIGITS FROM E
01B8  0E0A     341         MVI   C,0AH      ;LOAD C WITH 0AH = 10 DECIMAL
```

FIGURE 12-6 (continued)

356

CHAPTER TWELVE

```
LOC  OBJ      SEQ       SOURCE STATEMENT

01BA CDC701  342          CALL MSBCD      ;CONVERT UPPER BCD DIGIT IN E
01BD 0E01    343          MVI  C,01H      ;LOAD C WITH 01H = 01 DECIMAL
01BF FE00    344 LSBCD:   CPI  00H        ;CHECK IF A DEC TO ZERO
01C1 C8      345          RZ              ;RETURN IF ZERO
01C2 09      346          DAD  B          ;ADD BC TO HL
01C3 3D      347          DCR  A          ;DEC BCD IN A
01C4 C3BF01  348          JMP  LSBCD
01C7 FE0A    349 MSBCD:   CPI  0AH        ;CHECK IF MS BCD DIGIT COUNTED DOWN TO ZERO
01C9 D8      350          RC              ;CARRY SET IF A LESS THAN 10 DECIMAL
01CA 09      351          DAD  B          ;ADD BC TO HL
01CB D610    352          SUI  10H        ;SUBTRACT ONE FROM MS BCD
01CD C3C701  353          JMP  MSBCD
             354
             355 ;************************************************************
             356 ;MULTO-MULTIPLIES 16 BIT BINARY NUMBER IN DE X 16 BIT BINARY
             357 ;      NUMBER IN BC TO GIVE A 32 BIT RESULT IN DE AND HL.
             358 ;INPUTS:       BC - 16 BIT BINARY NUMBER
             359 ;              DE - 16 BIT BINARY NUMBER
             360 ;OUTPUTS:      DEHL - 32 BIT BINARY NUMBER
             361 ;CALLS:        NOTHING
             362 ;DESTROYS:     A, D, E, H, L, FF'S
             363 ;************************************************************
             364
01D0 210000  365 MULTO:   LXI  H,0000H    ;CLEAR HL
01D3 3E10    366          MVI  A,10H      ;INITIALIZE BIT COUNTER TO 16 DECIMAL
01D5 29      367 LOOP2:   DAD  H          ;SHIFT MSB OF PARTIAL PRODUCT IN HL
             368                          ;   LEFT INTO CARRY
01D6 EB      369          XCHG            ;EXCHANGE DE AND HL
01D7 DADE01  370          JC   MULT1      ;JUMP IF SHIFT OF HL PRODUCED CARRY
01DA 29      371          DAD  H          ;SHIFT DE NOW IN HL LEFT
01DB C3E001  372          JMP  MULT2      ;TO CHECK NEXT BIT OF MULTIPLIER
01DE 29      373 MULT1:   DAD  H          ;SHIFT DE NOW IN HL LEFT TO CHECK NEXT BIT
01DF 23      374          INX  H          ;ADD CARRY OUT OF HL ABOVE TO DE NOW IN HL
01E0 EB      375 MULT2:   XCHG            ;MOVE PARTIAL PRODUCT LOW BACK TO HL
01E1 D2E901  376          JNC  MULT3      ;NO CARRY - BIT OF MULTIPLIER = 0
             377                          ;LOOP TO TEST NEXT BIT
01E4 09      378          DAD  B          ;MULTIPLIER BIT = 1, ADD MULTIPLICAND
01E5 D2E901  379          JNC  MULT3      ;NO CARRY FROM ADD - JUMP TO MULT3.
01E8 13      380          INX  D          ;CARRY-ADD TO HIGH ORDER WORD OF RESULT IN DE
01E9 3D      381 MULT3:   DCR  A          ;DECREMENT BIT COUNTER
01EA C2D501  382          JNZ  LOOP2      ;LOOP UNTIL 16 BITS DONE
01ED C9      383          RET             ;RETURN
             384
             385 ;************************************************************
             386 ;BINCVT-CONVERTS A 24 BIT BINARY NUMBER IN E, H, L TO A PACKED
             387 ;      BCD EQUIVALENT IN H, L, B, C
             388 ;INPUTS:       E, H, L - BINARY NUMBER
             389 ;OUTPUTS:      HLBC - PACKED BCD RESULT
             390 ;CALLS:        CNVT1, CNVT2
```

FIGURE 12-6 *(continued)*

LOC   OBJ   SEQ   SOURCE STATEMENT

```
                  391 ;DESTROYS:       A, B, C, D, E, H, L, FF's
                  392 ;**********************************************************
                  393
01EE 1619         394 BINCVT:MVI  D,19H      ;SET BIT COUNTER TO 25 DECIMAL
01F0 CD0902       395        CALL CNVT1      ;PRODUCE 2 LS BCD DIGITS IN B
01F3 48           396        MOV  C,B        ;SAVE 2 BCD DIGITS IN C
01F4 1619         397        MVI  D,19H      ;SET BIT COUNTER TO 25 DECIMAL
01F6 CD0902       398        CALL CNVT1      ;PRODUCE NEXT 2 MORE SIG BCD DIGITS
01F9 C5           399        PUSH B          ;SAVE 4 LS BCD DIGITS ON STACK
01FA 1619         400        MVI  D,19H      ;SET BIT COUNTER TO 25 DECIMAL
01FC CD0902       401        CALL CNVT1      ;PRODUCE NEXT 2 BCD DIGITS IN B
01FF 48           402        MOV  C,B        ;SAVE 2 BCD DIGITS IN C
0200 1619         403        MVI  D,19H      ;SET BIT COUNTER TO 25 DECIMAL
0202 CD0902       404        CALL CNVT1      ;PRODUCE LAST 2 BCD DIGITS
0205 60           405        MOV  H,B        ;POSITION MS BCD FOR RETURN
0206 69           406        MOV  L,C        ;POSITION NEXT 2 BCD FOR RETURN
0207 C1           407        POP  B          ;RESTORE 4 BCD DIGITS FROM STACK
0208 C9           408        RET
0209 AF           409 CNVT1: XRA  A          ;CLEAR ACC AND CARRY
020A 47           410        MOV  B,A        ;SET B TO ZERO
020B AF           411 CNVT2: XRA  A          ;CLEAR ACC AND CARRY
020C 15           412        DCR  D          ;DECREMENT BIT COUNTER
020D C8           413        RZ              ;DONE IF ZERO
020E 29           414        DAD  H          ;SHIFT H LEFT 1 BIT - MSB TO CARRY
020F 7B           415        MOV  A,E        ;MOVE HIGH ORDER BYTE TO ACC
0210 17           416        RAL             ;ROTATE CARRY FROM H TO LSB
0211 5F           417        MOV  E,A        ;RESTORE E
0212 78           418        MOV  A,B        ;MOVE BCD DIGIT INTO A
0213 8F           419        ADC  A          ;DOUBLE A AND ADD CARRY FROM E
0214 27           420        DAA             ;MAINTAIN BCD FORMAT
0215 47           421        MOV  B,A        ;SAVE B
0216 D20B02       422        JNC  CNVT2      ;NO CARRY FROM DAA - CONTINUE
0219 3E00         423        MVI  A,00H      ;CLEAR ACC, KEEP CARRY
021B 8D           424        ADC  L          ;IF CARRY ADD TO L
021C 6F           425        MOV  L,A        ;RESTORE L
021D 3E00         426        MVI  A,00H      ;CLEAR ACC, KEEP CARRY
021F 8C           427        ADC  H          ;PROPAGATE CARRY TO H
0220 67           428        MOV  H,A        ;RESTORE H
0221 3E00         429        MVI  A,00H
0223 8B           430        ADC  E          ;PROPAGATE CARRY TO E
0224 5F           431        MOV  E,A        ;RESTORE E
0225 C30B02       432        JMP  CNVT2      ;CONTINUE
                  433
                  434 ;**********************************************************
                  435 ;DSPTB - TABLE FOR TRANSLATING BCD CHARACTERS TO 7 SEGMENT CODE
                  436 ;            FOR OUTPUT TO DISPLAY
                  437 ;**********************************************************
                  438
```

FIGURE 12-6 *(continued)*

LOC	OBJ	SEQ	SOURCE STATEMENT			
0228	F3	439	DSPTB:	DB	0F3H	;"0"
0229	60	440		DB	060H	;"1"
022A	B5	441		DB	0B5H	;"2"
022B	F4	442		DB	0F4H	;"3"
022C	66	443		DB	066H	;"4"
022D	D6	444		DB	0D6H	;"5"
022E	D7	445		DB	0D7H	;"6"
022F	70	446		DB	070H	;"7"
0230	F7	447		DB	0F7H	;"8"
0231	76	448		DB	076H	;"9"
0232	77	449		DB	077H	;"A"
0233	C7	450		DB	0C7H	;"B"
0234	93	451		DB	093H	;"C"
0235	E5	452		DB	0E5H	;"D"
0236	97	453		DB	097H	;"E"
0237	17	454		DB	017H	;"F"
0238	67	455		DB	067H	;"H"
0239	83	456		DB	083H	;"L"
023A	37	457		DB	037H	;"P"
023B	60	458		DB	060H	;"I"
023C	05	459		DB	005H	;"R"
023D	00	460		DB	000H	;BLANK
		461				
		462		END		

PUBLIC SYMBOLS

EXTERNAL SYMBOLS

USER SYMBOLS

BCDCVT	A	01A7	BINCVT	A	01EE	BLANK	A	002F	BUFFR1	A 20A0
CNVT1	A	0209	CNVT2	A	020B	DISP1	A	0146	DISP2	A 0148
DS1	A	0014	DS2	A	0040	DS3	A	004D	DS4	A 005A
EXPND	A	0178	GETWT	A	0011	INSRT	A	016F	KYBINT	A 0128
LSBCD	A	01BF	MSBCD	A	01C7	MULT1	A	01DE	MULT2	A 01E0
SAVBP	A	20AC	SHIFT	A	00D0	SHOWSP	A	00A6	SHRTLP	A 011C
BUFFR2	A	20A4	START	A	0000	LNGLP	A	0119	DISPLY	A 0139
DISP3	A	014B	BUFFR3	A	20A6	MULTO	A	01D0	EXIT2	A 01A3
DSPTB	A	0228	DISP4	A	0161	STORE	A	0031	LOOP2	A 01D5
KYBUF	A	20AE	EXIT1	A	008E	BUFFR4	A	20AA	RDKY	A 0197
MULT3	A	01E9								

ASSEMBLY COMPLETE.

FIGURE 12-6 (continued)

This is done by writing programs as a series of independent modules or subroutines linked together. As is shown in Chap. 13, the advantage of modular programs is that each module can be individually tested and debugged before linking it with others. Also modules or subroutines can be stored in a file on a floppy disk and pulled out as needed to write a new program. Symbolic addresses are used in these modules, so they can be assembled to run at any location in memory.

If a program has been written in a modular style, as

was the scale program here, you can work your way through it one module at a time. The flowchart should help you identify the modules.

EQUATES   The program listing starts with a table of equates. This table tells the assembler program what absolute address to use for each symbolic address in the program. For example, the statement BUFFR1 EQU 20A0H tells the assembler program to enter the address 20A0H each time it reads the symbolic address BUFFR1 in the program.

DUMB-SCALE MODULE   After initializing, the processor reads port 00. As shown in Figure 12-3, port 00 is connected to the A/D converter, BCD output lines, and digit-strobe lines. The DS1 loop reads port 00 over and over until a 1 is found on the digit-strobe line for the most significant digit. When the strobe is found, data for that digit is read in from the A/D converter. Since the A/D converter is 3½-digit unit, the most significant digit can be only 1 or 0. This 1 or 0 is output in bit 3 of the 4-bit data word for the most significant digit. If a 1 is read, it is displayed, but a 0 is blanked.

Note that the program jumps over the RST 5.5 and RST 7.5 addresses.

After a blank or one character has been loaded in the buffer for the most significant digit of the weight, the program enters a loop to wait for the next digit strobe. When the digit 2 strobe is found, data for that digit is read in and stored in the next address in the buffer. Data for this and the remaining digits is in standard BCD format in the lower 4 bits of the byte.

In the same manner, data for digits 3 and 4 is read in when a strobe is found. The data read in is stored in the next two addresses in the buffer. When all four digits have been read into the buffer, a display subroutine is called to convert the BCD characters to seven-segment code and send the characters to the 8279 for display.

Next characters for "L" and "b" are loaded into another buffer in memory. The display routine is called again to read these characters from the buffer, convert them to seven-segment code, and send the seven-segment code to the 8279 display. If the display routine is called with 00H in the accumulator, then characters in the buffer pointed to by HL are displayed in the four LED displays. If the accumulator contains 01H when display is called, the characters are displayed in the two LED displays. The display subroutine is discussed in detail later. The first time you read through a program, it is best to follow the main flow of the program and just take a quick look at the description, inputs, and outputs for subroutines called. Then, when you have an overall view of the program operation, you can study the details of the subroutines.

The final part of the dumb-scale program module involves checking whether a key has been pressed on the keyboard. The 8279 continuously scans the keyboard. When a key is pressed, the 8279 waits a debounce time and checks whether the key is still pressed. If it is, the character is loaded into the internal FIFO and an interrupt request signal is sent out. For this scale the interrupt request output is connected to the RESTART 5.5 input of the 8085A. If the RST 5.5 interrupt is not masked and interrupts are enabled, a signal on this input stores the return address on the stack and jumps the program to the RST 5.5 address of 2CH. A jump instruction stored at this address takes the program execution to a routine called KYBINT, which services the 5.5 interrupt. It reads the character from the 8279 FIFO and stores it in a buffer called KYBUF. The character can be read from KYBUF when needed. On reset or when the 8085A receives an interrupt, it disables all interrupts except TRAP. Therefore, interrupts must be unmasked and enabled before the processor can respond to the keyboard.

Once the weight is taken in and displayed, the program unmasks and enables interrupts and checks KYBUF to see if KYBINT has stored a character there. A 1 in the MSB of the byte read from KYBUF indicates no character from the keyboard was present. If no character is found, the program loops back to the start of dumb scale. Weight is taken in again and displayed. Then KYBUF is checked again for a character from the keyboard. This loop through dumb scale is continued until a key is pressed. When a character from the keyboard is found in KYBUF, the program exits the dumb-scale module.

GET PRICE/LB MODULE   The first digit of the price per pound is actually read in during the last loop through dumb scale before exit. This is stored in register D, and the units display is changed to SP., to indicate that selling price is being entered.

The first digit is then moved back to the accumulator, registers D and E are cleared, and another loop called SHOWSP is entered. The functions of this loop are to (1) create the price per pound as four packed BCD digits in the D and E registers, (2) expand the four BCD digits to 1 byte each in a buffer so they can be displayed, and (3) enter and display each new key pressed until an end key is pressed. Characters are entered in the display from the right, as is done in many calculators. When a new character is entered, the previously entered characters are moved left one character to make room.

The subroutine INSRT shifts the contents of the D and E registers 4 bits left to make room for a new BCD character in the lower 4 bits of the E register. EXPND, as the name implies, expands the four packed BCD digits in D and E to one BCD character per byte in BUFFR3 in memory. This is the format required for the display routine DSPLY.

After the character read during the end of the dumb-scale module is inserted, expanded, and displayed, a keyboard read routine RDKY is called to read the next character from KYBUF. RDKY waits until a key is pressed and then brings the character for the key to the accumulator. To determine whether the key pressed is a valid BCD digit or an end-of-price-per-pound indicator, it is compared with 0AH. The carry flag is set if the accumulator is less than 0AH (valid BCD). The JC instruction then loops the program back to SHOWSP to INSRT, EXPND, and DISPLY the new digit. Any character other

MEMORY ADDRESS	CONTENTS	LABEL
20A0H	DIGIT 1 WEIGHT	
20A1H	DIGIT 2 WEIGHT	BUFFR 1
20A2H	DIGIT 3 WEIGHT	
20A3H	DIGIT 4 WEIGHT	
20A4H	"L"	BUFFR 2
20A5H	"B"	
20A6H	DIGIT 1 PRICE/LB.	
20A7H	DIGIT 2 PRICE/LB.	BUFFR 3
20A8H	DIGIT 3 PRICE/LB.	LATER USED FOR TOTAL PRICE
20A9H	DIGIT 4 PRICE/LB.	
20AAH	"S"	BUFFR 4
20ABH	"P"	LATER USED FOR PR.

REGISTERS	CONTENTS	
A	0FH	TERMINATOR KEY OF PRICE/LB.
BC	XXXX	NOT EASILY PREDICTABLE
DE	PRICE/LB.	AS 4 PACKED BCD DIGITS
HL	KYBUF	

FIGURE 12-7 "Mini" RAM and register map.

than a valid BCD digit causes the processor to exit the SHOWSP loop and proceed to the next module.

Before we go on to the next module, it is a good idea to draw a "mini" RAM and register map to help keep track of where everything is at this point. Figure 12-7 shows such a map. Note that the weight is stored in four memory locations known as BUFFR1, the price per pound is in four memory locations known as BUFFR3, and the price per pound is *also* in registers D and E in packed form.

CONVERT PRICE PER POUND TO BINARY

Both the weight and price per pound must be converted to binary before the two are multiplied to get the total price. Since the price per pound is already in DE for the BCDCVT routine, it is converted first. The BCDCVT subroutine does the conversion and returns a 16-bit binary result in HL. This result is moved to a buffer in memory labeled SAVBP.

CONVERT WEIGHT TO BINARY MODULE

Before BCDCVT can be used to convert the weight to binary, the unpacked BCD digits in BUFFR1 must be packed and moved to D and E. This is done with a series of simple move-and-rotate-left operations. The task is simplified by the fact that the upper nibble of each location in BUFFR1 was set to all 0's as the BCD digits were stored there.

When the packing is done and the result is stored in D and E, BCDCVT is called to convert the weight value to a 16-bit binary equivalent in H and L. The binary weight is moved to DE by the XCHG instruction.

MULTIPLY BINARY WEIGHT TIMES BINARY PRICE PER POUND

The binary price per pound is returned from the SAVBP buffer to registers B and C. Then subroutine MULTO is called to do the actual multiplication. MULTO multiplies the 16-bit binary number in B and C by the 16-bit binary number in D and E to give a 32-bit binary result in DEHL. The routine does integer multiplication. In other words, all values are treated as positive integers. A weight of 10.00 lb is represented as 1000 decimal, or $3E8_{16}$, and a price per pound of $5.56 is represented as 556 decimal, or $22C_{16}$. As mentioned in Chap. 7 the programmer has to determine the correct place for the decimal point in the final result.

Since the maximum weight is 10.00 lb and the maximum total price displayable in four digits is $99.99, the maximum price per pound for a 10.00-lb weight is $9.99. Treating the weight and price per pound as integers gives a maximum product of 999,000 decimal. This number requires only about 20 binary bits to express it, so that in this case the upper byte of the binary product in the D register is all 0's. The useful result of the binary multiplication of weight and price per pound is the 24 bits in registers E, H, and L.

CONVERT TOTAL PRICE TO BCD

Next the binary total price must be converted to BCD so that it can be displayed. The BINCVT subroutine converts the 24-bit binary number in E, H, and L to eight packed BCD digits in registers H, L, B, and C. The maximum price of $99.99 comes back in these registers as 00999900. The decimal point obviously goes between the L register contents and the B register contents. In this case, the two least significant BCD characters are 0 and can just be dropped. However, this is not always the case, and so these digits must be rounded to the next higher BCD digit. Any carries produced must be propagated to the higher digits.

ROUND TOTAL PRICE TO NEAREST CENT AND POSITION FOR DISPLAY

The rounding is performed by comparing the lowest two BCD characters in C to 49_{16} in the accumulator. Carry is set if the contents of C are greater than 49_{16}. Any carry produced is added to the next digit in B. To keep BCD format in the result, a DAA is performed. The result is moved to the E register to position it for the expand routine. Any carry produced by the DAA operation on the B contents is added to L. To keep BCD format in the result a DAA operation is performed. The result is then moved to D for the expand routine.

EXPAND, DISPLAY, AND DELAY

EXPND converts the packed BCD in D and E to four unpacked BCD characters in BUFFR3. DISPLY is then called to send these four digits of the total price to the four LED displays.

BUFFR4 is loaded with the characters for "Pr." Display is called again to send these characters to the two-digit LED displays.

A nested delay loop is used to get the 3-s delay desired before jumping back to dumb scale. The number loaded into DE determines the number of loops through LNGLP, and the number loaded into B determines the number of times through SHRTLP. Loading 0000H into DE gives 65,536 times through LNGLP because DCX decrements DE to FFFFH before a 0 check is made. Each time through LNGLP the B register is loaded with 08_{16} and counted down to 0 with SHRTLP. When SHRTLP is exited, D and E are checked to see whether they have reached 0's. Note that a DCX instruction does not set any flags. Therefore, a common way to check a 16-bit register for 0 is to move 1 byte to the accumulator and OR the other byte with it. This will set the zero flag if both bytes are 0's.

This completes the first spiral through the assembly language program for the scale. The next spiral discusses in detail the four major subroutines: DISPLY, BCDCVT, MULTO, and BINCVT.

MAJOR SUBROUTINES

DISPLY

The main functions of the DISPLY subroutine are to convert the BCD or hex characters stored in a buffer to seven-segment code and to send the seven-segment code to the internal display buffer of the 8279.

After saving registers and setting the character counter, the program rotates the accumulator right to carry to check whether the four-digit or two-digit display is desired. For the four-digit field, a control word of 90_{16} is sent to the 8279 control port address of 1900H. If the two-digit field is desired, control word 94_{16} is sent to the control port address of 1900H. This control word sets a pointer to the internal display RAM of the 8279.

The next step is to convert the first BCD or hex characters in the main memory buffer to seven-segment code. The technique used for this conversion is important to note. In most cases this method is much faster and more efficient than the compare method used for the keyboard code in Chap. 11.

A table labeled DSPTB is set up in memory as shown at the end of the program. The desired output codes are put in order, as shown. The code desired for 00 is put at the top of the table, and the code for the highest character is put at the bottom.

To do a conversion, the pointer register HL is loaded with the address of the top of the table. The BCD or hex character is then simply added to the pointer. The result of the addition returned to HL points to the desired seven-segment code in DSPTB. MOV L,M then reads the seven-segment code from DSPTB to the L register for temporary storage.

Note that the seven-segment code shown on DSPTB is not the standard seven-segment code shown in Table 3-1. Figure 12-8 shows the seven-segment format required by the SDK-85. Values in DSPTB are the complements of the values required for each character using this format.

The weight display, price per pound, total price, and units all require a decimal point after the second digit from the left of the display field. This is inserted by ORing a 1 into the decimal point position of the seven-segment code.

After complementing, the character is sent to the 8279 write address, 1800H. The program loops back to DSP3 until four characters have been read from the buffer, converted, and sent to the 8279. For the units display only two characters actually need to be sent. However, no harm is done and the program is simplified by always sending four.

BCDCVT

This routine converts a four-digit BCD number in D and E to a 16-bit binary equivalent in HL. The method is quite simple. The binary equivalent of each decimal place is multiplied by the BCD digit in that place, and the result is added to the total in HL. Multipli-

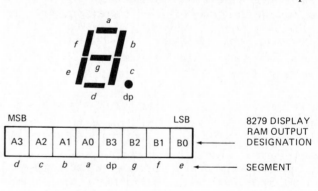

FIGURE 12-8 Segment format for 8279 output on SDK-85 board.

cation is done by successive addition. An example will help clarify this.

Suppose the BCD or decimal number 3421 is to be converted to binary. This number can be thought of as 3×1000 plus 4×100 plus 2×10 plus 1×1. Now, 1000_{10} is equal to 3E8 hex, 100_{10} is equal to 64 hex, 10_{10} is equal to 0A hex, and 01_{10} is equal to 01 hex. To get the binary equivalent of 3421_{10}, all that has to be done is to add 3E8 hex 3 times plus 64 hex 4 times plus 0A hex 2 times plus 01 hex once. This uses the successive-addition form for the multiplications.

The actual conversion of each BCD digit is done by one of two short subroutines. MSBCD operates on the BCD character if it is in the upper nibble, and LSBCD operates on the BCD character if it is in the lower nibble.

If you have trouble following routines such as these, take a specific BCD number such as 3421 and single-step your way through the routine with it. Steps for the most significant two digits follow.

HL is cleared and the binary equivalent of 1000_{10} is loaded into BC. The most significant two BCD digits are moved to A, and MSBCD is called. MSBCD adds 03E8 hex in BC to HL over and over until the most significant BCD character, 3 in this case, is counted down to 0, and then returns. BC is loaded with 0064 hex (100 decimal), and LSBCD is called. LSBCD adds 0064 hex in BC to the total in HL over and over until the least significant nibble, in this case 4, is counted down to 0, and then returns. The next two BCD digits in E are converted in a similar manner by using 0A hex and 01 hex. The routine can easily be altered to convert more or fewer BCD digits.

MULTO   MULTO multiplies a 16-bit binary number in BC by a 16-bit binary number in DE to give a 32-bit binary result in DE and HL. As mentioned in Chap. 7, there are several ways of doing binary multiplication. The BCDCVT routine above used successive addition. Figure 12-9 shows a common 8-bit × 8-bit routine using the shift-right-and-add algorithm shown in Chap. 7. A 16-bit × 16-bit multiply is easier to do on the 8080A/8085A with a shift-left-and-add method similar to the way pencil multiplication is often done. The ex-

```
;MULPLY:    MULTIPLIES THE 8 BIT BINARY NUMBER IN D BY THE 8 BIT
;           BINARY NUMBER IN C TO GIVE A 16 BIT BINARY NUMBER IN
;           B AND C
;INPUTS:    D - 8 BIT BINARY MULTIPLICAND
;           C - 8 BIT BINARY MULTIPLIER
;OUTPUTS:   BC - 16 BIT PRODUCT
;DESTROYS:  A, B, C, E, E/FS


MULPLY:    MVI    B,0        ;SET MS BYTE OF RESULT TO 0
           MVI    E,9        ;SET BIT COUNTER TO 9
MULP1:     MOV    A,C        ;MOVE MULTIPLIER TO ACC
           RAR               ;ROTATE LSB OF MULTIPLIER TO CARRY
           MOV    C,A        ;RESTORE SHIFTED MULTIPLIER TO C
           DCR    E          ;DECREMENT BIT COUNTER
           RZ                ;RETURN IF ZERO
           MOV    A,B        ;MOST SIG BYTE OF RESULT TO ACC
           JNC    MULP2      ;NO CARRY FROM RAR JUMP TO SHIFT
           ADD    D          ;CARRY FROM RAR, ADD MULTIPLICAND TO
                             ;HIGH ORDER BYTE OF RESULT IN B
                             ;REGISTER
MULP2:     RAR               ;SHIFT HIGH ORDER BYTE RIGHT
           MOV    B,A        ;RESTORE HIGH ORDER BYTE OF RESULT
           JMP    MULP1      ;GO CHECK NEXT BIT.
                             ;CARRY FROM THIS RAR IS SHIFTED
                             ;INTO THE MSB OF C BY THE RAR
                             ;AFTER MOV A,C THUS THE LOW ORDER
                             ;BYTE OF THE PRODUCT IS SHIFTED
                             ;INTO THE LEFT END OF C AS THE
                             ;MULTIPLIER IS SHIFTED OUT THE
                             ;RIGHT END OF C.
```

FIGURE 12-9 Assembly language routine for 8-bit × 8-bit binary multiplication using a shift-right-and-add algorithm.

1101	13$_D$ MULTIPLICAND
× 1011	11$_D$ MULTIPLIER
0000	SET PARTIAL PRODUCT REGISTER TO ZERO
00000	SHIFT LEFT
1101	ADD MULTIPLICAND BECAUSE *MSB* OF
01101	MULTIPLIER = 1
011010	SHIFT LEFT
0000	ADD ZERO BECAUSE NEXT DIGIT OF MULTIPLIER
011010	IS ZERO
011010	SHIFT LEFT
1101	ADD MULTIPLICAND BECAUSE NEXT DIGIT OF
1000001	MULTIPLICAND IS ONE
10000010	SHIFT LEFT
1101	ADD MULTIPLICAND BECAUSE *LSB* OF
10001111	MULTIPLICAND IS ONE

8 F$_H$ = 143$_D$

FIGURE 12-10 Short example of the MULTO algorithm.

ample in Figure 12-10, shown with fewer bits for simplicity, will help show how MULTO works.

Refer to Figure 12-10. The partial product is set to all 0's and shifted left 1 bit position. The most significant bit of the multiplier is tested. If it is 1, the multiplicand is added to the partial product and the result is shifted left. If the least significant bit was a 0, the partial product is just shifted one place to the left. Each bit of the multiplier is examined. If 1, the multiplicand is added to the shifted partial product; if 0, nothing is added to the partial product. You can see in Figure 12-10 that although the method is not immediately obvious, it does work.

The MULTO subroutine looks more complex than it is because of the shortage of 16-bit registers. The multiplicand is entered in B and C, and the multiplier is entered in DE. The partial product is started in HL, and then as the multiplier is shifted out of DE, the partial product expands also into DE. A 32-bit result fills all DE and HL.

There are two important points to recall when you are working your way through MULTO. First, DAD H adds the contents of HL to HL. This multiplies the contents of HL by 2, which is the same as shifting each bit of HL 1 bit position to the left. The most significant bit of H shifts into carry. Second, XCHG, JUMP, and all MOV instructions do not affect any flags. Therefore, a conditional jump instruction after XCHG refers to flags set by previous instructions. Using these hints, the comments in MULTO, and the example in Figure 12-10, you should be able to work your way through the routine. Don't be discouraged if it takes a while. Working through a program or routine almost always takes a lot of time.

BINCVT BINCVT converts a 24-bit binary number in registers E, H, and L to eight packed BCD digits in registers H, L, B, and C. It probably has the least obvious algorithm of all the routines of this scale program, but the actual operations are quite simple.

In a binary number each bit position represents a power of 2. An 8-bit binary number, for example, is equal to

$$b7 \times 2^7 + b6 \times 2^6 + b5 \times 2^5 + b4 \times 2^4$$
$$+ b3 \times 2^3 + b2 \times 2^2 + b1 \times 2^1 + b0$$

This can be shuffled around and expressed as

$$\text{Binary number} = (((((2b7 + b6)2 + b5)2 + b4)2$$
$$+ b3)2 + b2)2 + b1)2 + b0$$

where $b7$ through $b1$ are the binary bits. If each operation in the nested parentheses is done in BCD with the help of the DAA instruction, then the result is in BCD. This is how BINCVT produces a BCD equivalent.

The section of BINCVT up to CNVT1 sets the bit counter to 25 decimal and calls CNVT1 to produce two BCD digits at a time in B. These two BCD digits are saved, the bit counter set to 25 decimal again, and CNVT1 called to produce the next two BCD digits. These and the last two BCD digits are pushed on the stack to save them because all registers are in use. The four most significant BCD digits are produced by two more load-bit-counter-and-call CNVT1 cycles.

These four BCD characters are moved into H and L, and the four least significant BCD digits are popped off the stack to B and C.

The heart of this conversion is the CNVT1 and CNVT2 routine. Figure 12-11 shows a flowchart for it. After carry and the A and B registers are cleared, the 3-byte binary number in E, H, and L is shifted left 1 bit. The most significant bit of E moves into carry. The shift is done as follows. DAD H shifts the contents of HL 1 bit left. The MSB of H shifts into carry. Register E is moved to the accumulator. RAL shifts the contents of carry into the LSB of A and shifts the contents of A left 1 bit. The MSB of A is shifted left into carry. After shifting, the shifted result is returned to E.

The two BCD characters that are being built are kept in the B register. They are moved to the accumulator to be operated on. Here ADC A doubles their value and adds the bit shifted out of E and held in carry. This follows the procedure in the nested-parentheses equation. A DAA operation is done to keep the byte in A in BCD format, and the resultant two BCD characters are restored in B.

If no carry is produced by the DAA, then the program loops back to CNVT2 and continues the conversion. If the DAA prduces a carry because the byte in A was greater than 99 decimal, this overflow carry must be added to register L to be used in producing the next two more significant BCD digits. The effect of adding this carry to L is propagated up through H and E. Then the program loops back to CNVT2 to continue the conversion. Each pass through CNVT1 produces the next two more significant BCD digits in register B. After each run of CNVT1 (24 runs of CNVT2), EHL is left with a binary number equal to the original binary number minus the value of the two BCD digits produced. The bit counter in D must be loaded initially with 25 decimal, or 19$_{16}$, be-

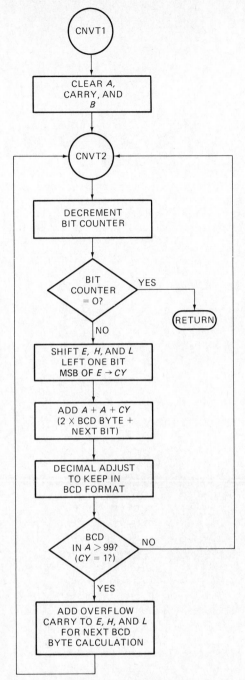

FIGURE 12-11 Flowchart for CNVT1 subroutine.

discussion, you may have thought of possible improvements and applications.

The entire digital portion of a simple scale such as this could be designed with a single-chip microcomputer such as the 8048 or 8748. However, an 8085A was used to give you more experience with 8080A/8085A assembly language.

More seven-segment displays could be added so that weight, price per pound, and total price could be displayed at the same time. The weight, price per pound, and total price could be sent to a printer to print labels to stick on packages of meat, etc. The total price could be sent to an electronic cash register and automatically entered.

A routine could be inserted in the program to subtract a tare or container weight. The scale could be used to count nuts, bolts, or money by simply entering a conversion factor of units per pound instead of the price per pound. Another application is as a postal scale. Adding lookup tables for each postal rate would allow the unit to compute postage required. Connecting the smart-scale output to a postage meter would allow automatic printout of the proper meter stamp.

The possibilities are nearly endless, and this is why microprocessors are having such an impact on our lives. Next in this chapter we discuss the hardware and software of an EPROM programmer based on the 6800 microprocessor.

AN MC6800-BASED EPROM PROGRAMMER

A very handy tool when you are working with microprocessors is an EPROM programmer. Experimental programs can be "burned" into an EPROM and stored or tried in a prototype instrument. If the program stored in the EPROM doesn't work in the instrument, it can be erased with an ultraviolet light. The EPROM can then be reprogrammed.

In this chapter we discuss a programmer for 2716 or equivalent MOS EPROMs. Moving one jumper wire allows the unit to also program 2732 EPROMs.

HARDWARE DESCRIPTION

With the exception of +25- and +5-V power supplies, the entire programmer is built on the two PC boards of the MEK6800D2. Figure 12-12 shows a photograph of the two PC boards complete with the programmer circuitry. The basic MEK6800D2 hardware is discussed first, and then the circuitry added for the programmer is explained.

BASIC MEK6800D2 BOARD   Figure 10-5 shows the schematic of the MEK6800D2 microcomputer board, and Figure 11-14 shows the schematic of the keyboard-display board. Refer to these as needed.

The basic CPU board comes with an MC6800 CPU, an

cause of the position of the DEC D instruction in the loop. If you want to really see the details of how this routine works, substitute some values in E, H, and L and work your way through the routine. A large piece of graph paper helps keep all the bits lined up.

SUMMARY

The preceding sections of this chapter have shown the hardware and software of a microprocessor-controlled, or smart, scale. The scale was designed to be simple enough to describe here. But as you read through the

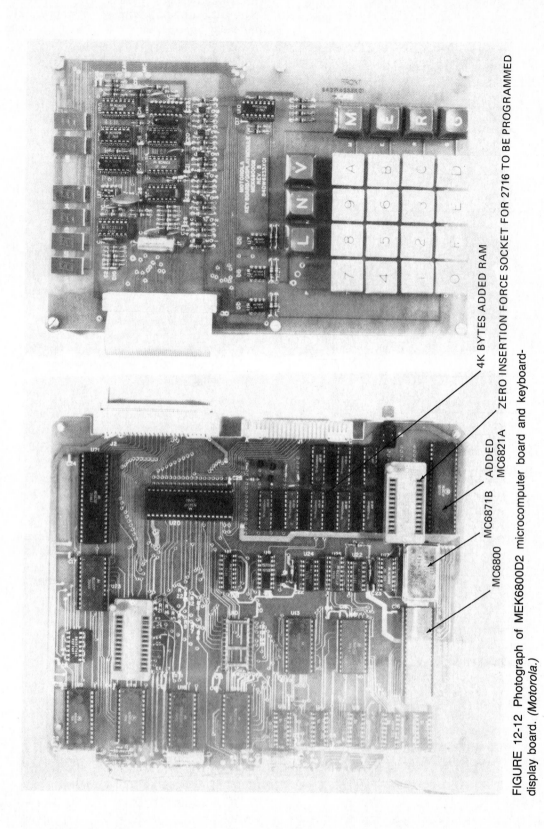

FIGURE 12-12 Photograph of MEK6800D2 microcomputer board and keyboard-display board. (*Motorola.*)

4K BYTES ADDED RAM

ZERO INSERTION FORCE SOCKET FOR 2716 TO BE PROGRAMMED

MC6821A

ADDED
MC6821A

MC6871B

MC6800

MC6871 clock generator, two MC6821 PIAs, an MC6850 ACIA, 256 bytes of RAM in two MC6810s, and 1 kilobyte of ROM in an MC6830. The ROM is programmed with the Motorola JBUG monitor program.

Socket spaces are present for three additional MC6830 RAMs, for two additional ROMs, and for address line and data line buffers. These buffers are needed if the board is interfaced to a large memory. A wire-wrap area is available for adding other ICs.

One of the MC6810 RAMs is used for a scratchpad by the JBUG monitor program, and one of the MC6821 PIAs is used to interface to the keyboard-display board. Therefore, the memory and I/O functions available to the user in the basic unit are 128 bytes of RAM at addresses 0000H to 007FH; two parallel I/O ports in an MC6821 at addresses 8004H and 8006H; a serial I/O port in the MC6850 ACIA at addresses 8008H and 8009H; and a 1-kilobyte monitor program at addresses E000H to E3FFH.

In addition to the keyboard and display hardware, the keyboard-display board contains circuitry to send data to and read data from magnetic tape in Kansas City standard format. This feature is handy for saving long programs.

HARDWARE ADDITIONS FOR THE 2716 EPROM PROGRAMMER

Hardware was added to the basic CPU board so data to be programmed into the 2716 can come from another 2716 or from RAM. (Programs on tape are first loaded into RAM and then used to program the EPROM.)

A 24-pin zero-insertion-force (ZIF) socket is soldered in IC position $U12$ for a 2716 to be copied from, while another is soldered in the wire-wrap area for the 2716 to be programmed. These ZIF sockets clamp onto the IC pins only when a small lever is pushed down. Therefore, ICs can be inserted and removed without danger of bending the pins. Figure 10-5 shows the connections to $U12$ for a 2716. As shown, the 2716 to be copied starts at address 6000H.

Eight MM5257N 4K × 1 static RAMs added in the wire-wrap area give 4 kilobytes of RAM storage. This allows the full program for a 2716 or 2732 to be loaded into RAM at once.

Figure 12-13 shows the schematic for the added RAM. The $\overline{2/3}$ line from the $U11$ decoder is used to enable this RAM block so that it occupies the 8-kilobit address space 2000H to 3FFFH. Since the memory actually uses 2000H to 2FFFH, 3000H to 3FFFH is foldback. In other words, either a 2 or a 3 in the leading digit will access the 4 kilobytes used.

Buffers were added in socket spaces $U1$ through $U5$ to drive the additional memory. A 7430 eight-input NAND gate in socket space $U7$ enables the data bus buffers whenever one of its inputs is low. Two additional MC6810 RAMs added to sockets $U14$ and $U16$ allow the program for the programmer to be kept on tape and loaded into this RAM to be run when needed.

Another MC6821 PIA ($U27$) added to the wire-wrap area is used to output and latch the addresses to the

2716 to be programmed. Figure 12-14 shows the connections of this PIA and the $U20$ PIA to the 2716 to be programmed. The added PIA occupies address space 8010H to 8013H.

Figure 12-14 also shows the switch circuitry to turn on and off the +25-V V_{PP} required by the 2716 during programming. Since the 25 V must be off except during programming or verifying of the 2716, the switch is designed to be in the off state when the processor is turned on. After a reset, all the outputs of the MC6821 are pulled high by an internal resistor. A high on the switch input turns it off. With the switch off, the diode to +5 V turns on and supplies the V_{PP} pin with the voltage required for a normal read operation. When the switch to +25 V is on, the diode is reverse-biased.

External power supplies are used for the +5 V, ±5 percent, and for the 26.5 V required for the V_{PP} switch. The specification for V_{PP} at the 2716 V_{PP} pin is +25 V, ±1 V, but because of the voltage drop across the transistor switch, a higher voltage must be used as the switch supply. A 0.1-μF capacitor on the V_{PP} pin helps prevent overshoot which might destroy the 2716 during switching.

SOFTWARE DESCRIPTION

JBUG MONITOR FUNCTIONS AND OPERATIONS

Since the JBUG monitor functions are a major part of the EPROM programmer software, these functions and their use are discussed first.

Note in the photograph of Figure 12-12 that in addition to the 16 keys for entering hexadecimal data, there are eight lettered keys. These keys correspond to monitor created functions as follows:

M—examine and change memory

E—escape from (abort) operation in progress

R—examine contents of MPU registers P, X, A, B, CC, S

G—go to specified program and begin execution of designated program

P—punch data from memory to magnetic tape

L—load memory from magnetic tape

N—trace one instruction

V—set (and remove) breakpoints

After a reset, the four-digit address field of the display shows a dash (—) in the leftmost digit as a *prompt* character. A prompt character tells you the microcomputer is sitting there waiting for you to enter a command. To use the M function or command, the desired four-digit hexadecimal address is entered with the keypad. This address is displayed in the address field. When the M key is pressed, the data field (two digits) of the display shows the data present at the displayed address. This data can be changed if desired by simply pressing the

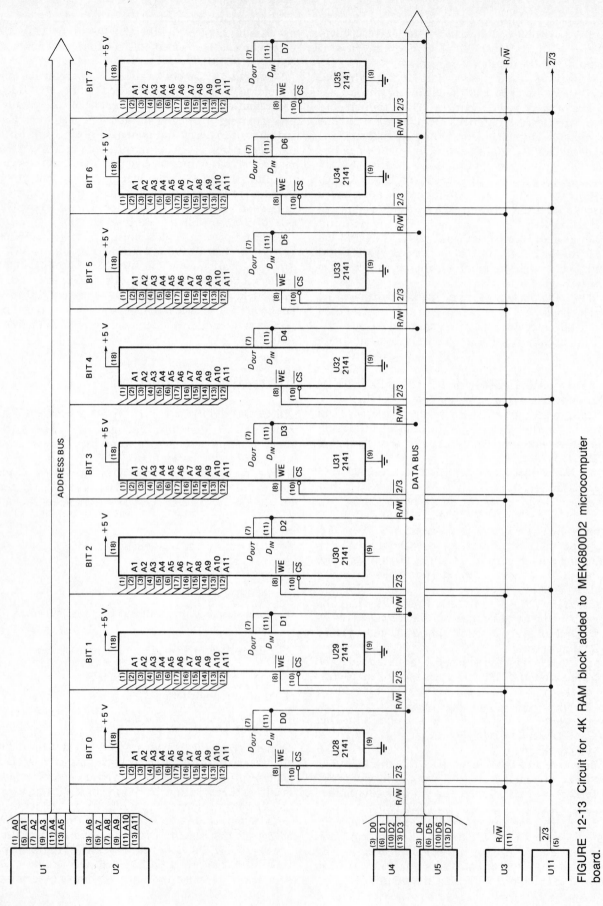

FIGURE 12-13 Circuit for 4K RAM block added to MEK6800D2 microcomputer board.

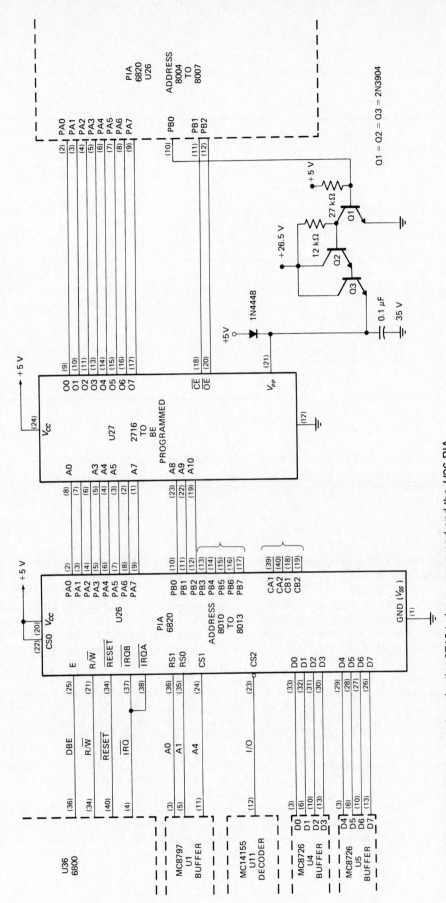

FIGURE 12-14 Circuit connections for the 2716 to be programmed and the *U26* PIA added to the MEK6800D2 microcomputer board.

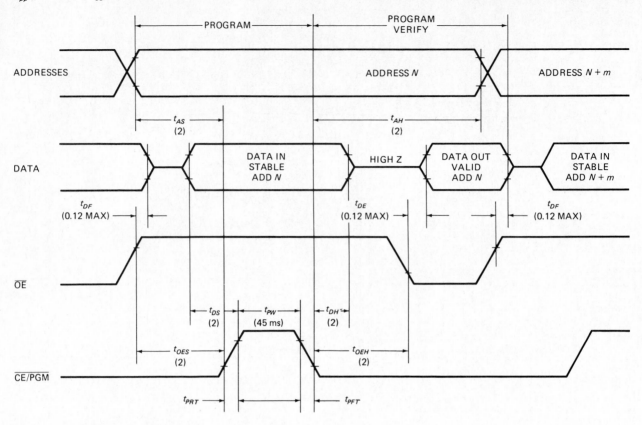

$V_{pp} = 25 \text{ V} \pm 1 \text{ V}, \ V_{CC} = 5 \text{ V} \pm 5\%$

NOTE: ALL TIMES SHOWN IN PARENTHESES ARE MINIMUM TIMES AND ARE μs UNLESS OTHERWISE NOTED.

FIGURE 12-15 *(a)* Programming waveforms of Intel 2716 EPROM.

two keys for the desired hex data. The data field display then shows the new data. Now there are two choices. Pressing the E (escape) key ends this command and returns the prompt character to indicate the processor is ready for a new command. Pressing the G key instead increments the address display by 1 and shows the contents of this address in the data display. The data at this address can be changed or left. Then pressing E gives back a prompt, or pressing G steps to the next address and displays its contents. The M command is used in this way to enter a program into RAM 1 byte at a time.

The next command of interest is the G (go) function used to execute programs. With the prompt character present, the starting address is entered on the keyboard. Pressing the G key starts execution of the program. Since the MEK6800D2 uses software display scanning, the display blanks while a program is running.

If a program gets stuck in a loop or in some other way fails, pressing the E key usually causes an escape from the program and gives back a prompt. The E key is connected almost directly to the $\overline{\text{NMI}}$ input of the processor. Therefore, pressing it vectors the processor to the $\overline{\text{NMI}}$ service routine, which gives control back to the monitor

program. Some program errors, however, write garbage in all available RAM, including an area used by the $\overline{\text{NMI}}$ routine. In this case, the only way to get back to the monitor control is to use the manual reset button on the CPU board!

The R, N, and V commands are explained in Chap. 13. The remaining commands of interest are P and L.

The P command is used to load a program onto magnetic tape from memory. With the use of the M command the starting address of the program to be sent is loaded into addresses A002H and A003H. The end address is loaded into addresses A004H and A005H, and E is pressed to get back a prompt. The tape output on the keyboard-display module is connected to the recorder microphone input. The recorder is started in the record mode, and then the P key is pressed. While the P command is executing, the display blanks. A prompt appears, to indicate that the program load to tape is done.

The L command is employed to read a program back from magnetic tape to RAM. First the earphone output from the tape recorder is connected to the tape input of the keyboard-display board. After the tape recorder is started in playback mode, the L key is pressed. A program read from the tape is loaded back into RAM at the

2716 and 2758 Program Characteristics[1]

$T_A = 25°C \pm 5°C$, V_{CC}[2] $= 5$ V $\pm 5\%$, V_{PP}[2,3] $= 25$ V ± 1 V

DC Programming Characteristics

Symbol	Parameter	Min	Typ	Max	Units	Test Conditions
I_{LI}	Input current (for any input)			10	μA	$V_{IN} = 5.25$ V$/0.45$
I_{PP1}	V_{PP} supply current			5	mA	$\overline{CE}/PGM = V_{IL}$
I_{PP2}	V_{PP} supply current during programming pulse			30	mA	$\overline{CE}/PGM = V_{IH}$
I_{CC}	V_{CC} supply current			100	mA	
V_{IL}	Input low level	−0.1		0.8	V	
V_{IH}	Input high level	2.0		$V_{CC} + 1$	V	

AC Programming Characteristics

Symbol	Parameter	Min	Typ	Max	Units	Test Conditions
t_{AS}	Address setup time	2			μs	
t_{OES}	$\overline{OE}$ setup time	2			μs	
t_{DS}	Data setup time	2			μs	
t_{AH}	Address hold time	2			μs	
t_{OEH}	$\overline{OE}$ hold time	2			μs	
t_{DH}	Data hold time	2			μs	
t_{DF}	Output enable to output float delay	0		120	ns	$\overline{CE}/PGM = V_{IL}$
t_{OE}	Output enable to output delay			120	ns	$\overline{CE}/PGM = V_{IL}$
t_{PW}	Program pulse width	45	50	55	ms	
t_{PRT}	Program pulse rise time	5			ns	
t_{PFT}	Program pulse fall time	5			ns	

1. Intel's standard product warranty applies only to devices programmed to specifications described herein.
2. V_{CC} must be applied simultaneously or before V_{PP} and removed simultaneously or after V_{PP}. The 2716/2758 must not be inserted into or removed from a board with V_{PP} at 25 ± 1 V to prevent damage to the device.
3. The maximum allowable voltage which may be applied to the V_{PP} pin during programming is ± 26 V. Care must be taken when switching the V_{PP} supply to prevent overshoot exceeding this 26 V maximum specification.

FIGURE 12-15 (continued) (b) Timing characteristics of Intel 2716 EPROM.

same address that it was recorded from. The starting address is written into the monitor scratchpad RAM at A002-3H and can be examined there if needed.

In summary, the EPROM programmer uses the M function to enter programs into RAM, the P function to store programs on tape, the L function to read them back from tape to RAM, and the G function to execute the programmer's program.

PROGRAMMING THE 2716 The 2716 MOS EPROM is discussed in Chap. 3. A complete data sheet for the device is shown in App. C. Initially and after being erased, all bits of a 2716 are in a 1 state; 0's are programmed into the desired locations. Once programmed to a 0, a bit location can be changed to a 1 only by erasing the entire memory.

Figure 12-15 shows the programming waveforms and characteristics of the Intel 2716. To program the device, first V_{PP} is raised to 25 V. The desired address is applied to the device, and $\overline{OE}$ is raised to a high state. The data byte to be programmed in is applied to the data *outputs* of the device, *O*0 through *O*7. At least 2 μs after the data is stable on these outputs, a 50-ms program pulse is applied to the $\overline{CE}/PGM$ input. Data must be held stable for at least 2 μs after the program pulse goes low. After a

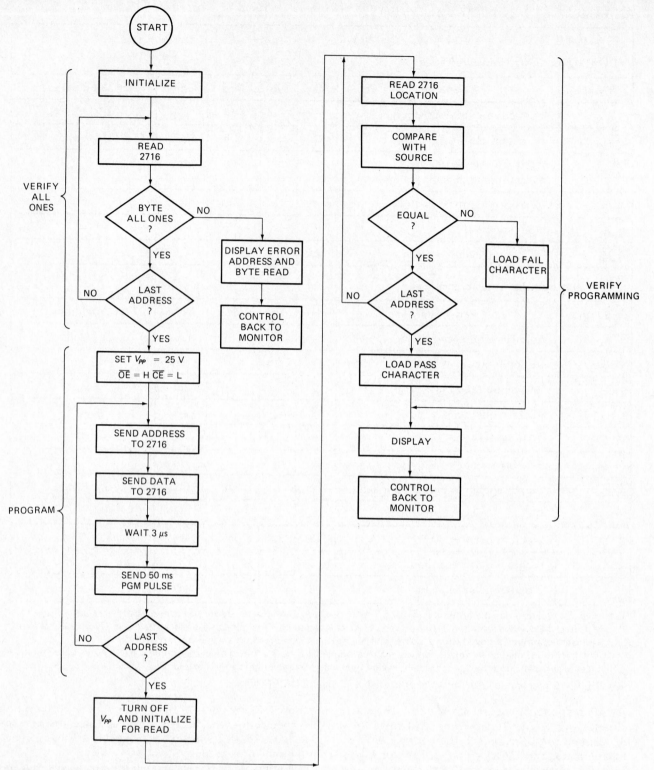

FIGURE 12-16 Flowchart for a 2716 EPROM programmer.

location is programmed, $\overline{OE}$ can be taken low and the location read to see if the programming was successful. If $\overline{OE}$ is left high, the data and address for the next location can be sent and then another 50-ms programming pulse applied to the $\overline{CE}$/PGM pin. Any number of or all the device locations can be programmed during a pro-

gramming session. They can be programmed in any order.

FLOWCHART FOR A 2716 PROGRAMMER Figure 12-16 shows the flowchart to program the 2716. The program consists of three main parts. First, the 2716 is

checked to see whether all locations to be programmed are 1's. In other words, were all the locations erased? Next, the desired locations are programmed. Finally, each location of the 2716 is read and compared with the data in the source locations. If any location doesn't verify, a fail sign of six Fs is loaded and displayed. If all locations verify, then six As are loaded and displayed.

ASSEMBLY LANGUAGE PROGRAM FOR A 2716 PROGRAMMER

Figure 12-17 shows the assembly language program to implement this flowchart on the modified MEK6800D2.

By using the M command of the JBUG monitor, the desired starting address in the 2716 to be programmed is entered in RAM addresses 0002H and 0003H. The ending address plus 1 is entered in RAM at addresses 0004H and 0005H. The starting address for the data source is entered in RAM addresses 0006H and 0007H, and the E key is pressed to get back to monitor. Since the start, end, and source addresses are entered by the user, program data can come from anywhere in the 65-kilobyte address space of the processor. Data can be programmed into any desired location of the 2716. Remember that the starting address for a 2716 to be copied is 6000H, and the starting address for the 4-kilobyte RAM storage is 2000H.

The starting address of the programmer program, 0010H, is then entered with the keyboard, and the G key is pressed. This causes the program to check for all 1's, program, and verify the 2716.

In the assembly language program in Figure 12-17, note the extensive use of the indexed addressing mode to initialize the PIAs. Since all the PIAs are mapped into memory at addresses near 8000H, this number is loaded into the index register as a base address. Then any of the PIA control or data ports can be accessed with a 2-byte indexed instruction such as STAA 10,X.

The code FFH is sent to three PIA data direction registers to initialize them as outputs. Then a control word is sent to all four PIA control registers to switch access from the data direction registers to the actual port registers.

After initializing the PIAs, the program saves the user-loaded 2716 start address in a buffer called WRKADD, or working address. The user-loaded source address is saved in a buffer called RDADD, or read address. The addresses in WRKADD and RDADD are incremented after each location is programmed. Setting these addresses aside, separate from the user-loaded addresses, preserves the user-loaded addresses for further use.

Next an address and a control signal to set $\overline{OE}$ low and $\overline{CE}$ low are sent to the 2716. The 2716 output word for this address is read and compared to FFH. If the word read is not equal to FFH, then the program branches to EXIT. EXIT loads the error address and the characters read there into a RAM buffer called DISBUF, or display buffer. A JMP to the OUTDS routine of the monitor program at address E0FEH displays the error address and character. OUTDS displays the address and data until a

key is pressed on the keyboard. In other words, jumping to OUTDS displays the address of the first error found and returns control to the monitor.

If the character read was equal to FFH, then the last read address is moved to the index register, incremented to point to the next location, and restored to WRKADD. The incremented address in X is compared with the user-loaded end address. If the desired end address has not been reached, the program loops back to read and test the next location. If the end address has been reached, then a branch to the program procedure at address 0070H is made.

Here indexed addressing is again used to reinitialize the PIAs. A control word to turn on V_{PP} to 25 V and to make $\overline{OE}$ high and $\overline{CE}$/PGM low is sent to the 2716 through the B port of U20. PA0 through PA7 of U20 are then changed to an output port to send data to the 2716.

After reinitialization, the user-loaded start address is again copied to WRKADD, and the user-loaded source address is copied to RDADD. Now PGMLUP does the actual programming. An address is sent to the 2716 through the A and B ports of U26. A data byte is read from the source address and sent to the outputs of the 2716, through U20 port A. After a 3.2-μs wait produced by the NOP, a control word making $\overline{CE}$/PGM high is sent to the 2716. DELAY loops for 50 ms, and then the control word to make $\overline{CE}$/PGM low is sent to the 2716.

The current programming address is brought from WRKADD, incremented, and restored to WRKADD. The incremented working address is compared with the user-loaded end address. If the end address has been reached, then the program branches to the verify section. If not, then the contents of RDADD (working source address) are brought to the index register, incremented, and restored to the RDADD buffer. After this, the program loops to program the next location. PGMLUP is repeated until all locations have been programmed and then exits to VERIFY.

VERIFY first returns V_{PP} to +5 V. $\overline{OE}$ is kept high so the outputs of the 2716 will be kept in a floating state. This is done so the outputs will not fight with the A port of U20, which was used as an output for programming. Next the program changes the A port of U20 to an input port. A control word is sent to make $\overline{OE}$ of the 2716 low. This enables the outputs so they can be read.

STRTAD is copied to the WRKADD buffer, and SRCSTR is copied to the RDADD buffer. The address in WRKADD is sent to the 2716. Data at this address is read and compared with the data pointed to by RDADD. If the byte read from the 2716 is not the same as the byte read from the source, then the program branches to FAIL. FAIL loads 0FH in all the locations of the display buffer, DISBUF, and jumps to OUTDS. As explained before, OUTDS displays the contents of DISBUF and returns control to the monitor. Therefore, a fail condition is indicated by a display of all Fs.

If the byte read from the 2716 is the same as the byte from the source address, then the last address read is brought from WRKADD to the index register and incre-

```
          ;********************************************************************
          ;
          ;                          2716 EPROM PROGRAMMER
          ;
          ;********************************************************************
          ;
          ;       THIS PROGRAM IS INTENDED TO RUN ON A MOTOROLA MEK6800D2
          ;MICROPROCESSOR DEVELOPMENT BOARD MODIFIED AS SHOWN IN
          ;FIG. 12-12.  THE 2716 TO BE PROGRAMMED IS PLACED IN THE
          ;ADDED SOCKET U27.  SOURCE OF PROGRAM DATA CAN BE FROM
          ;ANOTHER 2716 STARTING AT ADDRESS 6000H, A 4K BLOCK OF RAM
          ;STARTING AT 2000H, OR ANYWHERE ELSE IN THE ADDRESSING
          ;SPACE OF THE PROCESSOR.
          ;       THE MOTOROLA JBUG MONITOR PROGRAM
          ;IS USED AS PART OF THIS PROGRAM.  THE JBUG MONITOR IS
          ;IN ROM AT ADDRESSES E000 - E3FF.
          ;       THIS PROGRAM HAS THREE MAIN PARTS.  AFTER ENTERING
          ;THE START, END AND SOURCE ADDRESSES THE FIRST SECTION
          ;CHECKS TO SEE THAT THE DESIRED LOCATIONS ARE ALL
          ;ONES (FULLY ERASED).  THE NEXT SECTION DOES THE ACTUAL
          ;PROGRAMMING AND THE LAST SECTION CHECKS TO SEE
          ;IF THE PROGRAMMED EPROM AGREES WITH THE SOURCE.


                  ;EQUATES:
0002            STRTAD EQU  0002H              ;STARTING ADDRESS IN 2716
0004            ENDADD EQU  0004H              ;ENDING ADDRESS + 1 IN 2716
0006            SRCSTR EQU  0006H              ;STARTING ADDRESS OF SOURCE
0008            WRKADD EQU  0008H              ;TEMP STORAGE FOR 2716 LOC BEING PROG
000A            RDADD  EQU  000AH              ;TEMP STORAGE FOR SOURCE BEING READ
8000            PIABSE EQU  8000H              ;BASE ADDRESS OF PIAS
A00C            DISBUF EQU  0A00CH             ;START OF DISPLAY BUFFER
E0E0            DLY1   EQU  0E0E0H             ;DELAY IN MONITOR
E149            OUTDS3 EQU  0E149H             ;KEYBOARD START IN MONITOR
A020            XDSBUF EQU  0A020H
E3CA            DIGTBL EQU  0E3CAH
8022            SCNREG EQU  8022H
8020            DISREG EQU  8020H
A01C            SCNCNT EQU  0A01CH


0010                    ORG  0010H             ;ASSEMBLE TO RUN AT 0010


                  ;INITIALIZE PIAS

0010  CE8000            LDX   #PIABSE          ;#$8000
0013  86FF              LDA A #0FFH            ;CHARACTER FOR PORTS TO OUTPUTS
0015  A710              STA A 10H,X            ;SEND TO U26 PA0 - PA7
0017  A712              STA A 12H,X            ;SEND TO U26 PB0 - PB7
0019  A706              STA A 06H,X            ;SEND TO U20 PB0 - PB7
001B  8604              LDA A #04H             ;CONTROL WORD TO ACCESS 6820 PORTS
001D  A711              STA A 11H,X            ;SEND TO 4 PORTS
001F  A713              STA A 13H,X
0021  A705              STA A 05H,X
```

FIGURE 12-17 Assembly language program for 2716 EPROM programmer.

```
0023    A707              STA    A 07H,X

            ;CHECK LOCATIONS TO BE PROGRAMMED ARE ERASED

0025    DE02              LDX    STRTAD        ;COPY START ADD TO WRKADD
0027    DF08              STX    WRKADD
0029    DE06              LDX    SRCSTR        ;COPY SOURCE START ADD TO RDADD
002B    DF0A              STX    RDADD
002D    9608      CKNXT   LDA    A WRKADD      ;GET MS BYTE OF ADDRESS
002F    D609              LDA    B WRKADD+1
0031    CE8000            LDX    #PIABSE       ;#$8000
0034    E710              STA    B 10H,X       ;SEND LS BYTE OF ADDRESS TO 2716
0036    A712              STA    A 12H,X       ;SEND MS BYTE OF ADDRESS TO 2716
0038    8601              LDA    A #01H        ;CONTROL WORD VPP=5V, NOT OE=L, NOT CE=L
003A    A706              STA    A 06H,X       ;SEND TO 2716 THROUGH U20
003C    A604              LDA    A 04H,X       ;READ 2716
003E    81FF              CMP    A #0FFH       ;CHECK IF BYTE READ IS ALL ONES
0040    260B              BNE    EXIT          ;NO, EXIT WITH ERROR ADD
0042    DE08              LDX    WRKADD        ;GET WORKING PGM ADDRESS
0044    08                INX                  ;INCREMENT ADDRESS
0045    DF08              STX    WRKADD        ;RESTORE WORK ADDRESS
0047    9C04              CPX    ENDADD        ;CHECK IF LAST ADDRESS
0049    26E2              BNE    CKNXT         ;NO, CHECK NEXT ADDRESS
004B    2023              BRA    PGMSTR        ;YES, GO PROGRAM
004D    BDE29A    EXIT    JSR    MDIS2         ;2 HEX DIGITS IN A TO 2 BYTES IN A & B
0050    B7A010            STA    A 0A010H      ;SEND A TO DISBUF
0053    F7A011            STA    B 0A011H      ;SEND B TO DISBUF
0056    9608              LDA    A WRKADD      ;GET MS BYTE OF ERROR ADDRESS
0058    BDE29A            JSR    MDIS2         ;CONVERT TO ONE CHARACTER BYTE
005B    CEA00C            LDX    #0A00CH       ;DISBUF START
005E    A700              STA    A 00H,X       ;SEND MS CHAR OF ADDRESS TO DISBUF
0060    E701              STA    B 01H,X       ;SEND NEXT CHAR TO DISBUF
0062    9609              LDA    A WRKADD+1    ;GET LOWER BYTE OF ERROR ADD
0064    BDE29A            JSR    MDIS2         ;CONVERT TO ONE CHAR BYTE
0067    A702              STA    A 02H,X       ;SEND CHARACTER TO DISBUF
0069    E703              STA    B 03H,X
006B    7EE0FE            JMP    OUTDS         ;DISPLAY ERROR ADDRESS AND CHARACTER READ
006E    01                NOP
006F    01                NOP
0070    8604      PGMSTR  LDA    A #04H        ;VPP=25V, NOT OE=H, NOT CE=L
0072    CE8000            LDX    #8000H        ;LOAD BASE ADDRESS OF PIAS
0075    A706              STA    A 06H,X       ;SET VPP=25V, NOT CE=0, NOT OE=1
0077    4F                CLR    A             ;A=00 CONTROL WORD TO ACCESS DDR OF PIA
0078    A705              STA    A 05H,X       ;OPEN DATA DIRECTION REGISTER
007A    43                COM    A             ;A=FF DIRECTION WORD FOR OUTPUT
007B    A704              STA    A 04H,X
007C    8604              LDA    A #04H        ;CONTROL WORD FOR PORT ACCESS
007E    A705              STA    A 05H,X       ;SEND TO CONTROL PORT A OF U20
0081    DE02              LDX    STRTAD        ;COPY START ADDRESS
0083    DF08              STX    WRKADD        ;TO WORK ADDRESS BUFFER
0085    DE06              LDX    SRCSTR        ;COPY SOURCE START ADD
0087    DF0A              STX    RDADD         ;TO READ ADDRESS BUFFER
0089    CE8000    PGMLUP  LDX    #8000H        ;LOAD ADDRESS OF PIAS
008C    9608              LDA    A WRKADD      ;GET MSB OF ADDRESS FOR EPROM
008E    A712              STA    A 12H,X       ;SEND MSB OF ADDRESS TO EPROM
0090    9609              LDA    A WRKADD+1    ;GET LS BYTE OF WRKADD
0092    A710              STA    A 10H,X       ;SEND TO 2716
0094    DE0A              LDX    RDADD         ;LOAD CURRENT READ ADDRESS
```

FIGURE 12-17 *(continued)*

```
0096    A600                 LDA    A 00H,X        ;READ BYTE FROM SOURCE
0098    B78004               STA    A 8004H        ;SEND DATA TO 2716
009B    01                   NOP                   ;WAIT 3..2 MICRO SECOND
009C    8606                 LDA    A #06H         ;WORD FOR VPP=25V, NOT OE=1, NOT CE=1
009E    B78006               STA    A 8006H
00A1    CE0F00               LDX    #0F00H         ;LOAD DELAY CONSTANT
00A4    09          DELAY    DEX                   ;DELAY 50 MS
00A5    26FD                 BNE    DELAY
00A7    8604                 LDA    A #04H         ;TURN OFF NOT CE CHARACTER
00A9    B78006               STA    A 8006H        ;SEND TO CONTROL PORT
00AC    DE08                 LDX    WRKADD         ;LOAD CURRENT WORK ADDRESS
00AE    08                   INX
00AF    DF08                 STX    WRKADD         ;RESTORE
00B1    9C04                 CPX    ENDADD         ;CHECK IF LAST ADDRESS
00B3    2707                 BEQ    VERIFY         ;YES, VERIFY
00B5    DE0A                 LDX    RDADD          ;LOAD READ ADDRESS
00B7    08                   INX
00B8    DF0A                 STX    RDADD          ;RESTORE TO RDADD BUFFER
00BA    20CD                 BRA    PGMLUP         ;GO PROGRAM NEXT LOCATION
00BC    CE8000      VERIFY   LDX    #8000H         ;LOAD PIABSE
00BF    8605                 LDA    A #05H         ;CONTROL WORD VPP=5V, NOT OE=1, NOT CE=0
00C1    A706                 STA    A 06H,X        ;TURN OFF VPP KEEP NOT OE=1
00C3    4F                   CLR    A              ;A=00 CONTROL WORD TO ACCESS DDR
00C4    A705                 STA    A 05H,X        ;SEND TO CONTROL REGISTER
00C6    A704                 STA    A 04H,X        ;CHANGE DATA DIRECTION TO INPUT
00C8    8604                 LDA    A #04H         ;CONTROL WORD FOR PORT ACCESS
00CA    A705                 STA    A 05H,X        ;SEND TO CONTROL REGISTER
00CC    8601                 LDA    A #01H         ;WORD FOR VPP=5V, NOT OE=0, NOT CE=0
00CE    A706                 STA    A 06H,X        ;ENABLE 2716 OUTPUT
00D0    DE02                 LDX    STRTAD         ;COPY START ADDRESS TO WORK BUFFER
00D2    DF08                 STX    WRKADD
00D4    DE06                 LDX    SRCSTR         ;COPY SOURCE ADDRESS TO READ BUFFER
00D6    DF0A                 STX    RDADD
00D8    CE8000      CKMORE   LDX    #8000H
00DB    9608                 LDA    A WRKADD       ;GET MS BYTE OF ADDRESS
00DD    A712                 STA    A 12H,X        ;SEND TO 2716
00DF    9609                 LDA    A WRKADD+1     ;GET LS BYTE OF ADDRESS
00E1    A710                 STA    A 10H,X        ;SEND TO 2716
00E3    A604                 LDA    A 04H,X        ;READ 2716
00E5    DE0A                 LDX    RDADD          ;LOAD CURRENT READ ADDRESS
00E7    A100                 CMP    A 00H,X        ;COMPARE CHAR HERE WITH CHAR FROM 2716
00E9    2610                 BNE    FAIL           ;IF NOT SAME, FAIL EXIT
00EB    08                   INX                   ;RDADD IN X
00EC    DF0A                 STX    RDADD          ;RESTORE READ ADDRESS
00EE    DE08                 LDX    WRKADD         ;LOAD WRKADD
00F0    08                   INX
00F1    DF08                 STX    WRKADD         ;RESTORE TO WRKADD
00F3    9C04                 CPX    ENDADD         ;CHECK IF LAST ADDRESS
00F5    26E1                 BNE    CKMORE         ;NO, GO CHECK NEXT LOCATION
00F7    860A                 LDA    A #0AH         ;PASS CHAR = A
00F9    2002                 BRA    DISPLAY        ;DISPLAY AAAA - AA
00FB    860F        FAIL     LDA    A #0FH         ;FAIL CHARACTER = F
00FD    CEA00C      DISPLY   LDX    #0A00CH        ;DISPLAY BUFFER START ADDRESS
0100    A700                 STA    A 00H,X        ;SEND PASS OR FAIL
0102    A701                 STA    A 01H,X        ;CHARACTER TO ALL
0104    A702                 STA    A 02H,X        ;DISPLAY BUFFER LOCATIONS
0106    A703                 STA    A 03H,X
0108    A704                 STA    A 04H,X
```

FIGURE 12-17 *(continued)*

```
010A  A705              STA    A 05H,X
010C  BDE0FE            JSR    OUTDS           ;DISPLAY PASS OR FAIL
010F  3F                SWI
                      ;*
                      ;****ROUTINE TO DISPLAY 6 DIGITS IN DISBUF
                      ;*
      E0FE              ORG    0E0FEH
E0FE  CEA00C    OUTDS   LDX    #DISBUF         ;GET STARTING ADDRESS
E101  A600      OUTDS1  LDA    A 0,X           ;GET FIRST DIGIT
E103  4C                INC    A
E104  08                INX
E105  FFA020            STX    XDSBUF          ;SAVE POINTER
E108  CEE3C9            LDX    #DIGTBL-1
E10B  08        OUTDS2  INX
E10C  4A                DEC    A               ;POINT TO PATTERN
E10D  26FC              BNE    OUTDS2
E10F  7F8022            CLR    SCNREG          ;BLANK DISPLAY
E112  A600              LDA    A 0,X           ;GET PATTERN
E114  B78020            STA    A DISREG        ;SET UP SEGMENTS
E117  B6A01C            LDA    A SCNCNT
E11A  B78022            STA    A SCNREG        ;SELECT DIGIT
E11D  CE004D            LDX    #4DH            ;SETUP FOR 1MS DELAY
E120  8DBE              BSR    DLY1            ;DELAY 1MS
E122  FEA020            LDX    XDSBUF          ;RECOVER POINTER
E125  8CA012            CPX    #DISBUF+6
E128  271F              BEQ    OUTDS3
E12A  74A01C            LSR    SCNCNT          ;NO, MOVE TO NEXT DIGIT
E12D  20D2              BRA    OUTDS1
                      ;*
                      ;**SUBROUTINE TO MOVE LOW NIBBLE OF A TO B AND TO
                      ;****MOVE HIGH NIBBLE OF A TO LOW NIBBLE OF A
                      ;*
      E29A              ORG    0E29AH
E29A  16        MDIS2   TAB
E29B  C40F              AND    B #0FH          ;MASK LOW NIBBLE
E29D  84F0              AND    A #0F0H         ;MASK HIGH NIBBLE
E29F  44                LSR    A
E2A0  44                LSR    A
E2A1  44                LSR    A
E2A2  44                LSR    A               ;HIGH NIBBLE TO LOW NIBBLE
E2A3  39                RTS
                        END
```

FIGURE 12-17 (*continued*)

mented. The incremented address is compared with the user-loaded ENDADD. If the end address has not been reached, then the program branches back to CKMORE to check the next location in the 2716. If the last address has been reached, then 0AH is loaded into the accumulator and a branch to DISPLY is made. DISPLY loads this 0AH into all the locations in DISBUF. When this is done, a jump to OUTDS will display six As in the LEDs and return control to monitor.

This completes the discussion of the 2716 EPROM programmer program. The erasing procedure for 2716s is described in the data sheet in App. C.

In Chap. 13 we discuss troubleshooting techniques for microprocessor-based instruments such as the scale and EPROM programmer described here.

OVERVIEW OF INDUSTRIAL PROCESS CONTROL

An area in which microprocessors and microcomputers have had a major impact is *industrial process control*. Industrial process control involves first measuring system variables (such as motor speed, temperature, the flow of reactants, the level of liquid in a tank, or the thickness of a material) and then adjusting the system until the value of each variable is equal to a *set-point* value for that variable. The system controller must maintain each variable at its set-point value and compensate as quickly and accurately as possible for any change in the system, such as increased load on a motor. A simple example will show the traditional approach to control of a process variable and explain some terms used in control systems.

The circuit in Figure 12-18 shows one approach to controlling the speed of a dc motor. Attached to the shaft of the motor is a dc generator, or *tachometer*, which puts out a voltage proportional to the speed of the motor. The output voltage is typically a few volts per thousand rpm. A fraction of the output voltage from the tachometer is fed back into the inverting (−) input of the power amplifier. A positive voltage is applied to the noninverting (+) input of the power amplifier as a set point. When power is turned on, the motor accelerates until the voltage fed back from the tachometer to the inverting (−) input is nearly equal to the set-point voltage on the noninverting (+) input of the amplifier. This feedback scheme is called *proportional control*. A control loop of this type keeps the motor speed quite constant for applications in which the load does not change much. Some Winchester disk drives and high-quality phonograph turntables use this method of speed control.

For applications in which the load or set point changes drastically, however, a more complex feedback system is used. Here's why. Figure 12-19 shows two possible responses of a simple control loop, such as the motor-speed example in Figure 12-18, to a change in set point. Figure 12-19a shows an *underdamped* response. In this case, the variable, motor speed or temperature, for example, overshoots the new set point and bounces up and down for a while. The time it takes the variable to get within a given error range (band) is referred to as *settling time*. Even after the bouncing has settled, there is some *residual error* because of the finite gain of the amplifier. Figure 12-19b shows the effect of *overdamping* on the system response. For this case, the variable does not overshoot the set point. The variable gets within the error band and stays there. However, note that even after a long time there is still some residual error due again to the finite gain of the amplifier. Any feedback control system using only proportional

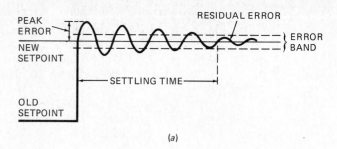

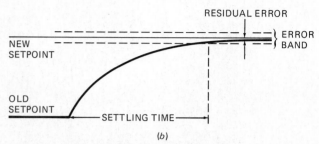

FIGURE 12-19 Response of a system to a change in set point. *(a)* Underdamped. *(b)* Overdamped.

control will have some residual error between the set point and the actual output.

The cure for residual error is to use *integral feedback*. Integral feedback is a signal that is proportional to the error voltage multiplied by the time that the error is present. Figure 12-20 shows how integral feedback can be produced and added to simple proportional control. Amplifier 1 compares the set-point voltage, V_S, with the feedback voltage from the tachometer, V_T. Note that V_S must be negative because of the inverting configuration in which the amplifiers are connected. If no error exists between the set point and the actual motor speed, the output of amplifier 1 is 0 V and amplifiers 2, 3, and 4 have no effect. If an error between V_S and V_T is present, a signal is applied to the inputs of amplifiers 2, 3, and 4. Amplifier 2 produces proportional feedback. Amplifier 3, an *integrator*, produces integral feedback. As long as any error is present, the output of the integrator ramps up or down to eliminate the error. The integrator, however, is slow to respond because the error signal must be present for some time before the integrator has much output.

The cure for slow response is to use *derivative feedback*. Derivative feedback is a signal proportional to the rate of change of the error between V_S and V_T. Amplifier 4 in Figure 12-20 is a *differentiator* which gives derivative feedback. As soon as an error voltage appears at its input, the differentiator gives a quick shot of feedback to try to correct the error. Because of the capacitor on its input, however, the differentiator has an effect for only a short time.

Process control loops that employ all three types of feedback are called *proportional-integral-derivative*, or PID, control loops. Because process variables change much more slowly than the microsecond operation of a microcomputer, a microcomputer with some simple I/O circuitry can perform all functions of the analog circuitry in Figure 12-20 for several PID loops.

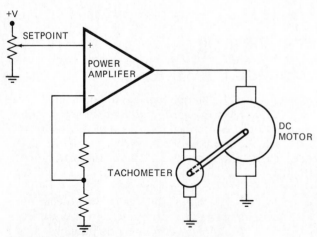

FIGURE 12-18 Controlling the speed of a dc motor with tachometer feedback.

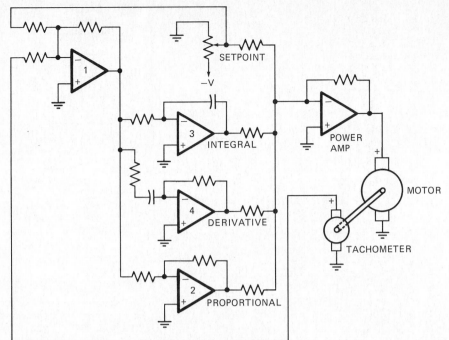

AN SDK-85-BASED PROCESS CONTROL SYSTEM

Figure 12-21 shows a block diagram of a microcomputer-based process control system. A keyboard, serviced on an interrupt basis, allows the user to enter set points and read the values of process variables. A programmable timer supplies clock "ticks." Clock ticks are used to tell the microcomputer when to service the next control loop. Data acquisition systems convert analog signals from pressure, temperature, and other types of sensors to digital signals that can be read in by the microcomputer input. The data acquisition systems are controlled by signals sent by the microcomputer. Other signals sent by the microcomputer from its output ports go to relays, solenoids, D/A converters, etc. These devices turn on heaters, flow valves, motors, and other actuators to control process variables.

Figure 12-22 shows in flowchart form one way in which the program for a microcomputer-based control system can operate. After power is turned on, the background program initializes ports, the timer, and process constants. It then sits in a loop, waiting for a user command from the keyboard or for a clock tick from the timer. When a clock tick occurs, the microcomputer calls an interrupt service routine. The service routine counts clock ticks. Every 20 clock ticks, the microcomputer calls a routine to service the next one of eight con-

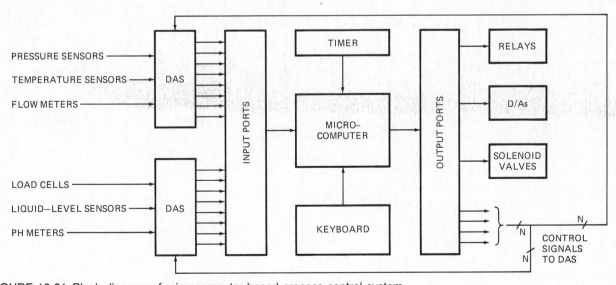

FIGURE 12-21 Block diagram of microcomputer-based process control system.

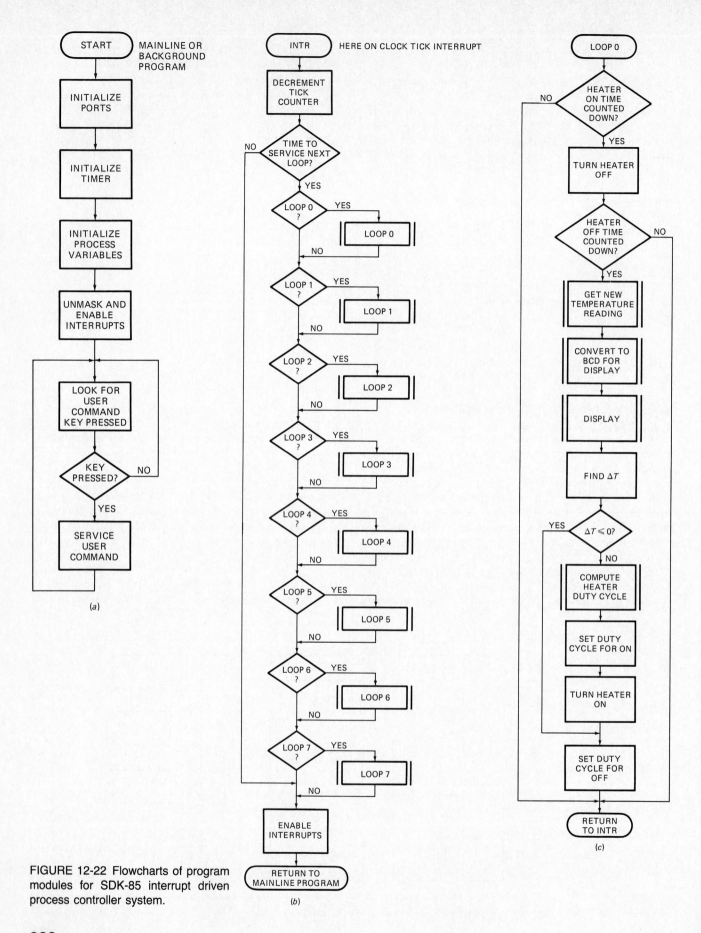

FIGURE 12-22 Flowcharts of program modules for SDK-85 interrupt driven process controller system.

trol loops. After a control loop is serviced, the microcomputer returns to the background program and waits for the next clock tick or a user command. The actual control routines are independent of one another, so each can be developed and tested individually.

LOOP0 in Figure 12-22c is an example of a control loop service routine that brings the temperature of a water bath up to a set-point temperature and maintains that temperature. LOOP0 calls a routine that does an A/D conversion, a routine to convert binary to BCD, a routine to display the current temperature, and a routine to calculate the feedback value based on the difference between the actual and set-point temperatures. Power sent to the heating element is controlled by varying the duty cycle of pulses sent to it.

In the following section we discuss the hardware added to an SDK-85 board to monitor and control the temperature of a water bath for set points within a range of 0 to 99°C. Then we show and discuss the program for an SDK-85-based control system.

HARDWARE FOR AN SDK-85 PROCESS CONTROL SYSTEM

HARDWARE FOR THE CLOCK TICK

The clock tick for this controller is generated on an interrupt basis, as described in Chap. 11. The timer in an 8155 is programmed to output a 1-kHz square-wave pulse train. A 74121 pulse-narrower circuit shown in Figure 12-23 reduces the high time of the pulses to about 15 μs, so one pulse cannot interrupt the processor more than once. The output of the pulse narrower is connected to the RST 6.5 input of the 8085A. A clock tick interrupt is produced every 1 ms.

HARDWARE FOR THE TEMPERATURE CONTROL LOOP

A common requirement in industries, such as in the manufacture of chemicals, is to heat a batch of reactants with a water bath to a precise temperature, accurately hold that temperature for a time, raise the temperature precisely to a new temperature, and hold that temperature accurately for a time. It is important that the temperature not overshoot the set point or bounce up and down around the set point too much. This is easily accomplished with microcomputer control. Figure 12-23 shows the parts that must be added to an SDK-85 for a temperature control loop. The major parts are a temperature sensor, an A/D converter, and a solid-state relay.

The *temperature sensor* is an LM335. The LM335 is a zener diode with a linear temperature coefficient of +10 mV/°C. The low-drift LM308 amplifier increases this to 40 mV/°C for input to the A/D converter. This allows the desired range of 0 to 99°C to use a greater portion of the A/D range. Applying a negative reference voltage to the noninverting input of the amplifier allows the output to be adjusted to 0.0 V at 0.0°C. The LM335 is rated for operation over −10 to +100°C. A temperature probe was made by sealing the LM335 in the end of a piece of ¼-in [6.3-mm] diameter glass tubing with high-temperature silicon sealant.

Analog-to-digital conversion is done by a National Semiconductor ADC0808 data acquisition system. The heart of the ADC0808 is an 8-bit binary output, successive-approximation A/D converter. On the input of the A/D converter is an eight-input multiplexer. The input channel being digitized by the A/D converter is determined by a 3-bit address on the ADD A, ADD B, and ADD C inputs. An eight-input data acquisition system was chosen so that eight variables could be monitored with the same A/D converter. A Schmitt trigger oscillator produces a 300-kHz clock for the A/D converter. The voltage drop of an LM329 low-drift zener is buffered by an LM308 amplifier to produce a V_{CC} and V_{REF} of 5.12 V for the A/D converter. With this reference voltage, the A/D converter will have 256 steps of 20 mV each.

Figure 12-24 shows the timing waveforms and parameters for the ADC0808. Note the sequence in which signals must be sent to the device first. The address of the desired channel is sent to the multiplexer address inputs. After 50 ns, address latch enable (ALE) is sent high. After another 2.5 μs, the start-conversion signal is made high and then low, and ALE is made low. When the end-of-conversion signal goes high, data can be read from the A/D converter into a port.

The *heater* can be either a 120-V ac hot plate or a 120-V ac immersion heater such as those used to heat a single cup of water for coffee. A solid-state relay switches the heater on and off. The microcomputer controls the heat output by pulsing the relay on and off for varying duty cycles.

The relay used is an optically isolated, zero-voltage-turn-on, zero-current-turn-off type. This type is a little more expensive, but it switches on and off rapidly, provides several thousand volts of electrical isolation between control input and power output, and produces very little electromagnetic interference (EMI) because the power is always on for some integral number of half cycles of the ac line voltage.

For very low-power applications, a D/A converter and power amplifier could be used to drive a heater. However, in high-power applications this is not very practical because the power amplifier dissipates as much as or more power than the load. For example, driving a 5000-W heater, the amplifier will dissipate 5000 W or more. The D/A converter and amplifier approach also has the disadvantage that it cannot easily use the ac line voltage. The pulse-width modulation technique has the added advantage that it can be used to control the speed of ac-dc (universal) motors.

A relay such as the Potter-Brumfield EOM1DB72 requires only 11 mA maximum at 5 V to turn it on, and with a proper heat sink it will control up to 25 A.

The driver transistor on the input of the relay serves three purposes: it supplies the drive for the relay, isolates the port pin from the relay, and holds the relay in the off position when power is first turned on. This is a good idea because the port pins are floating after a reset.

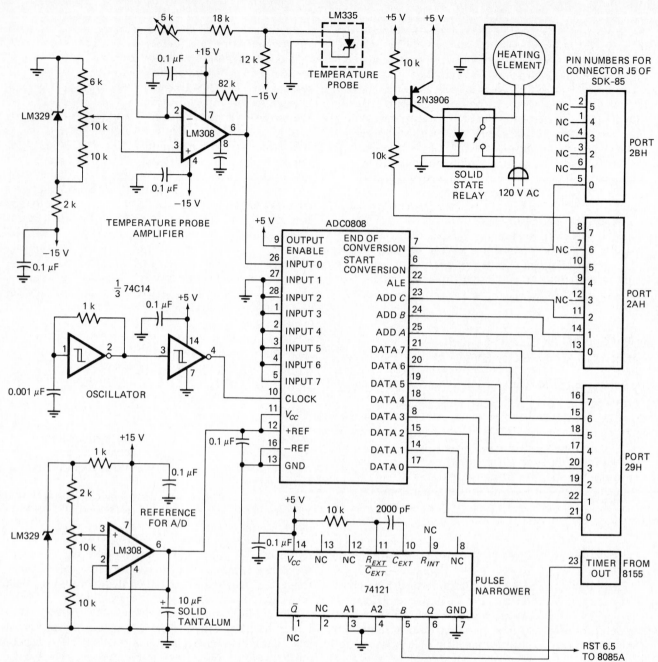

FIGURE 12-23 Circuitry added to an SDK-85 to measure and control the temperature of a water bath.

CONTROLLER PROGRAM

OVERVIEW   The program is written in a highly structured manner. Structured programming means that each task is done, or each problem solved, by breaking it down into smaller and smaller nested modules that can be individually written and debugged. Ideally, each module should have only one entry point and one exit point. Figure 12-22 shows flowcharts of how this program is structured. In the following sections we explain the operation of each major part of the program. Figure 12-25 shows the complete assembly language listing.

MAINLINE OR BACKGROUND PROGRAM   The mainline section of the program initializes port 29H as input, port 2AH as output, and port 2BH as input. It also initializes the 8155 timer for a 1-kHz square-wave output, sets the starting values for various control constants, unmasks interrupts, and enables interrupts. It then executes the JMP HERE instruction over and over while waiting for a clock tick interrupt. Normally the mainline program also would scan a keyboard to detect when the user desires to change a set point or read the current value of a variable such as temperature. To save space, a keyboard scan is omitted here.

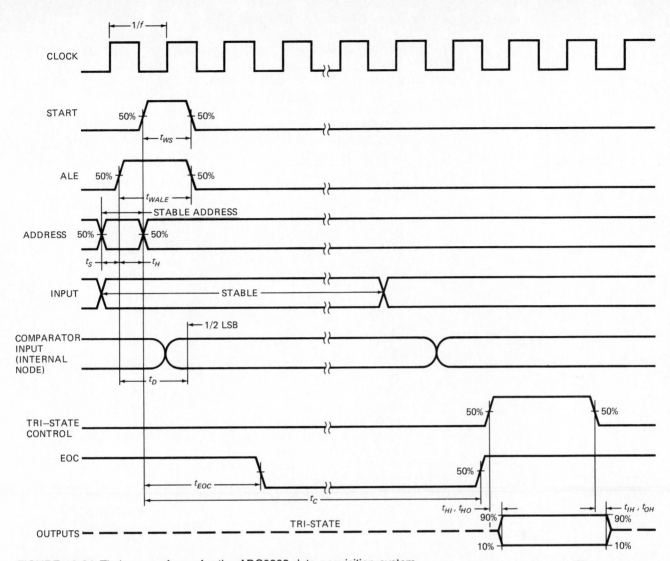

FIGURE 12-24 Timing waveforms for the ADC0808 data acquisition system.

INTR ROUTINE   INTR is the routine that counts clock ticks and services the next control loop after every 20 clock ticks. When the 8085A receives an RST 6.5 interrupt input from the timer, it disables interrupts and vectors to 0034H. The SDK-85 monitor program has, at this address, a jump to 20C8H. A jump to 203AH, the start of the INTR service routine, is loaded at address 20C8H. When the 8085A finally gets to the INTR routine at 2034H, it saves registers on the stack and gets the interrupt count from a memory location called INTCNT. [INTCNT is initialized to 14H (20_{10}) in the mainline program and again after each control loop is serviced.] The count is decremented and restored. If decrementing the count did not reduce it to 0, the routine jumps to EXIT2. It then restores registers, enables interrupts, and returns to the mainline program to wait for the next clock tick.

If 20_{10} (14H) interrupts have been counted, the INTR routine gets the number of the next control loop to be serviced from a memory location called LPCNT. This number is decremented to determine which control loop routine to call. Each control loop loads the number of the next control loop to be serviced into LPCNT and returns to the mainline program with a value of 0FFH in the accumulator. The 0FFH in the accumulator causes execution to fall through the remaining parts of the loop checker. If LPCNT somehow gets an erroneous number and no loop is found by the checker, then LPCNT is reloaded with 01H before returning to the mainline program. This causes loop 0 to be the next one serviced. If a loop was found and serviced, INTCNT is set to 20_{10} (14H), the registers are restored, interrupts are enabled, and execution is returned to the main program. Note that variables such as LPCNT and INTCNT are kept in named memory locations where they are easy to keep track of.

LOOP0   The routine that controls the temperature of the water bath is LOOP0. Figure 12-22 shows the flowchart for this routine. The time on for the output waveform to the solid-state relay is determined by counting down a number called DCHI. The time off is determined

```
00001                              ;THIS PROGRAM SERVICES EIGHT PROCESS CONTROL
00002                              ;LOOPS ON A ROTATING BASIS. IT IS WRITTEN
00003                              ;TO RUN ON AN INTEL SDK-85 WITH EXPANSION
00004                              ;RAM AT ADDRESS 2800H. TIMING FOR CONTROL
00005                              ;LOOPS IS GENERATED ON AN INTERRUPT BASIS
00006                              ;BY AN ON-BOARD 8155 TIMER. CONTROL LOOP 0
00007                              ;IN THE PROGRAM CONTROLS THE TEMPERATURE
00008                              ;OF A WATER BATH USING A PULSE WIDTH METHOD
00009                              ;
00010        2000       LPCNT   EQU     2000H      ;NEXT LOOP TO SERVICE
00011        2001       INTCNT  EQU     2001H      ;INTERRUPT COUNT
00012        2002       DCHI    EQU     2002H      ;HEATER ON COUNT
00013        2003       DCLO    EQU     2003H      ;HEATER OFF COUNT
00014        036E       UPDDT   EQU     036EH      ;MONITOR DISPLY ROUTINE
00015        2004       HOT     EQU     2004H      ;CURRENT TEMPERATURE
00016        2005       STPNT   EQU     2005H      ;SETPOINT TREMPERATURE
00017
00018                              ;START OF MAIN PROGRAM
00019
00020        2010    >          ORG     2010H      ;START OF MAIN PROGRAM
00021  2010  31C220             LXI     SP,20C2H;INIT STACK POINTER
00022  2013  3E00               MVI     A,00H      ;TIMER LOW COUNT
00023  2015  D32C               OUT     2CH
00024  2017  3E4C               MVI     A,4CH      ;TIMER HIGH AND MODE
00025  2019  D32D               OUT     2DH
00026  201B  3EC2               MVI     A,0C2H     ;START TIMER,INIT PORT
00027  201D  D328               OUT     28H        ;A IN,B OUT,C IN
00028  201F  3E14               MVI     A,014H     ;INITIALIZE INTERRUPT
00029  2021  320120             STA     INTCNT     ;COUNT TO 20 DECIMAL
00030  2024  3E01               MVI     A,01H
00031  2026  320020             STA     LPCNT      ;SET TO FIRST LOOP
00032  2029  320220             STA     DCHI       ;SET HEATER ON COUNT
00033  202C  320320             STA     DCLO       ;SET HEATER OFF COUNT
00034  202F  3E80               MVI     A,80H      ;TURN HEATER OFF
00035  2031  D32A               OUT     2AH
00036  2033  3E0D               MVI     A,0DH      ;UNMASK INTERRUPTS
00037  2035  30                 SIM
00038  2036  FB                 EI
00039  2037  C33720    >  HERE  JMP     HERE       ;WAIT HERE FOR INTERRUPT
00040                              ;SIMULATE MAIN PROGRAM
00041                              ;WAITING FOR USER ENTERED
00042                              ;COMMAND
00043
00044                              ;PROGRAM GOES HERE ON TIMER INTERRUPT
00045
00046  203A  C5         INTR   PUSH    B
00047  203B  D5                 PUSH    D
00048  203C  E5                 PUSH    H
00049  203D  F5                 PUSH    PSW        ;SAVE ALL
00050  203E  3A0120             LDA     INTCNT     ;GET INTERRUPT COUNT
00051  2041  3D                 DCR     A          ;DECREMENT
00052  2042  320120             STA     INTCNT     ;SAVE COUNT
00053  2045  C27920    >        JNZ     EXIT2      ;IF INTCNT DOWN TO ZERO,
```

FIGURE 12-25 Assembly language listing for SDK-85 process controller system with temperature control.

```
00054 2048 3A0020              LDA     LPCNT     ;GET NUMBER OF LOOP TO
00055 204B 3D                  DCR     A         ;SERVICE AND CALL ROUTINE
00056 204C CC0028   >          CZ      LOOP0     ;TO SERVICE THAT LOOP
00057 204F 3D                  DCR     A
00058 2050 CC6C28   >          CZ      LOOP1
00059 2053 3D                  DCR     A
00060 2054 CC7428   >          CZ      LOOP2
00061 2057 3D                  DCR     A
00062 2058 CC7C28   >          CZ      LOOP3
00063 205B 3D                  DCR     A
00064 205C CC8428   >          CZ      LOOP4
00065 205F 3D                  DCR     A
00066 2060 CC8C28   >          CZ      LOOP5
00067 2063 3D                  DCR     A
00068 2064 CC9428   >          CZ      LOOP6
00069 2067 3D                  DCR     A
00070 2068 CC9C28   >          CZ      LOOP7
00071 206B 17                  RAL               ; MSB TO CARRY
00072 206C DA7420   >          JC      EXIT1     ;LOOP FOUND IF CARRY,EXIT
00073 206F 3E01                MVI     A,01H     ;NOT FOUND, RESET LPCNT
00074 2071 320020              STA     LPCNT
00075 2074 3E14       EXIT1    MVI     A,14H     ;RESET INTCNT TO 20
00076 2076 320120              STA     INTCNT
00077 2079 F1         EXIT2    POP     PSW       ;RESTORE ALL REGISTERS
00078 207A E1                  POP     H
00079 207B D1                  POP     D
00080 207C C1                  POP     B
00081 207D FB                  EI
00082 207E C9                  RET               ;RETURN TO MAINLINE
00083                          ;
00084      20C8       >        ORG     20C8H     ;SETS INTERUPT 6.5
00085 20C8 C33A20     >        JMP     INTR      ;JUMP ADDRESS
00086                          ;
00087
00088                          ;ROUTINE TO SERVICE TEMPERATURE CONTROLLER
00089                          ;
00090      2800       >        ORG     2800H
00091 2800 3A0220     LOOP0    LDA     DCHI      ;GET HEATER ON COUNT
00092 2803 3D                  DCR     A
00093 2804 320220              STA     DCHI      ;SAVE DECREMENTED COUNT
00094 2807 C25F28     >        JNZ     EXIT      ;RETURN TO MAIN PROGRAM
00095 280A 3E01                MVI     A,01H     ;RESET DCHI TO
00096 280C 320220              STA     DCHI      ;FALL THROUGH VALUE
00097 280F 3E80                MVI     A,80H     ;TURN HEATER OFF
00098 2811 D32A                OUT     2AH
00099 2813 3A0320              LDA     DCLO      ;GET HEATER OFF COUNT
00100 2816 3D                  DCR     A
00101 2817 320320              STA     DCLO      ;SAVE DECREMENTED COUNT
00102 281A C25F28     >        JNZ     EXIT      ;RETURN TO MAIN PROGRAM
00103 281D 0600                MVI     B,00H     ;LOAD CHANNEL ADDRESS IN B
00104 281F CDA428     >        CALL    ADRD      ;DO A TO D CONVERT
00105 2822 E6FE                ANI     0FEH      ;DROP LOW BIT
00106 2824 0F                  RRC               ;DIVIDE BY 2
```

FIGURE 12-25 *(continued)*

385
PUTTING IT ALL TOGETHER

```
00107 2825 320420              STA       HOT      ;SAVE TEMPERATURE
00108 2828 4F                  MOV       C,A      ;LOAD FOR CONVERT
00109 2829 CDC728   >          CALL      CNVT1    ;CONVERT TO BCD FOR
00110 282C 78                  MOV       A,B      ;DISPLAY
00111 282D CD6E03              CALL      UPDDT    ;DISPLAY TEMPERATURE
00112 2830 3A0420              LDA       HOT      ;GET TEMPERATURE AGAIN
00113 2833 47                  MOV       B,A      ;SAVE IN B
00114 2834 3A0520              LDA       STPNT    ;LOAD SETPOINT TEMP
00115 2837 90                  SUB       B        ;FIND TEMP ERROR
00116 2838 57                  MOV       D,A      ;SAVE ERROR
00117 2839 CA5528   >          JZ        DONE     ;LEAVE HEATER OFF AND
00118 283C FA5528   >          JM        DONE     ;EXIT IF ERROR ZERO
00119 283F 016400              LXI       B,0064H  ;OTHERWISE COMPUTE DCLO,
00120 2842 CDD828   >          CALL      DIV1     ;DIVIDE 0064H BY ERROR,
00121 2845 79                  MOV       A,C      ;QUOTIENT IS VALUE FOR
00122 2846 320320              STA       DCLO     ;DCLO
00123 2849 3E04                MVI       A,04H    ;SET DCHI FOR 4 LOOPS ON
00124 284B 320220              STA       DCHI
00125 284E 3E00                MVI       A,00     ;TURN HEATER ON
00126 2850 D32A                OUT       2AH
00127 2852 C35F28   >          JMP       EXIT
00128 2855 3E01       DONE     MVI       A,01H    ;FALLTHROUGH
00129 2857 320220              STA       DCHI     ;VALUE FOR DCHI
00130 285A 3E7F                MVI       A,7FH    ;LONG OFF VALUE FOR
00131 285C 320320              STA       DCLO     ;DCLO
00132 285F 3E02       EXIT     MVI       A,02H    ;POINT TO NEXT LOOP
00133 2861 320020              STA       LPCNT
00134 2864 3E14                MVI       A,014H   ;RESET INTCNT TO 20
00135 2866 320120              STA       INTCNT
00136 2869 3EFF                MVI       A,0FFH   ;FALL THROUGH LOOP
00137 286B C9                  RET                ;CHECKER ON RETURN
00138                                             ;TO INTR ROUTINE
00139                 ;
00140                 ;DUMMY LOOPS HERE
00141                 ;
00142
00143 286C 3E03       LOOP1    MVI       A,03H    ;POINT TO NEXT LOOP
00144 286E 320020              STA       LPCNT
00145 2871 3EFF                MVI       A,0FFH   ;SET FOR FALL THROUGH
00146 2873 C9                  RET                ;LOOP DETECTOR
00147
00148 2874 3E04       LOOP2    MVI       A,04H    ;POINT TO NEXT LOOP
00149 2876 320020              STA       LPCNT
00150 2879 3EFF                MVI       A,0FFH
00151 287B C9                  RET
00152
00153 287C 3E05       LOOP3    MVI       A,05H    ;POINT TO NEXT LOOP
00154 287E 320020              STA       LPCNT
00155 2881 3EFF                MVI       A,0FFH
00156 2883 C9                  RET
00157
00158 2884 3E06       LOOP4    MVI       A,06H    ;POINT TO NEXT LOOP
00159 2886 320020              STA       LPCNT
```

FIGURE 12-25 *(continued)*

386

```
00160  2889  3EFF              MVI     A,0FFH
00161  288B  C9               RET
00162
00163  288C  3E07     LOOP5    MVI     A,07H    ;POINT TO NEXT LOOP
00164  288E  320020            STA     LPCNT
00165  2891  3EFF              MVI     A,0FFH
00166  2893  C9               RET
00167
00168  2894  3E08     LOOP6    MVI     A,08H    ;POINT TO NEXT LOOP
00169  2896  320020            STA     LPCNT
00170  2899  3EFF              MVI     A,0FFH
00171  289B  C9               RET
00172
00173  289C  3E01     LOOP7    MVI     A,01H    ;POINT TO NEXT LOOP
00174  289E  320020            STA     LPCNT
00175  28A1  3EFF              MVI     A,0FFH
00176  28A3  C9               RET
00177                   ;
00178                   ;ROUTINE TO CONTROL A TO D CONVERTER
00179                   ;       ;PORT 29H IS DATA FROM A TO D
00180                   ;PORT 2AH BIT 7 IS HEATER,BIT 5 IS START CONV
00181                   ;BIT 4 IS ALE,BITS 2,1,0 ARE CHANNEL ADDRESS
00182                   ;PORT 2BH BIT 0 IS END OF CONVERSION
00183  28A4  3E80     ADRD     MVI     A,80H    ;CONTROL FOR HEATER OFF
00184  28A6  B0               ORA     B        ;COMBINE WITH CHANNEL ADDRESS
00185  28A7  D32A              OUT     2AH
00186  28A9  3E90              MVI     A,90H    ;SEND ALE
00187  28AB  B0               ORA     B        ;KEEP CHANNEL ADDRESS ON
00188  28AC  D32A              OUT     2AH
00189  28AE  3EB0              MVI     A,0B0H   ;SEND START CONVERSION
00190  28B0  B0               ORA     B        ;KEEP CHANNEL ADDRESS ON
00191  28B1  D32A              OUT     2AH
00192  28B3  3E80              MVI     A,80H    ;TURN OFF ALE AND START
00193  28B5  B0               ORA     B        ;KEEP CHANNEL ADDRESS
00194  28B6  D32A              OUT     2AH
00195  28B8  DB2B     EOCL     IN      2BH      ;WAIT FOR END OF CONVERSION
00196  28BA  0F               RRC              ;TO GO LOW
00197  28BB  DAB828  >         JC      EOCL     ;
00198  28BE  DB2B     EOCH     IN      2BH      ;WAIT FOR END OF CONVERSION
00199  28C0  0F               RRC              ;TO GO HIGH
00200  28C1  D2BE28  >         JNC     EOCH
00201  28C4  DB29              IN      29H      ;READ DATA FROM A TO D
00202  28C6  C9               RET              ;RETURN WITH DATA IN ACC
00203
00204
00205                   ;ROUTINE TO CONVERT BINARY TO BCD
00206
00207                   ;THIS ROUTINE CONVERTS AN EIGHT BIT BINARY
00208                   ;NUMBER IN THE C REGISTER TO TWO BCD DIGITS
00209                   ;IN THE B REGISTER. BCD IN RANGE 00-99.
00210
00211  28C7  1609     CNVT1    MVI     D,09H    ;LOAD BIT COUNTER
00212  28C9  AF               XRA     A        ;CLEARS A
```

FIGURE 12-25 *(continued)*

```
00213  28CA  47                  MOV      B,A        ;CLEARS B
00214
00215  28CB  AF        CNVT2     XRA      A          ;CLEARS A
00216  28CC  15                  DCR      D          ;DECREMENT BIT COUNTER
00217  28CD  C8                  RZ                  ;RETURN IF DONE
00218  28CE  79                  MOV      A,C        ;GET BINARY
00219  28CF  17                  RAL                 ;SHIFT MSB TO CARRY
00220  28D0  4F                  MOV      C,A        ;SAVE SHIFTED IN C
00221  28D1  78                  MOV      A,B        ;BCD TO A
00222  28D2  8F                  ADC      A          ;DOUBLE A+ ADD CARRY FROM C
00223  28D3  27                  DAA                 ;KEEP IN BCD
00224  28D4  47                  MOV      B,A        ;SAVE BCD
00225  28D5  C3CB28   >          JMP      CNVT2      ;LOOP UNTIL DONE
00226
00227
00228                  ;ROUTINE TO DIVIDE 16 BIT NUMBER BY 8 BIT NUMBER
00229
00230                  ;THIS ROUTINE DIVIDES A 16 BIT NUMBER IN
00231                  ;THE B AND C REGISTER BY AN EIGHT BIT
00232                  ;NUMBER IN THE D REGISTER. THE QUOTIENT IS
00233                  ;RETURNED IN C AND THE REMAINDER IS IN B.
00234
00235  28D8  1E09      DIV1      MVI      E,09H      ;LOAD BIT COUNTER
00236  28DA  78                  MOV      A,B        ;
00237  28DB  47        DIV2      MOV      B,A        ;SAVE RESULT OF SUB D
00238  28DC  79                  MOV      A,C        ;MSB OF C INTO CARRY
00239  28DD  17                  RAL                 ;LATER MOVED TO LSB OF B
00240  28DF  4F                  MOV      C,A        ;SAVE SHIFTED RESULT
00241  28DF  1D                  DCR      E          ;DECREMENT BIT COUNTER
00242  28E0  CAED28   >          JZ       DIV3       ;EXIT IF DONE
00243  28E3  78                  MOV      A,B        ;MSB OF C STILL IN CARRY
00244  28E4  17                  RAL                 ;ROTATE IT INTO LSB OF B
00245                                                ;TO FINISH 16 BIT SHIFT
00246  28E5  92                  SUB      D          ;SUBTRACT DIVIDEND FROM
00247                                                ;SHIFTED RESULT
00248  28E6  D2DB28   >          JNC      DIV2       ;OK,GO FIND NEXT BIT
00249  28E9  82                  ADD      D          ;DIVISOR TOO BIG, ADD BACK
00250  28EA  C3DB28   >          JMP      DIV2       ;DO NEXT CYCLE
00251  28ED  3EFF      DIV3      MVI      A,0FFH     ;COMPLEMENT QUOTIENT
00252  28EF  A9                  XRA      C
00253  28F0  4F                  MOV      C,A        ;MOVE FOR EXIT
00254  28F1  C9                  RET                 ;RETURN TO INTR ROUTINE
00255                            END
```

FIGURE 12-25 *(continued)*

by a number called DCLO. At startup the mainline program initializes DCHI and DCLO as 01H. Therefore, the first time LOOP0 is called, execution falls through to an A/D conversion routine, called ADRD. The ADRD routine uses the ADC0808 to do an A/D conversion. It does this for the channel number passed to it in the B register. A binary number that is 2 times the actual binary value for the temperature is returned to LOOP0 in the accumulator. By masking the least significant bit and rotating the result right 1 bit position, the number is divided by 2 to give the correct binary value for the tem-

perature in degrees Celsius. This binary value is saved in memory.

Next a routine called CNVT1 is called to convert the binary value of the temperature to BCD so that it can be displayed on the SDK-85. CNVT1 is a reduced version of the BINCVT routine of the scale program in Figure 12-6. CNVT1 returns two BCD digits in register B. They are moved to the accumulator, and the SDK-85 monitor routine UPDDT is called to display these BCD digits on the data field of the SDK-85 display.

The next section of the LOOP0 control program com-

388

CHAPTER TWELVE

pares the actual temperature with the set-point temperature and sets the duty cycle constants accordingly. The binary temperature value is recalled and subtracted from the set-point temperature. If the difference (error) is 0 or negative, the output time on (DCLI) is set for 0, the output time off (DCLO) is set for 7FH, and the routine is exited. If the difference between the set-point and actual temperature is positive, this means the temperature is low and the heater should be turned on. In this case, the value of DCHI is made a constant of 04H to give a fixed on time for the heater. The value of DCLO for the off time is calculated in the following way. If ΔT is large, then DCLO should be small, so the heater is on for a long duty cycle. If ΔT is small, then DCLO should be large, so the heater is on for a short duty cycle. A first approximation to this relationship is $\Delta T \times DCLO = 100_{10} = 64H$. The value for DCLO, then, is just 64H divided by ΔT. For example, if the ΔT is 20°C, or 14H, then 64H ÷ 14H gives a value of 5 for DCLO. LXI B,0064H loads 64H into the B and C registers, and then the DIV1 routine divides this by the ΔT in the accumulator. The quotient is returned in the C register, moved into the accumulator, and stored in DCLO. The heater is turned on, and an exit is made from the control loop routine. Part of the exit process is to load the number of the next loop to be serviced into LPCNT and reset the interrupt counter to 14H. The accumulator also must be loaded with 0FFH, so execution falls through the loop checker on return to INTR.

When LOOP0 is called again after 160 ms, DCHI is decremented. If DCHI did not decrement to 0, then execution simply exits LOOP0 and returns to INTR. If DCHI is decremented to 0, the heater is turned off and DCLO is decremented. DCLO is decremented every time LOOP0 is serviced (160 ms) until it reaches 0. When DCLO reaches 0, a new A/D conversion is done and the feedback value for DCLO is recalculated. Feedback for this control loop is the value of DCLO calculated from the temperature error (ΔT).

An important point here is that the program sections that determine the feedback are small and easily altered without affecting the rest of the program. Routines could be added to produce integral and derivative feedback as well as the simple proportional feedback used. All that needs to be changed is the value of DCHI, the value of DCLO, and the rate at which these change as the set point is approached. The feedback section can be custom-tailored by experiment to give the best results for a particular system.

ADRD The subroutine that controls the ADC0808 data acquisition system is ADRD. The channel number to

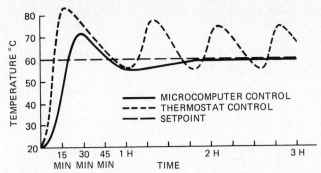

FIGURE 12-26 Temperature versus time responses of a microcomputer-controlled system and a thermostat-controlled system.

be digitized is passed to ADRD in the lower 3 bits of the B register. This is ORed with other control words sent to the device. The instructions in this routine create the control signals shown in Figure 12-24 and read in the binary data when a conversion is complete. The result is returned to the calling program in the accumulator.

CNVT1 A routine to convert an 8-bit binary number in C to two BCD digits in B is CNVT1. The operation of the routine is explained in the scale example earlier. The maximum value returned is 99. For temperature values greater than 99, a leading 1 is assumed but is not displayed.

DIV1 The routine DIV1 divides a 16-bit number in B and C by an 8-bit number in the accumulator. The routine uses a shift-left-and-subtract algorithm. The quotient is returned to the calling program in the C register, and the remainder in the B register.

SUMMARY AND CONCLUSION

Figure 12-26 shows a comparison of temperature-versus-time responses with microcomputer control and with traditional on-off thermostat control. The large initial overshoot of both systems and the large fluctuations of the thermostat curve are caused by the large thermal inertia of the hotplate used. The initial overshoot and the residual error of about 1°C in the microcomputer-controlled system can easily be eliminated by using a more complex feedback computation formula.

The advantages of computed feedback control are obvious. Because of these advantages, more and more industrial and research equipment is using computer power. The computers range from small programmable controllers that monitor 8 or 16 process control loops to room-size computers that control the operations in a whole factory.

REVIEW QUESTIONS AND PROBLEMS

1. List some logical steps for learning about a microprocessor-based instrument.

2. Define monitor program.

3. Describe the process used in the SDK-85-based scale to get weight data from the A/D converter.

4. Define modular programming. Give two advantages of modular programming.

5. Define equates, buffer, packed BCD, and integer multiplication.

6. Describe the function of the KYBINT routine in the scale program.

7. Why do you suppose that the multiplication in the scale program was not done by successive addition?

8. Calculate the exact delay for the "3-s" delay routine at the end of the scale program. Assume a 3-MHz clock frequency.

9. Describe how the lookup table technique used in the display routine of the scale could be used to convert ASCII characters to EBCDIC characters. Referring to Table 3-2 for these codes, show the first five entries in a lookup table for this conversion.

10. Explain the process used by BCDCVT to convert a BCD number to its binary equivalent.

11. Describe briefly three methods of multiplying two binary numbers with a microprocessor.

12. If the H and L registers contain 034A hex before a DAD H instruction is executed, what will they contain after the instruction? Show that the binary bit pattern has just been shifted one place left.

13. Draw a memory map for the MEK6800D2 microcomputer board used for the EPROM programmer.

14. Define ZIF socket and prompt character.

15. Why is the E key on the MEK6800D2 connected to the NMI input of the 6800 CPU?

16. Refer to Figure 12-15.
   a. What are the minimum and maximum widths for the PGM pulse?
   b. How long must data be kept valid after the PGM pulse goes low?
   c. How long must data be valid before the PGM pulse goes high?

17. Refer to the Figures 12-16 and 12-17.
   a. Describe the three major sections of the EPROM programmer program.
   b. Explain why the device is checked for all 1's before programming.
   c. Describe how a 2716 is erased.
   d. Why are the user-loaded starting address and source address copied into buffers for use?
   e. Define indexed addressing. Why is it more efficient than extended addressing for initializing the PIAs?
   f. Calculate the actual delay for the 50-ms delay loop, assuming a 614-kHz clock.
   g. Show in assembly language how a failure during a program verify could be indicated by a display of the failure address instead of six Fs.
   h. Describe how the OUTDS routine in the monitor passes control back to the monitor when a key is pressed.
   i. Why is a NOP inserted in the program before the PGM pulse is sent?

18. A solid-state relay is connected to the D0 line of an output port. A high on the output line turns on the relay and allows current to flow through a large solenoid which opens a door lock. Write an assembly language program for either the 8080A/8085A or 6800 that will open the lock only if a sequence of three hex characters is typed in within 3 s.

REFERENCES/BIBLIOGRAPHY

Intel Memory Design Handbook, Intel Corporation, Santa Clara, CA, 1981.

M6800 Programming Reference Manual, M68PRM(D), Motorola Semiconductor Products, Inc., Phoenix, AZ, 1976.

Intel MCS-80/85B User's Manual, Intel Corporation, Santa Clara, CA, January 1979.

8080/8085 Assembly Language Programming Manual, Intel Corporation, Santa Clara, CA, 1978.

M6800 Microcomputer System Design Data, Motorola Semiconductor Products, Inc. Phoenix, AZ, 1976.

Intel SDK-85 User's Manual, Intel Corporation, Santa Clara, CA, December 1977.

MEK6800D2 Manual, Motorola Semiconductor Products, Inc., Austin, Texas, 1976.

CHAPTER 13

PROTOTYPING AND TROUBLESHOOTING μP-BASED SYSTEMS

This chapter is made up of two parts. In the first part we discuss the procedures used for building and debugging hardware and software during the development of a microprocessor-based product. In the second part we describe the steps in manufacturing the product and the tools and techniques for troubleshooting microprocessor-based instruments in the product test section of production or in a field service situation.

At the conclusion of this chapter, you should be able to:

1. Describe steps for testing and debugging ROM, RAM, and I/O ports on development boards.

2. Describe steps for testing and debugging programs with development boards, using breakpoints and single-stepping.

3. Describe procedures for testing and debugging a microprocessor, ROM, RAM, and I/O ports with a logic timing/state analyzer.

4. Describe the use of a microprocessor development system and in-circuit emulator in debugging a microprocessor-based product.

5. Describe the flow of a product from design to finished product and the functions of the technicians who work at each stage.

6. Discuss how signature analysis is used to troubleshoot a microprocessor-based product.

BUILDING AND DEBUGGING A PROTOTYPE

GENERAL COMMENTS

Some of the points about construction discussed in Chap. 1 bear repeating and expanding here.

POWER SUPPLIES   Before you connect a power supply to LSI circuits, connect an oscilloscope (storage is best) to its outputs and check for turn-on transients. Some power supplies have turn-on transients which will destroy MOS LSI integrated circuits. If a supply is variable, make sure it is adjusted to the correct voltage before you connect it.

PHYSICAL CONSTRUCTION AND WIRING   In a prototype, use good-quality IC sockets that will withstand many insertions and removals. Poorly made sockets create more problems than they solve.

Be careful in handling the large ceramic-packaged ICs. They are expensive, and they break in half easily if dropped.

As discussed in Chap. 1, there is almost no such thing as too many bypass capacitors. Don't forget to connect a 15- to 50-μF electrolytic capacitor and a 0.1-μF disk capacitor in parallel from supply voltage leads to ground where the leads enter the board.

COLOR-CODE WIRING   If you are using wire wrap and extensive color coding is not feasible, code as much as possible. Especially when you are doing wire wrap, a wiring diagram such as that in Figure 13-1 produces fewer errors than using a normal schematic. Remember the trick of keeping track of which connections you have made by marking over each connection on the schematic with a transparent marking pen as you make it. Also identify ICs and other multiple-pin components by number on the drawing; then label the corresponding devices on the prototype.

When the wiring is completed, check the pin connections of each IC twice. Check once from the highest pin to the lowest, and once from the lowest pin to the highest. Make sure every pin is accounted for. All this may seem excessive, but aside from design errors, the most

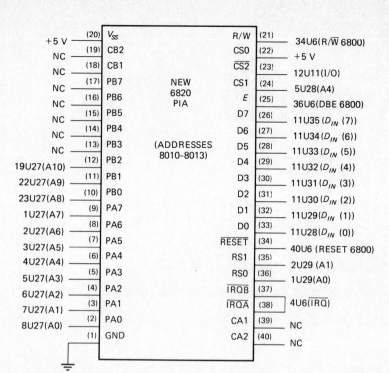

		NEW 6820 PIA			
+5 V	(20)	V_{SS}	R/W	(21)	34U6 (R/$\overline{W}$ 6800)
NC	(19)	CB2	CS0	(22)	+5 V
NC	(18)	CB1	$\overline{CS2}$	(23)	12U11 (I/O)
NC	(17)	PB7	CS1	(24)	5U28 (A4)
NC	(16)	PB6	E	(25)	36U6 (DBE 6800)
NC	(15)	PB5	D7	(26)	11U35 (D_{IN} (7))
NC	(14)	PB4	D6	(27)	11U34 (D_{IN} (6))
NC	(13)	PB3	D5	(28)	11U33 (D_{IN} (5))
		(ADDRESSES 8010–8013)			
NC	(12)	PB2	D4	(29)	11U32 (D_{IN} (4))
19U27 (A10)	(11)	PB1	D3	(30)	11U31 (D_{IN} (3))
22U27 (A9)	(10)	PB0	D2	(31)	11U30 (D_{IN} (2))
23U27 (A8)	(9)	PA7	D1	(32)	11U29 (D_{IN} (1))
1U27 (A7)	(8)	PA6	D0	(33)	11U28 (D_{IN} (0))
2U27 (A6)	(7)	PA5	$\overline{RESET}$	(34)	40U6 (RESET 6800)
3U27 (A5)	(6)	PA4	RS1	(35)	2U29 (A1)
4U27 (A4)	(5)	PA3	RS0	(36)	1U29 (A0)
5U27 (A3)	(4)	PA2	$\overline{IRQB}$	(37)	
6U27 (A2)	(3)	PA1	$\overline{IRQA}$	(38)	4U6 ($\overline{IRQ}$)
7U27 (A1)	(2)	PA0	CA1	(39)	NC
8U27 (A0)	(1)	GND	CA2	(40)	NC

FIGURE 13-1 Wiring diagram used to reduce errors during wiring.

common cause for a prototype circuit not working is wiring errors.

BUILDING AND TESTING THE PROTOTYPE HARDWARE

Often, in the rush to get a prototype built fast, there is a tendency to build all or a large section of the unit, turn on the power, and hope it works. Because errors or malfunctions tend to interact, this large section can be very difficult and time-consuming to troubleshoot. In the long run, the quickest way to get a prototype up and running is to build both hardware and software in small modules that can be built and tested one at a time. The smaller the sections or modules are built initially, the easier they are usually debugged. As sections are debugged, they can be linked together to form larger sections and these sections tested and debugged until the entire unit is assembled.

BASIC CPU MODULE   To simplify the prototyping of a microprocessor-based instrument, it may be built on a ready-made development board such as the SDK-85 or MEK6800D2 used for the examples in Chap. 12. Boards such as these are available for almost all processors. Most have on-board monitor programs in ROM that allow programs to be entered into RAM and run. Some have on-board keypads and seven-segment displays. Other boards require a teletype or CRT terminal to communicate with them. In addition to the basic clock, CPU, and monitor ROM ICs, most boards have a few hundred or thousand bytes of RAM, two or more parallel I/O ports ICs, and a serial I/O port IC. Socket space may be included on the board for more RAM, ROM, and I/O ports. Usually a wire-wrap area is included for user-added circuits.

Most microprocessor development boards come assembled and tested from the manufacturer. Those which come as kits such as the SDK-85 and the MEK6800D2 are quickly and easily assembled. If you build them carefully, usually they work the first time. The main point is that these boards get the prototype well along with few problems. Later we discuss how the CPU module is developed if a ready-made board is not used. Here we show how the prototype is built, tested, and debugged, using one of these development boards.

ADDING RAM AND ROM MODULES   When you are adding RAMs or ROMs, position the sockets so that address and data lines are easily daisy-chained from one to the next. Add a 0.1-μF bypass cap between V_{CC} and ground, next to each IC. Check all connections carefully. Then, before you plug in all those expensive memory ICs, check with an ohmmeter to make sure all wires go where they are supposed to go. Especially with wire wrap it is very easy to connect a wire one pin off when you are daisy-chaining address lines down a row of RAMs. All this checking may seem excessive, but it takes only a few minutes and often reveals errors that might take much longer to find with the memories operating.

Once the memories are powered up, the next job is to test them. For ROMs, a ROM with a known program can be used. It is read out with the monitor memory examine command and compared with a printout of the known program. Not every location has to be checked, but enough locations should be randomly checked to make sure all address lines are working correctly. If the ROM does not read out, the next section of the chapter on RAMs will discuss how to track down the problem.

For RAMs both read and write functions must be tested. An easy way to do this is to write a short program

```
START:  LXI     B,(# OF BYTES)              ;INITIALIZE BYTE COUNTER
        LXI     H,(START ADDRESS OF MONITOR)
        LXI     D,(START ADDRESS OF RAM)
LOOP:   MOV     A,M                        ;MOVE BYTE FROM MONITOR
        STAX    D                          ;WRITE TO RAM
        DCX     B                          ;DECREMENT BYTE COUNTER
        MOV     A,C                        ;CHECK IF ALL WRITTEN
        ORA     B                          ;
        JZ      TEST                       ;YES GO VERIFY
        INX     H                          ;POINT TO NEXT LOCATION IN
                                           ;MONITOR
        INX     D                          ;POINT TO NEXT LOCATION IN
                                           ;RAM
        JMP     LOOP                       ;GO WRITE NEXT BYTE
TEST:   LXI     B,(# OF BYTES)
        LXI     H,(START ADDRESS IN MONITOR)
        LXI     D,(START ADDRESS IN RAM)
LOOP2:  LDAX    D                          ;READ RAM
        CMP     M                          ;COMPARE WITH MONITOR
        JNZ     EXIT                       ;EXIT IF WRONG
        DCX     B                          ;DECREMENT BYTE COUNTER
        MOV     A,C                        ;CHECK IF BYTE
        ORA     B                          ;COUNTER EQUAL ZERO
        JZ      EXIT                       ;EXIT IF DONE
        INX     H                          ;ADVANCE TO NEXT LOCATION
        INX     D                          ;
        JMP     LOOP2                      ;GO READ NEXT RAM BYTE
EXIT:   RST1
```

FIGURE 13-2 8080A/8085A routine to test a section of RAM.

to copy all or some of the monitor program into the RAM to be tested. Figure 13-2 shows a routine to do this for an 8080A/8085A system. The routine reads a byte from the monitor program and writes it into a RAM location. The HL and DE pointers are incremented to point to the next location in the memories, and the byte counter is decremented. After all locations are written to, the pointers are reinitialized. Then each location is read from the RAM and compared with the corresponding byte from the monitor ROM. If the two bytes do not match, then the program jumps to the exit instruction, RST 1. If the bytes match, the program loops until all locations have been checked and returns control to the monitor program. For SDK-85, the RST 1 instruction saves all register contents on the stack and returns control to the monitor program. The examine register command can be then used to read out the contents of each register on the displays. For the example in Figure 13-2, after execution the H and L registers will show the last address read from the monitor program, and the D and E registers will show the last address read in RAM. If this last read address is the same as the last address of the added RAM, then a write to this RAM and a read from it were successful.

If the read-write operation was not successful, the next step is to find out why. How far the verify routine runs before exiting with an error will give you a clue to the problem. If the program runs most of the way through the RAM addresses before exiting, then a bad IC or an inactive high-order address line may be the problem. As with any IC, make sure you read the IC numbers carefully. A 2114, for example, is a 4K-bit static RAM organized as 1K × 4, and the 2141 is a 4K-bit static RAM organized as 4K × 1. They are obviously not interchangeable. If the program exits after a compare of the first bytes, then the problem could be the test program or signals not reaching the RAM ICs. The write-verify program of Figure 13-2 can be tested on some of the pretested RAMs that came with the board to eliminate this as a source of the problem. With this eliminated, the next step is to see whether the RAM is getting the proper signals and data. If the JZ TEST instruction in the test program of Figure 13-2 is replaced with an unconditional jump to start (JMP START), the program will be made into an endless loop. When run, it will attempt over and over to copy the monitor into the RAM. You can use an oscilloscope to check whether address, data, chip enables, write signals, etc. are getting to the device properly. Another endless loop routine can be written to check whether the proper signals are getting to a RAM or ROM for a read operation. In this way a combination of hardware and software is used to find the source of a problem.

I/O PORTS AND PROGRAMMABLE PERIPHERAL ICs   After the required ROM and RAM have been added

PROTOTYPING AND TROUBLESHOOTING µP-BASED SYSTEMS

and debugged, any additional I/O ports are wired in, and the connections are checked with an ohmmeter. Then a short program may be used to check each port. If a PIA is used for the ports, the program must first initialize the PIA. If it is initialized as an output port, the program can write a byte to it. Then you can use an oscilloscope or logic probe to find out whether the byte got to the output lines. If the written byte did not get to the output, then you can make the write-to-port program into an endless loop with a jump or branch instruction. As the program writes the byte over and over to the port, you can check with an oscilloscope or logic analyzer that the correct address, data, and enable signals are getting to the device. If the correct signals are getting to the device and it still doesn't work, recheck the initialization of the PIA in the program. Remember, some programmable peripheral ICs require that two or more control words be sent to initialize the device.

An input port can be tested in a similar manner. After being initialized, the port is read into the accumulator. The examine register command is used to display the accumulator contents. Resistors jumpered temporarily from the input lines to V_{CC} or ground can be used to put a data byte on the input port lines. Another endless-loop program that reads the port over and over allows an oscilloscope check that proper address, data, and enable signals are getting to the device.

Other programmable peripheral ICs are added and debugged in a similar manner. If the device is being used to output data, then a short program can be written to initialize the device and send data to it. An oscilloscope or logic analyzer can check whether the peripheral IC puts out the correct data. If the device is used to input data, then input signals can be simulated with resistors or pulse generators. Again, a short program can be written to initialize the device and read from it to the CPU. Here the data read can be examined to see whether it is correct.

EXTERNAL INTERFACE CIRCUITS The exact procedure used to build up and test external interface circuits depends on the specific circuit being built. The main point to remember is to, as much as possible, build and debug the circuits in sections.

In the microprocessor-based scale in Chap. 12, for example, the load-cell amplifiers were built and tested first. When these were working correctly and calibrated, the MC14433 A/D converter was added. An oscilloscope triggered on the digit-strobe output was used to check the BCD data out for each digit.

After the A/D converter was connected to an input port, the first section of the scale program was used to read the BCD weight data and store it in memory.

As another example, suppose you have to build the interface to an external display such as that on the MEK6800D2 display module in Figure 11-14. First wire the display segments in parallel. For common-cathode LED displays, connect all the segment lines temporarily to +5 V with 150-Ω resistors. Test each digit by grounding its two common cathodes. Each digit should show

an "8." Then remove the resistors on the segment leads and add and test the segment-driver transistors. When these are verified, add and test the digit-driver ICs. After all the display hardware is operating correctly, connect it to an output port and try to write data to the display with a program routine. A test routine might turn on all segments and cycle through, turning on one digit at a time for 1 ms. If the programmer has written the display scan program, it can be entered at this time and tested.

When the display is working satisfactorily, the keyboard can be wired and connected to the U10 encoder. After the hardware is debugged, the key scan routine can be entered and tested.

The above examples may seem simple-minded, but the need to build circuits in easily testable and debuggable sections cannot be overemphasized.

TESTING, DEBUGGING, AND LINKING PROGRAM MODULES

There are several methods to test program modules and debug them. One is to use breakpoints and the monitor single-step and examine register commands. This technique was used to develop the scale and EPROM programmer examples in Chap. 12. A second method makes extensive use of a logic analyzer. A third powerful, but very expensive, method uses an extension of a development system such as that in Figure 8-26. This technique is often referred to as *in-circuit emulation*. We discuss the three methods in the following sections. First we show how breakpoints and monitor commands were used to develop the programs for the first two examples in Chap. 12.

USING BREAKPOINTS AND MONITOR COMMANDS OF THE SDK-85 TO TEST AND DEBUG THE SCALE PROGRAM

The general steps for using this method are as follows. First, choose a logical and reasonable-size program module to start on. In the scale program, for example, a logical module to start with is the section to take in data from the A/D converter. Another suitable section might be the display routine DSPLY.

Second, look carefully at the section chosen to see what initialization is required. Determine the inputs required and the outputs produced. You may have to add a few program steps to the start of the module so that it can "stand by itself." For example, to test a multiplication routine, you have to load the input registers for the routine with known data so you can tell whether the output is correct. For example, the BCDCVT subroutine in the scale program converts four packed BCD digits in the D and E registers to a 16-bit binary equivalent. Therefore, an LXI D,1234H instruction should be added to the start of the routine for testing. The 1234 loaded into D and E should produce a result of $04D2_{16}$ in H and L at the conclusion of the routine.

An instruction must be put at the end of the module to return control to the monitor program when the module is finished. For the SDK-85 board the restart 1 (RST 1 or CF hex) instruction does this. The RST 1 instruction vectors the program execution to address 0008H. The SDK-85 monitor program has at this address a routine to save all registers in memory and return control to the monitor. The examine register command of the monitor can then be used to display the data that was contained in each register before the RST 1 instruction. This stop in program flow and return to monitor control with registers saved is often called a *breakpoint*. Remember, the object is to fix the program module so that it can stand by itself. It must have defined inputs and predictable outputs to be testable.

Third, enter the hex codes for the program module into the on-board RAM, using the substitute memory command of the monitor program. This is done on the SDK-85 as follows. After a reset, press the substitute memory key. Then enter the first hex address with the keypad. Press the NEXT key to display the data at that address. Enter the desired hex code by just pressing the desired keys. Press the NEXT key to step to the next address. When all the hex program bytes have been entere, press the EXEC key to return to monitor control.

Fourth, run the program. To do this on the SDK-85, press the GO key, enter the starting address of the program module, and press the execute (EXEC) key. When the program section finishes and control returns to the monitor, use the examine register command to find out whether the program produced the correct results in each register.

DEBUGGING A PROGRAM MODULE

Now, if the program module did not work correctly, as is often the case, start the debug procedure. If the program module was hand-assembled, check carefully to see that the correct hex code was entered for each instruction. Recheck the jump addresses. Remember, for the 8080A/8085A the low-order address byte is loaded into memory first and then the high-order byte. Make sure "immediate" instructions have the second or second and third bytes present with the immediate data. Make sure that push operations are followed by the same number of pop operations in the proper order. If a program has more pushes than pops, the stack will grow downward in RAM each time the routine is run and may write over the program in RAM. Then compare the program in RAM with the written version to make sure that a typing error did not put the wrong hex code in RAM and that no instructions were left out. If errors are found here, they can be corrected and the module retried. If no error was found or the program still doesn't work, the next step is to reduce the size of the section being tested.

The size of program module can be reduced by simply putting a breakpoint earlier in the program. To do this, just replace an instruction with CF, the hex code for RST 1. Refer to the "get data from A/D converter" section of the scale program in Figure 12-6. A good place to put a breakpoint might be in place of the ANI instruction at

address 001DH in the printout. If the section of the program up to this point operates correctly, then the breakpoint exit to the monitor should leave the MS digit from the A/D converter in the accumulator. At this point, three major outcomes are possible. First, the breakpoint may exit with the correct character in the accumulator. In this case, the ANI instruction can be replaced and the CF moved to a later place in the program to test the next small section. Second, the breakpoint may exit with the wrong character in the accumulator. In this case, check the order in which the BCD and strobe lines are connected to the input port.

A third possibility, which is quite common in programs containing loops, is that the program execution may get stuck in a loop and never reach the breakpoint. In this case, for the scale example leave the program running around the loop and check with an oscilloscope to make sure the digit strobe is getting to the proper pin of the I/O port. If *it is*, trigger the oscilloscope on the rising edge of the I/OR input of the 8755 input port, and see whether the strobe is getting through the 8755. To escape from the loop and get back to monitor control, press the reset button.

If inserting breakpoints with the RST 1 instruction does not help you find the trouble, then try the *single-step* command. As the name implies, this command allows you to step through execution of your program one instruction at a time. To use the single-step command on the SDK-85, the SINGLE STEP key is pressed and the starting address entered with the keyboard. When the NEXT key is pressed, the instruction at this address executes and the display increments to the next instruction address. If the EXEC key is pressed and then the EXAM REG key is pressed, you can check all the registers. Pressing SINGLE STEP again and then pressing NEXT execute the next instruction in the routine. This is an easy way to essentially put a breakpoint after each instruction. If the correct data is not appearing in a register at a given point, you can insert the correct data and step through the rest of a routine. This will tell you whether the rest of the routine works if it is given correct data. Then you can go back and find out why wrong data got into a register at a given step. Since each program routine is different, you have to think out the best approach for it. Try to think of experiments you can perform to narrow down the problem and fix it.

LINKING PROGRAM MODULES

After adjacent modules are entered in RAM and debugged, they can be linked into larger modules. For example, in the scale program, the section that takes in data from the A/D converter could be linked to the DSPLY routine. This displays the weight on the four LED displays. After the KYBINT routine is debugged, the next section of the program from line 82 to line 90 can be added. This completes the debugging of the dumb-scale module of the program.

The rest of the program is linked together in a similar manner. A routine is first debugged independently and then added onto the debugged main program.

USING BREAKPOINTS AND THE MONITOR COMMANDS OF THE MEK6800D2 TO TEST AND DEBUG THE EPROM PROGRAMMER PROGRAM

The general steps to test a program module or routine on the MEK6800D2 are very similar to those discussed above for the SDK-85, but owing to variations in the monitor program, specifics are different. Therefore, the procedure is repeated here for the EPROM programmer program with the MEK6800D2.

First, choose a logical and reasonable-size program module to start with. In the EPROM programmer program, for example, it might be good to start with the initialization and then the program module that checks whether the 2716 to be programmed is erased.

Second, look carefully at the section chosen to find out what initialization is required. Determine the required inputs and the expected outputs. You may have to add a few program steps to the start of the module so that it can stand by itself. For example, to test the delay routine, you will have to load the delay constant at the start. Since the purpose of this routine is to control the critical width of an output pulse, you may wish to add steps to make the routine put out a continuous train of pulses. You can then verify the 50-ms pulse width with an oscilloscope on PB1 of *U26*. Figure 13-3 shows the additions necessary to do this. Remember, anytime you are going to use a PIA in testing a module, you must make sure it is initialized correctly.

If a program module is not going to be made into an endless loop for testing, as was done with the delay loop above, then an instruction must be put at the end of the module to return control to the monitor. If this isn't done, the processor will finish the module and step on through RAM, trying to execute the random garbage found there.

For the MEK6800D2 a *software interrupt instruction* (SWI) placed at the end of a routine returns control to the JBUG monitor program. SWI also pushes the program counter contents, index register contents, and the *A* and *B* accumulator contents on the stack. The examine contents of MPU registers command, R, of the JBUG monitor can be used to display the contents of these registers.

Remember, to be testable a module must have defined inputs and predictable outputs. In addition, make sure that a test is nondestructive. For example, in the EPROM programmer, a V_{PP} of 25 V might overheat and destroy a 2716 if left on for too long. Since an empty socket will read into a 6821 PIA as all 1's, the first part of the program can be tested with the socket empty. As the first part of the program runs, the V_{PP} pin can be checked to make sure it isn't accidentally getting the 25 V at the wrong time.

Third, enter the program module into the on-board RAM, using the examine and change memory command, M, of the JBUG monitor. Enter the starting address and hex code desired in that location with the keypad. Pressing the G key increments the address and allows the next program byte to be entered. When all the program is entered, press the E key to return control to the monitor. Make sure to enter the required starting and ending addresses in RAM locations 0002H through 0007H.

Fourth, run the program module. To do this, enter the starting address and press the G key. If the module executes to the SWI instruction, a prompt will reappear to tell you the monitor is waiting for your next command. The R command can be used to look at registers. If the program gets stuck in a loop and doesn't reach the SWI, then press the E key to get back to monitor control.

DEBUGGING A PROGRAM MODULE If the program section did not work correctly, check that the right hex code was entered for each instruction and that no instructions were left out. If the program was hand-assembled, check that the written program has the correct hex code for each instruction. If no coding or typing errors are found, then reduce the size of the module being tested.

The size of the program module can be reduced by simply inserting the SWI instruction earlier in the program section. SWI produces a breakpoint in the program and allows you to check register and memory buffer contents with the monitor commands. In the EPROM programmer program, for example, suppose the section of the program to verify all 1's in the 2716 doesn't work. The SWI breakpoint can be moved back to address 003E in place of the CMPA #$FF instruction. At

```
START: LDX    #$8000    ;LOAD BASE ADDRESS OF PIA'S
       LDAA   #$FF      ;LOAD CODE FOR OUTPUT
       STAA   06,X      ;SEND TO PIA
       LDAA   #$04      ;LOAD CODE FOR PORT CONTROL
       STAA   07,X      ;SEND TO CONTROL PORT OF U26
       LDAA   #$06      ;LOAD CODE FOR PB1 HIGH
       STAA   06,X      ;SEND TO PB OF U26
       LDX    #$0F00    ;LOAD DELAY CONSTANT
DELAY: DEX              ;DECREMENT
       BNE    DELAY     ;LOOP IF NOT ZERO
       LDAA   #$04      ;LOAD CODE FOR PB1 LOW
       STAA   $8006     ;SEND TO PB OF U26
       JMP    START     ;REPEAT OVER AND OVER
```

FIGURE 13-3 6800 routine to test DELAY routine from EPROM programmer.

this point the program will have addressed and read only the first address in the 2716. The index register should contain 8000H, the A accumulator should contain FF, and the B accumulator should contain the least significant byte of the address sent to the 2716. Also, the pins of the 2716 can be checked with an oscilloscope to see whether the proper address and enable signals got to them. If the address and enable signals did not get to the 2716 socket, a possible cause is incorrect initializing of the PIAs. If signals got to the 2716 socket but are incorrect, then this is a clue to which part of the module has the problem.

To help in tracking down problems in a program module, the MEK6800D2 has two very useful commands: breakpoint insertion and removal, V, and trace one instruction, N.

Breakpoints inserted with the V command have the advantage that the instruction replaced by an SWI instruction is saved by the monitor. The SWI routine also saves all register contents on the stack. After registers have been checked, pressing the E key and then the G key causes the instruction to be reinserted in RAM. Then the program executes to the next breakpoint. This permits you to easily step through a program module from breakpoint to breakpoint. The JBUG monitor allows up to five breakpoints of this type to be inserted. Each is inserted by entering the desired address and pressing the V key. The V command is faster and gives less chance of error in inserting and removing breakpoints than substituting directly. The V command breakpoints are removed by simply pressing the V key when a prompt character is present.

The trace one instruction command, N, allows single-stepping through a program one instruction at a time. After the program has been run to a breakpoint, pressing the N key causes the processor to execute the next instruction and stop. Registers can be examined. Pressing the N key again causes the processor to execute the next instruction and stop again. In this way you can single-step through a program module, checking registers at each step to find a problem.

LINKING PROGRAM MODULES   After a program module is debugged, it can be linked to another module and the combination tested. In the EPROM programmer program in Figure 12-17, for example, after the section that initializes the PIAs and tests the 2716 for all 1's is debugged, add the EXIT routine. This routine is supposed to display the address of the first byte not all 1's read from the 2716 and the byte read there. Running the all-1's test on an empty socket will always read all 1's, and the program will never take the error exit. The program can be tricked into taking the error exit by jumpering one of the 2716 socket data pins to ground. The EXIT routine should then display the first address tested in the 2716 and the byte you programmed with the jumper to ground. If the section does not work, then breakpoints or trace one instruction can be used again to track down the problem.

When this section operates correctly, the next section to add is the actual 2716 programming module starting at address 0070H in Figure 12-17. This section can be tested initially with no 2716 in the programming socket. You might want to insert a breakpoint first at address 009BH and run to that point. You can then check that the proper address, data, and enable signals are getting to the socket. The next place to put a breakpoint might be at address 00A1H. After the program runs to this breakpoint, you can check whether the PGM pulse ($\overline{CE}$ equals 1) is getting to the 2716 socket. A next logical place to put a breakpoint is at address 00ACH. Here you can check that the PGM pulse gets turned off again. When all checks out to this point, remove the earlier breakpoints and insert one at address 00BCH. This allows the program to loop through PGMLUP. If you have not already checked the accuracy of the 50-ms delay, you can connect an oscilloscope to the $\overline{CE}$/PGM pin of the 2716 socket and measure the PGM pulse width as the program loops.

After this section is thoroughly tested, the VERIFY section of the program is entered into RAM and tested with the previous sections. If an empty 2716 socket is used again, then the VERIFY program takes the FAIL exit. The FAIL routine displays all Fs on the seven-segment LED displays. A good place to put a breakpoint in testing the VERIFY section is at address 00D0H. After the program runs to this breakpoint, you can check whether the V_{PP} of 25 V was turned off and the 2716 was enabled to output. This is important because if V_{PP} is left at 25 V too long with $\overline{CE}$ high, the 2716 will overheat and die.

If these signals are correct and the program takes the FAIL exit without problem, you can insert an erased 2716 in the programming socket and attempt to program it. Note that the relatively expensive 2716 is not inserted until everything that might destroy it has been thoroughly checked out! Now, instead of trying to program the entire 2 kilobytes of the 2716, try to copy 100 bytes or so from the monitor program into it. Since programming the entire 2716 takes about 2 min, you would have to wait that long to find out whether the program worked. Also you would have to erase the 2716 or use another one to make another test.

If the 2716 programs correctly, then after a brief time the display should show all As. If the display shows all Fs, then you have to find out whether the problem is in the programming section or the verify section. To find out whether the 2716 was programmed, move it to the ROM-to-be-copied socket at address 6000H. You can manually read it out, using the examine and change memory command, M, of the JBUG monitor, and compare the data read with a printout of the monitor program. If the 2716 contains the correct programming, then reinsert it in the programming socket and breakpoint or single-step your way through the verify program module.

All the checking and testing described may seem very laborious, but experience shows that it is the fastest way to debug and produce a working prototype.

BUILDING, TESTING, AND DEBUGGING A PROTOTYPE WITHOUT A DEVELOPMENT BOARD AND MONITOR PROGRAM

Developing the prototype of a microprocessor-based instrument without the benefit of a development board and monitor program can be quite time-consuming and frustrating unless a logic analyzer or emulator is available. In this section we show first how a logic analyzer is used to develop hardware and software and then how an emulator is used.

TESTING AND DEBUGGING HARDWARE BY USING A LOGIC ANALYZER

Before we discuss the use of a logic analyzer with microprocessors, it is a good idea to go back and review the discussion at the end of Chap. 5 on logic analyzers and their display formats. As was pointed out there, a logic analyzer is similar to a multichannel oscilloscope. Data on the parallel inputs is sampled, and the samples are stored in a memory. When a trigger occurs, the data samples stored in memory are displayed in the desired format on a CRT. The trigger can come from some external source or from a trigger qualifier circuit. A trigger qualifier circuit compares the input data word with a word programmed by the user with switches. When the two words match, the qualifier sends out a pulse which triggers the analyzer to display the contents of memory. Most trigger qualifiers also have an output to the front panel which can be used to trigger a standard oscilloscope.

Most analyzers have an adjustable trigger delay which permits a *pretrigger*, *center-trigger*, or *posttrigger* display of the data streams. For a unit with 1024 memory bits on each input channel, *pretrigger mode* shows the 1024 samples taken before the trigger. *Center-trigger mode* delays the trigger pulse to the display 512 counts, so the display shows 512 samples before the trigger and 512 bytes after the trigger. *Posttrigger mode* delays the trigger to the display 1024 counts, so the display shows only the 1024 samples after the trigger.

The clock that determines when samples are taken can come either from the system clock of the unit being tested or from an internal asynchronous oscillator. If the system clock is used, the instrument is often called a *logic-state analyzer* because each data sample corresponds to a state or machine cycle of the unit being tested. A logic-state analyzer is an excellent tool for debugging software and program flow. However, unless it has latch-mode capability, a logic-state analyzer is not very useful in detecting glitches.

To detect glitches, an asynchronous clock of 5 to 10 times the system clock frequency is used to take data samples. An instrument that can use this high-speed asynchronous clock mode is often called a *logic-timing analyzer*. As the name implies, this type is useful for finding timing problems and glitches.

BRINGING UP A MICROPROCESSOR WITH A LOGIC-STATE OR LOGIC-TIMING ANALYZER   The first step in "bringing up" is to wire and test the clock and reset circuitry. If the clock and reset circuits are contained in a separate IC, as is the case for either an 8080A or a 6800 system, then wire this and check its output before you connect the microprocessor. If the clock and reset circuitry is contained in the same IC as the microprocessor, then make sure that the proper logic level is connected to each pin of the device. Remember, most MOS inputs cannot just be left open! A standard oscilloscope is probably the best tool to check the clock pulse levels and timing.

When clock and reset circuits check out, wire the microprocessor IC. Make sure every pin is accounted for. Using 10-kΩ or larger resistors, hard-wire the data bus lines with the binary code for a NOP instruction. For the 8080A/8085A, the code for NOP is 00H; for the 6800 the code for NOP is 01H. After a reset/restart, the processor should show a counting sequence on the address lines. A logic-state analyzer using the system clock and triggered on address 0000H will show this. At this point, also check that a memory read signal is present.

In an 8080A system, the next step is to wire the 8228 system controller and bus driver if one is used. The NOP resistors are moved to the memory side of the 8228, and the above test is repeated.

In a system that does not have an 8228-type device, or after the 8228 is checked out, the next part to wire is ROM and RAM decoders, if used. Connect them to the address lines and power. Set the trigger qualifier switches of a state analyzer to trigger on the first address to be selected in the block of memory to be chosen by the first decoder output. As the processor continuously performs NOPs, the decoder output should show a pulse for each address in the selected block. After one chip-select output is tested, the trigger qualifier switches can be changed to produce a trigger on the first address of the next block selected by the decoder.

When the decoders are operating properly, you can wire a ROM socket. Connect address lines, enable lines, and power to the device, but leave the data outputs unconnected. Insert a ROM with a known program in the socket. As the processor performs NOPs, it will step through each location in the ROM. A state analyzer connected to the outputs of the ROM and set to trigger on the first word in the ROM allows you to see whether the words are being read out of the ROM in the proper order. If they aren't, then use the state analyzer to check that the addresses are arriving at the pins of the ROM in the proper order. A common problem in wiring memories is interchanged address lines. If the ROM readout sequences properly, then remove the NOP resistors and connect the ROM output to the processor data lines.

For the rest of the prototype development, program routines in an EPROM can be used to make tests. For example, to add RAM to the system, sockets are wired in. A program to write an increasing number to each successive location in the RAM is put in the EPROM and run. Checking the RAM socket with a state analyzer indicates whether the proper addresses, data, and enable signals are getting to it. If they are, then the RAM ICs can be inserted, and another routine can be run to write to RAM, read back, and verify. If the proper logic signals seem to be getting to the RAM and it still doesn't program or read, a standard two-channel oscilloscope can be used to accurately check timing and to make sure all address, data, and control signals are solidly within legal levels.

I/O ports are wired and tested in a similar manner. A short program loop in the EPROM writes a word to the port. A state analyzer triggered on the address of the port shows whether the data was output from the port at the correct time. Writing both all 1's and all 0's to the port should be tried.

So far, most of this prototype development has been done using a logic-state analyzer. An example will show why a logic timing analyzer is also a valuable tool. Suppose that all the hardware seems to be operating correctly except that an interrupt input line gets randomly pulsed when a test program is run. To find the problem, the interrupt input line is used as a trigger for a timing analyzer. The multiple inputs of the analyzer can show activity on the address bus, data bus, and control bus. Look for a transition on one of these signal lines that corresponds to the pulse on the interrupt line. A read or write pulse, for example, might be coupled into the interrupt line if the wires are wrapped around each other.

Once the hardware is built and debugged, your next task is software testing and debugging. A logic state analyzer can be very useful for this also.

TESTING AND DEBUGGING SOFTWARE BY USING A LOGIC-STATE ANALYZER

As with debugging programs by using a monitor program, the key to successful debugging with a logic-state analyzer is to test and debug small sections at a time.

If the unit being built contains a display, the routine to send data to the unit is a good place to start. The display routine, when debugged, can act as a window into the system. A program module can send register contents or the results of a calculation to the display for you to check.

The next module to work on should be the keyboard scan routine if a keyboard is used. When debugged, this routine with a few additions should allow you to enter data into RAM. Then you can test and debug other program modules in a logical order for the specific instrument being built.

As described earlier, the program module must be made so that it can stand alone. Make sure peripheral ICs are initialized. Make sure that a HALT or JMP instruction is put at the end of the module, to prevent the CPU from trying to execute the random words stored in the rest of memory. When the program module is set up to stand alone, "burn" the module into an EPROM. Then plug the EPROM into the prototype board and run the program module.

This is the point at which a logic-state analyzer enters the picture. Use the analyzer to monitor the data sent to ports or to memory. If the program doesn't work, you can, if you have enough inputs on the analyzer, look at the data bus and address bus flow as the program executes. If the analyzer has only 16 inputs, then look at the address bus to see whether the program module steps through in proper sequence. The analyzer may show the address line contents in a 1's and 0's form, as in Figure 5-23b, or in a more easily readable hexadecimal form, as in Figure 5-23c. Typically, 16 words plus the trigger word are shown at a time in these formats. If the address lines are sequencing properly, then connect the analyzer inputs to the data bus and see whether the correct sequence of instruction and data occurs on the data bus. If no errors are found in the first 16 samples, change the trigger word to show the next 16 addresses and run the program again. Then you can check the address bus sequence and data bus sequence for this block. You may find that the displacement for a relative branch instruction was miscalculated and the program branches to the wrong address. Perhaps the wrong hex code for an instruction was entered into the EPROM, or a data word is not being read in from an input port. The main point here is that the analyzer lets you see what the processor is doing on a step-by-step basis.

As each program bug is found, it is corrected. The EPROM is erased, reprogrammed with the corrected version, and retried. The cycle is repeated as many times as necessary to get the module working. When one module is working, another can be added to it in the EPROM and debugged with the aid of the analyzer.

DEVELOPING HARDWARE AND SOFTWARE WITH AN IN-CIRCUIT EMULATOR OR SIMULATOR AND A DEVELOPMENT SYSTEM

DEVELOPMENT SYSTEM REVIEW   Since in-circuit emulators or simulators are used with microprocessor development systems (Figure 8-26), a good place to start is with a review of the function of a development system. Development systems usually contain a control unit, a CRT terminal for entering programs, one or more floppy disk drive units for storing programs, and a line printer to make "hard copies" of programs. The control unit typically has a CPU and a monitor program in ROM. The monitor program allows programs to be entered and run. In addition, the control unit contains a large RAM, perhaps of 64 or 32 kilobytes, and several I/O ports.

Two of the main functions of a development system are to *edit* and *assemble* assembly language programs. To do this, first the *editor program* is loaded into RAM from a floppy disk or tape. The editor program allows you to enter an assembly language program with comments into RAM with the CRT terminal. The area in RAM where the program is entered is called the *editor workspace*. Various editor commands make it easy to add or remove instructions without retyping the entire program.

When the entire program or the desired program module is entered into the editor workspace, it is loaded onto a disk or tape. This is called a *source file*. Then an *assembler program* is loaded into RAM. If the development system has enough RAM, the editor and assembler can be in RAM at the same time. If not, the assembler replaces the editor in RAM.

The assembler program, when run, usually requires two passes to assemble a program. In the first pass, the assembler reads the source file and makes a *symbol table* in RAM. What this means is that the assembler program equates an *absolute address* with each *symbolic address* used in the program. The listing at the end of the scale program in Figure 12-6, for example, is the symbol table. On the second pass, the assembler produces the *machine*, or *object*, *code* for each instruction and address. According to directions you give it, the assembler can do several things with the object code produced. It can send the code to a disk or tape for storage. This is called *creating an object file*. The assembler can send the machine code and the assembly language program with comments to a printer to produce a printout such as that for the scale in Figure 12-6. This is called a *list file*. Another option is that the assembler can simply *load* the machine code for the program into a specified area of RAM. This last option is valuable for use with an emulator.

EMULATOR HARDWARE

The main concept of an emulator is to use the large RAM, I/O ports, and control functions of the development system in place of parts of the board or system being built. In many ways an emulator functions as a very powerful monitor program, like those described for the development boards earlier. To do this, the basic CPU group hardware of the prototype is assembled, and ohmmeter checked. A socket at the end of the umbilical cable from the in-circuit emulator is plugged into the prototype in place of the CPU IC. Figure 13-4 shows a photograph and block diagram of the Intel ICE-85 emulator hardware.

The two large PC boards plug directly into the Intel Intellec control chassis shown in the development system in Figure 8-26. One board is the *trace board*, and the other is the *processor board*. A section of ribbon cable connects the processor board to a module containing bidirectional buffers. A multiconductor cable from the other side of the buffer module connects to a 40-pin male IC connector, which plugs into the 8085A socket of the prototype. To the system being developed, the response of the emulator through this umbilical cable is exactly the same as if an 8085A were plugged into the socket. In other words, the signals sent out to the prototype and the responses to signals from it are the same as for an 8085A in the socket.

USING AN IN-CIRCUIT EMULATOR

Most emulators or simulators have a lengthy manual which describes their operation and use. Therefore, if you get to work with an emulator, refer to the specific manual for it. In this section we give an overview of emulator use and capabilities.

The overall principle is to at first use the RAM, ROM, and I/O ports in the development system in place of those on the prototype. As each part of the prototype hardware is added and debugged, it takes over its function from the part in the development system.

To start using the emulator, the basic CPU-group ICs of the prototype are wired, and in place of the CPU, the umbilical cable from the emulator is plugged in. The CPU group can then be tested with a program in the development system RAM. Here is where the emulator comes into the operation.

The emulator executes the test program and, in the case of the ICE-85, stores 4 bytes of information for each machine cycle executed in a RAM on the trace board. The 4 bytes stored are status, address byte high, address byte low, and data. They are sometimes called *snap data*. For the ICE-85, the snap data for the last 44 machine cycles before a breakpoint is saved in the trace RAM. These snap bytes can be read out or printed to trace the program flow on the address and data buses. Emulator commands allow breakpoints to be set at any point. Programs can be executed at full speed or one step at a time.

When one section is debugged, another is built. The program routine to test it is edited, assembled with the development system editor and assembler functions, and run with the emulator function. Having all the units connected in this manner speeds up the prototype development cycles. If an error is found in a program module, you can jump back to the editor function, correct the error in the source program, assemble the corrected source program, and retry the corrected version without leaving your chair.

In addition to the snap data, the emulator has some other very powerful features for helping you find bugs. When the trial program runs or "emulates" to a breakpoint, it enters the *interrogation mode*. In this mode you can ask the emulator program to print or display memory contents, register contents, flags, or even the signals on the HOLD, INTR, READY, and RESET pins of the 8085A. Other commands allow you to change any of these and continue the emulation of the program.

As mentioned before, an emulator functions in many ways as a very powerful version of the monitor programs on development boards.

PRODUCTION PROCESSES

We have discussed the tools and techniques used to test and debug the prototype of a microprocessor-based instrument during development stages. Now we examine the manufacturing steps from this prototype to a saleable instrument.

PRODUCTION PROTOTYPE RUN

After the prototype of an instrument is thoroughly tested and debugged in the *development engineering department*, the *production engineering department* designs a printed-circuit board, chassis, case, connec-

FIGURE 13-4 *(a)* Intellec in-circuit emulator/85 (ICE-85) hardware. *(b)* Block diagram of the ICE-85. *(Intel Corporation.)*

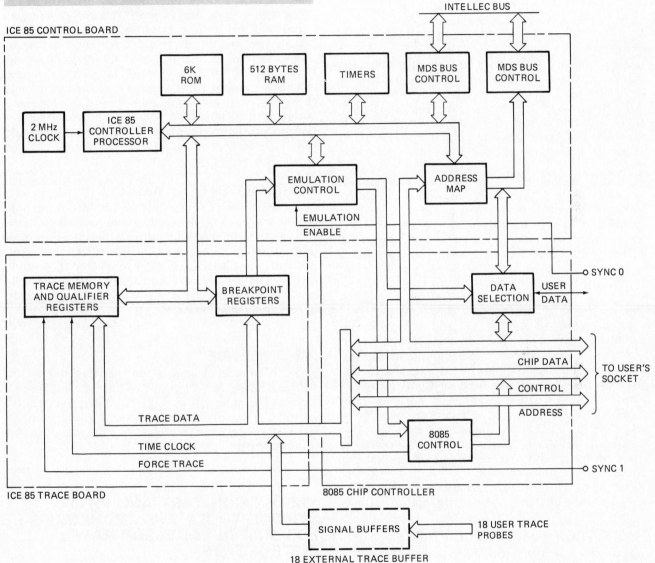

INTELLEC BUS

ICE 85 CONTROL BOARD

6K ROM

512 BYTES RAM

TIMERS

MDS BUS CONTROL

MDS BUS CONTROL

2 MHz CLOCK

ICE 85 CONTROLLER PROCESSOR

EMULATION CONTROL

ADDRESS MAP

EMULATION ENABLE

SYNC 0

TRACE MEMORY AND QUALIFIER REGISTERS

BREAKPOINT REGISTERS

DATA SELECTION

USER DATA

CHIP DATA

TO USER'S SOCKET

CONTROL

ADDRESS

TRACE DATA

TIME CLOCK

FORCE TRACE

8085 CONTROL

SYNC 1

ICE 85 TRACE BOARD

8085 CHIP CONTROLLER

SIGNAL BUFFERS

18 USER TRACE PROBES

18 EXTERNAL TRACE BUFFER

tors, knobs, labels, etc. for a production model. When this design is completed and the parts are gathered, the *production department* builds a prototype run of perhaps 10 units. These 10 units will probably be returned to the development engineering department to be tested and debugged. The reason is that some marginal timing or other factor may work by chance in the initial prototype. However, it may not work in 5 out of 10 of the production prototype units. When the 10 units have been debugged, heat-tested, and verified, the instru-

ment is released to full production. Schematics are finalized, and operator and service manuals are written.

PRODUCTION

In one section of production, the chassis is mechanically assembled with printed-circuit board edge connectors, transformers, switches, pots, connectors, etc. In another section, resistors, capacitors, ICs or IC sockets, etc. are "stuffed" into the printed-circuit boards. The

401

PROTOTYPING AND TROUBLESHOOTING μP-BASED SYSTEMS

loaded PC boards are passed through a wave soldering machine which quickly solders all the components at once. Any sensitive components may be hand-soldered. After soldering, the PC boards are cleaned with a solvent to remove the soldering flux.

BOARD TEST

Before the PC boards are plugged into the assembled chassis, they are sent to the board test section of production. Here technicians test and troubleshoot each board. A technician first does a careful visual inspection to check for solder bridges, components leads shorted together, broken or missing components, reversed ICs, diodes, or electrolytics, etc. After this, the board may be tested in several ways.

In large companies where a large number of boards may be produced, a computer-controlled board tester similar to the Sentry IC tester in Figure 5-20 may be used. These machines run diagnostic tests, and, in addition to telling whether the board passes or fails, they often indicate a possible source of a failure. The main disadvantage of these computer-controlled testers is their high cost.

A second method of checking boards is with a *test fixture*, or *test jig*. Test fixtures vary widely, but their main function is to supply power and input signals to the board and a readout of output signals from the board. The technician uses an oscilloscope or logic analyzer to troubleshoot the board. As problems are found and fixed, notes are kept so the problem can be more easily found if it occurs again. Also, if a particular problem occurs over and over, it can be referred to engineering.

Another method of checking boards is with an assembled instrument that is complete except for the board to be tested. An extender card is plugged into the chassis in place of the board to be tested. An extender card is just a PC board with traces from each pin on the end where it plugs into the chassis to each pin of an edge-connector socket on the other end. The board to be tested is plugged into the edge-connector socket. The extender board simply raises the board to be tested above the other boards in the instrument so that test probes can be connected easily.

After all the PC boards are tested and debugged, they are sent to the production test department.

PRODUCTION TEST PROCEDURES

A *production test technician* first plugs the power-supply regulator board into the chassis and checks the power-supply voltages before inserting memory or microprocessor boards. This is important because some semiconductor memories that require multiple supply voltages will be destroyed if, for example, +12 V is applied without +5 V present. When the ac power is first applied, a variable-voltage ac source may be used to gradually bring voltage up to 120 V ac. This gives time for the dielectric in large electrolytics to build up. After all power-supply voltages check out, the procedure var-

ies somewhat according to the size of the instrument. In a large instrument with a PC board for each of several functions such as clocks, processor, ROM, RAM, and I/O, the clock board is plugged in and tested first. The board was probably already tested in the board test, but now it must be tested and debugged in the chassis. The next boards to be inserted and tested are the processor board, the RAM, ROM, and peripheral controller boards. A small instrument probably has all three functions on one board, so after the power supply checks out, this one board is plugged in and debugged.

When the instrument is fully assembled and working, it is given a rough calibration, left running, and put on a shelf to "burn in" for 24 or 48 h. After the burn-in period, the instrument is rechecked and debugged again if necessary. Next it is left running for a second burn-in period, which may include some time in a heat chamber at a high temperature. The reason is that semiconductor parts may work initially, but fail for some reason after a few hours of operation. If the parts function properly for the first 10 to 100 h, they will probably work for several years without failure.

TEST, CALIBRATION, AND QC

After the instrument makes it through the burn-in procedure, it goes to the *test and calibration section*. Here calibration technicians fine-tune all the adjustments in the instrument for best performance. After calibration, the instrument goes to a *quality control*, or *QC*, *technician* who verifies that the instrument meets all specifications. When the instrument passes QC inspection, it is packaged for shipment.

FIELD SERVICE

If the instrument fails after the customer receives it, then a *field service engineer*, or *field service technician*, will be assigned to fix it. The instrument may be returned to a field service department in the plant, or the field service technician may go to the customer's location to fix it.

PRODUCTION TEST AND FIELD SERVICE TOOLS AND TECHNIQUES FOR MICROPROCESSOR-BASED INSTRUMENTS

GENERAL COMMENTS AND HINTS

The primary tools for a production test or field service technician are a multiple-trace oscilloscope, a DVM, the service manual for the instrument, and a systematic approach. Since technicians are valued on their ability to find and fix problems quickly, a systematic approach is particularly important for troubleshooting a complete instrument where the number of possible combinations of problems is very large.

1. Clearly identify the symptoms. It often helps to write them down.

2. Try to think of experiments or tests to perform to narrow down the problem.

3. Quickly verify that the power-supply voltages are within specifications so you don't spend hours tracing a problem caused by low V_{CC} around the microprocessor circuitry.

4. Check that clock signals are present and correct.

5. In a multiple-board system, remove all but the minimum number of boards required for basic operation. This may help narrow down the board with the problem. If you have spare boards, try exchanging them, one at a time, with those in the malfunctioning unit.

6. If you localize the problem to a specific board, follow the troubleshooting procedures described in Chaps. 9 and 10 to find the defective IC(s).

7. If DVM and oscilloscope checking do not reveal the problem, then more sophisticated instruments may be required. In a later section we discuss some of these.

Another useful hint, if you are in the position of a production test or field service technician, is to get a personal copy of the service manual for each instrument you have to work on. As you troubleshoot several units, you may want to write into the manual additional waveforms or voltages you found useful in tracing a problem. Also at the rear of the manual, list symptoms and the cure you found for each. This list and notes added to schematics are very valuable if you don't work on a particular instrument for a year and then return to it. The notes let you very quickly pick up where you left off.

Another technique that is handy in a field service situation is to carry a spare set of PC boards for each instrument you have to service. If a problem cannot be found quickly, then a problem board can just be replaced with a known good one. This keeps the customers happy because it means minimum down time for the instrument. Also it permits you to take the nonfunctioning board back to the plant, where you may have more sophisticated equipment to help you find the problem. Another point in favor of this approach is that your brain may work faster back in the plant, where you don't have an anxious customer peering over your shoulder. If this cannot be done, hopefully you can take one of the test instruments described next into the field with you for the tough problems.

TEST INSTRUMENTS FOR DIFFICULT PROBLEMS

When a problem in a microprocessor-based instrument or system cannot be found quickly with a standard multichannel oscilloscope, then more sophisticated tools must be used. We briefly discuss four common instruments for solving difficult problems.

LOGIC ANALYZERS   Logic-state analyzers and logic timing analyzers are very useful in debugging microprocessor-based systems. A powerful feature of these analyzers is the combinational trigger, which triggers the display only when the word on the parallel inputs matches a switch-selected word. Also the memory in these analyzers lets you see what happened on the buses before and after the trigger.

Use of a logic-state analyzer in the field is the same as discussed earlier. If an instrument, for example, runs part of its program properly but hangs up at a particular point, then connect the instrument address bus to the inputs of the analyzer and set the combinational trigger for the first address in the region of the program where it hangs up. Then, as the program runs, compare the sequence shown on the analyzer with the sequence in the program printout to find the specific problem point. With pretrigger recording you can look at the address sequence that led up to the problem. Instead of checking the sequence on the address bus, check the sequence on the data bus. Comparing long sequences can be tedious and time-consuming unless the analyzer has a built-in way of doing it. The Tektronix 7D01 analyzer, for example, records and displays parallel data from a working system on half the screen and parallel data from a nonfunctioning system on the other half. As shown in Figure 5-23b, any bits that disagree between the two halves are intensified in the display for the nonfunctioning system. This works very well if you have a functioning unit to compare with. However, in the field this is usually not the case. To overcome this problem, instruments such as the Fluke 9010A Microsystem Troubleshooter have been developed.

FLUKE 9010A MICROSYSTEM TROUBLESHOOTER

As you can see from the picture of the 9010A in Figure 13-5, it has a keyboard, a display, and an umbilical cable with an IC plug on the end of it. The unit also contains a minicassette tape recorder. For troubleshooting, the 9010A is used as follows. The microprocessor IC in a fully functioning unit is removed, and the plug at the end of the umbilical cord is inserted in its place. Then the learn function of the 9010A is executed. This function finds and maps RAM, ROM, and I/O registers that can be read into and written from. It also computes signatures for blocks of ROM. All these parameters are stored in the 9010A's RAM and/or on a minicassette tape. The microprocessor on a malfunctioning unit is removed, and the plug at the end of the umbilical cable is inserted in its place. An automatic test function is then executed. In this mode the 9010A automatically tests the buses, RAM, ROM, ports, power supply, and clock on the malfunctioning unit. Any problem found, such as stuck nodes or adjacent trace short-circuits, is

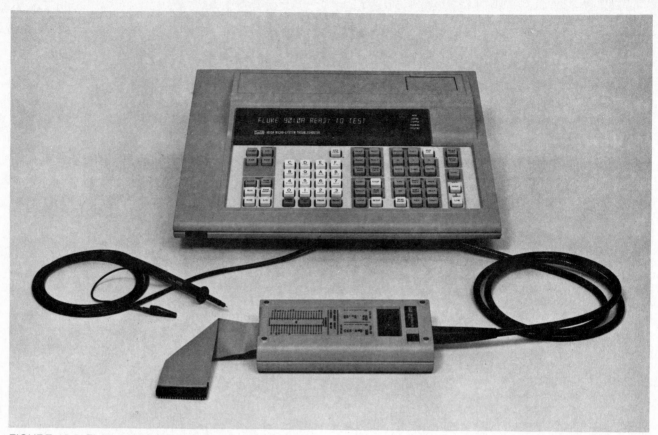

FIGURE 13-5 Fluke 9010A Microsystem Troubleshooter.

indicated on the display. Because of its built-in intelligence, the 9010A can be programmed to do other tests as well.

The point of an instrument such as the 9010A is that it moves the power of stimulus-response, automatic testing out into field service. Previously *automatic test equipment* (ATE) of this type was limited to the production test area within the factory because of its size and cost. The 9010A fits easily into a suitcase and requires only an umbilical cord, pod, and tape for each processor type that the field service engineer has to troubleshoot. Using the 9010A, the field service engineer does not have to be intimately familiar with the programming language of each type of malfunctioning unit to troubleshoot it.

DESIGNED-IN SELF-TEST AND SIGNATURE ANALYSIS

A strong trend in microprocessor-based products is to include *self-test* and diagnostic routines into their design. This simplifies the instruments needed to troubleshoot these products. However, it requires a little more effort at the design stage because the hardware must be created so that functional blocks can be isolated easily for testing. In most cases, it is well worth the time.

An operating microprocessor-controlled instrument is really a collection of digital feedback loops. The CPU sends out an address to ROM to get an instruction. The instruction fed back to the CPU determines the next operation of the CPU, etc. As long as all these feedback loops are connected, the instrument can be very difficult to troubleshoot. For *signature analysis*, some of the feedback connections are broken. Individual parts of the instrument are exercised with a loop test program. As the test program exercises the system part, a unique pattern of 1's and 0's appears at each point in the circuit. A *signature analyzer*, in a somewhat mathematically complex way, converts this pattern of 1's and 0's to a four-digit pseudohexadecimal display. This four-digit display is known as the *signature* for the tested point. Figure 13-6, for example, shows the test signatures that might be found with an analyzer around a ROM-RAM decoder in a 6800 microprocessor system. Figure 13-7 shows a photograph of the Hewlett-Packard 5004A signature analyzer. Note that it is only about the size of a bench-model DVM.

With a signature analyzer, a technician troubleshoots a system section by exercising it and comparing the signatures produced at each point with the correct signature listed on the schematic or a signature table or printed on the circuit board. The signature contains no diagnostic information as to what the fault is. In fact, the HP 5005A uses symbols 0 through 9, A, C, F, H, P, and U instead of standard hex symbols. An incorrect signature simply tells the technician that the signal at that point is wrong. The theory is that if the input signatures to a device are correct but one or more of the output signatures is wrong, then quite likely the device

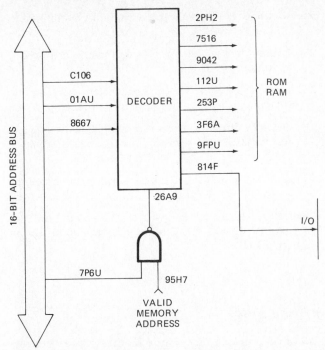

FIGURE 13-6 Signature pattern around a 6800 system ROM/RAM decoder.

FIGURE 13-7 Hewlett-Packard 5004 signature analyzer.

is bad and should be replaced.

With this approach the technician needs to know very little about the details of circuit operation for most troubleshooting. A signature analyzer allows a technician to know whether a part is operating correctly with a point-by-point comparison as a diagnostic routine is run. To further help the technician, most instruments designed for use with a signature analyzer have a detailed troubleshooting flowchart. This tells which test loop to run first for a particular symptom, which points to check with the signature analyzer, and which IC is bad if an incorrect signature is found at a particular point.

An example may clarify how this is used in practice. Suppose that the microprocessor IC itself in a system is to be tested. To do this, the microprocessor data bus must be disconnected from RAM, ROM, etc. If this was planned at the design stage, then perhaps just removing

a jumper plug from an IC socket will do it. Plugging in another jumper with resistors to V_{CC} or ground will hard-wire a NOP or some other instruction onto the data inputs of the processor. As the microprocessor runs, the signature at each pin of the microprocessor can be read and compared with that for the same pin of a known good unit. ROM and RAM decoders can also be checked at this point.

If the microprocessor IC signatures are correct, then the ROM section can be reconnected to it, and a ROM preprogrammed with test routines can be used to exercise other parts of the system. Front panel switches may be used to select the desired section of the diagnostic program. The signature analyzer is then used to check the signatures in each part of the instrument as its test routine is run.

Some instruments that have built-in displays may use these to indicate directly the results of an internal diagnostic test. The 5004A signature analyzer itself, for example, has diagnostic routines in its ROM, so the user can verify that the unit is working correctly at any time.

The logical extension of all this is instruments that constantly test themselves and notify the user of the type and source of any malfunction. A speech synthesizer describing the problem in a soothing voice might make troubleshooting more interesting!

REVIEW QUESTIONS AND PROBLEMS

1. What is the advantage of building a prototype one module at a time?

2. Describe a procedure for testing a 4-kilobyte block of RAM added to a development board.

3. Write a short assembly language routine for the 8085A or 6800 to test RAM at addresses 2000H through 27FFH.

4. Define breakpoint and single-stepping.

5. Describe a procedure for testing an I/O port on a development board.

6. List the steps in debugging a program module.

7. Show the assembly language additions you would have to make to test the BCDCVT and BINCVT routines in the scale program of Chap. 12.

8. Show the assembly language additions necessary to test the OUTDS routine in the EPROM programmer program of Chap. 12.

9. Define logic timing analyzer, logic-state analyzer, and pretrigger, center-trigger, and posttrigger recording.

10. Describe a procedure for testing and debugging a CPU or CPU group with a logic-state analyzer.

11. Describe procedures for testing and debugging ROM, RAM, and I/O ports by using a logic-state analyzer.

12. Describe the main functions of a microprocessor development system such as the Intel Intellec in Figure 8-26.

13. Define editor, assembler, source file, symbol table, and object file.

14. Explain how to use an in-circuit emulator to debug a program routine.

15. Draw a flowchart for the production process of an instrument from initial design to finished product.

16. Describe the task of each of the following types of technicians: engineering, board test, production test, calibration, QC, and field service.

17. Give the main advantage of test instruments such as the Fluke 9010A Microsystem Troubleshooter.

18. Define signature analysis, and describe how it is used to troubleshoot a microprocessor-based instrument.

REFERENCES/BIBLIOGRAPHY

M6800 MEK6800D2 Evaluation Kit II Manual, Motorola Semiconductor Products, Inc., Austin, TX, 1976.

Intel SDK-85 User's Manual, Intel Corporation, Santa Clara, CA, December 1977.

Intel In-Circuit-Emulator/85 Microcomputer Development System Operator's Manual, Intel Corporation, Santa Clara, CA, 1978.

Intel In-Circuit-Emulator/85 Microcomputer Development System Hardware Reference Manual, Intel Corporation, Santa Clara, CA, 1978.

Application Note 222—A Designer's Guide to Signature Analysis, Hewlett-Packard Company, Palo Alto, CA, April 1977.

Frohwerk, Robert A.: *Signature Analysis: A New Digital Field Service Method*, Hewlett-Packard Company, Palo Alto, CA, 1977.

Bechtold, Jim (ed.): "Signature Analysis: A New Concept in Digital Troubleshooting," *Hewlett-Packard Bench Briefs*, vol. 17, no. 2, pp. 1–4, March–April 1977; Hewlett-Packard Company, Palo Alto, CA.

Farly, Bruce G.: "The Role of Logic State Analyzers in Microprocessor Based Designs," *Application Note 167-13 Data Domain Measurement Series*, Hewlett-Packard Company, Colorado Springs, CO. Reprinted from 1975 *Wescon Professional Program*, September 16–19, 1975, Copyright 1975 Western Electronic Show and Convention.

"Troubleshooting a Microprocessor," *Logic Analyzer Application Note*, no. 57K1.0, Tektronix, Inc., 1977.

Fluke 9000 Series Micro System Troubleshooters Data Sheet, John Fluke Mfg. Co., Everett, WA, 1981.

CHAPTER 14

HIGHER-LEVEL LANGUAGES

In recent chapters we discussed the writing and debugging of assembly language programs. Here we talk about programming in high-level languages such as BASIC, FORTRAN, and Pascal. Also we show the relationship of these higher-level languages to assembly and machine languages.

At the conclusion of this chapter, you should be able to:

1. Describe the functions of a general-purpose operating system.

2. Explain the use of an interpreter.

3. Describe the use of a compiler.

4. List several high-level microcomputer languages.

5. Write or modify a simple BASIC program.

6. Write or modify a simple FORTRAN program.

7. Write or modify a simple Pascal program.

OVERVIEW

In Chap. 12 we said that it is a good idea to write assembly language programs wherever possible as a series of modules linked together. Most of the scale program, for example, is a collection of subroutines linked together. There are at least two advantages to this approach. First, each routine can be individually tested and debugged. Second, debugged routines can be kept in a file and pulled out for future programs. The BCD-to-binary, multiply, and binary-to-BCD routines of the scale pro-

gram could have been written right in the main program, but as independent subroutines they are stored more easily for future reference.

To help write assembly language programs faster, several microprocessor manufacturers maintain a library of useful assembly language programs and routines for their microprocessors. For example, Intel has an extensive library of 8080A/8085A assembly language routines and programs. For a yearly fee a subscriber gets a list of all the library programs. A printout of many short programs is contained right in the list. Longer programs can be obtained on paper tape, on floppy disks, or as a printout for a small fee.

The relationship of all this to higher-level languages is as follows. Programming in a higher-level language such as BASIC is very much like working with a preset library of routines. In a higher-language program, just the operation or function for the routine is written. Then a compiler program or an interpreter program converts the high-level statements to the equivalent machine language routines so they can be executed.

An ideal higher-level language might be a precisely defined form of written or spoken English. Except for simple one-word commands as described in Chap. 11, the technology has not reached this point yet. However, an example with a simple BASIC language statement will show how much more English-like it is than assembly language.

In the assembly language program for the scale in Chap. 12, a somewhat "messy" subroutine is used to multiply the selling price per pound by the weight to give the total price. In BASIC the single statement LET P=W*S specifies this multiplication. The asterisk indicates multiplication.

SYSTEMS AND PROGRAMS USED TO WRITE AND EXECUTE HIGH-LEVEL LANGUAGE PROGRAMS

HARDWARE

High-level language programs are usually developed on a large computer, a minicomputer, or a microcomputer system. In addition to a CPU, the system needs 64K or more words of RAM, a CRT terminal, a floppy or hard disk unit to store programs, and a printer to provide hard copy. Most microcomputer development systems such as the Intellec system in Figure 8-26 can be used to develop high-level language programs. Personal computers such as the Apple II, IBM, or TRS-80 also are used to develop high-level language programs.

SOFTWARE

To allow a user to develop and run programs, a system must have programs that it executes to interface the user to the hardware. The simplest is a monitor program.

MONITOR PROGRAM   The monitor program is most likely in ROM in the system. As with development boards such as the SDK-85 and MEK6800D2, the monitor ROM contains routines that initialize the system. The monitor ROM also contains routines that let the user load machine-code programs into memory, execute machine language programs, examine registers, insert breakpoints, etc., by using a teletype or CRT terminal. One major function of the monitor program, however, is to let the user load an operating system into the system RAM from a disk.

OPERATING SYSTEMS   A general-purpose operating system is a set of system programs or routines which, when run, allow a user to create high-level language programs without needing to know the hardware details of the system. For example, a user can write a program and save it on a floppy disk without knowing the format in which data is stored on the disk. In short, the operating system makes the system "user friendly."

The operating system routines may be in ROM, but usually they are loaded into RAM from a disk when the system is started up. A system loaded in this way is called a *disk operating system*, or DOS. To save RAM space, some systems load in only the basic part of the operating system at startup. Other routines are loaded into RAM from the disk as needed.

The operating system allows a user to:

1. Create, copy, delete, and manipulate disk files for data storage.

2. Enter programs, edit them, and store them on disks.

3. Link separately developed program modules.

4. Load linked programs into RAM and execute them.

5. Debug programs. The debugger routine lets you insert breakpoints and examine memory and registers. Some debuggers also contain *disassemblers*, which convert machine codes in memory to their equivalent assembly language instructions.

Operating systems also manage the system memory space and the transfer of data to printers and other peripherals.

EDITOR PROGRAMS   Editor programs allow a user to easily write the source program or text into a workspace in RAM. The editor has commands that make it easy to correct typing or program errors by simply moving around a cursor on the CRT screen and typing in the correction. Most editors let the user add or delete a line in the middle of a program by simply pressing one or two keys on the keyboard. A simple editor program may be part of the operating system. More powerful editors for specialized applications such as word processing are loaded in separately.

When a user's high-level source program is complete and edited in the editor workspace in RAM, the source program is copied into a file on a disk. The form of this source program is a long way from the binary machine codes that a microcomputer needs to execute the instruction. Translation of the program from a high-level form to executable code may be done in several ways. One way is to use an interpreter.

INTERPRETER PROGRAMS   When an interpreter runs, it reads a statement of the source program, translates each high-level statement to machine code, and executes that code immediately. It then reads the next high-level language statement, translates it, and executes it. Figure 14-1a shows this in flowchart form.

The advantage of an interpreter is that it is very user-friendly. Since an interpreter executes directly from the source program, it is easier to detect and correct errors in a program. If an error is found, the user can just correct the source program and rerun it.

The disadvantage of interpreters is that interpreted programs run 5 to 25 times slower than the same program translated by a compiler. The reason is that the interpreter does not produce a complete machine-code version of the program that can be run. The interpreter must translate each statement of the program each time it executes the program. In other words, with an interpreter, translation time is always part of the execution time. Remember this so that you can compare it with the operation of the compiler discussed next.

COMPILER PROGRAMS   A compiler program, when run, reads through a high-level language program and, in two or more passes through it, translates the entire program to *relocatable* machine code. Relocatable means that addresses such as jump addresses and call addresses are given symbolic names, so that the final program can be located anywhere in memory.

When compiling is complete the program must be *linked*. This means that any mathematical routines or

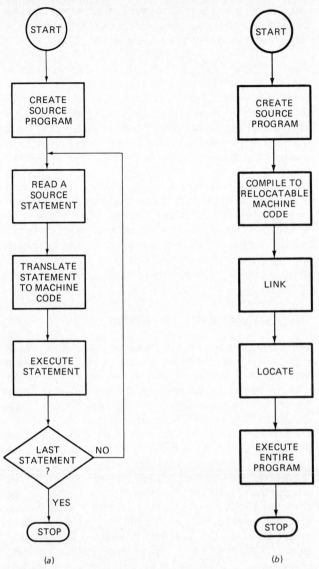

FIGURE 14-1 Comparison of compiler and interpreter operation. *(a)* Interpreter. *(b)* Compiler.

I/O routines needed by the user program are read from a library of routines and added to the code for the user program.

After all the program modules are linked, they must be *located.* All the symbolic addresses must be replaced with absolute addresses. The result of all these steps is a machine language program that can be *loaded* into RAM and executed.

Some compilers for the Pascal language translate the high-level language statements to an intermediate code called *P code.* The P code is then interpreted. Other compilers translate from high-level language to assembly language for the machine on which the program is to run. Compiling to assembly language has the advantage that debugged assembly language routines, which are not part of the high-level language vocabulary, can be added to the output from the compiler. The result is assembled to machine code with a standard assembler program.

Figure 14-1*b* shows the operation of a compiler in flowchart form for you to contrast with the operation of an interpreter. The advantage of the compiler approach is that the entire program is translated only once. The produced machine-code version is executed each time the program is run. Translation time is not part of the execution time as it is for an interpreter. Because of this compiled programs run 5 to 25 times faster than interpreted programs.

The disadvantage of compilers arises when an error is found. The error must be corrected in the source program, and then the entire program must be compiled, linked, located, and loaded again.

COMMON HIGH-LEVEL LANGUAGES

There are at least 25 common high-level languages, and some, such as BASIC, have countless dialects. Therefore, we give here a short overview of some of the most common.

FORTRAN   FORTRAN is an abbreviation for *formula translation,* and this gives a clue to its main use. Since this language contains many functions for complex mathematical operations, it is ideal for engineering calculations. Developed in the 1950s by IBM, it was one of the first common computer languages. Several versions of the language were developed. One of the most popular for about 10 years was FORTRAN IV. In 1977 the American National Standards Institute (ANSI) published a standard for FORTRAN. This was an attempt to get all manufacturers to use one form of the language so that programs would be "portable." *Portable* means that programs written on one machine can be run on several others. ANSI standard FORTRAN 77 is shown in a later section of this chapter. FORTRAN is usually used with a compiler.

Although it is an effective language for problem solving and large libraries of engineering programs are available, FORTRAN is somewhat difficult to teach to beginners because of the formatting required in writing the programs. Therefore in the early 1960s two professors at Dartmouth College developed BASIC.

BASIC   BASIC is an acronym for *B*eginner's *a*ll-purpose *s*ymbolic *i*nstruction *c*ode. While not as powerful as FORTRAN, BASIC is easier to use and powerful enough for many problems. Many versions of BASIC have been developed. They range from a very small version for home microcomputers that requires only 4 kilobytes of memory to extended versions that use 32 kilobytes of memory and run on computer systems. Later we show how programming is done in standard BASIC. BASIC is usually used with an interpreter, but BASIC compilers are available.

ALGOL   Another engineering-oriented language is ALGOL. The name is an acronym of *Algo*rithmetic *l*anguage. It is common in Europe and in some university computer centers, but not in industry because of the dominance of FORTRAN.

COBOL COBOL is an acronym for *Common business-oriented language*. As the name states, this language is structured to easily handle bookkeeping chores such as ledgers, accounts, and payroll. COBOL program statements use many words instead of symbols. For example, a COBOL statement might read SUBTRACT WITHDRAWALS FROM OLD BALANCE GIVING NEW BALANCE. English-like statements such as this make COBOL programs more understandable to nonprogrammers.

PL/1 AND PL/M Also developed by IBM, PL/1 is an extensive language more powerful than FORTRAN. Because it is such a large language, it requires a very large compiler. Therefore, it can be run only on sizable computer systems.

Since most users do not need all the power of the complete PL/1, several subset languages have been developed. One is PL/M, which was developed by Intel. An Intel compiler for PL/M is used with the Intellec system of Figure 8-26 to convert PL/M programs to 8080A/8085A machine code. PL/M allows a programmer to write programs for an 8080A faster than with assembly language directly. However, the disadvantage of PL/M for many applications is that the compiled PL/M program usually gives much more machine code than one written directly in assembly language and assembled. This means the compiled program will require more memory space.

APL Another language developed by IBM is APL, or *A programming language*. This is a very powerful language because it lets a user define instructions. Also, it has a very large library of predefined functions. However, it requires a large machine to run on and uses many symbols not found on standard typewriter terminals. APL statements are quite cryptic, and so many comments are needed to make APL programs followable.

PASCAL Pascal, named after the French mathematician Blaise Pascal, was developed by Nichlaus Wirth in Switzerland in the early 1970s. It is more modular than FORTRAN and more versatile than BASIC. Another reason for recent interest in Pascal is the fact that it is a compact language, so an interpreter or a compiler for it will fit in a relatively small system. Also Pascal produces very efficient machine-code programs when it is compiled. A compiled Pascal program runs several times faster than the same program compiled from FORTRAN or BASIC.

The organization of Pascal makes it easy to write *structured* programs. In structured programs, only certain well-defined structures or blocks are permitted. Ideally, each structure has only one entry point and one exit point. Therefore, each structure can be written and debugged individually. This is an extension of the modular approach described for assembly language pro-

grams. The rules for structured programming forbid the jumping all over the place which can turn a program into a hard-to-debug pile of "logical spaghetti."

ADA Ada is a highly structured language developed by the U.S. Department of Defense. It contains the best features of Pascal, ALGOL, and PL/1, plus real-time multitasking, which is required in control applications. Ada is based on the concept of software *components*, which are like complex subroutines. The components are separately compiled, cataloged, and stored in a program library. The library can be customized by including only those components needed for a specific application. Then the compiler calls these components as needed during compilation. When Ada is fully developed, it should allow programmers to develop complex programs from a catalog of software components, just as a hardware designer uses components from a catalog of LSI devices to develop hardware systems.

PROGRAMMING IN BASIC

BASIC has been one of the most popular high-level languages because programs are easily written in it. It has been made even more popular by the introduction of relatively inexpensive home computer systems such as the Apple II, TRS-80, and H-8, which contain full BASIC capability. Unfortunately, each company has developed its own version of the language. There are probably at least 100 different versions of BASIC, ranging from very tiny ones for use on memory-starved systems to extended versions with much more power than the original Dartmouth BASIC. Therefore, if you work on a machine in BASIC, make sure you read the manual to get the specifics for that machine. In this section we describe the use of a middle-of-the-road version of BASIC.

PROGRAM FORMAT

Figure 14-2 shows a simple BASIC program to add two numbers entered from a terminal and print the result on the terminal. When run, the program sits and waits for the user to enter two numbers, A and B, to be added. After the two numbers are entered, it adds them and prints out Y=(the sum). The program then loops back and waits for the user to enter the next two numbers to be added. You can see how much more English-like this program is than assembly language programs.

LINE NUMBER	STATEMENT
210	INPUT A,B
220	LET Y = A + B
230	PRINT "Y = ";Y
240	GOTO 210

FIGURE 14-2 Simple BASIC program to add two numbers and print the sum.

As shown in Figure 14-2, each BASIC statement has a line number to indicate the desired sequence of execution. Line numbers can be any numbers between 0 and 9999. For example, the line numbers could be 1, 2, 3, 4, etc. However, if consecutive numbers such as these are used and a statement is left out, then the line numbers for the whole program must be changed to insert it. So statements usually are assigned numbers at increments of 5 or 10, as shown in Figure 14-2. Statements can then be inserted with a minimum of shuffling.

The BASIC statements themselves start with the action to be performed, such as INPUT, LET, or PRINT. The second part of the statement gives the object of the action. Later we list and discuss the common BASIC statements.

CONSTANTS, VARIABLES, AND NUMBERS

Both constants and variables are given symbolic names in BASIC. The symbolic name can be any letter of the alphabet or a letter followed by a number from 0 to 9. The program in Figure 14-2, for example, used A to represent one variable input by the user and B to represent a second. The sum of these two is represented by the letter Y. A statement such as LET C=12 in a program gives the value 12 to the variable or constant named C. Other examples of legal names for variables or constants are P6, V9, X, R3, and H. One criticism of BASIC, is that longer, and therefore more mnemonic, names for constants and variables are not allowed.

BASIC recognizes and writes numbers in various formats. Numbers may be integers or real quantities. Integers are whole numbers that have no fractional part. Examples are 6, 6.00, and −9.0. Real quantities are numbers that have a fractional part. Examples are 5.26, −7.02, and 0.0136. Most machines running BASIC allow only six or seven significant figures. Commas to mark off thousands are not permitted. Very large or very small numbers are represented in scientific notation. For example, 120,000,000,000,000 is represented as 1.2E13, and .0000123 is represented as 1.23E−5. The E says that the number following it is an exponent of 10.

COMMON ARITHMETIC OPERATORS

Many statements in BASIC use arithmetic operators. Table 14-1 shows the common operators and the symbols used for them in BASIC programs. The symbols for addition and subtraction are the same as the arithmetic symbols. The symbol for multiplication is the asterisk because X might be confused with a variable or constant named X. The asterisk must always be used between two quantities to show multiplication in BASIC. Division is represented with a slash because most terminals do not have the ÷ symbol.

Exponentiation is the process of raising numbers to a power. The statement LET X=R↑2 means that X=R². The exponent does not have to be an integer, so this operator can be used to produce roots of a number. For example, LET X=R↑.5 will produce the square root of R, and LET X=R↑.3333 will produce the cube root of R for X.

Parentheses are used to group variables or constants so that the proper operation is performed. For example, LET Y=A/B+C has a different meaning than LET Y= A/(B+C).

If more than one operator is present in a BASIC statement, the operations are executed in a specific order. Table 14-1 shows this order. Operations within parentheses are always performed first. Y=A/(B+C) adds B to C and then divides the result into A. Other operations are performed in the order shown. In the statement LET Y=A+B*C, for example, B is multiplied by C, and the result is added to A. In the statement LET Y= C−R↑2*3.14159, the value of R is squared and then multiplied by 3.14159, and the result is subtracted from C to give the value of Y. If operators with the same priority are used in a statement, they are performed from left to right. The statement A/B*C divides A by B and multiplies the result by C. Putting parentheses around B*C causes this multiply operation to be performed before the division if desired.

COMMON RELATIONAL OPERATORS

In addition to the arithmetic operators, BASIC uses the relational operators shown in Table 14-2.

The equal symbol has two uses. One is the normal arithmetic meaning of equal. The other is an assignment meaning. An example of this second meaning is the statement LET A=A+1. In normal arithmetic this is impossible. But in BASIC this statement means to re-

TABLE 14-1
COMMON BASIC ARITHMETIC OPERATORS

PRIORITY	SYMBOL	OPERATION
1	( )	Grouping
2	↑	Exponentiation
3	*	Multiplication
3	/	Division
4	+	Addition
4	−	Subtraction

TABLE 14-2
RELATIONAL OPERATORS IN BASIC

SYMBOL	MEANING
=	Assignment or equal
>	Greater than
<	Less than
< > or #	Not equal
,	Separation

TABLE 14-3
COMMONLY AVAILABLE FUNCTIONS IN BASIC

FUNCTION	MATHEMATICAL FORM	DESCRIPTION
ABS (X)	$\lvert X \rvert$	Absolute value of X
ATN (X)	arctan X	Angle in radians with tangent = X
COS (X)	cos X	Cosine of X (X in radians)
EXP (X)	e^X	Base of natural log to power X
INT (X)	$[X]$	Next lower integer to X
LOG (X)	ln X	Natural logarithm of X
RND (X)		Random number between 0 and 1, not including 0 and 1
SIN (X)	sin X	Sine of X (X in radians)
SQR (X)	$\sqrt{X}$	Square root of X
TAN (X)	tan X	Tangent of X (X in radians)

place the stored value of A with A+1. In other words, it means to increment the stored value of A.

The other relational operators are pretty much self-explanatory. They are used in statements of the form IF A>B THEN (line number). If A is greater than B when this program step is executed, then program execution will jump to the line number at the end of the statement.

LIBRARY FUNCTIONS

Besides the simple arithmetic operators discussed, BASIC has several more complex mathematical functions already built in. Table 14-3 shows commonly available functions and their descriptions.

Note that SIN, COS, and TAN require the angle (X) to be in radians. Arctangent, ATN, returns the value of an angle in radians. EXP (X) computes the value of e, the base of natural logarithms, raised to the power X. LOG(X) computes the natural, or base e, logarithm of X. ABS(X) returns the absolute value of X. SQR(X) produces the square root of X much faster than by using a LET Y=X↑.5 approach. The remaining two library functions need a little more explanation.

INT(X) is a form of rounding off a number. It always rounds a number to the next lower (more negative) integer. For example, INT(6.7) gives 6 and INT(−6.7) gives −7. Rounding to the nearest integer can be produced by adding 0.5 to the quantity in parentheses. INT(6.7+0.5) gives 7, and INT(−6.7+0.5) gives −7.

RND(X) produces a random number between 0 and 1 such as 0.123456. However, values of exactly 0 or 1 are never produced. Most BASIC systems require a *seed*, or X value, of 0 for this function to work properly. In some machines the random numbers follow the same sequence after each system reset, so the numbers are not totally random. Random numbers between 0 and some number other than 1 can be produced by simply multiplying RND(X) by a constant. LET A=50*RND(0) pro-

duces a random number between 0 and 50. LET A= INT(6*RND(0))+1 produces a random number between 1 and 6, which is useful for a dice game.

BASIC STATEMENTS

The simple program in Figure 14-2 introduced three BASIC statements. Table 14-4 shows a list of these and other commonly used BASIC statements with examples. Here we discuss the meaning of each and show how they are used with operators to write programs.

REM, the first statement in Table 14-4, stands for remark. This is used to insert comments in a BASIC program. Anything written after a REM is printed out in a program but is ignored when the program is run. As with assembly language programs, a large number of comments help make a program understandable at a later date.

The LET statement is used to assign values to constants or variables in a program. It is probably the most used statement in BASIC programs.

PRINT causes the quantities following PRINT in the line to be printed out on a printer or terminal. PRINT N prints the numerical value of N. Anything enclosed in quotation marks is printed literally as written. For example, PRINT "CAT" prints the letters CAT. Commas and semicolons are used to format the printed quantities on a line. Each line on the printing device is divided into four zones of 15 spaces each. Printing starts with the leftmost space in a zone. PRINT A,B,C prints the value of A in the first zone, the value of B in the second zone, and the value of C in the third zone. This format is handy for producing data tables, etc. If a semicolon is used between quantities in a PRINT statement, the quantities are printed right next to each other in the same zone. PRINT A;",";B;",";C, for example, prints the value of A, the value of B, and the value of C just separated by commas.

TABLE 14-4
BASIC STATEMENTS WITH EXAMPLES

```
REM THIS IS A LIST OF BASIC STATEMENT
    EXAMPLES.
LET Y=A*X↑2+B*X+C
LET P=P+1
LET C=SQR(A↑2+B↑2)
PRINT N
PRINT "PLEASE TYPE IN THE NUMBER OF
    CONVERSIONS"
PRINT A, B, C
PRINT "THE ANSWER IS";N
INPUT A
INPUT A, B, C
GOTO 990
GOSUB 720
RETURN
STOP
END
IF C3=>Y4 THEN 210
IF Y=7 THEN 420
IF C3=>Y4 THEN 210 ELSE 990
IF SQR(X)=8 THEN 290
FOR A=1 TO 20
    .
    .
    .
NEXT A
FOR P=2 TO 20 STEP 2
    LET Y=5*P*(C+3)
    PRINT "P = "; P, "Y = "; Y
NEXT P
DATA 0.2, 3.1, 2.0, 1.8, 0.9
READ R₁, R₂, R₃, R₄, R₆
```

INPUT, when executed, causes the program to wait for a value to be entered on the terminal by the user. A question mark is printed on the terminal to indicate that the user must enter a value. If the input statement contains more than one quantity, for example INPUT A,B,C, then the program waits until values are entered for each quantity in the statement before continuing.

GOTO (line number) produces an unconditional jump to the indicated line number. GOSUB (line number) essentially calls a subroutine starting at the indicated line number. When the GOSUB statement is executed, a return address is stored. Therefore, if a RETURN statement is put at the end of the subroutine, the program returns to the main program after finishing the subroutine. As with assembly language programs, memory space is saved by writing often used procedures as subroutines.

A STOP statement and an END statement both terminate the execution of a program. The END statement is put as the last statement of a program to return control to the monitor and make sure that the computer does not try to execute random garbage in memory. The STOP statement can be put at any place in the program. It is used in place of a GOTO (END) statement when the line number of the END statement is not known. In a very long program this is quite possible.

The IF...THEN (line number) statement is a real workhorse because it produces conditional jumps based on a comparison of two or more quantities. Any of the relational operators in Table 14-2 can be used. IF C3=>Y4 THEN 210 jumps program execution to line 210 if the value of C3 is equal to or greater than the value of Y4. Some versions of BASIC have a statement of the form IF...THEN (line number) ELSE (line number) to specify one of two destinations. Some versions of BASIC also have the relational operators AND, OR, and NOT, which are very handy. For example, with these operators a statement can be used such as IF C3>7 AND Y2=3.2 THEN 320. This statement produces a jump to line 320 only if the value of C3 is greater than 7 *and* the value of Y2 is equal to 3.2. If the OR operator is used, then the jump to line 320 is made if either one condition or the other is true.

A FOR...NEXT loop is used to perform an operation over and over some number of times. For the first example of this in Table 14-4, the program executes the statements between the FOR statement and the NEXT statement 20 times and then proceeds to the statement after the NEXT A. Each time through the loop the value of A is incremented by 1.

The second example of the FOR...NEXT loop in Table 14-4 shows another use for it. Here the STEP 2 at the end of the FOR statement says that the variable P is incremented by 2 each time through the loop. The short program section shown prints out a table showing the values of P from 2 to 20 in steps of 2 as well as the values of Y produced for each value of P. Neither the step size nor the beginning and ending values for P have to be integers. For example, FOR statements could be FOR N=2.5 TO 9.6 STEP 0.01 or FOR N=9.6 TO 2.5 STEP −0.01.

DATA and READ statements are used together in place of multiple LET statements. Suppose a program is being written to find the average rainfall for a 5-day period. A LET statement could be used to enter the rainfall for each day, for example, LET R1=0.2, LET R2=3.1, LET R3=2.0, etc. A simpler way, however, is to use a READ statement and a DATA statement is shown in the example in Table 14-4. The data values are written in after DATA, with each value separated from the next by a comma. The symbolic names for each value are entered after READ in the desired order and separated by commas. READ and DATA statements must have the same number of entries. When executed, the READ statement gives the corresponding value from the DATA statement to each of its symbols.

This concludes an initial description of some common BASIC operators, functions, and statements. With this seemingly small set of tools a great many BASIC programs can be written. Next we show some program examples to help fix the basics of BASIC in your mind.

```
A.   10   REM COIN FLIPPING PROGRAM
     20   LET T = 0
     30   LET H = 0
     31   PRINT "TYPE THE NUMBER OF FLIPS DESIRED."
     40   INPUT F
     50   FOR N = 1 TO F
     60   LET R = RND (0)
     70   IF R < .5 THEN 100
     80   LET T = T + 1
     90   GOTO 110
    100   LET H = H + 1
    110   NEXT N
    120   PRINT "NUMBER OF TAILS =";T
    130   PRINT "NUMBER OF HEADS =";H
    140   GOTO 20

B.   TYPE THE NUMBER OF FLIPS DESIRED
     ?___(ENTER NUMBER FOR EXAMPLE 50)
     NUMBER OF TAILS = 27
     NUMBER OF HEADS = 23
     TYPE THE NUMBER OF FLIPS DESIRED
     ?___
```

PROGRAM EXAMPLES

At this point it would be appropriate to show how parts of the scale or EPROM programmer program could be done in BASIC. This is not easily done. In some cases, the program for a microprocessor-based smart instrument is written in BASIC and then compiled to machine code to run the instrument. However, most versions of BASIC don't have the I/O, interrupt, and bit manipulation capabilities for writing programs such as these. For example, standard BASIC does not allow inputting from or outputting to several ports.

Some extended versions of BASIC have solved this problem by including statements to read from or write to ports. Also several extended versions allow necessary assembly language routines to be included in the BASIC program. For some applications this can give the best of both programming worlds—assembly language for "bit fiddling" and BASIC for complex mathematical operations. The scale and EPROM programmer programs done in this way, however, end up all or mostly assembly language. Therefore, the examples in this section illustrate other uses of BASIC.

Most BASIC programs are run in an interactive manner. That is, the user sits down at a terminal connected to some type of computer or microcomputer and enters a program. The program may be to play a game, solve some mathematical or engineering problem, or perform financial bookkeeping chores. As the program runs, the user interacts with it by entering data, answering questions, or giving instructions as required by the program. Thus a dialog is carried on between the machine and the user. It is this type of BASIC program that is shown here.

COIN FLIPPING Figure 14-3 shows a simple program to toss a coin some number of times (typed in by the user) and print the number of heads and then the number of tails. The number of tails, T, and the number of heads, H, are initially set to 0 in lines 20 and 30. An instruction to the player is printed by the PRINT statement in line 31. Remember, anything enclosed in quotation marks after a PRINT statement is printed exactly as written. Note that using line numbers at intervals of 10 allowed the programmer to add this line without shifting the numbers of all the other lines.

When the INPUT F statement is executed, the machine simply sits and waits until the user enters a number. After the number, F, is entered, the program uses a FOR...NEXT loop to "flip the coin" F times. Each time through the loop, the LET R=RND(0) statement in line 60 produces a random number between 0 and 1. If the number is less than 0.5, then the program jumps to line 100 and increments the head count. If the number is greater than 0.5, then the program goes to line 80 and increments the tail counter. In either case, after the appropriate counter is incremented, the next flip is run. When all flips are done, the number of heads and number of tails are printed. As shown in the run of the program in Figure 14-3b, the material in quotes is printed as written. For a variable without quotes, the value of that variable is printed. Using a semicolon as the separator means the value of T, for example, is printed right after the equal sign. A comma here would cause the value of T to be printed in the next zone.

The GOTO 20 statement in line 140 makes this program into an endless loop. Once started, it will run over and over until the user hits an escape or break key,

A.
```
5    REM CONVERT FAHRENHEIT TO CELSIUS AND KELVIN
10   PRINT "FAHRENHEIT","CELSIUS","KELVIN"
15   DATA 7.2, 31.0, 46.3, 53.3, 64.7
20   DATA 73.6, 84.2, 94.7, 98.6, 99.2
25   READ F
30   LET C = (F - 32)*5/9
35   REM ROUND TO NEAREST TENTH OF A DEGREE
40   LET C = INT ((10*C) + .5)/10
45   LET K = C + 273.2
50   PRINT F, C, K
```

B.

FAHRENHEIT	CELSIUS	KELVIN
7.2	-13.8	259.4
31	-.6	272.6
46.3	7.9	281.1
53.3	11.8	285
64.7	18.2	291.4
73.6	23.1	296.3
84.2	29	302.2
94.7	34.8	308
98.6	37	310.2
99.2	37.3	310.5

?OD ERROR IN 25

FIGURE 14-4 BASIC program to convert Fahrenheit temperatures to Celsius and Kelvin. *(a)* Program listing. *(b)* Sample run.

which resets the machine. Each time the user is asked to enter the number of flips.

CONVERTING FAHRENHEIT TEMPERATURES TO CELSIUS AND KELVIN

Figure 14-4*a* shows a short program to convert a series of Fahrenheit temperatures to equivalent Celsius and Kelvin temperatures. Results of each conversion are rounded to the nearest tenth of a degree and printed as shown in Figure 14-4*b*.

The Fahrenheit temperatures to be converted are contained in the DATA statements in lines 15 and 20. *Note:* Line numbers at increments of 5 can be used if not too many additions are expected. The READ F statement reads out a value from the DATA statement, and the LET statement in line 30 converts it to the equivalent Celsius value. When executed, this LET statement first subtracts 32 from the F value, then multiplies the result by 5, and finally divides this result by 9. The Celsius result produced will probably have many more decimal places than justified by the number of significant figures in the data, so each result is rounded to the nearest tenth of a degree. This is done by the INTEGER function in the LET statement on line 40.

Since the result is to be rounded to the nearest tenth, the value for C from above is multiplied by 10. As shown earlier in the capsule summary of the INT(X) function, 0.5 is added, so positive and negative numbers are both rounded to the nearest integer when the INT function is performed. After rounding, the result is divided by 10 to put the decimal point back in the right place. An example may help to clarify this. A Fahrenheit temperature of

46.3° converts to a Celsius of 7.94444°. Multiplying by 10 gives 79.4444, and adding 0.5 gives 79.9444. The INT function rounds this to 79, and dividing by 10 gives 7.9°C.

The Kelvin temperature equivalent is produced by simply adding 273.2 to the rounded C value in line 45. The PRINT statement in line 50 prints the values of F, C, and K in three different zones, as shown in Figure 14-4*b*, because commas are used as separators. Note that leading and trailing zeros are left out.

Since there is no END statement after line 50, the program goes back and reads the next value from the DATA statement in line 15, converts it, and prints the results. After the values in the first DATA statement are used up, those in the second DATA statement are read, converted, and printed. When all data values are used, most machines terminate the program and print a message, as shown in Figure 14-4*b*, which says that the READ statement in line 25 ran out of data. A more elegant way to end the program might be to put an impossible value such as −500 at the end of the last DATA statement. A check for this value after each READ can be made with an IF...THEN (line number) statement. The included line number can produce a jump to an END statement if the −500 value is read.

This program is expandable in many directions. More values can be added to each DATA statement, and as many DATA statements as required can be added. Instead of the READ and DATA statements, a FOR...NEXT loop can be used to print a conversion table for each 0.1° over the required range. Using the statement FOR F=0

A. 70 REM CHECKBOOK TRANSACTION RECORD PROGRAM
    80 PRINT "ENTER LAST MONTH'S ENDING BALANCE"
    90 INPUT B
    100 PRINT "ENTER TRANSACTION NUMBER, DATE, AMOUNT"
    110 INPUT C,D$,A
    120 PRINT C,D$,A
    121 LET B = B + A
    122 IF B < 0 THEN 150
    130 PRINT "NEW BALANCE", ,B
    140 GOTO 110
    150 PRINT "WHOOPS, YOU HAVE OVERDRAWN, CHECK AGAIN"
    160 GOTO 80

B. ENTER LAST MONTH'S ENDING BALANCE
    ? 365.27
    ENTER TRANSACTION NUMBER, DATE, AMOUNT
    ? 25, 12/1/78, -65.92
    25   12/1/78   -65.92
    NEW BALANCE   299.35
    ? 26, 12/7/78, 221.73
    26   12/7/78   221.73
    NEW BALANCE   521.08
    ? 27, 12/14/78, -530.72
    27   12/14/78   -530.72
    WHOOPS, YOU HAVE OVERDRAWN, CHECK AGAIN
    ENTER LAST MONTH'S ENDING BALANCE
    ?

FIGURE 14-5 BASIC program for a checkbook record. *(a)* Program listing. *(b)* Sample run.

TO 100 STEP 0.1 with the program prints a table of F, C, and K for 1000 values between 0 and 100°F. This is the kind of job that humans find tedious and repetitious, but that computers do easily and accurately.

CHECKBOOK RECORD   As mentioned, BASIC can be used to perform financial bookkeeping chores. Figure 14-5 shows a program to produce a printout of checks written, deposits made, and current balance for an account. The format is similar to that of the record kept in a checkbook.

The user is first directed to enter last month's ending balance. Then the user is instructed to enter the check or deposit number, date, and amount deposited or withdrawn. A minus sign in front of the amount indicates a withdrawal. No sign is needed for a deposit. As each entry of number, date, and amount is made, the program prints these in table form, computes the new balance, and prints it in the proper column. Figure 14-5b shows a printout of a few entries for the program. If the balance goes negative, a warning message is printed out.

Note the dollar sign used after D in statements 110 and 120. This identifies the D as a *string* variable and allows the date to be entered in the familiar month/day/year format. This would not be possible with regular variables such as C or A.

STRINGS

BASIC can work with strings of letters or characters as well as with numbers. A *string* of characters is symbolically represented by a single letter with a $ after it, as shown above. This is then called a *string variable*. String variables are very useful because they let the programmer refer to a word or sentence with a simple symbol. For example, a program might at the start have the statements LET Y$="YES", LET N$="NO", LET E$= "YOU HAVE MADE AN ERROR". Then each time the programmer wants to refer to the word or sentence in the program, only the string variable name has to be used. For example, PRINT E$ prints the whole sentence "YOU HAVE MADE AN ERROR". String variables also allow users to enter commands as words rather than numbers. Figure 14-6 shows in a simple way how strings might be used at the end of a game to determine whether the user wishes to play another game.

Dimension statements, DIM, specify the amount of memory required for each string. Some machines limit the length of a string to 72 characters; other machines allow up to 255 characters per string. After each string variable is dimensioned and defined in the program of Figure 14-6, a user-entered string, A$, is compared with Y$. If the string entered is exactly equal to the defined Y$, then the program execution jumps to the start of the game program. If A$ is not equal to Y$, then it is com-

```
915    DIM A$(3)
920    DIM Y$(3)
925    DIM N$(2)
930    DIM E$(22)
935    LET Y$ = "YES"
940    LET N$ = "NO"
945    LET E$ = "YOU HAVE MADE AN ERROR"
950    PRINT "DO YOU WANT TO PLAY ANOTHER
       GAME?"
955    PRINT "TYPE YES OR NO."
965    INPUT A$
970    IF A$ = Y$ THEN (LINE NUMBER OF
       GAME START)
975    IF A$ = N$ THEN 995
980    PRINT E$
985    PRINT
990    GOTO 955
995    END
READY
```

FIGURE 14-6 Example of the use of strings in BASIC.

pared with N$. An exact match sends the program execution to the END statement at line 995. No match sends the program on to print the error message, E$, and repeat the question. A PRINT statement with nothing after it, as in line 985, simply leaves a blank line to make the printout look better.

Another use of strings in a program might be to allow access to a computer to only those who type in certain passwords. Again a string of characters typed in is compared with a previously defined string variable. If the two match, then the computer prints a "READY" message. If not, the computer prints a "SORRY, THAT DOESN'T TURN ME ON" message.

ARRAYS

Another useful feature in BASIC is the ability to work with arrays. In mathematics a list or column of numbers is called a *one-dimensional array*. The dimension of this array is the length of the list. A list with two or more columns is referred to as a *two-dimensional array*. A chessboard is a two-dimensional array. The table data in Figure 14-4b is also a two-dimensional array. A three-dimensional array consists of two or more pages of two-dimensional arrays, or a structure like that of a three-dimensional tic-tac-toe game.

A single letter is used to name an array in BASIC. Whenever an array is to have more than 10 values in any dimension, a DIM statement must be used to set aside memory space for the array. For example, DIM X(27) sets aside 27 locations for a one-dimensional array named X. Since BASIC cannot write subscripts in the normal way, the *elements*, or values, in the array are

referred to as X(1),X(2),X(3),...,X(27). DIM Y(12,4) defines a two-dimensional array with 12 rows and 4 columns. In a two-dimensional array each element is identified by a row number and a column number. For example, Y(1,2) identifies the element in the first row and the second column. Likewise, an element in a three-dimensional array is identified by a row number, column number, and a page number.

Arrays are a very powerful tool, and most examples of their use are too lengthy to show here. Figure 14-7 shows a program to input, sort, and print out in descending order up to 300 data values, or test scores. The program uses only a single, one-dimensional array, but it gives a small idea of what is possible.

The DIM X(300) statement at the start of the program sets the maximum size of the array at 300 elements, or entries. The dimension statement could be larger if needed, but there is no point in tying up memory unnecessarily.

After typing in the number of scores to be sorted, the user is directed to enter five scores, separated by commas, per line. When the program runs, these scores are stored sequentially in the memory locations set aside by the DIM(300) statements. After the N scores have been entered, the program proceeds to sort the scores into descending order.

The sort section works as follows. The first score in the array is compared with the second. If the first score is larger than the second score, then the first is left in place and compared with the third score. If the first score is smaller than or equal to the second, then it is moved to a temporary storage location called T. The second score is moved into the first position, and then the first score is moved from T to position 2 in the array. In other words, lines 230 to 250 just interchange the order of the two scores as you would do if you were hand-shuffling papers. Thus the program compares each score in the list with the score in the first location and either switches or not, depending on which is larger. After the first run of the I, FOR...NEXT loop, the largest number will have been moved to the first location in the array. The next run through the I, FOR...NEXT loop will put the second largest score in the array into the number 2 position. Each run through this loop will load the next-lower score into the next position of the array in memory. If you have trouble following the sort routine, make up an array of 5 or 10 numbers and work your way through it with them.

When all the scores are in descending order in the array, a title is printed. The scores are then printed in descending order at 10 scores per line. With 10 scores per line, the entire array prints out on one page for easy viewing.

The program could be easily modified to include a student identification number with each test score. In this section of the chapter, we have been able to show only a small part of what is possible with BASIC. For more information and examples, consult the references at the end of the chapter.

```
90    REM INPUTTING AN ARRAY OF TEST SCORES
100   DIM X(300)
110   PRINT "TYPE IN NUMBER OF TEST SCORES"
120   INPUT N
130   PRINT "TYPE IN 5 TEST SCORES PER LINE"
140   PRINT "SEPARATE SCORES WITH COMMAS"
150   FOR I = 1 TO N STEP 5
160   INPUT X(I), X(I+1), X(I+2), X(I+3), X(I+4)
170   NEXT I
180   PRINT
190   REM SORTING THE SCORES INTO DESCENDING ORDER
200   FOR I = 1 TO N - 1
210     FOR J = I + 1 TO N
220     IF X(I) > X(J) THEN  260
230     LET T = X(I)
240     LET X(I) = X(J)
250     LET X(J) = T
260     NEXT J
270   NEXT I
280   PRINT
310   PRINT "SCORES IN DESCENDING ORDER"
320   PRINT
330   FOR I = 1 TO N STEP 10
340   PRINT X(I); X(I+1); X(I+2); X(I+3); X(I+4); X(I+5)
350   PRINT X(I+6); X(I+7); X(I+8); X(I+9)
360   PRINT
370   NEXT I
380   END
```

FIGURE 14-7 BASIC program to input, sort, and print 300 data values or test scores in descending order.

PROGRAMMING IN FORTRAN

Since FORTRAN was to some extent the parent of BASIC, they are quite similar in many ways. FORTRAN is much more powerful and at first glance seems somewhat more difficult. However, if it is approached systematically, FORTRAN is not much harder to learn than BASIC.

Although FORTRAN is one of the oldest high level languages, it is still one of the most popular, for several reasons. There are huge libraries of engineering programs written in FORTRAN, and many computers are set up to run FORTRAN programs. Intel has even developed a powerful FORTRAN compiler, the FORTRAN-80, that will run in the 8080A-based development system shown in Figure 8-26. The Intel FORTRAN-80 is based on the ANSI FORTRAN 77, as is the following discussion.

DATA TYPES

INTEGER   *Integer*, or *fixed-point*, *constants* have no decimal point. They may be positive or negative. Examples are +3, −726, and 92. Most computers limit the size of an integer constant to less than about ±32000. When numbers larger than 999 are written in a program, commas are *not* used to mark off thousands.

An integer variable (one that can only have integer values) is usually given a name starting with one of the letters of the alphabet from I to N. The name can have up to six letters or letters and numbers, but it must start with a letter. Examples of integer variables names are INCOME, NUMBER, and MAX.

REAL NUMBERS   *Real*, or *floating-point*, *numbers*, contain an integer part and a fractional, or decimal, part. The numbers 3.7, .0003, and 3.14159 are examples. Each real number is stored in 4 memory bytes in one of the floating-point formats shown in Chap. 7. Very large and very small numbers may be written in scientific notation. For example, 3.7E6 means 3.7×10^6, and 2.15E−6 means 2.15×10^{-6}.

Real variables often are given names that start with a letter other than those reserved for integer variables. These names also can have up to six letters or letters and numbers, but they must start with a letter.

LOGICAL AND CHARACTER DATA   Variables and constants can be a logical type or character type. Logical data can have only the value "true" or "false."

Character data is a string of 1 to 255 letters of the alphabet or symbols. In a program, character data is set off by an apostrophe at each end. For example, 'DATA ENTERED IS OUTSIDE LEGAL RANGE'. Logical or character quantities can also be given symbolic names of up to six letters and numbers.

ARRAYS AND FILES FORTRAN allows data to be worked with in arrays and files. An array may have up to seven dimensions, and any value anywhere in the array can be accessed by referring to its coordinates in the array. For example, FARNHT(2,3) refers to the value in row 2 and column 3 of a two-dimensional array. An array, then, is a random-access storage.

Internal files are essentially serial storage. Data must be written in or read out sequentially. A way to remember this is to think of tape and disk files.

ARITHMETIC, RELATIONAL, AND LOGICAL OPERATORS

Table 14-5 shows a list of the FORTRAN operators and their priority. The operation with the highest priority is at the top of the list. As in BASIC, operations with highest priority in an expression are performed first. Note that two asterisks are used for exponentiation rather than the ↑ of BASIC. The statement $Z=X**2+Y**2$ in FORTRAN makes Z equal to X^2+Y^2. The other arithmetic symbols are the same as in BASIC.

The relational operators in FORTRAN use pairs of letters rather than the > symbol. If you remember the 6800 relative branch mnemonics, these should look familiar. .LT. represents "less than," .LE. is "less than or equal to," .EQ. is "equal to," .NE. is "not equal to," .GT. is

TABLE 14-5
FORTRAN OPERATORS IN ORDER OF DECREASING PRIORITY

PRIORITY	SYMBOL	DESCRIPTION
1	( )	Grouping
2	**	Exponentiation
3	*	Multiplication
3	/	Division
4	+	Addition
4	−	Subtraction
5	.LT.	Less than
5	.LE.	Less than or equal to
5	.EQ.	Equal to
5	.NE.	Not equal to
5	.GT.	Greater than
5	.GE.	Greater than or equal to
6	.NOT.	Logical not
7	.AND.	Logical and
8	.OR.	Logical or
9	.EQV.	Logical equivalent
9	.NEQV.	Logical not equivalent

"greater than," and .GE. is "greater than or equal to." Note the periods before and after each two letters.

The logical and boolean operators are used with logical variables, often with an IF statement. For example, IF(A.OR.B) [statement] means that if the logical variable A is true *or* the logical variable B is true, *then* execute the statement written after (A.OR.B). Here .EQV. stands for "equivalent to," and .NEQV. stands for "not equivalent to." A.EQV.B means that A and B have the same logical state, true or false.

SUBROUTINES AND FUNCTIONS

In addition to the arithmetic operators, FORTRAN has a built-in library of subroutines to perform functions such as taking the square root of a number. These are referred to as *intrinsic* functions. Table 14-6 shows some commonly used functions and their forms. A function from this library is called by simply using it in an expression. For example, the expression $C=SQRT(A**2+B**2)$ automatically calls the square root function during part of the compiling procedure.

PROGRAM FORM AND STATEMENTS

PROGRAM FORM Before we discuss FORTRAN statements, it is a good idea to look at the format of a FORTRAN program. Figure 14-8a shows a short FORTRAN program to convert entered Fahrenheit temperatures to Celsius and Kelvin and print a table of the results. FORTRAN programs often are written on special paper set up with 72 columns. Each column represents a space where a character can be entered; therefore, each line can have a maximum of 72 characters.

The first five columns are reserved for labels or a C. FORTRAN does not require a line number on each line as BASIC does. (The line numbers at the far left in Figure 14-8a are inserted by the compiler and are not required.) However, in FORTRAN any line referenced by a GO TO, DO, or other statement is given a label consisting of a number with one to five digits. A C in the first column indicates that the line is a comment.

The sixth column serves mostly as a separator. If a FORTRAN statement is longer than one line, a continuation character is put in column 6 to indicate that the contents of a line are a continuation of the previous line. Statements or comments are written in the last 66 spaces of each line.

An optional first statement of a FORTRAN program is the program statement. This statement gives an up-to-six-character symbolic name to the program. PROGRAM FCNVRT is an example. After this, a variety of statements are used. TYPE statements are often next in a program.

TYPE STATEMENTS TYPE statements either confirm or change the type of data referred to by a variable

```
   1            PROGRAM FCNVRT
      C         CONVERTS FAHRENHEIT TEMPERATURES TO CELSIUS AND KELVIN
      C         INPUT DEVICE IS CONSOLE = UNIT 5,
      C         OUTPUT DEVICES ARE LINE PRINTER = UNIT 4, CONSOLE = UNIT 6
      C         UNIT 4 = :LP: TO BE SPECIFIED AT RUN TIME


   2            REAL CLSIUS,KELVIN,FARNHT
   3            INTEGER NUMBER
   4            DIMENSION FARNHT(100)
   5            WRITE (6,9)
   6   9        FORMAT ('CONVERSION OF FAHRENHEIT TO CELSIUS AND KELVIN')
   7            WRITE (4,10)
   8  10        FORMAT ('1CONVERSION OF FAHRENHEIT TO CELSIUS AND KELVIN')
   9  11        WRITE (6,12)
  10  12        FORMAT ('PLEASE TYPE IN NUMBER OF TEMPERATURES TO BE CONVERTED')
  11  13        WRITE (6,14)
  12  14        FORMAT ('MAX NUMBER IS 100. FOR NUMBER LESS THAN 10 TYPE 00N.'
      1         ,/,'FOR NUMBER GREATER THAN 10 TYPE ONN')
  13            READ (5,20) NUMBER
  14  20        FORMAT (I3)
  15            IF (NUMBER. GT. 100) GOTO 11
      C         REPEAT DIRECTIONS IF NUMBER IS TOO LARGE
  16            WRITE (6,25)
  17  25        FORMAT ('PLEASE TYPE IN ALL FAHRENHEIT VALUES SEPARATED BY COMMAS')
  18            READ (5,*)(FARNHT(I),I=1,NUMBER)
      C         TAKE IN "NUMBER" OF VALUES AND STORE IN ARRAY "FARNHT"
  19            WRITE (6,40)
  20            WRITE (4,40)
  21  40        FORMAT (2X,'FAHRENHEIT',4X,'CELSIUS',4X,'KELVIN',/)
      C         PRINT HEADING FOR TABLE AND SKIP LINE
  22            DO 60,I=1,NUMBER
  23            CLSIUS=(ANINT((FARNHT(I)-32)*50/9))/10
      C         CALCULATE CELSIUS TEMP ROUNDED TO NEAREST .1 DEGREE
  24            KELVIN=CLSIUS+273.2
  25            WRITE (6,50)FARNHT(I),CLSIUS,KELVIN
  26            WRITE (4,50)FARNHT(I),CLSIUS,KELVIN
  27  50        FORMAT(4X,F6.1,6X,F6.1,4X,F6.1)
  28  60        CONTINUE
  29            END
```

MODULE INFORMATION:

```
    CODE AREA SIZE     = 0417H   1047D
    VARIABLE AREA SIZE = 01CAH    458D
    MAXIMUM STACK SIZE = 000AH     10D
    39 LINES READ
```

0 PROGRAM ERROR(S) IN PROGRAM UNIT FCNVRT

0 TOTAL PROGRAM ERROR(S)
END OF FORTRAN COMPILATION

(a)

FIGURE 14-8(a) Compiler listing of FORTRAN program to convert
Fahrenheit temperatures to Celsius and Kelvin.

```
CONVERSION OF FAHRENHEIT TO CELSIUS AND KELVIN
   FAHRENHEIT    CELSIUS    KELVIN

      -32.0       -35.6      237.6
        5.0       -15.0      258.2
      100.0        37.8      311.0
       98.6        37.0      310.2
        0.0       -17.8      255.4
```

(b)

FIGURE 14-8 (continued) (b) Sample run of FORTRAN program to convert Fahrenheit temperatures to Celsius and Kelvin.

or constant name. For example, the statement INTEGER NUMBER says that the variable NUMBER can have only integer values. Since NUMBER begins with an N, which automatically identifies it as an integer quantity, the statement only confirms this. However, suppose a real variable named KELVIN is desired. The K at the start of the name identifies KELVIN as an integer. However, the statement REAL KELVIN can be used to override the assignment to real or integer based on the first letter of the name.

Other TYPE statements identify logical or character types. For example, the statement of LOGICAL PASS indicates that the variable PASS is of the logical type. The statement CHARACTER∗15 NAME indicates that NAME is a character type variable of up to 15 characters. Even when not required, TYPE statements are often used to show all constants and variables with their types at the beginning of a program for documentation.

Arrays of data are defined by a DIMENSION statement such as DIMENSION CHESS (8,8). This defines CHESS

TABLE 14-6
FORTRAN "INTRINSIC" FUNCTIONS (INTEL)

FORM	FUNCTION	FORM	FUNCTION
SIGN(a1,a2)	Transfer sign of $a2$ to $a1$	COS(a)	Return cosine of a
IDIM(a1,a2)	Return $a1$-$a2$ if > 0; otherwise 0 (integer)	TAN(a)	Return tangent of a
DIM(a1,a2)	Return $a1$-$a2$ if > 0; otherwise 0 (real)	ASIN(a)	Return arcsine of a
		ACOS(a)	Return arccosine of a
MAXO (a1,...,an)	Select largest integer value from list	ATAN(a)	Return arctangent of a
AMAXI (a1,...,an)	Select largest real value from list	ATAN2(a1,a2)	Return arctangent of $a1/a2$
		SINH(a)	Return hyperbolic sine of a
AMAXO (a1,...,an)	Select largest integer value from list, returns as real	COSH(a)	Return hyperbolic cosine of a
MAXI (a1,...,an)	Select largest real value from list, returns as integer	TANH(a)	Return hyperbolic tangent of a
		INT(a)	Convert a to type integer
MINO (a1,...,an)	Select smallest integer value from list	IFIX(a)	Convert a to type integer
AMINI (a1,...,an)	Select smallest real value from list	REAL(a)	Convert a to type real
		FLOAT(a)	Convert a to type real
AMINO (a1,...,an)	Select smallest integer value from list, returns as real	ICHAR(a)	Convert a to type integer
MINI (a1,...,an)	Select smallest real value from list, returns as integer	AINT(a)	Truncate a to integer value
SQRT(a)	Return $\sqrt{a}$ for $a > 0$	ANINT(a)	Round a to nearest whole number (real)
EXP(a)	Return e∗∗a	NINT(a)	Round a to nearest integer
ALOG(a)	Return log (a) for $a > 0$	IABS(a)	Return absolute value of a (integer)
ALOG10(a)	Return log 10(a) for $a > 0$	ABS(a)	Return absolute value of a (real)
SIN(a)	Return sine of a	MOD(a1,a2)	Return remainder from $a1/a2$ (integer)
		AMOD(a1,a2)	Return remainder from $a1/a2$ (real)
		ISIGN(a1,a2)	Transfer sign of $a2$ to $a1$

as a two-dimensional array of eight rows and eight columns. DIMENSION FARNHT(100) defines a one-dimensional array, or list, of 100 elements.

ASSIGNMENT STATEMENTS Assignment statements are similar to the LET statements of BASIC. The first of these, arithmetic assignment statements, look very much like standard arithmetic statements. Examples are K=7, P=3.22, C=SQRT(X**2+Y**2).

Assignment statements are also used to give values to logical quantities, as in LUVTAX='FALSE', or to character quantities, as in NAME='HALL'.

DATA STATEMENTS DATA statements are used in place of multiple assignment statements to give initial values to variables, arrays, and array elements. For example, the statement DATA I,J,K/5,10,15/ will give initial values of 5 to I 10 to J, and 15 to K.

GO TO STATEMENTS GO TO statements produce unconditional jumps of program execution. The simplest GO TO statement has the form GO TO [statement label]. This causes the program execution to jump to the statement preceded by the label in the statement. Remember that in FORTRAN a label consists of a number between 1 and 9999.

Another form is GO TO (stl,stl,stl...)exp, where *stl* is a statement label and *exp* an integer expression. The statement GO TO (110,220,330)K causes program execution to jump to statement 110 if the value of K is 1, statement 220 if K is 2, and statement 330 if K is 3.

IF STATEMENTS FORTRAN has a variety of IF statements that produce conditional jumps of program execution. The simplest, similar to the multiple GO TO above, has the form IF(exp)stl1,stl2,stl3, where *exp* is an expression and *stl* stands for statement label. In the statement IF(B+C−A) 110,220,330, for example, program execution jumps to statement 110 if the value of the expression B+C−A is less than 0. If the value of the expression is equal to 0, the jump is to statement 220; if the value of the expression is greater than 0, the jump is to statement 330.

Another FORTRAN IF statement has the form IF(exp) [statement]. The statement can be any FORTRAN statement except a DO or an IF type of statement. An example of this type is IF(Y.GT.100)GO TO 10.

An IF statement also can be used with an END IF statement to form an IF block, as shown in Figure 14-9a. If the expression in parentheses is true, then the statements contained between the IF(exp)THEN and the ENDIF are executed. If the expression is not true, the IF block is skipped.

IF blocks can be nested as shown in Figure 14-9b. Also they can be combined with an ELSE IF(exp)THEN statement or an ELSE statement to provide jumps for several different conditions. Figure 14-9c shows an example of how these statements are used together. Note the ENDIF statement used to terminate the IF block. Identing as in Figure 14-9 is not necessary, but is done for clarity. The WRITE and FORMAT statements are explained later.

```
(a)    IF (EXP) THEN
          STATEMENT
          .
          .
          .
          STATEMENT
       END IF
```

```
(b)    IF (EXP) THEN
          STATEMENT
          STATEMENT
          IF(EXP) THEN
             STATEMENT
             STATEMENT
             .
             .
             .
          END IF
          STATEMENT
          .
          .
          .
       END IF
```

```
(c)       IF (LINE.EQ.1) THEN
             WRITE (6,20) NAME
      20     FORMAT...
          ELSE IF (LINE.EQ.2) THEN
             WRITE (6,40) FEDTAX
      40     FORMAT...
          ELSE IF (LINE.EQ.3) THEN
             WRITE (6,60) STATAX
      60     FORMAT...
          ELSE
             WRITE (6,80) FICA
      80     FORMAT...
          END IF
```

FIGURE 14-9 FORTRAN IF blocks. *(a)* Simple IF END IF. *(b)* Nested IF blocks.

DO STATEMENTS AND DO LOOPS The DO loop in FORTRAN is very similar to the FOR...NEXT loop of BASIC. It allows a group of statements to be done over and over for some number of times. The form of this statement is DO stl var=e1,e2,e3, where *stl* is the statement label of the last statement in the loop and *var* is an index variable. The starting value for the index is e1, the end value for the index is e2, and e3 is the step size for each run through the DO loop. For example, DO 200 K=1,50,1 loops through the statements between DO and statement 200 for 50 times. Another example of the DO loop is shown at the end of FCNVRT in Figure 14-8a.

CONTINUE STATEMENT A continue statement is a NOP. It performs no action. It is just used as a convenient place to hang a label for a DO loop because the last statement of a DO loop cannot be FORMAT, GO TO, IF, or DO. The DO loop in Figure 14-8a shows its use.

SUBROUTINE, CALL AND RETURN STATEMENTS FORTRAN uses subroutines to avoid writing an often used procedure over and over. The first statement in a routine must be SUBROUTINE, which gives the routine a symbolic name and lists the variables or arrays used. The form is SUBROUTINE name (var,var,var...). The subroutine may have a RETURN statement at any point to return control to the main program. However, an END statement is required at the end of the routine, and there it has the same effect as a RETURN.

The subroutine is called from the main program with a CALL statement of the form CALL [name] (var,var,var...). In this statement the actual values of the variables are written to be passed to the subroutine. The listed variable values must be in the same order as in the SUBROUTINE statement. Figure 14-10 shows an example are in which these statements are used to convert an IF block to a subroutine.

READ, WRITE, AND FORMAT STATEMENTS

These three statements are important because they are used to input data from a card reader, console, or some other device and to output data to a console or line printer. They are also probably the main reason BASIC was developed.

A typical READ statement has the form READ (UNIT #,stl) [var,var,...]. The unit number refers to the number given to an input device in the system in which the program is run. For example, the console in the Intel system on which the program in Figure 14-8a was run is called unit 5. The *stl* refers to the statement label of a FORMAT statement. READ (5,20) NUMBER means to input a value for the variable NUMBER from UNIT#5 and store it in memory in a location called NUMBER. The 20 in the statement says that the FORMAT statement with label 20 will show the form in which the data is to be read in. A FORMAT statement at label 20, for example, might be 20 FORMAT(I3). This means that the data to be read in is a three-digit integer.

At this point, you may wonder why a FORMAT statement is needed in FORTRAN, but not in BASIC. The answer is that FORTRAN was developed at a time when one of the main ways of inputting programs, data, or commands to a computer was with 80-column punched

```
          CALL PAYCHK (3)
          SUBROUTINE PAYCHK (LINE)
          IF (LINE.EQ.1) THEN
              WRITE (6,20) NAME
     20       FORMAT...
          ELSE IF (LINE.EQ.2) THEN
              WRITE (6,40) FEDTAX
     40       FORMAT...
          ELSE IF (LINE.EQ.3) THEN
              WRITE (6,60) STATAX
     60       FORMAT...
          ELSE
              WRITE (6,80) FICA
     80       FORMAT
          END IF
          RETURN
          END

     NEXT STATEMENT OF MAIN PROGRAM
```

FIGURE 14-10 FORTRAN IF block converted to a subroutine.

cards, such as shown in Chap. 3. Since several data values may be sent in on a single card, the READ statement has to know how many values are on the card and how many digits each has. BASIC, which is used mostly with CRT consoles or teletypes, is set up for a free-form input, so that values can just be typed in with separators.

WRITE statements in FORTRAN also usually have FORMAT statements with them. The simplest form is WRITE (UNIT#,stl) [variable,...,variable]. An example from Figure 14-8a is WRITE (4,50) FARNHT(I),-CLSIUS,KELVIN. This statement says to output to UNIT#4, which is the line printer, the values of variables FARNHT(I), CLSIUS, and KELVIN. The format in which the values are to be printed is shown in the FORMAT statement labeled 50.

A little more discussion of FORMAT statements will help you see what this one means. Table 14-7 shows some *repeatable edit descriptors* used in FORMAT statements. Some examples will help clarify these. I3 in a FORMAT statement indicates that a three-digit integer is to be input or output. F6.1 refers to a REAL variable quantity with six total characters including sign and decimal point. The 1 in the descriptor says that the number has one digit to the right of the decimal point. −237.3 is an example of a quantity that would require the F6.1 descriptor to read or write it. E and F descriptors are the same for input operations, but for output Ew.d.Ee allows real numbers to be printed in scientific

TABLE 14-7
REPEATABLE EDIT DESCRIPTORS USED IN FORTRAN FORMAT STATEMENTS

Iw	Integer descriptor
F$w.d$	Real number descriptor
E$w.d$	Real number descriptor
E$w.d$Ee	Real number descriptor
Lw	Logical descriptor
A	Variable-length alphanumeric descriptor
Aw	Fixed-length alphanumeric descriptor
Bw	Binary descriptor
Zw	Hexadecimal descriptor
where	
I,F,E,L, and A	Indicate the type of data being edited
B and Z	indicate the number base of data being edited
w =	nonzero, unsigned integer constant representing the width of entire edited field
d =	unsigned integer constant representing number of digits that should follow the decimal point
e =	nonzero, unsigned integer constant representing the width of the exponent field

Source: Intel.

notation. For example, E4.1E2 allows a number such as -7.6×10^{-6} to be output.

Logical variables can have only the values TRUE or FALSE. An L5 descriptor allows these to be input or output. For alphanumeric character input, an A20 descriptor allows a string of 20 characters to be input or output.

Several descriptors, separated by commas, can be written in a single FORMAT statement to indicate that several data values are to be read from a card or printed on a line. An X is used to specify a blank. The statement FORMAT (4X,F6.1,6X,F6.1,4X,F6.1) from the program in Figure 14-8a, for example, means that starting from the left margin print four blanks, the value of one real variable, six blanks, the value of the next real variable, four blanks, and the value of the last real variable.

Another type of FORMAT statement that simplifies the output of characters and words is used several times in Figure 14-8a. For this the WRITE statement simply gives the unit number and the label of the FORMAT statement as, for example, WRITE(6,9). The corresponding FORMAT statement then contains the words to be printed enclosed in apostrophes, as in 9 FORMAT ('CONVERSION OF FAHRENHEIT TO CELSIUS AND KELVIN'). If the message extends for more than one line, a continuation character is put in column 6 and the message continued on the next line, as shown for FORMAT statement 14 in Figure 14-8a. The apostrophe form can be used with other descriptors in a FORMAT statement, as shown in the one at line 40 in Figure 14-8a. Three column headings separated by blanks are printed by this statement. The slash at the end of the statement tells the printer to do a carriage return, line feed, or, in other words, skip a line before printing the next line.

When the only I/O device is a console, a technique called *list-directed formatting* simplifies the input of data by allowing it to be entered in free form. The READ statement for this has the form READ(UNIT #,*) [variable]. The asterisk in place of the FORMAT statement label indicates that no FORMAT statement is required and that a value typed in will be just read as written. Numbers are typed in with decimal points if they have a fractional part. Characters or words are typed in, set off by apostrophes. The READ statement at compiler line number 18 in Figure 14-8a is an example.

PROGRAM EXAMPLE

The previous sections have discussed FORTRAN statements and formatting. Many of the statements in the program of Figure 14-8a have been explained and used as examples. This section explains the operation of the complete program.

After a PROGRAM statement and some comments, real and integer variables are declared. Note how much more mnemonic these names are than BASIC names. The DIMENSION statement defines an array named FARNHT for 100 values.

The program title is written to the console and to the line printer. The 1 before the title in the message sent to the printer is a control character that tells the printer to skip to the top of the next page before printing the message. This puts the title at the top of the page, to make the printout look good.

The directions to the user then are sent only to the console. The READ (5,20) NUMBER and FORMAT(I3) statements wait for the user to enter the number of conversions to be run. Note that normal FORMAT statement, FORMAT(I3), requires the user to enter leading zeros if the number entered is less than 100. *List-directed formatting* could have been used here, but the normal formatting was left in to illustrate this type of problem.

If an illegal number is entered, the IF statement sends program execution back to the statement with label 11 to print the directions on the console again. If a legal number is entered, the user is directed on the console to enter the Fahrenheit values, separated by commas. READ(5,*) (FARNHT(I),I=1,NUMBER) takes in these values as they are typed and enters them in the array FARNHT in memory. This is a somewhat sneaky READ statement because it contains an implied DO loop. What the statement says is for the index I=1 to I=NUMBER, read a value from the input device and insert it in array FARNHT as element I. This is shown because it is a common shorthand for inputting multiple values.

When all the values are entered, the output table headings are printed on the console and the line printer. A DO loop is then used to calculate and print the equivalent Celsius and Kelvin values for each Fahrenheit temperature. The statement to calculate the Celsius equivalent first subtracts 32 from the Fahrenheit value and multiplies the result by 50/9. This gives a result that is 10 times the desired value. ANINT, one of the library functions shown in Table 14-6, rounds this result to the nearest integer. Dividing by 10 moves the decimal point back to the proper position and gives a result rounded to the nearest 0.1°. Kelvin is calculated by simply adding 273.2 to the Celsius value. Figure 14-8b shows a printout for a run of the program with five values entered.

COMPILING AND RUNNING A FORTRAN PROGRAM

Compiling a FORTRAN program so that it can be run is a multiple-step process.

CREATING A SOURCE FILE   The first step is to write and enter the basic program unit, such as that in Figure 14-8a, into RAM by using an editor program. When the program is all entered, it is loaded into a file on a floppy disk or tape. This is called a *source file*.

COMPILING   Second, a compiler program is used that translates the source file to relocatable object code. Relocatable means that symbolic addresses are used so that the module later can be located anywhere in memory. At this point, the compiler produces a printout of the program unit such as that in Figure 14-8a.

For its own reference the compiler assigns line numbers at the left of the printout to each statement. As shown in the printout, it also shows the number of bytes of memory required for the object module. Note that the short program in Figure 14-8*a* requires a total of 1515 bytes of memory. Since each REAL value in the array FARNHT requires 4 bytes, this array alone requires 400 bytes of variable storage.

The compiler also prints a listing of any syntax (grammatical) errors found in the program unit or indicates that no errors were found. If errors are found, the source file can be corrected and recompiled until the "no errors" message appears.

LINKING   Third, the corrected object module must be *linked* to the object modules for the library functions and the I/O routines. Remember, the "intrinsic" functions such as SQRT(X),ANINT(X), etc. are essentially subroutines. If they are used in a program unit, the code for them must be linked to the code for the unit. The I/O routines that are called by READ and WRITE statements also must be linked to the main module.

LOCATING   Fourth, after all the program object parts are linked together, they must be located. This means that all the symbolic addresses must be replaced with absolute addresses. The result is machine-code program that can be entered into RAM and executed. It may also be loaded in a disk file for storage.

PROGRAMMING IN PASCAL

Pascal is a newer language than FORTRAN or BASIC, so it does not have some of their problems. Pascal allows a programmer to easily write structured modular programs which are easy to debug. Pascal is easily adapted to run on relatively small microprocessor-based systems.

GENERAL STRUCTURE OF A PASCAL PROGRAM

A Pascal program has two main parts, a HEADING and a BLOCK. Figure 14-11*a* shows a complete program. The *heading* simply gives the program a title and lists any predeclared parameters, such as INPUT and OUTPUT. An example is PROGRAM FCNVRT (INPUT,OUTPUT).

The BLOCK part of the program is made up of two parts: a declaration part and the main body. In the declaration part, which must come first, all variables, arrays, functions, files, sets, etc. are given names and their types identified. For example, if an integer variable named K is going to be used in a program, it must be declared at the start of the BLOCK by the statement VAR K:INTEGER.

The main body of the program contains the statements that manipulate the declared data types. It starts with BEGIN and concludes with END . . Note the re-

quired period after END.. Statements are separated by semicolons. No line numbers are used.

DECLARATIONS

Figure 14-12 shows a *syntax diagram* for the five kinds of declarations that may be made at the start of a Pascal program. Only the kinds actually used in a program need to be declared, so in a particular program not all of these may be present. A syntax diagram simply shows the correct format and punctuation for each kind of declaration.

The five kinds of declarations are LABEL, CONST (constant), TYPE, VAR(variable), and PROCEDURE or FUNCTION. The bottom line of the syntax diagram represents the main body of the program, which starts with the word BEGIN, contains statements separated by semicolons, and ends with the word END..

LABEL   A LABEL in Pascal is used with a GOTO statement to unconditionally jump execution to some other statement. A label must be an integer of one to four digits. The syntax diagram shows that the correct format for a LABEL declaration is LABEL 33; or LABEL 26,78;.

CONST   A declared constant (CONST) can be a number or a string of characters. Examples of declared constants in the correct format are CONST PI = 3.14159; ADDON = −32; OK = 'YES';.

VAR   A variable name can be any length, but some compilers use only the first eight letters. Any variable used in a BLOCK must be declared at the start of the BLOCK and its type identified. The four simple types of variable are REAL, INTEGER, CHARACTER (abbreviated CHAR), and BOOLEAN. Some examples of declarations for these types are as follows: VAR INCREMENT: REAL; COUNT:INTEGER; JOE:CHAR; RESULT:BOOLEAN;.

Real variables contain an integer and a fractional part, and integer variables can have only an integer part. Character variables can have any letter of the alphabet or punctuation symbol as their value. Boolean variables can have only a value of TRUE or FALSE.

Variables declared at the start of a main block are *global* to the entire program. Variables declared only in a subBLOCK, such as a procedure or function, are *local* to that block. This is different from BASIC where all variables are global.

In addition to the four simple types of variables, Pascal permits more complex variable types known as *structured types*. These include arrays, records, sets, and files. An example of the declaration of the array variable is VAR A:ARRAY[1..100]OF INTEGER:. This means that the letter A represents an array of 100 integer values.

TYPE   Pascal also allows the user to declare new types of variables and the values allowed for them. For example, the declaration TYPE DAY = (SUNDAY, MONDAY,

```
PROGRAM FCNVRT;
(* CONVERTS FAHRENHEIT TO CELSIUS AND KELVIN *)
VAR
    LP : TEXT;
    DISK : FILE OF CHAR;
    FAREN : REAL;
FUNCTION CNVRT (F:REAL) : REAL;
    (* CONVERTS FAHRENHEIT TO CELSIUS *)
BEGIN (* CNVRT *)
 CNVRT : = (ROUND((F-32.0)*50.0/9.0))/10.0
END; (* CNVRT *)


BEGIN (* MAINLINE PROGRAM *)
  REWRITE (LP,'PRINTER:');   (* OPENS LINE PRINTER AS FILE *)
  REWRITE (DISK,'#4:DATA.TEXT');   (* OPENS DISK FILE NAMED DATA *)
  WRITELN ('CONVERSION OF FAHRENHEIT TO CELSIUS AND KELVIN'); (* TO CONSOLE *)
  WRITELN (LP,'CONVERSION OF FAHRENHEIT TO CELSIUS AND KELVIN'); (* PRINTER *)
  WRITELN ('ENTER FAHRENHEIT VALUES - AFTER LAST VALUE, TYPE -1000.0');
    REPEAT
        READ (FAREN);
        IF FAREN     -1000.0 THEN
            WRITELN (DISK,FAREN)
      UNTIL FAREN =  -1000.0;
  CLOSE (DISK,LOCK);   (* NON-STANDARD PROCEDURE TO CLOSE FILES *)
  RESET (DISK,'#4:DATA.TEXT');
  WRITELN (LP,' FAHRENHEIT  CELSIUS   KELVIN  ');
    WHILE NOT EOF(DISK) DO
        BEGIN
         READLN (DISK,FAREN);
         WRITELN (LP,FAREN:8:1,CNVRT (FAREN):12:1,CNVRT (FAREN)+273.2:8:1)
        END;   (* WHILE *)
    CLOSE (LP)
END. (* END OF MAINLINE PROGRAM *)
```

(a)


```
CONVERSION OF FAHRENHEIT TO CELSIUS AND KELVIN
 FAHRENHEIT  CELSIUS   KELVIN
     12.0       -11.1    262.1
     45.7         7.6    280.8
    345.5       174.2    447.4
   -200.7      -129.3    143.9
```

(b)

FIGURE 14-11 Pascal program to convert Fahrenheit temperatures to Celsius and Kelvin. (a) Program listing. (b) Sample run.

TUESDAY, WEDNESDAY, THURSDAY, FRIDAY, SATURDAY); defines a variable of type DAY that can have values of SUNDAY through SATURDAY. This is similar to the predefined type BOOLEAN which can have values of TRUE or FALSE. Another example of a TYPE declaration is TYPE COLOR = (RED,ORANGE,YELLOW, GREEN, BLUE,INDIGO,VIOLET);.

Once a new type is declared, it is used in a variable declaration just as any other type. For example, VAR X:COLOR states that X is a variable of type COLOR and

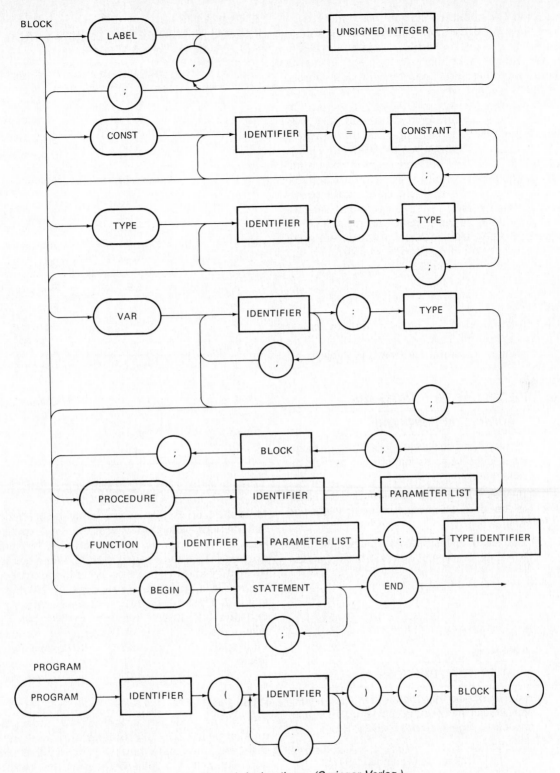

FIGURE 14-12 Syntax diagram for Pascal declarations. *(Springer-Verlag.)*

can have only the values declared in the corresponding TYPE declaration. Being able to declare new data types in this way is a very powerful feature of Pascal.

PROCEDURE AND FUNCTION   These subroutines

in Pascal are called by name as needed in a program. After the procedure or function is completed, execution is returned to the next statement in the main body of the program.

Procedures are general subroutines that are written

to do some job. *Functions* are subroutines that return a specific result such as the sine of an angle or the square root of a number. Table 14-8 shows a list and explanation of the predeclared, or intrinsic, functions in Pascal.

Unlike assembly language programs in which subroutines can be placed anywhere, a procedure or function in Pascal must occur in the program before the BLOCK that calls it. The procedure or function subBLOCK is first given a heading. The heading gives the routine a name, for example, PROCEDURE MINMAX; or FUNCTION FCNVRT;. Then any variables, types, constants, etc. not declared for the whole program are declared for the procedure. Values of variables declared in the procedure or function are local. After the required declarations, the statements of the subBLOCK are listed. This is shown for the FUNCTION FCNVRT at the top of Figure 14-11*a*. The listing for a procedure looks very much like that of a complete short program.

OPERATORS

Table 14-9 shows a list of the Pascal arithmetic logical and relational operators. Most are quite standard. The priority of arithmetic operators is shown in Table 14-9.

TABLE 14-8
PASCAL PREDECLARED FUNCTIONS

ARITHMETIC FUNCTIONS	
abs(x)	Computes the absolute value of x. The type of the result is the same as that of x, which must be either integer or real.
sqr(x)	Computes x∗x. The type of the result is the same as that of x, which must be either integer or real.
sin(x)	For the following, the type of x must be either real or integer; the type of the result is always real:
cos(x)	
arctan(x)	
exp(x)	
ln(x)	(natural logarithm)
sqrt(x)	(square root)

TRANSFER FUNCTIONS	
trunc(x)	x must be of type real; the result is the greatest integer less than or equal to x for $x \geq 0$, and the least integer greater or equal to x for $x < 0$.
round(x)	x must be of type real; the result, of type integer, is the value x rounded. That is, round(x) = trunc(x + 0.5) for $x \geq 0$ and round(x) = trunc(x − 0.5) for $x < 0$
ord(x)	The ordinal number of the argument x in the set of values defined by the type of x.
chr(x)	x must be of type integer, and the result is the character whose ordinal number is x (if it exists).

TABLE 14-9
PASCAL ARITHMETIC, LOGICAL, AND RELATIONAL OPERATORS

PRIORITY	OPERATOR	DESCRIPTION
1	NOT	Boolean not
2	*	Multiplication
2	/	Division (real)
2	DIV	Integer division
2	MOD	Modulus (remainder from integer division)
2	AND	Boolean and
3	+	Addition
3	−	Subtraction
3	OR	Boolean or
4	=	Equal
4	< >	Not equal
4	<	Less than
4	< =	Less than or equal to
4	> =	Greater than or equal to
4	>	Greater than
4	IN	In a set

Multiplication or division is done before addition or subtraction. Note that Pascal has no exponentiation operator. However, two new operators, DIV and MOD, are present. DIV is integer division. When one integer is divided by another in this way, only the integer part of the result is returned. For example, 5 DIV 2 gives a result of 2. The MOD operator returns the integer remainder after an integer division. For example, 5 MOD 2 gives a result of 1. This MOD operator is useful for determining whether one variable divides evenly into another.

STATEMENTS

After all the necessary declarations are made, the main-body statements of the program are written. At this point you may wonder how the programmer knows what to declare before the body of the program is written. The answer is that the programmer probably doesn't know at the start. Programs are written by successive approximation. As more variables, etc. are needed, they are added to the declaration lists. *Dummies* can be inserted in the declarations and removed later if not needed.

Figure 14-13 shows a syntax diagram for the common Pascal statements. The first is the assignment statement.

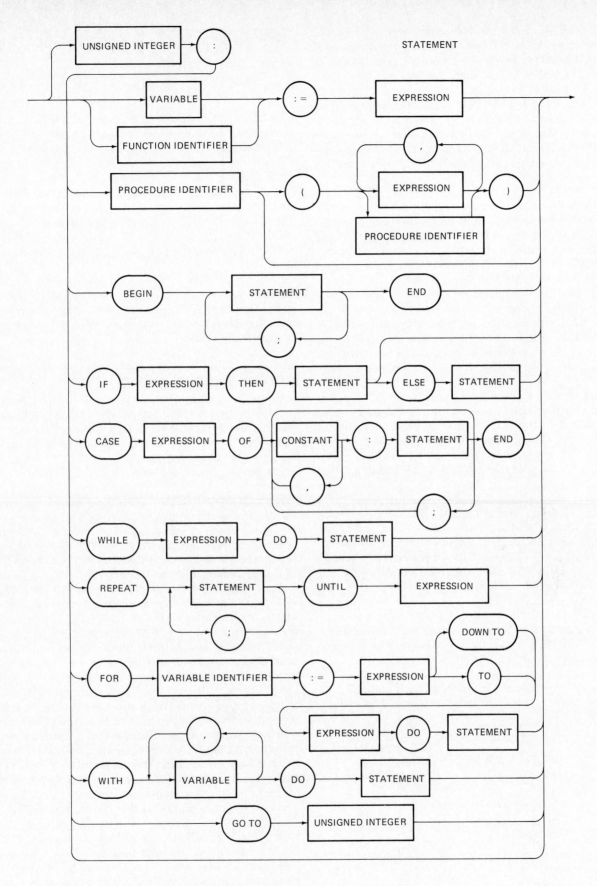

STATEMENT

FIGURE 14-13 Syntax diagram for Pascal statements. *(Springer-Verlag.)*

ASSIGNMENT STATEMENT This statement corresponds to the LET statement of BASIC. It uses a colon and equal sign together. Examples are X:=Y+2, X:=X+1, CUBE:=X*X*X, and Y:=SQRT(X).

PROCEDURE STATEMENT This statement is used to *call* a procedure. It starts with the name of the procedure and then contains in parentheses the values of parameters passed to the procedure. The statement MINMAX (3,7,9) calls procedure MINMAX and passes the values 3, 7, and 9 to it.

BEGIN AND END. BEGIN and END. are not statements. They are used like brackets to contain, or set off, a group of statements. A BEGIN and an END. are used at the start and finish of the main body of the programs. They also can be used to group several statements into a compound statement which can be treated as a single statement. The program example in Figure 14-11 shows this. Any place in a syntax diagram where a statement is indicated, a compound statement enclosed by BEGIN and END can be used. Note that this END does not have a period after it as the one at the end of the program does!

FOR...DO This statement allows a series of statements to be repeated for some number of times. It is similar to the FOR...NEXT loop met previously. Figure 14-14 shows how it is used to calculate and print the square root of each number from 1 to 16. The compound statement between BEGIN and END will be executed 16 times.

REPEAT...UNTIL By using this statement, a statement or group of statements is repeated until some condition is detected. In the program of Figure 14-11, for example, the REPEAT...UNTIL loop reads in Fahrenheit values typed by the user until the impossible value -1000 is detected.

WHILE...DO Another way of performing repetitive tasks is with this statement. An example is WHILE H>0 DO READ(TIME);. The program in Figure 14-11 shows another example of the WHILE...DO loop.

GOTO This produces an unconditional jump to a labeled statement. The label must be a one- to four-digit integer.

```
PROGRAM ROOT2

VAR I: INTEGER; X:REAL;
BEGIN
FOR I:=1 TO 16 DO
    BEGIN X:=I;
    WRITELN (I, SQRT(X));
    END;
WRITELN ('DONE');
END.
```

FIGURE 14-14 Example of a FOR . . . DO loop in Pascal.

(a) `IF Y = 1 THEN P:=7 ELSE P:=14;`

(b) `IF Y = 1 THEN P:=7`
`    ELSE IF Y = 2 THEN P:=14`
`        ELSE WRITELN ('ILLEGAL VALUE');`

FIGURE 14-15 Pascal examples of IF . . . THEN . . . ELSE statements. *(a)* Single-level IF . . . THEN . . . ELSE. *(b)* Nested IFs.

IF...THEN...ELSE If the expression in this statement is true, the statement after THEN is executed. If the expression is not true, execution continues with either the main program or a statement written after an ELSE, as shown in Figure 14-15a. IF statements can also be nested as shown in Figure 14-15b.

CASE AND WITH These are shown in Figure 14-13 because they are included in the standard syntax diagram, but there is not room here to explain them. Consult one of the references at the end of the chapter for further information.

CONSOLE INPUT AND OUTPUT

If input and output are from and to a console, then in most systems the four statements READ, READLN, WRITE, and WRITELN are all that is needed for I/O. The reason is that the console is *preconnected* as input and output. The parameter list (INPUT,OUTPUT) in the program heading indicates such.

WRITE AND WRITELN Both these statements print the indicated quantities on the system console, but WRITELN sends a carriage return and line feed after sending the message. Therefore, a WRITELN statement causes the material sent by the next WRITE or WRITELN to be printed on the next line.

Some examples will best show the form for the WRITE and WRITELN statements (see Figure 14-16). Quantities in parentheses after WRITE or WRITELN are printed on the same line. Values of REAL variables are printed in scientific notation, as in example c. In example d the first integer after the variable gives the desired total number of digits, and the second integer gives the number of digits to the right of the decimal point. If these values are given, then the printout will not be in scientific notation. In the program of Figure 14-11a, these numbers are also used to center the printed temperatures under their headings. If the specified field width is greater than the number, the number is printed at the right of the field. Character strings enclosed between single quotation marks are printed as written. Examples e and f contrast the effects of WRITE and WRITELN.

READ AND READLN The difference between these two is, again, a carriage return. The statement READLN(I), where I is an integer variable, causes the machine to sit and wait for the user to enter an appropriate value for I. A carriage return must be typed after

```
a. WRITE (1,2,3,4,)              1    2    3    4
b. WRITE ('YOU HAVE MADE AN ERROR')    YOU HAVE MADE AN ERROR
c. WRITE (SQRT (R))              1.50000000E + 00
                                     (for R = 2.25)
d. WRITE (SQRT (R):6:2)          1.50 (for R = 2.25)
e. WRITELN ('THE ANSWER =',A)    THE ANSWER = (value of A)
   WRITELN ('DID YOU GET IT?')   DID YOU GET IT?
f. WRITE ('THE ANSWER =',A)      THE ANSWER = (value of A) DID YOU GET IT?
   WRITELN ('DID YOU GET IT?')
```

FIGURE 14-16 Examples of WRITE and WRITELN statements.

each value is entered. If more than one value is typed in, only the last one typed before the carriage return is entered. With a READ the values for several variables can be typed in, separated by blanks, before a carriage return is typed. Figure 14-17 shows a few examples of READ and READLN statements.

INPUT AND OUTPUT TO OTHER DEVICES: FILES

In computer jargon the word *file* has two meanings. The first meaning is a sequence of memory locations in which data is stored. An internal file of this type is a series of RAM locations. Data is stored also on tapes and disks in sequential files.

The second meaning of *file* is any input or output device such as a line printer, disk drive, or paper-tape reader. This use of the term *file* is a holdover from the days when the main means of input to a computer were with a file of punched cards.

The program in Figure 14-11*a* converts Fahrenheit temperatures to Celsius and Kelvin and then prints a table of the results. The program could have been written in a very similar manner to that of the FORTRAN example in Figure 14-8*a*. However, a different approach was chosen to show the use of files.

This program reads a series of Fahrenheit values input on a console and stores them in a file named DATA on a floppy disk. When all the desired values have been entered, typing an impossible value of −1000 ends the REPEAT...UNTIL loop. After some housekeeping and printing of table headings, the program enters a DO...WHILE loop. This loop reads a stored value from the disk file DATA, converts it to Celsius and Kelvin, and prints the Fahrenheit, Celsius, and Kelvin values on the file line printer.

Before an external file can be read from or written to,

it must be symbolically connected to the system, or *opened*. This is done with a REWRITE statement. As shown in Figure 14-11*a*, REWRITE (LP,'PRINTER:') opens a line printer file named PRINTER. The statement REWRITE(DISK, '#4:DATA.TEXT') opens a disk file called DATA.TEXT. As values are read in from the console by the READ(FAREN) statement, the WRITEL N(DISK, FAREN) statement stores each in the disk file DATA.TEXT. No reference is needed to the file name DATA.TEXT in the WRITELN statement because the REWRITE statement says that anything sent to the external file DISK will be stored on the disk in the file DATA.TEXT.

Data written to a disk or line printer does not go directly to the device. It goes through a buffer. This is necessary because data cannot be written to, for example, a disk 1 byte at a time. A disk file consists of one or more records (sectors). A record is 256 or, for double density, 512 bytes. (See the section in Chap. 5 on disks.) Data to be stored on disk is first loaded into a 256- or 512-byte buffer. When the buffer becomes full, it is loaded into a sector on the disk. The CLOSE statement adds an end-of-file character (EOF) to the end of the data in the buffer and writes the buffer contents into a sector even if the buffer wasn't full. For a line printer the buffer holds enough data for one line of print, often 80 characters.

Data is then sequentially written into the buffer one location at a time and to the disk file in chunks. A *file pointer* keeps track of the location in the file of the current data byte. As each data value is entered, the file pointer increments 1.

When the file is closed, the file pointer is left pointing at the end of the file. Before the file can be read, the file pointer must be moved to the start of the file again. This is done by the RESET statement. In Figure 14-11*a* the RESET (DISK,'#4:DATA.TEXT') moves the file pointer

STATEMENT	DATA TYPED	RESULT
READ (FAREN)	−40.0 (CR)	FAREN = −40.0
READ (A,B,C)	5(BLANK) 7 (BLANK)9 (CR)	A=5, B=7, C=9
READLN (A,B,C)	5(BLANK) 7 (BLANK)9 (CR)	A=9
READLN (A,B,C)	5(CR) 7 (CR) 9 (CR)	A=5, B=7, C=9

FIGURE 14-17 Examples of READ and READLN statements.

to the first location in the disk file DATA.TEXT. The following WHILE...DO loop then reads a value from the file DATA.TEXT, increments the file pointer, converts the value to Celsius and Kelvin, and prints the Fahrenheit, Celsius, and Kelvin values. This process is continued until the end-of-file character is detected. The file PRINTER is then closed and the program terminated.

Note that the file DISK is declared as a *file of characters* at the start of the program. To simplify the program somewhat, the Fahrenheit values are stored in the disk file as ASCII characters rather than real numbers. Since FAREN is declared as a REAL variable, these are converted to numbers by the READLN(DISK,FAREN) statement for the conversion to Celsius and Kelvin values. The WRITELN statement converts the calculated values back to ASCII characters to go to the line printer. When the function CNVRT is called, as in the WRITELN(LP, FAREN:8:1, CNVRT(FAREN):12:1, CNVRT (FAREN) + 273.2:8:1); statement, the value of FAREN is automatically substituted for the dummy variable F in the CNVRT expression. In reality, the rounding did not need to be done in FUNCTION CNVRT because the statement WRITELN(FAREN:8:1); automatically rounds a result to one decimal place. However, ROUND was used to show how one of the predeclared functions can be used. The program as written will run on an Apple II computer with Pascal capability.

TRENDS IN THE DEVELOPMENT OF MICROCOMPUTER-BASED PRODUCTS

Development of the microprocessor has made it possible to build a great deal of intelligence into many instruments and systems. It is possible to do with software many tasks that are difficult, if not impossible, to do with hardware. Even a very small generalized system such as an SDK-85 or MEK6800D2 can be programmed to do many different jobs. However, the cost of software development can be very high. To combat this high cost, manufacturers use the following approaches:

1. Intelligent peripheral devices are used wherever possible to reduce the number of tasks that must be programmed or done with discrete hardware. Serial port chips such as the 8251A relieve the processor of the burden of sending and receiving serial data. Coprocessors such as the Intel 8087 mathematics processor can perform many mathematical operations that otherwise would have to be done with program routines. An important point is that the coprocessor can do this while the main processor is doing other tasks.

2. As much software as possible is developed in high-level languages. Programs developed in high-level languages are usually written more quickly and are easier to debug, document, and modify. For many programs the high-level language programmer does not need to know the intricate hardware details such as register structure that an assembly language programmer does.

Compiled higher-level language programs usually take more memory and run somewhat slower than carefully written assembly language programs. However, the decreasing per-bit cost of memory and the greater speed of product development with high-level languages usually outweigh these disadvantages.

Assembly language is used only when speed is critical or the program involves a great deal of individual bit testing and manipulating.

Operating systems, such as ISIS-II which runs on an Intellec system (see Figure 8-26), allow object-code modules developed from several languages to be linked. Using this feature, a programmer might write one section of a program in FORTRAN, which is strong in mathematical operations, a section in PL/M, which is strong in I/O operations, and another bit-fiddling section in assembly language.

3. Structured, modular programming techniques are used. Whatever language is used, programs should be written in blocks, or modules, which can be individually debugged and are often reusable. Each block should ideally have only one entry point and one exit point. (For further information on structured programming, consult the references at the end of this chapter.)

4. As much tested and debugged software as possible is purchased from software vendors, so that in-house efforts can focus on the specific application being developed. This avoids reinventing the wheel. New languages such as Ada enable programmers to develop programs by using a library catalog of complex software components. More and more "canned" software is becoming available, much of it as *firmware*, or *silicon software*. The name *firmware* is given to programs or routines stored in ROMs. Monitors, editors, and interpreters are commonly available as firmware.

Another example of silicon software is the Intel 80130. This device contains the basic software building blocks (called *primitives*) used for creating a real-time operating system. The device also has an 8259A-type priority interrupt controller and three 16-bit programmable timers. These are important parts of the hardware for a real-time operating system that might be used to control a series of processes. (The industrial control application in Chap. 12 contains some elements of a simple real-time system.)

Another example of silicon hardware-software is a five-chip set of ICs produced by Western Digital Corporation. This set of chips directly executes Pascal P code without the need for a compiler or interpreter.

The overall trend of microcomputer evolution is toward putting more and more hardware and software functions in single ICs. Because of this computers are becoming smaller, more powerful, and nearly as common as telephones. Computer intelligence is evolving at a very rapid rate. It is now possible to link computers all over the world in intelligent networks which help people share information and resources. It remains to be seen whether the power of computers to unite people will triumph over the destructive power of nuclear weapons.

REVIEW QUESTIONS AND PROBLEMS

1. Define and describe the use of the following:
   a. Monitor program
   b. Editor program
   c. Compiler program
   d. Interpreter program

2. Define source program and object program.

3. Explain the differences between a compiler and an interpreter. Give an advantage and disadvantage of each.

4. What do the following high-level language acronyms stand for:
   a. FORTRAN   b. BASIC
   c. ALGOL   d. COBOL
   e. APL

5. Why are BASIC statements usually not given closely consecutive numbers?

6. Identify the items in the following list which are valid names for BASIC variables:
   a. A   b. BET
   c. C6   d. D*
   e. 7F   f. A1
   g. ZZ   h. Z

7. List the BASIC arithmetic operators with their symbols in order of the priority in which they are performed in a statement.

8. a. Describe the order in which the operations in the following statements will be performed:
   LET Y=A+B*C
   LET X=A↑2+B*7
   LET P=SQR (A/7+C−B*3)
   b. If A = 7, B = 1, and C = 5, calculate the values for Y, X, and P.

9. Explain the following BASIC statements:
   a. LET X=X+1
   b. LET Y=R*SIN(A)
   c. PRINT "NOW ENTER A NUMBER"
   d. PRINT A,X,P
   e. GOSUB 290
   f. RETURN
   g. IF Z=7 THEN 310
   h. INPUT N

10. Add a statement to the coin-flipping program of Figure 14-3 which will cause a jump to an END statement if the user enters a negative number of flips.

11. Write a program to simulate the rolling of one die and then a program for two dice. Remember, the two dice are independent, and LET A=INT(6*RND(0))+1 gives a random number between 1 and 6.

12. In the program of Figure 14-4, rewrite the statement at line 40 to round the result to the nearest 0.01°.

13. Rewrite the program in Figure 14-4 to input the Fahrenheit temperatures in an array in the way the test scores are entered in the program of Figure 14-7.

14. Define floating-point number and fixed-point number.

15. Which are valid FORTRAN variable names?
    a. MODE   b. BYTE
    c. FAHRENHEIT   d. C3
    e. CATCH-22   f. TO FAR

16. Describe the meaning of the following FORTRAN expressions:
    a. Y=A−B/2
    b. C=SQRT(A**2+B**2)
    c. B=ANINT(A/7)
    d. Y .LE. X
    e. A=R*(B+6*(S/T)+9)
    f. Y=R*SIN(A)

17. Which are legal FORTRAN statement labels?
    a. 27   b. LOOP
    c. D3   d. 999
    e. C   f. −23
    g. 50

18. Explain the purpose of TYPE statements in FORTRAN.

19. Describe the use of the following FORTRAN statements:
    a. DIMENSION
    b. DATA
    c. GO TO
    d. IF (exp) [stl,stl,stl]
    e. If (exp) [statement]
    f. CALL [name] (var,var,var,...)
    g. READ (UNIT#, format stl) [var]
    h. WRITE (UNIT#, format stl) [var]
    i. FORMAT (   )

20. Describe the output of the following WRITE—FORMAT pairs of statements:
   a. WRITE (4,20) D3,NAME,WAGE
   b. 20 FORMAT (2X,15,10X,A20 10X,F4.2)
   c. WRITE (6,40)
   d. 40 FORMAT ('YOU HAVE MADE AN ERROR')

21. Write a FORTRAN program to print the sine of an angle whose value in radians is entered by a user.

22. Rewrite the program in Figure 14-8a to convert input Celsius temperatures to Fahrenheit and Kelvin.

23. Write a FORTRAN program to print a table of sines for every degree from 0 to 90°. Remember, for the intrinsic function SIN(X), X must be in radians.

24. Define the following for a Pascal program:
   a. HEADING   *b.* BLOCK
   c. Declarations

25. Which are valid declarations in Pascal?
   a. VAR NUMBER:REAL,
   b. VAR 7SEC:INTEGER:
   c. CONST e=2.71828, PASS='YES';
   d. VAR TEST2:INTEGER;
   e. VAR PASS;BOOLEAN

26. Explain the use of the TYPE declaration in Pascal.

27. What is the difference between a function and a procedure in Pascal?

28. List the Pascal arithmetic operators in order of their priority from highest to lowest.

29. Show the results produced by 26 DIV 4 and by 26 MOD 4.

30. Describe the effect of these Pascal statements:
   a. P:=P+1;
   b. CUBE(23)
   c. FOR K=1 TO 40 DO
          BEGIN Y:=K;
          WRITELN (Y,Y∗Y);
          END
   d. REPEAT
          READ(P);
          WRITELN SIN(P);
        UNTIL P>=1.57
   e. IF R>2 THEN GO TO 222
          ELSE IF R=2 THEN GO TO 444
            ELSE GO TO 888;

31. Describe the difference between the WRITE and WRITELN statements.

32. Define the two meanings of the term *file*.

33. Write a short Pascal program to compute the 6 percent sales tax for a price typed in and to print out the price, and tax, and the total.

34. Modify the program in Figure 14-11a to convert typed-in Celsius temperatures to Fahrenheit and Kelvin values and print the results. (F=9C/5+32.)

REFERENCES/BIBLIOGRAPHY

Jensen, Kathleen, and Niklaus Wirth: *Pascal User Manual and Report*, 2d ed., Springer-Verlag, New York, 1978.

Bowles, K. L.: *MICROCOMPUTER Problem Solving Using Pascal*, Springer-Verlag, New York, 1977.

Haag, James N.: *Comprehensive Standard FORTRAN Programming*, Hayden Book Company, Rochelle Park, NJ, 1969, 1976.

Kieburtz, Richard B.: *Structured Programming and Problem Solving with Pascal*, Prentice-Hall, Englewood Cliffs, NJ, 1978.

Spencer, Donald D.: *Game Playing with BASIC*, Hayden Book Company, Rochelle Park, NJ, 1977.

Mullish, Henry: *A Basic Approach to BASIC*, Wiley, New York, 1976.

Meissner, Loren P.: *Rudiments of FORTRAN*, Addison-Wesley, Reading, MA, 1971.

Preliminary ISIS-II FORTRAN-80 COMPILER OPERATOR'S MANUAL, Intel Corporation, Santa Clara, CA, 1978.

FORTRAN-80 Programming Manual, Intel Corporation, Santa Clara, CA, 1978.

Krouse, Tim (Design Engineering): "Pascal Language: Parts One through Five," *Electronic Design*, vols. 19, 20, 23, 25, September 13 and 27, November 8, December 6, 1978.

APPENDIX A

STATE DIAGRAMS AND ALGORITHMIC STATE MACHINE CHARTS

All but the simplest digital systems are synchronous. By *synchronous* we mean that the outputs change only when a clock pulse occurs. Figure A-1a shows a 2-bit binary synchronous counter made with JK flip-flops. An E signal into gate 1 gates a clock to the flip-flops. Whenever QA and QB are both 0's, the P output of gate 2 is asserted high. If the E signal is high, the QB and QA outputs step through the binary states 00, 01, 10, and 11 over and over.

The operation of this circuit can be represented by truth tables, timing waveforms, or boolean expressions. Two other common graphic representations are *state diagrams* and *algorithmic state machine charts*.

STATE DIAGRAMS

Figure A-1b shows a state diagram for the 2-bit binary counter in Figure A-1a. A state diagram shows the sequence of states through which the circuit steps. It also shows the conditions required for the circuit to go from one state to the next when a clock pulse occurs.

The large oval bubble contains a symbol, or code, identifying the state. Arrows show the direction of transition from one state to another. Symbols next to the transition arrows show the conditions required to change to the next state as well as the output levels for the previous state. For Figure A-1b, the format for these symbols is E/P, QB, QA. To change from state 0 to state 1, for example, E must be 1. In state 0, the P output is high, QB is low, and QA is low. The looped arrows to the left of the large oval bubbles indicate that if the conditions to the left of the slash are present, the system will hold in that state. For example, the loop next to state 1 indicates that the system will hold in state 1 as long as E is low. Clock pulses will have no effect.

Microprocessors are an example of complex state sys-

tems, or "machines." Figure 9-19 shows the transition state diagram for an 8085A microprocessor.

ALGORITHMIC STATE MACHINE CHARTS

Another common graphical way of representing the operation of a sequential system is with an *algorithmic state machine,* or ASM, *chart.* An algorithm is a formula or method for solving a problem. Digital systems, or "machines," are used to solve control problems by sequencing through a series of states. This, then, is the origin of the name *algorithmic state machine.*

Figure A-1c shows an ASM chart for the simple sequential "machine" in Figure A-1a. A state is represented by a rectangular box. The name (number) of the *state* is in a circle to the left of the box. The code for each state is written over the state box. If some condition must be present for the machine to change to the next state, then a diamond-shaped decision box is included in the representation for a state. In Figure A-1c, for example, the E signal must be a 1 for the machine to change from state 0 to state 1 when a clock pulse occurs. If an output signal is asserted during a state, then the name for that *signal* is written in the state box. The P signal in Figure A-1c is asserted in state 0, so P is written in the state 0 box.

SUMMARY

State diagrams and algorithmic state machine charts are very powerful tools for designing sequential digital systems. For further information regarding the use of state diagrams and ASM charts in design and analysis of more complex systems, refer to the following books.

435

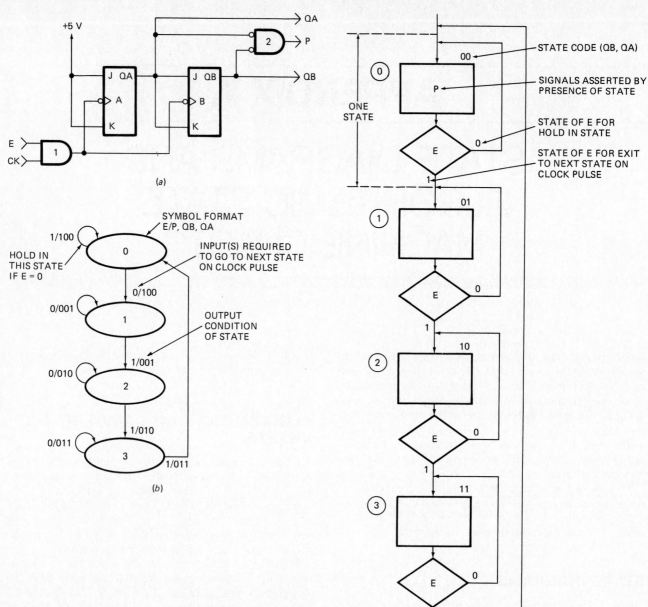

FIGURE A-1 Two-bit binary counter with clock enable and
"0" state detector. *(a)* Circuit. *(b)* State diagram. *(c)* Algorithmic state machine chart.

REFERENCES/BIBLIOGRAPHY

Fletcher, William I.: *An Engineering Approach to Digital Design*, Prentice-Hall, Englewood Cliffs, NJ, 1980.

Peatman, John B.: *Digital Hardware Design*, McGraw-Hill, New York, 1980.

Winkel, David, and Franklin Prosser: *The Art of Digital Design: An Introduction to Top-Down Design*, Prentice-Hall, Englewood Cliffs, NJ, 1980.

APPENDIX B

EXPLANATION OF NEW LOGIC SYMBOLS*

CONTENTS

	Title	Page
1.	INTRODUCTION	438
2.	SYMBOL COMPOSITION	438
3.	QUALIFYING SYMBOLS	439
	3.1 General Qualifying Symbols	439
	3.2 Qualifying Symbols for Inputs and Outputs	439
	3.3 Symbols Inside the Outline	441
4.	DEPENDENCY NOTATION	441
	4.1 General Explanation	441
	4.2 G, AND	441
	4.3 Conventions for the Application of Dependency Notation in General	442
	4.4 V, OR	443
	4.5 N, Negate (Exclusive OR)	443
	4.6 Z, Interconnection	443
	4.7 C, Control	444
	4.8 S, Set and R, Reset	444
	4.9 EN, Enable	444
	4.10 M, Mode	445
	4.11 A, Address	446
5.	BISTABLE ELEMENTS	447
6.	CODERS	448
7.	USE OF A CODER TO PRODUCE AFFECTING INPUTS	448
8.	USE OF BINARY GROUPING TO PRODUCE AFFECTING INPUTS	449
9.	SEQUENCE OF INPUT LABELS	449
10.	SEQUENCE OF OUTPUT LABELS	449

LIST OF TABLES

Table	Title	Page
I.	General Qualifying Symbols	439
II.	Qualifying Symbols for Inputs and Outputs	440
III.	Symbols Inside the Outline	440
IV.	Summary of Dependency Notation	447

If you have questions on this Explanation of New Logic Symbols, please contact:

F.A. Mann MS 84
Texas Instruments Incorporated
P.O. Box 225012
Dallas, Texas 75265
Telephone (214) 995-3746

IEEE Standards may be purchased from:

Institute of Electrical and Electronics Engine
345 East 47th Street
New York, N.Y. 10017

International Electrotechnical Commission (IEC)
publications may be purchased from:

American National Standards Institute, Inc.
1430 Broadway
New York, N.Y. 10018

---

*From *Bipolar Microcomputer Component Data Book for Design Engineers*, 3d ed., © 1981 by Texas Instruments.

by F. A. Mann

1 INTRODUCTION

The International Electrotechnical Commission (IEC) has been developing a very powerful symbolic language that can show the relationship of each input of a digital logic circuit to each output without showing explicitly the internal logic. At the heart of the system is dependency notation, which will be explained in Section 4.

The system was introduced in the USA in a rudimentary form in IEEE/ANSI Standard Y32.14-1973. Lacking at that time a complete development of dependency notation, it offered little more than a substitution of rectangular shapes for the familiar distinctive shapes for representing the basic functions of AND, OR, negation, etc. This is no longer the case.

Internationally, Working Group 2 of IEC Technical Committee TC-3 is preparing a new document (Publication 617-12) that will consolidate the original work started in the mid 1960's and published in 1972 (Publication 117-15) and the amendments and supplements that have followed. Similarly for the USA, IEEE Committee SCC 11.9 is revising the publication IEEE Std 91/ANSI Y32.14. Texas Instruments is participating in the work of both organizations and this Supplement to the TTL Data Book introduces new logic symbols in anticipation of the new standards. When changes are made as the standards develop, future editions of the TTL Data Book will take those changes into account.

The following explanation of the new symbolic language is necessarily brief and greatly condensed from what the standards publications will finally contain. This is not intended to be sufficient for those people who will be developing symbols for new devices. It is primarily intended to make possible the understanding of the symbols used in this book; comparing the symbols with functional block diagrams and/or function tables will further help that understanding.

2 SYMBOL COMPOSITION

A symbol comprises an outline or a combination of outlines together with one or more qualifying symbols. The shape of the symbols is not significant. As shown in Figure 1, general qualifying symbols are used to tell exactly what logical operation is performed by the elements. Table I shows the general qualifying symbols used in this data book. Input lines are placed on the left and output lines are placed on the right. When an exception is made to that convention, the direction of signal flow is indicated by an arrow as shown in Figure 11.

All outputs of a single, unsubdivided element always have identical internal logic states determined by the function of the element except when otherwise indicated by an associated qualifying symbol or label inside the element.

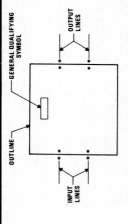

FIGURE 1 – SYMBOL COMPOSITION

*Possible positions for qualifying symbols relating to inputs and outputs

The outlines of elements may be abutted or embedded in which case the following conventions apply. There is no logic connection between the elements when the line common to their outlines is in the direction of signal flow. There is at least one logic connection between the elements when the line common to their outlines is perpendicular to the direction of signal flow. The number of logic connections between elements will be clarified by the use of qualifying symbols and this is discussed further under that topic. If no indications are shown on either side of the common line, it is assumed there is only one connection.

When a circuit has one or more inputs that are common to more than one element of the circuit, the common-control block may be used. This is the only distinctively shaped outline used in the IEC system. Figure 2 shows that unless otherwise qualified by dependency notation, an input to the common-control block is an input to each of the elements below the common-control block.

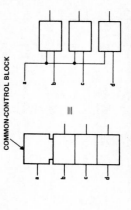

FIGURE 2 – ILLUSTRATION OF COMMON-CONTROL BLOCK

438
APPENDIX B

A common output depending on all elements of the array can be shown as the output of a common-output element. Its distinctive visual feature is the double line at its top. In addition the common-output element may have other inputs as shown in Figure 3. The function of the common-output element must be shown by use of a general qualifying symbol.

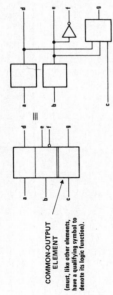

COMMON-OUTPUT ELEMENT

(must, like other elements, have a qualifying symbol to denote its logic function).

FIGURE 3 – ILLUSTRATION OF COMMON-OUTPUT ELEMENT

3 QUALIFYING SYMBOLS

3.1 General Qualifying Symbols

Table I shows the general qualifying symbols used in this data book. These characters are placed near the top center or the geometric center of a symbol or symbol element to define the basic function of the device represented by the symbol or of the element.

3.2 Qualifying Symbols for Inputs and Outputs

Qualifying symbols for inputs and outputs are shown in Table II and will be familiar to most users with the possible exception of the logic polarity and analog signal indicators. The older logic negation indicator means that the external 0 state produces the internal 1 state. The internal 1 state means the active state. Logic negation may be used in pure logic diagrams; in order to tie the external 1 and 0 logic states to the levels H (high) and L (low), a statement of whether positive logic (1 = H, 0 = L) or negative logic (1 = L, 0 = H) is being used is required or must be assumed. Logic polarity indicators eliminate the need for calling out the logic convention and are used in this data book in the symbology for actual devices. The presence of the triangular polarity indicator indicates that the L logic level will produce the internal 1 state (the active state) or that, in the case of an output, the internal 1 state will produce the external L level. Note how the active direction of transition for a dynamic input is indicated in positive logic, negative logic, and with polarity indication.

TABLE I – GENERAL QUALIFYING SYMBOLS

SYMBOL	DESCRIPTION	EXAMPLE
&	AND gate or function.	SN7400
≥1	OR gate or function. The symbol was chosen to indicate that at least one active input is needed to activate the output.	SN7402
=1	Exclusive OR. One and only one input must be active to activate the output.	SN7486
=	Logic identity. All inputs must stand at same state.	SN74180
2k	An even number of inputs must be active.	SN74180
2k+1	An odd number of inputs must be active.	*
1	The one input must be active.	SN7404
▷ or ▽	A buffer or element with more-than usual output capability (symbol is oriented in the direction of signal flow).	SN74S436
⊓	Schmitt trigger; element with hysteresis.	SN74LS18
X/Y	Coder, code converter (DEC/BCD, BIN/OUT, BIN/7-SEG, etc.).	SN74LS347
MUX	Multiplexer/data selector.	SN74150
DMUX or DX	Demultiplexer.	SN74138
Σ	Adder.	SN74LS385
P-Q	Subtracter.	SN74LS385
CPG	Look-ahead carry generator.	SN74182
π	Multiplier.	SN74LS384
COMP	Magnitude comparator.	SN74LS682
ALU	Arithmetic logic unit.	SN74LS381
⎍	Retriggerable monostable.	SN74LS422
⎍⎍	Non-retriggerable monostable (one-shot).	SN74121
G	Astable element. Showing waveform is optional.	SN74LS320
!G	Synchronously starting astable.	SN74LS624
G!	Astable element that stops with a completed pulse.	SN74LS595
SRGm	Shift register. m = number of bits.	SN54LS590
CTRm	Counter. m = number of bits; cycle length = 2^m.	SN74LS668
CTR DIVm	Counter with cycle length = m.	SN74187
ROM	Read-only memory.	SN74170
RAM	Random-access read/write memory.	SN74LS222
FIFO	First-in, first-out memory.	SN74AS877
1=0	Element powers up cleared to 0 state.	SN74LS608
Φ	Highly complex function; "gray box" symbol with limited detail shown under special rules.	

*Not all of the general qualifying symbols have been used in this book, but they are included here for the sake of completeness.

439

EXPLANATION OF NEW LOGIC SYMBOLS

TABLE II – QUALIFYING SYMBOLS FOR INPUTS AND OUTPUTS

Logic negation at input. External 0 produces internal 1.

Logic negation at output. Internal 1 produces external 0.

Active-low input. Equivalent to —◁ in positive logic.

Active-low output. Equivalent to ▷— in positive logic.

Active-low input in the case of right-to-left signal flow.

Active-low output in the case of right-to-left signal flow.

Signal flow from right to left. If not otherwise indicated, signal flow is from left to right.

Bidirectional signal flow.

Dynamic inputs active on indicated transition

	POSITIVE LOGIC	NEGATIVE LOGIC	POLARITY INDICATION
	1 / 0	0 / 1	H / L
	not used	not used	not used
	0 / 1	1 / 0	L / H

Nonlogic connection. A label inside the symbol will usually define the nature of this pin.

Input for analog signals.

Internal connection. 1 state on left produces 1 state on right.

Negated internal connection. 1 state on left produces 0 state on right.

Dynamic internal connection. Transition from 0 to 1 on left produces transitory 1 state on right.

Internal input (virtual input). It always stands at its internal 1 state unless affected by an overriding dependency relationship.

Internal output (virtual output). Its effect on an internal input to which it is connected is indicated by dependency notation.

The internal connections between logic elements abutted together in a symbol may be indicated by the symbols shown. Each logic connection may be shown by the presence of qualifying symbols at one or both sides of the common line and if confusion can arise about the numbers of connections, use can be made of one of the internal connection symbols.

The internal (virtual) input is an input originating somewhere else in the circuit and is not connected directly to a terminal. The internal (virtual) output is likewise not connected directly to a terminal.

TABLE III – SYMBOLS INSIDE THE OUTLINE

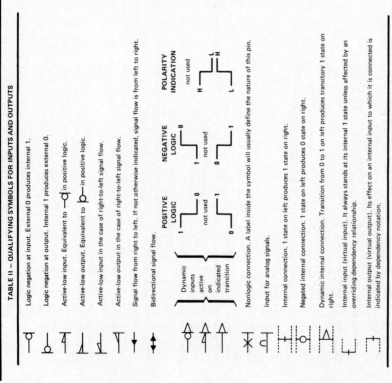

Postponed output (of a pulse-triggered flip-flop). The output changes when input initiating change (e.g., a C input) returns to its initial external state or level. See § 5.

Bi-threshold input (input with hysteresis)

NPN open-collector or similar output that can supply a relatively low-impedance L level when not turned off. Requires external pull-up. Capable of positive-logic wired-AND connection.

Passive-pull-up output is similar to NPN open-collector output but is supplemented with a built-in passive pull-up.

NPN open-emitter or similar output that can supply a relatively low-impedance H level when not turned off. Requires external pull-down. Capable of positive-logic wired-OR connection.

Passive-pull-down output is similar to NPN open-emitter output but is supplemented with a built-in passive pull-down.

3-state output

Enable input
When at its internal 1-state, all outputs are enabled.
When at its internal 0-state, open-collector and open-emitter outputs are off, three-state outputs are at normally defined internal logic states and at external high-impedance state, and all other outputs (e.g., totem-poles) are at the internal 0-state.

Usual meanings associated with flip-flops (e.g., R = reset, T = toggle)

J, K, R, S, T

Data input to a storage element equivalent to:

Shift right (left) inputs, m = 1, 2, 3 etc. If m = 1, it is usually not shown.

Counting up (down) inputs, m = 1, 2, 3 etc. If m = 1, it is usually not shown.

Binary grouping. m is highest power of 2.

The contents-setting input, when active, causes the content of a register to take on the indicated value.

The content output is active if the content of the register is as indicated.

Input line grouping . . . . indicates two or more terminals used to implement a single logic input.
e.g., The paired expander inputs of SN7450.

Fixed-state output always stands at its internal 1 state. For example, see SN74185.

The application of internal inputs and outputs requires an understanding of dependency notation, which is explained in Section 4.

In an array of elements, if the same general qualifying symbol and the same qualifying symbols associated with inputs and outputs would appear inside each of the elements of the array, these qualifying symbols are usually shown only in the first element. This is done to reduce clutter and to save time in recognition. Similarly, large identical elements that are subdivided into smaller elements may each be represented by an unsubdivided outline. The SN541LS440 symbol illustrates this principle.

3.3 Symbols Inside the Outline

Table III shows some symbols used inside the outline. Note particularly that open-collector, open-emitter, and three-state outputs have distinctive symbols. Also note that an EN input affects all of the outputs of the circuit and has no effect on inputs. When an enable input affects only certain outputs and/or affects one or more inputs, a form of dependency notation will indicate this (see 4.9). The effects of the EN input on the various types of outputs are shown.

It is particularly important to note that a D input is always the data input of a storage element. At its internal 1 state, the D input sets the storage element to its 1 state, and at its internal 0 state it resets the storage element to its 0 state.

The binary grouping symbol will be explained more fully in Section 8. Binary-weighted inputs are arranged in order and the binary weights of the least-significant and the most-significant lines are indicated by numbers. In this data book weights of input and output lines will be represented by powers of two usually only when the binary grouping symbol is used, otherwise, decimal numbers will be used. The grouped inputs generate an internal number on which a mathematical function can be performed or that can be an identifying number for dependency notation. See Figure 28. A frequent use is in addresses for memories.

Reversed in direction, the binary grouping symbol can be used with outputs. The concept is analogous to that for the inputs and the weighted outputs will indicate the internal number assumed to be developed within the circuit.

Other symbols are used inside the outlines in this data book in accordance with the IEC/IEEE standards but are not shown here. Generally these are associated with arithmetic operations and are self-explanatory.

When nonstandardized information is shown inside an outline, it is usually enclosed in square brackets [like these].

4 DEPENDENCY NOTATION

4.1 General Explanation

Dependency notation is the powerful tool that sets the IEC symbols apart from previous systems and makes compact, meaningful, symbols possible. It provides the means of denoting the relationship between inputs, outputs, or inputs and outputs without actually showing all the elements and interconnections involved. The information provided by dependency notation supplements that provided by the qualifying symbols for an element's function.

In the convention for the dependency notation, use will be made of the terms "affecting" and "affected". In cases where it is not evident which inputs must be considered as being the affecting or the affected ones (e.g., if they stand in an AND relationship), the choice may be made in any convenient way.

So far, ten types of dependency have been defined and all of these are used in this data book. They are listed below in the order in which they are presented and are summarized in Table IV following 4.11.

Section	Dependency Type or Other Subject
4.2	G, AND
4.3	General rules for dependency notation
4.4	V, OR
4.5	N, Negate, (Exclusive OR)
4.6	Z, Interconnection
4.7	C, Control
4.8	S, Set and R, Reset
4.9	EN, Enable
4.10	M, Mode
4.11	A, Address

4.2 G (AND) Dependency

A common relationship between two signals is to have them ANDed together. This has traditionally been shown by explicitly drawing an AND gate with the signals connected to the inputs of the gate. The 1972 IEC publication and the 1973 IEEE/ANSI standard showed several ways to show this AND relationship using dependency notation. While nine other forms of dependency have since been defined, the ways to invoke AND dependency are now reduced to one.

In Figure 4 input **b** is ANDed with input **a** and the complement of **b** is ANDed with **c**. The letter G has been chosen to indicate AND relationships and is placed at input **b**, inside the symbol. A number considered appropriate by the symbol designer (1 has been used here) is placed after the letter G and also at each affected input. Note the bar over the 1 at input **c**.

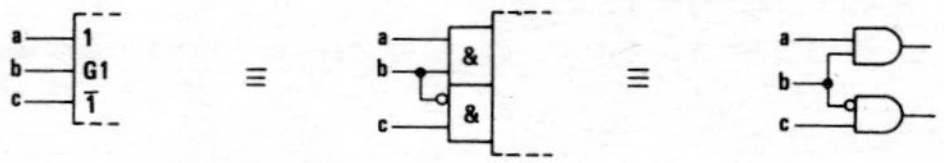

FIGURE 4 – G DEPENDENCY BETWEEN INPUTS

In Figure 5, output **b** affects input **a** with an AND relationship. The lower example shows that it is the internal logic state of **b**, unaffected by the negation sign, that is ANDed. Figure 6 shows input **a** to be ANDed with a dynamic input **b**.

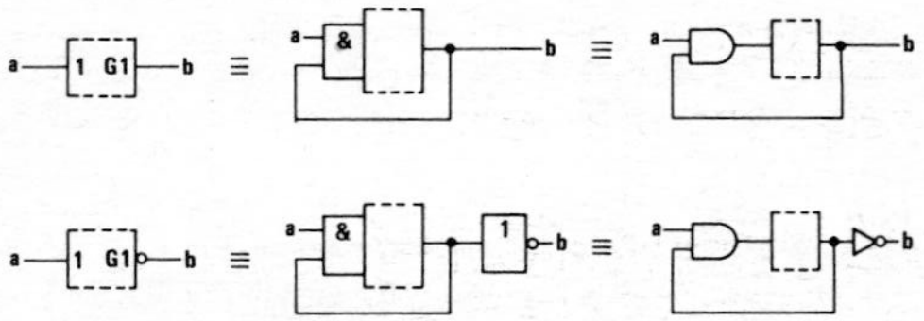

FIGURE 5 – G DEPENDENCY BETWEEN OUTPUTS AND INPUTS

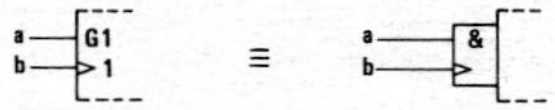

FIGURE 6 – G DEPENDENCY WITH A DYNAMIC INPUT

The rules for G dependency can be summarized thus:

When a Gm input or output (m is a number) stands at its internal 1 state, all inputs and outputs affected by Gm stand at their normally defined internal logic states. When the Gm input or output stands at its 0 state, all inputs and outputs affected by Gm stand at their internal 0 states.

4.3 Conventions for the Application of Dependency Notation in General

The rules for applying dependency relationships in general follow the same pattern as was illustrated for G dependency.

Application of dependency notation is accomplished by:

1) labeling the input (or output) *affecting* other inputs or outputs with the letter symbol indicating the relationship involved (e.g., G for AND) followed by an identifying number, appropriately chosen, and

2) labeling each input or output *affected* by that affecting input (or output) with that same number.

If it is the complement of the internal logic state of the affecting input or output that does the affecting, then a bar is placed over the identifying numbers at the affected inputs or outputs. See Figure 4.

If two affecting inputs or outputs have the same letter and same identifying number, they stand in an OR relationship to each other. See Figure 7.

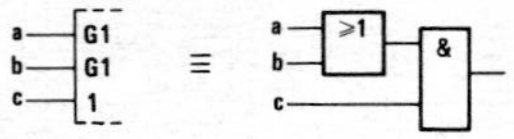

FIGURE 7 – OR'ED AFFECTING INPUTS

If the affected input or output requires a label to denote its function (e.g., "D"), this label will be *prefixed* by the identifying number of the affecting input. See Figure 12.

If an input or output is affected by more than one affecting input, the identifying numbers of each of the affecting inputs will appear in the label of the affected one, separated by commas. The normal reading order of these numbers is the same as the sequence of the affecting relationships. See Figure 12.

If the labels denoting the functions of affected inputs or outputs must be numbers, (e.g., outputs of a coder), the identifying numbers to be associated with both affecting inputs and affected inputs or outputs will be replaced by another character selected to avoid ambiguity, e.g., Greek letters. See Figure 8.

FIGURE 8 — SUBSTITUTION FOR NUMBERS

4.4 V (OR) Dependency

The symbol denoting OR dependency is the letter V. See Figure 9.

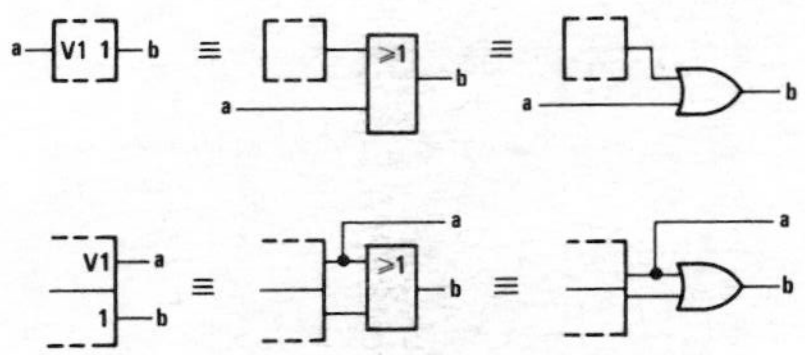

FIGURE 9 — V (OR) DEPENDENCY

When a Vm input or output stands at its internal 1 state, all inputs and outputs affected by Vm stand at their internal 1 states. When the Vm input or output stands at its internal 0 state, all inputs and outputs affected by Vm stand at their normally defined internal logic states.

4.5 N (Negate) (X-OR) Dependency

The symbol denoting negate dependency is the letter N. See Figure 10. Each input or output affected by an Nm input or output stands in an exclusive-OR relationship with the Nm input or output.

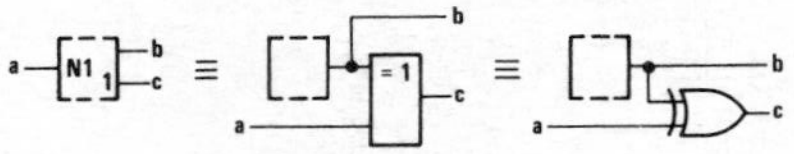

If $a = 0$, then $c = b$
If $a = 1$, then $c = \bar{b}$

FIGURE 10 — N (NEGATE) (X-OR) DEPENDENCY

When an Nm input or output stands at its internal 1 state, the internal logic state of each input and each output affected by Nm is the complement of what it would otherwise be. When an Nm input or output stands at its internal 0 state, all inputs and outputs affected by Nm stand at their normally defined internal logic states.

4.6 Z (Interconnection) Dependency

The symbol denoting interconnection dependency is the letter Z.

Interconnection dependency is used to indicate the existence of internal logic connections between inputs, outputs, internal inputs, and/or internal outputs.

The internal logic state of an input or output affected by a Zm input or output will be the same as the internal logic state of the Zm input or output, unless modified by additional dependency notation. See Figure 11.

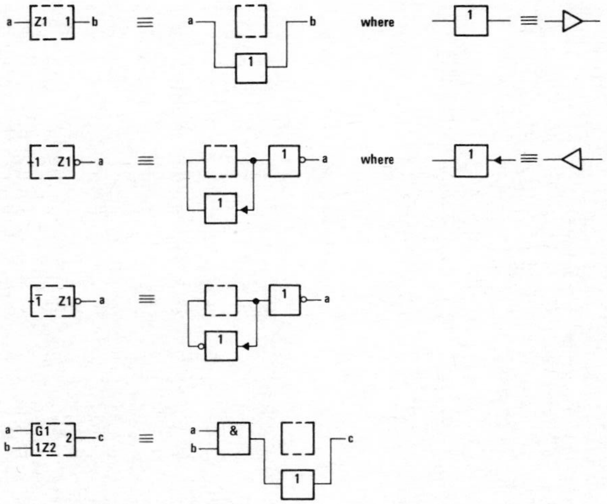

FIGURE 11 — Z (INTERCONNECTION) DEPENDENCY

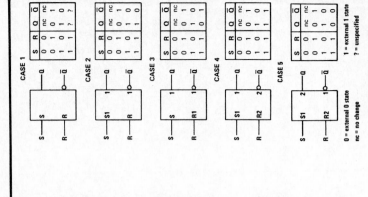

CASE 1

S	R	Q	Q̄
0	0	nc	nc
0	1	0	1
1	0	1	0
1	1	?	?

CASE 2

S	R	Q	Q̄
0	0	nc	nc
0	1	0	1
1	0	1	0
1	1	1	0

CASE 3

S	R	Q	Q̄
0	0	nc	nc
0	1	0	1
1	0	1	0
1	1	0	1

CASE 4

S	R	Q	Q̄
0	0	nc	nc
0	1	0	1
1	0	1	0
1	1	1	1

CASE 5

S	R	Q	Q̄
0	0	nc	nc
0	1	0	1
1	0	1	0
1	1	1	1

0 = external 0 state   1 = external 1 state
nc = no change   ? = unspecified

**FIGURE 13 – S (SET) AND
R (RESET) DEPENDENCIES**

Set and reset dependenies are used if it is necessary to specify the effect of the combination R=S=1 on a bistable element. Case 1 in Figure 13 does not use S or R dependency.

When an Sm input is at its internal 1 state, outputs affected by the Sm input will react, regardless of the state of an R input, as they normally would react to the combination S=1, R=0. See cases 2, 4, and 5 in Figure 13.

When an Rm input is at its internal 1 state, outputs affected by the Rm input will react, regardless of the state of an S input, as they normally would react to the combination S=0, R=1. See cases 3, 4, and 5 in Figure 13.

When an Sm or Rm input is at its internal 0 state, it has no effect.

Note that the noncomplementary output patterns in cases 4 and 5 are only pseudo stable. The simultaneous return of the inputs to S=R=0 produces an unforeseeable stable and complementary output pattern.

4.9 EN (Enable) Dependency

The symbol denoting enable dependency is the combination of letters EN.

An ENm input has the same effect on outputs as an EN input, see 3.1, but it effects only those outputs labeled with the identifying number m. It also affects those inputs labeled with the identifying number m. By contrast, an EN input affects all outputs and no inputs. The effect of an ENm input on an affected input is identical to that of a Cm input. See Figure 14.

4.7 C (Control) Dependency

The symbol denoting control dependency is the letter C.

Control inputs are usually used to enable or disable the data (D, J, K, R, or S) inputs of storage elements. They may take on their internal 1 states (be active) either statically or dynamically. In the latter case the dynamic input symbol is used as shown in the third example of Figure 12.

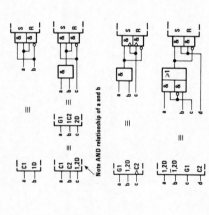

Note AND relationship of a and b

Input c selects which of a or b is stored when d goes low.

FIGURE 12 – C (CONTROL) DEPENDENCY

When a Cm input or output stands at its internal 1 state, the inputs affected by Cm have their normally defined effect on the function of the element, i.e., these inputs are enabled. When a Cm input or output stands at its internal 0 state, the inputs affected by Cm are disabled and have no effect on the function of the element.

4.8 S (Set) and R (Reset) Dependencies

The symbol denoting set dependency is the letter S. The symbol denoting reset dependency is the letter R.

When an ENm input stands at its internal 1 state, the inputs affected by ENm have their normally defined effect on the function of the element and the outputs affected by this input stand at their normally defined internal logic states, i.e., these inputs and outputs are enabled.

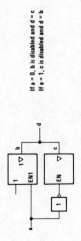

If a = 0, b is disabled and d = c
If a = 1, c is disabled and d = b

FIGURE 14 — EN (ENABLE) DEPENDENCY

When an ENm input stands at its internal 0 state, the inputs affected by ENm are disabled and have no effect on the function of the element, and the outputs affected by ENm are also disabled. Open-collector outputs are turned off, three-state outputs stand at their normally defined internal logic states but externally exhibit high impedance, and all other outputs (e.g., totem-pole outputs) stand at their internal 0 states.

4.10 M (Mode) Dependency

The symbol denoting mode dependency is the letter M.

Mode dependency is used to indicate that the effects of particular inputs and outputs of an element depend on the mode in which the element is operating.

If an input or output has the same effect in different modes of operation, the identifying numbers of the relevant affecting Mm inputs will appear in the label of that affected input or output between parentheses and separated by solidi. See Figure 19.

4.10.1 M Dependency Affecting Inputs

M dependency affects inputs the same as C dependency. When an Mm input or Mm output stands at its internal 1 state, the inputs affected by this Mm input or Mm output have their normally defined effect on the function of the element, i.e., the inputs are enabled.

When an Mm input or Mm output stands at its internal 0 state, the inputs affected by this Mm input or Mm output have no effect on the function of the element. When an affected input has several sets of labels separated by solidi (e.g., C4/2→/3+), any set in which the identifying number of the Mm input or Mm output appears has no effect and is to be ignored. This represents disabling of some of the functions of a multifunction input.

The circuit in Figure 15 has two inputs, b and c, that control which one of four modes (0, 1, 2, or 3) will exist at any time. Inputs d, e, and f are D inputs subject to dynamic control (clocking) by the a input. The numbers 1 and 2 are in the series chosen to indicate the modes so inputs e and f are only enabled in mode 1 (for parallel loading) and input d is only enabled in mode 2 (for serial loading). Note that input a has three functions. It is the clock for entering data. In mode 2, it causes right shifting of data, which means a shift away from the control block. In mode 3, it causes the contents of the register to be incremented by one count.

Note that all operations are synchronous.
In MODE 0 (b = 0, c = 0), the outputs remain at their existing states as none of the inputs has an effect.
In MODE 1 (b = 1, c = 0), parallel loading takes place thru inputs e and f.
In MODE 2 (b = 0, c = 1), shifting down and serial loading thru input d take place.
In MODE 3 (b = c = 1), counting up by increment of 1 per clock pulse takes place.

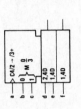

FIGURE 15 — M (MODE) DEPENDENCY AFFECTING INPUTS

4.10.2 M Dependency Affecting Outputs

When an Mm input or Mm output stands at its internal 1 state, the affected outputs stand at their normally defined internal logic states, i.e., the outputs are enabled.

When an Mm input or Mm output stands at its internal 0 state, at each affected output any set of labels containing the identifying number of that Mm input or Mm output has no effect and is to be ignored. When an output has several different sets of labels separated by solidi (e.g., 2,4/3,5), only those sets in which the identifying number of this Mm input or Mm output appears are to be ignored.

In Figure 16, mode 1 exists when the a input stands at its internal 1 state. The delayed output symbol is effective only in mode 1 (when input a = 1) in which case the device functions as a pulse-triggered flip-flop. See Section 5. When input a = 0, the device is not in mode 1 so the delayed output symbol has no effect and the device functions as a transparent latch.

FIGURE 16 — TYPE OF FLIP-FLOP DETERMINED BY MODE

Address dependency provides a clear representation of those elements, particularly memories, that use address control inputs to select specified sections of a multidimensional array. Such a section of a memory array is usually called a word. The purpose of address dependency is to allow a symbolic presentation of the entire array. An input of the array shown at a particular element of this general section is common to the corresponding elements of all selected sections of the array. An output of the array shown at a particular element of this general section is the result of the OR function of the outputs of the corresponding elements of selected sections. If the label of an output of the array shown at a particular element of this general section indicates that this output is an open-circuit output or a three-state output, then this indication refers to the output of the array and not to those of the sections of the array.

Inputs that are not affected by any affecting address input have their normally defined effect on all sections of the array, whereas inputs affected by an address input have their normally defined effect only on the section selected by that address input.

An affecting address input is labelled with the letter A followed by an identifying number that corresponds with the address of the particular section of the array selected by this input. Within the general section presented by the symbol, inputs and outputs affected by an Am input are labelled with the letter A, which stands for the identifying numbers, i.e., the addresses, of the particular sections.

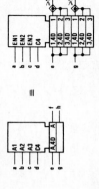

FIGURE 20 – A (ADDRESS) DEPENDENCY

Figure 20 shows a 3-word by 2-bit memory having a separate address line for each word and uses EN dependency to explain the operation. To select word 1, input a is taken to its 1 state, which establishes mode 1. Data can now be clocked into the inputs marked "1,4D". Unless words 2 and 3 are also selected, data cannot be clocked in at the inputs marked "2,4D" and "3,4D". The outputs will be the OR functions of the selected outputs, i.e., only those enabled by the active EN functions.

The identifying numbers of affecting address inputs correspond with the addresses of the sections selected by these inputs. They need not necessarily differ from those of other affecting dependency-inputs (e.g., G, V, N, . . .), because in the general section presented by the symbol they are replaced by the letter A.

In Figure 17, if input a stands at its internal 1 state establishing mode 1, output b will stand at its internal 1 state only when the content of the register equals 9. Since output b is located in the common-control block with no defined function outside of mode 1, this output will stand at its internal 0 state when input a stands at its internal 0 state, regardless of the register content.

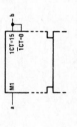

FIGURE 17 – DISABLING AN OUTPUT OF THE COMMON-CONTROL BLOCK

In Figure 18, if input a stands at its internal 1 state establishing mode 1, output b will stand at its internal 1 state only when the content of the register equals 15. If input a stands at its internal 0 state, output b will stand at its internal 1 state only when the content of the register equals 0.

FIGURE 18 – DETERMINING AN OUTPUT'S FUNCTION

In Figure 19 inputs a and b are binary weighted to generate the numbers 0, 1, 2, or 3. This determines which one of the four modes exists.

At output e the label set causing negation (if c = 1) is effective only in modes 2 and 3. In modes 0 and 1 this output stands at its normally defined state as if it had no labels. At output f the label set has effect when the mode is not 0 so output e is negated (if c = 1) in modes 1, 2, and 3. In mode 0 the label set has no effect so the output stands at its normally defined state. In this example 0,4 is equivalent to (1/2/3)4. At output g there are two label sets. The first set, causing negation (if c = 1), is effective only in mode 2. The second set, subjecting g to AND dependency on d, has effect only in mode 3.

Note that in mode 0 none of the dependency relationships has any effect on the outputs, so e, f, and g will all stand at the same state.

FIGURE 19 – DEPENDENT RELATIONSHIPS AFFECTED BY MODE

4.11   A (Address) Dependency

The symbol denoting address dependency is the letter A.

If there are several sets of affecting *Am* inputs for the purpose of independent and possibly simultaneous access to sections of the array, then the letter A is modified to 1A, 2A, ... Because they have access to the same sections of the array, these sets of A inputs may have the same identifying numbers.

Figure 21 is another illustration of the concept.

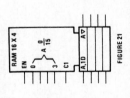

FIGURE 21

FIGURE 21 – ARRAY OF 16 SECTIONS OF FOUR TRANSPARENT LATCHES WITH 3-STATE OUTPUTS COMPRISING A 16-WORD X 4-BIT RANDOM-ACCESS MEMORY

TABLE IV – SUMMARY OF DEPENDENCY NOTATION

TYPE OF DEPENDENCY	LETTER SYMBOL*	AFFECTING INPUT AT ITS 1-STATE	AFFECTING INPUT AT ITS 0-STATE
Address	A	Permits action (address selected)	Prevents action (address not selected)
Control	C	Permits action	Prevents action
Enable	EN	Permits action	Prevents action of inputs. ◊ outputs off. ▽ outputs at external high impedance, no change in internal logic state. Other outputs at internal 0 state.
AND	G	Permits action	Imposes 0 state
Mode	M	Permits action (mode selected)	Prevents action (mode not selected)
Negate (X-OR)	N	Complements state	No effect
RESET	R	Affected output reacts as it would to S = 0, R = 1	No effect
SET	S	Affected output reacts as it would to S = 1, R = 0	No effect
OR	V	Imposes 1 state	Permits action
Interconnection	Z	Imposes 1 state	Imposes 0 state

*These letter symbols appear at the AFFECTING input (or output) and are followed by a number. Each input (or output) AFFECTED by that input is labeled with that same number. When the labels EN, R, and S appear at inputs without the following numbers, the descriptions above do not apply. The action of these inputs is described under "Symbols Inside The Outline", see 3.1.

BISTABLE ELEMENTS

The dynamic input symbol, the postponed output symbol, and dependency notation provide the tools to differentiate four main types of bistable elements and make synchronous and asynchronous inputs easily recognizable. See Figure 22. The first column shows the essential distinguishing features; the other columns show examples.

Transparent latches have a level-operated control input. The D input is active as long as the C input is at its internal 1 state. The outputs respond immediately. Edge-triggered elements accept data from D, J, K, R, or S inputs on the active transition of C. Pulse-triggered elements require the setup of data before the start of the control pulse; the C input is considered static since the data must be maintained as long as C is at its 1 state. The output is postponed until C returns to its 0 state. The data-lock-out element is similar to the pulse-triggered version except that the C input is considered dynamic in that shortly after C goes through its active transition, the data inputs are disabled and data does not have to be held. However, the output is still postponed until the C input returns to its initial external level.

Notice that synchronous inputs can be readily recognized by their dependency labels (1D, 1J, 1K, 1S, 1R) compared to the asynchronous inputs (S, R), which are not dependent on the C inputs.

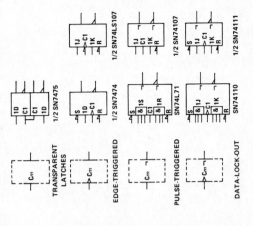

TRANSPARENT LATCHES 1/2 SN7475

EDGE-TRIGGERED 1/2 SN7474 1/2 SN74LS107

PULSE-TRIGGERED SN74L71 1/2 SN74107

DATA-LOCK-OUT SN74110 1/2 SN74111

FIGURE 22 – FOUR TYPES OF BISTABLE CIRCUITS

6 CODERS

The general symbol for a coder or code converter is shown in Figure 23. X and Y may be replaced by appropriate indications of the code used to represent the information at the inputs and at the outputs, respectively.

Indication of code conversion is based on the following rule:

Depending on the input code, the internal logic states of the inputs determine an internal value. This value is reproduced by the internal logic states of the outputs, depending on the output code.

The indication of the relationships between the internal logic states of the inputs and the internal value is accomplished by:

1) labelling the inputs with numbers. In this case the internal value equals the sum of the weights associated with those inputs that stand at their internal 1-state, or by

2) replacing X by an appropriate indication of the input code and labelling the inputs with characters that refer to this code.

The relationships between the internal value and the internal logic states of the outputs are indicated by:

1) labelling each output with a list of numbers representing those internal values that lead to the internal 1-state of that output. These numbers shall be separated by solidi as in Figure 24. This labelling may also be applied when Y is replaced by a letter denoting a type of dependency (see Section 7). If a continuous range of internal values produces the internal 1 state of an output, this can be indicated by two numbers that are inclusively the beginning and the end of the range, with these two numbers separated by three dots, e.g., 4 . . . 9 = 4/5/6/7/8/9, or by

2) replacing Y by an appropriate indication of the output code and labelling the outputs with characters that refer to this code as in Figure 25.

Alternatively, the general symbol may be used together with an appropriate reference to a table in which the relationship between the inputs and outputs is indicated. This is a recommended way to symbolize a PROM after it has been programmed.

X/Y

FIGURE 23 – CODER GENERAL SYMBOL

FIGURE 24 – AN X/Y CODE CONVERTER

INPUTS			OUTPUTS			
c	b	a	g	f	e	d
0	0	0	0	0	0	0
0	0	1	0	0	0	1
0	1	0	0	0	1	0
0	1	1	0	1	1	0
1	0	0	0	1	0	1
1	0	1	0	0	0	0
1	1	0	0	0	0	0
1	1	1	1	0	0	0

X/Y symbol — inputs a (1), b (2), c (4); outputs d (1/4), e (2/3), f (3/4), g (7).

FIGURE 25 – AN X/OCTAL CODE CONVERTER

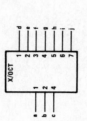

INPUTS			OUTPUTS						
c	b	a	j	i	h	g	f	e	d
0	0	0	0	0	0	0	0	0	0
0	0	1	0	0	0	0	0	0	1
0	1	0	0	0	0	0	0	1	0
0	1	1	0	0	0	0	1	0	0
1	0	0	0	0	0	1	0	0	0
1	0	1	0	0	1	0	0	0	0
1	1	0	0	1	0	0	0	0	0
1	1	1	1	0	0	0	0	0	0

X/OCT symbol — inputs a (1), b (2), c (4); outputs d, e, f, g, h, i, j labelled 1, 2, 3, 4, 5, 6, 7.

---

7 USE OF A CODER TO PRODUCE AFFECTING INPUTS

It often occurs that a set of affecting inputs for dependency notation is produced by decoding signals on certain inputs to an element. In such a case use can be made of the symbol for a coder as an embedded symbol. See Figure 26.

If all affecting inputs produced by a coder are of the same type and their identifying numbers correspond with the numbers shown at the outputs of the coder, Y (in the qualifying symbol X/Y) may be replaced by the letter denoting the type of dependency. The indications of the affecting inputs should then be omitted. See Figure 27.

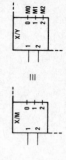

FIGURE 26 – PRODUCING VARIOUS TYPES OF DEPENDENCIES

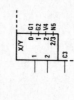

FIGURE 27 – PRODUCING ONE TYPE OF DEPENDENCY

8 USE OF BINARY GROUPING TO PRODUCE AFFECTING INPUTS

If all affecting inputs produced by a coder are of the same type and have consecutive identifying numbers not necessarily corresponding with the numbers that would have been shown at the outputs of the coder, use can be made of the binary grouping symbol (see 3.1). k external lines effectively generate 2^k internal inputs. The bracket is followed by the letter denoting the type of dependency followed by the $\frac{m1}{m2}$. The m1 is to be replaced by the smallest identifying number and the m2 by the largest one, as shown in Figure 28.

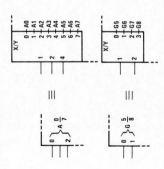

FIGURE 28 – USE OF THE BINARY GROUPING SYMBOL

9 SEQUENCE OF INPUT LABELS

If an input having a single functional effect is affected by other inputs, the qualifying symbol (if there is any) for that functional effect is preceded by the labels corresponding to the affecting inputs. The left-to-right order of these preceding labels is the order in which the effects or modifications must be applied. The affected input has no functional effect on the element if the logic state of any one of the affecting inputs, considered separately, would cause the affected input to have no effect, regardless of the logic states of other affecting inputs.

If an input has several different functional effects or has several different sets of affecting inputs, depending on the mode of action, the input may be shown as often as required. However, there are cases in which this method of presentation is not advantageous. In those cases the input may be shown once with the different sets of labels separated by solidi. No meaning is attached to the order of these sets of labels. If one of the functional effects of an input is that of an unlabelled input of the element, a solidus will precede the first set of labels shown.

If all inputs of a combinational element are disabled (caused to have no effect on the function of the element), the internal logic states of the outputs of the element are not specified by the symbol. If all inputs of a sequential element are disabled, the content of this element is not changed and the outputs remain at their existing internal logic states.

Labels may be factored using algebraic techniques.

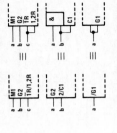

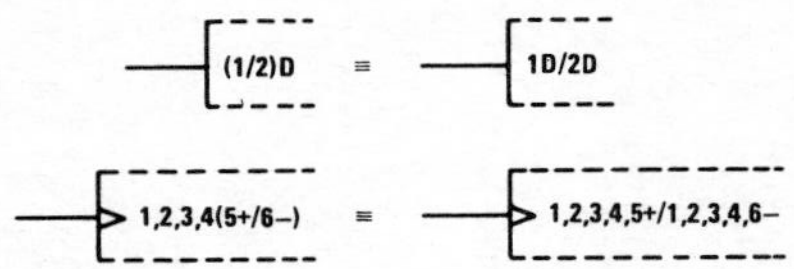

FIGURE 29 – INPUT LABELS

FIGURE 30 – FACTORING INPUT LABELS

10 SEQUENCE OF OUTPUT LABELS

If an output has a number of different labels, regardless of whether they are identifying numbers of affecting inputs or outputs or not, these labels are shown in the following order:

1) if the postponed output symbol has to be shown, this comes first, if necessary preceded by the indications of the inputs to which it must be applied;

2) followed by the labels indicating modifications of the internal logic state of the output, such that the left-to-right order of these labels corresponds with the order in which their effects must be applied;

3) followed by the label indicating the effect of the output on inputs and other outputs of the element.

Symbols for open-circuit or three-state outputs, where applicable, are placed just inside the outside boundary of the symbol adjacent to the output line. See Figure 31.

If an output needs several different sets of labels that represent alternative functions (e.g., depending on the mode of action), these sets may be shown on different output lines that must be connected outside the outline. However, there are cases in which this method of presentation is not advantageous. In those cases the output may be shown once with the different sets of labels separated by solidi. See Figure 32.

Two adjacent identifying numbers of affecting inputs in a set of labels that are not already separated by a nonnumeric character should be separated by a comma.

If a set of labels of an output not containing a solidus contains the identifying number of an affecting Mm input standing at its internal 0 state, this set of labels has no effect on that output.

Labels may be factored using algebraic techniques.

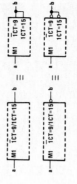

FIGURE 31 – PLACEMENT OF 3-STATE SYMBOLS

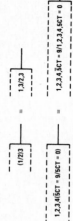

FIGURE 32 – OUTPUT LABELS

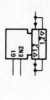

$$(1/2)3 \equiv 1,3/2,3$$

$$1,2,3,4(5CT = 9/5CT = 0) \equiv 1,2,3,4,5CT = 9/1,2,3,4,5CT = 0$$

FIGURE 33 – FACTORING OUTPUT LABELS

If you have questions on this Explanation of New Logic Symbols, please contact:

F.A. Mann MS 84
Texas Instruments Incorporated
P.O. Box 225012
Dallas, Texas 75265
Telephone (214) 995-3746

IEEE Standards may be purchased from:
Institute of Electrical and Electronics Engineers, Inc.
345 East 47th Street
New York, N.Y. 10017

International Electrotechnical Commission (IEC) publications may be purchased from:
American National Standards Institute, Inc.
1430 Broadway
New York, N.Y. 10018

APPENDIX C

DATA SHEETS:
INTEL 2716
ZILOG Z80

CONTENTS

INTEL 2716 16K (2K × 8) UV ERASABLE PROM	452
Pin Configuration	452
DC Characteristics	453
AC Characteristics	454
AC Waveforms	454
Erasure Characteristics	455
Device Operation	455
ZILOG Z80 AND Z80A	456
Pin Description	457
Timing Waveforms	458
Instruction Set	459
Z80 AC Characteristics	461
AC Timing Diagram	462
DC Characteristics	463
Z80A AC Characteristics	464

2716*
16K (2K × 8) UV ERASABLE PROM

- ■ **Fast Access Time**
  - — 350 ns Max. 2716-1
  - — 390 ns Max. 2716-2
  - — 450 ns Max. 2716

- ■ **Single +5V Power Supply**

- ■ **Low Power Dissipation**
  - — 525 mW Max. Active Power
  - — 132 mW Max. Standby Power

- ■ **Pin Compatible to Intel® 5V ROMs (2316E, 2332, and 2364) and 2732 EPROM**

- ■ **Simple Programming Requirements Single Location Programming Programs with One 50 ms Pulse**

- ■ **Inputs and Outputs TTL Compatible during Read and Program**

- ■ **Completely Static**

The Intel® 2716 is a 16,384-bit ultraviolet erasable and electrically programmable read-only memory (EPROM). The 2716 operates from a single 5-volt power supply, has a static standby mode, and features fast single address location programming. It makes designing with EPROMs faster, easier and more economical. For production quantities, the 2716 user can convert rapidly to Intel's pin-for-pin compatible 16K ROM (the 2316E) or the new 32K and 64K ROMs (the 2332 and 2364 respectively).

The 2716, with its single 5-volt supply and with an access time up to 350 ns, is ideal for use with the newer high performance +5V microprocessors such as Intel's 8085 and 8086. The 2716 is also the first EPROM with a static standby mode which reduces the power dissipation without increasing access time. The maximum active power dissipation is 525 mW while the maximum standby power dissipation is only 132 mW, a 75% savings.

The 2716 has the simplest and fastest method yet devised for programming EPROMs — single pulse TTL level programming. No need for high voltage pulsing because all programming controls are handled by TTL signals. Now, it is possible to program on-board, in the system, in the field. Program any location at any time — either individually, sequentially or at random, with the 2716's single address location programming. Total programming time for all 16,384 bits is only 100 seconds.

PIN CONFIGURATION*

```
        2716                      2732†
     A7 □ 1    24 □ Vcc       A7 □ 1    24 □ Vcc
     A6 □ 2    23 □ A8        A6 □ 2    23 □ A8
     A5 □ 3    22 □ A9        A5 □ 3    22 □ A9
     A4 □ 4    21 □ Vpp       A4 □ 4    21 □ A11
     A3 □ 5    20 □ OE        A3 □ 5    20 □ OE/Vpp
     A2 □ 6    19 □ A10       A2 □ 6    19 □ A10
     A1 □ 7 16K 18 □ CE       A1 □ 7 32K 18 □ CE
     A0 □ 8    17 □ O7        A0 □ 8    17 □ O7
     O0 □ 9    16 □ O6        O0 □ 9    16 □ O6
     O1 □ 10   15 □ O5        O1 □ 10   15 □ O5
     O2 □ 11   14 □ O4        O2 □ 11   14 □ O4
    GND □ 12   13 □ O3       GND □ 12   13 □ O3
```

†Refer to 2732 data sheet for specifications

PIN NAMES

A_0–A_9	ADDRESSES
$\overline{CE}$/PGM	CHIP ENABLE/PROGRAM
$\overline{OE}$	OUTPUT ENABLE
O_0–O_7	OUTPUTS

MODE SELECTION

PINS MODE	$\overline{CE}$/PGM (18)	$\overline{OE}$ (20)	V_{PP} (21)	V_{CC} (24)	OUTPUTS (9-11, 13-17)
Read	V_{IL}	V_{IL}	+5	+5	D_{OUT}
Standby	V_{IH}	Don't Care	+5	+5	High Z
Program	Pulsed V_{IL} to V_{IH}	V_{IH}	+25	+5	D_{IN}
Program Verify	V_{IL}	V_{IL}	+25	+5	D_{OUT}
Program Inhibit	V_{IL}	V_{IH}	+25	+5	High Z

BLOCK DIAGRAM

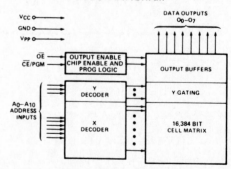

*Pin 18 and pin 20 have been renamed to conform with the entire family of 16K, 32K, and 64K EPROMs and ROMs. The die, fabrication process, and specifications remain the same and are totally uneffected by this change.

PROGRAMMING

The programming specifications are described in the Data Catalog PROM/ROM Programming Insutrctions on Page 4-83.

Absolute Maximum Ratings*

Temperature Under Bias . . . . –10°C to +80°C
Storage Temperature . . . . –65°C to +125°C
All Input or Output Voltages with
　Respect to Ground . . . . +6V to –0.3V
V_{PP} Supply Voltage with Respect
　to Ground During Program . . . . +26.5V to –0.3V

*COMMENT: Stresses above those listed under "Absolute Maximum Ratings" may cause permanent damage to the device. This is a stress rating only and functional operation of the device at these or any other conditions above those indicated in the operational sections of this specification is not implied. Exposure to absolute maximum rating conditions for extended periods may affect device reliability.

DC and AC Operating Conditions During Read

	2716	2716-1	2716-2
Temperature Range	0°C – 70°C	0°C – 70°C	0°C – 70°C
V_{CC} Power Supply [1,2]	5V ± 5%	5V ± 10%	5V ± 5%
V_{PP} Power Supply [2]	V_{CC} ± 0.6V [3]	V_{CC} ± 0.6V [3]	V_{CC} ± 0.6V [3]

READ OPERATION

D.C. and Operating Characteristics

Symbol	Parameter	Limits			Unit	Conditions
		Min.	Typ. [4]	Max.		
I_{LI}	Input Load Current			10	μA	V_{IN} = 5.25V
I_{LO}	Output Leakage Current			10	μA	V_{OUT} = 5.25V
I_{PP1} [2]	V_{PP} Current			5	mA	V_{PP} = 5.85V
I_{CC1} [2]	V_{CC} Current (Standby)		10	25	mA	$\overline{CE}$ = V_{IH}, $\overline{OE}$ = V_{IL}
I_{CC2} [2]	V_{CC} Current (Active)		57	100	mA	$\overline{OE}$ = $\overline{CE}$ = V_{IL}
V_{IL}	Input Low Voltage	–0.1		0.8	V	
V_{IH}	Input High Voltage	2.0		V_{CC}+1	V	
V_{OL}	Output Low Voltage			0.45	V	I_{OL} = 2.1 mA
V_{OH}	Output High Voltage	2.4			V	I_{OH} = –400 μA

NOTES: 1. V_{CC} must be applied simultaneously or before Vpp and removed simultaneously or after Vpp.

2. Vpp may be connected directly to V_{CC} except during programming. The supply current would then be the sum of I_{CC} and I_{PP1}.

3. The tolerance of 0.6V allows the use of a driver circuit for switching the Vpp supply pin from V_{CC} in read to 25V for programming.

4. Typical values are for T_A = 25°C and nominal supply voltages.

5. This parameter is only sampled and is not 100% tested.

Typical Characteristics

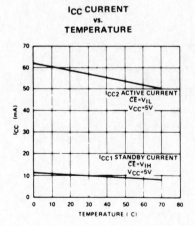

I_{CC} CURRENT
vs.
TEMPERATURE

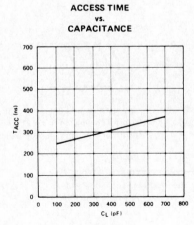

ACCESS TIME
vs.
CAPACITANCE

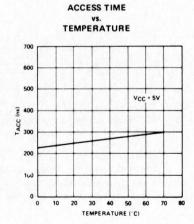

ACCESS TIME
vs.
TEMPERATURE

A.C. Characteristics

Symbol	Parameter	2716 Limits			2716-1 Limits			2716-2 Limits			Unit	Test Conditions
		Min	Typ[4]	Max	Min	Typ[4]	Max	Min	Typ[4]	Max		
t_{ACC}	Address to Output Delay			450			350			390	ns	$\overline{CE} = \overline{OE} = V_{IL}$
t_{CE}	$\overline{CE}$ to Output Delay			450			350			390	ns	$\overline{OE} = V_{IL}$
t_{OE}	Output Enable to Output Delay			120			120			120	ns	$\overline{CE} = V_{IL}$
t_{DF}	Output Enable High to Output Float	0		100	0		100	0		100	ns	$\overline{CE} = V_{IL}$
t_{OH}	Address to Output Hold	0			0			0			ns	$\overline{CE} = \overline{OE} = V_{IL}$

Capacitance[5] $T_A = 25°C$, f = 1 MHz

Symbol	Parameter	Typ.	Max.	Unit	Conditions
C_{IN}	Input Capacitance	4	6	pF	$V_{IN} = 0V$
C_{OUT}	Output Capacitance	8	12	pF	$V_{OUT} = 0V$

A.C. Test Conditions:

Output Load: 1 TTL gate and C_L = 100 pF
Input Rise and Fall Times: ≤20 ns
Input Pulse Levels: 0.8V to 2.2V
Timing Measurement Reference Level:
    Inputs   1V and 2V
    Outputs 0.8V and 2V

A. C. Waveforms (1)

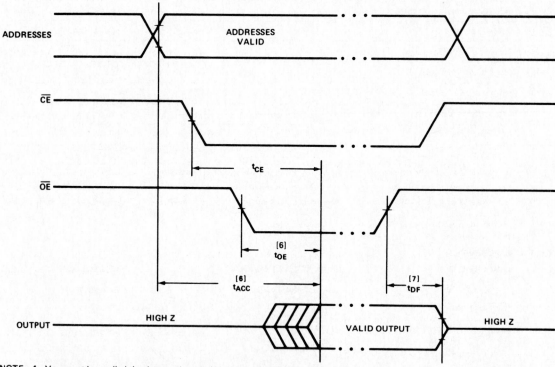

NOTE:
1. V_{CC} must be applied simultaneously or before Vpp and removed simultaneously or after Vpp.
2. Vpp may be connected directly to V_{CC} except during programming. The supply current would then be the sum of I_{CC} and I_{PP1}.
3. The tolerance of 0.6V allows the use of a driver circuit for switching the Vpp supply pin from V_{CC} in read to 25V for programming.
4. Typical values are for $T_A = 25°C$ and nominal supply voltages.
5. This parameter is only sampled and is not 100% tested.
6. $\overline{OE}$ may be delayed up to $t_{ACC} - t_{OE}$ after the falling edge of $\overline{CE}$ without impact on t_{ACC}.
7. t_{DF} is specified from $\overline{OE}$ or $\overline{CE}$, whichever occurs first.

ERASURE CHARACTERISTICS

The erasure characteristics of the 2716 are such that erasure begins to occur when exposed to light with wavelengths shorter than approximately 4000 Angstroms (Å). It should be noted that sunlight and certain types of fluorescent lamps have wavelengths in the 3000—4000Å range. Data show that constant exposure to room level fluorescent lighting could erase the typical 2716 in approximately 3 years, while it would take approximatley 1 week to cause erasure when exposed to direct sunlight. If the 2716 is to be exposed to these types of lighting conditions for extended periods of time, opaque labels are available from Intel which should be placed over the 2716 window to prevent unintentional erasure.

The recommended erasure procedure (see Data Catalog page 4-83) for the 2716 is exposure to shortwave ultraviolet light which has a wavelength of 2537 Angstroms (Å). The integrated dose (i.e., UV intensity $\times$ exposure time) for erasure should be a minimum of 15 W-sec/cm^2. The erasure time with this dosage is approximately 15 to 20 minutes using an ultraviolet lamp with a 12000 μW/cm^2 power rating. The 2716 should be placed within 1 inch of the lamp tubes during erasure. Some lamps have a filter on their tubes which should be removed before erasure.

DEVICE OPERATION

The five modes of operation of the 2716 are listed in Table I. It should be noted that all inputs for the five modes are at TTL levels. The power supplies required are a +5V V_{CC} and a V_{PP}. The V_{PP} power supply must be at 25V during the three programming modes, and must be at 5V in the other two modes.

TABLE I. MODE SELECTION

PINS MODE	CE/PGM (18)	OE (20)	Vpp (21)	Vcc (24)	OUTPUTS (9-11, 13-17)
Read	V_{IL}	V_{IL}	+5	+5	D_{OUT}
Standby	V_{IH}	Don't Care	+5	+5	High Z
Program	Pulsed V_{IL} to V_{IH}	V_{IH}	+25	+5	D_{IN}
Program Verify	V_{IL}	V_{IL}	+25	+5	D_{OUT}
Program Inhibit	V_{IL}	V_{IH}	+25	+5	High Z

READ MODE

The 2716 has two control functions, both of which must be logically satisfied in order to obtain data at the outputs. Chip Enable (CE) is the power control and should be used for device selection. Output Enable (OE) is the output control and should be used to gate data to the output pins, independent of device selection. Assuming that addresses are stable, address access time (t_{ACC}) is equal to the delay from CE to output (t_{CE}). Data is available at the outputs 120 ns (t_{OE}) after the falling edge of OE, assuming that CE has been low and addresses have been stable for at least $t_{ACC} - t_{OE}$.

STANDBY MODE

The 2716 has a standby mode which reduces the active power dissipation by 75%, from 525 mW to 132 mW. The 2716 is placed in the standby mode by applying a TTL high signal to the CE input. When in standby mode, the outputs are in a high impedence state, independent of the OE input.

OUTPUT DESELECTION

The outputs of two or more 2716s may be OR-tied together on the same data bus. Only one 2716 should have its output selected (OE low) to prevent data bus contention between 2716s in this configuration. The outputs of the other 2716s should be deselected by raising the OE input to a TTL high level.

PROGRAMMING

Initially, and after each erasure, all bits of the 2716 are in the "1" state. Data is introduced by selectively programming "0's" into the desired bit locations. Although only "0's" will be programmed, both "1's" and "0's" can be presented in the data word. The only way to change a "0" to a "1" is by ultraviolet light erasure.

The 2716 is in the programming mode when the V_{PP} power supply is at 25V and OE is at V_{IH}. The data to be programmed is applied 8 bits in parallel to the data output pins. The levels required for the address and data inputs are TTL.

When the address and data are stable, a 50 msec, active high, TTL program pulse is applied to the CE/PGM input. A program pulse must be applied at each address location to be programmed. You can program any location at any time — either individually, sequentially, or at random. The program pulse has a maximum width of 55 msec. The 2716 must not be programmed with a DC signal applied to the CE/PGM input.

Programming of multiple 2716s in parallel with the same data can be easily accomplished due to the simplicity of the programming requirements. Like inputs of the paralleled 2716s may be connected together when they are programmed with the same data. A high level TTL pulse applied to the CE/PGM input programs the paralleled 2716s.

PROGRAM INHIBIT

Programming of multiple 2716s in parallel with different data is also easily accomplished. Except for CE/PGM, all like inputs (including OE) of the parallel 2716s may be common. A TTL level program pulse applied to a 2716's CE/PGM input with V_{PP} at 25V will program that 2716. A low level CE/PGM input inhibits the other 2716 from being programmed.

PROGRAM VERIFY

A verify should be performed on the programmed bits to determine that they were correctly programmed. The verify may be performed wth V_{PP} at 25V. Except during programming and program verify, V_{PP} must be at 5V.

Z80®-CPU
Z80A-CPU

Product Specification
MARCH 1978

The Zilog Z80 product line is a complete set of micro-computer components, development systems and support software. The Z80 microcomputer component set includes all of the circuits necessary to build high-performance microcomputer systems with virtually no other logic and a minimum number of low cost standard memory elements.

The Z80 and Z80A CPU's are third generation single chip microprocessors with unrivaled computational power. This increased computational power results in higher system through-put and more efficient memory utilization when compared to second generation microprocessors. In addition, the Z80 and Z80A CPU's are very easy to implement into a system because of their single voltage requirement plus all output signals are fully decoded and timed to control standard memory or peripheral circuits. The circuit is implemented using an N-channel, ion implanted, silicon gate MOS process.

Figure 1 is a block diagram of the CPU, Figure 2 details the internal register configuration which contains 208 bits of Read/Write memory that are accessible to the programmer. The registers include two sets of six general purpose registers that may be used individually as 8-bit registers or as 16-bit register pairs. There are also two sets of accumulator and flag registers. The programmer has access to either set of main or alternate registers through a group of exchange instructions. This alternate set allows foreground/background mode of operation or may be reserved for very fast Interrupt response. Each CPU also contains a 16-bit stack pointer which permits simple implementation of

multiple level interrupts, unlimited subroutine nesting and simplification of many types of data handling.

The two 16-bit index registers allow tabular data manipulation and easy implementation of relocatable code. The Refresh register provides for automatic, totally transparent refresh of external dynamic memories. The I register is used in a powerful interrupt response mode to form the upper 8 bits of a pointer to a interrupt service address table, while the interrupting device supplies the lower 8 bits of the pointer. An indirect call is then made to this service address.

FEATURES

- Single chip, N-channel Silicon Gate CPU.
- 158 instructions—includes all 78 of the 8080A instructions with total software compatibility. New instructions include 4-, 8- and 16-bit operations with more useful addressing modes such as indexed, bit and relative.
- 17 internal registers.
- Three modes of fast interrupt response plus a non-maskable interrupt.
- Directly interfaces standard speed static or dynamic memories with virtually no external logic.
- 1.0 μs instruction execution speed.
- Single 5 VDC supply and single-phase 5 volt Clock.
- Out-performs any other single chip microcomputer in 4-, 8-, or 16-bit applications.
- All pins TTL Compatible
- Built-in dynamic RAM refresh circuitry.

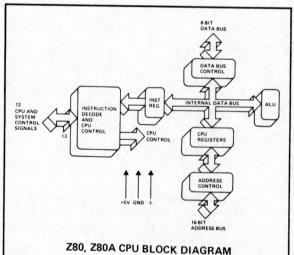

Z80, Z80A CPU BLOCK DIAGRAM

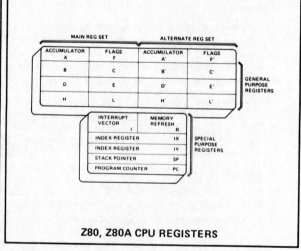

Z80, Z80A CPU REGISTERS

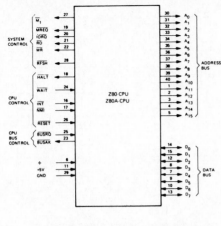

Z80, Z80A CPU PIN CONFIGURATION

A_0-A_{15}
(Address Bus)

Tri-state output, active high. A_0-A_{15} constitute a 16-bit address bus. The address bus provides the address for memory (up to 64K bytes) data exchanges and for I/O device data exchanges.

D_0-D_7
(Data Bus)

Tri-state input/output, active high. D_0-D_7 constitute an 8-bit bidirectional data bus. The data bus is used for data exchanges with memory and I/O devices.

$\overline{M_1}$
(Machine
Cycle one)

Output, active low. $\overline{M_1}$ indicates that the current machine cycle is the OP code fetch cycle of an instruction execution.

$\overline{MREQ}$
(Memory
Request)

Tri-state output, active low. The memory request signal indicates that the address bus holds a valid address for a memory read or memory write operation.

$\overline{IORQ}$
(Input/
Output
Request)

Tri-state output, active low. The $\overline{IORQ}$ signal indicates that the lower half of the address bus holds a valid I/O address for a I/O read or write operation. An $\overline{IORQ}$ signal is also generated when an interrupt is being acknowledged to indicate that an interrupt response vector can be placed on the data bus.

$\overline{RD}$
(Memory
Read)

Tri-state output, active low. $\overline{RD}$ indicates that the CPU wants to read data from memory or an I/O device. The addressed I/O device or memory should use this signal to gate data onto the CPU data bus.

$\overline{WR}$
(Memory
Write)

Tri-state output, active low. $\overline{WR}$ indicates that the CPU data bus holds valid data to be stored in the addressed memory or I/O device.

$\overline{RFSH}$
(Refresh)

Output, active low. $\overline{RFSH}$ indicates that the lower 7 bits of the address bus contain a refresh address for dynamic memories and the current $\overline{MREQ}$ signal should be used to do a refresh read to all dynamic memories.

$\overline{HALT}$
(Halt state)

Output, active low. $\overline{HALT}$ indicates that the CPU has executed a HALT software instruction and is awaiting either a non-maskable or a maskable interrupt (with the mask enabled) before operation can resume. While halted, the CPU executes NOP's to maintain memory refresh activity.

$\overline{WAIT}$
(Wait)

Input, active low. $\overline{WAIT}$ indicates to the Z-80 CPU that the addressed memory or I/O devices are not ready for a data transfer. The CPU continues to enter wait states for as long as this signal is active.

$\overline{INT}$
(Interrupt
Request)

Input, active low. The Interrupt Request signal is generated by I/O devices. A request will be honored at the end of the current instruction if the internal software controlled interrupt enable flip-flop (IFF) is enabled.

$\overline{NMI}$
(Non
Maskable
Interrupt)

Input, active low. The non-maskable interrupt request line has a higher priority than $\overline{INT}$ and is always recognized at the end of the current instruction, independent of the status of the interrupt enable flip-flop. $\overline{NMI}$ automatically forces the Z-80 CPU to restart to location 0066_H.

$\overline{RESET}$

Input, active low. $\overline{RESET}$ initializes the CPU as follows: reset interrupt enable flip-flop, clear PC and registers I and R and set interrupt to 8080A mode. During reset time, the address and data bus go to a high impedance state and all control output signals go to the inactive state.

$\overline{BUSRQ}$
(Bus
Request)

Input, active low. The bus request signal has a higher priority than $\overline{NMI}$ and is always recognized at the end of the current machine cycle and is used to request the CPU address bus, data bus and tri-state output control signals to go to a high impedance state so that other devices can control these busses.

$\overline{BUSAK}$
(Bus
Acknowledge)

Output, active low. Bus acknowledge is used to indicate to the requesting device that the CPU address bus, data bus and tri-state control bus signals have been set to their high impedance state and the external device can now control these signals.

INSTRUCTION OP CODE FETCH

The program counter content (PC) is placed on the address bus immediately at the start of the cycle. One half clock time later $\overline{MREQ}$ goes active. The falling edge of $\overline{MREQ}$ can be used directly as a chip enable to dynamic memories. $\overline{RD}$ when active indicates that the memory data should be enabled onto the CPU data bus. The CPU samples data with the rising edge of the clock state T_3. Clock states T_3 and T_4 of a fetch cycle are used to refresh dynamic memories while the CPU is internally decoding and executing the instruction. The refresh control signal $\overline{RFSH}$ indicates that a refresh read of all dynamic memories should be accomplished.

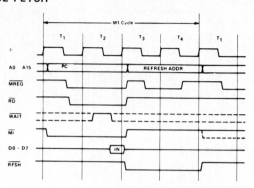

MEMORY READ OR WRITE CYCLES

Illustrated here is the timing of memory read or write cycles other than an OP code fetch (M_1 cycle). The $\overline{MREQ}$ and $\overline{RD}$ signals are used exactly as in the fetch cycle. In the case of a memory write cycle, the $\overline{MREQ}$ also becomes active when the address bus is stable so that it can be used directly as a chip enable for dynamic memories. The $\overline{WR}$ line is active when data on the data bus is stable so that it can be used directly as a R/W pulse to virtually any type of semiconductor memory.

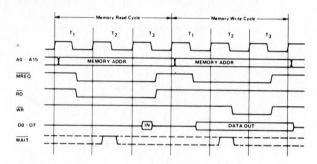

INPUT OR OUTPUT CYCLES

Illustrated here is the timing for an I/O read or I/O write operation. Notice that during I/O operations a single wait state is automatically inserted (Tw*). The reason for this is that during I/O operations this extra state allows sufficient time for an I/O port to decode its address and activate the $\overline{WAIT}$ line if a wait is required.

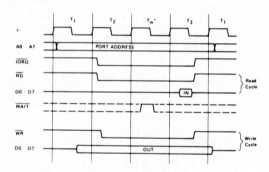

INTERRUPT REQUEST/ACKNOWLEDGE CYCLE

The interrupt signal is sampled by the CPU with the rising edge of the last clock at the end of any instruction. When an interrupt is accepted, a special M_1 cycle is generated. During this M_1 cycle, the $\overline{IORQ}$ signal becomes active (instead of $\overline{MREQ}$) to indicate that the interrupting device can place an 8-bit vector on the data bus. Two wait states (Tw*) are automatically added to this cycle so that a ripple priority interrupt scheme, such as the one used in the Z80 peripheral controllers, can be easily implemented.

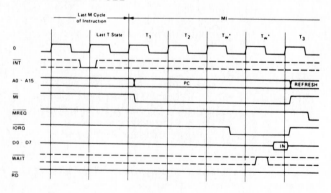

Z80, Z80A Instruction Set

The following is a summary of the Z80, Z80A instruction set showing the assembly language mnemonic and the symbolic operation performed by the instruction. A more detailed listing appears in the Z80-CPU technical manual, and assembly language programming manual. The instructions are divided into the following categories:

8-bit loads	Miscellaneous Group
16-bit loads	Rotates and Shifts
Exchanges	Bit Set, Reset and Test
Memory Block Moves	Input and Output
Memory Block Searches	Jumps
8-bit arithmetic and logic	Calls
16-bit arithmetic	Restarts
General purpose Accumulator	Returns
& Flag Operations	

In the table the following terminology is used.

b ≡ a bit number in any 8-bit register or memory location

cc ≡ flag condition code
- NZ ≡ non zero
- Z ≡ zero
- NC ≡ non carry
- C ≡ carry
- PO ≡ Parity odd or no over flow
- PE ≡ Parity even or over flow
- P ≡ Positive
- M ≡ Negative (minus)

d ≡ any 8-bit destination register or memory location

dd ≡ any 16-bit destination register or memory location

e ≡ 8-bit signed 2's complement displacement used in relative jumps and indexed addressing

L ≡ 8 special call locations in page zero. In decimal notation these are 0, 8, 16, 24, 32, 40, 48 and 56

n ≡ any 8-bit binary number

nn ≡ any 16-bit binary number

r ≡ any 8-bit general purpose register (A, B, C, D, E, H, or L)

s ≡ any 8-bit source register or memory location

s_b ≡ a bit in a specific 8-bit register or memory location

ss ≡ any 16-bit source register or memory location

subscript "L" ≡ the low order 8 bits of a 16-bit register

subscript "H" ≡ the high order 8 bits of a 16-bit register

( ) ≡ the contents within the ( ) are to be used as a pointer to a memory location or I/O port number

8-bit registers are A, B, C, D, E, H, L, I and R

16-bit register pairs are AF, BC, DE and HL

16-bit registers are SP, PC, IX and IY

Addressing Modes implemented include combinations of the following:

Immediate	Indexed
Immediate extended	Register
Modified Page Zero	Implied
Relative	Register Indirect
Extended	Bit

	Mnemonic	Symbolic Operation	Comments
8-BIT LOADS	LD r, s	r ← s	s ≡ r, n, (HL), (IX+e), (IY+e)
	LD d, r	d ← r	d ≡ (HL), r (IX+e), (IY+e)
	LD d, n	d ← n	d ≡ (HL), (IX+e), (IY+e)
	LD A, s	A ← s	s ≡ (BC), (DE), (nn), I, R
	LD d, A	d ← A	d ≡ (BC), (DE), (nn), I, R
16-BIT LOADS	LD dd, nn	dd ← nn	dd ≡ BC, DE, HL, SP, IX, IY
	LD dd, (nn)	dd ← (nn)	dd ≡ BC, DE, HL, SP, IX, IY
	LD (nn), ss	(nn) ← ss	ss ≡ BC, DE, HL, SP, IX, IY
	LD SP, ss	SP ← ss	ss ≡ HL, IX, IY
	PUSH ss	(SP−1) ← ss_H; (SP−2) ← ss_L	ss ≡ BC, DE, HL, AF, IX, IY
	POP dd	dd_L ← (SP); dd_H ← (SP+1)	dd ≡ BC, DE, HL, AF, IX, IY
EXCHANGES	EX DE, HL	DE ↔ HL	
	EX AF, AF′	AF ↔ AF′	
	EXX	$\begin{pmatrix} BC \\ DE \\ HL \end{pmatrix} ↔ \begin{pmatrix} BC' \\ DE' \\ HL' \end{pmatrix}$	
	EX (SP), ss	(SP) ·· ss_L, (SP+1) ·· ss_H	ss ≡ HL, IX, IY

	Mnemonic	Symbolic Operation	Comments
MEMORY BLOCK MOVES	LDI	(DE) ← (HL), DE ← DE+1 HL ← HL+1, BC ← BC−1	
	LDIR	(DE) ← (HL), DE ← DE+1 HL ← HL+1, BC ← BC−1 Repeat until BC = 0	
	LDD	(DE) ← (HL), DE ← DE−1 HL ← HL−1, BC ← BC−1	
	LDDR	(DE) ← (HL), DE ← DE−1 HL ← HL−1, BC ← BC−1 Repeat until BC = 0	
MEMORY BLOCK SEARCHES	CPI	A−(HL), HL ← HL+1 BC ← BC−1	
	CPIR	A−(HL), HL ← HL+1 BC ← BC−1, Repeat until BC = 0 or A = (HL)	A−(HL) sets the flags only. A is not affected
	CPD	A−(HL), HL ← HL−1 BC ← BC−1	
	CPDR	A−(HL), HL ← HL−1 BC ← BC−1, Repeat until BC= 0 or A = (HL)	
8-BIT ALU	ADD s	A ← A + s	CY is the carry flag
	ADC s	A ← A + s + CY	
	SUB s	A ← A − s	
	SBC s	A ← A − s − CY	s ≡ r, n, (HL) (IX+e), (IY+e)
	AND s	A ← A ∧ s	
	OR s	A ← A ∨ s	
	XOR s	A ← A ⊕ s	

	Mnemonic	Symbolic Operation	Comments
8-BIT ALU	CP s	$A - s$	$s = r, n$ (HL) (IX+e), (IY+e)
	INC d	$d \leftarrow d + 1$	$d = r$, (HL) (IX+e), (IY+e)
	DEC d	$d \leftarrow d - 1$	
16-BIT ARITHMETIC	ADD HL, ss	$HL \leftarrow HL + ss$	$ss \equiv$ BC, DE, HL, SP
	ADC HL, ss	$HL \leftarrow HL + ss + CY$	
	SBC HL, ss	$HL \leftarrow HL - ss - CY$	
	ADD IX, ss	$IX \leftarrow IX + ss$	$ss \equiv$ BC, DE, IX, SP
	ADD IY, ss	$IY \leftarrow IY + ss$	$ss \equiv$ BC, DE, IY, SP
	INC dd	$dd \leftarrow dd + 1$	$dd \equiv$ BC, DE, HL, SP, IX, IY
	DEC dd	$dd \leftarrow dd - 1$	$dd \equiv$ BC, DE, HL, SP, IX, IY
GP ACC. & FLAG	DAA	Converts A contents into packed BCD following add or subtract.	Operands must be in packed BCD format
	CPL	$A \leftarrow \bar{A}$	
	NEG	$A \leftarrow 00 - A$	
	CCF	$CY \leftarrow \overline{CY}$	
	SCF	$CY \leftarrow 1$	
MISCELLANEOUS	NOP	No operation	
	HALT	Halt CPU	
	DI	Disable Interrupts	
	EI	Enable Interrupts	
	IM 0	Set interrupt mode 0	8080A mode
	IM 1	Set interrupt mode 1	Call to 0038$_H$
	IM 2	Set interrupt mode 2	Indirect Call
ROTATES AND SHIFTS	RLC s	(diagram)	$s \equiv r$, (HL) (IX+e), (IY+e)
	RL s	(diagram)	
	RRC s	(diagram)	
	RR s	(diagram)	
	SLA s	(diagram)	
	SRA s	(diagram)	
	SRL s	(diagram)	
	RLD	(diagram)	
	RRD	(diagram)	

	Mnemonic	Symbolic Operation	Comments
BIT S. R. & T	BIT b, s	$Z \leftarrow \bar{s}_b$	Z is zero flag
	SET b, s	$s_b \leftarrow 1$	$s \equiv r$, (HL) (IX+e), (IY+e)
	RES b, s	$s_b \leftarrow 0$	
INPUT AND OUTPUT	IN A, (n)	$A \leftarrow (n)$	
	IN r, (C)	$r \leftarrow (C)$	Set flags
	INI	$(HL) \leftarrow (C), HL \leftarrow HL + 1$; $B \leftarrow B - 1$	
	INIR	$(HL) \leftarrow (C), HL \leftarrow HL + 1$; $B \leftarrow B - 1$; Repeat until B = 0	
	IND	$(HL) \leftarrow (C), HL \leftarrow HL - 1$; $B \leftarrow B - 1$	
	INDR	$(HL) \leftarrow (C), HL \leftarrow HL - 1$; $B \leftarrow B - 1$; Repeat until B = 0	
	OUT(n), A	$(n) \leftarrow A$	
	OUT(C), r	$(C) \leftarrow r$	
	OUTI	$(C) \leftarrow (HL), HL \leftarrow HL + 1$; $B \leftarrow B - 1$	
	OTIR	$(C) \leftarrow (HL), HL \leftarrow HL + 1$; $B \leftarrow B - 1$; Repeat until B = 0	
	OUTD	$(C) \leftarrow (HL), HL \leftarrow HL - 1$; $B \leftarrow B - 1$	
	OTDR	$(C) \leftarrow (HL), HL \leftarrow HL - 1$; $B \leftarrow B - 1$; Repeat until B = 0	
JUMPS	JP nn	$PC \leftarrow nn$	
	JP cc, nn	If condition cc is true $PC \leftarrow nn$, else continue	cc: NZ PO, Z PE, NC P, C M
	JR e	$PC \leftarrow PC + e$	
	JR kk, e	If condition kk is true $PC \leftarrow PC + e$, else continue	kk: NZ NC, Z C
	JP (ss)	$PC \leftarrow ss$	ss = HL, IX, IY
	DJNZ e	$B \leftarrow B - 1$, if B = 0 continue, else $PC \leftarrow PC + e$	
CALLS	CALL nn	$(SP-1) \leftarrow PC_H$; $(SP-2) \leftarrow PC_L, PC \leftarrow nn$	cc: NZ PO, Z PE, NC P, C M
	CALL cc, nn	If condition cc is false continue, else same as CALL nn	
RESTARTS	RST L	$(SP-1) \leftarrow PC_H$; $(SP-2) \leftarrow PC_L, PC_H \leftarrow 0$; $PC_L \leftarrow L$	
RETURNS	RET	$PC_L \leftarrow (SP)$, $PC_H \leftarrow (SP+1)$	
	RET cc	If condition cc is false continue, else same as RET	cc: NZ PO, Z PE, NC P, C M
	RETI	Return from interrupt, same as RET	
	RETN	Return from non-maskable interrupt	

$T_A = 0°C$ to $70°C$, $V_{CC} = +5V \pm 5\%$, Unless Otherwise Noted.

Signal	Symbol	Parameter	Min	Max	Unit	Test Condition
Φ	t_c	Clock Period	.4	[12]	µsec	
	$t_w(\Phi H)$	Clock Pulse Width, Clock High	180	[E]	nsec	
	$t_w(\Phi L)$	Clock Pulse Width, Clock Low	180	2000	nsec	
	$t_{r,f}$	Clock Rise and Fall Time		30	nsec	
A_{0-15}	$t_D(AD)$	Address Output Delay		145	nsec	
	$t_F(AD)$	Delay to Float		110	nsec	
	t_{acm}	Address Stable Prior to $\overline{MREQ}$ (Memory Cycle)	[1]		nsec	$C_L = 50pF$
	t_{aci}	Address Stable Prior to $\overline{IORQ}$, $\overline{RD}$ or $\overline{WR}$ (I/O Cycle)	[2]		nsec	
	t_{ca}	Address Stable from $\overline{RD}$, $\overline{WR}$, $\overline{IORQ}$ or $\overline{MREQ}$	[3]		nsec	
	t_{caf}	Address Stable From $\overline{RD}$ or $\overline{WR}$ During Float	[4]		nsec	
D_{0-7}	$t_D(D)$	Data Output Delay		230	nsec	
	$t_F(D)$	Delay to Float During Write Cycle		90	nsec	
	$t_{S\Phi}(D)$	Data Setup Time to Rising Edge of Clock During M1 Cycle	50		nsec	$C_L = 50pF$
	$t_{S\overline{\Phi}}(D)$	Data Setup Time to Falling Edge of Clock During M2 to M5	60		nsec	
	t_{dcm}	Data Stable Prior to $\overline{WR}$ (Memory Cycle)	[5]		nsec	
	t_{dci}	Data Stable Prior to $\overline{WR}$ (I/O Cycle)	[6]		nsec	
	t_{cdf}	Data Stable From $\overline{WR}$	[7]		nsec	
	t_H	Any Hold Time for Setup Time	0		nsec	
$\overline{MREQ}$	$t_{DL\Phi}(MR)$	$\overline{MREQ}$ Delay From Falling Edge of Clock, $\overline{MREQ}$ Low		100	nsec	
	$t_{DH\Phi}(MR)$	$\overline{MREQ}$ Delay From Rising Edge of Clock, $\overline{MREQ}$ High		100	nsec	
	$t_{DH\overline{\Phi}}(MR)$	$\overline{MREQ}$ Delay From Falling Edge of Clock, $\overline{MREQ}$ High		100	nsec	$C_L = 50pF$
	$t_w(\overline{MRL})$	Pulse Width, $\overline{MREQ}$ Low	[8]		nsec	
	$t_w(\overline{MRH})$	Pulse Width, $\overline{MREQ}$ High	[9]		nsec	
$\overline{IORQ}$	$t_{DL\Phi}(IR)$	$\overline{IORQ}$ Delay From Rising Edge of Clock, $\overline{IORQ}$ Low		90	nsec	
	$t_{DL\overline{\Phi}}(IR)$	$\overline{IORQ}$ Delay From Falling Edge of Clock, $\overline{IORQ}$ Low		110	nsec	
	$t_{DH\Phi}(IR)$	$\overline{IORQ}$ Delay From Rising Edge of Clock, $\overline{IORQ}$ High		100	nsec	$C_L = 50pF$
	$t_{DH\overline{\Phi}}(IR)$	$\overline{IORQ}$ Delay From Falling Edge of Clock, $\overline{IORQ}$ High		110	nsec	
$\overline{RD}$	$t_{DL\Phi}(RD)$	$\overline{RD}$ Delay From Rising Edge of Clock, $\overline{RD}$ Low		100	nsec	
	$t_{DL\overline{\Phi}}(RD)$	$\overline{RD}$ Delay From Falling Edge of Clock, $\overline{RD}$ Low		130	nsec	
	$t_{DH\Phi}(RD)$	$\overline{RD}$ Delay From Rising Edge of Clock, $\overline{RD}$ High		100	nsec	$C_L = 50pF$
	$t_{DH\overline{\Phi}}(RD)$	$\overline{RD}$ Delay From Falling Edge of Clock, $\overline{RD}$ High		110	nsec	
$\overline{WR}$	$t_{DL\Phi}(WR)$	$\overline{WR}$ Delay From Rising Edge of Clock, $\overline{WR}$ Low		80	nsec	
	$t_{DL\overline{\Phi}}(WR)$	$\overline{WR}$ Delay From Falling Edge of Clock, $\overline{WR}$ Low		90	nsec	
	$t_{DH\overline{\Phi}}(WR)$	$\overline{WR}$ Delay From Falling Edge of Clock, $\overline{WR}$ High		100	nsec	$C_L = 50pF$
	$t_w(\overline{WRL})$	Pulse Width, $\overline{WR}$ Low	[10]		nsec	
$\overline{M1}$	$t_{DL}(M1)$	$\overline{M1}$ Delay From Rising Edge of Clock, $\overline{M1}$ Low		130	nsec	$C_L = 50pF$
	$t_{DH}(M1)$	$\overline{M1}$ Delay From Rising Edge of Clock, $\overline{M1}$ High		130	nsec	
$\overline{RFSH}$	$t_{DL}(RF)$	$\overline{RFSH}$ Delay From Rising Edge of Clock, $\overline{RFSH}$ Low		180	nsec	$C_L = 50pF$
	$t_{DH}(RF)$	$\overline{RFSH}$ Delay From Rising Edge of Clock, $\overline{RFSH}$ High		150	nsec	
$\overline{WAIT}$	$t_s(WT)$	$\overline{WAIT}$ Setup Time to Falling Edge of Clock	70		nsec	
$\overline{HALT}$	$t_D(HT)$	$\overline{HALT}$ Delay Time From Falling Edge of Clock		300	nsec	$C_L = 50pF$
$\overline{INT}$	$t_s(IT)$	$\overline{INT}$ Setup Time to Rising Edge of Clock	80		nsec	
$\overline{NMI}$	$t_w(\overline{NML})$	Pulse Width, $\overline{NMI}$ Low	80		nsec	
$\overline{BUSRQ}$	$t_s(BQ)$	$\overline{BUSRQ}$ Setup Time to Rising Edge of Clock	80		nsec	
$\overline{BUSAK}$	$t_{DL}(BA)$	$\overline{BUSAK}$ Delay From Rising Edge of Clock, $\overline{BUSAK}$ Low		120	nsec	$C_L = 50pF$
	$t_{DH}(BA)$	$\overline{BUSAK}$ Delay From Falling Edge of Clock, $\overline{BUSAK}$ High		110	nsec	
$\overline{RESET}$	$t_s(RS)$	$\overline{RESET}$ Setup Time to Rising Edge of Clock	90		nsec	
	$t_F(C)$	Delay to Float ($\overline{MREQ}$, $\overline{IORQ}$, $\overline{RD}$ and $\overline{WR}$)		100	nsec	
	t_{mr}	$\overline{M1}$ Stable Prior to $\overline{IORQ}$ (Interrupt Ack.)	[11]		nsec	

[12] $t_c = t_w(\Phi H) + t_w(\Phi L) + t_r + t_f$

[1] $t_{acm} = t_w(\Phi H) + t_f - 75$

[2] $t_{aci} = t_c - 80$

[3] $t_{ca} = t_w(\Phi L) + t_r - 40$

[4] $t_{caf} = t_w(\Phi L) + t_r - 60$

[5] $t_{dcm} = t_c - 210$

[6] $t_{dci} = t_w(\Phi L) + t_r - 210$

[7] $t_{cdf} = t_w(\Phi L) + t_r - 80$

[8] $t_w(\overline{MRL}) = t_c - 40$

[9] $t_w(MRH) = t_w(\Phi H) + t_f - 30$

[10] $t_w(\overline{WRL}) = t_c - 40$

[11] $t_{mr} = 2t_c + t_w(\Phi H) + t_f - 80$

NOTES:

A. Data should be enabled onto the CPU data bus when $\overline{RD}$ is active. During interrupt acknowledge data should be enabled when $\overline{M1}$ and $\overline{IORQ}$ are both active.

B. All control signals are internally synchronized, so they may be totally asynchronous with respect to the clock.

C. The $\overline{RESET}$ signal must be active for a minimum of 3 clock cycles.

D. Output Delay vs. Loaded Capacitance
   $T_A = 70°C$ $V_{CC} = +5V \pm 5\%$
   Add 10nsec delay for each 50pf increase in load up to a maximum of 200pf for the data bus & 100pf for address & control lines

E. Although static by design, testing guarantees $t_w(\Phi H)$ of 200 µsec maximum

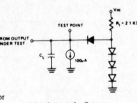

Load circuit for Output

A.C. Timing Diagram

Timing measurements are made at the following voltages, unless otherwise specified:

	"1"	"0"
CLOCK	$V_{CC} -.6V$	.45V
OUTPUT	2.0 V	.8 V
INPUT	2.0 V	.8 V
FLOAT	Δ V	± 0.5 V

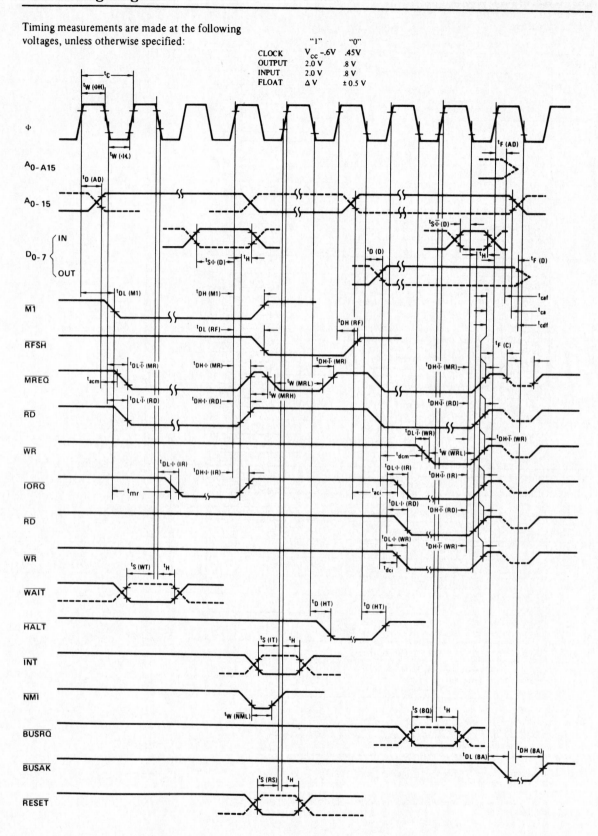

Absolute Maximum Ratings

Temperature Under Bias	Specified operating range.	
Storage Temperature	-65°C to +150°C	
Voltage On Any Pin with Respect to Ground	-0.3V to +7V	
Power Dissipation	1.5W	

*Comment

Stresses above those listed under "Absolute Maximum Rating" may cause permanent damage to the device. This is a stress rating only and functional operation of the device at these or any other condition above those indicated in the operational sections of this specification is not implied. Exposure to absolute maximum rating conditions for extended periods may affect device reliability.

Note: For Z80-CPU all AC and DC characteristics remain the same for the military grade parts except I_{cc}.

I_{cc} = 200 mA

Z80-CPU D.C. Characteristics

T_A = 0°C to 70°C. V_{cc} = 5V ± 5% unless otherwise specified

Symbol	Parameter	Min.	Typ.	Max.	Unit	Test Condition
V_{ILC}	Clock Input Low Voltage	-0.3		0.45	V	
V_{IHC}	Clock Input High Voltage	V_{cc} -.6		V_{cc}+.3	V	
V_{IL}	Input Low Voltage	-0.3		0.8	V	
V_{IH}	Input High Voltage	2.0		V_{cc}	V	
V_{OL}	Output Low Voltage			0.4	V	I_{OL}=1.8mA
V_{OH}	Output High Voltage	2.4			V	I_{OH} = -250μA
I_{CC}	Power Supply Current			150	mA	
I_{LI}	Input Leakage Current			10	μA	V_{IN}=0 to V_{cc}
I_{LOH}	Tri-State Output Leakage Current in Float			10	μA	V_{OUT}=2.4 to V_{cc}
I_{LOL}	Tri-State Output Leakage Current in Float			-10	μA	V_{OUT}=0.4V
I_{LD}	Data Bus Leakage Current in Input Mode			±10	μA	$0 \le V_{IN} \le V_{cc}$

Z80A-CPU D.C. Characteristics

T_A = 0°C to 70°C. V_{cc} = 5V ± 5% unless otherwise specified

Symbol	Parameter	Min.	Typ.	Max.	Unit	Test Condition
V_{ILC}	Clock Input Low Voltage	-0.3		0.45	V	
V_{IHC}	Clock Input High Voltage	V_{cc} -.6		V_{cc}+.3	V	
V_{IL}	Input Low Voltage	-0.3		0.8	V	
V_{IH}	Input High Voltage	2.0		V_{cc}	V	
V_{OL}	Output Low Voltage			0.4	V	I_{OL}=1.8mA
V_{OH}	Output High Voltage	2.4			V	I_{OH} = -250μA
I_{CC}	Power Supply Current		90	200	mA	
I_{LI}	Input Leakage Current			10	μA	V_{IN}=0 to V_{cc}
I_{LOH}	Tri-State Output Leakage Current in Float			10	μA	V_{OUT}=2.4 to V_{cc}
I_{LOL}	Tri-State Output Leakage Current in Float			-10	μA	V_{OUT}=0.4V
I_{LD}	Data Bus Leakage Current in Input Mode			±10	μA	$0 \le V_{IN} \le V_{cc}$

Capacitance

T_A = 25°C, f = 1 MHz, unmeasured pins returned to ground

Symbol	Parameter	Max.	Unit
C_Φ	Clock Capacitance	35	pF
C_{IN}	Input Capacitance	5	pF
C_{OUT}	Output Capacitance	10	pF

Z80-CPU Ordering Information

C — Ceramic
P — Plastic
S — Standard 5V ±5% 0° to 70°C
E — Extended 5V ±5% -40° to 85°C
M — Military 5V ±10% -55° to 125°C

Capacitance

T_A = 25°C, f = 1 MHz, unmeasured pins returned to ground

Symbol	Parameter	Max.	Unit
C_Φ	Clock Capacitance	35	pF
C_{IN}	Input Capacitance	5	pF
C_{OUT}	Output Capacitance	10	pF

Z80A-CPU Ordering Information

C - Ceramic
P - Plastic
S - Standard 5V ±5% 0° to 70°C

$T_A = 0°C$ to $70°C$, $V_{cc} = +5V \pm 5\%$, Unless Otherwise Noted.

Signal	Symbol	Parameter	Min	Max	Unit	Test Condition	
φ	t_c	Clock Period	.25	[12]	μsec		[12] $t_c = t_{w(ΦH)} + t_{w(ΦL)} + t_r + t_f$
	$t_{w(ΦH)}$	Clock Pulse Width, Clock High	110	[E]	nsec		
	$t_{w(ΦL)}$	Clock Pulse Width, Clock Low	110	2000	nsec		
	$t_{r,f}$	Clock Rise and Fall Time		30	nsec		
A_{0-15}	$t_{D(AD)}$	Address Output Delay		110	nsec		
	$t_{F(AD)}$	Delay to Float		90	nsec		
	t_{acm}	Address Stable Prior to MREQ (Memory Cycle)	[1]		nsec	$C_L = 50pF$	[1] $t_{acm} = t_{w(ΦH)} + t_f - 65$
	t_{aci}	Address Stable Prior to IORQ, RD or WR (I/O Cycle)	[2]		nsec		[2] $t_{aci} = t_c - 70$
	t_{ca}	Address Stable from RD, WR, IORQ or MREQ	[3]		nsec		[3] $t_{ca} = t_{w(ΦL)} + t_r - 50$
	t_{caf}	Address Stable From RD or WR During Float	[4]		nsec		[4] $t_{caf} = t_{w(ΦL)} + t_r - 45$
D_{0-7}	$t_{D(D)}$	Data Output Delay		150	nsec		
	$t_{F(D)}$	Delay to Float During Write Cycle		90	nsec		[5] $t_{dcm} = t_c - 170$
	$t_{SΦ(D)}$	Data Setup Time to Rising Edge of Clock During M1 Cycle	35		nsec		
	$t_{S\barΦ(D)}$	Data Setup Time to Falling Edge of Clock During M2 to M5	50		nsec	$C_L = 50pF$	[6] $t_{dci} = t_{w(ΦL)} + t_r - 170$
	t_{dcm}	Data Stable Prior to WR (Memory Cycle)	[5]		nsec		
	t_{dci}	Data Stable Prior to WR (I/O Cycle)	[6]		nsec		[7] $t_{cdf} = t_{w(ΦL)} + t_r - 70$
	t_{cdf}	Data Stable From WR	[7]				
	t_H	Any Hold Time for Setup Time		0	nsec		
MREQ	$t_{DLΦ(MR)}$	MREQ Delay From Falling Edge of Clock, MREQ Low		85	nsec		
	$t_{DHΦ(MR)}$	MREQ Delay From Rising Edge of Clock, MREQ High		85	nsec		
	$t_{DH\barΦ(MR)}$	MREQ Delay From Falling Edge of Clock, MREQ High		85	nsec	$C_L = 50pF$	
	$t_{w(MRL)}$	Pulse Width, MREQ Low	[8]		nsec		[8] $t_{w(MRL)} = t_c - 30$
	$t_{w(MRH)}$	Pulse Width, MREQ High	[9]		nsec		[9] $t_{w(MRH)} = t_{w(ΦH)} + t_f - 20$
IORQ	$t_{DLΦ(IR)}$	IORQ Delay From Rising Edge of Clock, IORQ Low		75	nsec		
	$t_{DL\barΦ(IR)}$	IORQ Delay From Falling Edge of Clock, IORQ Low		85	nsec		
	$t_{DHΦ(IR)}$	IORQ Delay From Rising Edge of Clock, IORQ High		85	nsec	$C_L = 50pF$	
	$t_{DH\barΦ(IR)}$	IORQ Delay From Falling Edge of Clock, IORQ High		85	nsec		
RD	$t_{DLΦ(RD)}$	RD Delay From Rising Edge of Clock, RD Low		85	nsec		
	$t_{DL\barΦ(RD)}$	RD Delay From Falling Edge of Clock, RD Low		95	nsec		
	$t_{DHΦ(RD)}$	RD Delay From Rising Edge of Clock, RD High		85	nsec	$C_L = 50pF$	
	$t_{DH\barΦ(RD)}$	RD Delay From Falling Edge of Clock, RD High		85	nsec		
WR	$t_{DLΦ(WR)}$	WR Delay From Rising Edge of Clock, WR Low		65	nsec		
	$t_{DL\barΦ(WR)}$	WR Delay From Falling Edge of Clock, WR Low		80	nsec		
	$t_{DH\barΦ(WR)}$	WR Delay From Falling Edge of Clock, WR High		80	nsec	$C_L = 50pF$	[10] $t_{w(WRL)} = t_c - 30$
	$t_{w(WRL)}$	Pulse Width, WR Low	[10]		nsec		
M1	$t_{DL(M1)}$	M1 Delay From Rising Edge of Clock, M1 Low		100	nsec	$C_L = 50pF$	
	$t_{DH(M1)}$	M1 Delay From Rising Edge of Clock, M1 High		100	nsec		
RFSH	$t_{DL(RF)}$	RFSH Delay From Rising Edge of Clock, RFSH Low		130	nsec	$C_L = 50pF$	
	$t_{DH(RF)}$	RFSH Delay From Rising Edge of Clock, RFSH High		120	nsec		
WAIT	$t_{s(WT)}$	WAIT Setup Time to Falling Edge of Clock	70		nsec		
HALT	$t_{D(HT)}$	HALT Delay Time From Falling Edge of Clock		300	nsec	$C_L = 50pF$	
INT	$t_{s(IT)}$	INT Setup Time to Rising Edge of Clock	80		nsec		
NMI	$t_{w(NML)}$	Pulse Width, NMI Low	80		nsec		
BUSRQ	$t_{s(BQ)}$	BUSRQ Setup Time to Rising Edge of Clock	50		nsec		
BUSAK	$t_{DL(BA)}$	BUSAK Delay From Rising Edge of Clock, BUSAK Low		100	nsec	$C_L = 50pF$	
	$t_{DH(BA)}$	BUSAK Delay From Falling Edge of Clock, BUSAK High		100	nsec		
RESET	$t_{s(RS)}$	RESET Setup Time to Rising Edge of Clock	60		nsec		
	$t_{F(C)}$	Delay to Float (MREQ, IORQ, RD and WR)		80	nsec		
	t_{mr}	M1 Stable Prior to IORQ (Interrupt Ack.)	[11]		nsec		[11] $t_{mr} = 2t_c + t_{w(ΦH)} + t_f - 65$

NOTES:

A. Data should be enabled onto the CPU data bus when RD is active. During interrupt acknowledge data should be enabled when M1 and IORQ are both active.

B. All control signals are internally synchronized, so they may be totally asynchronous with respect to the clock.

C. The RESET signal must be active for a minimum of 3 clock cycles.

D. Output Delay vs. Loaded Capacitance
    $TA = 70°C$ $V_{cc} = +5V \pm5\%$
    Add 10nsec delay for each 50pf increase in load up to maximum of 200pf for data bus and 100pf for address & control lines.

E. Although static by design, testing guarantees $t_{w(ΦH)}$ of 200 μsec maximum

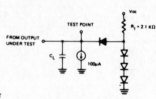

Load circuit for Output

APPENDIX D

VOTRAX SC-01 SPEECH SYNTHESIZER PHONEME CHART

TABLE 1.
PHONEME CHART

PHONEME CODE	PHONEME SYMBOL	DURATION (ms)	EXAMPLE WORD	PHONEME CODE	PHONEME SYMBOL	DURATION (ms)	EXAMPLE WORD
00	EH3	59	jacket	20	A	185	day
01	EH2	71	enlist	21	AY	65	day
02	EH1	121	heavy	22	Y1	80	yard
03	PA0	47	no sound	23	UH3	47	mission
04	DT	47	butter	24	AH	250	mop
05	A2	71	made	25	P	103	past
06	A1	103	made	26	O	185	cold
07	ZH	90	azure	27	I	185	pin
08	AH2	71	honest	28	U	185	move
09	I3	55	inhibit	29	Y	103	any
0A	I2	80	inhibit	2A	T	71	tap
0B	I1	121	inhibit	2B	R	90	red
0C	M	103	mat	2C	E	185	meet
0D	N	80	sun	2D	W	80	win
0E	B	71	bag	2E	AE	185	dad
0F	V	71	van	2F	AE1	103	after
10	CH*	71	chip	30	AW2	90	salty
11	SH	121	shop	31	UH2	71	about
12	Z	71	zoo	32	UH1	103	uncle
13	AWI	146	lawful	33	UH	185	cup
14	NG	121	thing	34	O2	80	for
15	AH1	146	father	35	O1	121	aboard
16	OO1	103	looking	36	IU	59	you
17	OO	185	book	37	U1	90	you
18	L	103	land	38	THV	80	the
19	K	80	trick	39	TH	71	thin
1A	J*	47	judge	3A	ER	146	bird
1B	H	71	hello	3B	EH	185	get
1C	G	71	get	3C	E1	121	be
1D	F	103	fast	3D	AW	250	call
1E	D	55	paid	3E	PA1	185	no sound
1F	S	90	pass	3F	STOP	47	no sound

/T/ must precede /CH/ to produce CH sound.
/D/ must precede /J/ to produce J sound.

TABLE 2. PHONEME CATEGORIES ACCORDING TO PRODUCTION FEATURES

VOICED					"VOICED" FRICATIVE	"VOICED" STOP	FRICATIVE STOP	FRICATIVE	NASAL	NO SOUND
E	EH	AE	UH	OO1	Z	B	T	S	M	PA0
E1	EH1	AE1	UH1	R	ZH	D	DT	SH	N	PA1
Y	EH2	AH	UH2	ER	J	G	K	CH	NG	STOP
Y1	EH3	AH1	UH3	L	V		P	TH		
I	A	AH2	O	IU	THV			F		
I1	A1	AW	O1	U				H		
I2	A2	AW1	O2	U1						
I3	AY	AW2	OO	W						

Votrax® reserves the right to alter its product line at any time, or change specifications or design without notice and without obligation.
Copyright Votrax® 1980

INDEX

Accumulator:
  Intel 8080A microprocessor, 197
  Motorola MC6800 μP, 256–257
Accumulator register, 187
ACIA (*see* Motorola, MC6850 ACIA)
Acquisition time, sample and hold, 172
A/D (analog-to-digital) converter:
  charge balance, 169
  in data acquisition sys., 317,
    381–383
  defined, 161
  in digital arithmetic, 178
  for DVMs, 168, 169, 172–174
  dual slope, 168–169
  "flash," 166–167
  in microprocessor-based scale,
    346–347
  parallel comparator, 166–167
  simultaneous, 166–167
  single counter, 169–170
  single-slope (single-ramp), 167–168
  specifications: accuracy, 169, 172
    conversion time, 172
    linearity, 172
    quantizing error, 172
    resolution, 166, 168, 169, 172
    speed, 167, 169, 170, 172
  in speech recognition units,
    317–318
  successive approximation, 171–172
  tracking, 170
Ada, 410, 432
Addition:
  BCD and excess-3 BCD, 179–180
  binary, 178–179
  hexadecimal, 180–181
  octal, 180
Address buffer, 225
  Intel 8212, 225, 227, 230
  Motorola MC8T96, 272
  72LS244, 225
Address bus: defined, 195
Address decoder: defined, 228
  Intel 3205, 228, 236, 237
  Intel 8205, 237, 238
  74LS138, 236
  74155, 272
Address decoding, 235–239
  for Intel 8080A system, 236–237
  for Intel 8085A system, 238, 239
  for Motorola MC6800 system,
    272–275
  worksheet, 236
Address latch: 8080A/8085A, 199
Addressing modes:
  absolute (6502), 282
  absolute X and absolute Y (6502),
    282
  accumulator (6800, 6502), 264, 282
  base page (6800, 6502), 258, 282
  direct (6800), 258
  direct page register (6809), 279

Addressing modes (*Cont.*):
  extended (6800), 264
  extended indirect (6809), 280
  immediate (6800, 6502), 258, 282
  implied (6800, 6502), 264, 282
  indexed (6800), 258, 264
    (Z80), 250
    (iAPX 86/10), 288
    (6502), 282
  indexed indirect (6809, 6502), 281,
    283
  indirect (6809), 279–280
    (iAPX 86/10, 6502), 288, 282
  indirect indexed (6502), 283
  MOS Technology 6502, 282–283
  Motorola MC6800, 258–264
  Motorola MC6809, 279–280
  post indexed (6502), 283
  relative (6800, 6502), 264, 282
  relative indirect (6809), 280
  zero page (6502), 282
  zero page X, zero page Y (6502), 282
  Z-80, 250
ALGOL, 409–410
Algorithm: defined, 187
Algorithmic state machine (ASM) chart,
    435–436
ALU (arithmetic logic unit)
  in microprocessor CPU, 194, 197,
    198
  74181, 185–188
Analog: definition, 17, 18, 161
Analog-to-digital-converter (*see* A/D
    converter)
AND function, 19
AND gate, 19–20
  boolean algebra expressions for,
    19–20
  schematic symbol illustrated, 19
  truth table, 19
Aperture time, for sample and hold,
    172
APL, 410
Apple II personal computer, 281, 408,
    432
Arithmetic logic unit (*see* ALU)
Arrays:
  BASIC, 417
  defined, 417
  FORTRAN, 419
ASCII code, 78–83
Assembler, 217, 219, 399–400
Assembler directives, 217, 219
Assembling a program:
  with an assembler, 217–219,
    399–400
  defined, 212
  by hand, 216–217
Assembly language: defined, 199
  grammar rules, 210–211
  program examples (*see* Program
    examples)

Astable multivibrator, 43, 45
Asynchronous counters, 104–111
Asynchronous data communication,
    325
Audio reverb block diagram, 123
Automatic test equipment (ATE),
    403–405
  (*See also* Test system)
Automobile seat belt warning system,
    66

BASIC, 409–418
  arithmetic operators, 411
  arrays, 417
  constants, variables, numbers, 411
  library functions, 412
  program examples, 410, 414–418
    add 2 numbers, 410
    checkbook record, 416
    coin flipping, 414
    sort (arrays), 417–418
    strings, 416–417
    temperature conversion, 415
  program format, 410–411
  relational operators, 411–412
  statements, 412–413
  strings, 416–417
Baud rate, 125, 325
BCD (binary coded decimal):
  codes, 75–77
  addition, 179–180
  subtraction, 184
  (*See also* Decimal adjust operation)
BCDIC, 79–82
Benchmark programs, 283, 286
Bidirectional bus, defined, 195
  driver, 222
Binary addition, 178–179
Binary division, 189–190
Binary multiplication, 187–189
Binary numbers (code), 17, 74–76
Binary point, 18
Binary subtraction, 181–184
BISYNC (binary synchronous protocol),
    335–337
Bit: definition, 18
Bit-oriented protocol, 337
Bit-slice processors, 288
Board test, 402
Boolean algebra, 55–58, 67–68
  for combinational logic circuits, 54
  for common logic gates, 19–21
  De Morgan's theorem, 56–58
  distributive properties, 55, 56
  equivalents, 55–58
  reductions, 55, 56
Breakpoint, 220
  defined, 395
  for debugging programs, 394–397
Bubble memory (*see* Magnetic bubble
    memory)
Buffer, 55

Buffer *(Cont.):*
  address, 225
  disk file, 431
  memory, in μP-based scale program,
    361
  testing of, 248
Bus standards, parallel, 318–325
  GPIB, HPIB, IEEE-488, 318, 321,
    323–325
  Intel multibus, 318, 322
  S-100, 318–321
Bus structure, 195
Bypass capacitors, 4, 31, 391, 392
Byte: defined, 91, 236
Byte-controlled protocol (BCP), 337

Calibration technician, 402
Carry flag:
  Intel 8080A/8085A, 197
  Motorola MC6800, 258
CCD (charge coupled device), 130–132,
    146–147
Central processing unit (*see* CPU)
Character generator: Motorola MC6571,
    338–340
Charge coupled device (*see* CCD)
Check-sum, 152–153
Clock cycles, microprocessor, 214
Clocks:
  crystal oscillators, 43–46
  digital, 107, 109, 110
  555 timer, multivibrators, 43–45
  real-time, μP, 294, 381, 383
CMOS logic (complimentary-MOS logic),
    4, 30, 37–41
  bypass capacitor values, 4
  crystal oscillator, 45, 46
  current parameters, 30, 38, 39
  gate structure (circuitry), 37–38
  guidelines for using, 40–41
  interfacing, 38–39, 49
  noise margin, 38, 39
  power dissipation, 39
  propagation delay times, 30, 39
  SOI, SOS, 40
  three-state output, 39–40
  threshold voltage, 38, 39
  transfer curve, 38, 39
  transmission gates, 39–40
  voltage parameters, 30, 38, 39
COBOL, 410
Code conversions (*see* Number system
    conversions)
Code converters, 77, 83–86, 91
CODEC (coder-decoder), 175–176
Codes:
  ASCII, 78–83
  BCD, 75–77
  BCDIC, 79–82
  binary, 17, 74–76
  EBCDIC, 79–82
  excess-3 BCD, 76, 77
  gray, 76, 77
  hexadecimal, 75, 76
  Hollerith, punched cards, 79–82
  octal, 74–76
  selectric, IBM, 79–84
  seven-segment, 76–77
Coding, assembly language, 212, 216
Color-coded wiring, 3, 391
Combinational logic, 54–70, 78, 83–85
Comments, in assembly language
    programs, 210–211

Comparators, 42, 166–167
  in A/D converters, 166–173
  as detectors, 42
  to produce logic level signals, 42
Compiler program, 408–409
  BASIC, 409
  FORTRAN, 424–425
Complement: defined, 20
  one's-, 181
  two's-, 182
Computer structure, 194, 195
Control bus: defined, 195
Conversion time, for A/D converter,
    172
Coprocessors, 432
Core memory (*see* Magnetic core
    memory)
Count widow, 113, 117
Counter:
  used in A/D converter, 169–170
  asynchronous, 104–111
    as digital clock, 107, 109, 110
    as frequency divider, 104–106
    as organ tone generator, 106–108
    problems of, 107, 111
  defined, 104
  modulo, 104
  ring, 122–125
  ripple, 104–111
  synchronous: defined, 111
    as MSI freq. counter, 113,
      116–119
    as LSI freq. counter, 120–121
    as modulo-*n* divider/counter,
      112–115
CPU (central processing unit), 194–199
  Intel 8080A/8085A, 197–199,
    222–223, 229–232
  Motorola MC6800, 256–257
"Crash," of magnetic disk head, 152,
    154
CRT:
  character generators, 338–340
  composite video signal, 338, 339
  field, 337
  frame, 338
  graphics, 340, 342
  of oscilloscope, 6
  programmable controllers, 339–342
    Intel 8275, 340, 341
    Motorola MC6845, 340, 342
  raster, 337
  scanning, 337–338
  of TV/video monitor, 337–339
Crystal oscillators, 43–46
Current sinking, defined, 23
Current sourcing, defined, 24
Current loop, 20 mA and 60 mA,
    325–326
Current-tracing probe, 5, 248
Cyclic redundancy check (CRC),
    152–153, 337

D flip-flop, 98–101
D latch, 98, 99
D/A (digital-to-analog) converter:
  in A/D converters, 169–172
  binary weighted resistor, 161–162
  characteristics, errors, specifications:
    accuracy, 164
    gain error, 164, 166
    linearity errors, 164, 166
    monotonicity, 166

D/A converter *(Cont.):*
  offset error, 164, 166
  output settling time, 166
  resolution, 164, 166
  in CODEC, 175–176
  defined, 161
  in digital arithmetic, 178
  in digital voice communication,
    174–176
  hybrid, 163–165
  interfacing to microprocessor,
    315–316
  monolithic, 163, 164
  for motor speed control, 316
  in programmable power supply, 316
  R/2R ladder, 162–163
  in speech synthesis, 316
"Daisy-chaining," 4
Data acquisition system, 317, 381–383
Data bus: defined, 195
Data communication standards,
    325–329
  20–60 mA current loop, 325
  RS-232C, 327, 328
  RS-422, 327
  RS-423, 327, 329
  teletype, 125, 325
DCE (data communication equipment),
    327
Debouncing:
  keyboard (key switch) software, 302
  hardware switch, 41–42, 97, 123
Debugging programs (*see*
    Troubleshooting, assembly
    language programs)
Decimal adjust operation, 180
  Intel 8080A/8085A, 204
Decode: defined for counters, 111
Decoder, 83–85
  defined, 83
  7447 BCD-to-7-segment, 83–85
    dependency notation logic symbol,
      85
  (*See also* Address decoders,
    decoding)
Decoupled wiring, 4
Delay constants, 214, 216, 362
De Morgan's theorem, 56–58
Demultiplexers, 68–70, 74
  dependency notation logic symbols,
    74
Dependency notation logic symbols:
  D flip-flop, 99, 100
  description, 437–450
  demultiplexer, 74
  JK flip-flop, 101–102
  multiplexer, 73–74
  NAND, 63–64
  7493 binary counter-divider, 105
  74193 binary up-down counter, 114
  74194 universal shift register,
    127–128
  TTL RAM (MCM93425), 136, 137
Derivative feedback control, 378, 379
Development system (*see*
    Microprocessor development
    system)
Differentiator, 378, 379
Digital: defined, 17, 18, 161
Digital arithmetic, 178–192
Digital clocks, 107, 109, 110
  LSI, using MM53103, 110
Digital-to-analog converter (*see* D/A)

Digital readout oscilloscopes, 15
Digital voice communication, 174–176
Digital voltmeter (DVM), 6, 168, 169, 172–174
DIP (dual in-line package), 21
Direct I/O, defined, 237
Direct memory access (*see* DMA)
Disk operating system (DOS), 408
Disks and disk files, 152–154, 431
Divider, frequency: defined, 104
Division: binary, 189–190
    assembly routines for, 388–389
    pencil method, 189
    shift-left-and-subtract method, 189–190
    successive subtraction method, 189
DMA (direct memory access), 242–243, 257
DMOS (double-diffused MOS), 37
Double-buffered, defined, 332
Double-level handshake I/O, 296
DTE (data terminal equipment), 327

EBAM, 155
EBCDIC code, 79–82
ECL (emitter-coupled logic), 4, 30, 31–33
    circuits, bypass capacitor values, 4
    crystal oscillator, 45, 46
    families, 33
    fanout, 30, 33
    internal circuitry, operation, 31, 32
    noise margins, 32, 33
    propagation delay, design tricks, 33
    termination, 32, 33
    transfer curves, 31–33
    voltage parameters, 30–32
Editor program, 399–400, 408
    (*See also* Text editor)
EEPROM, 88–89
Enable input:
    on D latch, 98–99
    on RS latch, 97, 98
    on three-state device, 28, 29
Encoder, 85–86
    defined, 83
    74148 8-line-to-3-line priority, 85–86
        dependency notation logic symbol, 86
Encoding, defined for keyboards, 123
End-around carry, 181, 182
EPROM, 88, 89
EPROM programmer, μP-based, 365–377
    assembly language program, 373–377
    flowchart, 372–373
    hardware, 365–369
        MEK6800D2, 365–367
    J-BUG monitor, 367, 370–371
    programming data, 2716 EPROM, 370–372
Equally tempered scale (ETS), 108
Equates, 217, 308, 350, 360
Error detection:
    check-sum, 152–153
    cyclic redundancy check, 152–153, 337
    framing error, 125, 330
    hamming code, 78
    overrun error, 125, 330
    parity, 78, 330
Excess-3 BCD, 76, 77

Excess-128 BCD number, 192
Exclusive NOR (XNOR) gate, 20–21
    boolean algebra expressions for, 21
    schematic symbol illustrated, 21
    truth table, 21
Exclusive OR (XOR) gate, 20–21
    boolean expressions for, 20–21
    schematic symbol illustrated, 21
    truth table, 21
Exclusive OR "tree" circuit, 78
Execution time, program instruction:
    comparison chart, 283
    Intel 8080A/8085A, 214–216
    MOS Technology R6502A, 283
    Motorola MC6800, 258–263, 267, 268, 275

Fall time, defined, 25
Fanout:
    CMOS, 30
    defined, 24
    ECL, 30, 33
    TTL, standard, 22, 24, 30
Field:
    defined for SDLC, 337
    defined for TV, CRT, 337
Field service engineer, 402–403
FIFO (first-in-first-out) RAM, 314
File:
    closing and opening, 431
    definitions, 150, 431
    FORTRAN, 419
    of characters, 432
    pointer, 431
Firmware, defined, 432
555 timer, 43–45
Fixed-point numbers, 190–191
Flag flip-flops, Intel 8080A/8085A, 197
Flags, Motorola MC6800, 258
Flip-flops, 98–104
    D type, 98–101
    JK type, 101–103
    problems of, 103–104
    T or toggle, 104
Floating-point numbers, 191–192, 418
Floppy disks, 152–154
    error detection, 152–153
    handling precautions, 154
    hard-sectored, 152–153
    soft-sectored, 152
Flowcharts, 209–210
Foldback, defined, 238, 348
Folding trick, multiplexers, 72, 73
FORTRAN, 409, 418–425
    compiling and running programs, 424–425
    data types, 418–419
    intrinsic functions, 419, 421
    operators, 419
    program example, temperature conversion, 420–421, 424
    program format, 419
    statements, 419, 421–424
FPGA (field programmable gate array), 91–92
FPLA (field programmable logic array), 91
Frame:
    defined, for SDLC, 337
    defined, for CRT, TV, 338
Framing error, defined, 125, 330
Frequency counter, 113, 116–121
    LSI, 120–121

Frequency counter (*Cont.*):
    MSI, 113, 116–119
    troubleshooting MSI, 119
FSK (frequency shift keying), 150, 151, 327
Full adder, 179
Full duplex mode, defined, 326
Full subtractor, 181, 182
Furnace controller, 66–68

Galpat, defined, 155
Gas hot-air furnace controller, 66–68
Glitch, 11–12, 96, 97, 158, 159
Glomper clip, 8
GPIB bus standard, 318, 321, 323–325
Graphics: computer, CRT, 340, 342
Gray code, 76, 77
Grounding problems, 48, 49

Half adder, 178–179
Half duplex mode, defined, 326
Half subtractor, truth table, 181
HALT state:
    Intel 8080A/8085A, 243
    Motorola MC6800, 257
Hamming code error detecting, 78
Handshake I/O, 296–301
    double handshake, 296
    simple strobe, 296
    single handshake, 296
Hard-sectored disk, 152
Hardware, defined for microprocessors, 220
Hexadecimal addition, 180–181
Hexadecimal numbers (code), 75, 76
Hexadecimal subtraction, 185
High-level languages:
    acronyms (abbreviations) for, 409–410
    development software, 408–409
    development systems, 408
    (*See also* BASIC; FORTRAN; Pascal)
HMOS (high-performance MOS), 37
HOLD state, Intel 8080A/8085A, 242–243
HPIB parallel bus standard, 318, 321, 323–325
Hybrid D/A converter, 163–165

IBM personal computer, 288, 408
IC (integrated circuit), 1
IEEE-488 bus standard, 318, 321, 323–325
I^2L (integrated injection logic), 33–34
In-circuit emulation, 394, 399–401
Indeterminate state:
    for latches, 97, 98
    for TTL voltage levels, 23
Industrial process control, 377–378
Input-output (I/O) ports, 194–195
    decoding, 237, 239
    direct, 237
    in Intel 8080A system, 225–228
    memory-mapped, 237
    in Motorola MC6800 system, 272–274
    testing, 249
Instruction cycle:
    defined, 8080A, 223, 224
    defined, 8085A, 229
    defined, Motorola MC6800, 275
Instruction decoder:
    Intel 8080A/8085A, 199

Instruction decoder *(Cont.):*
  Motorola MC6800, 257
Instruction fetch, 195–196
Instruction register:
  Intel 8080A/8085A, 199
  Motorola MC6800, 257
Instruction set:
  Intel 8048, 251–253
  Intel 8080A/8085A, 199–209
  MOS Technology 6502, 282–285
  Motorola MC6800, 258–266
  Zilog Z80, 249–250, 459–460
Integer numbers, FORTRAN, 418
  (See also Fixed-point numbers)
Integral feedback, 378, 379
Integrator:
  in A/D converters, 168
  in process control loops, 378, 379
Intel:
  2716 EPROM, 88, 89, 91, 225,
    370–372, 452–455
  3205 address decoder, 236
  8048 microprocessor and subfamilies
    (8021, 8022, 8031, 8035, 8049,
    8051, 8071), 250–254, 286
  8080A microprocessor
    address decoding, 236–237
    control signals, 222–225
    CPU, 197–199, 222–223
    halt state, 243
    hold state/DMA, 242–243
    instruction cycles, machine cycles,
      states, 223–224
    instruction set, 199–209, 215
    interrupts, 243
    port decoding, 237–238
    program execution time, 214–216
    programming examples, 211–216
    system, 225, 226
    troubleshooting, 246–249
    wait state, 241–242
  8085A microprocessor:
    address decoding, 238–239
    CPU, 197–199, 229
    control signals, 229–230
    halt state, 243
    hold state and DMA, 242–243
    instruction cycles, machine cycles,
      states, 229
    instruction set, 199–209, 215
    interrupt I/O, 292–294
    interrupts, 243–245
    microcomputer system, 230–235
    port decoding, 239
    timing parameters, 239–241
    transition state diagram, 245–246
    troubleshooting, 246–249
    wait state, 241–242
  8086 16-bit μP (iAPX 86/10),
    286–288
  8087 mathematics coprocessor, 432
  8088 16-bit μP (iAPX 88/10), 288
  8155/8156 RAM, I/O port, timer,
    230, 234–235, 239, 294,
    296–299
  8205 address decoder, 237–238
  8212 I/O port, 225, 227
  8224 clock generator, 222–223
  8228 system controller/driver,
    222–223
  8215A USART, 228, 231, 332–335
  8255A programmable peripheral
    interface, 228, 237–238

Intel *(Cont.):*
  8259 PICU (programmable interrupt
    controller unit), 294–295
  8275 CRT controller, 340, 341
  8279 keyboard-display interface,
    312–314
  8284 clock generator, 286
  8288 bus controller, 286
  8355 ROM, I/O port, 230, 234, 239
  8755 EPROM, I/O port, 230, 234
  80130 (silicon software), 432
  multibus, 318, 322
  SDK-85 microprocessor development
    board, 230–233, 310–313, 348
Interfacing logic families:
  NMOS-TTL-NMOS, 49
  TTL-CMOS-TTL, 31, 38–39, 49
  TTL-ECL-TTL, 49
  TTL-I L-TTL, 50
  TTL-PMOS-TTL, 49
  TTL subfamily interconnections, 31,
    48–49
Interfacing logic gates:
  to incandescent lamps, 50
  to LEDs, 50
  to mechanical relays, 51
  to neon lamps, 50
  to solid-state relays, 51
Interfacing microcomputer:
  to A/D converter, 316–318
  to cassette-tape (KCS), using
    MEK6800D2, 305, 307
  to CRT display monitor, 337–342
  to D/A converter, 315–316
  to high-power devices, 314–315
  to keyboards, 302–308, 310–313
    hardware method, 302–303
    MEK6800 keyboard display,
      305–308
    SDK-85, keyboard display,
      310–313
    software method, 303–305
  to liquid crystal displays, 314
  for motor speed control, 316
  to power supplies, 316
  to robots (using stepper motors), 315
  to seven-segment displays, 305–314
    hardware multiplexing, 312–314
    MEK6800D2 display routine,
      305–312
    SDK-85 display using 8279,
      310–313
    software multiplexing, 305–312
  to speech (voice) recognition unit,
    317–318
  to speech synthesizer, 297–299, 316
  to stepper motors, 315
  to tape reader: using Intel 8155, 296
    using Motorola 6831, 301
  to X-Y plotters (stepper motors), 315
Interpreter program, 408, 409
  BASIC, 409
Interrupt I/O, 292–295
  Intel 8085A example, 292–294
  Motorola MC6800 example, 294, 295
Interrupts:
  for generating clock-tick, 381, 383
  Intel 8080A, 243
  Intel 8085A, 243–245
  Motorola MC6800, 275–278
  priority control, 294–295
  for timing (real-time clock), 294
Inverter (NOT gate), 20

Inverter (NOT gate) *(Cont.):*
  boolean algebra expression for, 20
  dynamic MOS, 128–129
  PMOS, 128
  schematic symbol illustrated, 20
  truth table, 20

JBUG monitor *(see* Monitor program,
    JBUG)
JK flip-flops, 101–103
  master-slave, 101, 102
  negative-edge-triggered, 101, 102
  one's catching, 103
  positive-edge-triggered, 101, 102
  timing diagrams, 101–103
Josephson junction memories, 146

Kansas City standard (KCS), 150, 151
K-map (Karnaugh map):
  adjacencies, 63–64
  applications, automobile seat belt
      warning system, 66
    gas hot-air furnace controller, 68
    7-segment code converter, 83, 84
  don't-care states, 83, 84
  octets, 65
  quads, 65
  rules for constructing, 62–63
  rules for simplification, 63–66
Keyboard encoder, 123–125

Label (symbolic address), 212
Latch:
  D type, 98, 99
  NAND, 97, 98
  RS with enable, 97, 98
Latency time, defined, 147
LCD (liquid crystal display), 314
LIFO (last-in-first-out memory), 207
Line drivers and receivers, 48, 327
Linear select address decoding, 237
Linking:
  high-level language modules, 409
  assembly language modules, 395,
    397
Liquid crystal displays (LCDs), 314
List file, 400
Locating, defined for:
  high-level language modules, 409
  FORTRAN programs, 425
Logic analyzer, 15, 156–159
  center-trigger recording, 159
  display formats, 157–159
  used in field service, 403
  latch-mode storage, 158
  pretrigger recording, 158, 398
  used in troubleshooting
      microcomputers, 247, 248, 278
  used in testing, debugging μP
      prototype hardware, 398–399
  used in testing, debugging software,
    399
Logic functions:
  for common gates, 19–21
  table of, for 74181 ALU, 187
Logic gates, 19–21
  AND, 19–20
  defined, 19
  inverter (NOT), 20
  NAND, 20
  NOR, 20
  OR, 20
  as switches, 68–70

Logic gates (Cont.):
  XNOR (exclusive NOR), 20–21
  XOR (exclusive OR), 20–21
Logic probes, 6
Logic pulser, 5, 248
Logic state analyzer, defined, 398
  (See also Logic analyzer)
Logic state detector, LED, 50
Logic timing analyzer, defined, 398
  (See also Logic analyzer)
Lookup table, 91, 358–359, 362
Low-power Schottky TTL, 29–31
Low-threshold PMOS, 36
LSD (least significant digit), 19
LSI circuits, 21, 34, 40, 107, 110

Machine code, 400
Machine cycle:
  Intel 8080A, 223–224
  Intel 8085A, 229
  Motorola MC6800, 275
Machine language, defined, 199
Macros, 219
Magnetic bubble memory (MBM),
  147–148
Magnetic core memory, 148–149
Magnetic disk data storage, 150,
  152–155
  floppy disk, 152–154
  hard (rigid) disk, 154–155
Magnetic tape memory (storage),
  149–151
  recording modes, 150, 151
Mark, 152
  defined for teletype, 325
  defined for current loop, 325
  defined for RS232C, 327
Masking, 55, 205, 244–245
  interrupts: Intel 8085A, 244–245
    Motorola MC6800, 266, 276–278
Memories:
  CCD (charge coupled device),
    146–147
  electron beam accessed (EBAM), 155
  Josephson junction, 146
  Magnetic bubble, 147–148
  Magnetic core, 148–149
  Magnetic disk, 150–155
  Magnetic tape, 149–151
  RAM, 135–146
  ROM (read-only), 86–91
  read-write, 135–159
  shift register, 146–148
Memory map, 238
  for SDK-85 board, 348–349
Memory-mapped I/O, defined, 237
Memory testing, 155–159, 249, 278,
  392–393
Memory-to-memory architecture, 288
Microcomputer development boards:
  Intel SDK-85, 230–233, 310–313,
    348–349
  Motorola MEK6800D2, 270–274,
    305–310, 312, 365–367,
    370–371
Microcomputer development system
  (see Microprocessor development
  system)
Microcomputer interfacing (see
  Interfacing microcomputers)
Microcomputer, 16-bit:
  Intel iAPX 86/10 (8086), 286–288
  Intel iAPX 88/10 (8088), 286, 288

Microcomputer, 16-bit (Cont.):
  Motorola MC68000, 286, 288
  Texas Instruments TMS9900, 286,
    288
Microcomputer system (see
  Microprocessor system)
Microinstructions, 288
Microprocessor development boards
  (see Microcomputer development
  boards)
Microprocessor development system,
  217–219, 399–400
Microprocessor families:
  Intel 8048, 250–254
  Intel 8080A/8085A, 197–249
  Intel 8086/8088, 286–288
  MOS Technology 6500, 281–286
  Motorola MC6800-6809, 256–281
  Motorola 68000, 288
  Texas Instruments TMS 9900, 288
  Zilog Z80, 249–250,
    456–464
Microprocessor interfacing (see
  Interfacing microcomputers)
Microprocessor system:
  block diagram, 194–195
  Intel 8080A, 8085A, 197–219,
    222–249
  learning a new processor, 228–229
  Motorola MC6800, 270–275
Minterm expression, 64
Mixed logic, 60–62
Mnemonic, defined, 199, 210
MODEM (modulator-demodulator),
  325–328
  null MODEM, 327, 328
Modular programming, 359–360, 407,
  425, 432
  (See also Structured programming)
Modulo, of counters, dividers, 104
Monitor program,
  defined, 230, 348, 408
  Intel SDK-85, 8355, 348
    for debugging scale program,
      394–395
  JBUG of Motorola MEK6800D2, 308,
    309, 367, 370–371, 373, 374
    for debugging EPROM programmer
      program, 396–397
  used in troubleshooting RAM, 249, 278
Monolithic: defined for ICs, 163
Monostable multivibrator (555), 43, 45
Monotonicity, D/A converter, 166
MOS handling guidelines, 34–36,
  40–41
MOS logic, 30, 34–41
MOS logic families, 36–41
  CMOS, 4, 30, 37–41
  low-threshold PMOS, 36
  NMOS, 36–37
  PMOS, 36
  VMOS, DMOS, and HMOS, 37
MOS Technology:
  6502 microprocessor, 280–286
    addressing modes, 282–285
    architecture, 280, 281
    control signals, 281
    flags, 281–282
    instruction set, 282, 284–285
    system, 283
    timing, 283–286
  6522 PIA, 283
  6530 programmable interface, 283

MOS transistors, 34, 35
Motor speed control, 314–315, 378,
  381
Motorola:
  MC6800 microprocessor, 256–281
    addressing modes, 258–264
    clock, frequencies, 272
    CPU, 256–257
    flags, 258
    instruction set, 258–266
    interrupts, 275–278
    program examples, 266–269, 272
    signal descriptions, 257–258
    system, 270–275
    timing, 275, 276
  MC68B00, 283
  MC6801, single-chip μP with UART,
    279
  MC6802 microprocessor with clock,
    279
  MC6809 microprocessor, 279, 281
  MC6810 128X8 RAM, 272
  MC6821 PIA, 272–275, 298, 300–301
  MC6830 IK × 8 mask-programmed
    ROM, 272
  MC6845 CRT controller, 340, 342
  MC6846 ROM, I/O timer, 279
  MC6847 video display generator, 342
  MC6850 ACIA (UART), 272, 274,
    329–332
  MC6871B clock generator, 272, 275
  MC68000 16-bit microprocessor,
    286, 288
  MEK6800D2 keyboard display,
    305–307
MSD (most significant digit) 18, 19
Multiplexed displays, 117, 119,
  305–314
Multiplexer applications:
  address selector for ROMs, 87, 88, 90
  code conversion, 86, 87
  data transmission, 70
  oscilloscope 8-channel display, 70, 71
  synthesizing logic functions, 70–73
Multiplexers, 68–74, 86–88, 90
  dependency notation logic symbols,
    73–74
Multiplication, binary, 187–189
  add-and-shift-right method, 187–189
  assembly language routines for, 357,
    361, 363–364
  pencil method, 187, 188
  repeated addition method, 187
Multipliers, external to a computer, 188
Multivibrators, 43–45
  astable, 43, 44
  monostable, 43, 45

NAND function, derived, 20
NAND gate:
  CMOS, 37–38
  defined, 20
  NMOS, 36
  TTL, 21, 22, 26–27
  truth table, 20
NAND latch, 97, 98
Negative logic, 58–59
Nested subroutines, defined, 207
Nibble, defined, 236
N-key rollover, 313
NMOS, 36–37
Noise margin:
  CMOS, 38, 39

Noise margin *(Cont.)*:
  ECL, 31–33
  TTL, 23
Nondestructive turn-on, 314
NOR function, derived, 20
NOR gate:
  CMOS, 38
  defined, 20
  ECL, 32
  NMOS, 36
  truth table, 20
NOT function, 20
NOT gate (inverter), 20
Number system conversions:
  BCD to binary: assembly language
      routine, 356–357, 361–363
  binary to BCD: assembly language
      routine, 357–358, 361,
      364–365, 387–389
  binary to decimal, 18
  binary to hexadecimal, 75
  binary to octal, 74–75
  decimal to binary, 18–19
  decimal to BCD, 75–76
  decimal to hexadecimal, 75
  decimal to octal, 74
  hexadecimal to binary, 75
  hexadecimal to decimal, 75
  hexadecimal to octal, 75
  octal to binary, 74–75
  octal to decimal, 74
  octal to hexadecimal, 75
Number systems:
  base-2 (binary), 17–19, 76
  base-8 (octal), 74–76
  base-10 (decimal), 17–18, 76
  base-16 (hexadecimal), 75, 76
  binary, 17–19, 74, 76
  decimal, 17–18, 76
  fixed-point, 190–191
  floating-point, 191–192
  hexadecimal, 75, 76
  octal, 74–76

Object code or file, 219, 400,
    424–425
Octal addition, 180
Octal numbers (code), 74–76
Octal subtraction, 184
One's catching, 103
One's-complement, defined, 181
  subtraction, 181–182
Op amp (*see* Operational amplifier)
Op code (*see* Operation code)
Open collector TTL, 28
Operand, defined, 199
Operating system, 408, 432
Operation code, defined, 199, 210
Operational amplifier, 42, 161–162
  as comparator, 42
  as integrator, 168
  negative feedback, 161
  summing point, 162
  virtual ground, 162
Optical couplers, 51
OR gate, 20
  boolean algebra expressions for, 20
  schematic symbol illustrated, 20
  truth table, 20
Organ tone generator schematic, 108
OR-NOR gate (ECL), 31, 32
Oscillators, crystal, 43–45
Oscilloscope 8-input multiplexer, 70, 71

Oscilloscope probes, 8–9
  compensation, 8, 9
Oscilloscopes, 6–15
  ac coupling, 8
  alternate sweep mode, 7
  automatic triggering mode, 11
  beam finder, 6
  calibrator output, 8
  chopped sweep mode, 7
  dc coupling, 7–8
  delayed sweep modes, 11–12
  external trigger source, 10
  focus control, 6
  horizontal, vertical controls, 6
  how to get a trace on, 6
  intensity control, 6
  internal trigger source, 9
  line source, 10
  normal triggering mode, 11
  single-sweep mode, 11
  sweep mode controls, 6, 7, 11–13
  trace rotation control, 6
  trigger holdoff control, 11
  trigger input coupling controls, 10
  trigger slope, level controls, 10–11
  triggering controls, 6, 9–11
  for troubleshooting μC, 246–249, 278
  variable controls calibration, 7
  X-Y mode, 12, 13
  Z axis, 13
Overrun error, 125, 330
Overshoot, 8, 9, 46, 47

P-code, 409, 432
Page boundary crossing, 282
Page boundary error, 282
PAL (programmable array logic), 92
Parity, 77–78
  error, 330
  flag, 8080A/8085A, 197
  generator and checker system, 78
Pascal, 410, 425–432
  console input and output, 430–431
  constants, variables, 425–427
  declarations, 425
  operators, 428
  procedures and functions, 427–428
  program example, temperature
      conversion, 426, 431–432
  program structure, 425
  statements, 428–431
  syntax diagram, 425, 427, 429
PC (printed circuit) board, 3
Personal computer:
  Apple II, 281, 408
  IBM, 288, 408
  TRS-80, 408
PIA (*see* Motorola MC6821 PIA)
Phonemes, defined, 298
PICU (*see* Intel 8259 PICU)
Pipelining, defined, 283
PIPO shift register, 126–128
PISO shift register, 125–126
Pixels, defined, 340
PLA (programmable logic array), 91, 92
PL/1 and PL/M, 410
PMOS, 36
Polled I/O, 291
Port decoding, 237, 239
  Intel 8080A, 237
  Intel 8085A, 239
Power dissipation, logic families
    compared, 30, 39, 40

Power-on reset circuit, 123
  for 8080A system, 225–226
  for keyboard encoder, 124
Power supplies, in prototyping, 4, 391
Prescaler, 117, 120, 121
Printed circuit board, 3
Process control system, block diagram,
    379
  (*See also* Temperature controller)
Production department, 401–402
Production engineering dept., 400–401
Production process, for μP-based
    instrument, 400–402
Production test technician, 402–403
Program, microcomputer, defined, 194
Program counter, 8080A/8085A, 199
Program examples, assembly language:
  Intel 8085A interrupt I/O, 292–294
  Intel 8085A keyboard scan, 303–305
  Intel 8085A μP-based scale, 349–365
  Intel 8085A 1-ms delay, 212–214
  Intel 8085A take-in-data, 211–212
    when-data-ready-pulse, 212, 213
  Intel 8085A temperature/process
      controller, 382–389
  Motorola MC6800 EPROM
      programmer, 372–377
  Motorola MC6800 interrupt I/O, 294,
      295
  Motorola MC6800 multiple precision
      addition, subtraction, 267–269,
      272
  Motorola MC6800 take-in-data, at
      timed intervals, 267, 268
    when-data-ready signal, 266–267
  simple 3-step, 195–196
Program execution, example, 195–197
Programmable parallel ports, 234,
    296–301
  Intel 8155, 234, 296–299
  Intel 8755A, 234
  Motorola MC6821, 298, 300–301
Programmable peripheral interface IC
    (*see* Intel 8155, 8255, 8355,
    8755)
Programmable power supply, 316
Programmable serial port ICs:
  Intel 8251A, 332–335
  Motorola MC6850, 329–332
Programmer (*see* EPROM programmer)
PROM (programmable read-only
    memory), 88
Propagation delay time, defined, 24
  for different logic families, 30
  ECL, 30, 31, 33
  TTL, standard, 22, 24–25, 30
Proportional control, 378, 379
Proportional-integral-derivative (PID)
    control loops, 378, 379
Prototype circuit:
  for digital circuits, 1, 2
  for μP-based systems, 391–394
Prototyping, 1–4, 391–394
Prototyping boards, 1–3
Pseudoinstructions, 217, 219
PSW (program status word), 207
Pulse code modulation (PCM), 176
Pulse generators, 15
Pulse parameters, defined, 24–25
Pulse sources:
  crystal oscillators, 43–46
  555 timer, multivibrator, 43–45
Pulse width, defined, 25

Pulse width modulation, 381
Pulser probe (see Logic pulser)

Quality-control (QC) technician, 402
Quantizing error, 172

RAM (random access memory),
    135–146
  CMOS, 141
  defined, 135
  dynamic MOS, 141–146
  ECL, 136
  in 8080A system, 225
  nonvolatile, 141
  static MOS, 136–141
  timing parameters, waveforms,
    137–146
  TTL, 135–137
Ramp generator, 168
Random access memory (see RAM)
Raster, defined, 337
Read-only memory (see ROM)
Read-write memories, 135–159
Real numbers, 191, 418
  (See also Floating-point numbers)
Real-time clock, 294, 381, 383
Record, defined for magnetic tape, 150
Redundancy, defined, 146
Refresh:
  dynamic MOS shift register,
    128–130
  dynamic RAM, 141, 143–146
Register:
  accumulator, 187–188
  defined, 121–122, 135
  in full adder circuit, 179
  general-purpose, defined, 288
  shift, 122–132
Register bank:
  Intel 8048, 251
  Texas Instruments TMS9900, 288
Register file, defined, 135
Relays, 51
Reset, power-on, 123
Reset condition, defined, 97
Reset operation:
  Intel 8080A, 225, 243, 258
  Intel 8085A, 243, 245, 246
  Motorola MC6800, 258, 273, 276
Reverb, audio: block diagram, 123
RFI protection, 51
Ring counter, 122–125
  as control state counter, 123
  in keyboard encoder, 123–125
  for stepper motor control, 123
Ripple counters, 104–111
Risetime, defined, 25
ROM (read-only memory), 86–91
  applications, character generator, 91
    code conversion, 86–87, 91
    computer memory, 91
    keyboard encoder, 91
    logic gate replacement, 91
    look up table, 91
  data sheet parameters, 89, 91
  "homemade," 86–87
  mask-programmed, 87
  programmable, 88–89, 91
  programming procedure, 91
  structure and addressing, 89, 90
Routine, defined, 206
RS latch, 97, 98
RS-232-C, 327, 328

RS-422, 327
RS-423, 327, 329
S-100 parallel bus standard, 318–321
Sample and hold circuit 14, 172
Sampling oscilloscope, 13–14
SAR (successive approximation
    register), 171
Scale, microprocessor-based, 345–365
  A/D converter, 346–348
  assembly language program,
    350–359
  block diagram, 346
  flowchart, 349
  load cell and amplifier, 346–347
  program description, 360–365
Scanning, CRT
  interlaced, 337–338
  noninterlaced, 338
Schmitt trigger, 42, 43
Schottky diode, 29
Schottky diode-clamped transistor, 29,
    31
Schottky TTL, 4, 29–31
"Scratch pad" register, defined, 197,
    199
SDK-85 (see Microcomputer
    development boards, Intel
    SDK-85)
SDLC (synchronous data link control),
    337
Sector, defined for magnetic disk,
    152
Self-test diagnostics, 404–405
Serial access memory:
  CCD, 146–147
  magnetic bubble, 147–148
  magnetic tape, 149–150
Serial data communication, 125,
    325–337
SET, assembler directive, 217
Set condition, defined, 97
Set point, 377
Settling time:
  for D/A converter, 166
  for process control system, 378
Seven-segment display code, 76, 77
Shift register, 122–132
  defined, 122
  MOS, 128–132
    bucket brigade, 130, 131
    charge coupled device (CCD),
      130–132
    dynamic, 128–129
    static, 129–130
  parallel-in-parallel-out (PIPO),
    126–128
    for binary math operations, 128
  parallel-in-serial-out (PISO), 125–126
    for parallel-to-serial data conversion,
      123–126
  serial-in-parallel-out (SIPO), 122–126
    as ring counter, 122–125
    as serial-to-parallel data converter,
      125, 126
  serial-in-serial-out (SISO), 122, 123
    for delay, 123
  universal, 74194, 126–128
Shift register memory, 146
  (See also Serial access memory)
Sign-and-magnitude numbers,
    182–184
Sign bit, 182–184

Sign bit flag:
  Intel 8080A/8085A, 197
  Motorola MC6800, 258
Signal conditioning, 41–51
  input, 41–45
  interfacing, 48–51
  output, 45–48
Signature analysis, 404–405
Signed numbers, 182–184
Silicon software, 432
Simple strobe I/O, 296
Single-chip microcomputers, 250–254,
    286
Single-level handshake I/O, 296
Single-stepping, defined, 220, 395
  in debugging 8080A/8085A
    hardware, 247
  in debugging 8080A/8085A software,
    220
  in debugging 6800 software, 397
SIPO shift register, 122–126
SISO shift register, 122, 123
Skew, defined, 138
Snap data, 400
Soft-sectored floppy disks, 152
Software, defined, 220
Software components, 410
SOI CMOS (silicon-on-insulator), 40
Solderless prototyping boards, 1–3
SOS CMOS (silicon-on-sapphire), 40
Source program or file, 219, 400, 424
Space, defined for teletype, 325
Speech recognition unit, 317
Speech synthesizer:
  VOTRAX, interfacing to μC, 297–299
    phoneme chart, 465–466
  Texas Instrument TMC 0280, 316
Speed-power product, 29, 30
Stack, defined, 207
Stack operations, 207–209
Stack pointer, 8080A/8085A, 199
State:
  defined for microcomputer, 214
  defined for 8080A, 223, 224
  defined for 8085A, 229
  logic, defined, 19
State diagrams, 245–246, 435–436
Stepping (stepper) motor, 123, 152,
    315
Storage oscilloscopes, 13
Storage time, defined, 27
Strobe input, 98
Structured programming, 382, 410,
    425, 432
  (See also Modular programming)
Subroutines, 206–209, 212–214, 265,
    267, 268
Subtraction:
  BCD and excess-3 BCD, 184
  binary, 181–184
    standard, 181, 182
    one's-complement, 181–182
    two's-complement, 182–184
  hexadecimal, 185
  octal, 184
Successive approximation register, 171
Sum-of-products expression, 54–55,
    64–66
Summing point, 162
Switch debouncing, 41–42, 97, 123
Symbol table, 219, 400
Symbolic address (label), 212, 308, 400
Synchronous counters, 111–121

Synchronous serial data
    communication, 335–337
  BISYNC, 335–337
  SDLC, 337

Tape reader (interface to
    microcomputer):
  Intel 8155, 296–297
  Motorola 6821, 301
Teletypes:
  data format, 125, 325
  20 and 60 mA current loops,
    325–326
Temporary register, 8080A/8085A, 197
Temperature controller (SDK-85-based
    process control system), 379–389
  assembly language program,
    382–389
  flowcharts, 380
  hardware, 381, 382
    clock-tick, 381
    data acquisition system, 381–383
    temperature sensor, 381
Termination, for transmission lines:
  using diode clamps (TTL), 47–48
  using resistors (ECL), 32, 33, 47
Test fixture (test jig), 402
Test instruments:
  current tracing probe, 5, 248
  DVM, 6
  logic probe, 6
  logic pulser, 5, 248
  logic state analyzer, 15, 156–159
  microsystem troubleshooter,
    403–404
  ohmmeter, 5
  oscilloscopes, 6–15
  pulse generators, 15
  signature analyzer, 404–405
  time-mark generator, 15
Test system, IC, 155–156
Testing:
  assembly language programs,
    394–396, 399–400
  computer-controlled, 155
  galpat, 155
  prototype circuits, μP, 392–394,
    398–399
  prototype circuits, digital, 4–5
Texas Instruments:
  TMC 0280 speech synthesizer, 316
  TMS 9900, 286, 288
  TMS 9995, 288
Text editor, 217
T flip-flop, 104
Three-state output:
  CMOS, 39–40
  TTL, 28
Threshold voltage (switching
    threshold):
  CMOS, 38, 39
  defined, 23
  ECL, 31–32
  TTL, standard, 23
Time-mark generator, 15
Timing diagram:
  for data acquisition system
    (ADC0808), 383

Timing diagram (Cont.):
  defined, illustrated, 96, 97
  for D flip-flop, 100, 101
  for D latch, 99, 101
  for GPIB data transfer, 324
  for JK flip-flops, 101, 103
  for handshake data transfer:
    Intel 8155, 297
    Motorola 6821, 301
  logic analyzer format, 157–159
  for logic switches, multiplexers, 68, 69
  MC6800 read cycle, 276
  for NAND latch, 98
  for programming Intel 2716, EPROM,
    370
  RAM, read cycle, 139, 142
    write cycle, 140, 144
    refresh, 145
  for ring counter, 123
  7493 binary counter/divider, 105
  7493 modulo-10 counter/divider, 106
  7493 modulo-11 counter/divider, 107
  7493 showing unwanted count
    states, 111
  74193 as modulo-11
    counter/divider, 115
  74193 up-down counter, 114
  8080A cycles and states, 224
  8085A instruction fetch cycle, 229
  SISO shift register, 122
  Speech synthesizer (Votrax), 299
  T flip-flop, 104
Track, defined for magnetic disk,
    152
Transfer curves:
  CMOS, 38, 39
  ECL, 31, 32
  TTL, 12, 13, 23
Transistor switch, bipolar, 25–26
Transition state diagram, 8085A,
    245–246
Transition times, defined, 25
Transmission gate, 39–40
Transmission lines, 33, 45–48
Transparent refresh, defined, 249
Tree wiring, 4
Troubleshooting:
  assembly language program,
    219–220
    scale program, 395
    EPROM programmer, 396–397
  general techniques for, 403
  Intel 8080A/8085A μC system,
    246–249
  Motorola MC6800 μC system, 278
  MSI frequency counter, 119
Truncated, defined, 190
Truth tables:
  for common logic gates, 19–21
  for combinational logic circuit, 54,
    55
TTL, 4, 21–31
  crystal oscillator, 43–44, 46
  guidelines for using, 31
  interfacing, 31, 38–39, 48–50
  high power and low power, 29–30
  low-power Schottky, 29–31
  open collector, 28

TTL (Cont.):
  standard, data sheet parameters,
    21–25
    current specifications, 22–24, 30
    fanout, 22, 24, 30
    internal circuitry, 25–28
    logic voltage levels, 22, 23, 30
    maximum ratings, 21–23
    noise margin, 23
    propagation delay time, 22, 24–25,
      30
    threshold voltage, 23
  Schottky, 29–31
  three-state output, 28
  transfer curve, 12, 13, 23
TTL circuits, bypass capacitor values, 4
Two-key lockout, 312–313
Two-key rollover, 125, 303
Two's-complement, defined, 182
  subtraction, 182–184

UART:
  General Instruments AY5-1013,
    125–126, 325
  Intel 8251A, 332–335
  Motorola MC6850 ACIA, 274
Undershoot:
  oscilloscope waveform with, 8, 9
  in transmission line reflections, 46,
    47
USART, 126, 228
  Intel 8251A, 332–335

Video monitor, 337, 338
Virtual ground, 162
VMOS (vertical MOS), 37
Voice recognition (see Speech
    recognition unit)
Volatile memory, defined, 135
Voltage transients, suppression, 4
VOTRAX SC-01 speech synthesizer,
    297–299
  phoneme chart, 465–466

Wait for interrupt, MC6800, 265–266,
    278
Wait state, 8080A/8085A, 241–242
Waveforms (see Timing diagram)
Winchester disk drives, 154–155
Wire wrapping, 2, 3
Word recognizer, 10

XNOR (exclusive NOR) gate, 20–21
XOR (exclusive-OR) gate, 20–21
X-Y mode, for transfer curves, 12, 13

Zero flag:
  Intel 8080A/8085A, 197
  Motorola MC6800, 258
Zilog Z80 microprocessor, 249–250,
    286
  data sheets, 456–464
Zone coordinates on system
    schematics, 230, 232–233